CLINICAL GENETICS
A SOURCE BOOK FOR PHYSICIANS

Edited by

Laird G. Jackson, M.D.
Professor of Medicine, Pediatrics, Obstetrics, and Gynecology (Genetics)
Director, Division of Medical Genetics
Jefferson Medical College
Thomas Jefferson University
Philadelphia, Pennsylvania

and

R. Neil Schimke, M.D.
Professor of Medicine and Pediatrics
Director, Division of Metabolism, Endocrinology, and Genetics
Kansas University College of Health Sciences and Hospital
Kansas City, Kansas

A WILEY MEDICAL PUBLICATION
JOHN WILEY & SONS, INC.
New York • Chichester • Brisbane • Toronto

Copyright © 1979 by John Wiley & Sons, Inc.

All rights reserved. Published simultaneously in Canada.

Reproduction or translation of any part of this
work beyond that permitted by Sections 107 or 108
of the 1976 United States Copyright Act without the
permission of the copyright owner is unlawful. Requests
for permission or further information should be addressed
to the Permissions Department, John Wiley & Sons, Inc.

Library of Congress Cataloging in Publication Data

Main entry under title:

Clinical genetics.

   (A Wiley medical publication)
   Includes index.
   1.   Medical genetics.   I.   Jackson, Laird G.
II.   Schimke, R. Neil, 1935-   [DNLM: 1. Genetics, Human. 2. Hereditary diseases. QZ50.3 C641]
RB155.C572   616'.042   78-24414
ISBN 0-471-01943-7

Printed in the United States of America

10 9 8 7 6 5 4 3 2 1

Contributors

David E. Anderson, Ph.D.
Professor of Biology
Section of Medical Genetics
The University of Texas System Cancer Center
Texas Medical Center
Houston, Texas

Frederick R. Bieber, M.S.
Department of Human Genetics
Medical College of Virginia
Richmond, Virginia

Angelo M. DiGeorge, M.D.
Professor of Pediatrics
Temple School of Medicine
Director, Section of Endocrine
and Metabolic Disorders
St. Christopher's Hospital for Children
Philadelphia, Pennsylvania

Salvatore DiMauro, M.D.
Professor of Neurology
College of Physicians and Surgeons
Columbia University
New York, New York

Kenneth D. Gardner, Jr., M.D.
Professor of Medicine
Chief, Division of Renal Disease
Department of Medicine
University of New Mexico School of Medicine
Director, Hemodialysis Unit
Bernalillo County Medical Center
Albuquerque, New Mexico

Raymond Garrett, M.D.
Division of Renal Diseases
Department of Medicine
University of New Mexico School of Medicine
Albuquerque, New Mexico

Enid F. Gilbert, M.D.
Professor of Pathology and Pediatrics
University of Wisconsin Center for Health Sciences
Madison, Wisconsin

Peter Hathaway, M.D.
Division of Metabolism, Endocrinology, and Genetics
Kansas University College of Health Sciences and Hospital
Kansas City, Kansas

William A. Horton, M.D.
Assistant Professor of Medicine and Pediatrics
Department of Medicine
Division of Metabolism, Endocrinology, and Genetics
University of Kansas Medical Center
Kansas City, Kansas

Laird G. Jackson, M.D.
Professor of Medicine, Pediatrics,
Obstetrics, and Gynecology (Genetics)
Director, Division of Medical Genetics
Jefferson Medical College
Thomas Jefferson University
Philadelphia, Pennsylvania

Charles H. Kirkpatrick, M.D.
Senior Investigator
Head, Clinical Allergy
and Hypersensitivity Section
Laboratory of Clinical Investigation
National Institute of Allergy
and Infectious Diseases
Bethesda, Maryland

Irene H. Maumenee, M.D.
The Wilmer Ophthalmological Institute
Johns Hopkins University
School of Medicine
Baltimore, Maryland

Robert F. Murray, Jr., M.D., M.S., F.A.C.P.
Professor, Pediatrics and Medicine

Chief, Division of Medical Genetics
Department of Pediatrics and Child Health
Director, Division of Genetic Counseling
Center for Sickle Cell Disease
Howard University College of Medicine
Washington, D.C.

Walter E. Nance, M.D., Ph.D.
Professor and Chairman
Department of Human Genetics
Medical College of Virginia
Richmond, Virginia

Audrey Hart Nora, M.D.
Associate Clinical Professor of Pediatrics
University of Colorado Medical Center
Director of Genetics
Children's Hospital of Denver
Denver, Colorado

James J. Nora, M.D.
Professor of Pediatrics
Director of Pediatric Cardiology
University of Colorado Medical Center
Denver, Colorado

John M. Opitz, M.D.
Professor of Medical Genetics and Pediatrics
University of Wisconsin Center for Health Sciences
Madison, Wisconsin

Eberhard Passarge, M.D.
Professor and Chairman
Department of Human Genetics
University of Essen
Essen, West Germany

Mario C. Rattazzi, M.D.
Associate Professor of Pediatrics
Division of Human Genetics
Department of Pediatrics
State University of New York at Buffalo
Children's Hospital of Buffalo
Buffalo, New York

R. Neil Schimke, M.D., F.A.C.P.
Professor of Medicine and Pediatrics
Director, Division of Metabolism,

Endocrinology, and Genetics
Kansas University College of Health Sciences
and Hospital
Kansas City, Kansas

Charles I. Scott, Jr., M.D.
Professor of Pediatrics
Director, Medical Genetics Clinic
Department of Pediatrics
The University of Texas
Health Science Center at Houston
Houston, Texas

Robert L. Summitt, M.D.
Professor of Pediatrics, Anatomy,
and Child Development
University of Tennessee Center
for The Health Sciences
Memphis, Tennessee

Elliot S. Vesell, M.D.
Professor and Chairman
Department of Pharmacology
The Pennsylvania State University
College of Medicine
Hershey, Pennsylvania

Harry H. White, M.D.
Clinical Professor of Neurology
University of Kansas School of Medicine
Director of Neurological Education
Shawnee County Medical Foundation
Topeka, Kansas

Carl J. Witkop, Jr., D.D.S., M.S.
Professor and Chairman
Division of Human and Oral Genetics
School of Dentistry
University of Minnesota
Minneapolis, Minnesota

Preface

The contemporary practicing physician, no matter what his area of expertise, usually becomes uneasy when the conversational topic turns to inherited disease. This feeling is understandable when considered in its perspective because only the most recent medical school graduates have had much experience in the application of genetic principles to man. Also, geneticists have tended to publish their interesting case material in the genetics journals, which are hardly a place where they will be encountered by the ordinary physician. Those of us interested in clinical genetics have broadened our literary horizons over the past few years. However, this has not substantially improved the picture for the non-geneticists, since the various papers are now scattered throughout a host of both general and special journals. Standard genetics texts are usually very basic in their orientation and frequently contain little information of clinical relevance. More specialized texts are available, but unless a physician has a nearly unlimited budget (and a similar amount of time), he cannot hope to encompass the field.

With all the above in mind, we felt it would be useful to compile most of the current information in clinical medical genetics into a single volume emphasizing its clinical application. We hope that this book can serve as a focal source for inquiry into a specific genetic disease in a designated organ system. Obviously, "new" conditions are appearing on the medical horizon with such discouraging regularity that no one text can hope to cover the field. Our contributors have performed a yeoman effort to make available a tremendous amount of information in a readable, clinically-oriented form that we hope will be useful to students, to physicians, and to nurses and paramedical personnel who come in contact with patients with heritable disease. It was occasionally necessary for us as editors to make some changes for the sake of both uniformity and brevity. It is likely that we occasionally overdid it. The content of the various chapters is sound; we can only apologize to our colleagues and our readers for our editorial shortcomings.

We are indebted to our secretaries, Alice Algie, Patricia Cionci, Jane Gottlieb, Joan Glazerman, and Barbara Lawson for collating, typing, assisting with the editing process (mainly by finding the pages we misplaced), and generally putting up with our bad penmanship and vile humor. Support for this work was provided in part by the Fogarty Foundation and the National Foundation—March of Dimes.

Laird G. Jackson, M.D.
R. Neil Schimke, M.D.

Contents

Part One. Introduction   1

1. Fundamentals of Clinical Genetics   3
   R. Neil Schimke and Laird G. Jackson

Part Two. Multisystem Disease   33

2. Cytogenetic Disorders   35
   Robert L. Summitt

3. Genetics and Cancer   85
   Laird G. Jackson, David E. Anderson, and R. Neil Schimke

4. Genetic Disorders of the Immune System   121
   Charles H. Kirkpatrick

5. Genetic Metabolic Disorders   153

   Lysosomal Storage Disorders   153
   Mario C. Rattazzi

   Abnormalities in Carbohydrate Metabolism   185
   R. Neil Schimke

   Abnormalities in Amino Acid Metabolism   194
   R. Neil Schimke

   Disorders of Porphyrin Metabolism   208
   Laird G. Jackson

   Disorders of Purine Metabolism   217
   Laird G. Jackson

6. Heritable Connective Tissue Disorders   229
   William A. Horton

7. Pharmacogenetics   245
   Elliot S. Vesell

Part Three. Organ System Diseases   267

8. Genetics of Cardiovascular Diseases   269
   James J. Nora and Audrey H. Nora

9. Genetics of Endocrine Diseases   285
   Angelo M. DiGeorge

10. Genetics of the Gastrointestinal Tract   331
    Eberhard Passarge

11. Genetics and Hematology   349
    Peter Hathaway and R. Neil Schimke

12. Genetic Diseases of Skeletal Muscle   373
    Salvatore DiMauro

13. Heritable Neurologic Diseases   391
    Harry H. White

14. Heritable Eye Diseases   415
    Irene H. Maumenee and Laird G. Jackson

15. Hereditary Hearing Loss   443
    Frederick R. Bieber and Walter E. Nance

16. Heritable Diseases of the Kidney   463
    Kenneth D. Gardner, Jr. and Raymond Garrett

17. Genetic Disorders of the Respiratory System   495
    Enid F. Gilbert and John M. Opitz

18. Heritable Skeletal Dysplasias   519
    Charles I. Scott, Jr.

19. Genetic Disorders of the Skin   545
    Laird G. Jackson

20. Hereditary Defects of Teeth and Oral Structures   575
    Carl J. Witkop, Jr.

Part Four. Genetic Counseling   595

21. The Technique of Genetic Counseling   597
    Robert F. Murray, Jr.

22. Prenatal Diagnosis   613
    Laird G. Jackson

Glossary   617

Index   631

Part 1
INTRODUCTION

1
Fundamentals of Clinical Genetics

R. Neil Schimke
Laird G. Jackson

A BRIEF HISTORY OF GENETICS

Although sophisticated knowledge of the biologic and chemical bases of heredity is a fairly recent scientific acquisition, a thumbnail sketch of the salient historical developments is appropriate. Notions of genetics have been found among the stone carvings of the Chaldeans dating from 6,000 years ago, and the ancient Greeks advocated a stringent form of eugenics by recommending infanticide for the deformed. The Talmud contains proscriptions against circumcision for the brothers of males who were bleeders, thereby tacitly recognizing the X-linked inheritance of hemophilia. In the early eighteenth century, Pierre de Maupertuis studied certain single-gene disorders in man and developed a concept, albeit somewhat faulty, of the structural basis of heredity. Mendel, in the mid-nineteenth century, showed that the occurrence of simple physical characteristics in plants had a statistical predictability. At the turn of the present century, Johannsen coined the term gene and differentiated between an individual's genetic makeup, or genotype, and his external appearance, or phenotype.

Sutton and Boveri, in 1903, independently proposed the chromosomal theory of heredity. Garrod provided the basis for biochemical genetics through his study of alcaptonuria. His concept of heritable inborn errors of metabolism was confirmed and extended by Beadle and Tatum in the late 1930s and early 1940s, as they formulated the one gene:one enzyme concept* and, in essence, proved the correctness of Garrod's early twentieth-century conceptualization of an enzymatic error as being responsible for some of man's heritable diseases. Even prior to that time, a distinguished group of scientists including Morgan, Bridges and Sturtevant, working with the common fruit fly, *Drosophila,* established that genes were arranged in a linear sequence along chromosomes.

Avery, MacCleod and McCarty, in 1944, demonstrated that deoxyribonucleic acid (DNA) was the hereditary material, and a decade later the work of Wilkins, Franklin, Watson and Crick led to the proposal by Watson and Crick of the now

---

*Later it was shown that some proteins are composed of more than one polypeptide chain, e.g., hemoglobin, and are thus the product of more than one gene.

well-known double helical structure of DNA. In 1956, Tjio and Levan in Sweden and Ford and Hamerton in Great Britain reported that man had 46 rather than 48 chromosomes and three years later Lejeune demonstrated the chromosomal basis of the Down syndrome. In 1961, Nirenberg and Matthaei broke the genetic code, and elucidation of the translational mechanism followed shortly thereafter. At about the same time Jacob and Monod were reporting their now classic work on regulation and control of gene action.

These workers are but a few of those who have fostered interest in genetics over the centuries. Despite this background and the fact that every college-level biology text contains reference to genetics, the practical application of this discipline to the medical sciences did not occur before 1960. It is not surprising therefore that many clinicians know little about medical genetics. However, for many of them—not only pediatricians and obstetricians, but also those in the medicosurgical subspecialities—genetics is an important facet of clinical practice.

CHEMISTRY OF THE GENE

In this scientifically enlightened era, every schoolboy knows that deoxyribonucleic acid (DNA) is the genetic material. The DNA molecule is composed of two tightly coiled helical chains of nucleotides, or bases, the backbone of each chain being formed by sugar (deoxyribose)–phosphate groups. The two complementary chains of DNA are interconnected across the helical loops by hydrogen bonds between the bases, with adenine (A) being bonded to thymine (T), and guanine (G) to cytosine (C) (Fig. 1). During replication the two chains or strands come apart and each acts as a template for the synthesis of a complementary strand. The sequential attachment of new bases requires a specific enzyme, DNA-dependent DNA synthetase (replicase).

In this way each daughter cell contains a strand of DNA from the original cell plus a newly synthesized complementary molecule, a so-called semiconservative method of nucleic acid duplication. The genetic information is stored within the DNA in the form of a triplet code, i.e., each sequence of three bases codes for a single amino acid, and this triplet sequence is called a *codon*. A permutation of the four bases gives rise to 4^3 or 64 codons. Since there are only 20 amino acids, there appear to be excess codons. However, the code is degenerate, i.e., more than one triplet may code for a given amino acid; moreover certain codons act as starters or initiation points for translation of the genetic message, and others function as termination codons.

The genetic information in the DNA is then transcribed into a message of single-stranded ribonucleic acid (RNA) called *messenger RNA* (m-RNA). This process is achieved by a synthetic sequence similar to DNA duplication but requiring another specific enzyme, DNA-dependent RNA synthetase (transcriptase).† Each message is formed by a particular gene, or in some cases a number

---

†Some RNA viruses, the genetic information of which is stored in a double-stranded molecule of RNA, first transcribe this information to a molecule of DNA using an enzyme called reverse-transcriptase (RNA-dependent DNA synthetase). They are then able to use the cell's own enzyme to complete the process of information transfer as described herein.

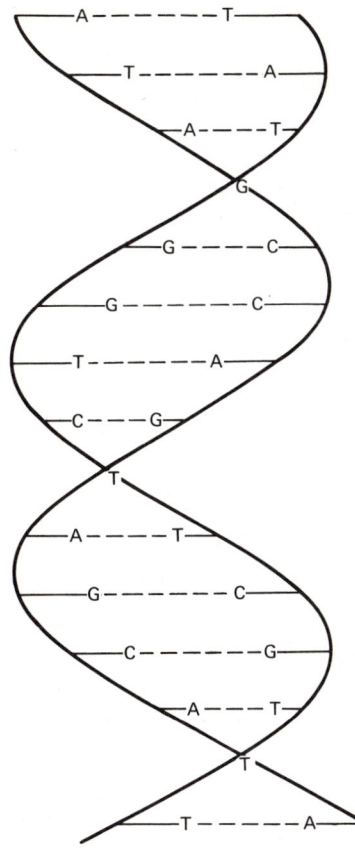

Figure 1. The DNA molecule. The two purines, adenine (A) and guanine (G), are joined to the pyrimidines, thymine (T) and cytosine (C) by hydrogen bonds to form the double helix. Chemical configuration of single strand of DNA molecule, showing sugar–phosphate "backbone" to which the four bases (A, C, G and T) are attached.

of genes, so that every base in the m-RNA is complementary to the corresponding base in DNA with the exception that adenine of DNA pairs rather with uracil (U) of RNA than with thymine (Fig. 2). The codons are usually expressed in m-RNA equivalents, i.e., UUU for phenylalanine rather than AAA (Table 1). The m-RNA then traverses the nuclear membrane and passes into the cytoplasm.

Here it becomes associated with another species of RNA, *ribosomal RNA* (r-RNA), aggregates of which are called *ribosomes*. Several ribosomes begin to attach to one end of the m-RNA and travel along it in a specific direction, "reading" the message as they go. This complex of m-RNA and ribosomes is the polysome, and the process—which results in protein synthesis—is *translation*. As the ribosome translates a codon, it calls for the proper amino acid to be inserted into the growing protein (polypeptide) chain, and the appropriate acid is brought into proper alignment by a third RNA, *transfer RNA* (t-RNA). The ribosome recognizes and pairs the m-RNA and t-RNA so that a peptide bond can be formed between the carboxy terminal of the growing polypeptide chain and the amino end of the amino acid brought into position by the t-RNA (Fig. 2).

The process continues until a chain-terminating codon is reached, at which juncture the completed polypeptide chain is released. This sequence continues

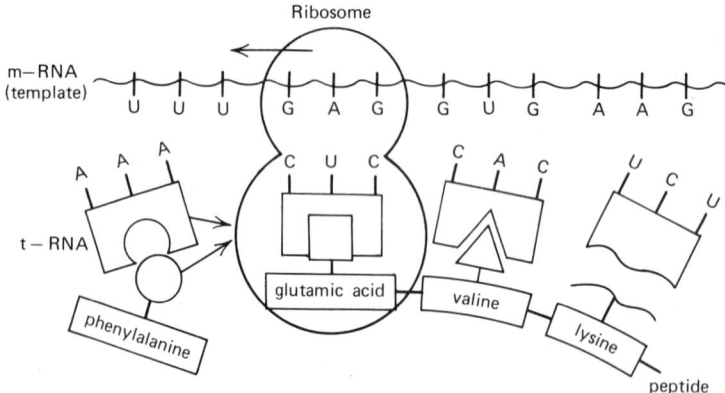

Figure 2. Protein synthesis. A strand of messenger RNA (m-RNA) is formed on DNA template in the nucleus. It then travels to the cytoplasm where it attaches to a ribosome. A t (transfer)-RNA–amino-acid complex the anticodon triplet of which matches the codon triplet being "read" (translated) on the m-RNA fits into place against the m-RNA. As m-RNA strand is being translated, t-RNA molecules are split from attached amino acids. The latter are joined by peptide bonds to form the protein molecule coded by the original DNA strand. T = thymine; U = uracil; C = cytosine; A = adenine; G = guanine.

until sufficient quantities of the polypeptide are produced. Control of this process ultimately may reside in the interaction of the specific gene with the polypeptide product (feedback inhibition), with nuclear protein or with other small molecules, the synthesis of which is under the control of still other genes. There are then *structural genes* that code for the structure of polypeptides, such as enzymes, hormones, hemoglobins and *regulatory genes,* or control genes, that control the extent of the translational process and hence the amount of structural genetic product.

The concept of regulatory, as distinct from structural, genes was first demonstrated by Jacob and Monod in *Escherichia coli.* They demonstrated the existence of a gene that directed or operated a set of structural genes to produce a series of proteins. These so-called operator genes were in turn regulated by molecules that could turn them on (*inducers*) or off (*repressors*). Although the original theory may have only limited, or perhaps no, direct applicability to a higher organism like man, it nonetheless supplies formal evidence for the existence of gene-control mechanisms. No regulatory devices like these have been rigorously demonstrated in higher animals, but they undoubtedly make up part of more complex genotypes.

Point Mutations

Abrupt heritable changes in the genetic makeup (genotype) of an organism are termed mutations, and they may be different types. The commonest is probably a point mutation in which a single base in the DNA is changed, perhaps by either

Table 1. The Genetic Code: Nucleotide Base Triplets and Corresponding Amino Acids

DNA Triplet	RNA Triplet	Amino Acid	DNA Triplet	RNA Triplet	Amino Acid
AAA*	UUU	Phenylalanine	ATA	UAU	Tyrosine
AAG	UUC	Phenylalanine	ATG	UAC	Tyrosine
AAT	UUA	Leucine	ATT	UAA	Chain end
AAC	UUG	Leucine	ATC	UAG	Chain end
GAA	CUU	Leucine	GTA	CAU	Histidine
GAG	CUC	Leucine	GTG	CAC	Histidine
GAT	CUA	Leucine	CTT	CAA	Glutamine
GAC	CUG	Leucine	GTC	CAG	Glutamine
TAA	AUU	Isoleucine	TTA	AAU	Asparagine
TAG	AUC	Isoleucine	TTG	AAC	Asparagine
TAT	AUA	Isoleucine	TTT	AAA	Lysine
TAC	AUG	Methionine	TTC	AAG	Lysine
CAA	GUU	Valine	CTA	GAU	Aspartic acid
CAG	GUC	Valine	CTG	GAC	Aspartic acid
CAT	GUA	Valine	CTT	GAA	Glutamic acid
CAC	GUG	Valine	CTC	GAG	Glutamic acid
AGA	UCU	Serine	ACA	UGU	Cysteine
AGG	UCC	Serine	ACG	UGC	Cysteine
AGT	UCA	Serine	ACT	UGA	Chain end
AGC	UCG	Serine	ACC	UGG	Tryptophan
GGA	CCU	Proline	GCA	CGU	Arginine
GGG	CCC	Proline	GCG	CGC	Arginine
GGT	CCA	Proline	GCT	CGA	Arginine
GGC	CCG	Proline	GCC	CGG	Arginine
TGA	ACU	Threonine	TCA	AGU	Serine
TGG	ACC	Threonine	TCG	AGC	Serine
TGT	ACA	Threonine	TCT	AGA	Arginine
TGC	ACG	Threonine	TCC	AGG	Arginine
CGA	GCU	Alanine	CCA	GGU	Glycine
CGG	GCC	Alanine	CCG	GGC	Glycine
CGT	GCA	Alanine	CCT	GGA	Glycine
CGC	GCG	Alanine	CCC	GGG	Glycine

*A = Adenine; C = Cytosine; G = Guanine; T = Thymine; U = Uracil.

ionizing radiation or certain chemicals or perhaps as a random event in the transcriptional or replicative process. The point mutation is said to be *missense* if the codon for one amino acid is changed for that of a different amino acid. The codon for glutamic acid, (e.g.), may be changed to the codon for lysine, a modification that requires only a single base change. This seemingly minor event may have implications for the individual organism, depending on whether the mutation took place in a position in the gene critical for the function of the resultant polypeptide. If this mutation occurred in that part of the beta chain gene for human hemoglobin coding for the sixth amino acid from the amino terminal of the molecule, hemoglobin C would result—with all the clinical consequences that that implies.

If the same mutation occurred rather at the 26th than at the sixth position, hemoglobin E would be synthesized. While each of these hemoglobin molecules is abnormal, the diseases resulting from these two changes are different. If the same mutational event took place in another polypeptide, more particularly in a part of another polypeptide that was not critical for function, perhaps no disease would develop. In this latter case the mutational event might be detected only by electrophoretic study of the normal and mutant polypeptides, revealing the charge difference occasioned by a change from the negative glutamic acid to the double-positively charged lysine residue.

Large numbers of serum protein and enzyme variants have been found in man, most of which reflect no apparent clinical consequences of the mutational event; they are useful, however, in the study of population genetics, because they can be used to examine, e.g., inbreeding patterns and early human migration (Table 2). Missense mutations that occur in a gene coding for an enzyme polypeptide might yield an enzyme with no, or perhaps only partial, ability to catalyze a given chemical reaction. The clinical result is an inborn error of metabolism, many examples of which are discussed in this work.

Some mutations are *nonsense mutations*, i.e., they code for no amino acid. When such a nonsense triplet is reached in the translational process, synthesis stops and the incomplete polypeptide chain falls off the ribosome. For nonsense mutations no peptide will be detected by immunologic means, whereas with missense mutations a peptide will be immunologically detectable but functionally inadequate. If a missense mutation occurs at a site important for *both* antigenicity and function,

Table 2. Polymorphic Blood Groups and Protein Loci Used as Genetic Markers

Blood Groups

ABO	RhCc
Duffy	RhDd
Kell	RhEe
Kidd	S
Lewis	Secretor
MN	

Red Cell Enzymes, Serum Proteins and Physical Characteristics

Acid phosphatase	Glyoxylase I
α_1-Acid glycoprotein	Haptoglobin α
Adenosine deaminase	Hemoglobin β
Adenylate kinase 1	Peptidase A
α_1-Antitrypsin	Peptidase C
Ceruloplasmin	Pepsinogen 5
Esterase D	Phosphoglucomutase 1
Glucose-6-phosphate dehydrogenase	6 - Phosphogluconate dehydrogenase
Glutamate-pyruvate transaminase	Third component of complement
Glutathione reductase	Transferrin
Group-specific component	

the peptide, while present, will not be detected by these techniques, and such mutations will simulate a nonsense mutation.

When a point mutation occurs in a chain-terminating codon, the resulting peptide will be lengthened, the extent depending on the duration of the message before the next nonsense or termination codon is reached. Lengthening presumably occurs because the m-RNA message contains excessive bases not normally translated. Hemoglobin Constant Spring is such a mutation, and persons with this conjugated protein have an elongated beta chain. Another type of mutation is a *deletion,* or loss of various number of bases. If the loss occurs "in register," i.e., in multiples of three bases, one or more amino acids will be deleted in the final peptide chain. If a deletion occurs "out of register," i.e., if anything less than a triplet is deleted, the mutation is a *frame shift.* The reader can assess for himself the possible consequences of changes like these and why they would be rare (see Chapter 11 for examples of chain-terminating and frame-shift mutations as they occur in the human hemoglobins).

Mutations can occur at any time in any cell. Those in gametes are heritable, and it is likely that the majority of human genetic disease results from mutational events such as those just described. Once in a gamete, they will be transmitted to subsequent generations with a finite probability. Mutations can occur in somatic cells as well and may be the direct cause of some forms of neoplasia in man.

Not all mutations are deleterious; presumably man has evolved because beneficial mutations occurred in an appropriate environment so that the mutation became fixed in the population, a process termed *natural selection.* Also, some mutations may have been selectively neutral, the mutation conferring neither advantage nor disadvantage but simply being "carried along" in the genotype, perhaps because other closely linked genes were being subjected to positive selection. Many geneticists feel that no mutation is neutral, that our present ignorance simply precludes identification of the selective process that favored the earlier retention of a given mutant in the population. In any case all investigators would agree that most mutations, past and present, are disadvantageous.

GENE AND CHROMOSOME

A gene, then, can be defined simply as a specific segment of DNA containing information for the production of a functional protein. The gene coding for a human hemoglobin beta chain, which contains 146 amino acids, is 146×3, or a minimum of 438 bases long, plus a few more necessary as "starters and stoppers." The m-RNA coding for this gene is at least as long, although as previously mentioned messengers translating more than one gene at a time exist. Each chromosome is composed of literally thousands of genes. The DNA composing the chromosomes may be somewhat dispersed in interphase (nondividing) nuclei, or tightly coiled and folded onto itself in metaphase. The chromosome also contains certain basic proteins (protamines and histones) as well as acidic ones that may in some fashion regulate gene function.

Man is a *diploid* organism, i.e., the chromosomes are paired except for the male X and Y sex chromosomes, the sizes of which are discrepant. Each gamete contains a *haploid* complement of chromosomes, and the diploid number is restored at fertilization in the zygote. The paired chromosomes are *homologous,* i.e., each number 21 chromosome is a homologue of the other. A meiotic accident leading to the addition of a whole set of chromosomes (*polyploidy*) is fortunately rare, for the consequences in man are lethal. The addition or loss of a specific chromosome (*aneuploidy*) is much commoner. Losses of parts of chromosomes (*deletions*) and exchange of chromosomal material (*translocations*) can also occur—with varying clinical consequences. The various syndromes resulting from chromosomal imbalance are discussed more fully in Chapter 2, "Cytogenetic Disorders."

One can think of an individual gene as occupying a specific site or *locus* on a given chromosome. As chromosomes exist in pairs, so do genes and if an individual bears two genes that are exactly alike on each of two homologous chromosomes, he is said to be *homozygous.* If the genes are unlike, e.g., if a mutation has occurred in one, the person is said to be *heterozygous.* A special notation is necessary for a male who harbors a gene mutation on his single X-chromosome and he is said to be *hemizygous* for the condition. The two genes, one normal and the other mutated, are *alleles* of each other.

An allele is operationally defined as an alternative form of the normal gene at that genetic locus. A genetic locus should be regarded as a position that can be occupied by a number of different alleles. A series of alleles, each present in such frequency in the general population that it could not be maintained by recurrent mutation alone, is termed a *polymorphism.* The beta chain locus of hemoglobin, may be occupied by the normal gene, the hemoglobin S mutation or the hemoglobin E mutation. All are alleles of each other; indeed there are more than 100 known beta chain alleles. Note that the mutation leading to the new allele can occur anywhere in the beta chain locus, the beta S mutation is in the sixth, the beta E in the 26th position yet both are alleles of the normal beta chain gene.

Since alleles are located on homologous chromosomes, they must separate from each other, or *segregate,* at meiosis; this is Mendel's first law. In other words each meiotic product receives one or the other allele, but never both and never neither. An individual who is a mixed heterozygote for hemoglobin S and hemoglobin E transmits one or the other gene to every offspring. If the other parent is homozygous normal, each child will be a heterozygote, for either hemoglobin S or hemoglobin E (Fig. 3a).

The alpha chain locus for hemoglobin is on another chromosome. Hence the alpha and beta chain genes are *nonalleles* and would be expected to *assort independently* of one another; this is Mendel's second law. If an individual, e.g., is doubly heterogyzous for an alpha chain mutation (hemoglobin G) and a beta chain mutation (hemoglobin S), he may have offspring with both G and S, neither G nor S or one or the other in a theoretical 1:1:1:1 ratio, providing of course that the other parent is normal (Fig. 3b).

LINKAGE

An exception to Mendel's second law of independent assortment is *linkage,* in which the two loci in question are in proximity on the same chromosome. The more closely linked two loci are, the likelier that deviations from the expected

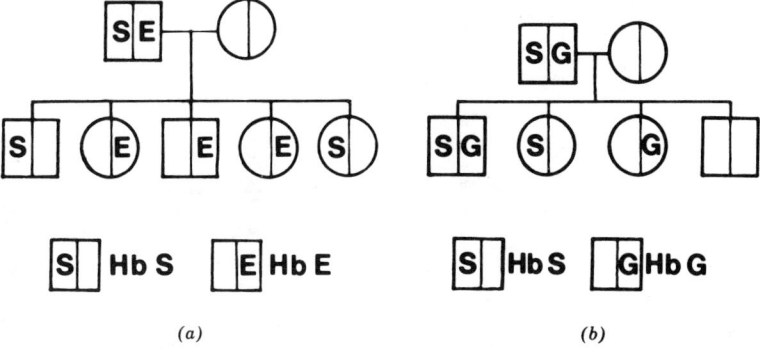

Figure 3. Alleles and nonalleles. *a.)* Beta chain genes of hemoglobins S and E are alleles. Note that offspring of a parent with both abnormal alleles receives either one or the other hemoglobin, but never both and never neither. *b.)* Nonalleles are independent of one another, as with S (beta chain variant) and G (alpha chain variant) hemoglobins. The theoretical 1:1:1:1 ratio in offspring of such a mating as shown here will rarely be evident in any single family, but will be attained when large numbers of such matings are totaled.

1:1:1:1 ratio will occur in offspring. The beta chain locus of hemoglobin A and the delta chain locus of hemoglobin A_2, are known to be contiguous. If an individual were doubly mutant for both beta and delta chain abnormalities only rarely would *crossing over* or recombination occur between homologous chromosomes during DNA synthesis and replication. In general the shorter the distance between two loci, the less frequently they will recombine.

The frequency with which recombination occurs between linked genes is a rough measure of how far apart they are, and linkage studies have led to the construction of genetic maps of man's chromosomes (Fig. 4). This type of study is of more than academic interest, since it may be possible to make an inferential diagnosis of an unfavorable condition (which may not be clinically apparent) by measuring the gene product of another, closely linked gene locus. The greater the distance between two gene loci, the likelier that recombination will occur. Even genes known to be on the same chromosomes, as X-linked genes, if sufficiently far apart behave as though they were on different chromosomes and, for practical purposes, assort independently.

Two factors about linkage are important. First, it is the gene loci that are linked, not the genes. If two mutant gene loci are present in the family on the same chromosome, they are in the coupled phase. The more closely linked, the more frequently the two mutants will stay together in the various offspring. On the other, if the two mutations have entered the pedigree on different chromosomes, the two gene loci (not the genes) are still linked, but in repulsion (Fig. 5). In this latter case the more closely linked the two loci, the likelier they will stay apart. This concept is sometimes difficult to understand (one of the classic texts of general genetics should be consulted for a more detailed treatment of the topic).

Second, linkage is often confused with *association*. Linkage implies a direct

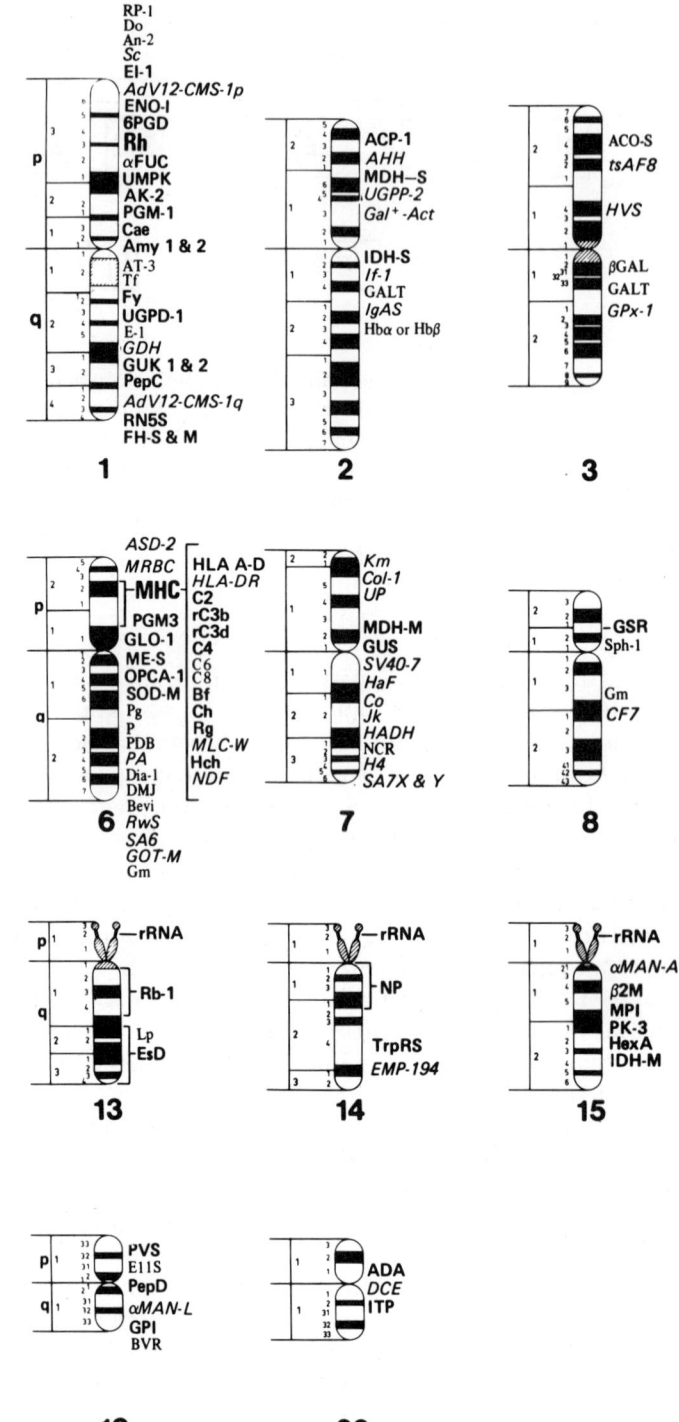

Figure 4. (Left)
(See legend on pages 14, 15.)
(Courtesy of V.A. McKusick, M.D.)

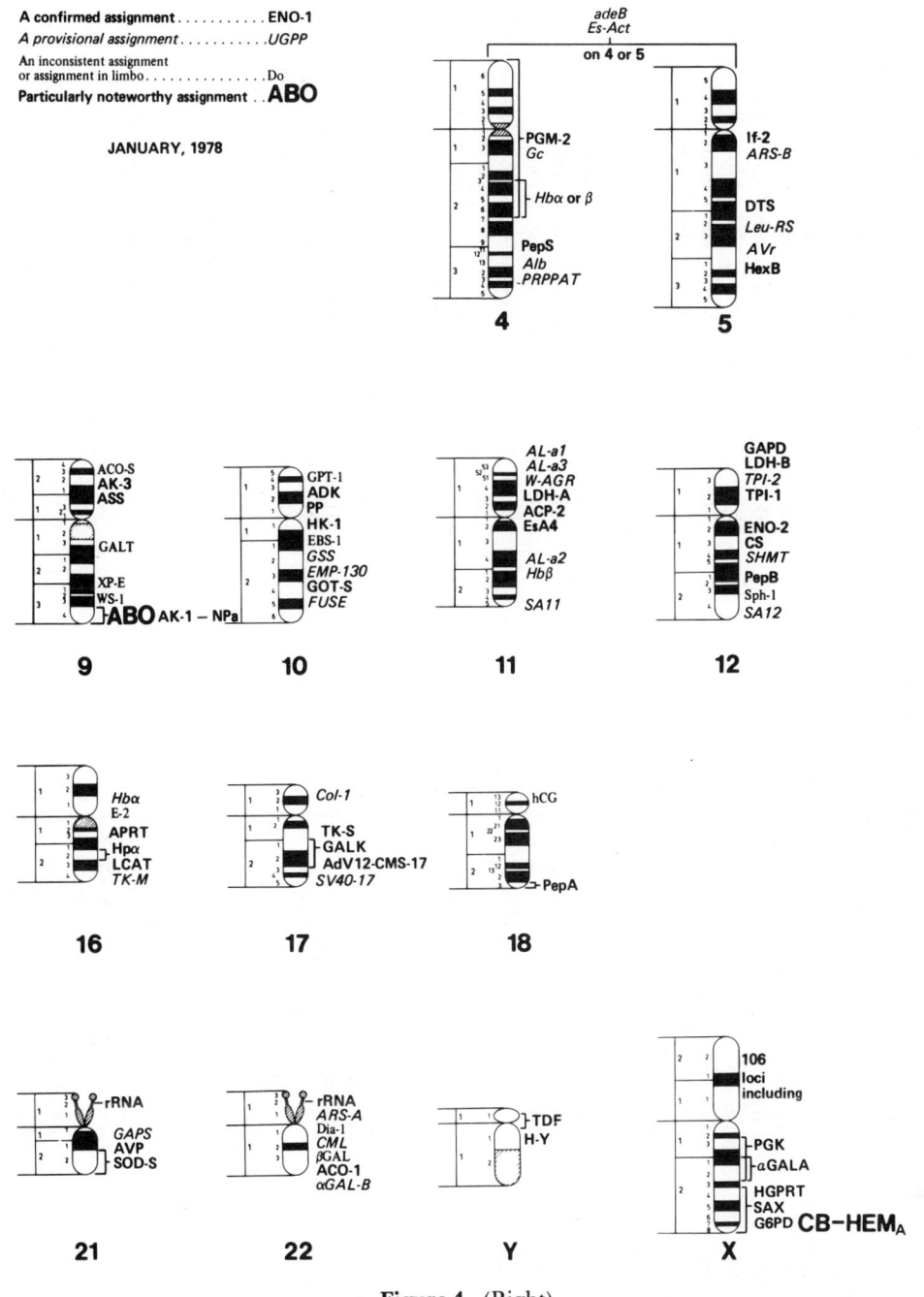

Figure 4. (Right)

Figure 4. The human gene map as currently known. *Key:*

ABO	ABO blood group (chr. 9)	Hex B	Hexosaminidase B (chr. 5)
ACO	Aconitase, mitochondrial (chr. 3)	HGPRT	Hypoxanthine-guanine phosphoribosyltransferase (X chr.)
ACO-S	Aconitase, soluble (chr. 9)		
ACP-1	Acid phosphatase-1 (chr. 2)	HK-1	Hexokinase-1 (chr. 10)
ACP-2	Acid phosphatase-2 (chr. 11)	HLA	Major histocompatibility complex (chr. 6)
ADA	Adenosine deaminase (chr. 20)	Hpα	Haptoglobin, alpha (chr. 16)
adeB	FGAR amidotransferase (chr. 4 or 5)	HVS	Herpes virus sensitivity (chr. 3)
ADK	Adenosine kinase (chr. 10)	H-Y	Y histocompatibility antigen (Y chr.)
AdV12-CMS-1p	Adenovirus-12 chromosome modification site-1p (chr. 1)	If-1	Interferon-1 (chr. 2)
AdV12-CMS-1q	Adenovirus-12 chromosome modification site-1q (chr. 1)	If-2	Interferon-2 (chr. 5)
		IDH-1	Isocitrate dehydrogenase-1 (chr. 2)
AdV12-CMS-17	Adenovirus-12 chromosome modification site-17 (chr. 17)	IDH_m	Isocitrate dehydrogenase, mitochondrial (chr. 15)
		ITP	Inosine triphosphatase (chr. 20)
AHH	Aryl hydrocarbon hydroxylase (chr. 2)	LCAT	Lecithin-cholesterol acyltransferase (chr. 16)
AK-1	Adenylate kinase-1 (chr. 9)	LDH-A	Lactate dehydrogenase A (chr. 11)
AK-2	Adenylate kinase-2 (chr. 1)	LDH-B	Lactate dehydrogenase B (chr. 12)
AK-3	Adenylate kinase-3 (chr. 9)	αMAN	Lysosomal α-D-mannosidase
AL	Lethal antigen: 3 loci (a1, a2, a3) (chr. 11)		
Amy-1	Amylase, salivary (chr. 1)	MDH-1	Malate dehydrogenase-1 (chr. 2)
Amy-2	Amylase, pancreatic (chr. 1)	MDH-2	Malate dehydrogenase, mitochondrial (chr. 7)
ASS	Argininosuccinate synthetase (chr. 9)	ME-1	Malic enzyme-1 (chr. 6)
APRT	Adenine phosphoribosyltransferase (chr. 16)	MHC	Major histocompatibility complex (chr. 6)
AVP	Antiviral protein (chr. 21)	MPI	Mannosephosphate isomerase (chr. 15)
		MRBC	B-cell receptor for monkey red cells (chr. 6)
Bf	Properdin factor B (chr. 6)		
β2M	β2-Microglobulin (chr. 15)	NP	Nucleoside phosphorylase (chr. 14)
		NPa	Nail-patella syndrome (chr. 9)
C2	Complement component-2 (chr. 6)		
C4	Complement component-4 (chr. 6)	OPCA-I	Olivopontocerebellar atrophy I (chr. 6)
C8	Complement component-8 (chr. 6)		
Cae	Cataract, zonular pulverulent (chr. 1)	P	P blood group (chr. 6)
CB	Color blindness (deutan and protan) (X chr.)	PepA	Peptidase A (chr. 18)
Ch	Chido blood group (chr. 6)	PepB	Peptidase B (chr. 12)
CS	Citrate synthase, mitochondrial (chr. 12)	PepC	Peptidase C (chr. 1)

14

DCE	Desmosterol-to-cholesterol enzyme (chr. 20)	PepD	Pepsinogen (chr. 6)
DTS	Diphtheria toxin sensitivity (chr. 5)	Pg	Peptidase D (chr. 19)
		PGK	Phosphoglycerate kinase (X chr.)
El-1	Elliptocytosis-1 (chr. 1)	PGM-1	Phosphoglucomutase-1 (chr. 1)
EllS	Echo 11 sensitivity (chr. 19)	PGM-2	Phosphoglucomutase-2 (chr. 4)
ENO-1	Enolase-1 (chr. 1)	PGM-3	Phosphoglucomutase-3 (chr. 6)
ENO-2	Enolase-2 (chr. 12)	6PGD	6-Phosphogluconate dehydrogenase (chr. 1)
Es-Act	Esterase activator (chr. 4 or 5)	PHI	Phosphohexose isomerase (chr. 19)
EsA4	Esterase-A4 (chr. 11)	PK3	Pyruvate kinase-3 (chr. 15)
ESD	Esterase D (chr. 13)	PP	Inorganic pyrophosphatase (chr. 10)
		PVS	Polio sensitivity (chr. 19)
FH-1 & 2	Fumarate hydratase-1 and 2 (S and M) (chr. 1)		
αFUC	Alpha-L-fucosidase (chr. 1)	Rg	Rodgers blood group (chr. 6)
Fy	Duffy blood group (chr. 1)	Rh	Rhesus blood group (chr. 1)
		rRNA	Ribosomal RNA (chr. 13, 14, 15, 21, 22)
Gal+-Act	Galactose + activator (chr. 2)	rC3b	Receptor for C3b (chr. 6)
αGAL	α-Galactosidase (Fabry disease) (X chr.)	rC3d	Receptor for C3d (chr. 6)
βGAL	β-Galactosidase (chr. 22)	RN5S	5S RNA gene(s) (chr. 1)
GALT	Galactose-1-phosphate uridyltransferase (chr. 3)		
GAPD	Glyceraldehyde-3-phosphate dehydrogenase (chr. 12)	SA7	Species antigen 7 (chr. 7)
		SAX	X-linked species (or surface) antigen (X chr.)
GAPS	Phosphoribosyl glycineamide synthetase (chr. 21)	Sc	Scianna blood group (chr. 1)
Gc	Group-specific component (chr. 4)	SHMT	Serine hydroxymethyltransferase (chr. 12)
GK	Galactokinase (chr. 17)	SOD-1	Superoxide dismutase-1 (chr. 21)
GLO-1	Glyoxylase 1 (chr. 6)	SOD-2	Superoxide dismutase-2 (chr. 6)
GOT-1	Glutamate oxaloacetic transaminase-1 (chr. 10)	SV40-T	SV40-T antigen (chr. 7)
G6PD	Glucose-6-phosphate dehydrogenase (X chr.)		
GSR	Glutathione reductase (chr. 8)	TDF	Testis determining factor (Y chr.)
GSS	Glutamate-γ-semialdehyde synthetase (chr. 10)	TK m	Thymidine kinase, mitochondrial (chr. 16)
GUK-1 & 2	Guanylate kinase-1 & 2 (S & M) (chr. 1)	TK s	Thymidine kinase, soluble (chr. 17)
GUS	Beta-glucuronidase (chr. 7)	TPI	Triosephosphate isomerase (chr. 12)
		TRPRS	Tryptophanyl-tRNA synthetase (chr. 14)
HADH	Hydroxyacyl-CoA dehydrogenase (chr. 7)	tsAF8	Temperature-sensitive (AF8) complementing (chr. 3)
HaF	Hageman factor (chr. 7)		
HEMA	Classic hemophilia (X chr.)	UGPP	Uridyl diphosphate glucose pyrophosphorylase (chr. 1)
Hex A	Hexosaminidase A (chr. 15)	UMPK	Uridine monophosphate kinase (chr. 1)

16   Introduction

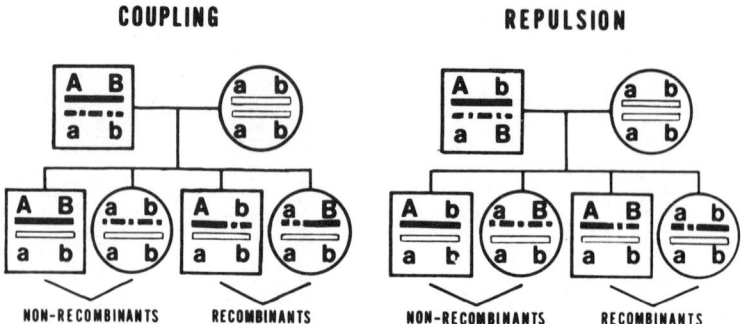

Figure 5. Genetic coupling and repulsion. The concepts of genes in coupling and repulsion are easy to understand if one remembers that gene *loci* are linked, not the genes per se.

physical connection; association is statistical only. Peculiarities in the distribution of ABO blood groups are associated with gastric cancer and peptic ulcer (see Chapter 10, "Genetics of the Gastrointestinal Tract"). The meaning of these associations is unclear, and it has nothing at all to do with linkage. Although the ABO blood groups are simply inherited, neither gastric cancer nor peptic ulcer shows any indication of resulting from a single-gene abnormality; hence linkage is impossible.

This process of homologous chromosomal pairing at meiosis (synapsis) with physical exchange of parts results in new gene combinations' being formed (Fig. 5). Occasionally there is a certain degree of mispairing or slippage between the homologous chromosomes. If recombination should take place within this area, the two resultant chromatids will be *unequal* with respect to the gene loci within the misaligned area (Fig. 6). One of them will contain a *gene duplication*, the other will bear a *gene deficiency*. If the former participates in fertilization, the duplication may become fixed in the population.

The newly duplicated gene and the original gene will ultimately undergo independent mutation, perhaps to the extent that the duplicated gene comes to subserve a new but usually related function. It is felt that the component chains of various human hemoglobins (A, A_2, F and so on) arose in this manner. The alpha chain locus is not located on the same chromosomes with the other

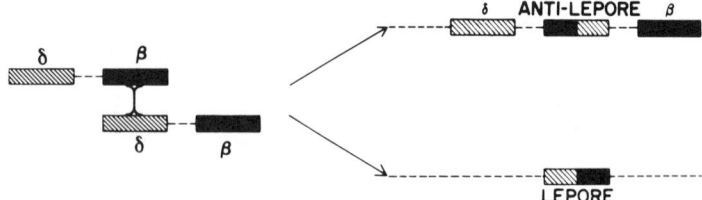

Figure 6. The origins of Lepore and anti-Lepore hemoglobins through nonhomologous pairing and unequal crossing-over, involving closely linked δ and β loci. See Chapter 11 for a discussion of these variants. (Reproduced with permission of V.A. McKusick and F.H. Ruckle, *Science*, 1977)

nonalpha loci (beta, delta, gamma); it likely has been separated from the others in earlier times by nonhomologous chromosomal exchange or translocation. The nonalpha loci are known to be closely linked and hence are the result of a fairly recent evolutionary event. The beta and delta chains are known by classic genetic studies to be immediately adjacent; they differ in structure by only a few amino acids, and hemoglobin A$_2$ occurs only in primates, not in lower animals.

All this evidence supports the contention that the delta chain gene locus has only recently been derived from the beta locus, presumably by the phenomenon of unequal crossover. The cytochrome enzyme system and the structural genes coding for antibodies are among the proteins also thought to be derived from this process. Recombination, with or without gene duplication, has obviously been and likely continues to be of considerable importance in man's evolution.

GENES IN FAMILIES

Familial diseases are not necessarily *hereditary* since they may be related to infectious or toxic vectors. Similarly the term *congenital* does not imply that a condition is genetic, e.g., rubella syndrome is congenital, but hardly hereditary. By the same token, genetic disease may not be congenital in the sense of its observable effect's, the *phenotype,* being evident at birth. Huntington's chorea is an established heritable entity with an average symptom-onset in the fourth decade of life. It is important to have these distinctions in mind before we begin a study of the transmission of familial genetic disease.

This section will be concerned with simple, unifactorial, single gene or Mendelian inheritance (the terms are used synonymously). A disorder may be dominant or recessive, depending on whether it is *clinically* evident in the heterozygote or homozygote, respectively. Use of the term dominant or recessive does not confer special properties on the gene; it simply reflects the way in which we look at the phenotypic effect of the gene. It is a tacit assumption in medical genetics that homozygosity for dominant conditions implies lethality. Obviously this line of reasoning applies only to diseases, since many protein polymorphisms can be regarded as dominant, or at least codominant, if considered in terms of electrophoretic, enzymatic or chemical detectability.

Use of the terms autosomal or X-linked (sex-linked) is obvious, indicating that the gene in question is located either on one of the autosomes (without specifying which one) or on the X-chromosome. The Y-chromosome contains only male-determining genes and perhaps a gene or genes coding for certain male antigens, but apparently nothing else. Figure 7 shows the common notations used in composing a pedigree. all of the single-gene disorders are rare, population frequencies being on the average of 1/50,000 births. In any one family, however, there may be a high risk of recurrence, the rate depending on the mode of inheritance.

Autosomal Dominant Traits

A pedigree showing typical autosomal dominant inheritance is shown in Figure 8. Note that transmission occurs from either sex to either sex, and males and

• PEDIGREE SYMBOLS •

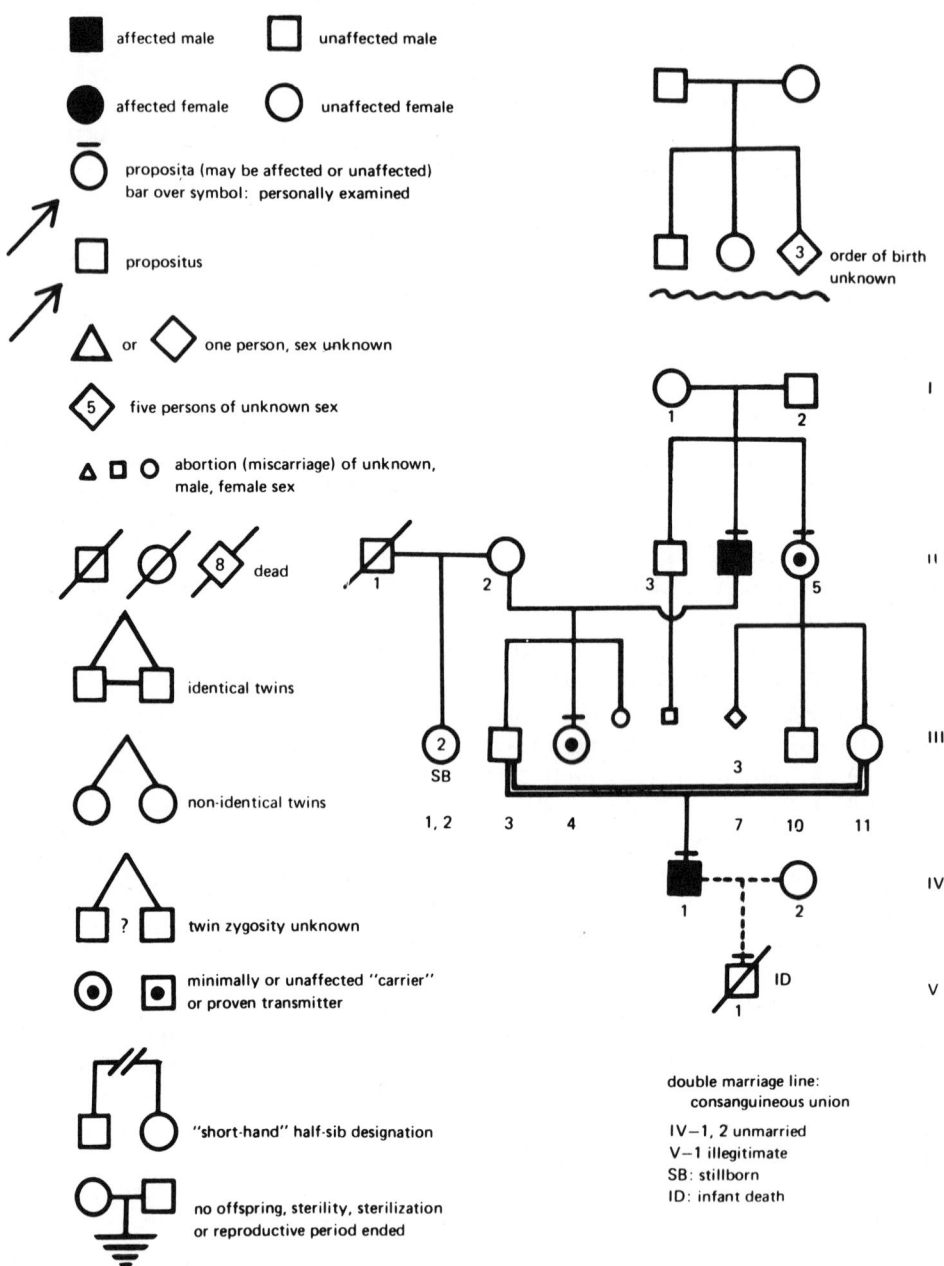

Figure 7. Symbols generally used when constructing a pedigree. (Courtesy of *Am. J. Med. Genetics*)

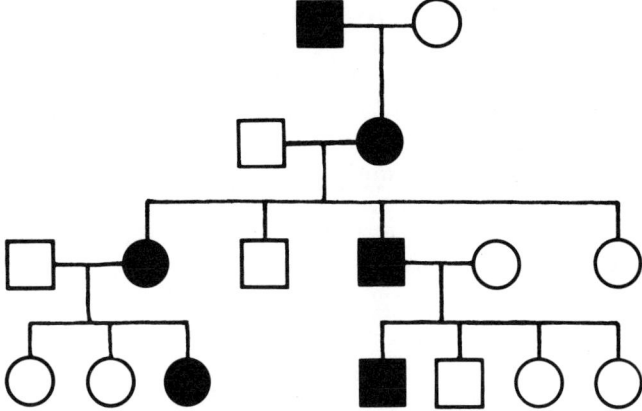

Figure 8. Typical autosomal dominant pedigree. Both sexes are affected, and transmission occurs through both sexes. The heterozygote is affected.

females are equally at risk. By definition, only one of the two alleles is abnormal and an affected individual has a 50% chance of transmitting the condition to an offspring. The pedigrees are "vertical" and multiple generations are characteristically affected. It is important to remember that chance has no memory and that each child born to an affected parent has an independent 50% risk of developing the condition. The likelihood, e.g., of having five children with an autosomal dominant condition is $(\frac{1}{2})^5$ or 1/32. This means that of all the families with at least five children at risk for the condition, one in 32 or about 3% of them will have five affected children. The 50% proportion may not hold in any one family, being based on the potential offspring of all affected individuals in the population. The presence of two of three or only one affected of five children in a family does not negate autosomal dominant inheritance.

Of all the terms in genetics *penetrance* and *expressivity* seem to cause the most difficulty. They are used when the exact biochemical basis of the gene defect is unknown, and the clinician must rely solely on external phenotype. Penetrance is an all-or-none phenomenon and refers to the detectability of the gene. If, e.g., both a man and his grandson had the Marfan syndrome—a well-known autosomal dominant condition—but the man's daughter (the boy's mother) had no clinical symptoms of the condition, the gene would be said to be nonpenetrant or incompletely penetrant (synonymous terms). The mother obviously transmitted the gene (to invoke a new mutation in her son would be absurd), but the clinician's ability to detect this gene is currently limited to physical examination, since the basic gene defect is unknown. It would appear, however, to reside in disordered connective tissue. The reason why the mother can harbor and yet not manifest the gene is unknown, but is presumably related to other modifying factors in her genome that protect her from the consequences of the Marfan gene.

One may read about a gene's being 50% or 80% penetrant. This means that, given all opportunities for the gene to express itself when it is identified by pedigree analysis, it is detectable by the usual means only 50% or 80% of the time. If the gene is penetrant, it may be *variably expressed*. Expressivity is analogous to clinical grade of severity. If in the foregoing example the mother showed only dislocation of her ocular lenses without any of the other stigmata of the Marfan syndrome, the gene would be said to be penetrant, but variably expressed. These two terms, penetrance and expressivity, are used most often with autosomal dominant conditions since, generally speaking, less is known about the basic gene defect in this class of conditions than in recessively inherited disorders. If the means were available to detect dominant disorders biochemically, the terms would fall into disuse.

Pleiotropism and Anticipation
Another frequent concept in descriptions of genetic disease is *pleiotropism*. It refers to the multiplicity of effects of a single gene, perhaps in multiple-organ systems. To use the Marfan syndrome again as an example, the gene defect, when fully expressed, comprises ectopia lentis, arachnodactyly and mitral or aortic insufficiency or both—often with aortic dissection, kyphoscoliosis, inguinal hernia and other connective-tissue defects. Multiple systems—notably eye, skeletal and cardiovascular—are involved. These then are pleiotropic manifestations of the gene. For the Marfan syndrome, the common systemic denominator is likely to be a connective tissue abnormality.

In other conditions (and this holds for recessives as well as dominants), the underlying basis for pleiotropism is obscure. It is important to note that secondary effects are not pleiotropic effects. Vascular thrombosis in sickle cell anemia is secondary to the deformed erythrocyte and not due to an action of the hemoglobin S gene outside the beta chain. Obviously it is not always possible to differentiate pleiotropism from secondary change.

Anticipation, another term used with autosomal dominant conditions, implies that with succeeding generations the disorder occurs at a progressively earlier age. Anticipation has no biologic basis, and the term should be removed from the genetics literature. If a condition is detected earlier in a later generation, it is usually because it is more readily diagnosed in younger than in older individuals, or because, having discovered an older affected individual in a family, the clinician is seeking additional cases and is more attuned to the possible diagnosis.

Autosomal Recessive Traits

Autosomal recessive conditions characteristically occur in equal proportions in multiple male and female siblings whose parents are normal (Fig. 9). The sibs are then homozygous for the defective gene and their parents are obviously heterozygous. Such traits are termed recessive for precisely that reason, namely, the heterozygote generally shows no evidence of the condition, although rarely a carrier of the abnormal gene in single dose may show a slight manifestation. Autosomal recessive disorders are often inborn errors of metabolism, resulting from defects in genes coding rather for enzymes than for structural proteins. Mutations in genes coding for the latter polypeptides are generally dominantly (or codominantly) inherited.

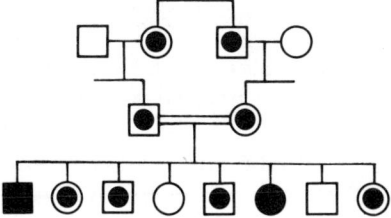

Figure 9. Autosomal recessive inheritance. The parents are unaffected and heterozygous: the offspring who are affected are homozygous for the mutant gene. Note consanguinity.

Parents with a child harboring an autosomal recessive condition run a 25% risk of having another affected child with each subsequent pregnancy. Moreover, on statistical grounds one-half their offspring will be asymptomatic heterozygotes like themselves, and the remaining 25% will be homozygous normal. Considerable effort is currently being expended on heterozygote detection, since it offers a means of identification of high-risk families prior to the birth of an affected child. This topic is considered in more detail in subsequent chapters as it relates to specific diseases.

Consanguinity is occasionally found in the families of persons with autosomal recessive traits. Generally, the rarer the condition, the likelier that the parents will be consanguineous or inbred. Inbreeding does not increase the frequency of the gene in the population, but it does increase the frequency of the homozygote and thus the number of clinically affected individuals. By the same token when evaluating a family history, if parental consanguinity is uncovered, the investigator should consider the possibility that the condition may be a rare autosomal recessive disorder. Extensive consanguinity can give rise to "pseudodominant inheritance", i.e., an affected individual who marries an unaffected but related and therefore heterozygous carrier may have an affected child, and the pedigree appears to show dominant inheritance.

There are many more heterozygous carriers of a given gene in the population than there are homozygotes. One means of determining the frequency of heterozygosis is the Hardy-Weinberg law. It states that given two alleles at a single locus, by convention termed p and q, the frequency of the three possible genotypes (homozygous p, heterozygous pq and homozygous q) will be related to one another according to the equation $p^2 + 2pq + q^2 = 1$. Whereas this relationship holds only under ideal circumstances—i.e., no mutation, no selective advantage of one genotype versus the others and so forth—it is a sound working means to arrive at the incidence of heterozygotes for counseling purposes. If, e.g., phenylketonuria (PKU) has a population incidence of 1/10,000 (q^2), then the gene frequency, q, is the square root of this figure, or 1/100. By definition $p + q = 1$, so that p = 99/100, or for practical purposes 1. The frequency of the PKU heterozygote is then 2 pq, or 2 (1) (1/100 or 1/50). Thus for every homozygote with PKU in the population, there are 200 heterozygotes. Obviously most families in which both parents are carriers have no affected children, an important fact to remember when reading about the feasibility of "eliminating" certain genes by various techniques of reproduction limitation.

X-Linked Recessive

These conditions are, for practical purposes, limited to males who are *hemizygous*. In any X-linked recessive pedigree (Fig. 10) only males are affected, and they are related to one another through the maternal side of the family. Any evidence of male-to-male transmission rules out X-linked inheritance because males transmit their Y-chromosome (and hence maleness) to their sons. The heterozygous females are *carriers* of the defective gene and transmit the disorder to one-half their sons. By the same token, one-half their daughters will be heterozygous like themselves. Because the condition is recessive, carriers usually show no clinical abnormality. There are exceptions, however. If the condition is not so severe as to preclude reproduction by affected males, e.g., then the daughter of an affected male and a carrier female could be affected since she would be homozygous.

Second, if the girl were simultaneously to have the Turner syndrome and only one X-chromosome, the latter bearing the defective gene, she would be affected. A third possibility relates to the so-called *Lyon hypothesis*. This interpretation attempts to account for the fact that females with two X-chromosomes do not have twice as much of some X-linked gene product as have males with only one X-chromosome. A system of *dosage compensation* has been evolutionarily derived whereby one of the two X-chromosomes in every female becomes essentially inactivated in early embryogenesis. The inactive X-chromosome is morphologically represented as the *sex-chromatin* (Barr body), study of which is useful in the diagnosis of sex chromosome aneuploidy (The general rule is that the number of sex chromosomes equals the number of Barr bodies minus one). Inactivation is random, i.e., it is as likely to happen to a woman's paternally derived, as to her maternally derived, X-chromosome.

In a small proportion of females heterozygous for an X-linked disorder, the normal X-chromosome may become inactive in enough of the cellular anlage that, for practical purposes, the woman behaves clinically like an affected male. Rare carriers of Duchenne type muscular dystrophy, e.g., may show considerable muscular disability. Women who are carriers for X-linked albinism may have a retina that appears speckled, with some areas showing normal pigmentation and others showing no pigmentation. Fortunately the women affected with

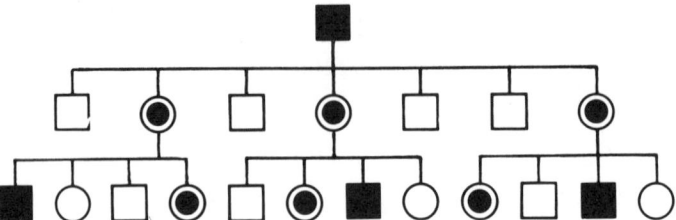

Figure 10. X-linked recessive pedigree showing, as expected, only affected males. Remember that in X-linked disorders where condition is not severe enough to preclude survival and reproduction, *all* daughters of affected males will be carriers. None of his sons, however, will be affected.

an X-linked recessive disorder by the Lyon hypothesis are very few, but the phenomenon itself may allow for carrier detection in families containing only one affected male in which the decision must be made whether the affected male resulted from a gene defect transmitted from his mother or from a new mutation. In the former instance the risk of recurrence is 50%; in the latter it is 0.

X-Linked Dominant Traits

Pedigrees of X-linked dominant traits resemble autosomal dominants except that there is no male-to-male transmission (Fig. 11). In large or in pooled pedigrees, twice as many affected females as males may be discerned. A little thought reveals why: affected females transmit the condition to one-half their daughters and to one-half their sons, whereas affected males transmit to all their daughters and to none of their sons. In keeping with the Lyon hypothesis, affected females are usually less severely affected than males, since it would be unusual for all the requisite cells to inactivate the normal X-chromosome.

BAYESIAN ANALYSIS

The foregoing methods of estimating the risk of recurrence of a dominant or recessive trait are based only on the statistics of allele segregation and transmission. In 1763 an English clergyman-scientist, T. Bayes, devised a method of comparing two probabilities (one that an event will occur, the other that it will not), using *all* available information (1). His method takes into account not only allele segregation but also all the remaining family history, whether negative or positive. The theorem states that the probability of some event a, when condition b applies, is equal to the probability of a × the probability of b.

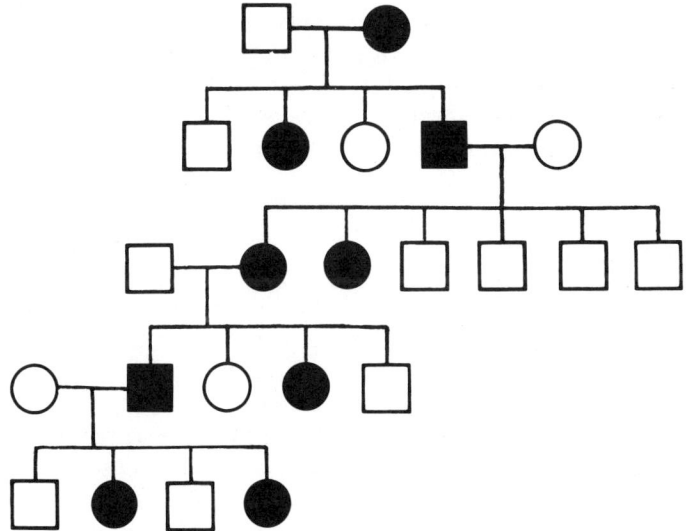

Figure 11. X-linked dominant pedigree.

Certain terms describe the use of the method, including 1) *prior probability,* which refers to the probability of the event in question disregarding any ancillary information; 2) *conditional probability,* which is the probability that the particular condition(s) will apply if the event in question is assumed to be true; 3) *joint probability,* which is the probability in absolute terms that both the event in question and the particular condition(s) are true (the joint probability is equal to the product of the prior and conditional probabilities); and 4) *posterior probability,* which is equal to the joint probability of one of two events compatible with some particular condition divided by the sum of the joint probabilities of the two events.

This type of calculation, although seemingly complex, can be reduced to simple mathematics with a little thought. Interested readers should consult the works of Emery (1) and Murphy and Chase (2) for a more detailed exposition of this topic. Bayesian analysis may be utilized to calculate risks in specific genetic situations:

1. Determination of the carrier status of a female relative of a male affected with an X-linked disorder in which normal first-degree male relatives also exist (see following paragraph).
2. Utilization of laboratory carrier test values for X-linked and autosomal recessive carrier probability determination (3).
3. Estimation of risk of being affected with a condition with incomplete or age-dependent "penetrance" (Huntington's chorea, e.g.).

The Bayes principle in X-linked disorders is illustrated in the pedigree shown in Figure 12a. The consultand, C, has two normal sibs, which reduces the possibility that her mother is a carrier even though her grandmother, with two affected sons, is almost certainly one. The consultand's mother then has a prior probability (a) of ½ of being a carrier. The conditional probability of both sibs, being normal if the mother were a carrier is ½ × ½ = ¼, and if she were not a carrier, it would be 1. The joint probability in the former case is ⅛ and in the latter instance ½. The posterior probability that the consultand's mother is heterozygous is one-fifth, and the consultand therefore has a ¹/₁₀ probability (½ × ⅕) of being a carrier and only a 1/20 (¹/₁₀ × ½) chance of having an affected male child. It is readily apparent that this risk is far less than the ½ chance expected when Mendelian ratios are used (Fig. 12b).

SPORADIC CASES

Sporadic or isolated examples of a given condition provide no information about its heritability. A sporadic case may be the result of a new dominant mutation, the first instance of a rare autosomal recessive in a family or, if it occurs in a male, an X-linked recessive trait that has been carried by unaffected females for many generations and by chance has not affected any at-risk males. Sporadic cases of PKU are obviously heritable because the genetic nature of the condition has been determined previously in family studies. Other sporadic disorders—e.g., congenital cataracts—may have resulted from an intrauterine insult and not be at all heritable. Conditions that are environmentally induced mimics of gene-

Fundamentals of Clinical Genetics 25

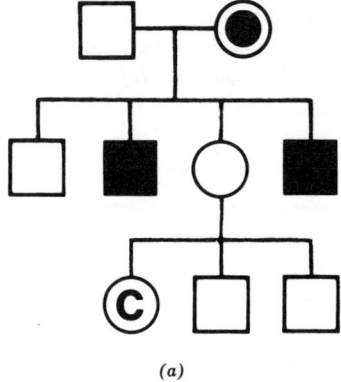

(a)

	Consultand's Mother Heterozygous	Consultand's Mother Not Heterozygous
Prior probability	½	½
Conditional probability	$(½)^2 = ¼$	1
Joint probability	⅛	½
Posterior probability	$\dfrac{⅛}{⅛ + ½} = ⅕$	$\dfrac{½}{⅛ + ½} = ⅘$

(b)

Figure 12. Example of Bayesian analysis as it applies to an X-linked disorder. *a.)* Pedigree (consultand female indicated by ©). *b.)* Risk of being carrier is considerably reduced, over Mendelian ratios. (Adapted from McKusick, V.A., *Human Genetics.* Prentice-Hall, Englewood, N.J.)

tic disease are called *phenocopies*. Patients with these abnormalities must be examined closely and compared to others harboring a similar constellation of defects if the true cause of the condition is to be determined.

Even if the sporadic case is a heritable entity, the mode of inheritance may be obscure. Some conditions like retinitis pigmentosa or peroneal muscular atrophy, e.g., are inherited in a number of ways. In general when such a situation obtains, the autosomal recessive form is the severest, the autosomal dominant the mildest and the X-linked form intermediate.

Careful evaluation of a series of sporadic cases (or for that matter even familial ones) frequently reveals *genetic heterogeneity*. This means that detailed evaluation of a disease category by clinical, genetic and biochemical means may reveal not one but several diseases, not all of which need be genetic. Each is a separate entity with a distinct risk of recurrence and often a characteristic clinical course. Multiple examples of genetic heterogeneity will be found throughout this textbook.

Chromosomal studies are of no value in single-gene disorders, since the basis for most of them is felt to be a point mutation. It is true that some complex chromosomal rearrangements may be transmitted in a manner simulating Mendelian inheritance and that some simple aneuploid states, such as trisomy 21, can theoretically reproduce in kind, but the recurrence risks in general are less than Mendelian risks. If in doubt as to whether a given condition has a chromosomal or a single-gene basis, perform chromosomal studies. To undertake such studies on patients with well-recognized diseases like the Marfan syndrome or PKU is a waste of time and money.

POLYGENIC INHERITANCE

In some conditions with a familial tendency the risk of the disorder is greater than the population incidence, but substantially less than that for single-factor inheritance. These conditions are termed *multifactorial* and probably result from the action of multiple genes (hence polygenic) of small effect, in contrast to the unifactorial disorders in which a single gene has a large effect. The term multifactorial inheritance properly includes the influence of the environment also, but the genetic component of the condition and that which is due to external forces are so intertwined as to be inseparable. The greater the genetic contribution to the cause of disease, the greater the *heritability* of the condition, a figure usually calculated from the incidence of the condition in relatives when compared to members of the general population. Normal traits, such as intelligence, stature and skin color, are felt to be multifactorial since they depend on both a person's genetic background and his environment.

There are pathologic conditions in man, of which the commonest are probably the usual nonspecific congenital malformations, such as congenital heart disease, cleft lip and palate, and club foot that follow a multifactorial mode of familial transmission. It is assumed that for some disease a person possesses a certain genetic liability that will be expressed only if environmental circumstances are appropriate. The environment may raise or lower the threshold of expression. If one plots the liability to a given disease (or for that matter a normal trait) in the population, the resultant curve is normally distributed. One tail of the curve represents the population incidence of the condition in question.

If one plots the same liability among relatives a normal curve is also generated, but the curve is shifted so that more relatives exceed the threshold and thus develop disease (Fig. 13). Such curves can be generated for liability (or risk) of disease in first or second generations or in relatives of more distant consanguinity. Genetic consultation is usually sought because the marital couple either has had a child with a multifactorial trait, or one of the prospective parents has the condition. For practical purposes, for conditions showing no sex predilection, the risk of recurrence for a subsequent affected child, given either an affected parent or a previously affected child, is roughly $1/20$, or 5%. This risk figure must be compared to the risk for the birth of an affected child, given a negative antecedent family history, i.e., the population incidence of the disease in question.

If we assume, e.g., that the incidence of congenital heart disease in the general population is about $1/1000$, then this is the risk that any person with a negative familial history will have a child with congenital heart disease. If a first-degree relative is affected (parent or previous child), the risk becomes about 50 times greater for a subsequent pregnancy ($1/1000 \times 50 = 1/20$, or 5%). It is important to remember that the *absolute* risk remains low in comparison to unifactorial

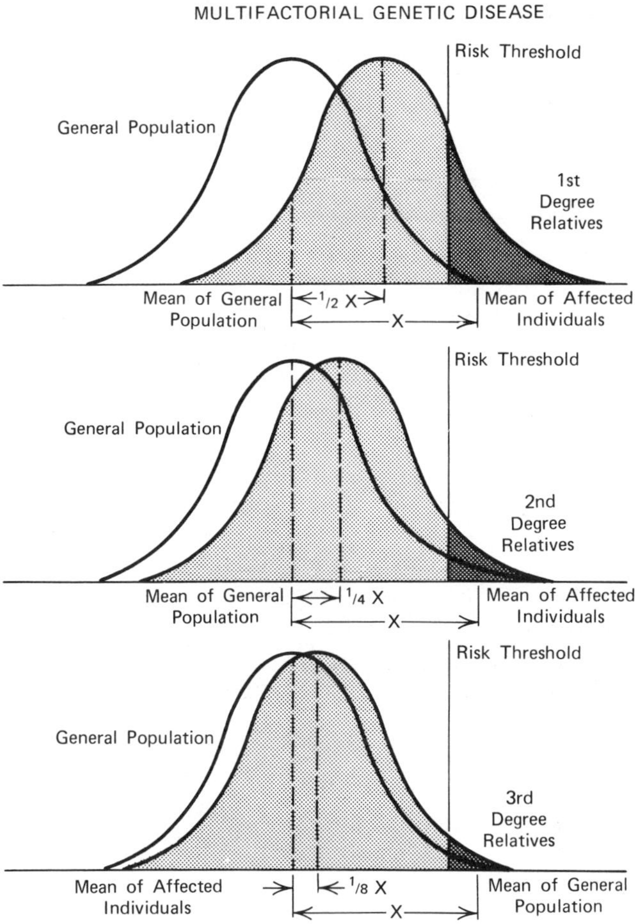

Figure 13. Increased risk to relatives of multifactorial genetic disease. The more distant the affected relative, the lower the risk. (Courtesy of C.O. Carter)

disorders in which 25–50% recurrence risks are encountered (Table 3). Unfortunately unlike the situation in single-gene disorders wherein the risk of recurrence remains the same with each succeeding birth, with multifactorial disorders the risk increases with each affected child. Thus, if both a parent and a child or two children were affected, the risk to a third individual in the immediate family would be roughly 15%, or three times what it would be if one person were affected.

The greater the number affected, the greater the subsequent risk. In rare families with great liability for a given disease, the recurrence of risk may approach that found in Mendelian disorders. In this situation one must beware of a misdiagnosis and be certain that the condition is truly multifactorial and not a single-gene disorder, the clinical features of which may be similar. The effect of affected second degree and more distant relations on risk of recurrence is finite but small and for practical purposes can probably be ignored. Computer pro-

Table 3. Approximate Recurrence Risks for Some Common Congenital Malformations

Abnormality	Male Proband (%)	Female Proband (%)
Cleft lip/Cleft palate*		
Pyloric Stenosis:		
Son	5	20
Brother	4	10
Daughter	3	8
Sister	3	4
Hirschsprung's Disease:		
Short (pelvic colon)		
Affected segment		
Brothers	5	8
Sisters	1	3
Long affected segment		
Brothers	7	18
Sisters	10	9
Entire intestine below duodenum affected	25	25
Neural tube defect:†		
Spina bifida	1	1
Anencephaly	1	1
Congenital dislocation of hip		
Brothers	2 approx	1
Sisters	10 approx	7
Club foot (Talipes equivaras)	2	2
Scoliosis		
Early onset cases	2	2
Adolescent onset		
Brother		2
Sister		10

*See Chapter 10, "Genetics of the Gastrointestinal Tract."

†Geographically variable; above are United States figures, London would be 2%; risk after an affected child is 10 × population risk.

grams have been generated to facilitate the calculations resulting from this additional information.

For disorders in which the affected individual tends to be more often of the same sex, the risk of recurrence is modified, depending on the sex of the proband. Pyloric stenosis is five times commoner in males. The risk that an affected father will have an affected son is about 5% and that he will have an affected daughter about one-half this figure. If the index case is female, however, the risk to her sons is nearly 20% and to her daughters about 8%. The usual explanation for this discrepancy is that the less frequently affected sex is more extreme from a genetic point of view (has more "risk" genes) when they are affected, compared to normal members of their sex; hence the recurrence risk to their offspring is greater. This so-called explanation begs the question; nonetheless the phenomenon is real and must be taken into account when counseling.

ANTENATAL DIAGNOSIS AND TERATOLOGY

Accurate genetic counseling requires first and foremost an accurate diagnosis. The relevant family history is often invaluable in computing accurate recurrence risks to enable the informed family to plan future pregnancies. Another approach to family planning is antenatal diagnosis. Various techniques may approach the fetus or conceptus directly by amniocentesis or visualization or may involve study of the fetal effect on maternal physiology and metabolism. All these measures are directed toward intrauterine diagnosis of a given deleterious condition, so that an affected fetus may be therapeutically aborted. Relatively few conditions can be diagnosed antenatally and most of them require invasive techniques, although current figures suggest that amniocentesis is relatively innocuous in experienced hands. Chromosomal disorders can be diagnosed as can some inborn errors of metabolism, although only those latter conditions can be recognized, the genotype of which is expressed in cultured fibroblasts. Some hemoglobinopathies can be detected by direct sampling of placental vessels. Antenatal diagnosis of neural-tube defects, such as anencephaly and open meningomyelocele, may be accomplished by measurement of amniotic fluid α-fetoprotein. Ultrasonic scanning and direct fetal visualization may be useful for some birth defects with gross derangement of body form. Skeletal X-rays probably have only limited utility because the age of ossification of most pertinent structures is late. Focused laser techniques offer the possibility of a three-dimensional picture of the fetus. The field is rapidly changing, and no attempt will be made here to delineate the various procedures, their application and their limitations. Each chapter in this textbook contains references to various conditions that can be diagnosed by antenatal maneuvers.

The previous sections emphasized the role of genetics in the etiology of disease. Obviously most diseases are acquired and far outnumber genetic disorders. Clinicians called on to explain the pathogenesis of complex congenital malformations are often in a diagnostic quandary, particularly if they are asked to give risks of recurrence. For those malformation syndromes known to follow Mendelian inheritance patterns, or to be multifactorial, genetic counseling usually poses no problem, although even the most experienced counselor may occasionally supply erroneous risk figures, since it is difficult for one to be fully informed of every "new" syndrome. It is important to be aware, however, that certain environmentally induced conditions have a recurrence risk, given that the teratogenic influence is in a subsequent pregnancy. Congenital syphilis is a case in point. The thalidomide problem graphically demonstrated the effect of a single drug ingested by the mother at a critical period of limb morphogenesis. Although no definite cause-and-effect relationship can be established between most drugs and fetal dysmorphogenesis, a few conditions are recognized (Table 4).

The *fetal alcohol* syndrome may occur in the offspring of alcoholic mothers and

Table 4. Fetal Teratologic Syndromes

Syndrome	Major Clinical Features	Risk to Exposed Fetus
Diphenylhydantoin	Mild-to-moderate growth deficiency; borderline-to-mild mental deficiency; unusual facies: wide anterior fontanelle, metopic ridging, ocular hypertelorism, broad depressed nasal ridge, low-set abnormal ears, broad alveolar ridge, cleft lip and palate; hypoplasia of distal phalanges with small nails.	10% for syndrome; perhaps as high as 33% for isolated features.
Trimethadione (tridione)	Growth deficiency, mental deficiency variable with speech disorder prominent, unusual facies: mild brachycephaly, midface hypoplasia, short upturned nose, broad and low nasal ridge, mild synophrys, strabismus, ptosis, epicanthus, cleft lip and cleft or high arched palate; cardiovascular defects may be severe with septal defects and tetralogy of Fallot.	67% of known exposed pregnancies significantly affected.
Warfarin (coumarin)	Variable mental deficiency, hypotonia, seizures, hypoplastic nose with low nasal bridge, stippled epiphyses, vertebral and nasal cartilage.	All mothers reported required anticoagulant treatment for prosthetic mitral valves.
Aminopterin (folic acid antagonists)	Growth deficiency, mild mental deficiency, severe hypoplasia of frontal bone and other cranial bones with synostosis of lambdoid or coronal sutures (microcephaly), broad nasal bridge, prominent eyes, epicanthal folds, micrognathia, cleft palate, low-set ears, limbs with mesomelic shortening, clubfoot and synostosis.	Most exposed fetuses expire prenatally or in neonatal period; survivors (2) showed marked growth deficiency.

Rubella	Growth deficiency, mental deficiency (microcephaly); deafness, cataract or glaucoma and other eye defect; patent with ductus arteriosus, septal defects or other cardiovascular anomaly may have hematologic, renal or other complications in infancy.	First trimester fetal exposure results in ⅓ abortion, ⅓ affected fetus; early second trimester exposure still carries modest risk for mental deficiency and less severe eye problems.
Alchohol	Growth deficiency, moderate mental deficiency; eye-hand coordination and fine motor function affected; short palpebral fissures, mild microcephaly, maxillary hypoplasia; mild joint anomalies, altered palmar creases, cardiac murmur disappearing at one year.	Maternal intake of 60 ml or more of ethyl alchohol/day correlated with production of syndrome; many workers feel that smaller amounts still carry some risk.

comprises pre- and postnatal growth deficiency, mental and motor retardation and characteristic facial anomalies. The *fetal warfarin* syndrome includes developmental and mental deficiency, seizures, hypotonia, optic atrophy and stippled epiphyses on radiographic evaluation. There is a significant association between ingestion of *trimethadione* by the mother and offspring with a syndrome of diverse mental and physical handicaps, congenital heart disease and facial and genital abnormalities. Another anticonvulsant agent, diphenyl hydantoin, has been incriminated in a syndrome of growth deficiency, craniofacial and other nonspecific abnormalities, and characteristic hypoplasia of the distal phalanges.

Amphetamines may cause placental dysfunction with repeated abortions; maternal narcotic abuse may lead to fetal distress or neonatal withdrawal symptoms. Even the *nicotine* inhaled by mothers who smoke cigarettes may be a cause of infant low-birth weight and an increased liability to spontaneous abortion. Other agents may have more organ-specific teratogenic effects; they are discussed in the relevant chapters. It is important to recognize that not all infants of all mothers ingesting the offending drug necessarily develop the full, or even the partial, syndrome. The reasons for this finding are unknown, but may in part be related to remote genetic factors. In any case the risk of the mother's giving birth to another affected child, given her contined use of the offending drug, is in general high (probably in the range of 10–100%, depending on the agent).

REFERENCES

1. Emery AEH: Methodology in Medical Genetics. Edinburgh, Churchill-Livingstone, 1976
2. Murphy EA, Chase GA: Principles of Genetic Counseling. Chicago, Yearbook, 1975.
3. Gold RJM, Maag UR, Neal JL, Scriver CR: The use of biochemical data in screening for mutant alleles and in genetic counseling. Ann Human Genet 37:315, 1974

Part 2
MULTISYSTEM DISEASE

2
Cytogenetic Disorders

Robert L. Summitt

The modern era of human cytogenetics began with the discovery in 1949 by Barr and Bertram (1) of the sex chromatin mass, a dark-staining mass of chromatin in the nuclei of somatic cells of female mammals. This was followed by the observation that cells of certain females with short stature, primary amenorrhea and sexual infantilism (the Turner syndrome) lacked the sex chromatin mass (2), whereas cells of certain males with hypogonadism (the Klinefelter syndrome) contained a sex chromatin mass (3). The most significant discovery presaging the modern era of human cytogenetics was the demonstration by Tjio and Levan (4) that the normal human somatic chromosome number is 46, not 48 as previously claimed. The first abnormality described in man was an extra chromosome 21 in cells of patients with the Down syndrome. This observation was made by Lejeune and his co-workers in 1959 (5), and was followed by the description in rapid succession of abnormalities in the Turner (6) and Klinefelter syndromes (7) and in diverse children with mental retardation and multiple congenital anomalies.

CHROMOSOME IDENTIFICATION

Chromosomal analysis is ordinarily conducted on cells arrested at the metaphase of mitosis. A metaphase chromosome consists of two *chromatids* joined at a constriction called a *centromere*. Figure 1 shows the three major types of chromosomes. The *metacentric* chromosome is one in which the centromere is nearly equidistant from the terminal ends of the chromosome. An *acrocentric* chromosome is one in which the centromere is very near but not at one end of the chromosome. In a *submetacentric* chromosome the centromere is between a central and a terminal position. In a chromosome in which the two arms are of unequal length, or in which the arms are of equal length but can be differentiated, the short arm is designated *p*, the long arm is designated *q* and the centromere is designated *c*. Attached to the short arms of the acrocentric auto-

---

Supported in part by Special Project No. 900, Division of Health Services, MCHS, HSMHA, DHEW, and by a grant from the National Foundation–March of Dimes.

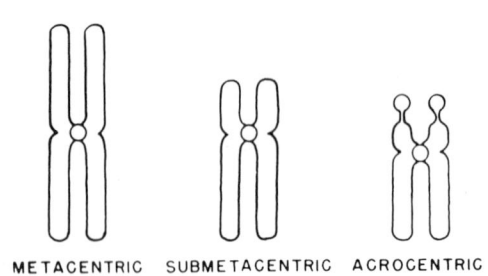

Figure 1. Three major morphologic types of chromosomes. On the left is a metacentric chromosome (centromeric constriction at or near center of chromosome length), on the right is an acrocentric chromosome (centromeric constriction near one end), and in the center is a submetacentric chromosome (centromeric constriction intermediate in position between those in metacentric and arocentric chromosome).

somes of man by a small thin segment of chromatin are highly condensed masses of chromatin called satellites and designated *s*.

During the early years of routine study of human chromosomes, chromosomal identification was based on size, position of centromere and features such as satellites and secondary constrictions. However, specific identification of every chromosome in the karyotype was not possible from the staining techniques available at that time. In 1962 autoradiography was introduced, allowing identification of the late replicating X-chromosome (the one that produces the X-chromatin mass), but was of limited value in identification of autosomes.

In 1970 Caspersson and his colleagues described a quinacrine staining pattern that was unique for each chromosome, making the identification of each chromosome possible (8). Specific pretreatment of chromosomal preparations followed by staining with a dilute Giemsa stain also produced specific banding patterns (9), and other differential staining techniques have subsequently been developed. Figure 2 shows the characteristic banding pattern diagrammed for each human chromosome. Figure 3 shows the characteristic Giemsa (G) and quinacrine (Q) banding patterns of a normal male cell. The Y-chromosome is the most brilliantly fluorescent chromosome in the Q-banded karyotype.

Table 1 summarizes nomenclature used in designating karyotypes (10). The general rule is that the total number of chromosomes is written first, followed by a comma and then the sex chromosome complement. Thus the normal female karyotype is written as 46,XX, whereas the normal male karyotype is 46,XY. The mapping of banded chromosomes is labeled by regions and component bands (Figs. 2 and 11). The long arm of chromosome 1, e.g., is divided into four regions, each of which includes 2 or more component bands. The terminal band of the long arm is designated q44, q designating the long arm, the first 4 designating region four and the second 4 designating band four within that region. The entire karyotype can thus be labeled.

The application of banding techniques allows the identification of many heretofore unrecognizable structural chromosomal aberrations such as small deletions from and additions to chromosomes not previously discernable, and of the sites of breakage and rearrangement in structurally altered chromosomes. All of these discoveries have added to our knowledge of the variations in the human phenotype which result from variations in the karyotype.

Cytogenetic Disorders 37

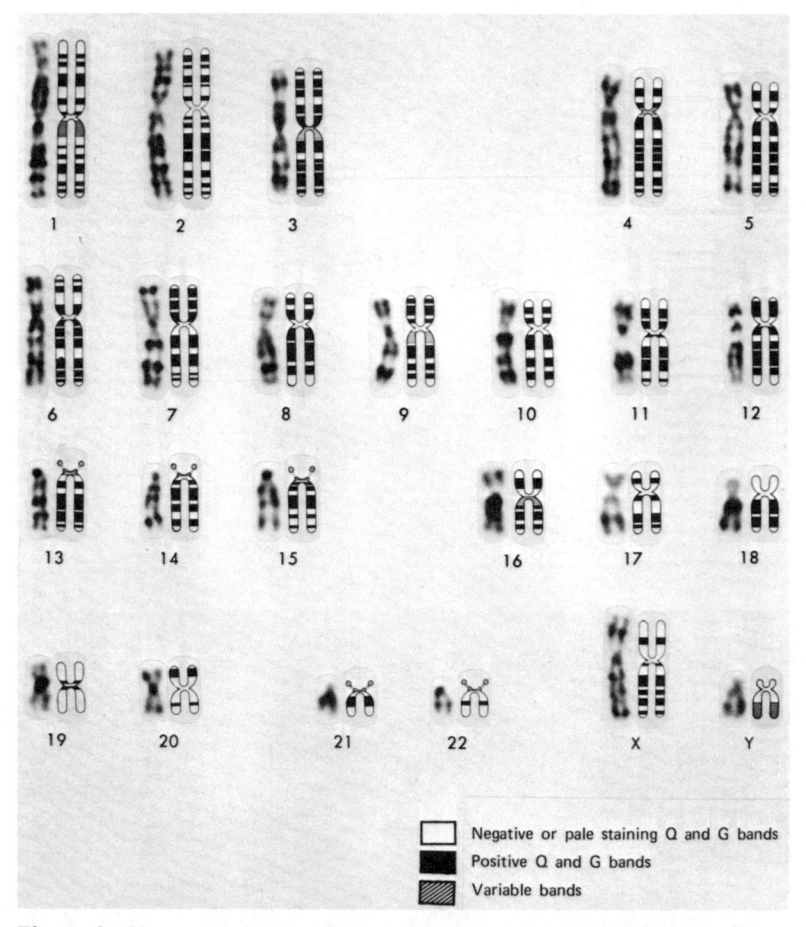

Figure 2. Human karyotype demonstrating schematically banding pattern (G-banding) of each chromosome in right-hand member of each pair and photographically banding pattern in Giemsa-stained chromosomes[2] in left-hand member of each pair. Q-banding patterns are in general similar but because they are demonstrated by fluorescent microscopy, bands that are dark in G-banded chromosomes are bright in Q-banded chromosomes and vice versa.

Cell-Culture Techniques

Study of human chromosomes requires the availability of cells in the metaphase of mitosis. To provide these, culture of the cells to be studied is usually necessary. Several cell sources are available, the commonest being peripheral blood lymphocytes. Solid tissues, such as skin, fascia and gonads, may be used, and bone marrow is studied in special situations. With bone marrow, culture is not necessary since cells in mitosis are already available. Amniotic fluid contains cells derived from the fetus, culture of which allows chromosomal analysis for prenatal cytogenetic diagnosis. Likewise various types of effusion contain cells the chromosomes of which may be analyzed. Cell-culture techniques and staining

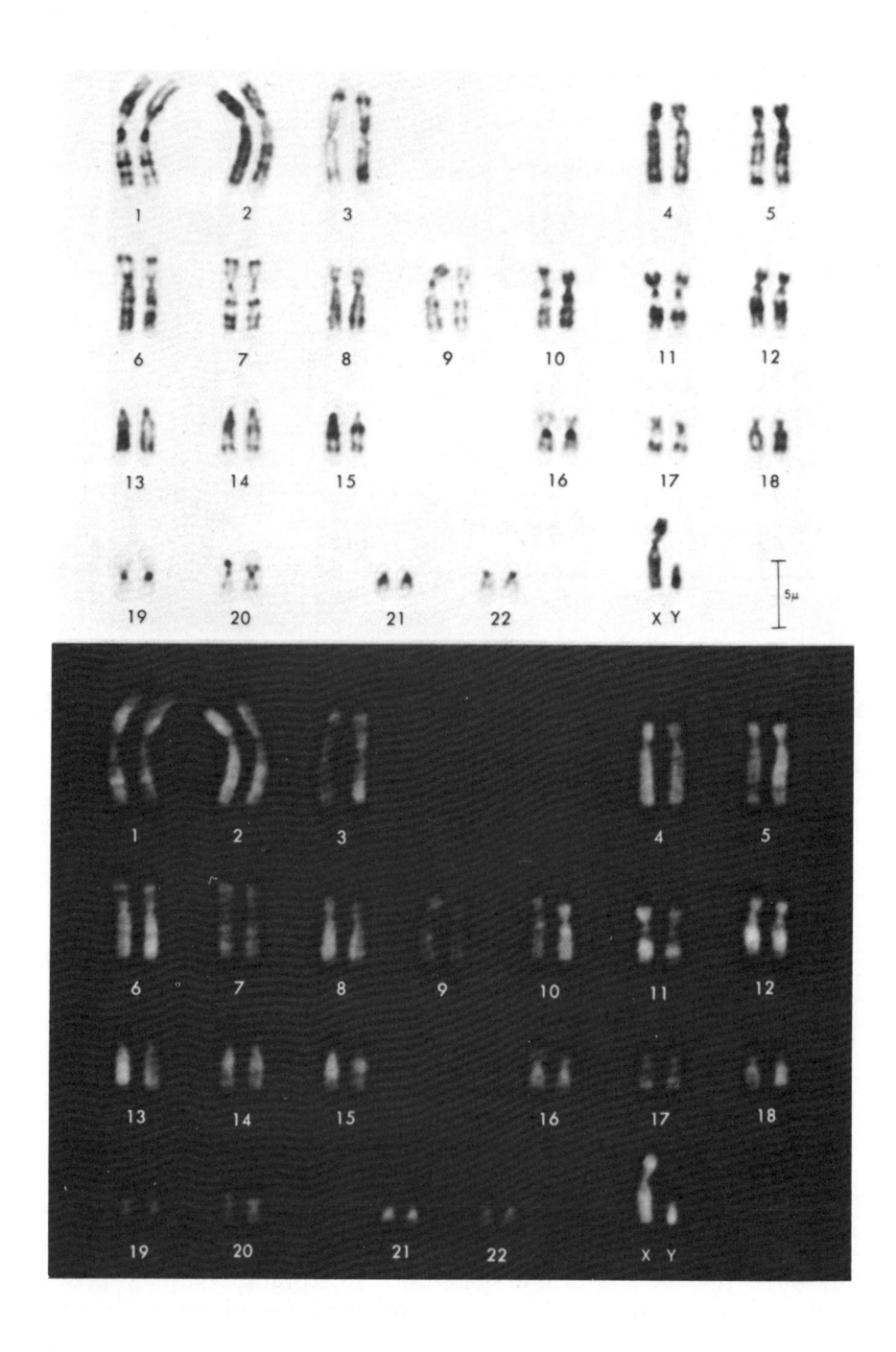

Figure 3. Dual G-banded/Q-banded karyotype of normal male cell.

Table 1. Nomenclature of Karyotype Description

Characteristic	Designation	Example
Normal female karyotype	46, XX	
Normal male karyotype	46, XY	
X-monosomy	45, X	
X-polysomy	Total chromosome number, sex chromosome complement	47, XXX 47, XXY
Y-polysomy	Total chromosome number, sex chromosome complement	47, XYY
Autosomal trisomy	47, XX or XY, + extra autosome	47, XX, +21
Polyploidy	Total chromosome number, sex chromosome complement	69, XXX
Ring chromosome of autosome	46, XX or XY, r (chromosome involved)	46, XX, r (9)
Ring X-chromosome	46, X, r (X)	
Isochromosome of long arm of X-chromosome	46, X, i (Xq)	
Isochromosome of autosome	46, XX or XY, i (chromosome and arm)	46, XX, i (21q)
Terminal deletion of short arm of autosome	46, XX or XY, del (chromosome and arm), *or* 46, XX or XY, del (chromosome) (breakpoint designation)	46, XX, del (5p) 46, XX, del (5) (p11)
Terminal deletion of long arm of autosome	46, XX or XY, del (chromosome and arm), *or* 46, XX or XY, del (chromosome) (breakpoint designation)	46, XX, del (12q) 46, XX, del (12) (q24)
Deletion of arm of sex chromosome	46, X, del (chromosome and arm)	46, X, del (Xq)
	or 46, X, del (chromosome) (breakpoint designation)	46, X, del (X) (q24)
Pericentric inversion (inv)	46, XX or XY, inv (chromosome) (breakpoint designation)	46, XX, inv (1) (p14; q24)
Robertsonian (rob) translocation, balanced	45, XX or XY, rob (chromosome and arms)	45, XX, rob (14q21q)
Robertsonian (rob) translocation, unbalanced	46, XX or XY, -chromosome not identifiable as such in karyotype, +rob (chromosomes involved)	46, XX, −14, +rob (14q21q)
Reciprocal (rcp) translocation, balanced	46, XX or XY, rcp (chromosome involved breakpoint designation)	46, XX, rcp (10; 21) (q26; q21)
Reciprocal translocation, unbalanced	46, XX or XY, der (translocated chromosome derivative), rcp (chromosome involved, breakpoint designation)	46, XX, der (10), rcp (10; 21) (q26; q21)
Mosaicism	First karyotype/second karyotype	46, XX/47, XX, +21

methodology are readily available (11). The culture of lymphocytes and amniotic cells requires several days, whereas fibroblasts from solid tissue must remain in culture for several weeks before sufficient cells for analysis are accumulated.

INCIDENCE

From 0.5–1.0% of all liveborn human infants bears a clinically significant chromosomal abnormality (Table 2) (12). Data acquired from the cytogenetic study of early spontaneous abortions indicate that roughly 50% of all spontaneously aborted fetuses harbor a chromosomal abnormality, the implication being that the severe effects of the chromosomal imbalance *cause* the abortions (13). Consideration of these data leads to the conclusion that at least 7% of human conceptions are karyotypically abnormal (14). However, 90% of them do not survive pregnancy (15). In the 10% that are liveborn, about one-half of the chromosomal abnormalities involve the sex chromosomes, half the autosomes.

Population surveys indicate that about 1 in 500 normal persons harbors a balanced structural rearrangement (12). The rearrangement usually has no apparent effect on the host but because of the chromosomal aberration, the host has an increased risk of producing offspring with an unbalanced form of the rearrangement, i.e., an *inherited* chromosomal abnormality. Those offspring may be mentally retarded and exhibit multiple anomalies. The application of contemporary banding techniques has led to the discovery of "normal variations" and heteromorphisms of the karyotype, all of which are compatible with a normal phenotype (16). With multiple-banding techniques the karyotype of each person includes an average of 5.08 heteromorphisms, or "minor variations" (17).

MECHANISMS OF CHROMOSOME IMBALANCE

Numerical chromosomal abnormalities include *trisomy* (an extra chromosome for a total of 47, the extra chromosome being one of the normally occurring ones), *monosomy* (absence of a chromosome) and *polyploidy* (one or more extra haploid sets of chromosomes for a total of 69—triploidy, or 92—tetraploidy). Structural aberrations include deletions, inversions and translocations. Structural abnormalities like these may lead to *partial trisomy* and *partial monosomy*.

Trisomy is caused by an abnormality of cell division, occurring usually in meiosis. Figure 4 depicts the phenomenon known as meiotic nondisjunction, specifically in oogenesis. Members of a chromosomal pair ordinarily separate or *disjoin* in the first meiotic division. Failure to do this, with inclusion of both chromosomes of the pair in the secondary oocyte, results in there being an extra chromosome in the mature ovum. At the second meiotic division the centromere of each chromosome splits, one chromatid of which then migrates to each pole of the spindle. Fertilization by a normal sperm then results in trisomy in the zygote.

The exclusion of both members of the chromosomal pair from the secondary oocyte leads to the lack of a chromosome in the mature ovum. Fertilization by a normal sperm then results in monosomy in the zygote. As indicated in Figure 4, nondisjunction may occur in the first or the second meiotic division. Utilization

Table 2. Incidence of Chromosome Abnormalities Among Liveborn Infants (12)

Abnormality	Incidence
Down syndrome (21 trisomy)	1/800
18-trisomy syndrome	1/8,000
13-trisomy syndrome	1/20,000
Turner syndrome (females)	1/10,000
Klinefelter syndrome (males)	1/1,000
Poly-X anomalies (females)	1/1,000
XYY karyotype (males)	1/1,000
Balanced structural rearrangement	1/520
Unbalanced structural rearrangement	1/1,700
Total	1/160

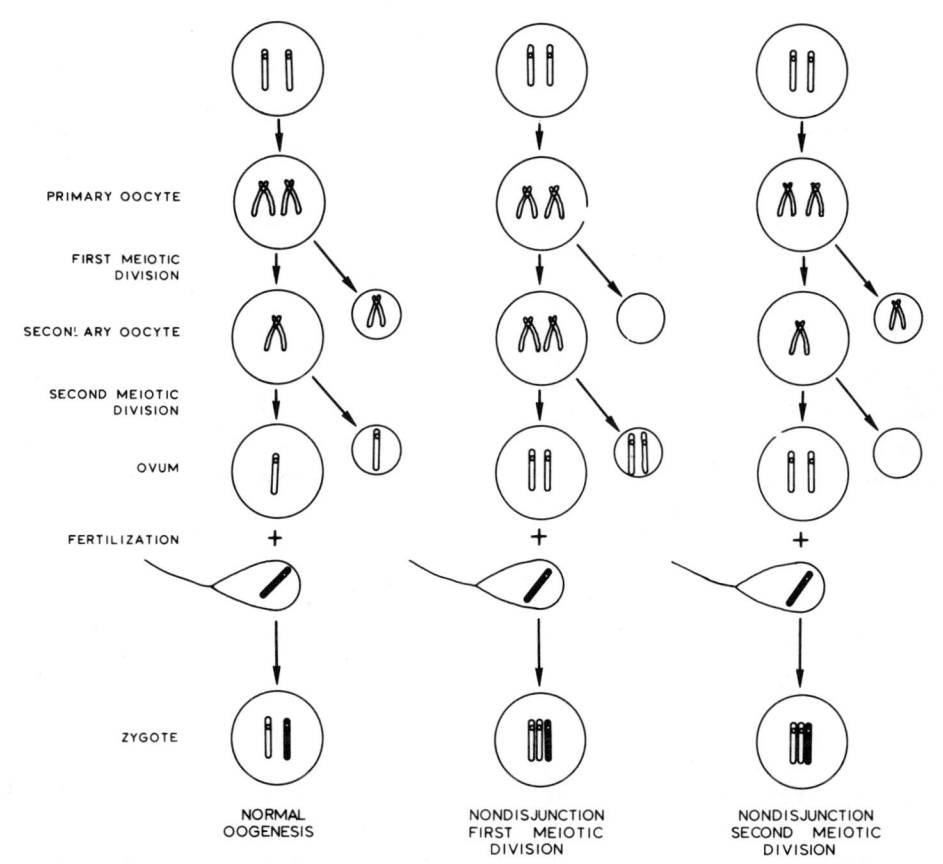

Figure 4. Meiotic nondisjunction. Figure depicts nondisjunction specifically in oogenesis, relative to theoretical ovum containing one pair of chromosomes. Left-hand column shows normal process of reduction division; center column shows nondisjunction at first meiotic division; right-hand column shows nondisjunction at second meiotic division, each case of nondisjunction resulting in trisomy in the zygote. (From Summitt RL: Autosomal abnormalities. A review. *G.P.* 36:96, 1967.)

of contemporary banding techniques in the analysis of chromosome variations and heteromorphisms makes it possible occasionally to determine the parental origin of an accessory chromosome, and if that accessory chromosome is the result of nondisjunction in first or second meiosis (18).

An abnormality in chromosome number may result from an error in mitosis as well as in meiosis. Normally, at the mitotic anaphase, the centromere of each chromosome splits, and the two separated chromatids of each migrate to opposite poles of the mitotic spindle, ultimately to be distributed to two daughter cells. Each of the latter will then contain an identical chromosome complement. An exception to this normal process is *mitotic nondisjunction,* in which the organism is made up of cells some of which contain one number of chromosomes, others a different number, a phenomenon known as *mosaicism* or *mixoploidy.* If nondisjunction occurs in the first mitosis after fertilization, the result is mosaicism (two cell lines, both abnormal). If nondisjunction occurs in any later mitotic stage, the resulting mosaicism includes three cell lines—two abnormal and one normal.

A second possible exception to normal mitosis is *anaphase lag.* If one member of a chromosome pair fails to migrate to the pole in mitotic anaphase, it will be excluded from the daughter cell, and again a mosaic results.

Polyploidy, one or more extra haploid sets (23) of chromosomes, is rare in liveborn human populations. Further, polyploidy almost always exists in a mosaic distribution, together with a normal cell line. For practical purposes the only two types of polyploidy that warrant consideration are triploidy (69 chromosomes) and tetraploidy (92 chromosomes). The former may result from one of several errors in meiosis, fertilization or mitosis (19).

Chromosomal Aberrations

Significant structural chromosomal aberrations include *deletions, ring chromosomes, inversions* and *translocations.* Figure 5 shows the mechanisms that may produce deletions, including ring chromosomes. A deletion may be terminal—and the entire segment distal to the break is lost—or interstitial—the result of two breaks—both on the same side of the centromere, in which the interstitial segment is broken out and lost while the proximodistal segments fuse to maintain continuity. A ring chromosome is also the result of two breaks, one on each side of the centromere followed by fusion of the two ends of the centric segment. This entails loss or deletion of the two terminal segments of the chromosome. In each of the situations caused by the chromosome breaks herein described, the result is *loss* of chromosomal mass, or *partial monosomy.*

Figure 6 shows *inversion,* the result of two breaks in a chromosome, followed by a 180° rotation of the interstitial selment and its reattachment to the terminal segments. Two breaks on the same side cause a *paracentric inversion;* if one break occurs on each side of the centromere, this is a *pericentric inversion.* Before contemporary chromosomal banding techniques were available, a paracentric inversion was not detectable, and a pericentric inversion was detectable only if the two breaks were not equidistant from the centromere, producing a chromosome with an altered arm ratio. Although it per se has no adverse effect on its host, an inversion, particularly a pericentric inversion, provides an opportunity for abnormalities to occur in meiosis, resulting in a gamete—and an infant—with a chromosomal imbalance.

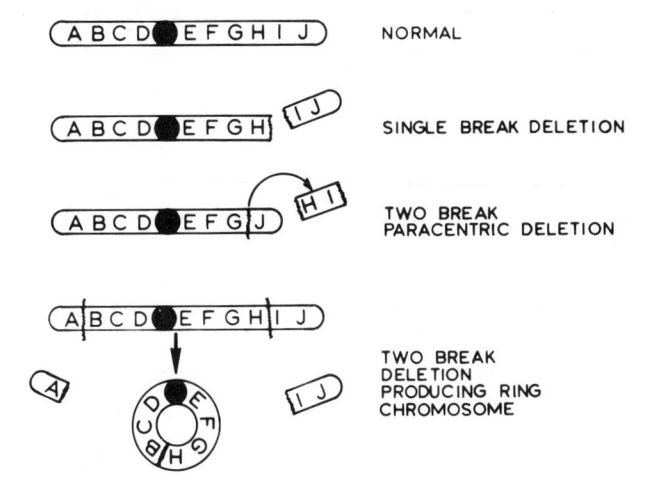

DELETIONS

Figure 5. Mechanism of chromosome deletion resulting in partial monosomy. Types include simple terminal deletion caused by single break, interstitial deletion resulting from two breaks on same side of centromere, and ring chromosome. (From Summitt RL: Autosomal abnormalities: A review. *G.P.* 36:96, 1967.)

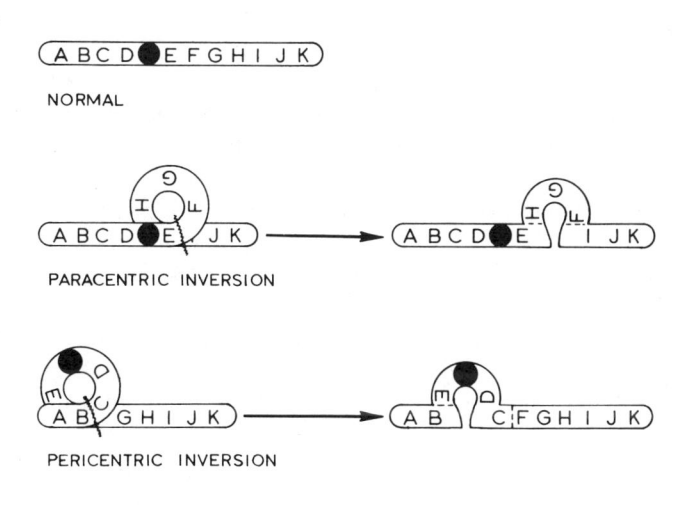

INVERSIONS

Figure 6. Mechanism of chromosome inversion. (From Summitt RL: Autosomal abnormalities. A review. *G.P.* 36:96, 1967.)

In a *translocation* a segment is lost by one chromosome and then attaches itself to another, usually nonhomologous, chromosome. Types of translocation in man include *simple* (the deletion of a segment from one, and its reattachment to another, chromosome without loss of mass from the recipient chromosome); *reciprocal* (a break in each of two chromosomes with exchange of terminal segments); and *Robertsonian* (the fusion of the long arms of two acrocentric chromosomes following a break in each at the centromere, a break in the short arm of one near the centromere and in the long arm of the other near the centromere, or, according to some authorities, a break in the short arm of each adjacent to the centromere). More complex translocations may involve the exchange of segments among more than two chromosomes, or the insertion of an interstitial segment of one chromosome (the result of two breaks in that chromosome) between the two segments of another chromosome separated by a single break. A translocation per se does not cause a phenotypic abnormality in its host so long as the translocation involves no net gain or loss of chromosomal mass. However, as with an inversion, abnormalities in segregation of the translocated chromosomes in meiosis may produce a chromosomal imbalance, such as partial trisomy or partial monosomy in the resulting infant.

ABNORMALITIES OF SEX CHROMOSOMES

Table 3 summarizes the major abnormalities of the sex chromosomes, including the Turner syndrome, the Klinefelter syndrome, X-polysomy and Y-polysomy. This discussion will not include those conditions in which the sex karyotype is not abnormal, although it is not in keeping with the sex phenotype.

Sex Chromatin

Consideration of abnormalities of the sex chromosomes would be incomplete without a discussion of X- and Y-chromatin. As mentioned in the introduction, the modern era of cytogenetics was ushered in by the discovery by Barr and Bertram of a small heteropyknotic mass of chromatin identifiable in the nuclei of somatic cells from female mammals, but not in cells from males. Subsequent investigation has identified this mass of chromatin, now termed the *X-chromatin mass,* as a highly condensed X-chromosome (Fig. 7). According to the concept proposed by Mary Lyon (the Lyon hypothesis) (20), so that cells of the normal female produce no more X-linked gene-derived products than do cells of the normal male, which contain only one X-chromosome, all X-chromosomes in a cell in excess of one are genetically inactivated. The cytogenetic manifestation of this genetically inactive X-chromosome is the X-chromatin mass, or Barr body.

In practice the commonest source of cells for determining the presence of the X-chromatin mass is the buccal mucosa. For various reasons an X-chromatin mass is ordinarily discernible in only 20–50% of the buccal mucosal cells of a person all of whose cells contain two X-chromosomes. In most laboratories, if less than 20% of cells of a buccal mucosal smear contain an X-chromatin mass, some of the cells are presumed to contain only one X-chromosome. Theoretically a buccal mucosal smear from a person whose cells contain three X-chromosomes

Table 3. Abnormalities of Sex Chromosomes

Abnormality	Cases (%)	Population Incidence
Turner Syndrome		1/10,000 females
45,X	57	
46,X, i (Xq) and mosaics including i (Xq) cell line	17	
46,X, del (Xq) and mosaics including del (Xq) cell line	1	
Mosaics 45, X/46, XX; 45, X/47, XXX	12	
Mosaics 45, X/46, XY	4	
Other [del (Xp), r (x), mosaics]	9	
Klinefelter Syndrome		1/1,000 males
47,XXY	82	
48,XXXY	3	
49,XXXXY	<1	
Mosaics	8	
Other	6	
Polysomy-X Females		1/1,000 females
47,XXX	98+	
48,XXXX	Rare	
49,XXXXX	Rare	
Mosaics	Rare	
Polysomy-Y		1/1,000 males
47, XYY	98+	
Other	Rare	

should reveal two X-chromatin masses in all cells. In reality some of the cells contain one X-chromatin mass, whereas a smaller number contain two. The resultant *maximum* of X-chromatin masses observable is one less than the number of X-chromosomes in cells of the tissue being evaluated.

Development of chromosomal fluorescent Q banding reveals that the Y-chromosome is the most brilliantly fluorescent of any in the karyotype. This bright fluorescence is confined to the long arm of the Y. Among normal males the size of the Y-chromosome varies widely. This characteristic is the result of variation in the size of the fluorescent segment, which is thought to be genetically inert (21). This portion of each Y-chromosome is tightly condensed in cells in interphase (similar to the condensation of all X-chromosomes more than one). When cells in interphase, such as those of the buccal mucosa, are stained with quinacrine, a bright fluorescent dot is visible within the nuclei of cells containing a Y-chromosome. This is known as the *Y-chromatin mass* (Fig. 8). In persons whose cells contain two Y-chromosomes, the buccal smear contains some nuclei with two Y-chromatin masses.

The practical implication of X- and Y-chromatin is that a study of the buccal-smear cells gives an indirect but accurate index of the number of X and Y chromosomes in the karyotype of those cells. Again, the maximum number of X-chromatin masses discernable is *one less than* the X-chromosomes in the cells examined, whereas the maximum of Y-chromatin masses is *equal to* the number of Y chromosomes. Results of a buccal smear examination should include the percentage of cells containing X- and Y-chromatin masses and the number of

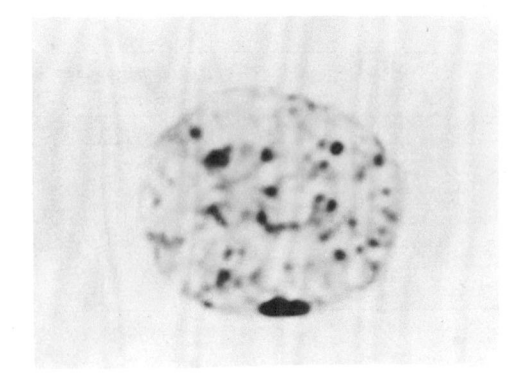

Figure 7. X-chromatin mass. (From Summitt RL: Abnormalities of sex determination-differentiation, in Hughes JG (ed): Synopsis of Pediatrics, 4th ed. St. Louis, Mosby, 1975, p. 586.)

such masses. In our laboratory a minimum of 100 cells is analyzed to establish percentages. Before a buccal smear is reported as being either X- or Y-chromatin negative, at least 200 cells are analyzed.

The Turner Syndrome

This syndrome may be defined as the spectrum of phenotypic features resulting from partial or complete monosomy-X, more specifically, partial or complete monosomy for the *short arm* of the X-chromosome. Complete monosomy-X involves the karyotype 45,X. Although the frequency of autosomal trisomy, resulting in all likelihood from meiotic nondisjunction, is correlated with advancing maternal age, the frequency of monosomy-X is not (23). Because of this, the frequency of mosaic karyotypes in the Turner syndrome, and the finding in testable 45,X cases that the paternally derived X is more often absent that the maternal-derived X, mitotic rather than meiotic errors are implicated in pathogenesis.

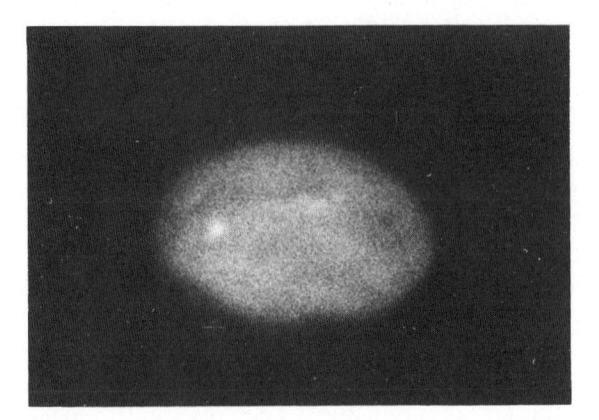

Figure 8. Y-chromatin mass. (From Summitt RL: Abnormalities of sex determination-differentiation, in Hughes JG (ed): Synopsis of Pediatrics, 4th ed. St. Louis, Mobsy, 1975, p. 586.)

Monosomy for the X-chromosome is the only type that is well documented in man. The phenotypic features of the Turner syndrome are shown in Table 4 and based on evaluation of *only* patients with the 45,X karyotype. The most constant features are shortness of stature and gonadal dysgenesis with its resultant sexual infantilism and sterility. In general the same phenotypic features are seen with other karyotypes with the exceptions subsequently noted. Webbing of the neck is rarely encountered with an i(Xq) karyotype. Cardiac lesions involve characteristically the left side of the heart and include aortic stenosis and coarctation of the aorta. However, ventricular septal defect (VSD), artrial septal defect (ASD) and other defects are reported.

Mental retardation is not ordinarily considered to be a feature of the Turner syndrome, but it was noted in 18% of cases making up Table 4. Although sterility is characteristic, a recent report notes 58 pregnancies in 25 women with 45,X (4), 45,X/46,XX (9), 45,X/47,XXX (4) and 45,X/46,XX/47,XXX (8) karyotypes (27). These 58 pregnancies produced 35 liveborn children, 12 of whom are abnormal. Eight of the latter had abnormal karyotypes: five with sex chromosomal abnormalities and three with the Down syndrome.

Partial Monosomy-X. As indicated in Table 3, only 57% of cases of the Turner syndrome are the result of complete monosomy-X, various other karyotypes making up the remainder. Partial monosomy-X may be the result of formation of an isochromosome of the long arm of X; 46,X,i(Xq). The mechanism producing this chromosome is shown in Figure 9 (28). Although this karyotype includes

Table 4. Clinical Features of Turner Syndrome (Percentages are based on incidence in 343 patients with 45, X karyotype [24–26])

Feature	*Percentage*
Short stature	97
Primary amenorrhea	96
Sterility	>99
Sexual infantilism	95
Peripheral lymphedema	41
Webbed neck	53
Cubitus valgus	58
Epicanthal folds	30
Low nuchal hairline	73
Shield chest	59
Cardiovascular anomaly	16
Hypertension	27
Urinary tract anomaly	43
Hypoplastic, hyperconvex nails	73
Short metacarpals or metatarsals	48
Pigmented nevi	60
Mental deficiency	18
Highly arched palate	45
Short neck	71
Defective vision	22
Defective hearing	53
Micrognathia	40
Pectus excavatum	38

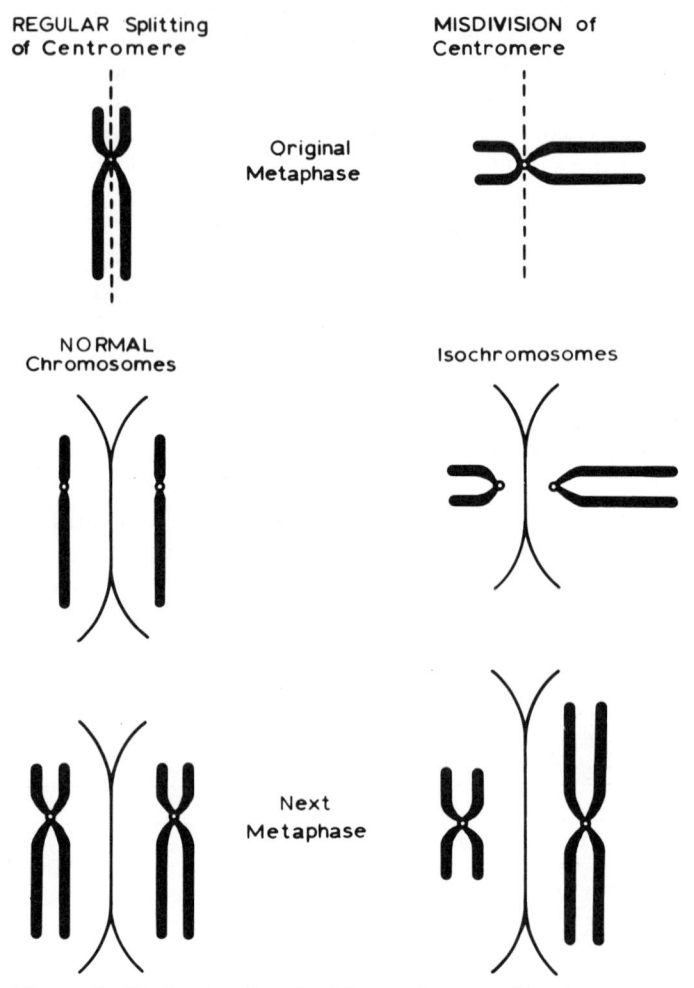

Figure 9. Mechanism involved in production of isochromosome, shown here specifically for long arm of X. (From Harnden DG [28].)

trisomy for the long arm of X, it also involves monosomy for the short arm. The buccal smear from patients with this anomaly is X-chromatin positive. Monosomy for the short arm of X seems to be critical in producing the gonadal dysgenesis and somatic abnormalities of the Turner syndrome.

Deletion of the short arm of X[46,X,del(Xp)] will also produce the Turner syndrome. The formation of a ring chromosome of X[46,X,r(X)] (Fig. 10) involves deletion of segments of both long and short arms (see Fig. 5). The short-arm deletion is in all likelihood responsible for the features of the Turner syndrome. Deletion of the long arm of X regularly produces gonadal dysgenesis, but the somatic features produced by this deletion are variable and the

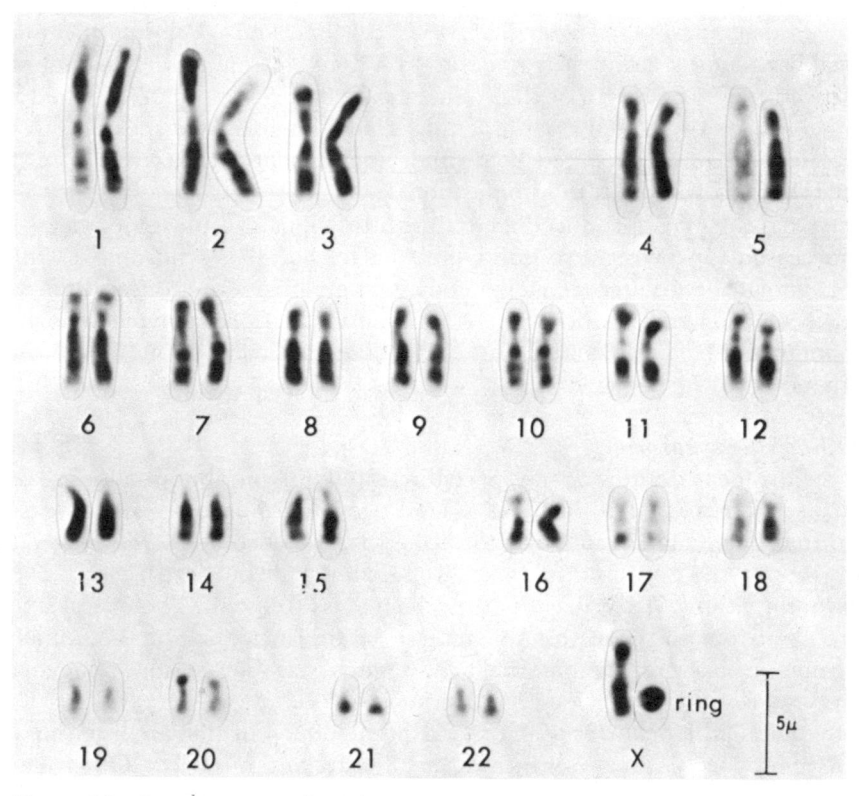

Figure 10. Karyotype revealing ring chromosome of X-chromosome: 46,X,r(X), resulting in monosomy for portion of long and short arms of X.

phenotype is not that which is ordinarily associated with the Turner syndrome. The buccal smears in patients with the Turner syndrome resulting from any form of deletion may be either X-chromatin positive or X-chromatin negative, depending apparently on whether the abnormal X is large enough to form a discernible X chromatin mass.

Mosaic Karyotypes. A significant percentage of cases of the Turner syndrome involves a mosaic karyotype, most commonly 45,X/46,XX. In general a normal 46,XX cell line mitigates the effect of the 45,X one, although this is a statistical generalization and cannot be applied to an individual case. Patients with a structural abnormality of the X-chromosome—such as 46,X,i(Xq), 46,X,r(X) or 46,X,del(Xp)—also often have a mosaic association with a 45,X cell line. An infrequent yet special type of mosaicism found in the Turner syndrome is 45,X/46,XY. This is caused by loss, as through anaphase lag, of a Y-chromosome from some cells of an originally 46,XY embryo. A Y-bearing cell line in the Turner syndrome predisposes the patient to neoplastic transformation in the dysgenetic gonads. This is one type of Turner syndrome in which such a predisposition exists and in which the gonads should be extirpated as soon as the diagnosis is made.

A second instance in which gonadal neoplasia occurs is the Turner syndrome with a mosaic karyotype involving a 45,X cell line and a cell line including a normal X and a small centric fragment: 45,X/46,X,+ fragment. The implication is that, although the origin of the centric fragment cannot be determined from its morphology or banding pattern, that it *does* involve predisposition to neoplasia means that it is a deleted Y-chromosome. Recent data derived from study of the HY antigen support this implication.

Once parents have produced a child with the Turner sundrome, their risk is not increased for having a second child so affected. This syndrome is a highly lethal abnormality in utero. Cytogenetic investigations of the conceptus of spontaneous abortion indicates that the commonest specific karyotypic abnormality encountered is 45,X (29), and at least 95% of the 45,X conceptions do not survive pregnancy.

The Klinefelter syndrome

This syndrome is defined as the spectrum of phenotypic abnormalities resulting from a sex karyotype that includes at least two X chromosomes and at least one Y-chromosome. This means that the Klinefelter syndrome is a *cytogenetic* diagnosis. Karyotypes encountered in the Klinefelter syndrome are listed in Table 3. The commonest is 47,XXY, with others being less frequent. The buccal smear is always X-chromatin positive. An increase in the incidence of the Klinefelter syndrome with advancing maternal age suggests that the causative abnormal sex karyotypes result from meiotic nondisjunction (23).

The phenotypic features of 47,XXY Klinefelter syndrome are listed in Table 5. The most constant features include small testes and infertility. Other features not tabulated include an average stature of affected men taller than the population mean, brachycephaly, minor ear anomalies, clinodactyly of the fifth finger, simian and other unusual palmar crease patterns, dermatoglyphic alterations—such as distal displacement of the axial triradius and decreased finger ridge count—body proportions showing an increased upper:lower body segment ratio, delayed adolescence, neurologic abnormalities and personality deviations (30–32).

The condition occurs with a frequency of 1 in 1000 live male births and accounts for 10–20% of males attending infertility clinics. Mental retardation is a feature in a small percentage of patients with the Klinefelter syndrome, accounting for its increased frequency in surveys of the residents of institutions for the mentally retarded. Characteristically the mental retardation is mild to moderate. As the number of X-chromosomes in the sex karyotypes of patients with the Klinefelter syndrome increases, the frequency of mental retardation in general also increases. Parents of a boy with the Klinefelter syndrome have no apparent increased risk of producing a second affected child.

A few patients with the Klinefelter syndrome have mosaic sex karyotypes, the commonest being 46,XY/47,XXY. As in the Turner syndrome, a normal karyotype in some cells mitigates the effect of the XXY cell line. Other mosaic karyotypes include 46,XX/47,XXY; 46,XY/47,XXY/48,XXYY; and 47,XXY/48,XXYY.

A few males are reported with a 46,XX karyotype. Although some authors include this as a variety of the syndrome, we do not. Some do not consider the

Table 5. Phenotypic Features of Klinefelter Syndrome with 47, XXY Karyotype (30)

Feature	Incidence (%)
Histologic evidence of impaired spermiogenesis	100
Small testes	99
Azoospermia	93
Gynecomastia	55
Scant facial hair	77
Scant pubic hair	61
Small penis	41
Decreased libido or potency	68
Decreased testosterone level	79
Increased gonadotropin level	75
Mental retardation	5

XX male to be the result of a chromosomal abnormality, but rather that it is a condition determined by a single-gene mutation in which the sex karyotype is not in keeping with the phenotype. This is similar to the situation in the androgen-insensitivity syndromes in which the sex karyotype is XY, whereas the phenotype is female. However, recent evidence gained from the study of the HY antigen (33) suggests that a small undetectable segment of the Y-chromosome is present, having been translocated to another chromosome. If this suspicion is confirmed, this would be by definition a variety of the Klinefelter syndrome.

The Poly-X Female
This condition involves females whose cells contain more than two X-chromosomes. Sex karyotypes with as many as five X-chromosomes have been described. The phenotypes of poly-X females are variable. Mental retardation is described and appears to be more frequent with increasing numbers of X chromosomes. Recent studies of 11 XXX females ascertained at birth and followed for nine years reveal that two-thirds of them are intellectually normal and well adjusted, whereas one-third of them have delayed early motor and speech development, mild intellectual deficit and disturbed interpersonal relationships (34). The approximate frequency of the abnormality is one in 1,000 liveborn females. Most affected females have normal gonadal function and are fertile. Further, their offspring are virtually always karyotypically normal.

XYY Male
The 47,XYY karyotype deserves consideration because of its unique effect on behavior. It was first brought to attention because of its increased frequency in males who were tall and manifested aggressive behavior. Although these phenotypic features are not encountered in all XYY males, the average stature is increased and aggressive behavior occurs more often than in the normal XY male. Some XYY males have had abnormalities of the external genitalia, but such features are uncommon. The XYY abnormality is relatively common, with a frequency of approximately 1 in 1,000 liveborn males. An extra Y-chromosome ordinarily has no effect on gonadal function. Also, virtually all offspring of XYY males have normal karyotypes.

AUTOSOMAL ABNORMALITIES

It is well known that specific abnormalities of the autosomes produce characteristic and recognizable spectra of phenotypic features.

Several points should be made about syndromes produced by autosomal abnormalities. Although each syndrome includes a recognizable pattern of phenotypic features, some of these individual features are common to all autosomal imbalances. Babies with these abnormalities show low birth weight for gestational age, a failure to thrive, significant mental retardation and major and minor congenital anomalies of multiple-organ systems. No anomaly is peculiar to a specific syndrome, and no phenotypic feature is a sine qua non of a specific syndrome.

The commonest autosomal abnormality is trisomy. The accessory chromosome is by definition one of the normally occurring ones. Three trisomy syndromes have been extensively investigated. Others are less well established but appear to be bona fide syndromes. As mentioned previously, the existence of complete autosomal monosomy remains in question. However, *partial monosomy* or *deletion* syndromes are known. Table 6 lists syndromes attributable to abnormalities of the autosomes, which are now to be considered.

Trisomy

21 Trisomy

The Down syndrome is the commonest autosomal abnormality in man, with a frequency of roughly 1 in 800 live births. Although known for more than 100 years, its chromosomal nature was first recognized in 1959. The clinical features are listed in Table 7. The phenotype of the affected neonate is ordinarily so typical that the clinical diagnosis may be made with confidence, but it must be made not on the basis of one or a few clinical features, but on that of the overall phenotype. Mental retardation, a constant feature, is significant. The average life expectancy of affected patients is reduced, although the syndrome is observable in adults 50–60 years old. Congenital cardiac lesions are frequent, and leukemia is 10–20 times as common in children with 21 trisomy as in normal children (36). Apparently the types of leukemia in the syndrome are the same as those in nontrisomic children of similar ages.

Although the incidence is 1 in 800 live births, it is apparently much commoner at conception. Most 21 trisomic conceptions do not survive pregnancy. The frequency of 21 trisomy, which is roughly 1 in 1,700 in infants born to mothers 20 years of age, doubles just about every four years until it reaches approximately 1 in 25 in infants born to mothers over 45 years of age (Table 8) (37). This correlation of the incidence with advancing maternal age suggests that 21 trisomy is the result of a nondisjunctional event in maternal meiosis, not in germ-cell formation in the father. However, recent evidence, gained from the study of heteromorphisms in banded chromosomes of affected patients and both parents (18), suggests that in as many as 30% of cases of 21 trisomy the accessory chromosome is paternal in origin.

46,XX or XY/47,XX or XY,+21 mosaicism occurs in approximately 1–2% of patients with the Down syndrome; in general, the phenotypic effects of 21 trisomy are mitigated by a normal cell line in some cells. On the one hand, those

Table 6. Established Syndromes Attributable to Autosomal Abnormalities

Syndrome
Numerical abnormalities
Trisomy
8 trisomy
9 trisomy
13 trisomy (Patau syndrome)
18 trisomy (Edwards syndrome)
21 trisomy (Down syndrome)
22 trisomy
Polyploidy
Triploidy
Tetraploidy
Structural abnormalities
Partial trisomy
1q trisomy
2q trisomy
3q trisomy
5p trisomy
7q trisomy
9p trisomy
10q trisomy
14q trisomy
22q trisomy
Deletion (partial monosomy)
4p- deletion syndrome
5p- deletion (cri du chat) syndrome
9p- deletion
13q- deletion
18p- deletion
18q- deletion
21q- deletion
22q- deletion

affected, who have been reported with normal or near-normal intelligence, are probably mosaics. On the other hand, it is very dangerous to assume that an infant with mosaic 21 trisomy will be less retarded or otherwise less affected than the one with nonmosaic trisomy (38). Since mosaic 21 trisomy is the result of a postfertilization error in cell division, it is not correlated with advancing maternal age.

In the vast majority of cases the 21 trisomy syndrome occurs as a sporadic event in a family. Observation reveals that when a couple produces an infant with uncomplicated 21 trisomy, the risk of their producing a second affected infant is about 1 in 100 when the mother's age is less than 40 years (39). After that age, the risk of recurrence is more a function of maternal age than of a previously affected infant (Table 8).

Translocation. Approximately 3.5% of cases of 21 trisomy are the result not of nondisjunction but of translocation, most commonly of the Robertsonian type. The long arm of chromosome 21 is attached to a chromosome of the D group

Table 7. Phenotypic Features of 21 Trisomy Syndrome (Down syndrome) (35)

Feature	Percentage
Mental retardation	99
Flat facial features	90
Oblique palpebral fissures	80
Epicanthal folds	40
Speckled iris (Brushfield spots)*	50
Ear anomalies, minor	50
Flat nasal bridge	60
Highly arched, narrow palate	70
Brachycephaly	75
Flat occiput	78
Dental abnormalities (small teeth)	65
Short, broad neck	45
Furrowed tongue (older patients)	50
Loose skin on posterior neck (infants)	81
Congenital heart disease	40–60
Duodenal obstruction	8
Short limbs	70
Short, broad hands; short fingers, especially 5th	70
Short middle phalanx of 5th finger	62
Single flexion crease, 5th finger	20
Incurved 5th finger	50
Simian or transverse palmar crease	48
Wide space between 1st and 2nd toes	45
Plantar furrow between 1st and 2nd toes	28
Dermatoglyphic features:	
10 ulnar loops or 8-9 with radial loop on 4th finger	90
Distal axial triradius	50
Loop distal in 3rd interdigital space	85
Arch tibial in hallucal area	>50
Hyperextensibility/hyperflexibility	47–77
Hypotonia	21–77
Absent Moro reflex in infants	81

*Virtually never seen in nonwhite patients.

(usually chromosome 14) in 44% of translocation cases, or to another chromosome of the G group in 56% of cases. In any such case the karyotype of the affected patient includes only 46 chromosomes, the extra chromosome 21 attached to another chromosome, thus not increasing the total chromosome count per se. The translocation producing 21 trisomy may occur as a sporadic or de novo event (a fresh mutation) or may be inherited from a phenotypically normal *balanced* translocation carrier-present. The karyotype of a person who carries a *balanced* Robertsonian translocation includes only 45 chromosomes, the 46th (a chromosome 21) being attached to another chromosome. The person is phenotypically normal because the second chromosome 21, although attached to another chromosome, nevertheless is normally functional. The translocation is inherited in 49% of cases of 21-D translocation but in only 5.6% of 21-G cases. One liveborn baby in 18,000 has translocation 21 trisomy.

Risk of recurrence is important in translocation 21 trisomy. Once a patient with the Down syndrome is clinically diagnosed, a chromosomal analysis is necessary, not only to confirm cytologically the diagnosis but also to establish if the

Table 8. Estimated Prevalence Rates of Down Syndrome Relative to Maternal Age by 5-Year Intervals (60)

Maternal Age	Prevalence
< 20	1/1,667
20–24	1/1,587
25–29	1/1,087
30–34	1/763
35–39	1/248
40–44	1/79
> 45	1/24

trisomy is the result of nondisjunction or of translocation. If the latter, then chromosomal analysis of the parents is mandatory to identify those translocations that are inherited from a carrier parent. If the translocation in the affected patient is a de novo occurrence, the risk of recurrence for the parents is negligible. On the other hand, if a phenotypically normal female is a carrier of a 21-D translocation as a result of having produced a 21-trisomic infant, or because she is identified in some type of family or population survey, then she must be advised that her risk of producing a 21-trisomic infant is 10 to 15% (40). If a male carries the 21-D translocation, his risk of producing a 21-trisomic offspring is less than 5%.

If a person carries a 21-G translocation, the risk of producing a 21-trisomic offspring depends on whether the translocation is 21–22 or 21–21. In the former case the risks are similar to those in 21-D translocation. However, a person who carries a 21-21 translocation has a risk of 100% of producing a 21-trisomic offspring. Whether male or female, the person can have *no* liveborn offspring other than those with 21 trisomy.

As with mosaic 21 trisomy, translocation trisomy is not correlated with advancing maternal age. According to both Wright and Day (41) and combined data, approximately 9% of 21-trisomic infants born to mothers less than 30 years old have *translocation* trisomy.

Any chromosome abnormality can be identified in the fetus from chromosomal analysis of cells from amniotic fluid obtained by transabdominal amniocentesis performed on a woman 14–16 weeks' pregnant. The availability of this technique must be made known to *any adult carrier* of a translocation involving chromosome 21, regardless of the carrier's age; to *any woman* who becomes pregnant at an age of 35 years or older; and to any couple, regardless of the wife's age, who has previously had an infant with 21 trisomy. If the fetus is found not to have 21 trisomy, the parents can be reassured; if it has, they can be offered a therapeutic abortion. By these means the selective prevention of 21 trisomy is possible.

18 Trisomy

The 18 trisomy syndrome (42) is much less frequent and much more severe than that resulting from 21 trisomy. It occurs with a frequency of about 1 in 8,000 live births, with a sex ratio of three females to one male. The phenotypic features are listed in Table 9. The abnormality is highly lethal, only 10% of affected infants surviving the first year of life and 50% dying by the second month. Affected

Table 9. Phenotypic Features of 18-Trisomy Syndrome (73)*

Feature	Percentage of patients
Severe mental and developmental retardation	100
Failure to thrive	100
Difficulty with feeding, poor suck	>95
Low birth weight for gestational age	
Hypertonia	50–80
Brain or spinal cord malformation	10–50
Meningomyelocele	10–20
Prominent occiput, elongated skull	>80
Low-set, malformed ears	>80
Ptosis of eyelids	10–50
Epicanthal folds	10–50
Corneal opacities	10–50
Microphthalmia	10–50
Short upper lip	10–50
Narrow palatal arch	>80
Cleft lip or palate	10–20
Micrognathia	>80
Short neck	50–80
Extra skin on posterior neck	40–60
Webbed neck	10–50
Short sternum	>80
Widely spaced nipples	10–50
Congenital heart disease (principally ventricular) septal defect and patent ductus arteriosus	>95
Eventration of diaphragm	10–50
Single umbilical artery	>80
Inguinal or umbilical hernia	50–80
Pyloric stenosis	10–50
Meckel diverticulum	40–60
Malrotation of bowel	10–20
Renal malformations, especially horseshoe kidney, hydronephrosis and hydroureter	50–80
Cryptorchidism	100
Small pelvis	>80
Limited hip abduction	>80
Clubfoot	40–60
Flexion deformity of fingers	>80
Overlapping fingers	>80
Distally placed, retroflexible thumb	40–60
Partial syndactyly	10–50
Ulnar or radial deviation of hands	10–50
Single palmar creases	10–50
Single crease on 5th finger	10–50
Hypoplastic fingernails and toenails	10–50
Prominent heels	50–80
Short, dorsiflexed hallux	50–80
Dermatoglyphic features:	
Low arches on 6 or more fingertips	>80
Distally displaced axial triradius	50–80

*Other features include prominent clitoris, bifid uterus, imperforate anus, dislocated hips, cleft deformities, of hands or feet, and phocomelia.

infants are so severely retarded that they require total nursing care. The incidence of 18 trisomy increases with advancing maternal age and thus is assumed generally to result from meiotic nondisjunction in maternal oogenesis. A few mosaic cases are known, and a few result from translocation or from crossing-over in an inversion loop. Those patients in whom the syndrome is caused by translocation or inversion ordinarily have only *partial trisomy,* are in general less severely affected and may survive to adulthood. In those persons with the mosaic and translocation forms, there is no correlation with advancing maternal age. Except for inherited translocation or inversion cases, risk of recurrence in the 18 trisomy syndrome is low.

13 Trisomy

As with 18 trisomy, an accessory chromosome 13 produces severe phenotypic effects. The 13-trisomy syndrome was first described by Patau and his associates in 1960 (43). It has a frequency of about 1 in 20,000 live births. As with the 21 and 18 trisomies, 13 trisomy is ordinarily assumed to result from meiotic nondisjunction, partly because of its association with advancing maternal age (44–45). When this is the responsible mechanism, the recurrence risk is low. However, some 13-D Robertsonian translocations have been found in infants with 13-trisomy syndrome.

Most cases of translocation 13 trisomy occur de novo, but a few are inherited from phenotypically normal parents who carry the translocation. Available information indicates that the risk of a 13-trisomic offspring from a 13-D translocation carrier is much lower than that from 21-D translocation carriers. A parent who carries a balanced 13-D translocation has a risk of less than 1% of producing a trisomic offspring. Rare mosaic 46/47,+13 infants are reported. Table 10 lists the phenotypic features of the 13 trisomy syndrome.

It is valuable to call attention to the fact that the 13-trisomic infant may have facial features including cleft lip and palate; hypotelorism with severe eye anomalies; and hypoplasia or aplasia of the crista galli, the median philtrum and other portions of the ethmoid bone and nasal septum. This craniofacial phenotype is termed category 1 by Snodgrass (46). Category 2, on the other hand, includes mild microcephaly and micrognathia, a large nose with a broad bridge and bulbous tip, redundant skin folds in the mandibular and periorbital regions and in some cases a large carp-shaped mouth. Severe eye defects and clefts are absent in category 2.

In the 13-trisomy syndrome fetal hemoglobin is detectable for variable intervals, much longer indeed than in unaffected infants (47). Also, the polymorphonuclear leukocytes of affected infants contain increased nuclear projections (47).

Cardiovascular defects in the 13-trisomy syndrome include VSD (50–60%), patent ductus arteriosus (50–60%), ASD (40–50%), dextrocardia (20–50%) and coarctation of the aorta (10–20%). Renal anomalies, present in more than 80% of cases, include multiple small renal cortical cysts, duplication of the renal pelvis, hydronephrosis and hydroureter. Anomalies of the central nervous system (CNS), found in 80% of cases at autopsy, include most commonly agenesis of the olfactory bulbs, agenesis of the corpus callosum, failure of hemispheric cleavage, cerebellar hypoplasia and hydrocephalus. Other internal malforma-

Table 10. Phenotypic Features of 13-Trisomy Syndrome (46)

Feature	Percentage of Patients
Profound mental and developmental retardation	100
Jitteriness and apneic episodes	50–80
Microcephaly	50–80
Presumptive deafness	50–80
Hypotonia	40–50
Seizures	20–30
Scalp defects	10–50
Hypotelorism	>80
Microphthalmia	50–80
Epicanthal folds	50–80
Absent eyebrows	10–50
Flat supraorbital ridges	10–50
Coloboma of iris	10–50
Low-set, malformed ears	>80
Cleft lip, or palate, or both	50–80
Micrognathia	50–80
Extra skin on posterior neck	50–80
Short neck	50–80
Congenital heart disease	50–80
Inguinal or umbilical hernia	10–50
Single umbilical artery	10–50
Pilonidal pit	10–50
Omphalocele	10–20
Limited hip abduction	10–20
Capillary hemangiomata	50–80
Polydactyly	50–80
Long hyperconvex nails	50–80
Retroflexible thumbs	50–80
Flexion deformity of fingers	50–80
Cleft deformities of hands	
Single palmar crease	50–80
Hypoplastic nails	10–50
Prominent heels	50–80
Short, dorsiflexed hallux	10–50
Clubfoot deformity	10–20
Dermatoglyphic features:	
Distal palmar axial triradius	>80
Hallucal tibial loop pattern	50–80
Three or more digital arches	10–50
Hallucal arch fibular S pattern	10–50

tions are malrotation of the gut, accessory spleens, Meckel diverticulum, abnormally attached mesentery, bicornuate uterus or vagina, or both, hypoplastic ovaries and abnormal Fallopian tubes. Mental retardation is, as a rule, profound. Prognosis for survival is poor; 95% of affected patients succumb in the first three years of life, and 50% survive less than one month.

Other Trisomies

Trisomy for some other autosomes are reported, for the most part sporadically. Several patients with complete or mosaic trisomy 8 have been described, to the extent that a trisomy 8 syndrome has been delineated (48, 49). Features of the syndrome are listed in Table 11. Cases of partial trisomy 8 are also described (50,

Table 11. Phenotypic Feature of 8-Trisomy Syndrome (48–51)*

Feature	Full Trisomy	Mosaic Trisomy	All
Mental retardation	4/4	18/20	22/24
Dysmorphic skull	2/2	10/12	12/14
Facial anomalies:			
Prominent forehead	2/2	7/10	9/12
Dysplastic ears	2/2	15/17	17/19
Strabismus	1/1	15/15	16/16
Plump nose with broad base	2/2	12/13	14/15
Low-set ears	1/1	4/11	5/12
Everted lower lip	0	5/6	5/6
High arched palate	0	7/10	7/10
Cleft soft palate	1/1	2/7	3/8
Micrognathia	3/3	13/14	16/17
Cardiovascular defect	2/3	5/13	7/16
Urinary tract anomaly	3/3	14/16	17/19
Short stature	0/2	5/17	5/19
Decreased weight	0/2	8/16	8/18
Vertebral anomalies	1/1	13/14	14/15
Narrow pelvis	1/1	9/10	10/11
Patellar dysplasia	0	11/12	11/12
Limited joint mobility	3/3	13/14	16/17
Deep flexion creases:			
Palms	0	6/8	6/8
Soles	1/1	7/8	7/8

*Incidence in 24 cases.

51) and are derived from translocations between chromosome 8 and another autosome. In general patients with partial trisomy have milder phenotypic abnormalities than those with the full-blown disorder. On the other hand, the description of a clinical syndrome resulting from trisomy for the short arm of chromosome 9 preceded delineation of the complete 9-trisomy syndrome (52, 54). Based on cases of trisomy for varying portions of the long arm of 9 in addition to trisomy for the short arm, various features of the full 9-trisomy syndrome are assigned to various portions of the long arm (Table 12; Fig. 11) (55).

The availability of banding techniques allows the assignment of specific phenotypic features to the excess or deficiency of specific chromosome segments, as in the 9p and complete 9 trisomy-syndrome just mentioned and the trisomy for different segments of chromosome 13 (56–59). Similarly the availability of these techniques allows the assignment, at least tentatively, of a clinical syndrome to trisomy for portions of chromosome 1q (Table 13) (60–62), chromosome 2p (Table 14) (63), chromosome 2q (Table 15) (64–67), chromosome 3q (Table 16) (68–69), chromosome 5p (Table 17) (70), chromosome 7q (Table 18) (71–76), chromosome 10q (Table 19) (77–81), chromosome 14q (Table 20) (82–93) and chromosome 22q (Table 21) (94–102). Because of the few cases reported and because of the obvious overlap of clinical features among these various partial trisomy syndromes, the exactness of their delineations varies markedly.

A phenotypically normal person may be identified as a carrier of a balanced

Table 12. Phenotypic Features of Syndrome Produced by Trisomy of Chromosome 9 (54, 55)

Short-arm Trisomy	Long-arm Trisomy
Mental retardation	Micrognathia
Microcephaly	Abnormal cranial sutures
Prominent forehead	Urinary tract anomalies
Deepset eyes	Cardiac defects
Protuberant ears	Congenital dislocation
Prominent nose	hip/knee
Carp-shaped mouth ("fish mouth")	
Clinodactyly	
Digital hypoplasia	
Nail hypoplasia	
Syndactyly	
Simian palmar crease	
Dermatoglyphic features:	
Absence of C triradius	
Absence of B triradius	

structural chromosome rearrangement because he or she has produced an unbalanced phenotypically abnormal offspring. This can be done by an investigation of couples who have had multiple spontaneous abortions or by some sort of population survey. In general the person has a risk of producing chromosomally abnormal offspring significantly higher than that of chromosomally normal parents. However, the actual risks depend on the types of rearrangement, the chromosome(s) and the sex of the carrier, and they may vary widely.

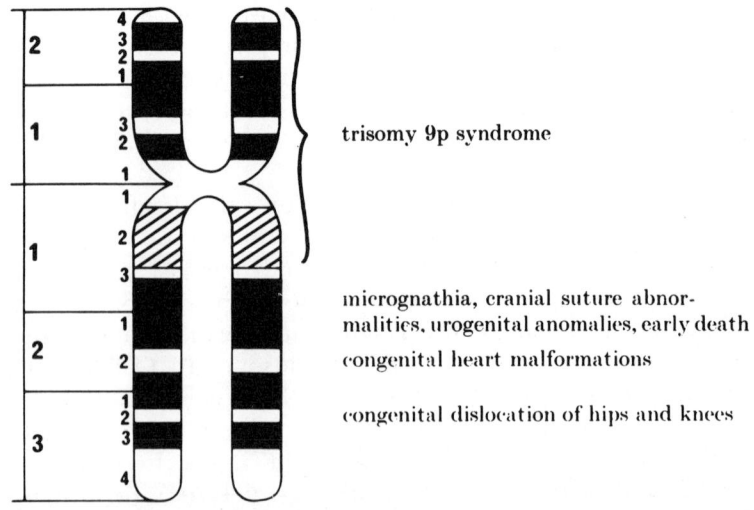

Figure 11. Map of G-banded chromosome 9, assigning features of 9-trisomy syndrome to various regions of accessory chromosome 9. (From Sutherland GR et al. [55].)

Table 13. Tentative Pattern of Anomalies Attributable to 1q Trisomy, Based on 3 Cases (60–62)

Anomaly
Death in newborn period or early infancy
Peaked nose
Micrognathia
Complex cardiac malformation
Absent or hypoplastic thymus
Absent gallbladder
Long fingers
Dysplastic kidneys
Hydrocephalus
Hirsutism

Table 14. Tentative Pattern of Anomalies Caused by 2p Trisomy, Based on 2 Cases (63)

Anomaly
Prenatal and postnatal growth deficiency
Profound mental retardation
High bulging forehead with frontal upsweep of hair
Flat, wide glabella and nasal bridge
Maxillary hypoplasia
Short nose with anteverted nares
Pointed chin
Secondary alveolar ridge
Myopia, astigmatism
Ptosis of eyelids
Dacryostenosis
Dolichostenomelia
Hyperextensible fingers
Subluxation of proximal interphalangeal joints
High interdigital pads and whorl patterns in dermatoglyphics
Flat feet, wide spaces between toes
Soft, thin skin with prolonged wound healing
Fine hair
Thin, fast-growing nails
Multiple fine creases on palms and soles
Long, narrow trunk
Hypoplastic external genitalia

Table 15. Tentative Pattern of Anomalies Caused by 2q Trisomy, Based on 8 Cases (65–67)

Anomaly
Severe mental retardation
Cleft palate
Platybasia
Hypertelorism
Long philtrum
Abnormalities of external ears
Micrognathia
Complex abnormalities in dermatoglyphics

Table 16. Pattern of Anomalies Caused by 3q Trisomy Based on 20 Cases (68–69)*

Anomaly	Percentage
Severe mental retardation	100
Failure to thrive	100
Short stature	100
Dolichocephaly with prominent forehead	50
Low-set ears	45
Cloudy corneas	25
Oblique palpebral fissures	55
Depressed nasal bridge	60
Short nose with anteverted nares	60
Hirsutism (persistent lanugo hair)	45
Cleft palate (posterior)	40
Micrognathia	50
Umbilical hernia or omphalocele	45
Clubfeet	60
Cryptorchidism (males)	58
Cardiac defect	45
Renal anomaly	40

*Fifteen of the 20 patients died in infancy.

Table 17. Tentative Pattern of Anomalies Caused by 5p Trisomy, Based on 6 Cases

Anomaly	Incidence
Death in infancy or early childhood	6/6
Failure to thrive	3/3
Severe mental retardation	3/3
Large head (CNS anomaly)	3/3
Dolichocephaly, promient occiput	3/3
Pupillary constriction	5/5
High, narrow palate	2/2
Respiratory problems	5/6
Hypotonia	5/6
Pectus excavatum	3/5
Hernia, umbilical or inguinal	2/4
Renal anomalies	3/3
Cardiac defect	2/6
Short first toes	2/2

POLYPLOIDY

Although not unusual in the products of spontaneous abortion, polyploidy is rare in liveborn infants, most of whom are mosaic with a normal as well as a polyploid cell line. A few triploid (69 chromosomes) infants, the majority of them mosaic, and six infants with tetraploidy (92 chromosomes) are reported (103). According to Neibuhr (104), in 1974 the literature included reports of approximately 230 triploid abortuses and 30 liveborn triploid infants. Of 255 in whom sex chromosome complements were determined, 84 were XXX, 139 were XXY

Table 18. Tentative Pattern of Anomalies Produced by 7q Trisomy, Based on 10 Patients*

Anomaly	Incidence
Mental retardation	7/7
Hypotonia	5/7
Abnormal head	8/8
Small palpebral fissures	7/10
Hypertelorism	6/8
Cleft palate	4/10
Large tongue	6/7
Micrognathia	5/7
Low-set ears	9/10
Ear anomalies	8/10

*Other anomalies include strabismus, epicanthal folds, skeletal anomalies, wide fontanelle and low birth weight for gestational age.

Table 19. Tentative Pattern of Anomalies Attributable to 10q Trisomy, Based on 10 Patients (77–80)

Anomaly	Incidence
Growth retardation	10/10
Mental retardation	10/10
Microcephaly	7/9
Hypotonia	7/10
Flat face	9/9
High, broad forehead	9/9
Small nose	8/9
Depressed nasal bridge	9/9
Arched, wide-spaced eyebrows	9/9
Blepharophimosis	10/10
Microphthalmia	6/7
Low-set ears	9/10
"Cupid-bow" mouth with prominent upper lip	9/10
Micrognathia	9/9
High-arched or cleft palate	9/9
Short neck	8/9
Eye anomaly	7/9
Cardiac defect	5/8
Hand and foot anomalies	7/7

and 10 (all abortuses) were XYY. Available data reveal a pattern of features probably constant enough to allow definition of a "triploidy syndrome" (Table 22).

No nonmosaic triploid infant has survived longer than seven days; mosaic diploid/triploid patients, however, may survive at least into the second decade of life. Polyploidy has accounted for 21% of 322 chromosomally abnormal abortuses among 1,291 studied in the review of Hamerton (23) and for 26% of 921 chromosomally abnormal abortuses among 1,500 reported by Boue (13). From these figures it may be deduced that as many as 1.2% of all conceptions have polyploid cell lines.

Table 20. Tentative Pattern of Anomalies Attributable to Partial 14q Trisomy, Based on 13 Cases*

Anomaly	Incidence
Mental deficiency	10/10
Short stature	9/10
Microcephaly	8/11
Downturned mouth	10/10
Long philtrum	3/4
Micrognathia	9/11
Cleft lip, palate, or both	6/12
Highly arched palate	2/3
Short neck	3/4
Convulsive seizures	4/6
Simian crease	4/8
Clinodactyly	6/7
Distal digital hypoplasia	3/7

*Other features include brachycephaly, hypotelorism, low-set ears, oxycephaly, cardiac defect and, in males, cryptorchidism.

MONOSOMY

In monosomy one complete chromosome is absent from the diploid set. Monosomy for the X-chromosome has been extensively described and, as mentioned previously, is the only type in man for which irrefutable evidence exists.

Deletions

Several partial monosomy or deletions of autosomes have been described, and syndromes attributable thereto are well known.

Table 21. Phenotypic Features Attributable to 22q Trisomy, Based on 30 Cases (94–102)*

Anomaly	Incidence
Mental retardation	23/23
Growth retardation	23/23
Microcephaly	22/25
Low-set, or malformed ears, or both	24/27
Preauricular appendages or sinuses	20/28
Cleft palate	15/27
Micrognathia	22/27
Cardiac defect	21/28
Abnormal, or low-set nipples, or both	5/10
Cubitus valgus	4/7
Finger-like or malapposed thumbs	9/14
Deformed lower limbs	10/18

*Other features include hypotonia, atretic ear canals, deafness, epicanthal folds, narrow bitemporal width, flat nasal bridge, long philtrum, downturned mouth, hypertelorism, downward slanting palpebral fissures, 13 pairs of ribs, renal aplasia, imperforate anus and dislocated hips.

Table 22. Phenotypic Features of Triploidy, Based on 31 Cases (104)

Feature	22 Cases Nonmosaic	9 Cases Mosaic
Large placenta	11/13	3/3
Hydatid degeneration of placenta	12/15	2/2
Hypotonia	6/9	3/5
Asymmetry	2/3	5/9
Hydrocephalus or large head	7/12	2/9
Agenesis of corpus callosum	5/13	
Dilated ventricles	3/8	1/1
Large posterior fontanelle	9/10	
Meningomyelocele	5/21	0/9
Hypertelorism	11/16	2/9
Microphthalmia	6/14	1/8
Colobomata	7/16	1/9
Low-set ears	9/14	3/7
Malformed ears	8/15	3/7
Palate/lip defects	5/21	2/7
Syndactyly	13/18	6/9
Simian crease	10/12	3/7
Malformations of feet	5/12	1/6
Cardiac defect	9/17	3/7
Cystic renal degeneration	5/16	
Adrenal aplasia	8/12	
Abnormal genitalia	12/13	2/4
Mental retardation		7/7
Low birth weight for gestational age	9/12 (36+ weeks' gestation)	6/8

Short Arm of Chromosome 4 (4p− syndrome; Wolf-Hirschhorn syndrome)

This abnormality was first described in 1965 by Wolf and associates (105), and 43 cases were recently summarized by Johnson and co-workers (106). The extent of deletion varies from case to case, with some degree of correlation between the percentage of 4p, which was absent, and the severity of the phenotype. The features of the 4p−syndrome are shown in Table 23. Attention has been called to the similarities between the 4p− and 5p− phenotypes (107). However, features peculiar to 4p− include coloboma of the iris; scalp defects; preauricular sinus or appendage; cleft lip, or palate, or both; convulsive seizures; delay in ossification in pelvic and carpal bones; hypoplastic dermatoglyphics; hypospadias; and *absence* of the characteristic cat-like cry of the 5p−syndrome.

Mental retardation is more profound in the 4p− than 5p−syndrome, and the prognosis for mental development in a 4p− infant is grave. Every reported case of 4p− has been a sporadic occurrence, and unlike the situation in the 5p− syndrome, no case has ever resulted from an inherited structural rearrangement. Thus the risk of recurrence is negligible. Nevertheless chromosomal analysis should be carried out on the parents of all affected patients to exclude an inherited deletion.

Short Arm of Chromosome 5 (5p− syndrome; cri du chat syndrome)

The 5p− syndrome was first reported in 1963 by Lejeune and his co-workers (108) in an infant with developmental delay, microcephaly, several somatic abnormalities and a peculiar high pitched, cat-like cry—thus the term, cri du chat,

Table 23. Phenotypic Features of 4p Deletion Syndrome (107)

Feature	Percentage
Profound mental retardation	100
Failure to thrive	76
Microcephaly	91
Midline scalp defect	14
Low birth weight	89
Convulsive seizures	47
Hypertelorism	74
Downward slanting palpebral fissures	31
Strabismus	36
Ptosis of eyelids	28
Epicanthal folds	26
Coloboma of iris	31
Broad base of nose	64
Prominent glabella	47
Beak nose	64
Low-set, large and "simple" ears	69
Preauricular dimples or appendages	33
Downturned mouth	36
Short philtrum	17
Cleft lip, or palate, or both	57
Micrognathia	69
Cardiac defect	55
Sacral dimple	33
Hypospadias or cryptorchidism in males, hypoplastic uterus in females	64
Foot deformity	>70
Simian crease (unilateral or bilateral)	>60
Dermatoglyphic features:	
Hypoplastic dermal ridge patterns	40
Low finger ridge count	>80
Double loop on thumb	>50
Premature pubarche	40

or cat-cry syndrome. This has been followed by numerous other descriptions, and a characteristic 5p− phenotype has emerged, consistently associated with the cat-like cry (109). The extent of deletion varies and although some deletions appear to be terminal, others are interstitial (109–111). Patients with 5p− frequently survive to adulthood, and estimates indicate that the 5p− syndrome may account for as many as 1% of low-grade institutionalized mental retardates.

Approximately 85% of cases are sporadic, whereas 15% are inherited from a phenotypically normal parent who carries a balanced structural chromosomal rearrangement, either translocation or inversion. In the former situation, the recurrence risk is low whereas a balanced structural rearrangement in a parent indicates an increased risk of unbalanced offspring for that parent. Parents of all patients should have a chromosomal analysis performed. The clinical features of the 5p− syndrome are enumerated in Table 24. Although the phenotypes in infancy and later are different, both are characteristic of 5p−. The characteristic cat-like cry disappears in late infancy. Even so, the voice characteristics of older affected patients continue to be unusual.

As noted earlier, the 4p− and 5p− syndromes share some features (Tables 23

Table 24. Phenotypic Features of the 5p Deletion (cri du chat) Syndrome (109)

Feature	Percentage
Infancy:	
Low birth weight for gestational age	
Mental retardation	
Failure to thrive	20–40
High-pitched, cat-like cry	90–100
Respiratory stridor	
Laryngomalacia	
Microcephaly	90
Round face	80
Hypertelorism	70
Strabismus	50
Broad nasal root	50
Downward slanting palpebral fissures	50
Epicanthal folds	70
"Almond-shaped" palpebral fissures	
Low-set, malformed ears	60
Micrognathia	60
Simian crease	70
Dermatoglyphic features:	
Distal axial triradius	
Reduced finger ridge count	
Older patients:	
Thin face	
High nasal bridge	
Facial asymmetry	
Dental malocclusion	
Short metacarpals and metatarsals	
Scoliosis	
Pes planus	
Premature graying of hair	
Poor muscular development	
Short stature	50

and 24). Chromosome banding techniques allow the clear differentiation of the chromosome involved in a case and also define the extent of deletion and whether it is terminal or interstitial.

Short Arm of Chromosome 9 (9p− syndrome)
The 9p− syndrome was first identified in 1972. Alfi has recently summarized the findings in six cases (Table 25) (112).

Long Arm of Chromosome 13 (13p− syndrome; r(13) syndrome)
Deletion of the short arm of chromosome 13 (and for that matter of the short arm of any acrocentric autosome) is not clinically significant according to current evidence. Deletion of the long arm of chromosome 13—occurring as a simple deletion or resulting from either ring chromosome formation or the segregation of a structural rearrangement in a parent—produces a definable phenotype

Table 25. Phenotypic Features of 9p Deletion Syndrome, Based on 6 Cases (112)

Feature	Incidence
Mental retardation	6/6
Trigonocephaly	6/6
Upward slanting palpebral fissures	5/6
Epicanthal folds	5/6
Flat nasal bridge	6/6
Anteverted nares	6/6
Long philtrum	6/6
Low-set ears	6/6
Abnormal ear lobules	4/6
Highly arched palate	6/6
Micrognathia	6/6
Low hairline	4/5
Short neck	6/6
Webbed neck	4/6
Widely spaced nipples	6/6
Cardiac murmur	4/6
Long fingers	6/6
Square nails	6/6

(Table 26) (113). Clinical severity varies according to the extent of the region deleted. Attention should be called to the unusual association of retinoblastoma to 13q− (114). We recently had the opportunity to study a fetus with 13q− determined by analysis of cultured amniotic cells (56). Major abnormalities included oligodactyly, agenesis of the corpus callosum, common aorticopulmonary trunk, VSD, atresia of the duodenum and severe renal hypoplasia-dysplasia.

The abnormalities of 13q− may be contrasted ("countertype") with those of 13 trisomy. This differentiation, in a general way, can be seen in numerous chromosomal deletions relative to trisomy for the same chromosome.

Short Arm of Chromosome 18 (18p− syndrome)

This deletion was first described by de Grouchy et al, in 1963 (115). Other cases were described subsequently, and in 1974 Schinzel and co-workers reported the findings in 82 cases (Table 27) (116). Unlike deletion of the short arm of an acrocentric autosome, 18p− produces significant abnormalities. While the reported cases of 18p− ordinarily involve deletion of virtually the entire short arm, the severity of the phenotype varies widely. In some patients mental retardation may be relatively mild, with defects particularly in speech, whereas in others 18p− may produce a severe abnormality of the forebrain. Evidence exists that patients are prone to connective-tissue disease, and a few of them have had a deficiency of immunoglobulin A.

In Schinzel's report (116), the parental karyotypes were known in 58 cases. Most of the deletions were de novo whereas in *six families,* there was a heritable structural rearrangement. In sporadic cases recurrence risk is low; in families in which segregation of a structural rearrangement produces the 18p− syndrome, risk of recurrence may be significantly increased.

Table 26. Phenotypic Features of 13q Deletion Syndrome (113)

Feature or Anomaly	
Mental retardation	Micrognathia
Microcephaly	Large, malrotated, low-set ears
Trigonocephaly	Short neck with lateral skin folds
Facial asymmetry	Congenital heart disease
Epicanthal folds	Hypospadias
Hypertelorism	Bifid scrotum
Ptosis of eyelids	Imperforate anus
Microphthalmia	Perineal fistula
Coloboma of iris	Absent or hypoplastic thumbs
Retinoblastoma	Short 5th finger
Wide, prominent nasal bridge	Clubfoot
Protruding upper incisors	Overriding toes
Cleft or highly arched palate	Short great toes
	Syndactyly of 4th and 5th toes

Table 27. Phenotypic Features of 18p Deletion Syndrome (42, 116)*

Feature	Percentage
Low birth weight for gestational age	>50
Mental retardation	100
Short stature	80
Round, flat face	97
Hypertelorism	60
Epicanthal folds	68
Strabismus	42
Ptosis of eyelids	65
Flat or broad nasal bridge	92
Wide mouth	90
Dental caries	40
Micrognathia	84
Large poorly differentiated, low-set ears	93
Short, broad neck	90
Webbed neck	15
Pectus excavatum	63

*Other anomalies include stubby hands with short digits, partial syndactyly of toes, microcephaly, cebocephaly or arrhinencephaly, cataract, clinodactyly, muscle hypotonia, cyclops deformity, lymphedema and shield chest.

Long Arm of Chromosome 18 (18q− syndrome)

The deletion of a portion of the long arm of chromosome 18 was first reported by Summitt and Patau in 1964 (117), and the findings of the 18q− syndrome were summarized in 54 cases by Schinzel et al (Table 28) (118). The phenotype of patients with 18q− is distinct from that of 18p− and to some extent involves a countertype to the features of 18 trisomy, particularly to partial 18q− trisomy. In addition to reported cases of 18q−, several other patients are reported with ring-18 chromosomes. The phenotypes of those with ring-18 chromosomes are variable, perhaps because of different degrees of long *and* short arm deletion.

Of the 54 cases of 18q− summarized by Schinzel, in which parental chromo-

Table 28. Phenotypic Features of 18q Deletion Syndrome (42, 118)

Feature	Percentage
Low birth weight for gestational age	
Short stature	78
Microcephaly	68
Severe mental retardation	100
Muscular hypotonia	91
Midfacial dysplasia	85
Carp-shaped mouth ("fish mouth")	87
Hypertelorism	81
Epicanthal folds	42
Strabismus	34
Prominent or deformed ears, or both	84
Atretic external auditory canals	50
Cleft lip	9
Cleft palate	29
Short neck	37
Widely spaced nipples	79
Congenital heart disease	35
Hypoplasia of labia minora	47
Cryptorchidism	52
Hypospadias	27
Hypoplastic scrotum	42
Prominent subacromial dimples	50
Epitrochlear, parapatellar and dorsal hand dimples	>50
Proximally placed thumbs	92
Long, tapering fingers	90
Clubfoot	21
Abnormal toe implantation	84
Dermatoglyphic features:	
High frequency of digital whorl patterns	>80
IgA deficiency	44
	(of 26 cases)

somes were studied, 35 were de novo whereas 12 were inherited from a parental carrier of a structural rearrangement. The midfacial dysplasia includes deepset eyes, depressed nasal bridge and malar flatness. The chin appears relatively prominent. The ears may be severely malformed or normally differentiated but large. Conductive deafness with or without atresia of the external auditory canals may be present. Female genital abnormalities include hypoplastic labia and clitoris; male abnomalities include small penis, hypospadias and occasionally cryptorchidism. Eye abnormalities include strabismus, nystagmus, glaucoma, tapetoretinal degeneration, colobomata and optic atrophy. A deficiency of immunoglobulin A was noted in 11 of 26 cases in which information was available. These accumulated data indicate a real but still undefined relationship between chromosome 18 and the synthesis of immunoglobulin A.

Long Arm of Chromosomes 21 and 22 (21q− and 22q−syndromes)

Simple deletions and those in the form of ring chromosomes are known for chromosomes 21 and 22. One of the first cases was reported by Reisman et al, in 1966 (119). The morphologic similarity between chromosomes 21 and 22 led to problems in defining the G-deletion syndromes prior to the advent of chromosomal banding techniques. Even so, on the basis of phenotypic differences, Warren and Rimoin (120) were able to identify two distinct syndromes, which they designated G-deletion syndromes I and II. The validity of their distinction was borne out by banding studies in two of their patients, and in others. The G-deletion syndrome I was shown to be due to 21q−, whereas G-deletion syndrome II was attributable to 22q−. Although the two syndromes share some features, yet, as shown in Table 29, each has distinctive features. The 21q− syndrome has features that to some extent seem the countertype of those in 21 trisomy.

Other Autosomal Deletions

Deletions of other autosomes, including 8p− (121), 10p− (122), 11q− (123–124) and 12p− (125–127), have been described. Not enough data are available at this time to allow sufficiently clear delineation of syndromes resulting from these deletions.

CYTOGENETICS AND NEOPLASIA

The association between chromosomal abnormalities and neoplasia has been a subject of investigation since the inception of modern cytogenetic techniques (see also Chapter 3).

Congenital Chromosomal Abnormalities

The increased incidence of leukemia in the Down syndrome is the best documented example of an association between a congenital chromosomal abnormality and neoplasia; this was discussed earlier (128). Also, transient hematopoietic aberrations resembling leukemia may occur in infants with the Down syndrome. The occurrence of leukemia and leukemia-like states suggests some form of general aberration in the control of hematopoiesis resulting from 21 trisomy.

Table 29. Phenotypic Features of Long Arm 21 Deletion and Long Arm 22 Deletion Syndromes (42)

21 Deletion*	22 Deletion*
Hypertonia	Hypotonia
Downward slanting eyes	Epicanthal folds
Prominent nasal bridge	Syndactyly of toes
Micrognathia	Ptosis of eyelids
Skeletal malformations	Bifid uvula
Growth retardation	Clinodactyly
Nail anomalies	
Hypospadias	
Inguinal hernia	
Cryptorchidism	
Pyloric stenosis	

*Features common to both syndromes include mental retardation, microcephaly, highly arched palate and large or low-set ears, or both.

An association between the eye tumor, retinoblastoma, and deletion of the long arm of chromosome 13 has also been mentioned earlier (see section on "13 trisomy"). Although the basis of this association is not firmly established, the possibility of retinoblastoma should be kept in mind in patients with 13 long-arm deletion.

Increased incidence of gonadal neoplasia is noted in some patients with the Turner syndrome and gonadal dysgenesis. This increased incidence of gonadoblastoma, seminoma or dysgerminoma seems to occur primarily in patients with mosaicism of the sex karyotype in which one cell line includes a Y-chromosome, i.e., 45,X/46,XY mosaicism, or a portion of a Y (129). Simpson estimates that 10–15% of 45,X/46,XY patients develop gonadal neoplasms. Because of this finding, any patient with a 45,X/46,XY or 45,X/46,X,−X,+mar (the "mar" might be a Y-derived fragment) karyotype should have gonadal tissue removed surgically. Although an occasional patient with 45,X Turner syndrome and a gonadal neoplasm is reported, the association of the 45,X karyotype and predisposition to gonadal neoplasia remains unproven. A patient with this abnormality may have an undiscovered Y–chromosome-bearing cell line.

Specific Chromosomal Aberrations
The first and most extensively documented association of a specific chromosomal abnormality in a specific neoplasm is that between chronic myelogenous leukemia (CML) and the Philadelphia (Ph[1]) chromosome. In 1960, Nowell and Hungerford described a deletion of a small acrocentric chromosome in bone-marrow cells from a patient with chronic myelogenous leukemia (130). Subsequent investigations established that a Ph[1] chromosome can be found in about 90% of cases of chronic myelogenous leukemia (131). The Ph[1] chromosome is not congenital, but current evidence indicates that it can be demonstrated in cells from patients prior to the development of frank leukemia, in some instances by several years (132). It is limited to the nonlymphocytic hematopoietic tissue and may not be demonstrable during drug-induced remission, only to reemerge when the patient relapses.

With the introduction of chromosomal banding techniques, it became apparent that all cases of the Ph[1] chromosome involved a deletion of a major portion of chromosome 22, not 21, as had been claimed previously. However, only in 1973 was it found the Ph[1] chromosomal abnormality involved not a simple deletion of the long arm of chromosome 22, but a translocation of that segment to another chromosome. Rowley (133) reported nine cases of CML in which the deleted segment of chromosome 22 was translocated to the terminal long arm of chromosome 9. Although this translocation is the commonest in CML, the translocation of the long arm of 22 to several other chromosomes has also been reported (134–138). The mechanism whereby a translocation might lead to a loss of control of leukopoiesis is still unknown. In those cases of CML with no demonstrable Ph[1] chromosome, no evidence of elongation of another chromosome has been found.

Although no consistent specific chromosomal abnormality has been found in other types of leukemia, aneuploidy has been found in hematopoietic cells in acute myelogenous leukemia (AML). Although early findings suggested that the chromosomal abnormalities in AML were random, recent evidence indicates a nonrandom distribution, with selective involvement of chromosomes 8 and 21, and to a less extent 7 and 9 (139-140). The commonest aberration appears to be trisomy.

Trisomy 21 has been reported in the bone-marrow cells of a few cases of chronic lymphocytic leukemia (141), and in three of four cases of plasma-cell leukemia a 14q+ marker chromosome replaced a normal chromsome 14 (142). Among other aberrations a deleted chromosome 20 was described in eight of 22 cases of polycythemia vera (143). Burkitt lymphoma, a neoplasm described only in the past few years, is the subject of extensive cytogenetic investigation, partly because of its association with the Epstein-Barr virus. The most frequent chromosomal abnormality was the absence of the distal band of the long-arm of chromosome 8 and its apparent translocation to the terminal long arm of chromosome 14.

Chromosomal abnormalities have also been found in solid tumors (144). In 28 of 29 benign meningiomas, 22-monosomy or deletion of the long arm of chromosome 22 was seen. Trisomy 8 or 14 has been observed in polyps of the colon. Although malignant solid tumors are not associated with specific chromosomal aberrations, most of them are aneuploid, with marker chromosomes peculiar to each individual case. In general the chromosome number is greater than 46 and during the natural history of the malignant process, evolution of the karyotypic abnormality occurs.

Genetic Diseases Predisposing to Neoplasia

Several rare inherited diseases predisposing their victims to certain malignant neoplasms are also associated with increased frequency of chromosome changes.

The Bloom syndrome, an autosomal recessively inherited condition, is characterized by low birth weight dwarfism, a predisposition to respiratory and gastrointestinal infections and telangiectatic lesions of the face and other exposed skin areas, which are progressive with age and made worse by exposure to sunlight. Affected children have a greatly increased risk of developing a malignant neoplasm. Although the most frequent neoplastic process is leukemia, other types of neoplasia involve the alimentary tract. As reported by German, of 16 patients who reached 20 years of age, six developed malignant neoplasm (145).

Another feature of the syndrome is chromosomal instability. Increased incidence of various types of chromosomal rearrangements in short-term cultured lymphocytes and, to a less extent, in cultured fibroblasts is demonstrated. The commonest rearrangement encountered is the quadriradial (Fig. 12), formed by the exchange of chromatid segments by two chromosomes. Certain chromosomes seem to be preferentially involved, including chromosome 1, one or more of the C group and the F group—in almost all cases at homologous sites in homologous chromosomes. Also noted are acentric fragments, dicentrics and other monocentric but structurally abnormal chromosomes (146).

The more recent development of techniques to allow observation of exchanges of segments between the two sister chromatids of a single replicated chromosome has led to the discovery that the rate of such "sister chromatid exchange" is greatly increased in cultured cells from children with the Bloom

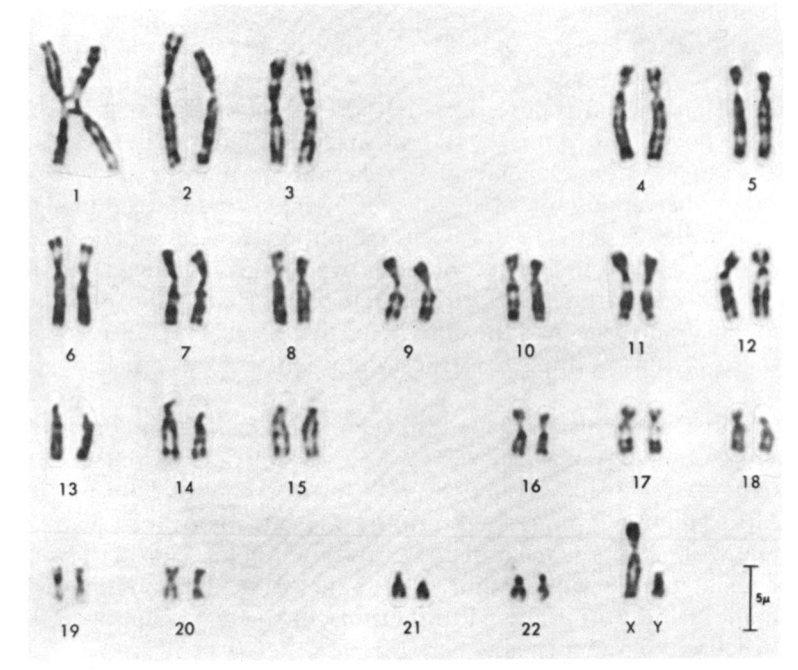

Figure 12. Karyotype showing quadriradial produced by exchange of chromosome segments between two chromosomes 1.

syndrome, as compared to cells from homozygous unaffected persons *or* persons heterogygous for the gene for the syndrome (147–148). The rate of DNA chain growth in cells from victims of the syndrome is significantly slower than that in normal cells (150). It appears from the findings in the syndrome, as well as in the other conditions still to be mentioned, that some correlation exists between chromosome instability and neoplastic potential.

The Fanconi syndrome (Fanconi pancytopenia) was described first in 1927. It appears to be inherited in an autosomal recessive manner and is characterized by pancytopenia; shortness of stature; abnormal skin pigmentation; and congenital limb, cardiac and renal defects. The pancytopenia is not congenital. Its appearance occurs at various ages, from 17 months to 22 years. Anemia is ordinarily the presenting complaint. Skin hyperpigmentation, which may appear before or after the onset of hematologic abnormalities, preferentially involves the trunk, skin folds of the axillae and groin and the neck. Limb anomalies include absent, hypoplastic or supernumerary thumbs; hypoplastic or absent radius; and other less frequent defects (Table 30). Death may result from the progressive panmyelopathy or from leukemia, which occurs with significantly increased frequency. Also, other neoplasms, such as carcinoma of the esophagus and of the anal skin, are reported, and it appears that the incidence of leukemia is increased in heterozygotes for the Fanconi pancytopenia gene.

In 1966, Bloom and co-workers described a variety of gaps, breaks, chromatid exchanges (not only between homologs as occurs in the Bloom syndrome) and endoreduplication (151). The increase in sister chromatid exchanges found in

Table 30. Phenotypic Features of Fanconi Pancytopenia (149)*

Feature	Percentage
Skeletal anomalies:	68
Absent, hypoplastic or supernumerary thumb	50
Hypoplastic or absent radius	13
Congenital hip dislocation	6
Retarded growth	54
Microcephaly	28
Microphthalmia	19
Strabismus	16
Deafness	6
External ear anomalies	4
Abnormal pigmentation	74
Gynecomastia	3
Cardiac or vascular defect	6
Renal anomalies:	29
Unilateral aplasia	21
Horseshoe kidney	10
Inguinal hernia	4
Hypoplasia or aplasia of adrenals	4
Hypogonadism	40
Hypospadias	5
Cryptorchidism	16
Hydrocephalus	3
Obesity	4

*Other anomalies include pes planus, Klippel-Feil anomaly, clubfoot, Sprengel deformity, mental retardation, hyperreflexia, septate bladder, Meckel's diverticulum and pilonidal sinus.

cells from patients with Bloom syndrome is not observed in cells from patients with the Fanconi syndrome (147–148). The commoner occurrence of various types of rearrangements and especially of gaps and breaks is now accepted by many workers as a constant feature of the syndrome.

Ataxia-telangiectasia (Louis-Bar syndrome) is an autosomal recessive condition characterized by, in early childhood, cerebellar ataxia, other neurologic manifestations, oculocutaneous telangiectasia, frequent and severe respiratory infections, abnormalities of lymphoid tissue and an immunodeficiency state (152–153). Increased incidence of neoplasia of the reticuloendothelial system, and other malignancies, has been reported in this condition.

Studies of the chromosomes of the lymphocytes of patients with the syndrome reveal an inconstant increase in gaps and breaks, a relatively constant increase in chromosomal rearrangements, such as dicentrics and abnormal monocentrics, and a very interesting appearance of clones (distinct, stable subpopulations) of chromosomally abnormal cells. These clones in some cases constitute the majority of cells analyzed and involve cells with translocations, preferentially involving the chromosomes of the D group, particularly chromosome 14. The mechanisms involving the occurrence of these clones and their significance, are poorly understood. Abnormalities are also detectable in cultured fibroblasts. However, no increase in sister chromatid exchanges has been demonstrated in cultured lymphocytes (154).

Xeroderma pigmentosum, a rare defect, is inherited in an autosomal recessive manner and characterized by profound sensitivity to ultraviolet light. Homozygous affected persons appear normal at birth, but exposure to sunlight leads to progressively severe injury to the skin. Changes include the appearance of freckles of various size and pigmentation, progressive dryness, telangiectasia, atrophy and keratoses. Photophobia may be intense, corneal ulceration and scarring may lead to blindness, and the eyelids may become scarred and ectropic. Skin changes in the patient progress relentlessly to the formation of large numbers of squamous cell and basal cell carcinomas, leading to his or her death before age 30. Other ectodermal and mesodermal neoplasms also occur with increased frequency. Genetic heterogeneity apparently exists in xeroderma pigmentosum, as demonstrated by mental retardation in some patients in addition to the features already mentioned (De Sanctis-Cacchione syndrome). The vast majority of patients with xeroderma pigmentosum are mentally normal.

Ultraviolet (UV) light routinely induces DNA damage in any exposed cells. Normally, however, an efficient repair mechanism reconstitutes the DNA without adverse effect. Evidence is convincing that the DNA-repair mechanism is defective in xeroderma pigmentosum cells (154). The defect is corrected when xeroderma pigmentosum cells are hybridized with cells of the normal golden hamster (156). In addition, the cellular defect in patients with both classic xeroderma pigmentosum and the De Sanctis-Cacchione syndrome is corrected when the two cell types form heterokaryons (157).

Other apparently distinct forms of xeroderma pigmentosum have been demonstrated by similar complementation experiments (158). This is another indication of the genetic heterogeneity in xeroderma pigmentosum. Available evidence points to a deficiency of an ultraviolet-specific endonuclease as the basis for the defective DNA repair, indicates that the "unrepaired" DNA is responsible for the exquisite sensitivity of patients to ultraviolet light, and suggests that these factors may lead directly or indirectly to the characteristic malignant neoplasms to which most patients succumb.

Chromosomal gaps, breaks and rearrangements of the types seen in the Bloom syndrome, Fanconi pancytopenia and ataxia-telangiectasia are not found in cultured lymphocytes or fibroblasts of patients with xeroderma pigmentosum. However, German et al (159) have noted chromosomally aberrant clones in cultured skin fibroblasts and suggested the existence of a predisposition in xeroderma pigmentosum cells to the formation of these clones of cells. Bartram and colleagues (147) have demonstrated an increased number of UV-induced sister chromatid exchanges in cultured xeroderma pigmentosum lymphocytes. All of these changes probably can be related to the abnormality in the DNA-repair mechanism, which seems to be the constant feature among all types of xeroderma pigmentosum. The possible role of the chromosomally aberrant cell clones in carcinogenesis is of great interest, but awaits further elucidation.

PRENATAL CYTOGENETIC DIAGNOSIS

Amniotic fluid contains cells derived from the fetus. These cells can be grown in culture, and chromosomal analysis performed on them. Transabdominal amniocentesis can be performed on the pregnant woman as early as the 13th or

14th week of pregnancy. These factors now make possible the prenatal diagnosis of chromosomal abnormalities in pregnancies at risk (160). Those pregnancies in which prenatal diagnostic studies specifically directed toward the detection of chromosomal abnormalities may be appropriate include:

1. A pregnancy in a woman who is, or whose mate is, a carrier of a balanced chromosomal rearrangement, either an inversion or a translocation. As noted in a previous section, a female who carries a *Robertsonian* translocation between chromosomes 14 and 21 has a risk in any pregnancy of 10–15% of producing an infant with the Down syndrome. For parents who carry balanced *reciprocal* translocations, the risk may be higher.

2. A pregnancy in a woman age 35 years or older. The risk for older mothers relative to the Down syndrome has been discussed. Accumulated data indicate that a woman who becomes pregnant between age 35 and 39 years has a 2.2% probability of producing a child with *some* variety of chromosomal abnormality. The risk for a woman 40 years old approximates 3.4%, rising to about 10% by 45 years.

3. A pregnancy in a woman of whatever age who has delivered a child with a chromosomal abnormality. Relative to the Down syndrome, accumulated experience from prenatal diagnostic programs reveals that about 1.5% of these pregnancies produce another affected child.

Maternal age seems to be the commonest reason for prenatal cytogenetic diagnostic procedures. Prenatal diagnostic procedures also are conducted for several noncytogenetic purposes. Probably chromosomal analysis should be conducted on cells from any specimen of amniotic fluid, whatever the reason for obtaining that specimen. This is routine in our laboratory.

Recent studies conducted in the United States (161) and in Canada (162) indicate that amniocentesis is a safe procedure, with no increase in overall fetal loss in women undergoing the procedure compared with controls on whom amniocentesis was not performed. Although actual risk may exceed the level suggested by these studies, it is certainly low enough to justify the procedure for the indications cited. Probably it is still not safe enough to be advocated in all pregnancies; neither are laboratory facilities available to handle such a load.

On the other hand, Milunksy (160) estimates that, in Massachusetts, only 4.1% of women older than 35 years have had diagnostic amniocentesis. Thus, even in those pregnancies in which the indication is definite, the procedure is grossly underutilized. This may be due to clinicians' lack of awareness of the procedure, lack of knowledge by laymen or the cost of the procedure. In general, if prenatal genetic diagnosis is to become significantly more widely utilized, support by governmental or other agencies will be necessary.

REFERENCES

1. Barr ML, Bertram EG: A morphological distinction between neurones of the male and female and the behavior of the nucleolar satellites during accelerated nucleoprotein synthesis. *Nature* 163:676, 1949
2. Polani PE, Hunter WF, Lennox B: Chromosomal sex in Turner's syndrome with coarctation of aorta. *Lancet* 2:120, 1954

3. Bradbury JT, Bunge RG, Boccabella RA: Chromatin test in Klinefelter's syndrome. *J Clin Endocrinol Metabol* 16:689, 1956
4. Tjio JH, Levan A: The chromosome number of man. *Hereditas* 42:1, 1956
5. Lejeune J: Le mongolisme. Premier exemple d'aberration autosomique humaine. *Ann Genet Sem Hop* 1:41, 1959
6. Ford CE, Jones KW, Polani PE, de Almeida JC, et al: A sex-chromosome anomaly in a case of gonadal dyslenesis (Turner's syndrome). *Lancet* 1:711, 1959
7. Jacobs PA, Strong JA: A case of human intersexuality having a possible XXY sex determining mechanism. *Nature* 183:302, 1959
8. Caspersson T, Zech L, Johansson C: Analysis of the human metaphase chromosome set by aid of DNA-binding fluorescent agents. *Exp Cell Res* 62:490, 1970
9. Sumner AT, Evans HJ, Buckland RA: A new technic for distinguishing between human chromosomes. *Nature New Biol* 232:31, 1972
10. Paris Conference: Standardization in human cytogenetics. Birth Defects 8:7, 1972
11. Lubs HA, McKenzie WH, Patil SR, Merrick S: New staining methods for chromosomes, in Prescott DM (ed): Methods in Cell Biology, 6th ed. New York, Academic, 1973, p. 345
12. Hook EB, Hamerton JL: The frequency of chromosome abnormalities detected in consecutive newborn studies—differences between studies—results by sex and by severity of phenotypic involvement, in Hook EB, Porter IH (eds): Population Cytogenetics: Studies in Humans. New York, Academic, 1977, p. 63
13. Boue JG, Boue A, Lazar P: Retrospective and prospective epidemiological studies of 1500 karyotyped spontaneous human abortions. *Teratology* 12:11, 1975
14. Jacobs PA: Human population cytogenetics, in de Grouchy J, Ebling FJG, Henderson IW (eds): Human Genetics. Proceeding of the Fourth International Congress of Human Genetics. Amsterdam, Excerpta Medica, 1972, p. 232
15. Polani PE: Autosomal imbalance and its syndromes, excluding Down's *Brit Med Bull* 25:81, 1969
16. Lubs HA, Patil SR, Kimberling WJ, Brown J, et al: Q and C banding polymorphisms in 7 and 8 year old children: Racial differences and clinical significance, in Hook EB, Porter IH (eds): Population Cytogenetics: Studies in Humans. New York, Academic, 1977, p. 133
17. McKenzie WH, Lubs HA: Human Q and C chromosomal variations: Distribution and incidence. *Cytogen Cell Genet* 14:97, 1975
18. Langenbeck U, Hansmann I, Hinney B, Hönig V: On the origin of the supernumerary chromosome in autosomal trisomies—with special reference to Down's syndrome. *Hum Genet* 33:89, 1976
19. Hamerton JL: Human Cytogenetics, Vol 1, in General Cytogenetics. New York, Academic, 1971
20. Lyon MF: Sex chromatin and gene action in the mammalian X-chromosome. *Amer J Hum Genet* 14:135, 1962
21. Bobrow M, Pearson PL, Pike MC, El-Alfi OS: Length variation in the quinacrine-binding segment of human Y chromosomes of different sizes. *Cytogenetics* 10:190, 1971
22. Caspersson T, Zech L, Johansson C, Lindsten J, et al: Fluorescent staining of heteropyknotic chromosome regions in human interphase nuclei. *Exp Cell Res* 61:472, 1970
23. Hamerton JL: Human Cytogenetics, Vol II. Clinical Cytogenetics. New York, Academic, 1971
24. Wilroy RS, Summitt RL, Martens PR, Tharapel AT, et al: Phenotype-karyotype correlations in 81 patients with the Turner syndrome. *Clin Res* 25:74A, 1977
25. Palmer CG, Reichmann A: Chromosomal and clinical findings in 110 females with Turner syndrome. *Hum Genet* 35:35, 1976
26. Simpson JL: Gonadal dysgenesis and abnormalities of the human sex chromosomes: Current status of phenotypic-karyotypic correlations. *Birth Defects* 11:23, 1975
27. Reyes FI, Koh KS, Faiman C: Fertility in women with gonadal dysgenesis. *Amer J Obstet Gynecol* 126:668–670, 1976

28. Harnden DG: Normal and abnormal cell division, in Hamerton JL (ed):   Chromosomes in Medicine. London, Heinemann, 1961, p. 17
29. Creasy MR, Crolla JA, Alberman ED: A cytogenetic study of human spontaneous abortions using banding techniques. *Hum Genet* 31:177, 1976
30. Gordon DL, Krmpotic E, Thomas W, Gandy HM, et al: Pathologic testicular findings in Klinefelter's syndrome. *Arch Intern Med* 130:726, 1972
31. Opitz JM: Klinefelter syndrome, in Bergsma D (ed):   Birth Defects Atlas and Compendium. Baltimore, Williams & Wilkins, 1973, p. 554
32. Becker KL, Hoffman DL, Albert A, Underdahl LO, et al: Klinefelter's syndrome. Clinical and laboratory findings in 50 patients. *Arch Intern Med* 118:314, 1966
33. Dosik H, Wachtel SS, Khan F, Spergel G, et al: Y-chromosomal genes in a phenotypic male with 46XX karyotype. *JAMA* 236:2505, 1976
34. Tennes K, Puck M, Bryant K, Frankenburg W, et al: A developmental study of girls with trisomy X. *Amer J Hum Genet* 27:71, 1975
35. Miller OJ: Chromosome twenty-one trisomy syndrome, in Bergsma D (ed):   Birth Defects Atlas and Compendium. Baltimore, Williams & Wilkins, 1973, p. 252
36. Miller RW: Neoplasia and Down's syndrome. *Ann NY Acad Sci* 171:637, 1970
37. Hook EB: Estimate of maternal age-specific risks of a Down syndrome birth in women aged 34–41. *Lancet* 2:33, 1976
38. Fishler K, Koch R, Donnell GN: Comparison of mental development in individuals with mosaic and trisomy 21 Down's syndrome. *Pediatrics* 58:744, 1976
39. Golbus MS: The antenatal detection of genetic disorders. Current status and future prospects. *Obstet Gynecol* 48:497, 1976
40. Hamerton JL: Robertsonian translocation in man: Evidence for prezygotic selection. *Cytogenetics* 7:260, 1968
41. Wright SW, Day R.W., Muller H, Weinhouse R: The frequency of trisomy and translocation in Down's syndrome. *J Pediatr* 70:420, 1967
42. Summitt RL: Abnormalities of the autosomes and their resultant syndromes. *Pediatr Ann* 2:40, 1973
43. Patau KA, Smith DW, Therman EM, Inhorn SL, et al: Multiple congenital anomaly caused by an extra autosome. *Lancet* 1:790, 1960
44. Magenis RE, Hecht F, Milham S Jr: Trisomy 13 (D$_1$) syndrome. Studies on parental age, sex ratio, and survival. *J Pediatr* 73:222, 1968
45. Magenis E, Hecht F: Chromosome thirteen trisomy syndrome, in Bergsma D (ed):   Birth Defects Atlas and Compendium. Baltimore, Williams & Wilkins, 1973, p. 249
46. Snodgrass GJAI, Butler LJ, France NW, Crome L, et al: The "D" (13–15) trisomy syndrome: An analysis of 7 examples. *Arch Dis Childh* 41:250, 1966
47. Powars D, Rohde RA, Graves D: Foetal haemoglobin and neutrophil anomaly in the D$_1$-trisomy syndrome. *Lancet* 1:1363, 1964
48. Caspersson T, Lindsten J, Zech L, Buckton KE, et al: Four patients with trisomy 8 identified by the fluorescence and giemsa banding techniques. *J Med Genet* 9:1, 1972
49. Fineman RM, Ablow RC, Howard RO, Albright J, et al: Trisomy 8 mosaicism syndrome. *Pediatrics* 56:762, 1975
50. Fryns JP, Vevessen H, Van den Berghe H, Van Kerckvoorde J, et al: Partial trisomy 8: Trisomy of the distal part of the long arm of chromosome number 8+(8q2) in a severely retarded and malformed girl. *Humangenetik* 24:241, 1974
51. Sanchez O, Yunis JJ: Partial trisomy 8 (8q24) and the trisomy-8 syndrome. *Humangenetik* 23:297, 1974
52. Rethore M-O, Larget-Piet L, Abonyi D, Boeswillwald M, et al: Sur quatre cas de trisomie pour le bras court du chromosome 9. Individualisation d'une nouvelle entite morbide. *Ann Genet* 13:217, 1970
53. Podruch PE, Weisskopf B: Trisomy for the short arms of chromosome 9 in two generations, with balanced translocations t(15p+;9q−) in three generations. *J Pediatr* 85:92, 1974

54. Centerwall WR, Miller KS, Reeves LM: Familial "partial 9p" trisomy: Six cases and four carriers in three generations. *J Med Genet* 13:57, 1976
55. Sutherland GR, Carter RF, Morris LL: Partial and complete trisomy 9: Delineation of a trisomy 9 syndrome. *Hum Genet* 32:133, 1976
56. Wilroy RS Jr, Summitt RL, Martens P, Gooch WM III, et al: Partial monosomy and partial trisomy for different segments of chromosome 13 in several individuals of the same family. *Birth Defects* 12:161, 1976
57. Noel B, Quack B, Rethore M-O: Partial deletions and trisomies of chromosome 13: Mapping of bands associated with particular malformations. *Clin Genet* 9:593, 1976
58. Escobar JI, Sanchez O, Yunis JJ: Trisomy for the distal segment of chromosome 13. *Amer J Dis Child* 128:217, 1974
59. Lewandowski RC Jr, Yunis JJ: New chromosome syndromes. *Amer J Dis Child* 129:515, 1975
60. Van den Berghe H, Van Eygen M, Fryns JP, Tanghe W, et al: Partial trisomy 1, karyotype 46, XY,12−,t(1q,12p)+. *Humangenetik* 18:225, 1973
61. Neu RL, Gardner LI: A partial trisomy of chromosome 1 in a family with a t(1q−;4q+) translocation. *Clin Genet* 4:474, 1973
62. Norwood TH, Hoehn H: Trisomy of the long arm of human chromosome 1. *Humangenetik* 25:79, 1974
63. Francke U, Jones KL: The 2p partial trisomy syndrome. *Amer J Dis Child* 130:1244, 1976
64. Bijlsma JB, deFrance H, Bleeker-Wagemakers EM, Wijffels JCHM: Duplication deficiency syndrome in familial translocation (2q−:5p+). *Humangenetik* 12:110, 1971
65. Forabosco A, Dutrillaux B, Toni G, Tamborino G, et al: Translocation equilibree t(2;13)(q32;q33) familiale et trisomie 2q partielle. *Ann Genet* 16:255, 1973
66. Rosenthal IM, Beligere N, Thompson F, Pruzansky S, et al: Trisomy of the distal portion of the long arm of chromosome 2, a new familial syndrome associated with mental retardation and a characteristic facies. *Amer J Hum Genet* 26:73a, 1974
67. Warren RJ, Panizales EG, Cantwell RJ: Inherited partial trisomy 2: 46,XX,1p+;t(1:2)(p36;q31). *Birth Defects* 11:177, 1975
68. Summitt RL: Familial ⅔ translocation. *Amer J Hum Genet* 18:172, 1966
69. Allderdice PW, Browne N, Murphy DP: Chromosome 3 duplication q21→qter deletion p25→pter syndrome in children of carriers of a pericentric inversion inv(3) (p25q21). *Amer J Hum Genet* 27:699, 1975
70. Opitz JM, Patau KA: A partial trisomy 5p syndrome. *Birth Defects* 11:191, 1975
71. Vogel W, Siebers JW, Reinwein H: Partial trisomy 7q. *Ann Genet* 16:277, 1973
72. Bass HN, Crandall BF, Marcy SM: Two different chromosome abnormalities resulting from a translocation carrier father. *J Pediatr* 83:1034, 1973
73. Berger R, Derre J, Ortiz MA: Les trisomies partielles du bras long du chromosome 7. *Nouv Presse Med* 3:1801, 1974
74. Serville F, Broustet A, Sandler B, Bourdeau M-J, et al: Trisomie 7q partielle. *Ann Genet* 18:67, 1975
75. Al Saadi A, Moghadam HA: Partial trisomy of the long arm of chromosome 7. *Clin Genet* 9:250, 1976
76. Slavka M-P, Lala Z, Kraigher A: Translocation reciproque dans la famille de deux proposants avec une trisomie partielle du chromosome 7q. *Ann Genet* 19:133, 1976
77. Yunis JJ, Sanchez O: A new syndrome resulting from partial trisomy for the distal third of the long arm of chromosome 10. *J Pediatr* 84:567, 1974
78. Roux C, Taillemite J-L, Baheux-Morlier G: Trisomie partielle 10q par translocation familiale t(t10q−;22p+). *Ann Genet* 17:59, 1974
79. Forabosco A, Bernasconi S, Giovannelli G, Dutrillaux B: Trisomy of the distal third of the long arm of chromosome 10. *Helv Paediatr Acta* 30:289, 1975
80. Prieur M, Forabosco A, Dutrillaux B, Laurent C, et al: La trisomie 10q24→10qter. *Ann Genet* 18:217, 1975

81. Moreno-Fuenmayor H, Zackai EH, Mellman WJ, Aronson M, et al: Familial partial trisomy of the long arm of chromosome 10 (q24–26). *Pediatrics* 56:756, 1975
82. Reiss JA, Wyandt HE, Magenis RE, Lovrien EW, et al: Mosaicism with translocation: Autoradiographic and fluorescent studies of an inherited reciprocal translocation t(2qt;14q−). *J Med Genet* 9:280, 1972
83. Short EM, Solitare GB, Breg WR: A case of partial 14 trisomy 47, XY,(14q⁻)+ and translocation t(9p+;14q−) in mother and brother. *J Med Genet* 9:367, 1972
84. Muldal S, Enoch BA, Ahmed A, Harris R: Partial trisomy 14q− and pseudoxanthoma elasticum. *Clin Genet* 4:480, 1973
85. Fryns JP, Cassiman JJ, Van den Berghe H: Tertiary partial 14 trisomy 47,XX,+14q−. *Humangenetik* 24:71, 1974
86. Raoul O, Rethore M-O, Dutrillaux B, Michon L, et al: Trisomie 14q partielle I.—Trisomie 14q partielle par translocation maternelle t(10:14) (p15.2;q22). *Ann Genet* 18:35, 1975
87. Turleau C, de Grouchy J, Bocquentin F, Roubin M, et al: Trisomie 14q partielle II.—Trisomie 14q partielle par translocation maternelle t(12;14) (q24.4:q21). *Ann Genet* 18:41, 1975
88. Laurent C, Dutrillaux B, Biemont M-Cl, Genoud J, et al: Translocation t(14q−;21q+) chez le pere. Trisomie 14 et monosomie 21 partielles chez la fille. *Ann Genet* 16:281, 1973
89. Pfeiffer RA, Büttinghaus K, Struck H: Partial trisomy 14− following a balanced reciprocal translocation t(14q−:21q+). *Humangenetik* 20:187, 1973
90. Lo Curto F, Maraschio P, Milanesi P, Severi F, et al: The syndrome of partial trisomy 14q. *Eur J Pediatr* 123:237, 1976
91. Fawcett WA, McCord WK, Francke U: Trisomy 14q−. *Birth Defects* 11:223, 1975
92. Pena SDJ, Ray M, McAlpine PJ, Ducasse C, et al: Tertiary trisomy 14: Is there a syndrome? *Birth Defects* 12:113, 1976
93. Yeatman GW, Riccardi VM: Partial trisomy of chromosome 14: (+14q−). *Birth Defects* 12:119, 1976
94. Hsu LYF, Shapiro LR, Gertner M, Lieber E, et al: Trisomy 22: A clinical entity. *J Pediatr* 79:12, 1971
95. Bass HN, Crandall BF, Sparkes RS: Probable trisomy 22 identified by fluorescent and trypsin-Giemsa banding. *Ann Genet* 16:189, 1973
96. Zellweger H, Ionasescu V, Simpson J: Trisomy 22. *J Genet Hum* 23:65, 1975
97. Begleiter ML, Kulkarni P, Harris DJ: Confirmation of trisomy 22 by trypsin-Giemsa banding. *J Med Genet* 13:517, 1976
98. Perez-Castillo A, Abrisqueta JA, Martin-Lucas MA, Goday C, et al: A new contribution to the study of 22 trisomy. *Humangenetik* 30:265, 1975
99. Emanuel B, Zackai EH, Aronson MM, Mellman WJ, et al: Abnormal chromosome 22 and recurrence of trisomy-22 syndrome. *J Med Genet* 13:501, 1976
100. Alfi OS, Sanger RG, Donnell GM: Trisomy 22: A clinically identifiable syndrome. *Birth Defects* 11:241, 1975
101. Uchida IA, Ray M, McRae KN, Besant DF: Familial occurrence of trisomy 22. *Amer J Hum Genet* 20:107, 1968
102. Punnett HH, Kistenmacher ML, Toro-Sola M, Kohn G: Quinacrine fluorescence and giemsa banding in trisomy 22. *Theor Appl Genet* 43:134, 1973
103. Golbus MS, Bachman R, Wiltse S, Hall BD: Tetraploidy in a liveborn infant. *J Med Genet* 13:329, 1976
104. Niebuhr E: Triploidy in man. Cytogenetical and clinical aspects. *Humangenetik* 21:103, 1974
105. Wolf W, Porsh R, Baitsch H, Reinwein H: Deletion on short arms of a B-chromosome without "cri du chat" syndrome. *Lancet* 1:769, 1965
106. Johnson VP, Mulder RD, Husen R: The Wolf-Hirschhorn (4p−) syndrome. *Clin Genet* 10:104, 1976
107. Miller OJ, Breg WR, Warburton D, Miller DA, et al: Partial deletion of the short arm of chromosome No. 4(4p−): Clinical studies in five unrelated patients. *J Pediatr* 77:792, 1970

108. Lejeune J, LaFourcade J, Berger R, Vialatte J, et al: Trois cas de deletion bras court d'un chromosome 5. *CR Acad Sci (Paris)* 257:3098, 1963
109. Breg WR, Ward PH: Chromosome five p− syndrome, in Bergsma D (ed): Birth Defects Atlas and Compendium. Baltimore, Williams & Wilkins, 1973, p. 248
110. Miller DA, Warburton D, Miller OJ: Clustering in deleted short arm length among 25 cases with the Bp− chromosome. *Cytogenetics* 8:109, 1969
111. Niebuhr E: Localization of the deleted segment in the cri-du-chat syndrome. *Humangenetik* 16:357, 1972
112. Alfi OS, Donnell GN, Allderdice PW, Derencsenyi A: The 9p− syndrome. *Ann Genet* 19:11, 1976
113. Allderdice PW: Chromosome thirteen q− syndrome, in Bergsma D (ed): Birth Defects Atlas and Compendium. Baltimore, Williams & Wilkins, 1973, p. 250
114. Knudson AG, Meadows AT, Nichols WW, Hill R: Chromosomal deletion and retinoblastoma. *N Engl J Med* 295:1120, 1976
115. de Grouchy J, Lamy M, Thieffry S, Arthuis M, et al: Dysmorphie complexe avec oligophrenie: Deletion des bras iourt d'un chromosome 17−18. *CR Acad Sci (Paris)* 256:1028, 1963
116. Schinzel A, Schmid W, Luscher U, Nater M, et al: Structural aberrations of chromosome 18, 1. The 18p− syndrome. *Arch Genet* 47:1, 1974
117. Summitt RL, Patau KA: Cytogenetics in mental defectives with anomalies. *J Pediatr* 65:1097, 1964
118. Schinzel A, Hayashi K, Schmid W: Structural aberrations of chromosome 18. II. The 18q− syndrome. Report of three cases. *Humangenetik* 26:123, 1975
119. Reisman LE, Kasahara S, Chung C-Y, Darnell A, et al: Antimongolism. Studies in an infant with a partial monosomy of the 21 chromosome. *Lancet* 1:394, 1966
120. Warren RJ, Rimoin DL, Summitt RL: Identification by fluorescent microscopy of the abnormal chromosomes associated with the G deletion syndromes. *Amer J Hum Genet* 25:77, 1973
121. Orye E, Craen M: A new chromosome deletion syndrome. Report of a patient with a 46,XY,8p− chromosome constitution. *Clin Genet* 9:289, 1976
122. Francke U, Kernahan C, Bradshaw C: Del (10)p autosomal deletion syndrome: Clinical, cytogenetic and gene marker studies. *Humangenetik* 26:343, 1975
123. Taillemite J-L, Baheux-Morlier G, Roux Ch: Délétion interstitielle du bras long d'un chromosome 11. *Ann Génét* 18:61, 1975
124. Engel E, Hirshberg CS, Cassidy SB, McGee BJ: Chromosome 11 long arm partial deletion: A new syndrome. *Amer J Mental Def* 80:473, 1976
125. Tenconi R, Baccichetti C, Anglani F, Pellegrino P-A, et al: Partial deletion of the short arm of chromosome 12(p11;p13). *Ann Genet* 18:95, 1975
126. Magnelli N, Therman E: Partial 12p deletion: A cause for a mental retardation, multiple congenital abnormality syndrome. *J Med Genet* 12:105, 1975
127. Orye E, Craen M: Short arm deletion of chromosome 12. *Humangenetik* 28:335, 1975
128. Cervenka J, Koulischer L: Chromosomes and Human Cancer. Springfield, Ill, CC Thomas, 1973
129. Simpson JL, Photopulos G: The relationship of neoplasia to disorders of abnormal sexual differentiation. *Birth Defects* 12:15, 1976
130. Nowell PC, Hungerford DA: A minute chromosome in human chronic granulocytic leukemia. *Science* 132:1197, 1960
131. Whang-Peng J, Canellos GP, Carbone PP, Tjio JH: Clinical implications of cytogenetic variants in chronic myelocytic leukemia(CML). *Blood* 32:755, 1968
132. Canellos GP, Whang-Peng J: Philadelphia-chromosome-positive preleukemic state. *Lancet* 2:1227, 1972
133. Rowley JD: A new consistent chromosomal abnormality in chronic myelogenous leukemia identified by quinacrine fluorescence and Giemsa staining. *Nature* 243:290, 1973
134. Gahrton G, Zech L, Lindsten J: A new variant translocation (19q+;22q−) in chronic myelocy-

tic leukemia. *Exp Cell Res* 86:214, 1974
135. Harland AA, Wolman SR, Distenfeld A, Cohen T: Another variant translocation in chronic myelogenous leukemia. *N Engl J Med* 294:164, 1976
136. Matsunaga M, Sadamori N, Tomonaga Y, Tagawa M, et al: Chronic myelogenous leukemia with an unusual karyotype: 46,XY,t (17q+;22q−). *N Engl J Med* 295:1537, 1976
137. Mammon Z, Grinblat J, Joshua H: Philadelphia chromosome with t(6;22) (p25,q12). *N Engl J Med* 294:827, 1976
138. Hayata I, Kakati S, Sandberg AA: Another translocation related to the Ph[1] chromosome. *Lancet* 1:1300, 1975
139. Rowley JD: Nonrandom chromosomal abnormalities in hematologic disorders in man. *Proc Nat Acad Sci USA.* 72:152, 1975
140. Yamada K, Furusawa S: Preferential involvement of chromosomes No. 8 and No. 21 in acute leukemia and preleukemia. *Blood* 47:679, 1976
141. Crossen PE: Giemsa banding patterns in chromic lymphocytic leukemia. *Humangenetik* 27:151, 1975
142. Philip P: Marker chromosome 14q+ in multiple myeloma. *Hereditas* 80:155, 1975
143. Mitelman F, Levan G: Clustering of aberrations to specific chromosomes in human neoplasms. *Hereditas* 82:167, 1976
144. Atkin NB: Chromosomes in human malignant tumors: A review and assessment, in German J (ed): Chromosomes and Cancer. New York, Wiley, 1974, p. 375
145. German J: Bloom's syndrome. II. The prototype of human genetic disorders predisposing to chromosome instability and cancer, in German J (ed): Chromosomes and Cancer. New York, Wiley, 1974, p. 601
146. German J, Crippa LP, Bloom D: Bloom's syndrome III. Analysis of the chromosome aberration characteristic of this disorder. *Chromosoma* 48:361, 1974
147. Bartram CR, Koske-Westphal T, Passarge E: Chromatid exchanges in ataxia telangiectasia, Bloom syndrome, Werner syndrome, and xeroderma pigmentosum. *Ann Hum Genet* 40:79, 1976
148. Chaganti RSK, Schonberg S, German J: A manyfold increase in sister chromatid exchanges in Bloom's syndrome lymphocytes. *Proc. Nat. Acad. Sci. USA* 71:4508, 1974
149. Minagi, H, Steinback HL: Roentgen appearance of anomalies associated with hypoplastic anemias of childhood: Fanconi's anemia and congenital hypoplastic anemia (erythrogenesis imperfecta). *Am. J. Roentgen., Rad. Ther. and Nuclear Med.* 97:100, 1966
150. Hand R, German J: A retarded rate of DNA chain growth in Bloom's syndrome. *Proc Nat Acad Sci USA* 72:758, 1975
151. Bloom GE, Warner S, Gerald PS, Diamond LK: Chromosome abnormalities in constitutional aplastic anemia. *N Engl J Med* 274:8, 1966
152. Harnden DG: Ataxia telangiectasia syndrome: Cytogenetic and cancer aspects, in German J (ed): Chromosomes and Cancer. New York, Wiley, 1974, p. 619
153. McFarlin DE, Strober W, Waldmann TA: Ataxia-telangiectasia. *Medicine* 51:281, 1972
154. Hatcher NH, Brinson PS, Hook EB: Sister chromatid exchanges in ataxia-telangiectasia. *Mutat Research* 35:333, 1976
155. Robbins JH, Kraemer KH, Lutzner MA, Festoff BW, et al: Xeroderma pigmentosum. An inherited disease with sun sensitivity, multiple cutaneous neoplasms, and abnormal DNA repair. *Ann Int Med* 80:221, 1974
156. Miller OJ. Cell hybridization in the study of the malignant process, including cytogenetic aspects, in German J (ed): Chromosomes and Cancer. New York, Wiley, 1974, p. 521
157. Day RS III, Kraemer KH, Robbins JH: Complementing xeroderma pigmentosum fibroblasts restore biological activity to UV-damaged DNA. *Mutat Res* 28:251, 1975
158. deWeerd-Kastelein EA, Keijzer W, Bootsma D: A third complementation group in xeroderma pigmentosum. *Mutat Res* 22:87, 1974
159. German J, Gilleran TG, Setlow RB, Regan JD: Mutant karyotypes in a culture of cells from a

man with xeroderma pigmentosum. *Ann Genet* 16:23, 1973
160. Milunsky A: Current concepts in genetics. Prenatal diagnosis of genetic disorders. *N Engl J Med* 295:377, 1976
161. NICHHD National Registry for Amniocentesis Study Group: Midtrimester amniocentesis for prenatal diagnosis. Safety and accuracy. *JAMA* 236:1471, 1976
162. Simpson NE, Dallaire L, Miller JR, Siminovich L, et al: Prenatal diagnosis of genetic disease in Canada: Report of a collaborative study. *Can Med Assoc J* 115:739, 1976

3
Genetics and Cancer

Laird G. Jackson
David E. Anderson
R. Neil Schimke

At present the question of genetic factors in neoplasia is no longer "*Is* cancer hereditary," but "*How* is cancer hereditary?" The notion that abnormal cellular events could produce intracellular imbalance leading to cancer was first enunciated by the German cytologist, Von Hansemann (1). In the early part of the present century the Flemish cytologist, Theodor Boveri, described mitotic irregularities in double-fertilized sea urchin eggs that led to the formation of disorganized cell groups and abnormal cellular chromosomal content (2). Boveri was convinced that this was the precursor of neoplastic growth and represented a general phenomenon that would explain all cancer. Although Boveri was unable to prove the theory, his work and publications became the stimulus for increasing effort by other workers to relate the genesis of neoplastic growth to hereditary factors. Many subsequent investigators, demonstrated the generational transmission of tumors in inbred strains of mice (3).

Although true that no murine tumor appears to be transmitted by a single gene, study of several mouse neoplasms, notably mammary tumor and leukemia, demonstrates that several genes influence the susceptibility of the mouse to tumor development. The principal causative agent in both tumors was a virus, transmitted in the mother's milk when the pups were immunologically susceptible. This finding did not negate the importance of genetics, as subsequent evidence demonstrated. First the mice in question were all of highly inbred strains and the transmissibility of the virus was influenced strongly by the genetic characteristics of the mouse. Newborn mice of another strain, when nursed by mothers carrying the virus, did not show the same susceptibility as did the pups of the original mother.

Second it was known that tumors were uncommon in the wild mouse; only in the inbred laboratory strains did they reach high proportions. The third piece of experimental evidence came from tumor transplantation from which it was observed that tumors arising in one strain could be transplanted to other mice of the same strain, but not to animals of a different strain. By observation of the number of "takes" in hybrid mice the genotype of which was a mixture derived from one susceptible and one resistant parent, an estimate was established of the number of genetic factors in the acceptance or rejection of the tumor. The genes responsible were called histocompatibility genes, and many such genes and their chromosomal location (the H2 locus) have been mapped in the mouse. Observation of the role of the histocompatibility genes in transplantability of the mouse tumors showed that these genes occasionally changed suddenly. This phenomenon was seen as resulting from somatic mutation, an interpretation that further supported the genetic theory of tumor susceptibility. Thus, although mouse leukemia was directly caused by a virus, the susceptibility of the mouse to the virus was controlled by the major histocompatibility locus.

MALIGNANCY AND GENETIC CONTROL

Recently the major histocompatibility complex (MHC) in man has been similarly investigated. Unfortunately the same type of strong association between the MHC, located on chromosome 6, and human neoplasia, has not been found (4). The best-studied genes in the MHC are those for antigens found in human leukocytes, the HLA system. No HLA association has been found in patients with acute leukemia. Patients with chronic myelogenous leukemia (CML), however, have the HLA-A3 antigen more frequently and HLA-A12 less frequently than controls (5). The only other positive associations include a scattered set of HLA relationships with Hodgkin's disease. At present none of these associations is convincing.

That human malignancy should be at all influenced by the genetic control of susceptibility to viruses is strengthened by the peculiar malignancy known as Burkitt's lymphoma (6). Burkitt described a lymphomatous tumor in the mandible in children of the central African plain. Because this geographic area corresponded to the yellow fever belt, a search for a mosquito-vectored virus ensued. Subsequently Epstein and Barr detected virus particles (E-B virus) in lymphoblastoid cell lines in tissue culture obtained from these tumors, and patients were shown to have serum antibodies to the antigen (7). Although the virus itself is ubiquitous, support for its oncogenic effect comes from the finding of a cytohistologically identical disease in the United States. This latter differs principally from Burkitt's lymphoma by a later age of onset (11 vs 7 years) and location of the tumors in the peripheral lymph nodes, gastrointestinal (GI) tract and bone marrow. In the United States the E-B virus itself is found in certain forms of nasopharyngeal carcinoma. The virus appears to have oncogenic potential although it apparently requires a coincident cellular change, such as a chromosomal alteration or immunologic deficit. The E-B viral lymphoma story remains the best-documented example of probable virus-associated malignancy in man.

Other work with viruses in human beings demonstrates increased susceptibility of cultured cells from certain "premalignant" conditions to transformation by oncogenic viruses, such as SV-40. Initially demonstrated in the cells of patients with Fanconi's pancytopenia, the work has been extended to other conditions, particularly those with chromosomal imbalance (8). These studies indicate that the cells of both the affected patients and their heterozygous parents show increased transformation. Further, Swift demonstrated a significant increase in early onset cancer in the heterozygote for the Fanconi syndrome gene (9). A similar investigation of ataxia-telangiectasia heterozygotes revealed similar but less marked increases to the extent that more than 5% of all persons dying from cancer less than age 45 are estimated to carry the ataxia-telangiectasia gene (10).

An important group of observations has been made on the cellular origin of cancer, utilizing genetic markers like glucose-6-phosphate dehydrogenase (G-6-PD) to distinguish single- from multicelled origins for tumors (11). The Burkitt lymphoma, e.g., has a single-cell origin and it is interesting that in all studies to date of early relapses of the tumor, the same cell clone returned, whereas in recurrences after long remission the emergence of new clones was seen (12). Genetic marker studies reveal similar findings in patients with acute lymphatic leukemia following radical therapy and bone-marrow transplantation. The findings suggest either the activation of latent tumor virus in donor cells or the transfer of virus from host to donor cell. In contrast, studies of G-6-PD markers in chronic myelogenous leukemia indicated that all leukemia cells had a common enzyme phenotype, whereas nonleukemic granulocytes and skin cells had mixed heterozygous phenotypes. This finding virtually eliminates cell recruitment as an explanation for the disorder and strongly supports a single-cell origin of CML (13).

Immunodeficiency Disorders

Because tumors occur with increasing frequency in certain immunodeficiency disorders (14), this and other observations prompted speculation on the role of the immune system in oncogenesis (15, 16). Burnet in particular suggested that malignant cells were a common event, but ordinarily the abnormal cells were destroyed by the host's immunologic surveillance mechanisms. This theory required two assumptions: 1) that the tumor cells must possess unique surface antigens permitting them to be recognized as foreign; and 2) that the host must be capable of developing an appropriate and effective immune response against these antigens. An obvious corollary to the second assumption predicts that cancer will occur more frequently in patients or in animals with immunodeficient responses. Accumulating evidence in recent years supports each of these two points. Both chemically and physically induced tumors possess specific antigens (17, 18).

Furthermore there are tumors in man, such as acute leukemia and the Burkitt lymphoma, that possess cell-specific antigens (19). Additional evidence for the immune role in cancer is suggested by experiments in which either transplanted immune cells from immune donors caused tumor regression or tumor regression occurred after activation of the immune system by BCG. This relationship between the immune system and the development of cancer may take several courses (20). First the immunodeficiency syndromes may render the host susceptible to infection with oncogenic agents. Second the same disorders may impair the host's ability to destroy spontaneously generated malignant cells. Third certain oncogenic agents, including chemicals and viruses, may have additional immunosuppressive properties facilitating or enhancing their oncogenic potential. Fourth the neoplastic process itself may suppress the immune response and perpetuate tumor growth. That some or all of these mechanisms work to increase the risk of development of cancer in persons with immunodeficiency diseases is clearly established (21).

Figures show that the incidence of malignancy in these patients may be as much as 10,000 times higher than in a comparable control population. In addition, unlike the control population, approximately 60% of cancers in the immunodeficient patient involve the lymphoreticular system. Each type of immunodeficient disease has a characteristic group of malignancies associated with it (see Chapter 4 for further description of clinical features of the immunodeficiency diseases). Undoubtedly antibiotics and stringent therapy for the immunodeficiency itself will expose more of the neoplastic potential of these disorders. Risks for the development of neoplasm in these diseases vary from 2% in subacute combined immunodeficiency to 8-10% in IgM deficiency, common variable hypogammaglobulinemia, Wiskott-Aldrich syndrome and ataxia-telangiectasia. Therapy for these tumors has not been successful although initial attempts at bone-marrow transplantation were encouraging in the Wiskott-Aldrich syndrome (22).

In an attempt to synthesize a single hypothesis accommodating most of the preceding information as it relates to neoplasms in man, Knudson developed what is now known as the "two-hit" model for human oncogenesis (23) and the occurrence of heritary and nonheritary tumors of identical form. His work was based on observations originally made on retinoblastoma in man and precipitated by the work of Mole in radiation carcinogenesis (24). As applied to tumors in man, Knudson's hypothesis states that when the first mutation occurs in a germ cell, the mutation will be present in all cells of the resultant individual and will be hereditary. The gene carrier may develop none, one or multiple tumors; the number will follow a Poisson distribution. If the first mutation is somatic, it will be confined to a single cell and the cancer will be nonhereditary.

The second mutation is always somatic in both the hereditary and nonhereditary types. Thus all tumors require two mutational events. In the hereditary form the first event has already occurred and is in all cells; therefore only a single second event is necessary for tumor development. Hereditary tumors will both occur at an earlier average age than nonhereditary ones and likely be multiple. The nonhereditary type will be late and occur singly, since tumor development in a single cell requires two infrequent mutational events.

When applied to existing data on retinoblastomas (25), Wilms' tumors (26), neuroblastomas and pheochromocytomas (27), the model appears to be valid. The hereditary forms of these tumors occur earlier and there are more bilateral or multiple site tumors compared with the nonhereditary forms. The estimated incidence of germinal mutations ranges from 22% for neuroblastomas and pheochromocytomas to 40% for retinoblastomas and Wilms' tumor. The model also shows close correlation with its prediction of distribution of multiple tumors in retinoblastoma. It predicts an average of three retinoblastomas per gene carrier, following a Poisson distribution. Of carriers 5% should have no tumor, 35% unilateral tumor and 60% bilateral tumors. The observed frequencies show that 10% of carriers develop no tumor, 25–40% have unilateral tumors and 60–75% develop bilateral tumors. Knudson has gone further with these data to suggest that perhaps all childhood cancers will fit this model, that they are therefore probably not preventable if they are the result of germinal mutation, and that only surveillance and early detection will be effective in reducing their toll.

On the other hand a recent critical review of retinoblastoma in Japan suggests that host resistance and other factors, rather than delayed mutation or a second mutational event, may explain the "penetrance" of the tumor gene (27a).

FAMILIAL CANCER AND CANCER FAMILIES

Many observations over the years indicate that the mutational model just described may also apply to tumors in adult human beings. Hereditary cancer characterized by early age of onset as compared to sporadic cases and by multiple tumors may develop in family members (28). Families with an excess of either a specific neoplasm or different neoplasms are referred to as *cancer fraternities* or *cancer families*. Warthin was one of the first workers to use these terms (29). He applied them to families in which affected relatives died of gastrointestinal, but no other kind of, cancer; or to families in which affected members developed cancer only of the thyroid or uterus or lip and no other. In his most publicized family (Family G), the male members showed an excess of gastrointestinal carcinoma, and the female members, carcinoma of the uterus. He attributes these aggregations to an inherited tendency to cancer.

Other investigators have used cancer family in referring to aggregations of any cancer among relatives, hereditary or not. The medical history of Family G has been repeatedly updated, the latest being the summary of Lynch et al (30). This family now includes 82 members with cancer either of the colon, endometrium, stomach or breast. The patient's age at diagnosis of these neoplasms averages about 49 years, a decade or two earlier than that observed for these neoplasms in the population at large. Of the affected family, 14% of the members have developed multiple primaries, mostly of the colon or of colon and uterus.

Lynch now applies the term cancer family syndrome not only to Warthin's family but to any family with increased incidence of adenocarcinoma primarily of colon and uterus (endometrium), increased frequency of multiple primary malignancies, early age at onset and vertical transmission of cancer consistent with an autosomal dominant mode of inheritance. The term cancer family, therefore, has no specific meaning, referring as it may to hereditary as well as to nonhereditary cancers. If hereditary, it may refer to various distinct clinicogenetic entities. Cancer family syndrome may also encompass several different clinicogenetic entities.

The study of familial cancers is important because of its potential for providing clues concerning carcinogenesis and its relevance to early detection and treatment. It is important, therefore, that the different types of cancer families be identified and distinguished. A first step in this direction is to purify the diagnosis in such families according to not only the site(s) of involvement but also the histologic type(s) of the primary and associated neoplasms. The diagnosis—and hence appellation—of any associated genetic condition could also be clarified.

A second step is to conduct pedigree studies of these carefully documented families. Attention to such variables as the cancer distribution pattern, age at diagnosis, tumor localization, associated tumors of the same or different sites or stigmatizing anomalies should lead to the identification of specific types of hereditary cancers and their mode of inheritance. The utility of this approach will be illustrated for some cancers found in adults specific hereditary types of which have previously been neither fully appreciated nor recognized (Table 1). Discussions of other hereditary cancers are in reviews by Knudson (31), Knudson et al (32) and Mulvihill (33), and in five recent monographs (34–39).

EMBRYONAL TUMORS

Malignant tumors are uncommon in the neonate; those that appear probably originate in maldeveloped fetal tissue. These embryonal neoplasms are usually solid, of mixed histologic types or sarcomatous. The overall incidence declines rapidly with advancing age and the majority develop before age 5. Recent evidence supports the conclusion that a significant proportion of "pure" embryonal neoplasms are heritable.

Wilms' Tumor

The world-wide incidence of Wilms tumor (nephroblastoma) is remarkably uniform (0.4–1/10,000 live births) with no racial or ethnic predilection. It comprises nearly 15% of all tumors diagnosed under age 15 with more than 50% occurring before age 3 and more than 90% by age 10. The tumor is probably derived from the metanephric blastema, which is normally fully differentiated by 34 weeks' gestation.

Until recently reports of familial Wilms tumor have been sparse. More aggressive therapy has led to a survival rate of more than 80%; consequently many descriptions of both sibs and two-generation transmission have been recorded. As in other reports of familial tumors of paired organs, bilaterality is frequent and an earlier average age of onset is the rule. A study of familial aggregates has led Knudson and Strong to the following conclusions (26): 1) approximately one-third of all cases are hereditary; 2) when hereditary, the tumor is transmitted as an autosomal dominant trait with about 63% penetrance, i.e., 37% of gene carriers manifest no overt malignancy; 3) about 15% of gene carriers have bilateral tumors and 43% have unilateral lesions; and 4) roughly 30% of all unilateral sporadic cases actually harbor the genetic trait. These data provide a basis for calculation of approximate risk for counseling families with Wilms' tumor (Table 2). More accurate calculations can be obtained by the application of Bayesian statistics as described in the introduction.

Other congenital anomalies are reported in Wilms' tumor patients; the association is significant, e.g., between sporadic congenital aniridia and Wilms' tumor, with nearly one-third of the patients with the eye lesion also having the kidney tumor (39). Conversely only 2% of Wilms' tumor patients have aniridia. The dominant genetic form of aniridia is not associated with Wilms' tumor. Patients with a syndrome of Wilms' tumor, aniridia, pseudohermaphroditism (in the male) and nephrotic syndrome have occasionally been described, although only once in sibs (40). Affected males usually have ambiguous genitalia, and the

Table 1. Simply Inherited Human Neoplasms*

Neoplasm or Disorder	Inheritance	Other Neoplasms
Phacomatoses		
von Recklinghausen's neurofibromatosis	AD†	Fibrosarcoma, neuroma, schwannoma, meningioma, polyps, optic glioma, pheochromocytoma
Tuberous sclerosis	AD	Adenoma sebaceum, periungual fibroma, glial tumors, rhabdomyoma of heart, renal tumor
von Hippel-Lindau syndrome	AD	Retinal angioma, cerebellar hemangioblastoma, other hemangiomas, pheochromocytoma, hypernephroma
Sturge-Weber syndrome	AD	Angioma of numerous organs
Nervous System		
Retinoblastomas, bilateral	AD	Sarcoma
Acoustic neuroma, bilateral	AD	
Neuroblastoma	AD	
Megalencephaly	AD	Ganglioneuroblastoma
Endocrine		
Multiple endocrine neoplasia I (Wermer's syndrome, MEN I)	AD	Adenomas of islet cells, parathyroid, pituitary and adrenal glands, malignant schwannoma, nonappendiceal carcinoid
Multiple endocrine neoplasia II (Sipple's syndrome, MEN II)	AD	Medullary carcinoma of thyroid gland, parathyroid adenoma, pheochromocytoma
Mucosal neuromas and endocrine adenomatosis (MEN III)	AD	Pheochromocytoma, medullary carcinoma of thyroid, neurofibroma, submucosal neuromas of tongue, lips and eyelids
Paraganglioma (chemodectoma)	AD	
Pheochromocytoma	AD	
Hyperparathyroidism	AD	
Mesoderm (Soft Tissue)		
Nevoid basal-cell carcinoma syndrome	AD	Basal-cell carcinoma, medulloblastoma, ovarian fibroma and carcinoma
Multiple hamartoma syndrome (Cowden syndrome)	AD	Papillomatosis of lip and mouth, hypertrophic and cystic breast with early cancer, thyroid adenoma and carcinoma, bone and liver cysts, lipoma, polyps and meningioma
Leopard syndrome	AD	multiple lentigines
Gingival fibromatosis ± hypertrichosis or other anomalies	AD AD	
Juvenile fibromatosis	AR	Multiple subcutaneous tumors
Familial cutaneous collagenoma	AR	Multiple skin nodules
Multiple leiomyoma	AD	Cutaneous, uterine, or esophageal leiomyoma, or both
Multiple lipomatosis, some times site-specific, neck or conjunctiva	AD	Skin cancer
Goldenhar syndrome	AR	Lipodermoid of conjunctiva, hemangioma
Macrosomia adiposa	AR	Adrenocortical adenoma
Multiple eruptive milia	AD	Colonic cancer
Universal melanosis	AD	
Nevi (pigmented and halo)	AD	Malignant melanoma
Albinism	AR	Skin cancer

Table 1. (Continued)

Neoplasm or Disorder	Inheritance	Other Neoplasms
Mesoderm (Soft Tissue)		
Xeroderma pigmentosum, xerodermoid pigmentosum (including De Sanctis-Cacchione syndrome)	AR	Skin cancer
Epidermolysis bullosa dystrophica	AD AR	Skin cancer arising in scars
Disseminated superficial actinic porokeratosis	AR	Skin cancer
Multiple sebaceous gland tumors and visceral carcinoma (Torre's syndrome)	AR	Diverse GI and UG cancers
Acrokeratosis verruciformis	AD	Warty hyperkeratosis
van den Bosch syndrome	XR	
Darier-White disease	AD	Keratotic papules
Scleroatrophy and keratosis of limbs	AD	Skin and bowel cancer
Pachyonychia congenita	AD	Hyperkeratosis, cutaneous horns, leukoplakia
Multiple trichoepithelioma (Spiegler-Brooke tumors; cylindromatosis)	AD	
Self-healing squamous epithelioma	AD	
Hyperkeratosis lenticularis perstans	AD	Skin cancer
Steatocystoma multiplex ± pachyonychia congenita	AD	
Lymphatic and Hematopoietic		
Histiocytic reticulosis, generalized or neural only (Letterer-Siwe disease)	AR	
Familial lipochrome histiocytosis	AR	
Kostmann's infantile genetic agranulocytosis	AR	Acute monocytic leukemia (chromosomal breaks)
Glutathione reductase deficiency	AR	Leukemia (chromosomal breaks)
Immunodeficiency		
Bruton's agammaglobulinemia	XR	Leukemia, lymphoreticular
Wiskott-Aldrich syndrome	XR	Lymphoreticular
Ataxia-telangiectasia	AR	Lymphoreticular, leukemia, carcinoma of stomach, brain tumors (chromosomal breaks)
Chediak-Higashi syndrome	AR	Pseudolymphoma
Multiple System		
Bloom's syndrome	AR	Leukemia, intestinal cancer (chromosomal breaks)
Fanconi's pancytopenia	AR	Acute monomyelogenous leukemia, squamous-cell carcinoma of mucocutaneous junctions, hepatic carcinoma and adenoma (chromosomal breaks)
Dyskeratosis congenital (Zinnsser-Cole-Engman syndrome)	XR AD	Leukoplakia with squamous-cell carcinoma, including cervical
Beckwith-Wiedemann syndrome	AR	Visceromegaly cytomegaly, macroglossia, adrenal cortical neoplasia, Wilms' tumor, hepatoma
Rothmund-Thomson syndrome	AR	Squamous-cell carcinoma
Werner's syndrome	AR	Sarcoma
Osteopoikilosis	AD	Nevi

Table 1. (Continued)

Neoplasm or Disorder	Inheritance	Other Neoplasms
Alimentary Tract		
Familial polyposis coli	AD	Intestinal polyps, carcinoma of colon
Gardner syndrome	AD	Intestinal polyps; osteomas; fibromas; sebaceous cysts; carcinoma of colon, ampulla of Vater, pancreas, thyroid and adrenal gland
Peutz-Jeghers syndrome	AD	Intestinal polyps, ovarian (granulosa cell) tumor
Colorectal carcinoma	AD	
Hereditary pancreatitis	AD	Carcinoma of pancreas
Tylosis with esophageal cancer	AD	Carcinoma of esophagus
Hepatocellular carcinoma	AD AR	
Familial, juvenile, and neonatal cirrhosis	AD AR	Hepatocellular carcinoma
Hemochromatosis	AD AR	Hepatocellular carcinoma
Urogenital		
Pure gonadal dysgenesis (XY), Reifenstein syndrome	XR	Gonadoblastoma, dysgerminoma
Testicular feminization	XR	
True hermaphroditism	AR?	
Ovarian tumors	AD	
Testicular tumors	AD	
Wilms' tumors	AD	
Hydronephrosis, familial	AD	Congenital sarcoma of kidney
Pulmonary		
Fibrocystic pulmonary dysplasia	AD	Bronchial adenocarcinoma
Aryl hydrocarbon hydroxylase inducibility	AD?	Brochogenic carcinoma
Vascular		
Multiple glomus tumors	AD	
Herediatry hemorrhagic telangiectasis of Rendu-Osler-Weber	AD	Angioma
Lymphedema with distichiasis	AD	Lymphangiosarcoma of edematous limb
Skeletal		
Multiple exostosis	AD	Osteosarcoma, chrondrosarcoma
Cherubism	AD	Fibrous dysplasia of jaws, giant cell tumor
Fibroosseous dysplasia	AD	Osteosarcoma, medullary fibrosarcoma
Paget's disease of bone	AD	Osteosarcoma
Skin and Appendages		
Breast cancer	AD	
Mastocytosis	AD	
Malignant melanoma	AD	
Neurocutaneous melanosis	AR	Malignant melanoma of skin and meninges
Noonan syndrome	AD	Schwannoma
Focal dermal hypoplasia (Goltz syndrome)	XD	Mucocutaneous papillomas

Table 1. (Continued)

Neoplasm or Disorder	Inheritance	Other Neoplasms
Inborn Errors of Metabolism		
Angiokeratoma diffusa (Fabry syndrome)	XR	
Tyrosinemia, hypermethioninemia, galactosemia	AR	Postcirrhotic hepatoma
Wilson disease, glycogen storage disease-IV		
α_1-antitrypsin deficiency	AR	Hepatoma
Vitamin D-resistant rickets	XR	Parathyroid adenoma

*Includes tumors reported in an inherited pattern; modified from Mulvihill (37) with permission.
†Abbreviations: AD, autosomal dominant; AR, autosomal recessive; XD, X-linked dominant; XR, X-linked recessive; AR?, mode of inheritance suspected but not proven.

gonadal tissue present usually contains only testicular elements (41). Several children with the syndrome of Aniridia-Wilms' tumor have now been described with interstitial deletions of the short arm of chromosome 11 (41a).

Hemihypertrophy is known to develop before and after the discovery of Wilms' tumor, and it may be partial, asymmetrical, or both. A family is described in which a mother with hemihypertrophy had three children with Wilms' tumor and an additional child with renal malformation (42). Another family of normal parents had one child with Wilms' tumor and hemihypertrophy and another child with hemihypertrophy alone (43). Isolated hemihypertrophy may be inherited as an autosomal dominant trait, raising an obvious parallel between the situation noted earlier with dominant aniridia. Hemihypertrophy in children is associated with hepatoblastoma and adrenocortical carcinoma as well as Wilms' tumor. Conceivably some hemihyperthrophic adults may be at risk for having children with congenital tumors, but firm data are lacking.

Wilms' tumor is reported in the Beckwith-Wiedemann syndrome, which may also feature adrenocortical carcinoma, hepatoblastoma and cardiac hamartomas (44). A related condition is reported from a consanguineous Yemenite Jewish family, in which two sibs died perinatally with metanephric blastomatoses and a third sib developed Wilms' tumor (45). The sibs also had gigantism and microphthalmia, but lacked other features of the Beckwith-Wiedemann syndrome.

Table 2. Approximate Risk of Subsequent Child Developing Embryonal Neoplasm with Various Family Histories

Family History	Wilms' tumor (%)*	Retinoblastoma (%)*
Parent with bilateral tumors, parent with unilateral and positive family history, unaffected parent with either 2 children affected, or child and near collateral relative affected	22	48
Parent with unilateral tumor and negative family history	9.5	7.5
Single child with bilateral tumors	10	1.2
Single child with unilateral tumor	<5	1.2

*Assumes 63% penetrance for Wilms' tumor and 95% for retinoblastoma.

Urinary tract abnormalities occur in some patients with Wilms' tumor and on occasion in their otherwise normal sibs (46). Nodular renal blastema and focal renal hamartomas are described in patients with trisomy 18.

Leukemia is reported in survivors of Wilms' tumor after a 5- to 15-year latent period (47). This complication may be a late effect of bone-marrow irradiation, or it may be analogous to what is seen in some patients with retinoblastoma who develop osteogenic sarcoma at a site distant from the radiation field (see next section). To date, none of the Wilms' tumor–leukemia patients has had a positive familial history for either disease.

Retinoblastoma

Retinoblastoma develops in about 1/18,000 liveborn infants. Roughly 30% of all cases are hereditary and transmitted as an autosomal dominant trait with penetrance in excess of 90% (25). Whereas nearly 60% of all gene carriers have bilateral, frequently multifocal, disease, nearly one-third have unilateral findings only. In fact, perhaps 20% of all unilateral cases may actually be the heritable tumor. The various recurrence risks shown in Table 2 make empiric allowance for this estimate.

Some children with sporadic retinoblastoma and other anomalies, including retardation, have a 13q-karyotype, suggesting that the retinoblastoma gene may reside on the number 13 chromosome (48). Continual monitoring of children with retinoblastoma shows that some them develop a second malignancy, particularly in the head and neck, presumably as a direct result of exposure to therapeutic irradiation or intracarotid chemotherapy. However, a number of second tumors develop outside the cranial area in patients treated only surgically, particularly those with bilateral, hence familial retinoblastoma. These second malignancies are most often long bone sarcomas, and they develop after a latency of some years (49).

Neuroblastoma

Neuroblastomas are derived from the ubiquitous neural crest tissue, and thus the tumors are frequently extraadrenal, mostly in the paravertebral area in conjunction with autonomic ganglia (50). Although patients commonly have as the presenting symptom an abdominal mass, more unusual initial symptoms are also described, including the Horner syndrome, polymyoclonia and hypoglycemia. In one series of nearly 500 cases, less than 25% survived three years (51). Thereafter mortality tended to remain unchanged, largely because of the natural tendency of the tumor to undergo spontaneous regression or maturation to ganglioneuromas or ganglioneurofibromas, particularly in women.

Although relatively few familial cases of neuroblastoma are reported, the lesion is described in twins, sibs and two generations (29). Some relatives of affected patients have had elevated urinary catecholamine levels, but this finding is not sufficiently consistent. Unlike the situation with Wilms' tumor, other associated congenital anomalies are not frequent. Second malignancies are not a significant feature so far in surviving patients. The assumption that all cases with multiple tumors represent the heritable form of the disease may not be warranted in view of the paucity of familial reports.

The available information is consistent when familial neuroblastoma occurs as an autosomal dominant trait with intermediate pentrance. Pentrance for a gene like this one may be difficult to determine when prognosis is generally poor and, alternatively, tumor maturation (and hence undetectability) is not uncommon. Familial risks of recurrence are probably similar to those shown for Wilms' tumor (Table 2).

Hepatoblastoma and Sarcomas

There is no evidence favoring genetic factors in the etiology of hepatoblastoma, since only one case of affected infant sisters is described (52). The tumor is associated with hemihypertrophy and the Beckwith-Wiedemann syndrome. More differentiated tumors (hepatomas) can secondarily complicate several genetic syndromes, but the genetic implications are those of the primary disease.

Sarcomas of bone, muscle or connective tissue are likewise rare in early childhood. Rhabdomyosarcomas are probably the commonest of the embryonal sarcomas, but familial aggregation is rare. Osteogenic sarcoma cannot properly be regarded as embryonal since it generally develops in older individuals. Sib–sib correlations among osteogenic sarcomas, rhabdomyosarcomas, brain tumors of various histologic types and other unusual sarcomas were found in a 1971 survey of U.S. death certificates (53). These sibling occurrences and the associated breast tumors in mothers of such children make it seem likely that on occasion sarcomas of early childhood are a manifestation of the cancer family syndrome.

ENDOCRINE GLAND NEOPLASIA

Endocrine tumors are seen in diverse ways since they may secrete excessive quantities of an appropriate hormone, elaborate ectopic hormones or show space-occupying interference with normal secretions. The various components of the endocrine system are diffuse, but recent investigation suggests that all the various tissues originate from the neural crest. Although the evidence for this contention is only tentative, the concept has stimulated considerable thought, especially as it relates to the commonest familial tumors of the endocrine system, multiple endocrine neoplasia (MEN) syndromes.

Multiple Endocrine Neoplasia, Type I

This heritable syndrome includes tumors of the pituitary and parathyroid glands, pancreas and less frequently thyroid epithelial cell and adrenal cortex (54). The parathyroid gland is the structure most commonly involved and the tumors are frequently asymptomatic, being detected by routine screening of serum calcium and phosphorus levels. Pancreatic involvement is interesting, since it is associated with production of a host of hormone hypersecretory states. Excessive gastrin secretion can produce the Zollinger-Ellison (ZE) syndrome, and hyperinsulinism produces the classic symptoms of hypoglycemia.

Glucagonomas may be clinically silent, but can produce an unusual symptom cluster characterized by a necrotizing skin rash, stomatitis and diabetes mellitus. Hyperplasia of the islet delta cells may result in the pancreatic cholera syndrome. Achlorhydria and low serum gastrin levels differentiate this diarrhea from that seen in the ZE syndrome. The commonest probable cause of the syndrome is excess vasoactive intestinal peptide, or VIP. Adrenocortical symptoms may be those of hypercortism or hyperaldosteronism or related to excessive sex steroids.

The thyroid may show simple adenomas, colloid goiter or a profile resembling thyroiditis. Medullary carcinoma is never seen in MEN I. A patient with MEN I syndrome may present with the carcinoid syndrome, produced by bronchial, thymic, pancreatic or intestinal carcinoid tumors; epithelial thymomas, schwannomas and multiple lipomas have also been described.

Multiple endocrine neoplasia, type I (MEN I) is inherited as an autosomal dominant trait with high penetrance, if the patient's entire lifespan is taken into account. Symptoms most commonly develop in the third decade, but some patients may not develop endocrine dysfunction until much later. Rarely are more than one or two endocrine glands cancerous. The spontaneous mutation rate is probably low, and all "sporadic" cases should be assumed to have affected relatives. At-risk family members should be screened at least annually if representative symptoms appear (Table 3). The suggestion that MEN I results from a genetic fault in differentiation of neural crest tissue, although attractive, appears to be an oversimplification.

Multiple Endocrine Neoplasia, Type II

Sipple first noted pheochromocytomas in patients with thyroid carcinoma, and the syndrome bears his name (55). Later workers discovered that the thyroid tumor was invariably a medullary carcinoma derived rather from the parafollicular than from the follicular cell and was responsible for secretion of large quantities of calcitonin (50). MEN II is inherited as an autosomal dominant trait, and its clinical features include signs of catecholamine excess, although some pheo-

Table 3. Screening Studies on Patients at Risk for MEN I

Endocrine Gland	Frequency of Involvement (%)	Screening Studies*
Parathyroid	90	Serum calcium and phosphorus
Pancreas	80	Glucose tolerance test with plasma insulin levels if possible; fasting serum gastrin
Pituitary	65	Plasma growth hormone response to insulin-induced hypoglycemia; metapyrone for ACTH, plasma (or urinary) gonadotropin levels
Adrenal cortex	35	AM and PM plasma cortisol; urinary steroids, less reliable
Thyroid	20	True thyroxine; I^{131}-uptake and scan if gland palpably abnormal

*Other screening studies include: urinary 5-HIAA, gastric analysis with measurement of basal and maximal acid content; upper GI series, biopsy if ulcer symptoms are present; X-rays of sella turcica; visual field examination; and selective stimlatory and suppression tests (dexamethasone and calcium infusion).

chromocytomas are asymptomatic. The medullary thyroid carcinoma may also be asymptomatic and undetectable except by the plasma calcitonin assay. The thyroid tumor is multifocal and invariably malignant. The pheochromocytoma may be unilateral, bilateral or extra-adrenal and is usually benign. As with MEN I, months or years commonly elapse between the appearance of one tumor and the development of another.

Medullary thyroid carcinoma secretes various ectopic hormones, such as ACTH, serotonin, VIP and prostaglandins.

Parathyroid hyperplasia, which is as frequent in MEN II as it is in MEN I, may be either secondary to calcitonin-induced hypocalcemia or a primary manifestation of the pleiotropic mutant gene. Other tumors described with MEN II include gliomas, glioblastomas and meningiomas. All of the tumors in MEN II, except possibly the parathyroids, could theoretically result from abnormal neural crest differentiation.

Plasma calcitonin assay is important for early detection of medullary thyroid tumor and is particularly useful in the periodic evaluation of potentially affected members of families in whom the proband harbors the MEN II gene. It may also be valuable in patients with known disease to assess completeness of surgical removal or later metastases.

Multiple Endocrine Neoplasia, Type III

This disorder is also called the multiple mucosal neuroma syndrome because, in addition to pheochromocytoma and medullary thyroid carcinoma, affected individuals have striking physical findings including neuromas of the conjuctival, labial and buccal mucosa, the tongue, the larynx and the GI tract (56). Enlarged corneal nerves; soft, "blubbery" lips; and pseudoprognathism due to soft-tissue hypertrophy of the chin are striking as is the tall, Marfanoid habitus with hypotonia, lax joints, kyphoscoliosis, genu valgus and pes cavus. Café au lait spots and diffuse lentiginous pigmentation may be present. Megacolon has also been seen. The phenotype is virtually diagnostic of underlying thyroid malignancy. Other symptoms are similar to those of MEN II, except that both parathyroid hyperplasia and ectopic hormone production are less common.

Autosomal dominant inheritance of the entity is strongly suggested, but the degree of penetrance and range of clinical expression are not fully known. Evaluation of the few families so far described suggests that penetrance is high and that most affected individuals have at least mucosal neuromas.

Other Familial Endocrine Tumors

There are a host of reports in the literature of combinations of endocrine tumors. Some of the latter may represent true tumor syndromes; others may be coincidental associations. Pituitary tumors are not uncommon neoplasms, but familial examples outside the diagnostic confines of MEN I are rare.

Familial hyperparathyroidism may be inherited as an autosomal dominant trait with nearly full penetrance and onset from adolescence to late adulthood. Multiple adenomas or diffuse hyperplasia occur also in nearly 90% of patients with MEN I and II, and a number of families with hyperparathyroidism may represent partial expression of these two endocrine neoplastic syndromes. Certainly affected families should have a full endocrinologic evaluation to exclude these conditions. Primary thyroid adenocarcinoma (not medullary) is reported in sibs, parent and child both as a complication of familial goiter and in association with nonendocrine tumors. At present the reports have no obvious common denominator and do not constitute sufficient evidence for familial thyroid cancer, distinct from the medullary type. A familial risk of recurrence would be negligible, unless other factors (irradiation, nonfamilial goiter) were operational. It seems likelier that the bulk of familial thyroid tumors so far described represent a partial expression of other genetic disorders, namely, those of the primary disease.

Aside from the malignant islet cell tumors found in MEN I, pancreatic endocrine tumors do not occur in a genetic form. Neither is there evidence supporting a heritable form of simple adrenocortical tumor, especially in adults. Malignant involvement of the adrenal glands in childhood is uncommon, but evidence exists for genetic factors. The frequent association of other neoplasms in the affected child or his first-degree relatives has led to the presumption that at least some of the early onset cases represent limited expression of the cancer family syndrome. Adrenocortical neoplasia occurs in children with Wilms' tumor, hepatoblastoma and hemihypertrophy, as mentioned earlier.

Familial neuroblastomas and embryonal tumors originating at least in part from the adrenal medulla have been discussed previously. Pheochromocytoma, the other major tumor of the adrenal medulla, occurs more commonly in adults than in children. Probably about one-fourth of all cases are hereditary, occurring with equal frequency as a unilateral or bilateral lesion and occasionally outside the adrenal as well. When inherited, pheochromocytoma follows an autosomal dominant mode of inheritance with nearly complete penetrance. Pheochromocytoma may be part of neurofibromatosis and von Hippel-Lindau disease. Recent ultrastructural and histochemical studies suggest that the paraganglia (carotid and aortic bodies, glomus jugulare, ganglion nodosum and other related structures) are both derived from the neural crest and rather intimately related to pheochromocytoma. A father with a glomus jugulare tumor, e.g., had two daughters with bilateral carotid body tumors, one of whom subsequently developed a pheochromocytoma (57).

Another reported patient developed pheochromocytoma, paraganglioma and bilateral carotid tumors at ages 11, 13 and 36, respectively (58). Although paragangliomas (or chromodectomas) have occurred without pheochromocytoma in family aggregates consistent with autosomal dominant inheritance, the structures are closely related embryologically.

BREAST CANCER

There is little question that breast cancer indeed has a hereditary basis. Obviously not all breast cancers are the consequence of an inherited genetic defect; rather among all patients there will be some fraction, between 15 and 30%, whose disease is genetically caused (32). Genetic studies from both high- and low-incidence countries convincingly demonstrate a consistent two-to threefold higher risk for the disease in close relatives of a patient than in controls or in comparison with expected rates (59). Additional studies indicate that the risks vary with age at the time of diagnosis and bilaterality of disease (60). Premenopausal onset increased the risk threefold to relatives, bilateral disease increased the risk fivefold, and there was a nearly 10-fold increase with both premenopausal and bilateral disease. Relatives of patients with postmenopausal and unilateral disease had no appreciably increased risk over controls.

In addition the risk varied according to the pattern of breast cancer occurrence in families (61). When the mother of the affected patient had prior disease (mother pedigrees), the average risk in the patient's sisters was fivefold higher than in similar-aged control sisters. If this increased risk is applied to the general probability (6-7%) of developing breast cancer by age 72 years, the lifetime probability for these sisters with affected mothers would be 30-35%. The actual lifetime probability through age 79 was calculated at 32 percent (62). The disease in this pedigree group usually developed in the premenopause and at similar ages among the affected sisters with 15% bilateral primary disease compared with a 3% frequency among breast cancer patients in general. The frequency of bilateral disease in the affected mothers was also higher than expected. The enhancing effects of early diagnosis and bilaterality on risk noted in the earlier studies were confined solely to this one pedigree group.

In pedigrees wherein the mothers were unaffected but prior disease involved a sister of the patient (sister pedigrees), the risks were lower and the disease characteristics also differed. The average risk for the remaining sisters of the two affected sisters was 2.7 times higher than in controls, or a 12 to 16% lifetime probability of breast cancer development (61, 62). The disease usually developed in the para- or postmenopausal period and often in a single breast. The disease in this pedigree group thus appeared to be more heterogeneous than that in the mother pedigrees, and its origin was probably multigenic. Wang et al (63) recently demonstrated that unaffected sisters of patients had significantly lower urinary etiocholanolone and androsterone and plasma dehydroepiandrosterone sulphate levels than did controls, particularly in the age interval from 30 to 40 years. No differences were noted between daughters of patients and controls.

In pedigrees in which prior disease involved a second-degree relative, the risks for sisters of patients were little different from control values (61). Further evidence of a genetic basis and heterogeneity in breast cancer is suggested by pedigree studies. At least seven potentially distinct familial or hereditary types of the disease are now identified by this approach (Table 4) (64-70).

Since many of the familial and hereditary forms of breast cancer have early onset and multicentric or bilateral occurrence, or both, women belonging to such families should be counseled about their high risk of breast cancer and taught breast self-examination from an early age. Examination by a physician should begin when the subject is between 25 and 30 years of age and include a physical breast examination, thermography and a baseline xeromammogram. Subsequent yearly examinations should be confined to physical examination of the breast plus thermography to reduce radiation exposure since the examinations are begun at a relatively early age. Mammography should of course be used to evaluate suspicious lesions.

Once breast cancer has developed, it should be treated as a systemic process,

Table 4. Familial Forms of Breast Cancer

Condition	Characteristics	Inheritance*	Reference
Breast cancer syndrome	Early onset and bilateral breast cancer in association with soft-tissue sarcoma, leukemia and other neoplasms	AD	64
Familial breast and ovarian cancer	Breast carcinoma and ovarian cyst-adenocarcinoma	AD	65
Familial breast and colonic cancer	Breast carcinoma and colonic cancer	AD	66
Cowden's disease (multiple hamartoma syndrome)	Breast cysts and carcinoma, thyroid adenoma and ADC, colonic polyps and ADC, papillomatosis of lips and mucosa	AD	67
Premenopausal breast cancer	Early onset and bilateral breast cancer	AD	68
Postmenopausal breast cancer	Late onset and unilateral breast cancer	AD	69
Lobular carcinoma in situ	Lobular carcinoma in situ, often bilateral and multicentric; invasive lobular and ductal and intraductal carcinoma	AD	70

*Abbreviations: AD, autosomal dominant mode of inheritance.

and any remaining breast tissue should be regarded as premalignant. The remaining breast should be examined at least every six months. Surgical management should be considered for patients in whom regular followup is difficult or in whom an underlying disease, such as fibrocystic disease, hinders adequate evaluation. The patient with unilateral breast cancer has a risk increase for cancer in the contralateral breast four times normal if she has a relative with unilateral breast cancer; and 36 times normal if she has a relative with bilateral breast cancer (61).

GASTROINTESTINAL TRACT CANCER

Epidemiologic studies suggest that cancers of the GI tract are influenced by an important but variable environmental effect. The incidence of esophageal cancer varies 100- to 400-fold between high- and low-incidence countries, whereas stomach cancer varies about 30-fold, and colon cancer only about 10-fold (71). Ratios by gender show a similar gradient, namely, a highly disparate 20:1 male-to-female ratio for esophageal cancer, decreasing to a constant 1:1 ratio, or one with a slightly higher incidence of females to males for cancer of the colon. Because these gradients seem to reflect the relative environmental contribution to each of these cancers, it is reasonable to assume that the converse of the gradients reflects the relative contribution of genetic factors. Thus the genetic contribution is minimal for esophageal, larger for gastric and largest for colonic, cancer (Table 5).

Esophagus

There is little convincing evidence to suggest a genetic basis for susceptibility to esophageal cancer. Only two pedigree studies are recorded. Freytes and Carri (72) described a neoplasm in three sibs and stomach cancer in another whose mother had died of an unspecified intestinal cancer. A large family was reported by Pour and Ghadirian (73) in which 13 descendants of the same parents developed the neoplasm and at much earlier ages than other patients from the same geographic area. This dearth of evidence points to the rarity of a direct genetic effect on the development of this neoplasm.

Increasing evidence, however, shows that esophageal cancer may occur with some rare genodermatoses. The best documented study describes the association with a late-onset type of tylosis (keratosis palmaris et plantaris), a dominantly inherited trait. Harper et al (74) described esophageal cancer in four tylotic families. In one of them, 17 members—all with tylosis—developed esophageal cancer. It was estimated that 95% of the remaining members would develop esophageal cancer by age 65. The age of tumor occurrence in this family was much earlier than it was for the usual esophageal cancer, averaging about 48 years (75). Six members of another tylotic family died of esophageal cancer, and in two others esophageal cancer was confined to only one tylotic family member. A different presentation involved a family in which tylosis was associated with esophageal stricture in two or possibly three generations, and esophageal cancer only developed in the proband (76). This family may carry a different gene or allele than that in the previous families mentioned.

Other neoplasms are also reported in patients with tylosis, such as bronchogenic carcinoma (75) and laryngeal and gastric cancer (77). A report on three families described an unusual form of skin lesion called sclerotylosis, in which the hyperkeratosis of hands and feet was accompanied by both atrophic changes of skin and nails and frequent development of cutaneous malignancy (78). Among 44 individuals with the condition, seven developed malignancies in the affected skin, another seven developed carcinoma of the tongue, tonsil, breast and uterus, but there were no cases of esophageal cancer. Based on these type associations, Ritter proposed that the association between tylosis and malignancy may be more generalized than previously suspected, as evidenced by the development of esophageal carcinoma in a 25-year-old woman with hyperkeratosis of the knuckle pads (tylositas articularis) and hyperkeratosis (keratosis pilaris) of the arms (79). The patient also had leukoplakia of the buccal mucosa. Further support is provided by Table 6, in which several other genodermatoses also appear to be associated with squamous-cell carcinomas of the same organs implicated in tylotic patients and their families.

Stomach

Numerous studies of gastric cancer are surprisingly consistent in indicating a risk 2–2.5 times higher for the neoplasm among the relatives of gastric cancer patients than in controls (80). Macklin (81) simultaneously studied families of two groups of patients, one with gastric cancer and the other with cancer of the large intestine. First-degree relatives of gastric cancer patients exhibited a rate of gastric cancer 2.6 times higher than expected and no excess of cancer of the large intestine, whereas the first-degree relatives of intestinal cancer patients exhibited a rate of colon cancer 3.2 times higher than expected with no excess of gastric cancer. These results indicated that each neoplasm had specific induction factors, i.e., the genetic mechanism for each was specific for that neoplasm only, not to neoplasm in general.

The most definitive evidence of a genetic component in gastric cancer comes from pedigree studies. A classic pedigree documents the neoplasm in direct descent through three generations and in four of the seven sibs of Napoleon Bonaparte, who was also affected (82). The age at death among the affected family members averaged 49 years, much earlier than that usually encountered for this neoplasm. Other familial aggregations are reported (83, 84), including a family with 12 affected individuals (85). Several family members, with or without gastric cancer, manifested parietal cell antibodies and cell-mediated immune defects, suggesting that mechanisms of autoimmunity and immunodeficiency, consistent with a genetic defect of T-lymphocytes, were involved in this family's susceptibility to gastric cancer. The role of the immune mechanism has also been implicated by the occurrence of this neoplasm in association with an inherited form of IgA deficiency and ataxia-telangiectasia.

Table 5. Genetic Gastrointestinal Disease–Cancer Syndromes

Disorder	Major Intestinal Manifestation	Extraintestinal	Predisposition to GI Cancer	Inheritance* (ref. 67–71)
Familial polyposis:				
Type I (Juvenile adenomatosis)	Adenomatous polyps (100–over 1000) in colon & rectum mean onset 24 yrs. of age.	None	Multiple adenocarcinoma in 80% of untreated patients by age 40	AD
Type II (Peutz-Jeghers)	Generalized polyposis, especially small bowel	Mucosal and digital pigmentation thecal cell ovarian tumors	2-3% risk for GI cancer	AD
Type III (Gardner)	Adenomatous polyps in colon and rectum	Subcutaneous and osseous tumors, epidermoid cysts, hypodontia	Carcinoma	AD
Polyposis with CNS tumors (Turcot)	Colonic polyps	Glioma	Adenomatous polyps only	AR
Generalized juvenile polyposis	Hamartomatous polyps (10—100) in colon, rectum, small bowel and stomach: cysts.	None	Carcinoma, not closely related to polyps, increased polyps and carcinoma in relatives	AD
Generalized GI adenomatous polyps	Adenomatous polyps in stomach, small and large bowel	Desmoid reported	Polyps only	AD?
Sporadic colonic polyps (solitary polyps)	Occasional colonic polyps	None	Adenocardinoma	AD
Hereditary gastrocolonic polyps	Gastric and colonic polyps	None	Adenocarcinoma of stomach and colon	AD

Hereditary colon cancer (site specific)	Adenocarcinoma	None	Adenocarcinoma 3X as frequent in relatives	AD
Ulcerative colitis	Colitis	None	Adenocarcinoma in 37% of patients in first 10 years, 20% in next decade	
Crohn's disease	Inflammatory bowel disease	None	Increased risk of small bowel adenocarcinoma	
Cancer family syndrome II (Torre)	Adenocarcinoma of small bowel and stomach	Sebaceous adenomas visceral cancer	Primary lesion	AD
Familial combined breast and colon	Adenocarcinoma of colon	Breast adenocarcinoma	Primary lesion	AD?
Cancer family syndrome I	Adenocarcinoma of colon, occasionally stomach	Adenocarcinoma of uterus, occasionally of breast	Primary lesion	AD
Cronkite-Canada syndrome	Diffuse polyposis	Alopecia, hypopigmentation, malabsorption	Adenocarcinoma is usual	Unknown (probably not genetic)

*Abbreviations: AD, autosomal dominant mode of inheritance; AR, autosomal recessive mode of inheritance; AD? mode of inheritance not proven.

Table 6. Genodermatoses Related to Squamous-Cell Carcinoma (SCC) of Skin and Other Sites

Condition	Cutaneous Neoplasm	Other Sites	Inheritance*
Dyskeratosis congenita	not reported	SCC oral mucosa, tongue nasopharynx, esophagus, lung, cervix	AR
Ectodermal dysplasia	SCC nailbeds	SCC tongue, cervix	AD
Epidermolysis bullosa dystrophica (several types)	SCC extremities	SCC oral mucosa, tongue	AD +AR
Hereditary keratoacanthoma	Keratoacanthomas	SCC larynx	AD
Hyperkeratosis lenticularis perstans (lower limbs)	SCC head, neck	SCC lung	AD
Sclerotylosis†	SCC nailbeds, hands	SCC tongue, tonsil; adenocarcinoma of breast and uterus	AD
Tylosis		SCC esophagus	AD

*Abbreviations: AD, autosomal dominant trait; AR, autosomal recessive trait.
†Linked with MNS blood group system.

Colorectum

The commonest sites of familial GI cancer are the colon and rectum. The majority of these cancers involve the various hereditary polypoid diseases (Table 5). The clinicogenetic attributes of these polypoid and other large intestinal diseases predisposing to cancer of this site have been eminently discussed (86–90) and need not be repeated here, except to emphasize the control aspects, since these diseases generally develop early in life. A diagnosis of multiple polyps should lead to examination of the patient's children and if there is a family history, to the patient's sibs as well. If these diseases can be diagnosed and treated before the onset of symptoms, prognosis is improved since two-thirds of symptomatic patients may already have carcinoma (89).

Familial and heritable neoplasms of the large intestine may also develop independently of any evidence of multiple polyps or predisposing disease. These heritable disorders are associated with a high risk for malignancy, approaching 50% (Table 5). They are neither so well known nor so well documented as the polypoid disease, perhaps because they are more difficult to recognize. A clear example is the cancer family syndrome referred to earlier. Colonic cancer is a relatively constant feature of this syndrome, which occurs in a dominant hereditary pattern. A report on another large group of families states that the neoplasms are apparently confined exclusively to the colon (91). In one such family, cancer of the colon developed in 42 of 94 members at risk, including some who may still have been too young to develop cancer of the colon. Colonic cancer led to 22 of 50 deaths in the family. Age at diagnosis of this neoplasm is slightly earlier than that in generalized hereditary adenocarcinomatosis. The mortality rate of nearly 50%, absence of similar findings in control patients from the same geographic area and rate of occurrence are consistent with an autosomal dominant gene effect.

Some important differences occur in the age of onset of these various syndromes. Torre's syndrome has an average age of diagnosis of 50 years, generalized hereditary adenocardionma of 45 years, hereditary colonic adenocarcinoma of 40 years and hereditary gastrocolonic adenocarcinoma of 35 to 40 years (92, 93). These averages are one or two decades earlier than that for diagnosis of colonic adenocarcinoma in the general population, a difference also evident in the polypoid diseases. The penetrance in hereditary adenocarcinomatosis is about 73% (94), about 85% in hereditary colonic adenocarcinoma and Torre's syndrome and 90% in hereditary gastrocolonic adenocarcinoma (93).

The heritable cancers of the large intestine developing independently of polyposis differ in one important respect from colon cancers developing either with polyposis or from colonic cancer patients in general. Colonic cancer is most frequent in the rectosigmoid and cecal areas, the sites involved in patients with polyposis. In the hereditary forms occurring independently of polyposis, the neoplasms are distributed more randomly throughout the colon, so that about 50% of them involve the transverse and right side of the colon, compared with only 20 to 25% in polyposis patients, or colonic cancer patients in general. This is important for the examination procedure; detection must visualize the entire colon, not merely the distal portion thereof. Heritable carcinomas of the colon, whether or not associated with polyposis, have much earlier average ages at diagnosis than those observed in colonic cancer in general; further they are characterized by high frequencies of multiple primaries (at least 14%) compared with about a 6% frequency among colonic cancer patients generally.

Skin

Cutaneous neoplasms are particularly informative from a genetic point of view, for they are more readily detected, diagnosed, treated and traced through families than are most other neoplasms. They also furnish excellent examples of the varying degrees of interaction that may exist between environment and genotype in the development of neoplasms. Two clear examples of such an interaction are provided by albinism (see Chapter 19, "Genetic Disorders of the Skin") and xeroderma pigmentosum. Both conditions are inherited as autosomal recessive disorders, but cutaneous malignancies develop only in areas of the bodies of homozygotes exposed to ultraviolet radiation.

Xeroderma pigmentosum is a rare disease in which the skin is extremely sensitive to radiation at the shortwave length of the sun's spectrum, i.e., below 3100 Å (95). Typical progression of the disease consists of an early stage in the first years of life when the erythemal response to sunlight may be abnormal. After several years other skin changes—including excessive freckling, keratosis and cancers—develop. The skin cancers involve both the ectodermal and mesodermal layers in the production of basal-cell carcinoma, squamous-cell carcinoma, melanoma, angiosarcoma, fibrosarcoma and keratoacanthoma (Table 7). The lesions are progressive, and there is no cure; only preventive steps will help, such as minimizing exposure to sunlight and regular dermatosurgical care. Still, death may intervene early in life.

A more complex form of the disorder is sometimes found in children in whom there is associated microcephaly, mental deficiency, bony abnormalities, cerebellar ataxia and endocrine deficiency (De Sanctis-Cacchione syndrome). The two forms are apparently clinically distinct and have not been seen in the same family. Both are inherited as autosomal recessive disorders. Xeroderma pigmentosum is associated with a deficiency of skin cell DNA repair after solar irradiation (96). It can be detected in tissue culture, and genetic heterogeneity has been suggested both by complementation studies in vitro and by wide variation in residual DNA repair rates in studied individuals. Successful prenatal diagnosis of the condition has recently been achieved through such studies (97).

The other hereditary neoplasms of the skin generally are little, if at all, influenced by sunlight or other environmental agents (Table 7). But increasing evidence points to the strong likelihood of an environmental influence in these conditions as well. *Porokeratosis of Mibelli* is primarily a benign disease, but about 7% of patients developed squamous-cell carcinomas, primarily of the extremities (98). Because of this location, it is conceivable that exposure to ultraviolet (UV) light may have played a role in the transformation of these lesions into carcinomas. Perhaps it is more significant that X-rays and radium were used as therapeutic modalities before the onset of carcinomas in some patients, a fact that may have played a role in the induction of carcinoma.

The hallmarks of the *nevoid basal-cell carcinoma syndrome* (NBCCS) include characteristic early-onset multiple basal-cell carcinomas, pit-like lesions of the

Table 7. Hereditary Forms of Cutaneous Malignancies

Condition	Primary Cutaneous Neoplasm	Other Tumors or Neoplasm	Inheritance*
Albinism	Squamous- and basal-cell carcinoma	—†	AR
Giant hairy pigmented nevus	Malignant melanoma	—†	AD
Malignant melanoma	Malignant melanoma	—†	AD?
Nevoid basal-cell carcinoma syndrome	Multiple basal-cell carcinomas	Squamous-cell carcinoma of antrum; medulloblastoma; ovarian fibroma, carcinoma or sarcoma	AD
Porakeratosis of Mibelli	Squamous-cell carcinoma	—†	AD
Xeroderma pigmentosum	Squamous- and basal-cell carcinoma, melanoma	—†	AR
Neurofibromatosis	Cafe-au-lait spots, cutaneous, neuro-fibromatosis of all sites	Sarcoma, acoustic neuroma(?), pheochromocytoma	AD
Tuberous sclerosis	Intracranial astrocytoma glioblastoma	Hamartomas in brain, kidney and heart	AD
von-Hippel Lindau syndrome	Vascular nevi of face	Hemangioblastoma of cerebellum, renal cell carcinoma (20%)	AD

*Abbreviations: AD, autosomal dominant trait; AR, autosomal recessive.
†None known or reported.

palms and soles, cysts of the jaws, ectopic calcification in soft tissues and skeletal anomalies. The syndrome follows an autosomal dominant mode of inheritance with high penetrance in both sexes. Medulloblastoma may also occur in small but significant fractions of children with the syndrome (99). Treatment for medulloblastoma includes surgery, followed by radiation to the cranium and spinal axis. Twelve patients who underwent such therapy have now been reported (100). They developed basal-cell carcinomas of the neck, shoulders, spinal areas, axilla and scalp. This distribution, localized primarily to the irradiated areas, was unlike that usually encountered in patients with the syndrome. Further the latent period for the development of these tumors following radiation therapy was relatively short (six months to three years), and the patients were very young, age two to six years. These observations suggest that radiation therapy acts as a carcinogen to accelerate tumor formation in gene carriers. Exposure to UV light may also enhance the development of tumors in patients with this syndrome.

Sunlight has also long been implicated in cutaneous *malignant melanoma*. Incidence increases with a decrease in latitude in the northern hemisphere and is most frequent on the head, neck and trunk of men and on the legs of women, trends that reflect differences in the sexes' exposure to sunlight. Further following sunlight exposure, malignant melanoma may develop in patients with albinism and xeroderma pigmentosum. Since the sites of involvement in men and women with the hereditary forms of cutaneous melanoma are the same as they are in the sporadic form of melanoma (101), sunlight conceivably is also involved in the hereditary form. The latter is characterized by an earlier average age at diagnosis and a higher frequency of multiple primaries than that seen in the sporadic form of malignant melanoma.

Interestingly the hereditary form also exhibits a significantly higher survival rate than does the nonfamilial form. The genetic mechanism underlying this neoplasm is undoubtedly complex, possibly involving a major gene with a degree of dominance that influences the development of functional nevi or pigmented lesions. Whether these abnormalities transmute into melanomas may depend on modifying genes that influence complexion, the expression of which in turn may be influenced by exposure to UV light. Evidence also exists of a cytoplasmic factor in the inheritance of melanoma since children from affected or carrier mothers have significantly higher frequencies of melanoma than have children of affected or carrier fathers (102). Thus the inheritance of melanoma involves a nuclear component transmitted by either the affected or carrier male or female parent plus a cytoplasmic component transmitted by the female parent. The magnitude of the effect of the nuclear and cytoplasmic components is similar and additive, but the basis or mode of action of the cytoplasmic component is as yet not known.

Genitourinary Cancer

Aside from the increased occurrence of uterine endometrial cancer in the cancer family syndrome already described, genetic factors in genitourinary (GU) tract tumors are most obvious in the gonad. Gonadal tumors are frequently associated with dysgenetic gonads in which a negative sex-chromatin pattern and a Y-chromosome are found (Table 8) (103). The occurrence of a dysgenetic gonad

Genetics and Cancer 111

Table 8. Tumors in Dysgenetic Gonads

Condition	Clinical Features	Gonad	Tumor
Male, normal (XY)	Normal	Undescended testes	Seminoma
Male, pseudohermaphrodite (XY or mosaic with Y)	Undescended testes Hypospadias, Variable internal genitalia	Undescended testis	Seminoma, gonadoblastoma
Testicular feminization syndrome	Normal female habitus, vagina No uterus or tubes Undescended testes	Dysgenetic testes	Sertoli cell or tubular adenoma
Pure gonadal dysgenesis (Xy or mosaic)	Eunuchoid female habitus Uterus and fallopian tubes	Dysgenetic testes	Gonadoblastoma, dysgerminoma,
Female pseudohermaphrodite nonadrenal (XY or Mosaic)	Female habitus with variable masculinization	Streak ovaries or indeterminate gonads	Gonadoblastoma, dysgerminoma
Mixed gonadal dysgenesis (XO/XY or other mosaic with Y)	Asymmetric gonads and internal genitalia	Streak or indeterminate gonad on one side	Gonadoblastoma
	Ambiguous external genitalia	Contralateral testis	Seminoma, sertoli cell adenoma
True hermaphrodite (XX or Mosaic)	Variable external and internal genitalia and secondary characteristics Usually male habitus	Bilateral ovotestis asymmetrical testes and ovary Ovotestis and contralateral ovary or testis	Dysgerminoma

in an individual with a sex chromosomal abnormality frequently causes a gonadal tumor, principally a dysgerminoma. The risk of gonadal tumor in a patient with a form of XY gonadal dysgenesis may be as high as 20-30% (104). The tumors usually develop in the second or third decade of life, but may occur earlier so that all patients with Y-chromosome–bearing dysgenetic gonads should be examined, and gonadal tissue removed at an appropriate time. In those individuals with dysgenetic gonads who have a chromosomal mosaic disorder and a 45X/46XY chromosomal constitution, the incidence of similar tumors is 15-20% with appearance at the same time of life.

In contrast, gonadal tumors are not a feature of the Turner syndrome in which there is no Y-chromosome, i.e., 45X or 45X/46XX mosaics, (deleted X). Patients with the Klinefelter syndrome (47XXY) show no increase in incidence of gonadal tumors, but are 20 times likelier to develop breast carcinoma than are normal males. Cryptorchidism does predispose the patient to testicular neoplasia and is a finding in 4-11% of males with testicular carcinoma. The undescended testis is at least 14 times likelier to undergo neoplastic transformation than is a normal one. The risk of neoplastic change is decreased only if orchiopexy is performed when the patient is between age 6-10.

Even in those gonadal tumors not associated with errors in sexual development, genetic factors are probably more important than is generally suspected (105). Reports have described 17 families with germinal-cell testicular tumors, including affected twins, sibs of nonconsanguineous parents and transmission vertically through successive generations. Several cell types are included in these reports, suggesting that genetic factors are not peculiar to any one cell type. Similar aggregations of ovarian tumors are reported within families, the data suggesting either that dominant genetic factors are responsible or that there are multiple genes operating. A high frequency of bilaterality and earlier age of onset in these familial ovarian tumors support the single-gene etiology.

Finally renal cell carcinoma is also reported in isolated family studies and known to occur in approximately 20% of patients with the dominantly inherited von Hippel-Lindau disease.

Skeletal Cancer

Few examples of hereditary skeletal neoplasia exist. Some skeletal conditions predispose to malignant change, and one example, hereditary multiple exostosis, is a systemic disorder in which knobby protrusions appear either from the metaphyseal ends of long bones or (occasionally) from flat bones of cartilaginous origin. It occurs as an autosomal dominant trait with nearly 100% penetrance in males but less than that in females. There appears to be no completely satisfactory explanation for this difference. Sporadic cases also appear, but many of these likely represent fresh mutations. The exostoses usually arise near the tendon or ligamentous attachment of a long bone and initially are shaped like a hook or horn; they subsequently become mushroom shaped. They apparently arise from cartilaginous nests embedded in the periosteum, later ossifying without disturbing the epiphyseal ossification centers.

In one-third of patients lesions are noticeable at birth and in the another two-thirds they appear during childhood. Their size increases prepubertally, but usually stops after puberty. If growth resumes later, the possibility of malignancy should be considered. The subject is frequently short because of reduced limb growth and curvature of the spine. Multiple exostoses are to be differentiated from enchondromatoses and solitary exostoses. Occasionally osteochondromas form and enlarge and—in a relatively high percentage of cases—become chondrosarcomas; this may occur in as many as 10% of cases. The tumors, however, grow slowly, and long-term survival is not unusual (106, 107).

Polyostotic fibrous dysplasia and Paget's disease of bone also result in osteogenic sarcoma in some cases. This is unusual in the former, but commoner in the latter, disease.

LEUKEMIA AND LYMPHOMA

Although the incidence of leukemia and lymphoma in certain genetic syndromes has certainly increased, especially in those with an immunodeficiency component, the question of a simply inherited form of either of these malignancies remains moot (Table 9). There is a high rate of concordance for childhood leukemia in monozygotic twins (53, 108). In these studies the age at diagnosis of the affected twins was earlier than for childhood leukemia generally, a feature observed in other inherited tumors. Frequently too the interval between onset of the disease in the two members of the twin pair was brief, raising the question of in utero transfer of leukemic cells between twins. In at least one report, however, a divergence in chromosomal markers seemed to militate against this possibility.

The reported patterns of familial leukemia aggregation seem generally to be like those of other childhood tumors, with many sibs being affected without further familial incidence, although exceptions are noted and pedigrees consistent with dominant inheritance seen. A report from Japan revealed that, of 20 sib pairs with leukemia, 30% had first-cousin parents and an additional 20% had second-cousin or first cousin-once removed parent sets (108). The population frequency of related marriages was only 4.5%, and the age of onset of disease was lower in the sibs of consanguineous, than of nonconsanguineous, parents. These and other pedigree reports are growing exceptions to the general lack of any clearly established Mendelian pattern. The overall evidence at present seems to favor the existence of a small but significant subgroup of simply inherited childhood leukemia or lymphoma, mainly in a dominant pattern.

Cytogenetic Changes in Leukemia

Chromosomal studies may help significantly in managing two specific forms of leukemia, chronic and acute myelogenous leukemia. Although the genetic significance of these diagnostic and prognostic cytogenetic changes may be argued, they seem to have some practical ability (see also Chapter 2).

Table 9. Heritable Conditions Predisposing to Leukemia or Lymphoma

Disorder	Lymphoma (%)*	Leukemia (%)*	Total Patients with Tumor (%)†
Severe combined Immunodeficiency	67	35	2
XL hypogammaglubulinemia	17	83	6
IgM deficiency	83		8
IgA deficiency		15	
Common variable immunodeficiency	56	10	8
Wiskott-Aldrich syndrome	79	13	8
Ataxia-telangiectasia	62	21	10
Fanconi's pancytopenia		+ ‡	
Xeroderma pigmentosum		+	
Werner syndrome		+	
Bloom syndrome		+	
Chédiak-Higashi	+		

*Percentage of tumors in these classes.
†Percentage of patients with some form of tumor.
‡Disease occurs but % of patients not known.

Chronic Myeloid Leukemia

Since the original description of an abnormally small G group chromosome in leukemic cells from chronic myeloid leukemic (CML) patients in 1960 (109), over 1,000 total cases have been studied, almost 200 of which have been studied with the banding techniques. The original finding of a deletion of part of the G long arms has been refined and identified as a 22q-or Philadelphia (Ph[1]) chromosome. The absent material from 22q is not deleted (110), but rather is translocated to another chromosome, usually 9q (111). The Ph[1] chromosome is in roughly 85% of all patients with CML and more than 90% of the time is due to a t(9;22) (q34;q11) rearrangement (112), a finding virtually diagnostic of CML. In about 16% of the untreated Ph[1] positive CML cases, additional chromosomal changes are found in the initial examination, including an extra C, missing Y in males (113), an isochromosome of the long arms of chromosome 17 or two Ph[1] chromosomes.

In CML, change in the original karyotype usually indicates onset of the acute phase of the disease, and in roughly 70% of patients additional chromosomal abnormalities are superimposed on the Ph[1]-positive cell line. The most frequent change is an additional Ph[1] chromosome (50%), but other aberrations similar to those already described may occur. In cases with two Ph[1] chromosomes, there is usually only one 9q+ so that the second Ph[1] appears to arise rather by duplication of the first than by a new translocation. When an extra C chromosome is added there appears to be no consistent abnormality, although in 15 of 21 cases in which this was found it was an additional number 8.

Of those Ph[1]-negative patients studied by banding methods, all show normal karyotypes in the chronic phase. No evidence of translocation or rearrangement of chromosomal material has been detected (114, 115). It is important to note the observed difference in survival and response to therapy in Ph[1]-positive and Ph[1]-negative cases of CML. Mean survival time of the former is 42 months, whereas the latter succumb an average of 15 months after diagnosis. Recently aggressive chemotherapeutic treatment in the acute phase of Ph[1]-positive cases shows some increase in remission rate so that the difference in survival may become even more significant (114).

Acute Leukemia

In contrast to the early finding of a specific chromosomal abnormality in one form of chronic leukemia, chromosomal changes described in acute leukemia have until recently been without a meaningful pattern. Although about 50% of all acute leukemias studied will show chromosomal abnormalities, these are regarded as variable (114). Early studies suggest that acute lymphoblastic leukemia (ALL) patients were likelier to have a hyperdiploid chromosomal karyotype, whereas acute myelogenous leukemic (AML) cells were likelier to be diploid or hypodiploid. Newer studies have partially clarified these features and suggest that prebanding AML studies have frequently underestimated the number of chromosomally abnormal patients (116). Of particular interest is the clarification of a previously confusing pattern as an 8;21 translocation frequently accompanied by loss of a Y- or X-chromosome. This translocation, described as t(8;21) (q22;q22), was found in 21% of cases in one series (117) and 17% of cases in another (118). It appears to be of diagnostic value and of particular prognostic significance described below. Other frequent changes are addition of a number 8 or loss of a number 7 chromosome.

Current studies of chromosomes in AML furnish data suggesting differences in survival and response to therapy similar to those seen in CML, which are correlated with specific chromosomal pictures (117-120). In one study of 50 cases of acute nonlymphocytic leukemia, the 25 cases with normal karyotypes at diagnosis had a mean survival time of 10 months as opposed to 2 months in those cases with abnormal karyotypes (119). Examination of only those cases with AML revealed a difference even more significant: 18 months for those with normal karyotypes against the two months for the karyotypically abnormal cases. Therapeutic remission was obtained in 87% of cases with normal karyotypes, 20% in those with some normal and some abnormal karyotypes and in none of the cases with only karyotypically abnormal cells. However, two groups of aneuploids had even better survival times.

Those with a C chromosome abnormality (probably +8) had a mean survival of 16 months, whereas those with the 8;21 translocation survived an average of 18 months, one patient being alive five years and three months after diagnosis. The translocation patients represented 16.7% of the total aneuploid patients, ranged in age from 20 to 50 years and responded to therapy with disappearance of the translocation chromosome. The 8;21 translocation cases are frequently missing a sex chromosome as previously described, and this chromosomal profile predicts a healthy response to therapy and relatively long survival. Those patients under age 70 years with other chromosomal abnormalities but with some karyotypically normal bone-marrow cells have a reasonable chance of responding therapeutically. Patients either over 70 years or with no normal marrow cells and abnormalities other than the 8;21 translocation or an extra C chromosome do poorly and in fact may be made worse by chemotherapeutic treatment.

LUNG

Tumors of the lung, especially bronchiogenic carcinomas, were relatively rare before the present century and most epidemiologic studies have thus focused on environmental factors. Many agents have been implicated, and major sources of occupational or industrial inhalant toxicity successfully approached. The common carcinogenic agent, tobacco, has not been so successfully studied or controlled. Because of this fact, Mulvihill in a recent review (121) suggested a search for genetic factors controlling the response of the host to an environmental carcinogenic challenge and then concentrating on the high-risk group as has been done with other cancers. In the past most such attempts revealed little or nothing to support the existence of significant genetic factors in lung carcinoma.

Family studies have been somewhat more revealing. Tokuhata found an increase in lung cancer about 2.5 times higher in the relatives of lung cancer patients than in control relatives (122). Other pedigree studies elsewhere support this conclusion and also demonstrate a synergistic effect of heredity and smoking. Some cancer family syndrome pedigrees show a moderate number of lung tumor cases—usually adenocarcinoma or alveolar cell carcinoma. Lung cancers are frequently reported as second primary lesions, and they may also be associated with decreased immune capacity. Both these features are compatible with genetic influences. Scleroderma and dominant heritable disorders of the lung featuring interstitial pulmonary fibrosis are associated with lung cancer in a few cases. In contrast no clear examples of genetic disease are associated with pulmonary tumors.

Recent studies attempt to connect the inducible enzyme system, aryl hydrocarbon hydroxylase (AHH), with lung cancer (123). This enzyme can convert some of the hydrocarbons in tobacco smoke to potent carcinogens. It can be found in human lung alveolar macrophages and is induced by cigarette smoke. Its inducibility was reported to be controlled by a single genetic locus, and a greater proportion of lung cancer patients were initially reported to have higher inducibility of AHH than were controls. Unfortunately these reports are not substantiated, and it appears that AHH is not yet ready to serve as a biochemical marker for lung cancer susceptibility.

Currently then there is little to suggest that important genetic factors operate in bronchiogenic carcinoma. Adenocarcinoma and alveolar cell carcinoma are less related to smoking than to other precipitating factors and themselves are frequently found in familial aggregations, in conjunction with immunodeficiency disease, other tumors or heritable lung disease. Perhaps the study of relationships between susceptible genotypes and environmental carcinogens will lead to new findings in the genetics of lung cancer.

REFERENCES

1. Triolo VA: Nineteenth century foundations of cancer research. Advances in tumor pathology, nomenclature and theories of oncogenesis. *Canc Res* 25:75, 1965
2. Boveri T: The origin of malignant tumors. Baltimore, *Williams & Wilkins,* 1929
3. Strong LC: Genetic concept for the origin of cancer. Historical review. *Ann NY Acad Sci* 71:819, 1958
4. Lawler SD: The HLA system and Neoplasia, in Lynch HT (ed): Cancer Genetics. Springfield, Ill., C C Thomas, 1976
5. Degos L, Drolet Y, Dausset J: HLA antigens in chronic myeloid leukemia (CML) chronic lymphoid leukemia (CLL). *Trans Proc* 3:309, 1971
6. Burkitt DP: A sarcoma involving the jaws of African children. *Br J Surg* 46:218, 1958
7. Ziegler JL, Magrath ET, Gerber P, Levine PH: Epstein-Barr virus and human malignancy. *Ann Intern Med* 86:323, 1977
8. Todaro GJ, Green H, Swift MR: Susceptibility of human diploid fibroblast strains to transformation by SV-40 virus. *Science* 153:1252, 1966
9. Swift M: Fanconi's anemia in the genetics of neoplasia. *Nature* 230:370, 1971
10. Swift M: Malignant disease in heterozygous carriers. *Birth Defects* 12:133, 1976
11. Fialkow PJ: Human tumors studied with genetic markers. *Birth Defects* 12:123, 1976
12. Fialkow PJ: Clonal origin and stem cell evaluation of human tumors in Mulvihill JJ, Miller RW, Fraumeni JF Jr. (eds): *Genetics of Human Cancer*. New York, Raven, 1977, p. 439
13. Fialkow PJ, Jacobson RJ, Papayannopoulou N: Chronic myelocytic leukemia: Clonal origin in a stem cell common to granulocyte, erythrocyte, platelet and monocyte/macrophage. *Am. J Med* 63:125, 1977
14. Kirpatrick C: Cancer and immunodeficiency diseases. *Birth Defects* 12:61, 1976

15. Thomas L: Open discussion, in Lawrence HS (ed): Cellular and Humeral Aspects of Hypersensitivity States. New York, Hoeber, 1959, p. 529
16. Burnet FM: Immunological Surveillance. Oxford: Pergamon, 1970
17. Klein G: Tumor-specific immune mechanisms. *N Engl J Med* 278:326, 1207, 1968
18. Allen DW, Coal WP: Viruses in human cancer. *N Engl J Med* 286:70, 1972
19. Fass L, Herberman RV, Ziegler J: Delayed cutaneous hypersensitivity to autologous extracts of Burkitt-lymphoma cells. *N Engl J Med* 282:776, 1970
20. Good RA: Relations between immunity and malignancy. *Proc Nat Acad Sci USA* 69:1026, 1972
21. Kersey JH, Spector BD, Good RA: Primary immunodeficiencies in cancer: The immunodeficiency cancer registry. *Int J Cancer* 12:333, 1973
22. Bock FH, Albertini RJ, Joo P, Anderson JL, et al: Bone marrow transplantation in a patient with the Wiscott-Aldrich syndrome. *Lancet* ii:1364, 1968
23. Knudson AG: Genetics and the etiology of childhood cancer. *Pediatr Res* 10:513, 1976
24. Mole RH: Cancer production by chronic exposure to penetrating gamma irradiation. *Natl Canc Inst Monogr* 14:217, 1964
25. Knudson AG Jr: Mutation and cancer; statistical study of retinoblastoma. *Proc Natl Acad Sci* 68:820, 1971
26. Knudson AG Jr, Strong LC: Mutation and cancer: A model for Wilms tumor of the kidney. *J Natl Canc Inst* 48:313, 1972
27. Knudson AG Jr, Strong LC: Mutation and cancer: Neuroblastoma and pheochromocytoma. *Am J Hum Genet* 24:514, 1972
27a. Matsunaga AE. Heretary retinoblastoma: Delayed mutation or host resistance? *Am J Hum Genet* 30:406, 1978
28. Lynch HT, Guirgis HA, Lynch PM, Lynch JF, et al: Familial cancer syndromes, a survey. *Cancer* 39:1867, 1977
29. Warthin AS: Heredity with reference to carcinoma as shown by the study of the cases examined in the pathological laboratory of the University of Michigan, 1895-1913. *Arch Intern Med* 12:546, 1913
30. Lynch HT, Krush AJ, Thomas, RJ, Lynch, J: Cancer family syndrome, in Lynch HT (ed): Cancer Genetics. Springfield, Ill., C C Thomas, 1976, p. 355
31. Knudson AG Jr: Genetics and etiology of human cancer, in Harris H, Hirschhorn K (eds): Advances in Human Genetics. New York, Plenum, 1977
32. Knudson AG Jr, Strong LC, Anderson DE: Heredity and cancer in man. *Prog Med Genet* 9:113, 1973
33. Mulvihill JJ: Congenital and genetic diseases, in Fraumeni JF (ed): Persons at High Risk of Cancer. New York, Academic, 1975, p. 3
34. Bergsma D (ed): Cancer and genetics. *Birth Defects* 12(1): 1-200 1976
35. Fraumeni JF Jr (ed): Persons at High Risk of Cancer. An Approach to Cancer Etiology and Control. New York, Academic, 1975
36. Lynch HT (ed): Cancer Genetics. Springfield, Ill., C C Thomas, 1976
37. Mulvihill JJ, Miller RW, Fraumeni JF Jr: The Genetics of Human Cancer. New York, Raven, 1977
38. Schimke RN: Genetics and Cancer in Man. Edinburgh, Churchill-Livingston, in press,
39. Bond, JV: Bilateral Wilms' tumor. *Lancet* 2:482, 1975
40. Barakat AY, Papadopoulou ZL, Chandra RS, Hollerman CE, et al: Pseudohermaphroditism, nephron disorder and Wilms' tumor, a unifying concept. *Pediatrics* 54:366, 1974
41. Denys P, Malvaux P, Van den Berghe, H, Tanghe, W, et al: Association d'un syndrome anatomopathologigue de pseudoharmaphrodisme masculin, d'unee tumor de Wilms, d'une nephropathie parenchymateuse et d'un mosaicisme XX/XY. *Arch Franc Pediatr* 24:729, 1967
41a. Riccardi M, Sujansky E, Smith AC, Franke U: Chromosomal imbalance in the Aniridia-Wilms Tumor association: 11p interstitial deletion. *Ped* 61:416, 1978

42. Meadows AT, Lichtfeld JL, Koop CE: Wilms' tumor in three children of a woman with congenital hemihypertrophy. *N Engl J Med* 291:23, 1974
43. Fraumeni JF Jr, Geiser CF, Manning MD: Wilms' tumor and congenital hemihypertrophy, report of five new cases and review of the literature. *Pediatrics* 40:886, 1967
44. Reddy JK, Schimke RN, Chang CHJ, Svoboda DJ, et al: Beckwith-Wiedemann syndrome. *Arch Pathol* 94:523, 1972
45. Perlman M, Levin M, Wittels B: Syndrome of fetal gigantism, renal hamartomas and nephroblastomatosis with Wilms' tumor. *Cancer* 35:1212, 1975
46. Miller RW, Fraumeni JF Jr, Manning MD: Association of Wilms' tumor with aniridia, hemihypertrophy and other congenital malformations. *N Engl J Med* 270:922, 1964
47. Schwartz AD, Lee H, Baum ES: Leukemia in children with Wilms' tumor. *J Pediatr* 87:374, 1975
48. Knudson AG Jr, Meadows AT, Nichols WW, Hill R: Chromosomal deletion and retinoblastoma. *N Engl J Med* 295:1121, 1976
49. Kitchin FO, Ellsworth RM: Pleiotropic effect of the gene for retinoblastoma. *J Med Genet* 11:244, 1974
50. Schimke RN: Tumors of the neural crest system, in Mulvihill JJ, Miller RW, Fraumeni JF Jr (eds): Genetics of Human Cancer. New York, Raven, 1977, p. 179
51. Wilson LM, Draper GJ: Neuroblastoma, its natural history and prognosis: A study of 487 cases. *Br Med J* 3:301, 1974
52. Fraumeni JF Jr, Rosen PJ, Hull EW, Barth RF, et al: Hepatoblastoma in infant sisters. *Cancer* 24:1086, 1969
53. Miller RW: Deaths from childhood leukemia and solid tumors among twins and other sibs in the United States, 1960–1967, *J Natl Cancer Inst* 46:203, 1971
54. Ballard HS, Frame B, Hartsock RJ: Familial multiple endocrine adenoma-peptic ulcer complex. *Medicine* 43:481, 1964
55. Sipple JH: The association of pheochromocytoma with carcinoma of the thyroid gland. *Am J Med* 31:163, 1961
56. Schimke RN, Hartmann WH, Prout TE, Rimoin DL: Syndrome of bilateral pheochromocytoma, medullary thyroid carcinoma and multiple neuromas. *N Engl J Med* 279:1, 1968
57. Pollack RS: Carotid body tumor-idiosyncracies. *Oncology* 27:81, 1973
58. Revak CS, Morris SW, Alexander GH: Pheochromocytoma and recurrent chemodectomas over a twenty-five year period. *Radiology* 100:53, 1971
59. Post RH: Breast cancer, lactation, and genetics. *Eugen Q* 13:1, 1966
60. Anderson DE: A genetic study of human breast cancer. *J Natl Cancer Inst* 48:1209, 1972
61. Anderson DE: Genetic study of breast cancer: Identification of a high risk group. *Cancer* 34:1090, 1974
62. Anderson DE: Genetic predisposition to breast cancer, *Proc Natl Conf Breast Cancer*, Montreal, 1976
63. Wang DY, Bulbrood RD, Hayward JL: Urinary and plasma adrogens and their relation to familial risk of breast cancer. *Eur J Cancer* 11:873, 1975
64. Li FP, Fraumeni JF: Familial breast cancer, soft tissue sarcomas, and other neoplasma. *Ann Intern Med* 83:833, 1975
65. Fraumeni JF Jr, Grundy GW, Creagan ET, et al: Six families prone to ovarian cancer. *Cancer* 36:364, 1975
66. Lynch HT, Krush AJ, Guirgia H: Genetic factors in families with combined gastrointestinal and breast cancer. *Am J Gastroenterol* 59:31, 1973
67. Burnett JW, Goldner R, Calton GJ: Cowden disease. Report of two additional cases. *Br J Dermatol* 93:329, 1975
68. Stephens FE, Gardner EJ, Woolf CM: A recheck of Kindred 107, which has shown a high frequency of breast cancer. *Cancer* 11:967, 1958

69. Woolf CM, Gardner EJ: The familial distribution of breast cancer in a Utah kindred. *Cancer* 4:515, 1951
70. Wheeler JE, Enterline HT, Roseman JM, et al: Lobular carcinoma in situ of the breast. *Cancer* 34:554, 1974
71. Doll R: Cancer in five continents. *Proc Soc Med* 65:49, 1972
72. Freytes MA, Carri J: Familial esophageal carcinoma, *Fac Cienc Med Cordoba* (Sp.) 26:215, 1968
73. Pour P, Ghadirian P: Familial cancer of the esophagus in Iran. *Cancer* 33w;1649, 1974
74. Harper PS, Harper RMJ, Howel-Evans AW: Carcinoma of the oesophagus with tylosis. *Q J Med* 39:317, 1970
75. Parnell DD, Johnson SAM: Tylosis palmaris et plantaris. Its occurrence with internal malignancy. *Arch Dermatol* 100:7, 1969
76. Shine I, Allison PR: Carcinoma of the oesophagus with tylosis (keratosis palmaris et plantaris). *Lancet* 1:951, 1966
77. Haines D: Primary carcinoma associated with tylosis. *J R Nav Med Serv* 53:75, 1967
78. Huriez C, Derminatti M, Agache P, et al: Une genodysplasie non encore individualisee: La genodermatose scleroatrophiante et kerrutodermique des extremites frequemment degenerative. *Sem Hop Paris* 4:481, 1968
79. Ritter SB: Esphageal cancer, hyperkeratosis, and oral leukoplakia. Occurrence in a 25-year-old woman. *JAMA* 235:1723, 1976
80. Graham S, Lilienfeld AM: Genetic studies of gastric cancer in humans: An appraisal. *Cancer* 11:945, 1958
81. Macklin MT: Inheritance of cancer of the stomach and large intestine in man. *J Natl Cancer Inst* 24:551, 1960
82. Sokoloff B: Predisposition to cancer in the Bonaparte family. *Am J Surg* 40:673, 1938
83. Cruze K, Clarke JS, Farra SE: Familial aspects of gastric adenocarcinoma. Report of 13 patients. *Am J Dig Dis* 6:7, 1961
84. Woolf CM, Isaacson EA: An analysis of 5 "stomach cancer families" in the State of Utah. *Cancer* 14:1005, 1961
85. Creagan ET, Fraumeni JF Jr: Familial gastric cancer and immunologic abnormalities. *Cancer* 32:1325, 1973
86. McConnell RB: The Genetics of Gastrointestinal Disorders. London, Oxford University Press, 1966
87. Harper PS: Heredity and gastrointestinal tumors. *Clin* Gastroenterol 2:675, 1973
88. Alm T, Licznerski G: The intestinal polyposes. *Clin Gastroenterol* 2:577, 1973
89. Bussey HJR: Familial Polyposis Coli. Baltimore, Johns Hopkins University Press, 1975
90. Erbe RW: Inherited gastrointestinal-polyposis syndromes. *N Engl J Med* 294:1101, 1976
91. Anderson DE, Strong LC: Genetics of gastrointestinal tumors, in Cancer Epidemiology, Environmental Factors. Proceedings of the XI International Cancer Congress, Florence, 1974
92. Anderson DE, Strong LC: Genetics of gastrointestinal tumors, in Excerpta Medica International Congress Series No. 351, Vol 3. Amsterdam, 1974, p. 267
93. Anderson DE: Unpublished data, 1976
94. Lynch HT, Kaplan AR: Cancer concordance and the hypothesis of autosomal dominant transmission of cancer diathesis in a remarkable kindred. *Oncology* 30:210, 1974
95. Cleaver JE: Xeroderma pigmentosum, DNA repair and carcinogenesis, in Lynch HT (ed): Cancer Genetics. Springfield Ill., C C Thomas, 1976
96. Cleaver JE, Bootsma D: Xeroderma pigmentosum: Biochemical and genetic characteristics. *Ann Rev Genet* p. 19, 1975
97. Ramsay CA, Coltart TM, Blunt S, Pawsey SA, et al: Prenatal diagnosis of xeroderma pigmentosum. Report of first successful case. *Lancet* 2:1109, 1974
98. Goerttler EA, Jung EG: Porokeratosis Mibelli and skin carcinoma. A critical review. *Humangenetik* 26:291, 1975

99. Neblett CR, Waltz TA, Anderson DE: Neurological involvement in the nevoid basal cell carcinoma syndrome. *J Neurosurg* 35:577, 1971
100. Strong LC: Theories of pathogenesis, in Mulvihill JJ, Miller RW, Fraumeni JF Jr (eds): Mutation and Cancer, New York, Raven, 1977
101. Anderson DE, Smith L Jr, McBride CM: Hereditary aspects of malignant melanoma. *JAMA* 200:741, 1967
102. Anderson DE: Inheritance of a genetic type of melanoma in man, in Rily V (ed): Pigmentation: Its Genesis and Biological Control. New York, Appleton-Century Crofts, 1972, p. 401
103. Taylor H, Barter R, Jacobson C: Neoplasms of dysgenetic gonads. *Am J Gynecol* 96:816, 1966
104. Simpson JL Photopulos G: The relationship of neoplasia to disorders of abnormal sexual differentiation. *Birth Defects* 12:15, 1976
105. Simpson JL, Photopolous G: Hereditary aspects of ovarian and testicular neoplasia, *Birth Defects* 12:51, 1976
106. Krooth RS, Macklin MT, Hilbish TF: Diaphyseal aclasis (multiple exostosis) on Guam. *Am J Hum Genet* 13:340, 1961
107. Dahlin DC: *Bone Tumors.* Springfield, Ill., C C Thomas, 1957
108. Kurita S, Kamei Y, Ota K: Genetic studies on familial leukemia. *Cancer* 34:1098, 1974
109. Nowell RC, Hungerford DA: A minute chromosome in human chronic myelogenous leukemia. *Science* 132:1197, 1960
110. Mayall BH, Carrano AV, Rowley JD: DNA cytophotometry of chromosomes in a case of chronic myelogenous leukemia. *Clin Chem* 20:1080, 1974
111. Rowley JD: A new consistent chromosome abnormality in chronic myelogenous leukemia identified by quinacrine fluorescence and giemsa staining. *Nature* 243:290, 1973
112. Gahrton G, Zech H, Lindsten J: A new variant translocation ($19q^+$, $22q^-$) in chronic myelocytic leukemic. *Exp Cell Res* 86:214, 1974
113. Lawler SD, Loss DS, Willshaw E: Philadelphia-chromosome positive bone-marrow cells showing loss of the Y in males with chronic myeloid leukemia. *Br J Haematol* 27:247, 1974
114. Rowley JD: Population cytogenetics of leukemia, in Hook EB, Porter IH (eds): Population Cytogenetics. New York, Academic, 189, 1977
115. Mitelman F, Brandt L, Nilsson PG: The banding pattern in Philadelphia-chromosomes negative chronic myeloid leukemia. *Hereditas* 78:302, 1974
116. Rowley JD, Porter D: Chromosomal banding patterns in acute nonlymphocytic leukemia. *Blood* 47:705, 1976
117. Sakurai M, Sandberg AA: Chromosomes and causation of human cancer and leukemia XI correlations of karyotypes with clinical features of acute myeloblastic leukemia. *Cancer* 37:285, 1976
118. Cork A, Trujillo JM, McCredie KB: Specific chromosomal changes in acute granulocytic leukemia. *Proc 15th Som Cell Gen Congr*, 1976
119. Colomb HM, Vardiman J, Rowley JD: Acute nonlymphocytic leukemia in adults: Correlation with Q-banded chromosomes. *Blood* 48:9, 1976
120. Trujillo JM, Cork A, Hart JS, George SL, et al: Clinical implications of aneuploid cytogenetic profiles in adults acute leukemia. *Cancer* 33:824, 1974
121. Mulvihill JJ: Host factors in human lung tumors. *J Natl Canc Inst* 57:3, 1977
122. Tokuhata GK, Lilienfeld AM: Familial aggregation of lung cancer in humans. *J Natl Cancer Inst* 30:289, 1963
123. Kellerman Luyten-Kellerman M, Shaw CR: Genetic variability of aryl hydrocarbon hydroxylase in human lymphocytes. *Am J Hum Genet* 25:327, 1973

4
Genetic Disorders of the Immune System

Charles H. Kirkpatrick

To survive in an environment laden with pathogenic and opportunistic microorganisms, man and other animals have evolved mechanisms that enable them to identify and to eliminate or confine harmful invaders. Many components of these defense systems are now identified and defined, and clinical observations show the need for normal function and integration of all systems. The situations are numerous, however, in which failure of a single component has rendered the subject unusually susceptible to infectious diseases.

DEVELOPMENT OF IMMUNE SYSTEM

In some respects the study of the immunodeficiency syndromes began in 1952 when Bruton reported on a patient with recurrent infections who was deficient in serum (γ) globulin (1). Other similar cases were soon recognized, and by 1954 certain morphologic counterparts of this disease, such as the absence of plasma cells and germinal centers in lymphoid organs, were known. During the 1960s, experiments with immunologically manipulated chickens led to formulation of the two-component model for development of the immune system (2), the components and cellular interactions of which are illustrated in Figure 1.

Pluripotent stem cells arise in the fetal yolk sac or liver and postnatally in the bone marrow. Studies conducted in chickens indicate that marrow-derived stem cells may receive differentiative influences from one of two sources—the thymus or the bursa of Fabricius, a lymphoid organ in the hindgut.

"Bursa-equivalent" Lymphocyte System

Although a morphologic counterpart of the bursa of Fabricius has not been identified in mammals, in birds the bursa directs differentiation of cells that migrate to the peripheral lymphoid tissues and populate both the lymphoid follicles and medullary cords of lymph nodes and the red pulp and germinal centers of the spleen; there they participate in antibody synthesis. Maturation of antibody-forming cells from stem cells probably occurs in two stages (3). The

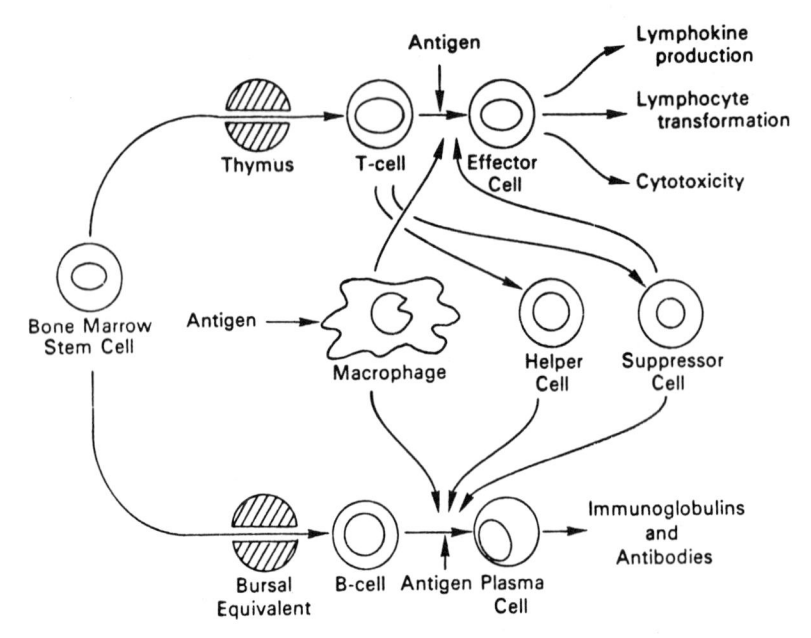

Figure 1. The two-component model for development of the cellular (thymus-dependent) and humoral (bursal-dependent) immune systems.

first stage takes place during embryonic life (4) and results in formation of "bursa-equivalent" lymphocytes (B-lymphocytes), the precursors of antibody-forming cells. These B-lymphocytes, or B-cells, can synthesize immunoglobulins but unless appropriately stimulated, they do not become immunoglobulin-secreting cells. They may be identified by easily detectable immunoglobulins on the cell membrane (5) and by other membrane-associated properties, such as receptors for complement components (6) and the Fc portion of aggregated IgG (7).

Not every B-cell has all these receptors. The first B-cells in the embryo bear IgM, and it is postulated that these cells are the precursors of IgG- and IgA-bearing cells (3). Differentiation of B-cells from stem cells does not require antigenic stimulation.

The second stage of B-lymphocyte maturation involves differentiation of lymphocytes and plasma cells capable of both synthesizing and secreting immunoglobulins. This response occurs as a consequence of antigenic stimulation and involves only the subpopulation of B-cells that can respond to that antigen. Ordinarily the fetus neither receives antigenic stimulation nor makes antibodies. However, observations show that the fetus infected in utero can synthesize specific antibodies as early as the 20th week of gestation. These antibodies are predominantly of the IgM class, but lesser amounts of IgA and IgG have also been detected (8).

IgG molecules can cross the placenta and the serum concentration of IgG in the newborn is essentially the same as the mother's. Synthesis of new IgG by the infant does not begin until the second or third month of life and during this time the maternal immunoglobulin is being catabolized. If the onset of IgG synthesis by the newborn is delayed, there may be a "physiologic" period during which the infant is hypogammaglobulinemic. Other maternal immunoglobulins do not cross the placenta. The newborn infant can synthesize IgM and by the end of the first year of life the serum IgM reaches near-adult levels; IgA synthesis begins about a month after birth, but adult serum levels may not be achieved until age 6-8 years even by normal children.

In vitro stimulation of normal B-lymphocytes with pokeweed mitogen (PWM) causes them to secrete newly synthesized immunoglobulins into the cytoplasm and the culture media (9). As will be described in more detail below, this phenomenon has provided a useful probe for study of the differentiative potential of lymphocytes of patients with a variety of hypogammaglobulinemic syndromes.

Thymus-dependent Lymphocyte System

The other major pathway of lymphocyte differentiation involves maturation of stem cells into thymus-derived lymphocytes (T-lymphocytes) and then into the effector cells of cell-mediated immunity (Fig. 1). In this two-stage process, the first stage is maturation of stem cells into T-cells. It probably requires both cell-cell interactions within the thymus gland and a thymic humoral factor. Thymus-derived lymphoid cells then migrate to peripheral lymphoid organs where they populate both the paracortical zones of lymph nodes and the periarteriolar lymphoid sheaths of the spleen and comprise the majority of circulating blood lymphocytes.

The T-lymphocytes have unique surface receptors enabling them to form rosettes with sheep erythrocytes (4). In addition T-cells respond with enhanced DNA, RNA and protein synthesis to diverse stimulants, including allogeneic cells and plant lectins, such as phytohemagglutinin (PHA), concanavalin A (Con-A) and pokeweed mitogen (PWM), a substance that is also mitogenic for B-cells. These responses are routinely used to assess T-cell functions in patients with possible immunodeficiency syndromes.

Like the first stage of B-cell maturation, T-cells form during embryonic life (10, 11) and apparently do not require stimulation by antigen. The role of the thymus in this process is shown by the in vitro development of T-cells from bone-marrow cells treated with thymus extracts (12, 13) and by the appearance of rosette-forming T-cells in an athymic patient after thymus transplantation (14). Presumably exposure to antigen triggers the second stage of differentiation in which an antigen-dependent subpopulation of T-cells matures to effector cells that then exert their activities directly as T-cell–mediated cytotoxicity or indirectly through effector molecules known as lymphokines.

Other cells also exert important modulating effects on immune responses and illustrate the complexities of cellular interactions (Fig. 1). They include 1) macrophages, which apparently function in antigen processing (15, 16), 2) two subpopulations of T-lymphocytes, the "helper" cells (17), necessary for optimal antibody responses to haptenic antigens and 3) suppressor cells (18), which exert negative influences on both humoral and cell-mediated responses.

GENETIC COMPONENTS AND IMMUNE RESPONSE

A detailed discussion of the contribution of genetic factors to normal immune responses, although beyond the scope of this chapter, is considered in recent reviews (19, 20). Among the crucial early observations were the discoveries that certain strains of inbred guinea pigs and mice respond to polymeric amino acid antigens by producing antibodies, whereas other strains do not. Breeding experiments showed that this trait was genetically controlled and transmitted in an autosomal dominant manner. Subsequently it was shown that in mice the genes regulating immune responses (Ir genes) are clustered within the chromosomal complex known as H-2, the latter which also determines the structure of the major transplantation antigens. Currently these genetic controls appear to be limited to immune responses requiring T-lymphocytes, and it is postulated, but not proven, that the products of the Ir genes may serve as antigen receptors on T-cells.

In man a functionally analogous genetic complex has been identified on chromosome 6. This genetic region determines the structure of the serologically defined (SD) leukocyte antigens HLA-A and HLA-B and the HLA-D locus that controls mixed leukocytic reactions (20). It also contains the gene(s) controlling serum levels of the second component of complement (C2) (21). The current evidence for immune response genes within this genetic region in man is indirect and is inferred from two kinds of observations.

First there are several instances in which specific antibody responses have been associated with certain HL-A antigens or haplotypes. In patients with celiac disease there is a significant association between high antigluten antibody titers and HL-A8 (22). It is reported that in normal subjects reduced secretory antibody responses to intranasal influenza vaccine are associated with antigen W-16 (23), and that in vitro lymphocyte transformation reactions to streptococcal antigens are associated with HL-A5 (24). In patients with ragweed pollinosis, Marsh et al (25) have noted a statistically significant association between IgE and IgG antibody responses to ragweed antigen Ra5 and HL-A7. Levine et al (26) have reported that clinical ragweed hay fever and IgE antibody responses to ragweed antigen E tend to be associated with certain HL-A antigens or haplotypes in families. In their report, of 46 family members 26 possessed the "hay fever haplotype" and 20 of the latter had symptomatic hay fever and positive immediate-type skin tests. Similar findings in hay-fever patients have been described by Blumenthal and co-workers (27).

The second general line of evidence for genetic control of immune responses in man derives from studies of leukocyte antigen frequencies in patients with various diseases in which immunologic factors are important (28). For example, HL-A8 occurs in over 80% both of patients with gluten enteropathy (29) and those with dermatitis herpetiformis who also have intestinal symptoms (30). The antigen W-27 is found in 6-8% of the normal population (28), but occurs in 90% of patients with ankylosing spondylitis (31, 32), 75% of patients with Reiter's disease (33) and over 50% of patients with acute anterior uveitis (34). In psoriasis, W-17 and HL-A13 are more, whereas HL-A12 is less, frequent (35); Dw2 (formerly HL-A7a) is found in roughly 60% of multiple sclerosis patients, but in only 20% of controls (36).

HOST-DEFENSE MECHANISMS

The most important incentive for evaluating host-defense systems is a clinical history of unusual susceptibility to repeated acute infections with common high-grade pathogens, or persistent or recurring infections with opportunists or organisms regarded as low-grade pathogens. It has recently become common to examine host-defense systems in patients with other diseases, especially malignancies and "autoimmune" disorders to identify and characterize secondary immunodeficiencies.

A systematic appraisal of the immunologic components of host defense may be conducted along the guidelines of the two-component model in Figure 1, but one must also consider functions of the nonlymphoid components, such as the phagocytic-microbicidal cells and the complement system. The clinical history is important. Problems beginning in infancy or early childhood strongly suggest congenital diseases, whereas disorders that become apparent later may result from either defects in regulatory processes or other diseases. The physical examination may provide important clues to immunologic defects (Table 1), and even simple tissue studies, such as blood leukocyte and bone-marrow examinations, may be helpful. Biopsies of lymph nodes are useful, especially if taken seven to 10 days after local antigenic stimulation.

The B-lymphocyte system should be evaluated to see if the patient can produce B-cells and if the cells are functional (Table 2) (5-7). In all these assays one must be aware of artifacts. Winchester et al (37), e.g., defined a potential technical pitfall in the assay for Fc-receptor–bearing cells. Preparations contaminated with monocytes or promonocytes may give falsely elevated values because the cells bind antigen-antibody complexes or aggregated IgG in the fluoresceinated reagent through their own surface Fc receptors. The products of B-cells may be measured quantitatively, but such assays do not supply information about functional activities.

Some hypogammaglobulinemic patients have, on the one hand, immunoelectrophoretic patterns showing qualitatively restricted heterogeneity and suggesting an inability to synthesize immunoglobulin molecules belonging to one or more of the IgG subclasses (38). On the other hand, immunoelectrophoresis may show peaks consisting of a single immunoglobulin class (paraproteins) in dysgammaglobulinemic syndromes. Functional assessment of the B-cell system involves measurement of antibody activities. Nearly every child has received immunizations with diphtheria-pertussis-tetanus (DPT) and poliomyelitis vaccines,

Table 1. Evaluation of Anatomic Components of Host-Defense Systems

Lymphoid Tissues:
   Tonsils and adenoids (by physical examination and lateral, soft tissue X-ray examination of neck)
   Lymph nodes (by physical examination and [rarely] by lymphangiography)
   Spleen (by physical examination and [rarely] by radiologic techniques)
   Thymus (by chest X-ray)

Histologic Studies:
   Total leukocyte and differential counts
   Bone-marrow examination
   Lymph node biopsy

Table 2. Evaluation of B-Lymphocyte System

Quantitation of B-Lymphocytes:
　Cells with easily detectable surface immunoglobulin (SMIg)
　Cells with complement receptors (form rosettes with EAC-coated erythrocytes)
　Cells with receptors for Fc portion of IgG (form rosettes with EA-coated erythrocytes)

Quantitative and Qualitative Assays of B-Lymphocyte Products:
　Quantitation of IgG, IgA, IgM, IgD and IgE
　Immunoelectrophoresis
　Titers of "natural" antibodies
　Humoral responses to immunization
　Analysis of IgA for secretory piece
　Immediate-type skin responses

Responses to B-Lymphocyte Mitogens:
　Immunoglobulin secretion
　Lymphocyte transformation

and antibody responses to these antigens usually indicate normal B-cell function. In doubtful cases "booster" immunizations may be used to produce anamnestic responses. Evaluation of primary antibody responses to previously unencountered antigens can be done with typhoid vaccine, keyhole limpet hemocyanin—a product of an inedible mollusk to which incidental exposure is presumably rare (39)—or ϕX-174, a bacteriophage that does not infect human coliform organisms (40), although clinical use of the latter two substances in man is currently under investigation and regulated by the U.S. government.

In man, immediate type wheal and flare skin responses are mediated through immunoglobulin E (IgE). An estimate of the functional integrity of this immunoglobulin class in sensitive subjects can be obtained by intradermal testing with common environmental antigens, such as *Candida*; streptokinase-streptodornase; or extracts of pollens, molds, house dust or animal danders. As mentioned earlier IgE antibody responses to some antigens are under genetic control, and positive responses by normal subjects depend on prior sensitizing exposures. Thus positive responses indicate functional IgE molecules, whereas negative responses may simply indicate that exposure has not occurred. Wu et al (9) reported that in vitro stimulation with PWM caused normal B-lymphocytes to secrete immunoglobulin into the culture fluid and cell cytoplasm. Lymphocytes from most panhypogammaglobulinemic patients fail to show this response.

T-Lympocytes: In-Vitro Stimulation

An analogous approach may be used for assessment of T-cell functions. The commonest method for quantitating T-cells depends on their unique property of forming spontaneous rosettes with sheep erythrocytes (5) (Table 3). Peripheral blood of normal subjects contains 60-80% T-lymphocytes, of which one-fourth to one-half form "active" rosettes. The T-lymphocytes have also been quantitated with cytotoxic anti-T–cell antisera, but antisera of this specificity are difficult to prepare and the assay is not widely used. Thymus-dependent lymphocytes respond to in vitro stimulation with mitogens and antigens in one of three ways, including 1) synthesis and secretion of lymphokines; 2) increased synthesis

Table 3. Evaluation of T-lymphocyte System

Quantitation of T-lymphocytes cells with receptors for sheep erythrocytes [E-rosette forming cells]; cells reacting with anti-T–cell antisera
Assays for soluble T-lymphocyte products (lymphokines)
Responses to T-lymphocyte mitogens and antigens (lymphocyte transformation)
Cytotoxicity
Delayed cutaneous hypersensitivity

of proteins, RNA and DNA (lymphocyte transformation); and 3) cytotoxicity (Fig. 1). The lymphokines are believed to function as the local mediators of the inflammatory responses of cellular immunity (42). Those that have been characterized chemically are proteins, but little is known about their mechanisms of action (Table 4) (43).

Some lymphokines, notably macrophage migration inhibitory factor (MIF), may be secreted in vitro by either T or B cells (44), a fact that explains some of the clinical reports in which patients with profound abnormalities in T-cell responses—but normal B-cells—produced MIF in vitro. Optimal lymphokine production in vitro also requires macrophages. All T-lymphocytes respond to stimulation with plant lectins, such as PHA, glutinin, Con-A and PWM, with increased synthesis of DNA, RNA and protein. The magnitude of these responses may be quantitated by measuring the incorporation of radioactive precursors into cell products. In other patients the numbers of T-cells are normal, but immunodeficiency is demonstrated by impaired cellular immune responses to specific antigens. There is also evidence that the lymphocytes responding to antigens with increased DNA synthesis belong to a subpopulation of cells distinct from those producing lymphokines (44).

In normal subjects the antigen-dependent lymphocyte transformation re-

Table 4. Products of Antigen and Mitogen-Stimulated Lymphocytes (Lymphokines)

Mediators Affecting:

Macrophages
  Migration inhibitory factor (MIF)
  Macrophage activating factor (MAF, indistinguishable from MIF)
  Chemotactic factor for macrophages

Polymorphonuclear Leukocytes
  Chemotactic factors for neutrophils, eosinophils and basophils
  Leukocyte inhibitory factor (LIF)

Lymphocytes
  Mitogenic factors
  Factors affecting antibody production
  Transfer factor (enhances E-rosette formation)

Other Cell Types
  Cytotoxic factors
  Growth-inhibitory factors, including clonal-inhibitory and proliferation-inhibitory factors
  Osteoclast-activating factor (OAF)
  Interferon
  Tissue factor
  Colony-stimulating activity

sponses have close correlations with delayed cutaneous hypersensitivity responses to the same antigens. However, in patients with cellular immunodeficiency diseases, there are numerous instances in which subjects with negative skin tests have antigen-responsive lymphocytes in their blood. A response similar to the lymphocytic transformation reaction occurs when leukocytes from two genetically dissimilar persons are cultured together. This response, the mixed leukocyte reaction, is genetically controlled by the HLA-D locus. Remember that in evaluating abnormalities of lymphocyte transformation responses, one must consider the possibility that serum factors may suppress cells that would otherwise respond normally. Numerous serum factors may modify in vitro lymphocyte responses, and no single substance has been implicated in all cases.

The third class of responses by antigen or mitogen-stimulated lymphocytes is cytotoxicity. These responses involve direct contact between cytotoxic and target cells; no role for antibody or complement has been detected. Presumably the cytotoxic response is involved in reactions against foreign cells, such as grafts or tumors.

The in vivo expression of cellular immune reactions is the delayed-type hypersensitivity skin test. These responses are observed in subjects sensitized by natural or intentional exposure to the antigens. Positive responses require intact function of the entire arc of this immune response, including 1) antigen processing by macrophages, 2) lymphokine production and replication by antigen-sensitive T-cells and 3) responses of inflammatory cells to the humoral mediators. The same immunologic processes are seen in contact allergic reactions, rejection of allogeneic skin grafts and the graft vs host reaction. So far there are no dependable methods for localizing T-cells in tissues, although studies with cells extracted from cutaneous infiltrations show that the lymphoid cells in certain malignancies are thymus-derived (45).

In addition to the antigen-specific immunologic components of host-defenses, several systems may be activated by immunospecific interactions, but function by amplification of inflammation in a nonspecific way (Table 5). They include the complement system, the mechanisms for generation of and responses to chemotactic molecules, and the phagocytic cell system with its bactericidal function as measured by reduction with nitroblue tetrazolium (NBT) dye.

IMMUNODEFICIENCY SYNDROMES

From the general model shown in Figure 1, it is possible to construct a classification of immunodeficiencies based on defective functions of either T- or B-lymphocytes, or both (Table 6). In most cases the underlying biologic defects or errors are not yet completely characterized.

Table 5. Non-Specific Components of the Host Defense System

Complement
Chemotaxis
Phagocytosis
Bactericidal activity
—NBT dye reduction

Table 6. Lymphocyte Defects and Genetic Aspects of Selected Immunodeficiency Syndromes

Disorder	T-Cells Stage 1*	T-Cells Stage 2*	B-Cells Stage 1*	B-Cells Stage 2*	Mode of Inheritance
Disorders affecting stem cells:					
Reticular dysgenesis	Y†	Y	Y	Y	Unknown
SCID‡ (thymic alymphoplasia)	Y	Y	(Y)§	(Y)	X-linked
SCID (Swiss-type)	Y	Y	(Y)	(Y)	Autosomal recessive
SCID with ADA deficiency	Y	Y	(Y)	(Y)	Autosomal recessive
SCID with ectodermal dysplasia and dwarfism	Y	Y	Y	Y	? Autosomal recessive
SCID (sporadic)	Y	Y	Y	Y	Unknown
Disorders affecting B-cells:					
Congenital hypogammaglobulinemia (Bruton-type)	N†	N	Y	Y	X-linked
Congenital hypogammaglobulinemia	N	N	Y	Y	Autosomal recessive
Common variable hypogammaglobulinemia	N	N	N	Y∥	? Autosomal recessive
IgA deficiency	N	(N)	N	N	Variable
Immunodeficiency with elevated IgM	N	N	N	(Y)	X-linked
X-linked immunodeficiency with normal or hyperglobulinemia	N	N	(N)	(Y)	X-linked
Hypogammaglobulinemia with thymoma	N	N	N	Y∥	Unknown
Disorders affecting T-cells:					
Thymus hypoplasia	Y	Y	N	N	Variable
DiGeorge syndrome	Y	Y	N	N	Variable
Chronic mucocutaneous candidiasis with endocrinopathy	N	Y	N	N	? Autosomal recessive
Hyper-IgE syndrome	N	Y	N	N	Unknown
Complex immunodeficiencies:					
Ataxia-telangiectasia	(Y)	Y	N	(N)	Autosomal recessive
Wiskott-Aldrich syndrome	Y	Y	N	(Y)	X-linked
Cartilage-hair hypoplasia	?	Y	N	(N)	Autosomal recessive

*indicates first or second stages of lymphoid cell differentiation (see text).
†Y, yes; N, no.
‡SCID = severe combined immunodeficiency.
§Statements in parentheses indicate defects of variable severity or expression.
∥Recent evidence indicates a role for suppressor T-cells (see text).

Stem-Cell Disorders

The rare syndrome, *reticular dysgenesis*, was first recognized in a pair of male twins (49) and is characterized by the absence of all types of leukocytes from the bone marrow and peripheral blood; megakaryocytes and erythrocyte precursors, however, are spared. The immunologic features of the disease include lymphopenia, hypoplasia of the thymus and peripheral lymphoid tissues and hypogammaglobulinemia (50). Death from infectious diseases occurs early.

The name severe combined immunodeficiency (SCID) is applied to a common immunodeficiency syndrome in which development of both B- and T-cell functions is affected. Several patients have been partially or completely restored to immunocompetence by grafts of stem cells from bone marrow (51) or fetal liver (52), suggesting that the underlying defect is production of precursors of immunocompetent cells. As more patients are studied, however, it becomes clearer that the disease shows considerable heterogeneity in both clinical severity and immunologic abnormalities (53, 54). Some patients with functional abnormalities of both cellular and humoral immune responses, e.g., have normal numbers of blood lymphocytes, circulating B-cells and immunoglobulins.

An especially interesting case is reported by Pyke et al (55). Blood lymphocytes and bone-marrow cells from the patient described, when cultured on monolayers of human thymus epithelium, gave rise to cells that both formed rosettes with sheep erythrocytes (T-cells) and produced hemolytic antibodies when stimulated with sheep erythrocytes in vitro. The data are suggestive of a defect in maturation of the thymus in a patient who has competent stem cells; yet by conventional criteria this patient had SCID.

Both autosomal recessive and X-linked recessive forms of SCID have been identified (53). In general the former has lower blood lymphocytic counts and a more fulminant course than the latter. In both forms clinical evidence of immunodeficiency appears in the first few months of life and is expressed clinically as recurrent infections of the respiratory tract, diarrhea, candidiasis, skin rashes and failure to thrive. Unless the defect is corrected, survival beyond the second year is rare.

When cellular and humoral immunodeficiencies are fully expressed, the diagnosis is not difficult. The patient has lymphopenia, hypoplasia of the lymphoid tissues, hypogammaglobulinemia and (usually) severe deficiency of both T- and B-lymphocytes. When the disorder is only partially expressed, as in the patient with normal numbers of blood lymphocytes or serum immunoglobulins, the diagnosis may be more difficult. The cells in such a patient are usually effete in that they do not respond to stimulation with antigens or mitogens, and the immunoglobulins do not possess antibody activity. O'Reilly and Dupont have devised a scheme that assigns diagnostic weight to various clinical and immunologic features and emphasizes the importance of functional studies (Table 7).

It has recently been recognized that one autosomal recessive form of SCID may be associated with deficiency of adenosine deaminase (ADA), the enzyme that converts adenosine to inosine (56). Several lines of evidence indicate that the enzymatic defect is closely related to the immunologic dysfunction (57-59). A patient with the ADA-deficient form of SCID who was treated with transfusions of ADA-containing erythrocytes showed 1) improvement in the lymphocytic responses to mitogens and allogeneic cells, 2) development of a thymus shadow on the chest X-ray, 3) increases in the blood lymphocyte count and 4) restoration of antibody responses as well as clinical improvement (60).

Disorders of B-Lymphocyte System

The striking immunologic component of all these syndromes is the subject's inability either to synthesize and secrete immunoglobulins or to produce immunoglobulin molecules with antibody activity.

Table 7. Diagnostic Factors in Severe Combined Immunodeficiency*

Diagnostic Feature	Age When Significantly Abnormal	Diagnostic Importance†
Family history	Birth	++
Absent response to PHA	Birth	++++
Absent response to allogeneic cells	Birth	++++
Positive response to allogeneic cells	Limited significance	±
Lymphopenia (<1200 cells/mm^3)	Birth	++
Absent thymus shadow	Birth, if no distress	++
Absent lymph nodes	4–6 months	++
Absent tonsils	6 months	++
Absent isoagglutinins	>6 months	++
Humoral unresponsiveness to protein antigens	>6 months	+++
Humoral unresponsiveness to poly saccharide antigens	>2 years	±
Failure to sensitize with CDNB‡	~2–4 months	+++
Negative skin tests to *Candida*, SKSD§	Depends on exposure	±
Failure to reject skin graft	Birth	++++

*From (51).

†*Key:* ±, ++, +++, ++++, relative scale with ± and ++ being relatively unimportant, and +++ being of diagnostic importance.
‡CDNB, chloro dinitro benzene.
§SKSD, streptokinase-streptodornase.

Congenital Hypogammaglobulinemia

The patients usually do well during the first few months of life because of the protective effects of maternal antibodies. By the end of the first year, however, they have repeated pyogenic infections, most frequently of the sinuses, bronchi and lungs and less often of the skin, GI tract and bones. Most common viral infections are handled normally, although viral hepatitis may be a progressive, lethal infection. Other infrequent complications include malabsorption, diarrhea, rheumatoid-like arthritis, dermatomyositis, hemolytic anemias, neutropenias and thrombocytopenias. Malignancies, especially of the lymphoreticular system occur with unusual frequency (61). The disorder occurs in both X-linked recessive and autosomal recessive forms.

The serum shows panhypogammaglobulinemia with IgG levels of 100 mg/dl or less and less than 1% of the normal concentrations of the other immunoglobulins. Natural antibodies to blood-group antigens or sheep erythrocytes are not detectable, and there are minimal or no antibody responses to immunizations. In the X-linked form the patients often lack B-lymphocytes in the peripheral blood (3, 7, 62-64) and the lymphoid organs and bone marrow are devoid of plasma cells. In-vitro stimulation of these lymphocytes with PWM does not induce immunoglobulin synthesis or secretion (9). The patients have normal thymus glands and T-cells are present in normal numbers and functionally intact, indications that they have stem cells but lack the organ responsible for differentiation of stem cells into B-lymphocytes. A recent report suggests that some of these males have deficient lymphocyte ecto-5′-nucleotidase, another of the purine nucleotide degradation enzymes similar to adenosine deaminase (64a).

Some patients with child-onset hypogammaglobulinemia have normal numbers of circulating B-lymphocytes (62, 65). In addition their lymphocytes may respond to antigens by cell division, but they fail to differentiate into plasma cells. The genetic basis of this disorder, which occurs in male and female patients, has not been determined.

Common Variable Hypogammaglobulinemia
This syndrome is observed in patients in whom there are no clinical symptoms during childhood, but who develop recurrent infections and immunodeficiency in adult life. The disorder occurs in both male and female patients. Although the inheritance patterns are not well defined, the detailed pedigree analyses reported by Wollheim and co-workers (66) suggest a genetic component. In addition Kamin et al (67) report that patients with common variable hypogammaglobulinemia and their relatives have abnormal lymphocyte responses to in vitro stimulation with mitogens, suggesting a genetic abnormality in T-cell function.

The serum IgG levels are usually less than 5 mg per ml; IgA is usually undetectable, but IgM and IgE levels may be low, normal or elevated. In most cases antibody production is markedly deficient; however, some patients can produce antibodies and may have Coombs'-positive hemolytic anemias and immediate-type allergic reactions to γ-globulin. Normal numbers of T-cell are usually present, although their functions may be deficient (68, 69).

Considerable heterogeneity also exists in the anatomic expressions of the disease. Most patients have lymphocytes with B-cell membrane properties although the numbers vary from patient to patient. Dickler et al (7) have offered a classification of variable immunodeficiency based on blood lymphocyte membrane properties. Four immunologic groups of patients were identified, including 1) those with normal numbers of Ig-bearing cells in whom the presumed defect was an inability to differentiate Ig-secreting plasma cells; 2) those with B-lymphocytes by some criteria, but whose cells lacked surface Ig and in whom the defect presumably involved both synthesis and secretion of immunoglobulins; 3) those patients with large numbers of "null cells" lacking all membrane-associated markers and in whom precursor cells could not differentiate into B-cells; and 4) those whose cells bore both B- and T-cell markers. A fifth possibility, in which all lymphocytes were T-cells, was not found in this syndrome. Also, patients with common variable immunodeficiency may have hyperplastic lymphoid tissues, especially in the intestinal tract, mucosa as well as splenomegaly and lymphadenopathy.

Thus most patients with common variable hypogammaglobulinemia can differentiate B-cells, but are unable to elaborate them into antibody-secreting plasma cells. Further, their lymphocytes, even though they can synthesize immunoglobulins, do not secrete immunoglobulins when they are stimulated with PWM (9). An underlying defect in this disease has been defined by Waldmann et al (70). They found that peripheral blood lymphocytes from patients with common variable hypogammaglobulinemia—when cocultured with pokeweed mitogen-stimulated cells from normal donors—suppressed secretion of immunoglobulins by the normal cells. The cells with the suppressor activity were T-cells. Indeed, if the suppressor T-cells were removed from the cultures, the residual B-cells from hypogammaglobulinemic patients could secrete immunoglobulins in response to stimulation with PWM. This indicates that in some instances common variable hypogammaglobulinemia is a disease with defective regulation of T-cell function in which the B-cells are affected secondarily.

Selective Immunoglobulin Deficiencies

A commonly recognized laboratory abnormality, *isolated deficiency of IgA* occurs in about 1 in 500 normal people (71). In most cases the abnormality occurs sporadically, although there are familial studies indicating that it may be transmitted as either an autosomal dominant or an autosomal recessive trait. The patients have normal numbers of IgA-bearing B-cells (62, 65) and when their cells are stimulated with PWM, they synthesize cytoplasmic immunoglobulin in a normal manner (3). According to our present understanding the synthetic and secretory processes in these cells are intact, but ordinarily fail to secrete IgA.

The clinical significance of IgA deficiency also varies. It is noted in patients with recurrent sinopulmonary infections, steatorrhea, gluten-enteropathy (nontropical sprue) and diverse autoimmune disorders (72). However, most persons with IgA deficiency are normal. Perhaps the most serious risk of the disorder is the possibility of developing antibodies against IgA after blood transfusions (73).

Immunodeficiency with hyper-IgM is characterized by low serum levels of IgG and IgA and high concentrations (15 to 150 mg per ml) of electrophoretically heterogeneous IgM. Serum IgD and IgE levels may also be elevated. The few cases studied have had normal numbers of immunoglobulin-bearing cells (65). The disorder is transmitted as an X-linked recessive trait and is also associated with congenital rubella. In addition to susceptibility to infection, these patients may have hemolytic anemias, thrombocytopenias, neutropenias or a malignant infiltrative process with widespread deposition of IgM-bearing lymphoid cells.

There are several other immunodeficiencies in which selective immunoglobulin abnormalities are recognized, most of which are sporadic and unaccompanied by predisposing genetic factors. Exceptions include a woman with the syndrome of immunodeficiency with hyper-IgM in whom immunoglobulin abnormalities were found in 10 of 37 family members (74), elevated levels of IgM in paternal relatives of male children with hypogammaglobulinemia (75) and families in which various immunoglobulin abnormalities occurred in the pedigrees (76, 77).

The *Lesch-Nyhan syndrome* is an X-linked disorder with retardation, self-mutilation, hyperuricemia and virtual absence of the enzyme hypoxanthine-guanine phosphoribosyltransferase. Allison et al (78) recently reported three patients in whom the percentages of B-cells, IgG levels and lymphocytic responses to B-cell mitogens were subnormal; they suggested that the purine salvage pathway was less important for T-cell than for B-cell responses. This report, as yet unconfirmed, is important because it indicates another disorder in which an enzymatic defect in purine metabolism may be accompanied by lymphocyte dysfunction.

Disorders of T-lymphocyte System

Patients with developmental defects of the T-cell system are especially susceptible to recurrent or chronic infections with certain viruses and fungi. In fact chronic candidiasis of the skin and mucous membranes is often a hallmark of

T-cell deficiency (79). Other complications include progressive infections with vaccinia or BCG vaccine following vaccination, and graft vs host reactions after blood transfusions.

The best-defined disorders with isolated T-cell deficiency are the *DiGeorge syndrome* of congenital absence of the thymus and parathyroid glands (80) and the *Nezelof syndrome* of cellular immunodeficiency with normal immunoglobulins (81). Both groups of patients lack thymus glands and are unable to differentiate T-cells from stem cells. Therefore they are unable to produce T-cell–dependent immune responses, including delayed cutaneous hypersensitivity reactions to common environmental antigens or after intentional sensitizations. Rejection of skin allografts is also delayed. In vitro studies show that T-cells are markedly diminished. Antigenic or mitogenic stimulation of their lymphocytes in vitro does not cause lymphocyte transformation or lymphokine production. In contrast antibody responses, serum immunoglobulins and the numbers of circulating B-lymphocytes are usually normal.

The hereditary aspects of these diseases are variable. A DiGeorge-like syndrome has been recorded in a mother and child (82). Lawlor and co-workers (83) collected 34 cases of cellular immunodeficiency with normal immunoglobulins and concluded that the mode of inheritance was autosomal recessive in 11 cases and X-linked recessive in five; in the remaining cases the information was inadequate to make a determination.

A female patient with cellular immunodeficiency lacked the enzyme *nucleoside phosphorylase* in her erythrocytes (84). The case is especially interesting because it represents an additional instance of enzymatic defect in purine metabolism in an immunodeficient patient.

In other patients with cellular immunodeficiency, the number of circulating T-cells and the responses of these cells to T-cell mitogens were normal; the functional defect involves maturation of effector cells in response to antigenic stimulation. These patients are often infected with *Candida albicans* in a unique syndrome *(chronic mucocutaneous candidiasis)* involving candidiasis of the skin, nails and mucous membranes, which is resistant to chemotherapy (79). Delayed skin responses to *Candida* and other naturally encountered antigens are usually absent. In addition blood lymphocytes from many of the patients do not respond to in-vitro stimulation with antigens, especially *Candida*, either by producing lymphokines or by replication (79, 85-87). In some respects these patients are functionally analogous to those with the common variable type of hypogammaglobulinemia; they have achieved the first stage of T-cell differentiation, but are unable to respond in the antigen-dependent stage. The role of suppressor cells in these patients is currently under investigation.

One form of *chronic mucocutaneous candidiasis* is associated with hypofunction of the endocrine glands; the parathyroids, adrenals and thyroid are most frequently affected. The syndrome occurs in sibships, and an autosomal recessive mode of inheritance is proposed (88).

In 1972, Buckley et al (89) described two children with recurrent pyogenic infections, defective cell-mediated immunity and extreme elevation of serum IgE levels *(hyper-IgE syndrome)*. Subsequently Clark and co-workers (90) and Hill and Quie (91) recorded similar cases and showed that defective cellular chemotactic responses were an additional feature of the syndrome. In most instances the disease occurs sporadically; however, Van Scoy et al (92) recently described a family in which elevated serum IgE levels and deficient cellular chemotaxis occurred in three generations. The etiology and mechanisms of the abnormalities in this disorder are still under study.

Complex Genetic Disorders and Immunodeficiency

The syndromes considered in this section are "complex" in that they are composed of abnormalities affecting organ systems other than the host defenses.

Ataxia-Telangiectasia

This disorder is transmitted as an autosomal recessive trait. Ataxia and choreoathetosis may become apparent during infancy, whereas oculocutaneous telangiectasia may not appear until the fifth or sixth year (93). Infectious complications are usually limited to the sinopulmonary system. Patients often succumb from respiratory failure, and lymphoreticular neoplasms are also common (61).

The immunologic dysfunction involves both the humoral and cellular components. The patients usually have dysplastic or hypoplastic thymus glands, a depletion of thymus-dependent regions in the lymphoid organs and deficient circulating T-cells (3, 94). The deficiency is functionally expressed as cutaneous anergy and prolonged survival of allografts. Most patients with ataxia-telangiectasia have severe IgA deficiency, even though the numbers of IgA-bearing B-cells in their bloodstream are normal, and these cells secrete immunoglobulin into the cytoplasm when stimulated with PWM. In some patients antibodies to IgA and hypercatabolism of IgA have been found.

Wiskott-Aldrich Syndrome

This is an X-linked recessive disorder marked by eczema, thrombocytopenia and recurrent infections, and the patients are susceptible to bacterial, viral and fungal infections. Infection, hemorrhage or the lymphoreticular neoplasma that occur in roughly 10% of the cases (61) can all lead to death.

There is progressive deterioration of thymus-dependent immunity with depletion of paracortical areas of lymph nodes, cutaneous anergy and loss of responses to T-cell mitogens, although the severity of defects varies with the patients (95, 96). Immunoglobulin abnormalities are common and include extreme elevations of IgE levels, moderate increases in IgA levels and low levels of IgM (96). Other unique features are the marked impairment of antibody responses to polysaccharide antigens (97, 98) and the frequent absence of monocyte receptors for IgG (95) and defective chemotactic responses by monocytes. An immunologic survey of clinically normal relatives of Wiskott-Aldrich patients disclosed hyper-IgM levels in five of eight persons; no abnormalities in cell-mediated immunity were detected (95).

Cartilage-hair Hypoplasia

A variety of immunologic defects has been observed in some patients with short-limbed dwarfism (Table 6) (99). A discrete, autosomal recessive disorder, cartilage-hair hypoplasia, is associated with repeated pulmonary infections and marked sensitivity to life-threatening infections with the varicella virus (100,

101). Studies of the defense systems disclose chronic neutropenia, peripheral lymphopenia and impaired cellular immune responses in vivo and in vitro (100). Antibody responses and immunoglobulins are either normal (100) or deficient (101).

Miscellaneous Disorders

In several other complex familial disorders immunologic defects have also been noted. In most instances the cases studied are few and the mechanisms of immunodeficiency not well defined. Examples include the *exomphalos-macroglossia-gigantism syndrome* (in which markedly depressed cellular immunity and moderately severe immunoglobulin abnormalities have been described in a female patient). Two deceased male siblings were also suspected of having a similar immunologic disorder (102). The term *immunologic anamnesis* designates a familial disorder characterized by repeated infections, eczema and lymphopenia in two siblings, a boy and a girl. The immunologic defect involved both the cellular and humoral responses with cutaneous anergy and inability to develop anamnestic responses to repeated immunizations. The serum contained a complement-dependent, lymphocytotoxic substance that was demonstrable only during periods of severe lymphopenia (103).

Duncan's disease (104) is a progressive disorder characterized by immunoglobulin abnormalities and lymphocytic infiltrations of tissues. In the terminal stage, the patients' thymus glands and thymus-dependent lymphoid tissues atrophied. The disorder affected only males, and there were several instances in which it was preceded by infectious mononucleosis, suggesting that the disorder is genetically controlled and due to an aberrant response to either the Epstein-Barr or some other virus. Observations of this sort may define the relationship between neoplastic disorders and immunodeficiency diseases.

GENETIC DISORDERS OF PHAGOCYTIC-BACTERICIDAL SYSTEM

The cellular components of this system include the fixed phagocytic cells of the reticuloendothelial system and the circulating neutrophils, monocytes and eosinophils. Neutrophils are present in the greatest number and have highly efficient microbicidal functions. Optimal activity of this defense mechanism requires contributions from four sources. Chemotactic factors cause directed leukocytic migration into areas of inflammation. Opsonins react with microorganisms and increase their susceptibility to ingestion (phagocytosis). Finally the organisms are ingested and either sequestered within or killed and digested by the phagocytic cells.

Disorders of Chemotaxis

Neutrophil and other circulating phagocytic cells may respond to diverse stimuli with increased ameboid movement. Nondirectional cellular movement is called random migration. If the stimulus reaches the cells across a concentration gradient, the cells respond with migration toward the stimulus. This phenomenon, *chemotaxis,* involves several incompletely understood events at the cell membrane that are accompanied by changes in calcium flux and the net surface charge (105, 106). Cellular chemotaxis can be studied in vitro (47) and in vivo (107), and a classification of abnormalities of chemotaxis has been presented by Gallin and Wolff (47). Abnormalities of production of chemotactic factors were divided into disorders with diminished production of chemoattractants, disorders due to inhibitors of chemotactic factors and disorders with inappropriate inactivation of chemoattractants. Abnormalities in which cellular responses to chemotactic stimuli are deficient were classified as defects in cellular adherence, defects in cellular deformability (a requisite for cells to permeate freely vascular and tissue spaces), defects in random migration and defects in directed migration. Although only a few genetic defects of chemotaxis have been recognized, clinical studies of abnormal chemotaxis are just beginning and one should anticipate identification of additional syndromes in the future.

Chemotactic Factors

Perhaps the most intensively studied and easily demonstrated chemotactic factor is the cleavage product of the fifth component of complement (C5a) (Fig. 2). The cleavage product of C3—C3a—and the trimolecular complex of C567 are also reported to have chemotactic activity. Deficient generation of serum chemotactic activity has been observed in patients with recurrent infections and isolated deficiencies of C3 or C5 (Table 8). In addition patients with deficiencies of the early complement components C1r or C2 generate serum chemotactic activity at abnormally slow rates, although normal levels of activity are eventually achieved.

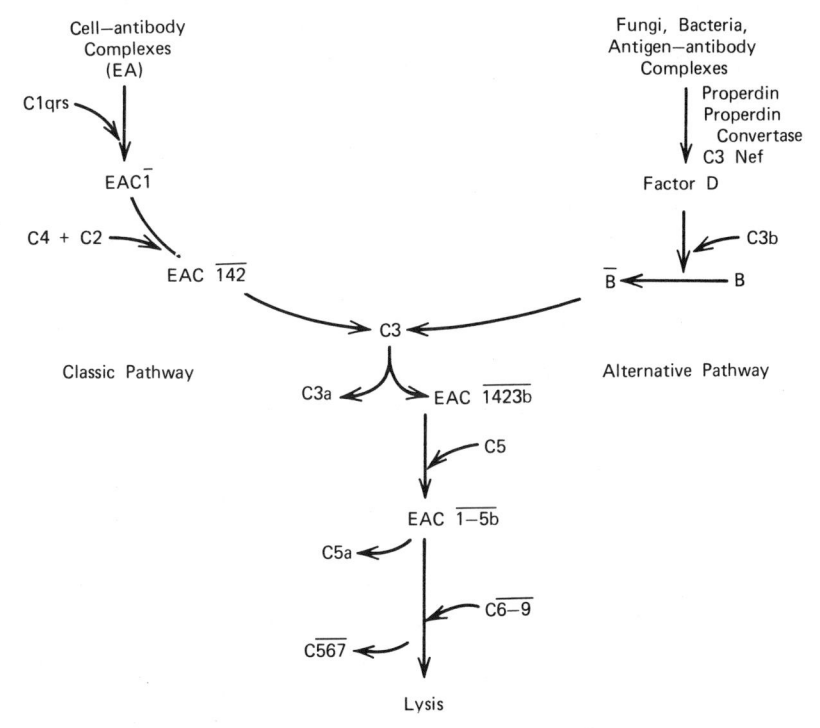

Figure 2. The components and reactions of the complement system.

Table 8. Genetic Disorders of Phagocytic-Bactericidal System

Disease	Deficiency	Inheritance*
Disorders of Chemotaxis		
Complement component deficiencies:		
C3	Impaired generation of chemotactic factors	AR
C5	Impaired generation of chemotactic factors	AR
C1r	Delayed generation of chemotactic factors	AR
C2	Delayed generation of chemotactic factors	AR
Antibody deficiency syndrome	Impaired generation of chemotactic factors	See Table 5
Cellular immunodeficiencies		
Wiskott-Aldrich syndrome	Impaired generation of chemotactic lymphokines, abnormal monocyte chemotaxis	X-linked
Chronic mucocutaneous candidiasis	Impaired generation of chemotactic factors lymphokines	AR
Leukocyte Defects:		
Neutropenia	Inadequate numbers of responsive cells	V
Chédiak-Higashi syndrome	Defective migration, possibly due to reduced deformability	AR
Lazy leukocyte syndrome	Deficient random migration and chemotactic responses	U
Hyper-IgE syndrome	Deficiency cellular chemotaxis	U
Impaired Opsonization		
Newborn	Impaired opsonic capacity of serum	
Antibody deficiency syndromes	Impaired opsonic capacity of serum	See Table 5
Complement deficiencies	Impaired opsonic capacity of serum	See Table 9
Sickle cell disease	Impaired opsonic capacity of serum	AI
Tuftsin deficiency	Impaired opsonic capacity of serum	U
Disorders of Phagocytosis		
Occur in diverse syndromes	Opsonic properties of complement or serum immunoglobulin systems are impaired	V
Disorders of Microbicidal Activity		
Chronic granulomatous disease	Complex disorder with impaired activity of hexose monophosphate shunt, O_2 utilization and H_2O_2 generation	Usually X-linked, but females may be affected
Myeloperoxidase deficiency	Impaired halogenation of ingested organisms	AR
Chédiak-Higashi syndrome	Impaired degranulation	AR
Glucose-6-phosphate dehydrogenase deficiency	Probably similar to CGD	X-linked

*AR, autosomal recessive; V, variable; U, unknown; AI, autosomal intermediate.

Serum from patients with hypogammaglobulinemia does not normally generate chemotactic activity after endotoxin treatment. Presumably this defect is due to deficiency of antiendotoxin antibody required for activation of the complement system and cleavage of C5 into C5a.

A second category of well-studied chemotactic factors are the chemotactic lymphokines, released from antigen-sensitive lymphocytes after stimulation with the appropriate antigens (Table 4). Deficient production of the chemotactic, as well as other, lymphokines is observed in various genetic disorders in which cell-mediated immunity is deficient. This is especially true of the Wiskott-Aldrich syndrome and certain forms of chronic mucocutaneous candidiasis.

Inhibitors of cellular chemotaxis and serum chemotactic factors are reported in association with other diseases too (47); however, at this time there is no evidence that production of these inhibitors is genetically determined.

Cellular Defects

The commonest cause of inability to accumulate leukocytes at inflammatory sites is probably *neutropenia* (108, 109). A second type of abnormal cellular chemotaxis is seen in the *Chédiak-Higashi* syndrome (110). The serum from these patients can be activated to produce chemotactic factors, but the leukocytes from these patients and from animals with the disease respond poorly to normal chemotactic stimuli. The disorder is transmitted in an autosomal recessive manner, and clinical features include loss of pigmentation of skin and hair, ocular albinism, neutropenia, recurrent infections and abnormally large lysosomes. The chemotactic defect may be due to reduced cellular deformability.

The *"lazy leukocyte" syndrome* is characterized by recurrent infections and by deficiencies of both random migration and chemotaxis (111) and in some cases by abnormalities of phagocytosis or bacterial killing. Abnormalities of cellular chemotaxis are found in the *"hyper-IgE" syndrome* (82-92) previously mentioned. This is a complex disorder characterized by recurrent pyogenic infections, eczema, defective cellular immunity, defective cellular chemotaxis and extreme elevation of serum IgE levels. The genetic components are unclear, although one report (92) described several affected members in a family. The mechanism of defective cellular chemotaxis in both the lazy leukocyte and hyper-IgE syndromes has not been defined.

Disorders of Opsonization

Opsonins are serum factors that react with the surface of microorganisms, rendering them more susceptible to ingestion by phagocytic cells. They are especially important for efficient phagocytosis of encapsulated organisms, such as pneumococci, *Hemophilus influenzae*, streptococci, and *Klebsiella pneumoniae*. *Heat-stable opsonins* are specific antibodies of the IgG_1 and IgG_3 subclasses and are found only in immunized subjects. The Fab portion of the immunoglobulin molecule reacts with antigenic determinants on the bacterial surface, and the Fc portion is free to interact with receptors on the phagocytic cells.

The *heat-labile* opsonic activity of nonimmune serum may be attributed to the complement system. Antibody-coated microorganisms may react with complement by the classical (C1-C4-C2-C3) pathway to form C3b, which will interact with receptors on neutrophils (Fig. 2). In addition some organisms may activate the alternative pathway directly by a process that does not require antibody, but which also leads to coating of the organisms with C3.

Impaired opsonization has been recognized in numerous conditions. The serum of newborn infants is deficient in both heat-stable and heat-labile opsonins. Sera from patients with many antibody-deficiency syndromes have impaired opsonic capacity, and this can be improved by replacement therapy with γ-globulin. Serum from patients with sickle cell disease is deficient in heat-labile opsonizing activity for the pneumococcus; opsonizing activity for *Salmonella,* however, is normal (112). Tuftsin is a γ-globulin–derived tetrapeptide composed of threonine-lysine-proline-arginine; it reportedly stimulates phagocytosis (113). *Deficiency of tuftsin* is described in splenectomized patients and in the members of three families.

Disorders of Phagocytosis

Following opsonization, potentially infectious organisms may be ingested by neutrophils or other phagocytic cells; however, little is known about the mechanism of recognition of an opsonized particle. Presumably ingestion of IgG but not IgM-opsonized particles is related to the surface receptors for IgG on neutrophils and macrophages. The divalent cations, Ca^{2+} and Mg^{2+}, are essential for optimal ingestion of particles just as they are for migration and spreading of the phagocytic cells. Particles opsonized by C3 may be ingested at lower concentrations of cations than are unopsonized particles. Substances that increase cyclic AMP, inhibit glycolysis, bind sulfhydryl groups, chelate divalent cations or have surface-active properties will impede phagocytosis (48).

Deficient phagocytic functions have been observed in patients with recurrent infections. The underlying defects in these subjects are diverse and include abnormalities of the complement system and deficiencies of opsonization, such as those noted in the newborn period and in patients with sickle cell anemia or hypogammaglobulinemia. Defective phagocytosis of properly opsonized particles is described in some patients with chronic granulomatous disease of childhood and in diabetic ketoacidosis. In contrast, phagocytosis of droplets of oil-red O occurs at a supernormal rate in the Chédiak-Higashi syndrome (114).

Disorders of Microbicidal Activity

Closely associated with phagocytosis by neutrophils is degranulation, the process by which the primary and secondary granules of the phagocytes discharge their enzymes into the phagosomes. By this process the cell can deliver the lysosomal contents to the ingested particles without exposing its own components to the potentially injurious enzymatic activities. Degranulation is associated with increased hexose monophosphate shunt-activity, increased oxygen consumption and production of hydrogen peroxide (H_2O_2), which agent is probably the most important of the bactericidal substances.

The enzymatic processes involved in formation of H_2O_2 are unknown. An intermediate step is formation of superoxide ion (O_2^-), which may then be converted to H_2O_2 by superoxide dismutase or react with preformed H_2O_2 to form hydroxyl radicals ($OH\cdot$). Two mechanisms are known by which neutrophils and macrophages protect themselves against potentially injurious effects of H_2O_2. One involves its degradation by catalase, and the other requires reduced glutathione, which is generated through reactions associated with the hexose monophosphate shunt.

There are several components of the bactericidal environment of the phagocytic vacuole. The internal pH is 3.5 to 4, which range itself is bactericidal or bacteriostatic for many organisms. Moreover the low pH provides optimal conditions for both conversion of superoxide to peroxide and activation of the hydrolytic enzymes contained in lysosomal granules. Other agents include lactoferrin, which inhibits microbial growth by chelating iron, and lysozyme, which digests cell walls and is especially effective against organisms either opsonized with complement components or exposed to ascorbate and H_2O_2. Although the latter compound is itself bactericidal, its most pronounced bactericidal effect is mediated through the enzyme myeloperoxidase and results in covalent binding of iodide to proteins of the ingested organisms.

The bactericidal activities of mononuclear phagocytes are generally similar to those of neutrophils. Blood monocytes have less myeloperoxidase activity than have neutrophils, and alveolar macrophages apparently lack this enzyme. Moreover mononuclear cells do not contain lactoferrin or cationic bactericidal proteins. Macrophages stimulated by either attachment to glass surfaces or exposure to lymphokines can digest more particles and have more hydrolytic enzymes, greater oxygen utilization (suggesting greater H_2O_2 production) and greater bactericidal activity. Presumably some of the bactericidal effect is due to iodination in which catalase instead of myeloperoxidase serves as the essential enzyme, although in most comparative studies mononuclear phagocytes have less efficient bactericidal activities than have neutrophils. The relatively long lifespan of mononuclear phagocytes, especially those of the reticuloendothelial system, may enable them to contain organisms in secondary lysosomes for prolonged periods.

Chronic granulomatous disease of childhood (CGD) is an especially striking example of the infectious complications that may occur in patients with impaired bactericidal mechanisms. The disorder exists in several forms; the commonest is transmitted as an X-linked recessive trait (48); however, variants of CGD are described in female patients and in male and female siblings (115). The patients have multiple recurrent infections of the lungs, bones and parenchymal organs. Granuloma formation causes enlargement of lymph nodes, liver and spleen.

A basic defect in CGD patients is that their neutrophils are unable to produce normal amounts of H_2O_2. Accordingly the patients are susceptible to infection by catalase and superoxide dismutase-producing organisms, such as staphylococci, gram-negative enteric bacteria, *Candida* and *Aspergillus*. Catalase-negative microorganisms, such as streptococci and pneumococci that produce their own H_2O_2, are only rarely causes of serious infections in CGD patients. However, certain catalase-negative, superoxide dismutase-positive organisms are killed by neutrophils from CGD patients, presumably through a synergistic action between a small amount of H_2O_2, myeloperoxidase (from lysosomal granules) and iodide. This observation further illustrates the importance of the peroxide-myeloperoxidase-halide system in the bactericidal function of neutrophils.

The nature of the defect in CGD patients has not been completely defined. Their cells have very active phagocytosis, but do not show concomitant increases in hexose monophosphate shunt-activity, oxygen consumption or H_2O_2 generation, all of which may be due to a deficiency of one of the nucleotide oxidases, a defect that would explain why neutrophils and monocytes from CGD patients fail to show reduction of NBT dye. This hypothesis has received support from the reports that neutrophils of CGD patients after ingesting latex beads coated with glucose oxidase showed improved metabolic and bactericidal activities (116, 117).

Deficiency of myeloperoxidase is an autosomal recessive disorder in which leukocytes have impaired bactericidal activity for *Candida,* and some patients are susceptible to deep infections with *Candida* (118). Myeloperoxidase is normally found in the primary granules and, as previously mentioned, is an important component of the hydrogen peroxide-myeloperoxidase-halide bactericidal system. Not all persons who lack the enzyme have infections, probably because they accumulate peroxide in neutrophil phagosomes.

In addition to the chemotactic defect already described, neutrophils and monocytes from patients with the *Chédiak-Higashi syndrome* contain giant lysosomal granules that fail to fuse with phagosomes, thereby causing a functional deficiency of degranulation and lysosomal enzyme release (114).

Deficiency of glucose-6-phosphate dehydrogenase is X-linked, and in some patients absence of the enzyme is associated with impaired bactericidal activity and a chronic granulomatous disease similar to, but less severe than, CGD (119).

Lipochrome histiocytosis syndrome includes lipochrome pigmentation of histiocytes, pulmonary infiltrates, hyperglobulinemia, splenomegaly and recurrent infections. A recent report describes sisters with this disease in whom a bactericidal defect similar to CGD was found (120).

DISORDERS OF COMPLEMENT SYSTEM

It has been known for years that this series of biochemical reactions provides an important system for amplification of immunoglobulin-dependent immunologic reactions. Although the exact nature of some of the reactions is still ill defined, existing data have permitted assigning functions to certain complement components and recognizing several diseases in which disorders of the complement system are critical (Table 9).

The "classic" complement pathway consists of nine components and 11 separate proteins. The alternative pathway contains five proteins and merges with the classic pathway at C3, the third component (Fig. 2) (46). The classic complement pathway is activated when antigenic determinants (E) react with IgG or IgM antibodies (A) to form immune complexes (EA). A complex containing a single molecule of IgM can initiate the complement sequence, whereas at least two molecules of IgG antibody appropriately spaced on the antigenic determinants of the membrane are required. Only the IgG_1, IgG_2 and IgG_3 subclasses of IgG can fix complement.

The sequence begins with the interaction of the $C1_q$ portion of C1, the first complement component, with the Fc portion of the immunoglobulin to form EAC1 complexes. This reaction also activates an esterase. The EAC1 complex reacts with C4 molecules, which are cleaved, and $EAC\overline{14}$ complexes are found. (The bar over the numbers indicates a component protein or complex of proteins in an activated form capable of continuing the sequence of reactions.) These complexes, in turn, react with C2 to form additional cleavage products and $EAC\overline{142}$ complexes. This product, in the presence of C3 molecules, mediates the formation of new cleavage products, C3a and C3b. The C3a is

Table 9. Deficiencies of Complement System in Man

Component	Associated Abnormalities	Mode of Inheritance*
C1q	SCID and other immunodeficiency diseases	U
C1r	SLE-like skin lesions, glomerulonephritis and arthralgia	AR
C1s	SLE-like syndrome	U
C4	SLE-like syndrome	U (AR in guinea pigs)
C2	SLE-like syndrome, glomerulonephritis and dermatomyositis may occur in normal persons	AR
C3	Recurrent infections	AR
C5	Recurrent infections, SLE-like syndrome	AR
C6	Probably none	AR
C7	None	Unknown
C8	Deficient serum bactericidal activity for *N. gonorrhoeae*	AR
C1 esterase inhibitor	Hereditary angioedema	AD
C3 inhibitor	Recurrent infections	U

*U, unknown; AR, autosomal recessive; AD, autosomal dominant.

released from the complex into the fluid phase where it exerts chemotactic and anaphylatoxic activities; C3b remains attached to the membrane as part of the complex, EAC$\overline{1423}$b, and confers opsonic activity. As the reaction proceeds, the C$\overline{1423}$b complex binds to and cleaves C5. One cleavage product, C5a, is released into the fluid phase where it has potent chemotactic activity. The C5b portion remains attached to the immune complex.

The late components C6, C7, C8 and C9 react sequentially in the process that typically terminates with disruption of the cell membrane and lysis of the cell; C$\overline{567}$ complexes have chemotactic activity in vitro, but their biologic significance is unknown. Fixation of C8 or C9 is required for cell lysis.

The alternative complement, or *properdin* pathway, differs from the classic pathway in that it may be activated either by antigen-antibody complexes, including complexes containing IgA, or by antibody-independent mechanisms involving interaction of properdin with complex polysaccharides, such as the cell walls or capsules of many bacteria and fungi (Fig. 2). The result of activation of this pathway is generation of a C3-cleaving enzyme, which is analogous to that produced by the classic pathway.

In addition there are inhibitory substances that apparently regulate the activity of the reaction cascade. They include an inhibitor of the C1 esterase, inactivators of C3 and C4 and an inhibitor of C3a.

Genetic Defects of Complement System

There are numerous reports of subjects in whom a single component of the complement system is deficient because the patient either cannot synthesize the protein or produces molecules devoid of function. Detailed studies of complement-dependent reactions in these patients have defined the relative contributions of individual components to certain in vitro phenomena and to the host-defense system. In general defects affecting the early components of the classic pathway (C1, C4, C2) (Fig. 2) are of relatively little clinical significance, presumably because the patients have intact alternative pathways and therefore are able to opsonize bacteria and generate chemotactic factors. On the other hand, Hurley et al (121) showed that guinea pigs depleted of alternative pathway components by administration of cobra venom showed impaired ability to clear *C. albicans* from the bloodstream and were killed by doses of the organisms that were not fatal to animals with either intact complement function or a genetic deficiency of C4, an early component of the classic pathway.

Deficiency of C1 q is reported in a few patients with hypogammaglobulinemia and the lymphopenic form of SCID. It is uncertain if the lack represents a direct genetically controlled abnormality in C1 q synthesis or is secondary to the lymphocyte abnormalities.

C1r deficiency has occurred in two siblings and an unrelated subject. The sibs suffered from cutaneous ulcerations that deteriorated to severe scarring, arthralgias without evidence arthritis, febrile episodes and focal glomerulonephritis (122). Immunofluorescent studies of the kidney biopsies showed coarse deposits containing IgG and C3. The renal disease in the sibs was not progressive, but the third patient developed renal failure.

C1s deficiency is described in one patient, a 6-year-old with a clinical syndrome suggestive of systemic lupus erythematosus (SLE). The LE cell preparation was negative, but the patient had an anti-DNA antibody.

C4 deficiency is established as an autosomal recessive trait in guinea pigs; the affected animals have no increased susceptibility to infections. Normally activation of their serum with endotoxin produces chemotactic factor (C5a), but generation of chemotactic factors by immune complexes is delayed; no other complement-dependent functions are abnormal. The disorder has been observed in a woman with a lupus-like syndrome. She did not have LE cells, antinuclear antibodies or IgG deposits in the skin. Recently a child with C4 deficiency and a lupus-like disorder has been reported.

C2 deficiency has been discovered in several pedigrees (124). It is compatible with health, presumably because functions of the alternative pathway are intact. In some cases, however, C2 deficiency has been associated with connective tissue diseases, such as SLE, dermatomyositis and chronic glomerulonephritis.

The rare disorder, *C3 deficiency,* has been identified in at least three patients and is transmitted as an autosomal recessive trait (125). Each patient was markedly susceptible to pyogenic infections. The one patient studied in depth was unable to mobilize neutrophils during infectious episodes (in vivo chemotaxis), could not opsonize bacteria and was unable to lyse antibody-coated bacteria or erythrocytes.

Deficiency of C5 is implicated in two clinical situations. In one the serum from an infant girl with recurrent infections, diarrhea and seborrheic dermatitis failed to opsonize yeast and staphylococci (126). The defect was corrected in vitro by addition of C5 to her serum, and infusions of fresh plasma led to her improvement. The second case involved two members of a family, one of whom had an SLE-like syndrome with mild nephritis, antinuclear antibodies and positive LE cell preparations (127). The patient's serum showed deficient capacity for generating chemotactic factors, but was not deficient in opsonins.

The *C6 deficiency* was serendipitously discovered in a young woman who was essentially well except for mild Raynaud's phenomenon and trace amounts of cryoglobulin in her serum (128), which latter had normal opsonic and chemotactic activities, but would not lyse antibody-coated bacteria or erythrocytes. The disorder was transmitted as an autosomal recessive trait. Subsequently other patients with C6 deficiency have been identified, one being a child who suffered from repeated infections with *Neiserria* organisms.

Two cases of *C7 deficiency* have been described. One, a healthy male, was incidentally found to have deficient hemolytic complement activity, which was attributed to C7 deficiency. The other was a patient with scleroderma. Neither one had recurrent infections (129, 130).

C8 deficiency has been observed in a woman with two or three episodes of disseminated gonococcal infections (131). Her serum contained no C8 and no lytic activity for erythrocytes or *N. gonorrhoeae;* all other complement components were present. In addition her serum showed normal opsonic activity for yeast and staphylococci, the opsonins generated chemotactic activity when treated with heat-aggregated IgG. The familial study indicated autosomal recessive inheritance.

These deficiencies furnish evidence for the relative importance of certain complement components in resisting infections. Defects affecting the early components of the classic pathway are not associated with infectious complications. The most profound defects occur in C3-deficient patients who are unable to opsonize microorganisms and therefore have deficient phagocytosis and bacterial killing. One would anticipate that C5-deficient patients would have inordinate numbers of infections because they are unable to generate an important chemotactic factor. However, with the exception of *Neisseria* infections, deficiencies of the other late components apparently do not predispose a person to infectious diseases.

C1 esterase inhibitor deficiency is accompanied by hereditary angioedema, an autosomal dominant disorder characterized by episodic massive, nonpitting edema of the extremities, genitalia, face or torso (132). Involvement of the tongue, larynx or pharynx may lead to death by asphyxiation. Edema of the intestinal mucosa may cause episodes of abdominal pain with evidence of obstruction. The pathogenesis of the disease is unclear. Affected persons have low levels of C1-esterase inhibitor and C4 even during asymptomatic periods. The disorder exists in two forms, in one of which patients are deficient in inhibitor molecules, whereas in the other, nonfunctional molecules are present.

C3 inhibitor deficiency has been observed in one patient with recurrent infections and temperature-sensitive urticaria (133). The low levels of C3 and properdin factor B were due to continuous activation of the alternative pathway and hypercatabolism of the components.

IMMUNODEFICIENCY AND ATOPIC DISEASES

Familial studies demonstrate that the predisposition to atopic disorders, such as asthma, allergic rhinitis and atopic dermatitis, is genetically influenced (134). The possible association of these disorders with certain serologically defined leukocytic antigens or haplotypes has already been described. From the immunologic viewpoint these disorders have in common the excessive production of reagenic (IgE) antibodies to antigens that are most frequently encountered on the mucous membranes of the respiratory and GI tracts, and this predisposition to production of IgE antibodies is often reflected in elevated levels of IgE in the serum.

Recently several investigators found hypofunction of the T-cell system in patients with atopic dermatitis (135). This finding is expressed as 1) decreased numbers of T-lymphocytes in the blood, 2) subnormal lymphocyte transformation responses to T-cell mitogens and 3) impaired delayed skin responses to common environmental antigens. The defects are more profound in patients with rather more than less severe forms of disease and do not appear to be secondary to treatment with antiinflammatory drugs. The pathogenesis of this immunodeficiency and its relationship to production of cutaneous lesions have not been defined.

THERAPEUTIC CONSIDERATIONS

The natural consequences of delineation of the underlying defects in immunologic deficiency diseases have been the attempts to tailor treatment programs for specific lesions. Plasma transfusion and γ-globulin therapy are well established for treatment of antibody-deficiency syndromes, and fresh plasma transfusions have been used to treat some patients with complement-component deficiencies (126). Several patients with SCID have received transplants of bone-marrow or fetal liver cells in attempts to promote differentiation of stem cells into immunocompetent cells (51, 52). Although there have been many failures because of genetic disparities between the grafted cells and the recipient or because of infectious complications, the successes clearly show the potential for this therapeutic approach. The correction of the immunologic defect in a patient with the ADA-deficient form of SCID with transfusions of ADA-containing erythrocytes represents a novel and apparently low-risk approach to treatment of this disorder (60). Obviously this observation is an incentive to identification of additional enzymatic defects and exploration of other forms of replacement therapy.

Thymus-dependent lymphocyte immunodeficiencies due to thymic hypoplasia have been corrected by thymus grafts (14, 136) and by thymosin (137)—an extract from thymus glands. Transfer factor has been used to correct defective lymphokine production in patients with chronic fungal infections and has been of therapeutic benefit (138). It is now under study in patients with various neoplastic disorders.

Finally recognition that the immunodeficiency in some adult-onset diseases—such as common variable hypogammaglobulinemia (70) and some chronic fungal infections (139)—may be secondary to active suppression of normal immune processes opens new avenues for therapeutic intervention in these diseases.

REFERENCES

1. Bruton OC: Agammaglobulinemia. *Pediatrics* 9:722, 1952
2. Cooper MD, Perey DY, Peterson RDA, et al: The two-component concept of the lymphoid system. *Birth Defects* 4:7, 1968

3. Cooper MD, Keightley RG, Wu L-YF, et al: Developmental defects of T and B cell lines in humans. *Transplant Rev* 16:51, 1973
4. Lawton AR, Self KS, Royal SA, et al: Ontogeny of B-lymphocytes in the human fetus. *Clin Immunol Immunopath* 1:84, 1972
5. Jondal M, Wigzell H, Aiuti F: Human lymphocyte subpopulations: Classification according to surface markers and/or functional characteristics. *Transplant Rev* 16:163, 1973
6. Shevach EM, Jaffe ES, Green I: Receptors for complement and immunoglobulin on human and animal lymphoid cells. *Transplant Rev* 16:3, 1973
7. Dickler HB, Adkinson NF, Fisher RI, et al: Lymphocytes in patients with variable immunodeficiency and panhypogammaglobulinemia. Evaluation of B and T cell surface markers and a proposed classification. *J Clin Invest* 53:834, 1974
8. Bellanti JA, Hurtado RC: Immunology of the fetus and newborn, in Avery GB (ed): *Neonatology*. Philadelphia, Lippincott, 1975, p. 521
9. Wu L-YF, Lawton AR, Cooper MD: Differentiation capacity of cultured B lymphocytes from immunodeficient patients. *J Clin Invest* 52:3180, 1973
10. Stites DP, Carr MC, Fudenberg HH: Development of cellular immunity in the human fetus: Dichotomy of proliferative and cytotoxic responses of lymphoid cells to phytohemagglutinin. *Proc Nat Acad Sci USA* 69:1440, 1972
11. Wybran J, Carr MC, Fudenberg HH: The human rosette-forming cell as a marker of a population of thymus-derived cells. *J Clin Invest* 51:2537, 1972
12. Komuro K, Boyse EA: In-vitro demonstration of thymic hormone in the mouse by conversion of precursor cells into lymphocytes. *Lancet* 1:740, 1973
13. Incefy GS, L'Esperance P, Good RA: *In vitro* differentiation of human marrow cells into T lymphocytes by thymic extracts using the rosette technique. *Clin Exp Immunol* 19:475, 1975
14. Kirkpatrick CH, Ottesen EA, Smith TK, et al: Reconstitution of defective cellular immunity with fetal thymus and dialyzable transfer factor: Long-term studies in a patient with chronic mucocutaneous candidiasis. *Clin Exp Immunol* 23:414, 1976
15. Waldron JA, Horn RG, Rosenthal AS: Antigen-induced proliferation of guinea pig lymphocytes *in vitro*: Obligatory role of macrophages in the recognition of antigen by immune T-lymphocytes. *J Immunol* 111:58, 1973
16. Rosenthal AS, Shevach EM: Macrophage-T lymphocyte interaction: The cellular basis for genetic control of antigen recognition, in Katz D, Benacerraf B (eds): The Role of Products of the Histocompatibility Gene Complex in Immune Responses. New York, Academic., 1976, p. 335
17. Katz DH, Benacerraf B: The regulatory influence of activated T cells on B cell responses to antigen. *Adv Immunol* 15:1, 1972
18. Gershon RK: A disquisition on suppressor T cells. *Transplant Rev* 26:170, 1975
19. Benacerraf B, McDevitt HO: Histocompatibility-linked immune response genes. *Science* 175:273, 1972
20. Bach FH, Bach ML, Sondel PM: Differential function of major histocompatibility complex antigens in T-lymphocyte activation. *Nature* 259:273, 1976
21. Wolski KP, Schmid FR, Mittal KK: Genetic linkage between the HL-A system and a defecit in the second component (C2) of complement. *Science* 188:1020, 1975
22. Scott BB, Swinburne ML, Rajah SM, et al: HL-A8 and the immune response to gluten. *Lancet* 2:374, 1974
23. Spencer MJ, Cherry JD, Terasaki PI: HL-A antigens and antibody response after influenza A vaccination. Decreased response associated with HL-A type W-16. *N Engl J Med* 294:13, 1976
24. Greenberg LJ, Gray ED, Yunis EJ: Association of HL-A5 and immune responsiveness in vitro to streptococcal antigens. *J Exp Med* 141:935, 1975
25. Marsh DG, Bias WB, Hsu SH: Association of the HL-A7 cross-reacting group with a specific antibody response in allergic man. *Science* 179:691, 1973
26. Levine BB, Stember RH, Fotino M: Ragweed hay fever: Genetic control and inkage to HL-A haplotypes. *Science* 178:1201, 1972
27. Blumenthal MN, Amos DB, Noreen H, et al: Genetic mapping of Ir locus in man: Linkage to second locus of HL-A. *Science* 184:1301, 1974

28. McDevitt HO, Bodmer WF: HL-A, immune-response genes and disease. *Lancet* 1:1269, 1974
29. Falchuk ZM, Rogentine GN, Strober W: Predominance of histocompatibility antigen HL-A8 in patients with gluten-sensitive enteropathy. *J Clin Invest* 51:1601, 1972
30. Gebhard RL, Katz SI, Marks J, et al: HL-A antigen types and small intestinal disease in dermatitis herpetiformis. *Lancet* 2:760, 1973
31. Schlosstein L, Terasaki PI, Bluestone R, et al: High association of an HL-A antigen, W27, with ankylosing spondylitis. *N Engl J Med* 288:704, 1973
32. Calin A, Fries JF: Striking prevalence of ankylosing spondylitis in "healthy" W27-positive males and females. A controlled study. *N Engl J Med* 293:835, 1975
33. Morris R, Metzger AL, Bluestone R, et al: HL-A antigen W27—a genetic link between ankylosing spondylitis and Reiter's syndrome? *N Engl J Med* 290:572, 1974
34. Brewerton DA, Caffrey M, Nicholls A, et al: Acute anterior uveitis and HL-A27. *Lancet* 2:994, 1973
35. White SH, Newcomer VD, Mickey MR, et al: Disturbance of HL-A antigen frequency in psoriasis. *N Engl J Med* 287:740, 1972
36. Jersild C, Hansen GS, Svejgaard A, et al: Histocompatibility determinants in multiple sclerosis, with special reference to clinical course. *Lancet* 2:1221, 1973
37. Winchester RJ, Fu SM, Hoffman T, et al: IgG on lymphocyte surfaces; technical problems and the significance of a third cell population. *J Immunol* 114:1210, 1975
38. Schur PH, Borel H, Gelfand EW, et al: Selective gamma-G globulin deficiencies in patients with recurrent pyogenic infections. *N Engl J Med* 283:631, 1970
39. Curtis JE, Hersh EM, Harris JE, et al: The human primary immune response to keyhole limpet hemocyanin: Interrelationships of delayed hypersensitivity, antibody response and *in vitro* blast transformation. *Clin Exp Immunol* 6:473, 1970
40. Wedgwood RJ, Ochs HD, Davis SD: The recognition and classification of immunodeficiency diseases with bacteriophage ØX 174, in Bergsma D (ed): Immunodeficiency in Man and Animals. Sunderland, MA, Sinauer Assoc., Inc., 1975, p. 331
41. Edelson RL, Smith RW, Frank MM, et al: Identification of subpopulations of mononuclear cells in cutaneous infiltrates. I. Differentiation between B cells, T cells and histiocytes. *J Invest Dermatol* 61:82, 1973
42. Bennett B, Bloom BR: Reactions *in vivo* and *in vitro* produced by a soluble substance associated with delayed-type hypersensitivity. *Proc Nat Acad Sci USA* 59:756, 1968
43. Remold HG: Purification and characterization of lymphocyte mediators in cellular immunity. *Transplant Rev* 10:152, 1972
44. Rocklin RE: Mediators of cellular immunity, their nature and assay. *J Invest Derm* 67: 372, 1976
45. Lutzner M, Edelson R, Schein P, et al: Cutaneous T-cell lymphomas: The Sézary syndrome, mycosis fungoides, and related disorders. *Ann Int Med* 83:534, 1975
46. Frank MM, Atkinson JP: Complement in clinical medicine. *Disease-a-Month* Jan 1975, p. 3
47. Gallin JI, Wolff SM: Leukocyte chemotaxis: Physiological considerations and abnormalities. *Clin Haematol* 4:567, 1975
48. Stossel TP: Phagocytosis. *N Engl J Med* 290:717, 774, 833, 1974
49. DeVall O, Seynhaeve V: Reticular dysgenesis. *Lancet* 2:1123, 1959
50. Alonzo K, Dew JM, Starke WR: Thymic alymphoplasia and congenital aleukocytosis (reticular dysgenesia). *Arch Pathol* 94:179, 1972
51. Dupont B, O'Reilly RJ, Jersild C, et al: Transplantation of immunocompetent cells, in Brent L, Holborow J (eds): Progress in Immunology, Proceedings of the Second International Congress of Immunology, Vol 5. New York, American Elsevier, 1974, p. 203
52. Buckley RH, Whisnant JK, Schiff RI, et al: Correction of severe combined immunodeficiency by fetal liver cells. *N Engl J Med* 294:1076, 1976
53. Miller, ME, Schieken RM: Thymus dysplasia. A separable entity from "Swiss agammaglobulinemia." *Am J Med Sci* 253:741, 1967
54. Hitzig WH, Landolt R, Müller G, et al: Heterogeneity of phenotypic expression in a family

with Swiss-type agammaglobulinemia: Observations on the acquisition of agammaglobulinemia. *J Pediatr* 78:968, 1971
55. Pyke KW, Dosch H-M, Ipp MM, et al: Demonstration of an intrathymic defect in a case of severe combined immunodeficiency disease. *N Engl J Med* 293:424, 1975
56. Meuwissen HJ, Pickering RJ, Pollara B, et al: Combined Immunodeficiency Disease and Adenosine Deaminase Deficiency. A Molecular Defect. New York, Academic, 1975
57. Fox IH, Keystone EC, Gladman DD, et al: Inhibition of mitogen mediated lymphocyte blastogenesis by adenosine. *Immunol Comm* 4:419, 1975
58. Hovi T, Smyth JF, Allison AC, et al: Role of adenosine deaminase in lymphocyte proliferation. *Clin Exp Immunol* 23:395, 1976
59. Polmar SH, Wetzler EM, Stern RC, et al: Restoration of in vitro lymphocytes responses with exogenous adenosine deaminase in a patient with severe combined immunodeficiency. *Lancet* 2:743, 1975
60. Polmar SH, Stern RC, Schwartz AL, et al: Enzyme replacement therapy for adenosine deaminase deficiency and severe combined immunodeficiency disease. *Pediatr Res* 10:392, 1976
61. Kersey JH, Spector BD, Good RA: Immunodeficiency and cancer. *Adv Cancer Res* 18:211, 1973
62. Preud'Homme JL, Griscelli C, Seligmann M: Immunoglobulins on the surface of lymphocytes in fifty patients with primary immunodeficiency diseases. *Clin Immunol Immunopathol* 1:241, 1973
63. Hayward AR, Greaves MF: Central failure of B-lymphocyte induction in panhypogammaglobulinemia. *Clin Immunol Immunopathol* 3:461, 1975
64. Rachelefsky GS, McConnachie PR, Ammann AJ, et al: Antibody-dependent lymphocyte killer function in human immunodeficiency diseases. *Clin Exp Immunol* 19:1, 1975
64a. Edwards ML, Magilvay DB, Cassidy JT, Fox IH: Lymphocyte ecto-5'-nucleotidase deficiency in agammaglobulinemia. *Science* 201:628, 1978
65. Cooper MD, Lawton AR: Circulating B-cells in patients with immunodeficiency. *Am J Pathol* 69:513, 1972
66. Wollheim FA, Belfrage S, Coster C, et al: Primary "acquired" hypogammaglobulinemia. *Acta Med Scand* 176:1, 1964
67. Kamin RM, Fudenberg HH, Douglas SD: A genetic defect in "acquired" agammaglobulinemia. *Proc Nat Acad Sci USA* 60:881, 1968
68. Geha RS, Schneeberger E, Merler E, et al: Heterogeneity of "acquired" or common variable agammaglobulinemia. *N Engl J Med* 291:1, 1974
69. Webster ADB, Asherson GL: Identification and function of T cells in the peripheral blood of patients with hypogammaglobulinemia. *Clin Exp Immunol* 18:499, 1974
70. Waldmann TA, Durm M, Broder S, et al: Role of suppressor T cells in pathogenesis of common variable hypogammaglobulinemia. *Lancet* 2:609, 1974
71. Collins-Williams C, Kokubu HL, Lamenza C, et al: Incidence of isolated deficiency of IgA in the serum of Canadian children. *Ann Allergy* 30:11, 1972
72. Hong R, Ammann AJ: Selective absence of IgA. Autoimmune phenomena and autoimmune disease. *Am J Pathol* 69:491, 1972
73. Vyas GN, Holmdahl L, Perkins HA, et al: Serologic specificity of human anti-IgA and its significance in transfusion. *Blood* 34:573, 1969
74. Feldman G, Koziner B, Talamo R, et al: Familial variable immunodeficiency: Autosomal dominant pattern with variable expression of the defect(s). *J Pediatr* 87:534, 1975
75. Kirkpatrick CH, Schimke RN: Paternal immunoglobulin abnormalities in congenital hypogammaglobulinemia. *JAMA* 200:105, 1967
76. Buckley RH, Sidbury JB: Hereditary alterations in the immune response: Coexistence of "agammaglobulinemia," acquired hypogammaglobulinemia and selective immunoglobulin deficiency in a sibship. *Pediatr Res* 2:72, 1968
77. Goldblum RM, Lord RA, Cooper MD, et al: X-linked B lymphocyte deficiency. I. Panhypo-γ-globulinemia and dys-γ-globulinemia in siblings. *J Pediatr* 85:188, 1974

78. Allison AC, Hovi T, Watts RWE, et al: Immunological observations on patients with Lesch-Nyhan syndrome, and on the role of de novo purine synthesis in lymphocyte transformation. *Lancet* 2:1179, 1975
79. Kirkpatrick CH, Rich RR, Bennett JE: Chronic mucocutaneous candidiasis: Model-building in cellular immunity. *Ann Int Med* 74:955, 1971
80. DiGeorge AM: Congenital absence of the thymus and its immunologic consequences: Concurrence with congenital hypoparathyroidism, in Bergsma D (ed): Immunologic Deficiency Diseases in Man. New York, National Foundation, 1968, p. 116
81. Nezelof C: Thymic dysplasia with normal immunoglobulins and immunologic deficiency: Pure alymphocytosis, in Bergsma D (ed): Immunologic Deficiency Diseases in Man. New York, National Foundation, 1968, p. 104
82. Steele RW, Limas C, Thurman GB, et al: Familial thymic aplasia. *N Engl J Med* 287:787, 1972
83. Lawlor GJ, Ammann AJ, Wright WC, et al: The syndrome of cellular immunodeficiency with immunoglobulins. *J Pediatr* 84:183, 1974
84. Giblett ER, Ammann AJ, Wara DW, et al: Nucleoside-phosphorylase deficiency in a child with severely defective T-cell immunity and normal B-cell immunity. *Lancet* 1:1010, 1975
85. Lehner T, Wilton JMA, Ivanyi L: Immunodeficiencies in chronic mucocutaneous candidosis. *Immunology* 22:775, 1972
86. Valdimarrson H, Higgs JM, Wells RS, et al: Immune abnormalities associated with chronic mucocutaneous candidiasis. *Cell Immunol* 6:348, 1973
87. Kirkpatrick CH, Montes LF: Chronic mucocutaneous candidasis. *J Cutan Pathol* 1:211, 1974
88. Sotos JF: The endocrine system, in Goodman RM (ed): Genetic Disorders of Man. Boston, Little, Brown, 1970, p. 720
89. Buckley RH, Wray BB, Belmaker EZ: Extreme hyperimmunoglobulinemia E and undue susceptibility to infection. *Pediatrics* 49:59, 1972
90. Clark RA, Root RK, Kimball HR, et al: Defective neutrophil chemotaxis and cellular immunity in a child with recurrent infections. *Ann Int Med* 78:515, 1973
91. Hill HR, Quie PG: Raised serum IgE levels and defective neutrophil chemotaxis in three children with eczema and recurrent bacterial infections. *Lancet* 1:183, 1974
92. Van Scoy RE, Hill HR, Ritts RE, et al: Familial neutrophil chemotaxis defect, recurrent bacterial infections, mucocutaneous candidiasis and hyperimmunoglobulinemia E. *Ann Int Med* 82:766, 1975
93. Boder E: Ataxia-telangiectasia: Some historic, clinical and pathologic observations, in Bergsma D (ed): Immunodeficiency in Man and Animals. Sunderland, MA, Sinauer Assoc., 1975, p. 255
94. Biggar WD, Good RA: Immunodeficiency in ataxia-telangiectasia, in Bergsma D (ed): Immunodeficiency in Man and Animals. Sunderland, MA, Sinauer Assoc., 1975, p. 271
95. Spitler LE, Levin AS, Stites DP, et al: The Wiskott-Aldrich syndrome. Immunologic studies in nine patients and selected family members. *Cell Immunol* 19:201, 1975
96. Blaese RM, Strober W, Waldmann T: Immunodeficiency in the Wiskott-Aldrich syndrome, in Bergsma D (ed): Immunodeficiency in Man and Animals. Sunderland, MA, Sinauer Assoc., 1975, p. 250
97. Cooper MD, Chase HP, Lowman JT, et al: Wiskott-Aldrich Syndrome. An immunologic deficiency disease involving the afferent limb of immunity. *Am J Med* 44:499, 1968
98. Blaese RM, Strober W, Brown RS, et al: The Wiskott-Aldrich syndrome. A disorder with a possible defect in antigen processing or recognition. *Lancet* 1:1056, 1968
99. Ammann AJ, Sutliff W, Millinchick E: Antibody mediated immunodeficiency in short limbed dwarfism. *J Pediatr* 84:200, 1974
100. Lux SE, Johnston RB, August CS, et al: Chronic neutropenia and abnormal cellular immunity in cartilage-hair hypoplasia. *N Engl J Med* 282:231, 1970
101. Saulsbury FT, Winkelstein JA, David LE, et al: Combined immunodeficiency and vaccine-related poliomyelitis in a child with cartilage-hair hypoplasia. *J Pediatr* 86:868, 1975
102. Greene RJ, Gilbert EF, Huand S-W, et al: Immunodeficiency associated with exomphalos-

macroglossia-gigantism syndrome. *J Pediatr* 82:814, 1973
103. Kretschmer R, August CS, Rosen FS, et al: Recurrent infections, episodic lymphopenia and impaired cellular immunity. *N Engl J Med* 281:285, 1969
104. Purtilo D, Cassel CK, Yang JPS, et al: X-linked recessive progressive combined variable immunodeficiency (Duncan's disease). *Lancet* 1:935, 1975
105. Gallin JI, Rosenthal AS: The regulatory role of divalent cations in human granulocyte chemotaxis: Evidence for an association between calcium exchanges and microtubule assembly. *J Cell Biol* 62:594, 1974
106. Gallin JI, Durocher JR, Kaplan AP: Interaction of chemotactic factors with the cell surface. I. Chemotactic factor induced changes in human granulocyte surface charge. *J Clin invest* 55:967, 1975
107. Rebuck JW, Crowley JH: A method for studying leukocyte functions *in vivo*. *Ann NY Acad Sci* 59:757, 1955
108. Patten E, Gallin JI, Clark RA, et al: Effects of cell concentration and various anticoagulants on neutrophil migration. *Blood* 41:711, 1973
109. Dale DC, Wolff SM: Skin window studies of the acute inflammatory responses of neutropenic patients. *Blood* 38:138, 1971
110. Clark RA, Kimball HR: Defective granulocyte chemotaxis in the Chediak-Higashi syndrome. *J Clin Invest* 50:2645, 1971
111. Miller ME, Oski FA, Harris MB: Lazy leukocyte syndrome. A new disorder of neutrophil function. *Lancet* 1:665, 1971
112. Winkelstein JA, Drachman RH: Deficiency of pneumococcal serum opsonizing activity in sickle cell disease. *N Engl J Med* 279:459, 1968
113. Najjar VA, Constantopoulos A: A new phagocytosis-stimulating tetrapeptide hormone, tuftsin, and its role in disease. *J Reticuloendothel Soc* 12:197, 1972
114. Stossel TP, Root RK, Vaughan M: Phagocytosis in chronic granulomatous disease and the Chédiak-Higashi syndrome. *N Engl J Med* 286:120, 1972
115. Holmes B, Park BH, Malawista SE, et al: Chronic granulomatous disease in females. A deficiency of leukocyte glutathione peroxidase. *N Engl J Med* 283:217, 1970
116. Johnston RB, Baehner RL: Improvement of leukocyte bactericidal activity in chronic granulomatous disease. *Blood* 35:350, 1970
117. Baehner RL, Nathan DG, Karnovsky ML: Correction of metabolic deficiencies in the leukocytes of patients with chronic granulomatous disease. *J Clin Invest* 49:865, 1970
118. Salmon SE, Cline MJ, Schultz J, et al: Myeloperoxidase deficiency: Immunologic study of a genetic leukocyte defect. *N Engl J Med* 282:250, 1970
119. Gray GR, Klebanoff SJ, Stomatoyannopoulos G, et al: Neutrophil dysfunction, chronic granulomatous disease, and non-spherocytic hemolytic anemia caused by complete deficiency of glucose 6-phosphate dehydrogenase. *Lancet* 2:530, 1973
120. Rodey GE, Park BH, Ford DK, et al: Defective bactericidal activity of peripheral blood leukocytes in lipochrome histiocytosis. *Am J Med* 49:322, 1970
121. Hurley DL, Gelfand JA, Fauci AS, et al: Complement deficiency and experimental candidiasis. Fifteenth Interscience Conference on Anti-microbial Agents and Chemotherapy, Washington, 1975, p. 203
122. Day NK, Geiger H, Stroud R: C1r deficiency: An inborn error associated with cutaneous and renal disease. *J Clin Invest* 51:1102, 1972
123. Clark RA, Kimball HR, Frank MM: Serum chemotactic factors in an animal model of a complement deficiency state, in Braun W, Ungar J (eds): Non-specific Factors Influencing Host Resistance. Basel, Karger, 1973, p. 205
124. Kemperer MR, Woodworth HC, Rosen RS, et al: Hereditary deficiency of the second component of complement (C'2) in man. *J Clin Invest* 45:880, 1966
125. Alper CA, Colten HR, Gear JSS, et al: Homozygous human C3 deficiency. The role of C3 in antibody production, C1s induced vasopermeability, and cobra venom-induced passive hemolysis. *J Clin Invest* 57:222, 1976

126. Miller ME, Nilsson UR: A familial deficiency of the phagocytosis-enhancing activity of serum related to a dysfunction of the fifth component of complement (C5). *N Engl J Med* 282:354, 1970
127. Rosenfeld SI, Leddy JP: Hereditary deficiency of fifth component of complement (C5) in man. *J Clin Invest* 53:67a, 1974
128. Leddy JP, Frank MM, Gaither T, et al: Hereditary deficiency of the sixth component of complement in man. I. Immunochemical, biologic and family studies. *J Clin Invest* 53:544, 1974
129. Wellek B, Opferkuch W: A case of deficiency of the seventh component of complement in man. Biological properties of a C7-deficient serum and description of a C7-inactivating principle. *Clin Exp Immunol* 19:223, 1975
130. Boyer JT, Gall EP, Norman ME, et al: Hereditary deficiency of the seventh component of complement. *J Clin Invest* 56:905, 1975
131. Petersen BH, Graham JA, Brooks GF: Human deficiency of the eighth component of complement. The requirement of C8 for serum *Neisseria gonorrhoeae* bactericidal activity. *J Clin Invest* 57:283, 1976
132. Frank MM, Gelfand JA, Atkinson JP: Hereditary angioedema: The clinical syndrome and its management. *Ann Int Med* 84:580, 1976
133. Alper CA, Abramson N, Johnston RB, et al: Increased susceptibility to infection associated with abnormalities of complement mediated functions and of the third component of complement (C3). *N Engl J Med* 282:349, 1970
134. Levine BB: Genetic factors in atopic allergic disease, in Kirkpatrick CH, Reynolds HY (eds): Immunologic and Infectious Reactions in the Lung. New York, Dekker, 1976, p. 419
135. Rachelefsky GS, Opelz G, Mickey MR, et al: Defective T cell function in atopic dermatitis. *J Allergy Clin Immunol* 57:569, 1976
136. Biggar WD, Park BH, Good RA: Immunologic reconstitution. *Ann Rev Med* 24:135, 1973
137. Wara DW, Goldstein AL, Doyle NE, et al: Thymosin activity in patients with cellular immunodeficiency. *N Engl J Med* 292:70, 1975
138. Kirkpatrick CH, Gallin JI: Treatment of infectious and neoplastic diseases with transfer factor. *Oncology* 29:46, 1974
139. Stobo JD, Paul S, Van Seoy RE, et al: Suppressor thymus-derived lymphocytes in fungal infection. *J Clin Invest* 57:319, 1976

5
Genetic Metabolic Disorders

Mario C. Rattazzi
R. Neil Schimke
Laird G. Jackson

LYSOSOMAL STORAGE DISEASES
Mario C. Rattazzi

GENERAL ASPECTS

The Concept of Lysosomal Storage Diseases

The concept of inborn lysosomal storage diseases was proposed in 1965 by Hers (1). He defined these metabolic disorders as progressive diseases, in which (not necessarily) homogenous material is stored within membrane-bound cytoplasmic inclusions in different organs as a result of impaired function of degradative enzymes. In the intervening years this hypothesis has been verified in several inherited disorders (Tables 1 and 2). The recognition of lysosomes as cellular organelles responsible for catabolism (2), the biochemical characterization of numerous lysosomal hydrolases and the elucidation of the chemical structure of the stored materials were critical factors in the identification of the enzyme deficiencies in these diseases. Studies on the genetically determined storage disorders have evolved from clinical description to investigation of lysosomal physiopathology and provide an insight into the molecular pathology of some of these diseases.

Function and Biochemical Features (3, 4)

Lysosomes are cellular organelles in which a number of hydrolytic enzymes are kept separate from the surrounding cytoplasm by a unit membrane (2). Small vesicles containing the lysosomal enzymes are detached from the (mature) face of the Golgi apparatus and are usually referred to as primary lysosomes. In

Table 1. Glycosphingolipidoses: Biochemical Aspects

Disorder and Eponym	Compound Primarily Stored	Enzyme Deficiency	Key to Steps in Figure 2
Gm1 gangliosidosis (Landing's disease)	Gm1 ganglioside	β-Galactosidase	1
Gm2 gangliosidosis I, (Tay-Sachs disease)	Gm2 ganglioside	β-D-N-Acetyl hexosaminidase A	2
Gm2 gangliosidosis II (Sandhoff disease)	Gm2 ganglioside, globoside	β-D-N-Acetyl hexosaminidase A,B	2, 3
Trihexosylceramidosis (Fabry's disease)*	Trihexosylceramide	α-Galactosidase	4
Lactosylceramidosis	Lactosylceramide	? β-Galactosidase	5
Glucosylceramidosis (Gaucher's disease)	Glucosylceramide	β-Glucosidase	6
Sphingomyelinosis (Niemann-Pick disease)	Sphingomyelin	Sphingomyelinase	7
Sulfatidosis (Metachromatic leukodystrophy)	Sulfatide	Arylsulfatase A	8
Globoid cell leukodystrophy (Krabbe's disease)	Galactosylceramide	Specific β-galactosidase	9
Lipogranulomatosis (Farber's disease)	Ceramide	Ceramidase	10

*X-linked recessive inheritance.

Table 2. Mucopolysaccharidoses: Biochemical Aspects

Disorder and Eponym	Compound Primarily Affected	Enzyme Deficiency	Key to Steps in Figure 3
MPS IH (Hurler syndrome); MPS IS (Scheie syndrome)	Dermatan and heparan sulfates	α-L-iduronidase	2
MPS II (Hunter syndrome)*	Dermatan and heparan sulfates	L-Iduronosulfate sulfatase*	1
MPS III A (Sanfilippo A syndrome)	Heparan sulfate	Heparan-N-sulfatase (sulfamidase)	3
MPS III B (Sanfilippo B syndrome)	Heparan sulfate	α-N-Acetyl glucosaminidase	5
MPS IV (Morquio syndrome)	Keratan and chondroitin sulfates	N-acetyl hexosamine-6-sulfate sulfatase	7
MPS VI (Maroteaux-Lamy syndrome)	Dermatan sulfate	N-acetyl galactosamine-4-sulfate sulfatase (arylsulfatase B)	6
MPS VII (Sly syndrome)	Dermatan and heparan sulfates	β-Glucuronidase	4

*X-linked recessive inheritance.

endocytosis, vacuoles containing extracellular material are formed by invagination of the plasma membrane. These endocytic vacuoles fuse with the primary lysosomes, thereby acquiring lysosomal enzymes that degrade the material within the vacuole and become digestive vacuoles, or secondary lysosomes. The cycle is often repeated several times. Undigestible residues accumulate within the vacuole, the activity of lysosomal hydrolases decreases and the resulting residual bodies usually become postlysosomes, which are practically devoid of enzymatic activity, and at times may discharge their content into the extracellular spaces.

Endogenous cellular materials are catabolized by a similar process. These mechanisms are present in most animal cells, and additional degradative mechanisms have been observed in some specialized cell types in specific organs and tissues. It is thought that lysosomal enzymes, e.g., are excreted by osteoclasts into the surrounding medium through direct fusion of primary lysosomes with the plasma membrane, a process known as *exocytosis*. Degradation of the bone matrix is thus initiated outside the cell and completed by endocytosis of partially degraded material.

Special centrifugation of cell and tissue homogenates allow the isolation of intact lysosomes. About 60 degradative enzymes are associated with this lysosomal fraction (3, 4), and with few exceptions all of them have an acidic optimum pH. In-vitro assay of intact lysosomal preparations yields only a fraction ($< 10\%$) of the enzymatic activity that can be measured after disruption of the lysosomes. This characteristic of lysosomal enzymes (latency) is the result of the relatively impermeable lysosomal membrane, which allows diffusion only of small molecules (< 250 daltons) with certain chemical properties. The same permeability characteristics of the membrane insure that a given complex biologic substrate will be degraded to its smallest constituents, which then escape for use into the cytoplasm. This degradation is generally due to *exoenzymes,* which sequentially cleave residues in the terminal position on the complex molecule.

Impaired cleavage results in a block of the stepwise degradation; the molecular weight of the undegraded substrate does not allow diffusion through the lysosomal membrane, and progressive accumulation ensues. The resulting numerous, abnormally distended cytoplasmic vacuoles (inclusion- or storage bodies), visible under the light microscope, often occupy the entire cytoplasmic space of the cell. The physicochemical properties of the accumulated compound may result in a characteristic ultrastructure of the storage bodies, from precipitation or aggregation of the compound in diverse configurations, i.e., "cytoplasmic membraneous bodies," "zebra bodies" or "myelin figures", which entities are often helpful in the identification of a given storage disorder.

A lysosomal enzyme may recognize as substrates different compounds having a common residue in the appropriate configuration. Thus in a given disease in addition to the substance that is predominantly stored, concentrations of the other compounds are often abnormally elevated in the same or different tissues. This phenomenon may also result from secondary inhibition of other, unrelated, lysosomal hydrolases by physicochemical interactions with the storage material. In some cases lysosomal enzyme deficiencies are accompanied by increased enzymatic activity of unrelated lysosomal hydrolases. This is regarded as the result of overproduction of primary lysosomes, presumably because of continued fusion of these vesicles with the storage vacuoles, the membrane of which may be abnormally unstable and hence favor fusion.

Impaired Functioning

The relationship between lysosomal storage and impaired cellular function is obscure. The intuitive notion that a cell completely filled with huge inclusion bodies cannot carry out its normal metabolic function is not substantiated by biochemical evidence. Observations of phagocytic cells suggest that the capacity of cells to engage in continued endocytosis is limited by the availability of membrane material necessary for the formation of new digestive vacuoles. Materials not removed from the cytoplasm could also result in synthesis of structures with abnormal composition and, therefore, abnormal function. Finally impaired degradation of small amounts of a toxic compound—also normally recognized as substrate by the enzyme affected by the mutation—may be overshadowed by a more dramatic accumulation of another compound, but it may be the primary cause of impaired cellular function.

A lysosomal enzyme may be present in several isozymes coded for by different genes, and one or more forms may be deficient in a given disease. Study of these isozymes is complicated by the widespread use of artificial, chromogenic or fluorogenic substrates that, because of their relatively simple molecular structure, can at times be recognized by different enzymes with strict specificities for natural compounds. In some instances these compounds can be cleaved at nearly normal rate by an abnormal enzyme that is inactive toward its natural substrate. Most lysosomal hydrolases are in almost all cells except erythrocytes, thereby allowing the potential use of leukocytes, cultured skin fibroblasts and amniotic fluid cells, tears, plasma and urine for biochemical diagnosis. However, this does not imply that all organs and tissues are equally affected in terms of storage by a given enzyme deficiency.

Electron microscopy and chemical studies often reveal some degree of storage in several tissues; massive storage, however, is generally limited to those tissues in which the turnover rate of a given compound or class of compounds is normally high. Consequently diseases resulting from deficient activity of different enzymes involved in the degradation of the same or related compounds may be clinically indistinguishable, and a precise diagnosis may be possible only by biochemical identification of the enzyme deficiency. Further, the relatively broad substrate specificity of some hydrolases may result in similar clinicopathologic features for mutations of enzymes primarily active in different degradative pathways. Finally the same enzyme deficiency may be associated with a spectrum of diseases, ranging from the early-onset, rapidly fatal one to the late-onset, mild one with near-normal life expectancy. This clinical variability may be due to different allelic mutations affecting the same enzyme, resulting in different degrees of impairment of the activity toward natural substrates in vivo, not detectable in vitro, especially when artificial substrates are used.

The known lysosomal storage disorders are described in greater detail in several recent books and reviews (5-14).

GLYCOSPHINGOLIPIDOSES

These diseases (Table 1) result from defects in the catabolism of glycosphingolipids, compounds with a common ceramide (N-acyl sphingosine) backbone, but with different residues (hexoses, phosphorylcholine) replacing the terminal hydroxyl group. The two main branches of this pathway, schematically shown in Figure 1, involve stepwise degradation of *cerebrosides* (Fig. 1, left) and *gangliosides* (Fig. 1, right) to ceramide. Degradation of sphingomyelin (phosphorylcholineceramide) and sulfatide (sulfuric acid ester of galactosyl ceramide) also has ceramide as the end product. Ceramide is further degraded to sphingosine and fatty acids. The conditions listed in Table 1 are all transmitted with an autosomal recessive mode of inheritance, with the exception of trihexosylceramidosis (Fabry's disease), which has an X-linked recessive inheritance.

Gm$_1$ Gangliosidosis (Landing's Disease, Generalized Gangliosidosis)

The disease exists in two main clinical forms. The early-onset (infantile) *type I* is characterized by progressive psychomotor deterioration, evident at age 6 months or earlier, and often accompanied by seizures. The patient has a coarse facies similar to that found in the mucopolysaccharidoses, corneal opacity, frequent retinal cherry-red macula, prominent hepatosplenomegaly and skeletal abnormalities (also similar to those found in mucopolysaccharidoses [dysostosis multiplex]). Vacuolated peripheral lymphocytes are present. Histopathologic findings include neuronal lipidosis with membranous cytoplasmic bodies, prominent visceral histiocytosis (foam cells) and cytoplasmic ballooning of the renal glomerular epithelium. The disease advances rapidly to blindness, deafness and spastic quadriplegia; death from pulmonary infections usually intervenes by 2 years of age.

In the late-onset (juvenile) *type II*, which appears to be the less common of the two, psychomotor development is normal during the first year. Ataxia, loss of coordination and progressive psychomotor degeneration, often with seizures, then ensue. The striking facial features of the early-onset type are absent in these patients as is hepatomegaly (usually), and bony lesions are less severe. Vacuolated peripheral lymphocytes and foamy histiocytes in the bone marrow are seen. The histopathologic profile is similar to that of type I, but visceral involvement is minimal. The disease advances to a decerebrate state, leading to death between age 5 and 10 years.

Massive quantities of Gm$_1$ ganglioside, and to a less extent of its asialo-derivative, accumulate in the brains of these patients; corresponding storage in visceral organs is found only in type I. Acidic glycosaminoglycans (formerly, mucopolysaccharides) of the keratan sulfate type (Fig. 2) are stored in the visceral organs in both infantile and juvenile forms, and increased urinary excretion of these compounds is reported. Visceral storage of partially degraded glycoprotein material, presumably derived from erythrocyte stroma, is also described. All these substances have in common a terminal β-D-galactose residue: their impaired degradation results from a profound deficiency of lysosomal β-galactosidase (Step 1, Fig. 1; Step 9, Fig. 2).

Two main forms of β-galactosidase, A and B, are present in normal tissues, but in the disease the activity of both is deficient. Immunologically cross-reactive material is found in the patients' tissues, in amounts comparable to those of the enzyme in normal tissues. This indicates that a structural mutation affects the catalytic properties of both enzymes. The structural relationships between

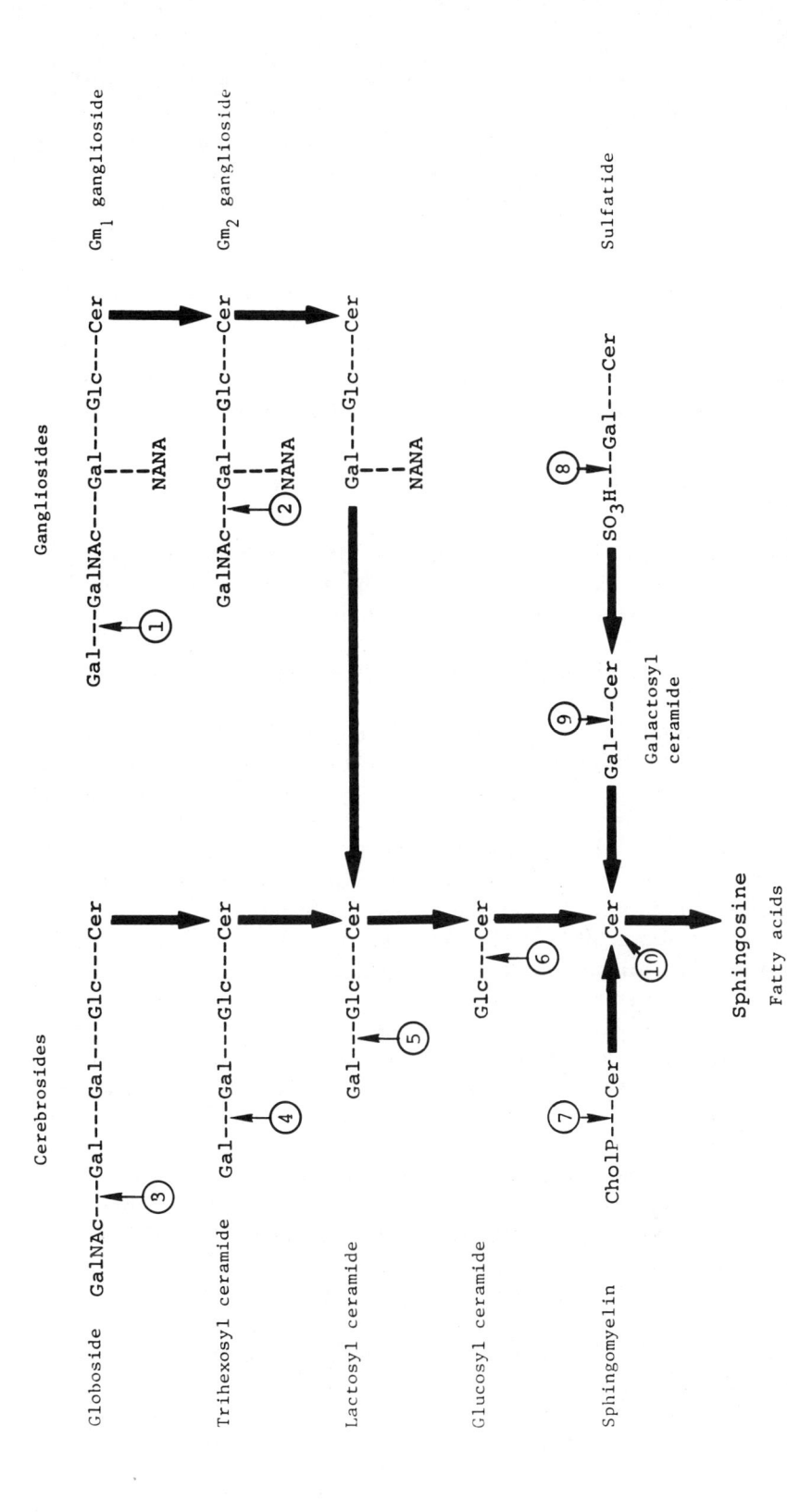

Figure 1. Chemical structure and catabolic pathways of major sphingolipids in men and sites of inborn errors of metabolism. Circled numbers: enzymatic steps listed in Table 1. Abbreviations: Cer, ceramide; CholP, phosphorylcholine; Gal, galactose; GalNac, N-acetyl galactosamine; Glc, glucose.

β-galactosidase A and B, however, have not been fully clarified. The different clinical features in type I and type II Gm_1 gangliosidosis, in the cases with intermediate phenotypes and in the rare milder forms with adult onset might result from allelic mutations. These changes may affect the activities of β-galactosidase toward different substrates (Gm_1 ganglioside, keratan sulfate and glycoproteins) differently in various combinations (15). Complementation—i.e., correction of β-galactosidase deficiency by fusion of fibroblasts from patients with type I or II disease with fibroblasts from a patient with a mild form of the disease has been observed; the β-galactosidase deficiency in this case, however, was probably secondary to a neuraminidase deficiency. Reduced enzymatic activity in leukocytes from parents of patients has also been found. Prenatal biochemical diagnosis was made in a few instances and confirmed in the aborted fetuses.

Gm_2 Gangliosidoses

In *Gm_2 gangliosidosis type I* (Tay-Sachs disease, infantile amaurotic idiocy), the first symptoms are noticeable at age 5-6 months. Psychomotor deterioration is evident before 1 year and usually advances rapidly in the second year, resulting in blindness, deafness, seizures and a decerebrate state. A doll-like appearance, with translucent skin and long eyelashes, is often noticeable. Macrocephaly is common in the second year, and a retinal cherry-red macula is often present as is optic-nerve degeneration. No visceral or skeletal involvement is detectable. The main histopathologic finding is neuronal lipidosis with characteristic "membranous cytoplasmic bodies" consisting of membrane-enclosed concentrical lamellar arrays visible by electron microscopy. Death, usually caused by bronchopneumonia, intervenes between age 3 and 4 years.

The disease is caused by massive accumulation of Gm_2 ganglioside in neurons. This substance can reach levels 300-fold higher in affected than in normal brains and accounts for about 90% of total ganglioside content in this organ. The asialo-derivative of Gm_2 ganglioside is also stored, but to a less extent. Impaired degradation of these substances is owing to generalized deficiency of β-D-N acetyl hexosaminidase A (Hex A), one of the two main forms of this enzyme normally in human tissues. Since the other major form of β-hexosaminidase, Hex B, is in the patient's tissues, the finding supports the contention that Hex A is the enzyme primarily responsible for Gm_2-ganglioside degradation in normal tissues, (Step 2, Fig. 1).

The storage of the asialo-derivative of the ganglioside can be explained by competitive inhibition of the normal Hex B activity against this compound by excess Gm_2 ganglioside, but it is not clear if Hex B participates in the catabolism of Gm_2 ganglioside in vivo. Several reports indicate that in vitro the purified enzyme does not cleave this substrate, but a recent report states that in certain experimental conditions Hex B can cleave Gm_2 ganglioside (15a).

The two enzymes share certain characteristics and a common antigenic determinant, although Hex A possesses a unique antigenic determinant, not detectable in Hex B. Somatic cell genetic studies show that Hex A is controlled by a gene on chromosome 15, and Hex B by a gene on chromosome 5.

The expression of human Hex A in man–rodent cell hybrids, however, requires Hex B. It seems clear then that the structure of the two enzymes is closely related, although the precise nature of that relationship has not been fully elucidated. Of the various structural models proposed, the one with the best experimental support postulates common subunits in Hex A and Hex B and unique subunits in Hex A (16, 17).

Biochemical confirmation of the diagnosis is easily obtained by assay of Hex A activity. The deficiency is detectable in utero by amniocentesis and enzymatic assay of cultured amniotic fluid cells. Ultrastructural and biochemical observations show that storage of Gm$_2$ ganglioside in fetal neural tissue is detectable as early as 20 weeks' gestation; heterozygotes are readily detectable. An important implication of heterozygote identification is the possibility of preventing the disease in Ashkenazic Jews, in which population Gm$_2$ gangliosidosis occurs with high frequency (1 in 3,600 live births).

Accurate detection of heterozygosity in pregnant women is possible by enzymatic assay of leukocytes. In contrast to plasma, the Hex A: Hex B ratio in these materials is not affected by pregnancy. Occurrence of Gm$_2$ gangliosidosis type I in high frequency, most probably an effect of genetic drift and inbreeding, is reported in some small isolates (Pennsylvania Dutch, French-Canadian populations).

Gm$_2$ gangliosidosis type II (Sandhoff disease) is an early-onset form of Gm$_2$ gangliosidosis, with clinicopathologic features virtually identical to those of Tay-Sachs disease. They include accumulation of Gm$_2$ ganglioside, its asialo-derivative, globoside (tetrahexosyl ceramide) in neural tissue and visceral organs and a generalized deficient activity of *both* Hex A and Hex B. The residual activity (1% of normal) is due to a minor form of β-hexosaminidase, Hex S, which is also detectable in normal tissues. Both Hex A and Hex B can cleave the terminal N-acetyl galactosaminyl residue of globoside. Thus the additional deficient activity of Hex B in Gm$_2$ gangliosidosis type II can account for the storage of this compound (Step 3, Fig. 1), which is not observed in type I disease.

Also attributable to Hex B deficiency is the defective catabolism of dermatan and chondroitin sulfates observed in cultured fibroblasts from affected patients (Step 8, Fig. 2). The genetic defect is detectable in heterozygotes as a reduction in total β-hexosaminidase activity. Gm$_2$ gangliosidosis type II is not particularly frequent in any ethnic group; its relative frequency in the general population, however, may be higher than was previously suspected. This is because when the enzymatic defect was not known a number of patients, clinically diagnosed as having Tay-Sachs disease, may in fact have had Sandhoff disease.

Gm$_2$ gangliosidosis type III (juvenile Gm$_2$ gangliosidosis) is described in a small number of patients. The symptoms more resemble those of neuronal ceroid lipofuscinosis than they do those of the type I disease, with onset between ages 2 and 6 years. Death usually intervenes between 5 and 15 years. No cherry-red macula was seen, but blindness from optic atrophy occurred in some cases. Neuronal lipidosis with the histopathologic and ultrastructural features of type I disease were present as were also "lamellar bodies." Levels of Gm$_2$ ganglioside and its asialo-derivative were elevated in the brain, but were lower than those in Tay-Sachs disease. A generalized deficiency of Hex A (~50% of normal) was detected in these patients by use of artificial substrates. Enzymatic assay with Gm₂ ganglioside as substrate showed a much severer impairment of activity, consistent with the degree of ganglioside storage. Hex B activity was normal.

The spectrum of Gm₂ gangliosidosis includes the rare variant AB, in which the activity of both Hex A and Hex B toward artificial substrates, globoside and asialo-Gm₂ ganglioside is normal or elevated. Activity of Hex A toward Gm₂ ganglioside, however, is about 10% of normal. Finally a few adults are described as having severe deficiency of Hex A, but without the clinical signs of Gm₂ gangliosidosis (18).

The nature of the biochemical defect in Gm₂ gangliosidosis is not completely clear. According to the attractive "unlike subunits" structural model (16), the subunit composition of Hex A is $(\alpha \beta)_2$; the structure of Hex B is $(\beta \beta)_2$ (17). The structural gene coding for the α subunits of Hex A is mutated in Tay-Sachs disease, and the gene coding for the β subunits common to both enzymes and carrying the active site is mutated in Sandhoff disease.

Trihexosyl Ceramidosis (Fabry's Disease; Angiokeratoma Corporis Diffusum)

The only known sphingolipidosis transmitted with an X-linked recessive inheritance, this disorder may appear in early childhood, with burning acral pains and typical corneal opacities visible by slit-lamp examination. In hemizygous males characteristic skin lesions (angiokeratomas) appear between the first and second decade as purple, raised angiectasis on abdomen, scrotum and thighs. Anhydrosis may be present. In later life renal deterioration and its complications develop, and death usually intervenes in the fifth decade. In heterozygous female, the symptoms are variable—as would be expected from the Lyon hypothesis. Corneal opacities are the commonest symptoms followed, in decreasing frequency, by skin lesions, acral burning pains, renal symptoms and cerebrovascular accidents.

The skin lesions, kidney glomerular and tubular epithelium, myocardium, central and peripheral nervous system, blood vessel endothelium and smooth muscle cells show extensive vacuolation. Vasa nervorum are also affected, which finding probably accounts for the characteristic "burning pain." Myelin-like lamellar cytoplasmic inclusion bodies, often accompanied by less typical pleiomorphic storage bodies, can be observed by electron microscopy in practically all tissues and organs in affected individuals. Trihexosylceramide is increased 5-10 times over normal levels in most tissues and is elevated in plasma, tears and urine. It is derived from globoside, a ubiquitous component of cell membranes, abundant in erythrocytes and present in plasma.

It is now well established that impaired activity of a specific trihexosyl ceramide α-galactosidase is responsible for storage of these compounds (Step 4, Fig. 1). The generalized enzymatic deficiency is detectable with artificial substrate, which procedure promotes demonstration in normal tissues of two electrophoretic forms of α-galactosidase (α-gal A, α-gal B). Only the more thermally labile α-gal A is deficient in the patients; thus the molecular abnormalities may be similar to those of Gm₂ gangliosidosis type I. It is not clear if the two enzyme forms are structurally related, for residual α-gal A activity has been detected in cells of some patients, suggesting genetic heterogeneity (19). This finding is reinforced by biochemical characteristics of the residual α-gal A, which suggest a structural alteration in some, but not all, cases (19). Somatic cell genetic studies conclusively show the existence of a gene for α-galactosidase on the X-chromosome.

Lactosylceramidosis

One patient with progressive neurologic deterioration; seizures; hepatosplenomegaly; foamy histocytes in bone marrow; and storage of lactosylceramide, glucosylceramide and Gm_3 ganglioside in liver tissue has been described. (12) Assay of lactosylceramide β-galactosidase showed reduction (15-20% of normal) of enzymatic activity in a liver biopsy and in cultured skin fibroblasts, compatible with the chemical findings (Step 5, Fig. 1). However, a recent reexamination of cultured skin fibroblasts from the patient, using newly developed assay systems, showed that the two forms of lactosylceramide β-galactosidase, recently identified, are both apparently normal in the patient's cells (20). In contrast sphingomyelinase was reduced to about 15% of normal, and the question is whether this case is an unusual type of sphingomyelinosis (Niemann-Pick disease).

Glucosylceramidosis (Gaucher's Disease)

This disorder occurs in three clinical forms, differing in age of onset and degree of CNS involvement: adult, chronic (*type I*); infantile, acute (*type II*); and juvenile, subacute (*type III*). The symptoms common to all forms includes marked splenomegaly, hepatomegaly, osteoporosis of long bones and pelvis with tendency to fractures, anemia and thrombocytopenia with tendency to episodic bleeding. Autofluorescent PAS-positive histocytes with characteristic cytoplasmic striations ("Gaucher cells") can be demonstrated in bone-marrow aspirates in all forms with the variable degree of involvement usually closely correlated with variability of clinical manifestations.

The onset of *type I* is in late childhood–early adulthood; patients do not have CNS signs of involvement, and the disease is usually compatible with normal life. With the exception of Gm_2 gangliosidosis in Ashkenazic Jews, this form of glucosylceramidosis is probably the commonest lipid storage disease; it too is observed more frequently in Ashkenazic Jews than in the general population, in which form it has about the same frequency as Tay-Sachs disease. In the rarer *type II*, onset is in early infancy with signs of CNS involvement. The disease rapidly advances to a decerebrate state, death usually intervening by 2 years of age. Patients with the rare *type III* have variable but less severe neural involvement (seizures, mental retardation and behavioral problems). Onset is in the first year of life, the patient usually dying either late in the first decade or early in the second decade.

The widespread organ histiocytosis reflects the generalized deposition of glucosylceramide, presumably derived from catabolism of the globoside derivatives in the cell membrane of RBCs and WBCs. Neuronal deposition in patients with type II and III diseases presumably derives from ganglioside degradation (Step 6, Fig. 1). Glucocerebroside β-glucosidase activity is reduced to 0-9% of normal in tissues from patients with type II disease; to about 15% of normal in type III; and to about 40% of normal in patients with the adult, type I form. Thus the degree of enzyme deficiency appears to be correlated with the severity of the disease, an unusual finding in the glycosphingolipidoses. Enzymatic activity can be measured with synthetic substrates, and assay of leukocyte extracts provides a convenient biochemical diagnostic test.

Heterozygotes can be identified by this method, and as expected the reduction of enzymatic activity is less pronounced for type I heterozygotes, with inconclusive results being possible. Genetic heterogeneity may exist even within each type. Prenatal diagnosis has been performed in a few cases, and storage of glucosylceramide in fetal liver has been detected as early as 18 weeks' gestation.

Sphingomyelin Lipidosis (Niemann-Pick Disease)

At least five subtypes of this disorder have been identified. The most frequent (infantile) *type A* (>80% of all cases) develops in the first months of life. Symptoms include hepatosplenomegaly, pulmonary involvement, osteoporosis and severe progressive psychomotor degeneration. Anemia and thrombocytopenia may also be present, and the patient's skin often has a yellow-green cast. A cherry-red macula is visible in about one-third of cases. Foamy cells are present in blood smears as are lipid-laden mulberry-shaped histiocytes in bone-marrow aspirates.

The disease advances to generalized cachexia, which in turn leads to death before or in the third year of life. Histiocytosis is generalized in all visceral organs except possibly in kidneys and bone marrow. The brain shows cortical atrophy, vacuolated neurons and cerebellar demyelination. In patients with *type B* disease the symptoms usually develop later, and the clinical course is protracted. Absence or reduced neurologic involvement is considered to be the distinctive feature of this clinical form. Histiocytosis of visceral organs is similar to that found in type A, but brain changes are minimal. *Type C* is similar to type A in that neurologic involvement is present, although its onset is delayed; the symptoms of visceral organ involvement also develop later than they do in type A. *Type D* patients, all originating from a single community in Nova Scotia, have symptoms similar to those of type C patients. Finally *type E* disease is demonstrated by a few patients and is clinically characterized by late onset and mild, chronic symptoms, devoid of neurologic signs.

The common biochemical feature in all types is reticuloendothelial storage of sphingomyelin, presumably derived from catabolism of plasma membranes and cellular organelles. A moderate increase in sphingomyelin is found in cerebral gray matter in type A. In addition, the cholesterol content of visceral organs is uniformly increased, and phospholipids are often elevated (21, 22). Severe sphingomyelinase deficiency has been demonstrated in type A (approximately 1% of normal) and type B (approximately 10% of normal) (Step 7, Fig. 1). In contrast, normal levels of total sphingomyelinase activity have been reported in type C, and in type E.

A correlation between clinicopathologic heterogeneity and enzymatic defect, however, is beginning to emerge (23, 24). Sphingomyelinase activity from normal liver and brain has been resolved in three major forms (activity peaks I, II and III) and two minor forms (peaks IV and V) by isoelectric focusing. A substantial reduction of peak II was demonstrated in one patient with type C disease. Further, electrofocusing of fibroblast extracts—which normally shows mainly peaks I, II and III—shows that in cells from two patients with type A disease the three peaks are barely detectable. In a patient with type B disease the same peaks are present at levels twice as high as in type A. Finally in fibroblasts from a patient with type E disease, the findings were similar to those obtained with tissues from the type C patient, i.e., absence of peak II; peak I was increased about five-fold over control values, suggesting a common moiety in enzyme peaks I, II and III and a specific moiety in peak II.

The recent development of an artificial chromogenic substrate may greatly simplify enzymatic assays for diagnosis and analysis (25). The enzymatic diagnosis of type A and B, using (^{14}C)-sphingomyelin as substrate and leukocytes or fibroblasts as enzyme source, has been extended to the identification of heterozygotes, confirming the autosomal recessive mode of transmission. Prenatal diagnosis has been made and confirmed by chemical and enzymatic studies of the fetal tissues.

Sulfatide Lipidosis (Metachromatic Leukodystrophy)

Three clinical forms are recognized, differing in age of onset. In the commonest, the *late infantile type*, progressive loss of locomotor ability occurs between age 1 and 2 years. Dysarthria, ataxia, loss of tendon reflexes and loss of bladder control gradually develop. Spastic quadriplegia, frequent seizures, progressive dementia, optic atrophy are characteristic of the terminal phase. Death usually occurs before the age of ten.

The less common, *juvenile form* has its onset in the second decade, often with personality changes and deterioration of scholastic performance. Gradually a cluster of symptoms similar to that of the infantile type develops; death occurs in the third decade. In the *adult form*, with onset in the third decade, mental and behavioral symptoms predominate; neurologic signs, such as incoordination and ataxia, may be evident only very late in the disease. These patients have often been diagnosed as schizophrenic, the real nature of the disease being recognized only at autopsy.

Despite the variable age of onset and duration of this lipidosis, the neuropathologic picture is fairly uniform, consisting of diffuse demyelination of central and peripheral nervous systems, and diffuse gliosis. Sulfatide storage is evident histologically as characteristic cytoplasmic inclusions, which stain metachromatically with a number of dyes. At electron microscopy, abundant membrane-bounded inclusions with variable lamellar aspect are visible in cells of the central nervous system, kidney, gallbladder and liver. The sulfatide levels are increased several times above normal, especially in white matter and kidney, and to a degree also in peripheral nerves, gallbladder and adrenal cortex. Sulfatide is also present in relatively large quantities in the urinary sediment and cells; this has been routinely exploited for diagnostic purposes together with demonstration of metachromatic inclusions in peripheral nerve biopsy material.

There are two forms of lysosomal arylsulfatase (AS-A, AS-B) in normal tissues; the former is profoundly deficient in tissues, leukocytes, fibroblasts, plasma and urine of patients with all three forms of sulfatide lipidosis when assayed with synthetic or sulfatide substrates (Step 8, Fig. 1) (26, 27). Normal levels of enzymatically inactive protein that cross-reacts with AS-A but not AS-B have been detected in several patients. The mechanisms by which accumulated undegraded sulfatide results in the extensive demyelination is not clear. Excess sulfatide (or deficiency of galactosylceramide) may interfere with membrane and organelle function of glial and Schwann cells. Alternatively myelin may have an abnormal, less stable composition because of excess sulfatide or defect of galactosyl ceramide, or both.

The degree of generalized AS-A deficiency seems to be the same in the three clinical types of sulfatide lipidosis. At present it is not possible to reconcile the severe deficiency with the clinical variability. In patients with a rare variant of sulfatide lipidosis, *mucosulfatidosis,* both AS-A and AS-B are deficient, but the inference that these enzymes may be under common genetic control is not supported by the absence of immunologic cross-reactivity between them. Interestingly deficient activity of AS-B with apparently normal activity of AS-A is present in the Maroteaux-Lamy syndrome (see "Mucopolysaccharidosis VI").

Enzymatic assay of AS-A permits unambiguous biochemical diagnosis even in the absence of clinical signs. The various tests are not as useful, however, in heterozygote detection. Qualitative techniques are essential for the recognition of exceptional cases of metachromatic leukodystrophy (MLD) in which enzyme variants may be present (28). Prenatal diagnosis of MLD has been made and confirmed in a few cases. The arylsulfatase pattern of cell-free amniotic fluid does not correspond to that of cells, and in two cases of confirmed fetal MLD the enzymatic pattern was indistinguishable from that in normal pregnancy (29).

Globoid Cell Leukodystrophy (Krabbe's Disease)

This rare metabolic disorder usually appears when the infant is between 4 and 6 months old and is characterized by chronic crying and excessive irritability. Progressive muscular rigidity develops rapidly and is often accompanied by seizures and optic atrophy. The patient usually succumbs before two years of age, but a few patients with delayed onset and more protracted course have been reported.

Pathologic findings are mainly limited to cerebral white matter, with profound demyelination, loss of oligodendroglia and severe astrocytic glyosis. "Globoid cells," presumably mesenchymal, are present in large numbers in white matter, often around blood vessels, and are pathognomonic. The cytoplasm of these peculiar cells is PAS-positive and shows intense acid phosphatase activity. Diverse cytoplasmic inclusions can be seen by electron microscopy although they are not membrane-limited, and no evidence of "classic" lysosomal storage bodies has been found in globoid or other cells. This morphologic finding has its counterpart in the absence of gross elevation of concentrations of any specific glycolipid in the nervous tissue. It is only by comparing the chemical composition of white matter from these patients with material from patients with other demyelinating disorders that a significant increase in galactosylceramide can be detected.

The lysosomal nature of the disease is reflected by the profound deficiency of the lysosomal enzyme, ceramide β-galactosidase, detected in tissues, leukocytes, cultured fibroblasts and serum from all patients examined. Decreased activity is found in parents of patients. Deficient activity of this enzyme likely results in impaired degradation of galactosylceramide (Step 9, Fig. 1) and in the formation of globoid cells. As with metachromatic leukodystrophy, the pathogenetic steps linking the enzyme deficiency or the relative excess of galactosylceramide, or both, to loss of oligodendroglia and demyelination are not understood. Thus, although genetically determined enzymatic defects are demonstrated and useful in diagnostic studies, including heterozygote detection and prenatal diagnosis, the exact pathogenetic mechanism of the disease is still unclear (30).

Lipogranulomatosis (Farber's Disease)

A rare lipid-storage disorder characterized by 1) infiltrative nodules in subcutaneous tissue and tendons at the level of the joints, 2) growth retardation and 3) neurologic handicap, has its onset in the first postnatal months and leads to death usually before age 3. The histopathologic features include generalized granulomatous infiltration of dermis, synovia, lymph nodes and visceral organs; CNS neurons become swollen; and storage vacuoles have been detected by electron microscopy in Kupfer cells, hepatocytes and histiocytes from dermal infiltrates. Ceramide is accumulated 10-60 times in different tissues, notably liver, and moderate accumulation of dermatan sulfate and glycolipids is also reported. A profound lack of ceramidase activity, with radioactive synthetic oleoylsphingosine as substrate, has been demonstrated in patients, with lesser degrees of deficiency in obligate heterozygotes (31). The deficiency is most probably responsible for ceramide accumulation in the patient's tissues (Step 10, Fig. 1). It is not known if the same enzymatic defect is present in the few patients who are described as less severely affected and as surviving longer.

Gm$_3$ Gangliosidosis

One patient is described, displaying most of the clinical features of Gm$_1$ gangliosidosis type I, but without skeletal dysplasia. The symptoms were present soon after birth, and the patient died at age 3 1/2 months. Excessive amounts of Gm$_3$ ganglioside and virtual absence of higher ganglioside homologues were found in brain and liver extracts. A biosynthetic rather than a degradative enzymatic defect was found in this case in that the activity of a glycosyltransferase —normally catalyzing the transfer of uridine diphosphate-N-acetyl galactosamine to Gm$_3$ ganglioside to give Gm$_2$ ganglioside and higher homologues—was virtually absent (32). Although Gm$_3$ gangliosidosis is not a lysosomal-storage disease (33), it is mentioned because it may pose problems of differential diagnosis with Gm$_1$ gangliosidosis and mucolipidosis II.

MUCOPOLYSACCHARIDOSES

This group of storage disorders encompasses defects in the catabolism of acidic mucopolysaccharides, more correctly called glycosaminoglycans (GAG). These

168 Multisystem Disease

HEPARAN SULFATE

```
         (2)              (4)          (5)
          ↓                ↓            ↓
L-IdUA ──── GlcN ──── GlcUA ──── GlcNAc ──── L-IdUA ····
  │  α  (3)→│    α          β           α
(1)→│       │
  │         │                    │
 SO4       SO3                  SO4
```

DERMATAN SULFATE

```
         (2)          (8)          (2)                    (4)
          ↓            ↓            ↓                      ↓
L-IdUA ──── GalNAc ──── L-IdUA ──── GalNAc ──────── GlcUA ────
  │  α      β            α           β                β
(6)→│                (1)→│
  │                      │            │
 SO4                    SO4          SO4
```

CHONDROITIN SULFATE

```
              (8)            (4)
               ↓              ↓
GalNAc ──── GlcUA ──── GalNAc ──── GlcUA ····
  │   β    6   β         β    4    β
(7)→│             (6)→│
  │                    │
 SO4                  SO4
```

KERATAN SULFATE

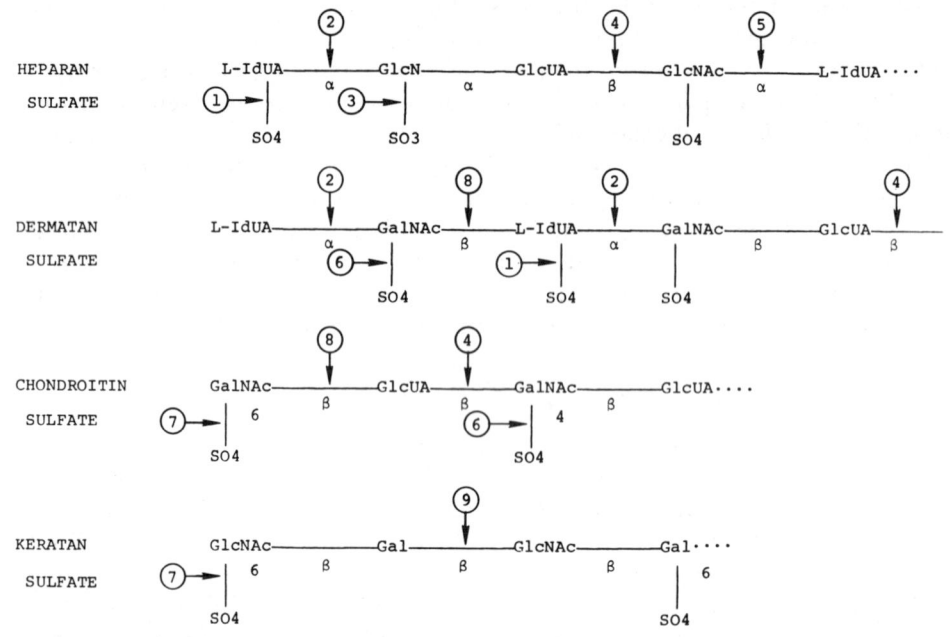

```
                        (9)
                         ↓
GlcNAc ──── Gal ──── GlcNAc ──── Gal ····
  │   6  β      β         β        │ 6
(7)→│                               │
  │                                 │
 SO4                               SO4
```

Figure 2. Chemical composition of glycosaminoglycans involved in mucopolysaccharidoses and sites of inborn errors of metabolism. Circled numbers (1–7): enzymatic steps listed in Table 2. Steps 8 (β-hexosaminidase) and 9 (β-galactosidase) are impaired in Gm$_2$ gangliosidosis and Gm$_1$ gangliosidosis, respectively. Abbreviations: Gal, galactose; GalNAc, N-acetyl galactosamine; GlcNAc, N-acetyl glucosamine; GlcUA, glucuronic acid; L-IdUA, L-iduronic acid.

complex polysaccharides of large molecular weight are linked to "core" protein as essential components of connective tissue. Four types of glycosaminoglycans are involved in storage diseases, including heparan sulfate, dermatan sulfate, chondroitin sulfate and keratan sulfate; the basic structure of these compounds is shown in Figure 2. In normal tissues, degradation proceeds sequentially by the action of lysosomal exoenzymes, glycosidases and sulfatases. Deficiency of any one of these enzymes results in a block of the degradative process and in intracellular accumulation.

The cleavage points and enzyme deficiencies in the mucopolysaccharidoses are also schematically shown in Figure 2. The widespread occurrence of the different GAG in tissues and organs contributes to the similarity of signs and symptoms of the mucopolysaccharidoses, the differences being attributed to the prevalence of a given compound in a given organ. Partially degraded fragments of average molecular weight smaller than that in normal tissue are usually found in large amounts in the urine of patients with most of the known mucopolysaccharidoses, an observation useful in screening studies. These fragments are thought to result from the action of *endoglycosidases*, such as lysosomal hyaluronidase (endo-β-N-acetyl hexosaminidase), which partially bypass the metabolic block; it is not known if alternative catabolic pathways are active only in abnormal storage. The catabolic block is expressed in fibroblasts from patients, a determinant in the clarification of genetic and biochemical aspects of these diseases.

The commonest mucopolysaccharidoses (MPS), classified by a combination of clinical and enzymatic criteria (34), are listed in Table 2. Excepting MPS II (Hunter syndrome), which is transmitted as an X-linked recessive trait, all these disorders have an autosomal recessive mode of inheritance. The salient clinical features are summarized in Table 3.

Mucopolysaccharidoses IH and IS (MPS IH; MPS IS)

This disorder can be divided into two main types the clinical manifestations of which differ widely. Infants with MPS IH (Hurler syndrome) are physically normal for the first 6 months of life, but thereafter rapidly develop typical features. The head is large with grotesque facies; corneal clouding is prominent; and the skin is thick and coarse, with hypertrichosis. Impaired joint mobility, thoracolumbar kyphosis, claw-like contracted hands with stubby digits and dwarfism are evident by age 3 years. Hydrocephalus, hepatosplenomegaly, cardiac murmurs and umbilical hernias are usual. Progressive dementia and physical deterioration usually lead to death from pneumonia or cardiac failure before age 10.

Typical radiologic signs in MPS IH include a J-shaped sella owing to a prominent optic sulcus; a "prowing" or "lipping" of the anteroinferior aspect of the vertebral bodies, primarily in the dorsolumbar area; kyphosis; flaring of iliac bones with shallow acetabula; and valgus of the femoral neck. Radioulnar growth disturbances, overconstriction of metacarpals and phalanges with proximal pointing are also typical. Histologic findings include extensive fibrosis of myocardium, and the reduction of proliferating cartilage, with large vacuolated cells, and periosteal and perichondreal involvement. Electron microscopic examination shows numerous "zebra bodies" in neurons, and "clear vacuoles," containing a fine granular reticulum are found in other affected tissues. Excess dermatan sulfate and heparan sulfate are excreted in the urine, with usually more of the former than of the latter compound.

The materials stored in tissues are predominantly dermatan sulfate and heparan sulfate, compounds that are not normally degraded because of a general deficiency of the enzyme α-L-iduronidase (Step 2, Fig. 2). Fibroblasts, leukocytes, serum and urine are markedly deficient in this enzyme. Decreased activity of β-galactosidase in patients' tissues, for some time regarded as the primary defect in MPS I and II, is conclusively shown to be from secondary inhibition by stored GAG. The activity of α-L-iduronidase is reduced in cultured skin fibroblasts and leukocytes from obligatory heterozygotes (35). Prenatal diagnosis has been successful in a few cases by the study of ^{35}S-GAG metabolism in cultured amniotic fluid cells. Metachromasia in amniotic fluid cells, because of stored GAG in amniotic fluid, is unreliable for prenatal diagnosis. Assay of α-L-iduronidase with phenyl-α-L-iduronide as substrate in amniotic fluid cells may be inaccurate since the enzyme's activity in normal amniotic fluid cells is low.

In contrast to those of patients with MPS IH, the clinical features of patients

Table 3. Mucopolysaccharidoses: Clinical Aspects

Disorder and Eponym	Main Clinical Features	Radiologic Signs	Intelligence	Age of Onset	Life Expectancy (yrs)
MPS IH (Hurler syndrome)	Coarse facies, corneal opacities, hypertelorism, rhinorrhea; (deafness), claw hands, stubby fingers, stiff, knobby joints, dwarfism, coarse skin, hypertrichosis, hepatosplenomegaly, and umbilical hernias	Dolichoscaphocephaly, J-shaped sella, prowing of dorsolumbar vertebral bodies, kyphosis, flared iliac bones, shallow acetebula, spatulated ribs, V-shaped radioulnar pointing and proximal constriction of metacarpals	Severe mental retardation	1st year of life	< 10
MPS IS	Corneal clouding, mild hypertrichosis, claw hands, aortic regurgitation and mild hepatomegaly	Minimal abnormalities	Normal	~ 5 years	Adulthood
MPS II (Hunter syndrome)	Same as in MPS IH, usually no corneal opacities, deafness and X-linked recessive inheritance	Same as in MPS IH	Severe mental retardation	1st year of life	10–15

MPS III A and B (Sanfilippo syndrome, types A and B)	Coarse facies, hypertrichosis, mild hepatomegaly and progressive spastic quadriparesis	Minimal abnormalities; increased thickness of calvarium and ovoid dorsolumbar vertebral bodies	Severe mental retardation	2–3 years	15–20
MPS IV (Morquio syndrome)	Short neck, wide mouth, broad maxilla, corneal opacities, pectus carinatum, flaring of ribs, kyphosis; dwarfism, flat-footed waddle, genu valgum, knobby joints, not stiff and frequent aortic regurgitation	Cervicodorsolumbar platyspondyly, odontoid hypoplasia, anterior subluxation of C1 on C2 and prow-shaped dorsolumbar vertebral bodies	Usually normal, or mild mental retardation	1–2	10–15
MPS VI (Maroteaux-Lamy syndrome)	Similar to MPS IH, with dwarfism, corneal opacities, hearing loss, hepatomegaly and progressive skeletal dysplasia	Similar to MPS IH; odontoid hypoplasia may be present	Usually normal	2–3	15–20
MPS VII (Sly syndrome)	Similar to MPS IH (corneal clouding can be absent), hepatosplenomegaly, umbilical and inguinal hernias and short stature	J-shaped sella, prowing of dorsolumbar vertebrae, spatulated ribs and metacarpal pointing	Mental retardation	< 1	?

with *MPS IS (Scheie syndrome)* are limited to corneal clouding, claw hands, carpal tunnel syndrome and aortic regurgitation. Normal height, intelligence and life expectancy and absence of visceroskeletal abnormalities are usually found. The urine contains dermatan sulfate and heparan sulfate in amounts and ratios comparable to that of patients with MPS IH; ^{35}S-GAG turnover in cultured skin fibroblasts is also similarly impaired. Deficient activity of α-L-iduronidase is demonstrable in fibroblasts and leukocytes from these patients, and reduced activity is also found in cells from obligatory heterozygotes. No cross-correction can be demonstrated between MPS IH and MPS IS cultured fibroblasts.

The reason for the milder clinical expression of MPS IS over MPS IH despite what appears to be the same enzymatic deficiency is not known, but conceivably the allelic mutations in the two diseases affect the structure and enzymatic activity in vivo of α-L-iduronidase in different ways, and the difference is not detectable in vitro. The postulated allelic mutations may give rise to "genetic compounds," with one MPS IH gene and one MPS IS gene, and the phenotypic expression being intermediate between the extremes. Patients with these intermediate clinical features have indeed been reported (34). Genetic or epigenetic variations in the efficiency of alternative catabolic routes may also modulate the phenotypic expression of the same primary enzyme defect (36).

Mucopolysaccharidosis II (MPS II)

The clinical, radiologic and pathologic features of MPS II (Hunter syndrome) are similar to those in patients with MPS IH. Corneal clouding, however, is generally absent. Deafness is more frequent, and survival time is generally longer than in MPS IH. Urinary excretion of dermatan sulfate and heparan sulfate is copious, and the amounts of both metabolites are about the same. An important distinctive feature of MPS II is the X-linked recessive mode of inheritance. Enzymatic studies established L-iduronosulfate sulfatase as the enzyme the deficiency of which is responsible for the X-linked MPS II (Step 1, Fig. 2) (37). A precise biochemical or genetic diagnosis, or both, differentiating MPS II from MPS IH is of great practical importance for proper genetic counseling, since the risks of affected offspring are considerably different in the two diseases, especially for offspring of female sibs of an affected male. Assay of L-iduronosulfate sulfatase will identify the affected male and the male fetus in utero. Heterozygote detection is much less reliable, and sisters of affected males should be regarded as carriers for counseling purposes. A mild form of MPS II, biochemically indistinguishable from the common severe form, is described in a few patients, with minimal or no neurologic involvement and apparently normal life expectancy.

Mucopolysaccharidoses III A and III B (MPS III)

The clinical hallmark of MPS III (Sanfilippo syndrome) is the severe mental retardation in contrast to the relatively mild somatovisceral signs and minimal radiologic findings (Table 3). Urinary GAG excretion may be normal, thus compounding the diagnostic difficulties; excessive excretion is due predominantly to heparan sulfate. Two types of MPS III, A and B, are now biochemically recognized, and in both types there is defective degradation and tissue accumulation of heparan sulfate, but this results from deficient activity of different enzymes. Heparan-N-sulfatase (sulfamidase) is deficient in MPS III type A (Step 3, Fig. 2); α-N-acetyl glucosaminidase is deficient in MPS III type B (Step 5, Fig. 2).

The recent finding of antigenically reactive, enzymatically inactive α-N-acetyl glucosaminidase in the urine of three patients with MPS III B in amounts exceeding those in normal persons indicates that a structural mutation is responsible for the defect (38). Enzymatic assays have also demonstrated decreased activity of sulfamidase and α-N-acetyl hexosaminidase in heterozygous parents of patients with MPS III A and MPS III B. Biochemical identification by enzymatic methods has also resulted in the realization that MPS III A may be five or six times more frequent than MPS III B (39). Finally, both enzyme assays are applicable to prenatal diagnosis.

Mucopolysaccharidosis IV (MPS IV)

In MPS IV (Morquio syndrome) the initial clinical profile is not dissimilar to that in MPS I and II, but as the disease progresses the typical features of the disorder become apparent. They include a wide mouth and a broad maxilla; corneal clouding; striking dwarfism; broad, flared rib cage and protruding sternum; knobby but mobile joints; genu valgum; and flat feet are usual toward the end of the first decade. Radiologic signs of severe platyspondyly are present. A characteristic hypoplasia of the odontoid process is noteworthy, accompanied by laxity of the ligaments between C-1 and C-2, abnormal mobility of the cervical spine and possible dislocation with spinal cord compression. The thoracic and spinal skeletal dysplasias account for the frequent pulmonary complications in MPS IV. These patients, who have normal or near-normal intelligence, may survive well into their forties, usually succumbing from cardio pulmonary complications.

Also characteristic of MPS IV is excess excretion of keratan sulfate and chondroitin sulfate, especially evident in the early stages. Fibroblasts from MPS IV patients are unable to cleave the 6-sulfate residues of either sulfate moiety; this indicates deficient activity of an N-acetyl hexosamine-6-sulfate sulfatase (14) (Step 7, Fig 2). Although heparan sulfate also has 6-sulfated hexosamine residues, there is no abnormal excretion of the compound in MPS IV. This finding suggests that hydrolysis of this sulfate group may be catalyzed by an enzyme other than the sulfatase deficient in MPS IV. Patients with the clinical and radiologic signs of MPS IV, but devoid of keratan sulfaturia, have been described, suggesting genetic heterogeneity. Heterozygote detection is not yet reliable, however.

Mucopolysaccharidosis VI (MPS VI)

The clinical profile of MPS VI (Maroteaux-Lamy syndrome) resembles that of MPS IH, but onset is delayed, survival longer and intelligence normal. Radiologic signs are similar to but milder than those in MPS IH. As in MPS IV, odontoid hypoplasia and subluxation of C1 or C2 are often found. Large amounts of dermatan sulfate are found in the urine of these patients, allowing a clear distinction between MPS IV and MPS VI. Deficiency (roughly 10% of normal) of an N-acetyl galactosamine 4-sulfate sulfatase, also called arylsulfatase B (AS-B), has been found by assay of tissue extracts, fibroblasts, leukocytes and long-term lymphoid cell lines from affected patients (Step 6, Fig. 2).

It appears likely that the natural substrate of AS-B consists of the sulfate groups linked in position 4 to N-acetyl galactosamine residues in dermatan sulfate and chondroitin sulfate. Although urinary excretion of the last-named compound is not excessive, the marked skeletal deformities in the syndrome suggest impaired degradation of chondroitin sulfate. AS-B activity was decreased in leukocytes and skin fibroblasts from two obligate heterozygotes and one of two potential heterozygotes. Considerable heterogeneity exists in this rare disorder, which finding may be due to different allelic forms of enzymatic deficiency.

Excessive storage of dermatan sulfate, from AS-B deficiency, is also found in *mucosulfatidosis,* a rare variant of MLD characterized by multiple sulfatase deficiency (AS-A, AS-B, AS-C deficiency). Increased levels of cerebroside sulfate (sulfatide) correlate closely with deficient AS-A activity, as in classic metachromatic leukodystrophy; excess steroid sulfate levels reflect deficient AS-C activity. The clinicopathologic features in mucosulfatidosis appear to include those of MLD and MPS VI. A distinctive feature is ichthyosis, found in 12 of 15 patients studied (40). The disorder is transmitted with autosomal recessive inheritance.

Mucopolysaccharidosis VII (Sly Syndrome)

The most recent addition to the mucopolysaccharidoses is MPS VII (Sly syndrome). Only a few patients have so far been reported, and heterogeneity in clinical features is evident. In three patients viscero skeletal and psychomotor involvement resembled that of MPS IH, although severity varied with the patient. In a fourth, much older (14 years) subject, cardiovascular involvement was prominent; radiologic signs of skeletal abnormalities were minimal and mental retardation was mild. A common feature in all reported patients is increased urinary excretion of dermatan and heparan sulfates; the levels varied in different patients and at different times in the same patient. In all subjects a deficiency of β-glucuronidase was found in fibroblast and leukocyte extracts, the deficiency being less pronounced in the older than in the younger patients (Step 4, Fig. 2). Reduced β-glucuronidase activity was found in serum or leukocytes, or both, of the 7 obligatory heterozygotes examined. Some relatives and sibs of the patients also had decreased enzymatic activity, consistent with the proposed autosomal recessive mode of inheritance.

Further studies on the qualitative aspects of the enzymatic deficiency are needed to substantiate the contention that different mutations are responsible for the apparent clinical and biochemical heterogeneity of MPS VII. A gene controlling the expression of β-glucuronidase in man has recently been assigned to chromosome 7, using man-rodent somatic cell hybrid (41, 42).

MUCOLIPIDOSES

This group of metabolic disorders is characterized by storage of acidic GAG, together with sphingo- and glycolipids in visceral and mesenchymal cells. Five diseases are classified as mucolipidoses. Of these, mucolipidosis I is an ill-defined clinical entity, with features of mucolipidosis II and III, psychomotor retardation and mild clinical and radiologic signs of mucopolysaccharidosis. Although the primary genetic defect in ML I is known, it is likely that different enzyme deficiencies exist in different patients. Of the remaining four conditions, fucosidosis and mannosidosis are caused by single enzyme deficiencies. Mucolipidosis II (ML II, I-cell disease) and mucolipidosis III (ML III, pseudo-Hurler polydystrophy) are characterized by abnormally low cellular levels of most lysosomal enzymes and correspondingly elevated levels in plasma and cell culture fluids. Mucolipidosis IV has recently been recognized as a distinct clinical entity and unlike ML II and ML III has the biochemical and ultrastructural features of a "classic" storage disease. Features of these disorders are summarized in Tables 4 and 5; the primary genetic defect has not yet been clarified.

Fucosidosis

Two types of fucosidosis are clinically distinguished. The typical features of *type I* are early onset, coarse facies, corneal opacities, hepatosplenomegaly, increased excretion of Na$^+$ and Cl$^-$ in sweat, kyphoscoliosis and radiologic signs similar to those in MPS IH, vacuolated lymphocytes, recurrent respiratory infections, retarded growth and early progressive severe psychomotor deterioration, culminating in a decerebrate state. Death usually intervenes by 4-6 years of age.

In *type II* a similar cluster of symptoms obtains, but neurologic involvement is less severe, survival is usually longer (up to the second or third decade), levels of Na$^+$ and Cl$^-$ in perspiration are not elevated. Typical features are angiokeratomas, the appearance and distribution of which closely resemble those in patients with Fabry disease. In both types of fucosidosis EM examination shows storage material in hepatocytes, Kupfer cells, neurons, endothelial cells and fibroblasts. Chemical studies show polysaccharides and glycolipids containing excess fucose in these same tissues. The lysosomal enzyme α-L-fucosidase is markedly deficient in visceral organs, brain, serum, cultured fibroblasts and leukocytes of both types I and patients (43). Assay of α-L-fucosidase in blood leukocytes and serum of obligatory heterozygotes shows that the predicted reduction of enzymatic activity can be detected in cell extracts; the low enzyme activity values found in these patients' plasma however, are not diagnostically helpful, since α-L-fucosidase activity can be reduced to similarly low levels in the plasma of normal persons.

Table 4. Mucolipidoses: Biochemical Aspects

Disorder and Eponym	Compound Affected	Enzyme Deficiency
ML I (sialidosis)	Sialyloligosaccharides	Neuraminidase
ML II (I-cell disease)	Glycolipids and Mucopolysaccharides	Lysosomal enzymes deficient in mesenchymal cells, excessive in plasma and urine; ? deficit in recognition factor
ML III (pseudo-Hurler polydystrophy)	Similar to ML II	Similar to ML II
ML IV	Mucopolysaccharides containing hyaluronic acid	? Hyaluronidase

Table 5. Mucolipidoses: Clinical Aspects

Disorder and Eponym	Main Clinical Features	Radiologic Signs	Intelligence	Age of Onset	Life Expectancy (yrs)
ML I (sialidosis)	Mild coarseness of facies suggesting Hurler syndrome; short trunk with kyphosis. progressive neurodegenerative disorder; muscular hypotonia; fine corneal opacities and decreased visual acuity; macular red spot	Mild signs of dysostosis multiplex	Variable—mild to severe mental retardation	Infancy to adolescence	10–25
ML II (I-cell disease)	Severe features, coarse facies with small orbits, prominent eyes with partly closed lids but clear cornea, full-flushed cheeks with telangiectatic vessels, prominent maxillary gingival hypertrophy, short neck and trunk with total height at 80 cm or less, inguinal hernia, stiff joints and frequent upper respiratory tract infections with ultimate cardiorespiratory death	Severe dysostosis multiplex; early bowing of long bones, club feet or hip dislocation	Severe mental retardation	Birth	4
ML III (pseudo-Hurler polydystrophy)	Early sign is joint stiffness with coarse facies and short stature, fine corneal clouding and frequent aortic insufficiency otherwise like mild Hurler syndrome	Moderate signs of dysostosis multiplex	Moderate mental retardation	3 years	Adolescence
ML IV	Congenital corneal opacities, Hurler-like features and psychomotor retardation; most patients are Ashkenazic Jews	Mild skeletal changes	Moderate to severe mental retardation	Birth	?

About one-half of the relatively few patients so far reported in the literature were of Italian descent; this suggests that the frequency of the gene(s) for fucosidosis may be elevated in Italians.

Mannosidosis

From the number of reported cases, this disorder appears to be more frequent than fucosidosis, which it resembles in clinical features. Coarse facies, corneal clouding and deafness, frequent upper respiratory infections, hepatomegaly, radiologic signs of dysostosis multiplex, craniosynostosis, vacuolized lymphocytes, often normal early growth followed by growth arrest and psychomotor retardation are the usual findings. Age of onset, duration of the disorder, severity of symptoms and degree of mental retardation all vary and suggest the existence of genetic heterogeneity. Although urinary excretion of GAG is normal, large quantities of mannose-rich oligosaccharides are present. In keeping with the ultrastructural appearance of lysosomal storage, greatly increased amounts of oligosaccharides rich in mannose can be extracted from brain and liver, compounds that most likely result from impaired degradation of the carbohydrate moiety of glycoproteins.

Abnormal activity of the enzyme α-D-mannosidase has been found in tissues, leukocytes, fibroblasts and body fluids of affected individuals. The deficient activity is limited, however, to two forms, A and B, which have an acid pH optimum. The third form, C, with an optimal pH of 7, is normally active. Biochemical and immunologic studies reflect a close structural relationship between forms A and B, both of which differ from form C in biochemical, kinetic and immunologic characteristics. Zinc ions appear essential for optimal activity and stability of the former. Recent observations suggest that, for some patients at least, the mutation resulting in enzymatic deficiency may affect the binding of Zn^{2+} to the enzyme molecule; and that, at least in vitro, excess Zn^{2+} may partially restore enzyme activity (44). These findings have therapeutic implications. Reduced activity of acidic α-D-mannosidase in obligatory heterozygotes is also reported. For accurate diagnosis the presence of normal levels of neutral α-D-mannosidase in serum, leukocytes and cultured fibroblasts in patients and heterozygotes has to be taken into account.

A disease closely resembling the clinical, ultrastructural, biochemical and enzymatic features of the disorder in man has been described in Angus cattle. This may furnish useful experimental information for future therapeutic attempts in man.

Mucolipidosis I (ML I)

A clinical entity comprising patients with various degrees of dysostosis and psychomotor impairment, normal GAG excretion, and storage of lipids and GAG in different tissues (44a), ML I should be considered a provisional classification. Among these patients, however, several have been found with a generalized deficiency of neuraminidase activity, and excretion of sialic acid-rich oligosaccharides (44b). In some patients a macular cherry-red spot and myoclonus were present, and the term cherry-red spot—myoclonus syndrome has been coined (44b). It remains to be determined how the neuraminidase deficiency, which appears genetically transmitted, can account for the heterogeneous clinical manifestations.

Mucolipidosis II (ML II)

Mucolipidosis II (inclusion-cell disease; I-cell disease) develops early, sometimes shortly after birth, and is characterized by coarse facies, hypertrichosis, rhinorrhea, hepatosplenomegaly, joint contractures, claw hands, severe psychomotor retardation and radiologic signs of severe dysostosis multiplex; the overall picture is similar to that of MPS IH. Diagnostically important are the absence of abnormal urinary excretion of GAG and a striking elevation of plasma levels of several lysosomal hydrolases, a typical feature of ML II. Cardiopulmonary failure at age 4-6 years is the usual cause of death. Abnormalities include cytoplasmic lysosomal inclusions (hence the name, Inclusion-cell disease) in cultured skin fibroblasts, which are visible by light or phase contrast microscopy. Morphologic evidence of lysosomal storage in many other cell types is also found by electron microscopy. Glycolipids and GAG are stored in pleomorphic membrane-bound cytoplasmic vacuoles in visceral organs and brain.

The activity of lysosomal hydrolases in these tissues is normal, however, except possibly for decreased β-galactosidase activity in liver and brain. In contrast, marked deficiency of several lysosomal hydrolases is found in extracts of cultured skin fibroblasts, although the activity of lysosomal enzymes in the culture medium from these cells is increased several fold over normal values. The discrepancy is not from leakage of enzymes into extracellular spaces as a result of abnormal lysosomal permeability, since ML II fibroblasts can on the one hand endocytose and retain lysosomal enzymes excreted by or extracted from normal or non-ML II cells. Lysosomal enzymes excreted by ML II cells, on the other hand, are not taken up by normal or non-ML II cells. Thus it is hypothesized that in ML II fibroblasts an alteration in the carbohydrate chain of most lysosomal enzymes (known to be glycoproteins) impairs the recognition and uptake that are normally responsible for reentry of excreted enzymes. This, if valid, implies—at least in mesenchymal tissues—a pathway for enzyme packaging in the lysosomes, which functions by excretion and selective endocytosis by neighboring cells.

Support for the hypothesis of an alteration of the carbohydrate chain in ML II-excreted lysosomal enzymes is the finding that periodate oxidation of normal lysosomal enzymes prevents their uptake by cells. Presumably this alters the carbohydrate "recognition signal" needed for binding to cell membrane receptors (45, 46). Furthermore, enzymes excreted by ML II fibroblasts appear to contain excess neuraminic acid. The abnormal carbohydrate composition may be a consequence of abnormal excretion (46a). However, the basic defect is still unknown. Whatever the metabolic error, it appears to be expressed in obligatory heterozygotes for the ML II gene: on average, plasma levels of lysosomal enzymes are significantly elevated in these subjects, although there is considerable overlap with normal values. The biochemical and morphologic characteristics of cultured fibroblasts from ML II patients have been found in cultured amniotic fluid cells from a few pregnancies at risk, making it possible to diagnose the disease in utero.

Mucolipidosis III (ML III)

Mucolipidosis III (Pseudo-Hurler polydystrophy), closely related to ML II, is clinically distinguishable from it by later onset, milder clinical symptoms, mild or absent mental retardation and survival to adulthood. Excessive urinary excretion of GAG is absent, plasma levels of lysosomal hydrolases are moderately elevated. The multiple lysosomal enzyme deficiency in cultured skin fibroblast and the numerous cytoplasmic inclusions already described, as well as the increased enzyme levels in cell culture fluids, are also found in ML III. Since, however, the primary biochemical defect in both conditions is unknown it is not possible to infer from these striking similarities that the two diseases are due to allelic mutations. More than one gene may be involved in the assembly of a carbohydrate moiety necessary for normal recognition, or in the control of exocytosis. Thus, mutations at unrelated loci could result in similar biochemical alterations and clinical characteristics in the two diseases.

Mucolipidosis IV (ML IV)

Several patients have been described, the majority *Ashkenazi Jews,* with this storage disease. The salient features are corneal opacities, a coarse facies, and psychomotor retardation. Corneal opacities are detectable in the first months of life (46b). Electron microscopic examination reveals characteristic storage bodies in biopsy tissues, including liver, in which hepatocytes show lamellar structures, and the Kupffer cells, clear vacuoles. Abnormal storage bodies are also evident in cultured skin fibroblasts and cultured amniotic fluid cells, a finding that has made prenatal diagnosis possible. The stored material apparently consists of gangliosides and sulfated mucopolysaccharides of which hyaluronic acid is the major component (46c). All lysosomal hydrolases tested so far have normal activities in patients' cells, and the primary enzyme defect is not clear. From family studies, the disorder apparently is inherited as an autosomal recessive trait.

OTHER STORAGE DISEASES*

Simple Lipid Storage Disorders

Two clinically different lysosomal-storage diseases result from impaired catabolism of cholesteryl esters and triglycerides, both of which are inherited as autosomal recessive traits. *Wolman's syndrome* (WS; familial xanthomatosis with adrenal calcification) has its onset in the first few weeks of life. Continued vomiting and diarrhea, failure of the patient to gain weight, progressive hepatosplenomegaly, fever and anemia rapidly lead to cachexia and death, usually by age 4-5 months. Peripheral lymphocytes with large lipid-laden vacuoles, large foamy histiocytes in bone marrow and striking enlargement and calcification of the adrenal glands are pathognomonic.

Cellular storage of lipids or infiltration of lipid-laden histiocytes, or both, are

---

*Glycogen storage disease, type II (Pompe's disease), currently held to be a lysosomal-storage disease, is discussed in conjunction with the other glycogen storage diseases.

in virtually every visceral organ, particularly in the liver, spleen and small intestine. The CNS appears less involved. Large excesses of triglycerides and cholesteryl esters are stored in the affected organs and tissues, especially the liver and spleen. In contrast, plasma lipid levels are usually normal, although alterations of lipoprotein patterns have been reported. Cultured skin fibroblasts also show massive lipid accumulation, and there is a profound deficit of lysosomal acid lipase–cholesteryl esterase in tissues, leukocytes and cultured fibroblasts. This enzyme complex seems to have both cholesteryl ester hydrolase activity and triglyceride hydrolase activity.

The second lysosomal-storage disease caused by defective catabolism of cholesteryl esters is *cholesteryl ester storage disease* (CESD). A relatively mild disorder of delayed onset and protracted course, its salient features include hepatosplenomegaly and hyperlipidemia related to cholesterol, triglycerides and phospholipids. It is now recognized that tissue storage of lipids, which in CESD patients involves liver, spleen and intestine, is not related exclusively to cholesteryl esters, but is also affected by triglycerides, similar to what is found in WS. Impaired activity of lysosomal acid lipase–cholesteryl esterase is also found in tissues and cells from CESD patients, although the deficiency does not seem complete. The biochemical relationship between he two disorders is unclear.

The same autosomal recessive mode of transmission, the same apparent enzyme deficiency and the existence of milder clinical forms of WS with comparable deficiency of acid lipase are on the one hand compatible with a spectrum of (allelic?) mutations affecting the same enzyme to different degrees. On the other hand, recent evidence shows that this enzyme system is heterogeneous (47-49). The metabolic defects in WS and CESD fibroblasts are corrected by cultivation with fibroblasts from patients with familial hypercholesterolemia, which lack specific membrane receptors for low-density lipoproteins (LDL) and therefore have impaired cholesterol metabolism, but normal lysosomal acid lipase–cholesteryl esterase activity. A net decrease in the content of cholesteryl esters was observed in the treated cells, in agreement with previous histochemical findings of storage correction by exposure of WS fibroblasts to culture medium from normal fibroblasts. Correction seems to be related to the release of lysosomal acid lipase in the medium by the nonlipase-deficient cells, a finding of possible therapeutic value.

The study of WS and CESD is important to the overall understanding of atherosclerosis. Arterial lipid deposition was described in WS infants who died, and it was particularly severe in one adult case of CESD. Further, decreased activity of cholesteryl esterase accompanies cholesteryl ester storage in lysosomes of smooth muscle cells from atheromatous aortas, both in experimental animals and in man, indicating that this lysosomal enzyme participates in the pathogenesis of the atherosclerotic lesion.

Cerebrotendinous xanthomatosis (CTX), a third rare disorder of neutral lipid storage, is characterized by tendon xanthomas, xanthelasmas, cataracts, progressive cerebellar ataxia and dementia (50, 51). Probably fewer than 20 patients have been reported. Clinical onset of symptoms is insidious, mental dullness or retardation occurring as early as age 10 or as late as the fourth or fifth decade. In adolescence or young adulthood the patient becomes progressively spastic, and ataxia usually accompanies the symptom. Cataracts also develop at this time.

Although tendon xanthomas may occur as early as the second decade, they commonly appear in the third or fourth decade, with the achilles, triceps, tibial tuberosities and extensor tendons of the fingers the sites in decreasing order of predilection. Although xanthomatous enlargement is slow and not painful, it and neurologic deterioration become severe and are attended by increased spasticity, tremors, and muscle wasting. Death occurs in middle age, usually in the 40s or 50s.

Usually cholesterol and other plasma lipid levels are normal. The cerebellum shows the most striking histologic changes, with foamy granulomatous lesions in the white matter, which is demyelinated. The cerebral cortex is spared, but the forebrain, brainstem and spinal cord are involved (50). The tendon xanthomas contain fat-staining material, and foamy cells have also been seen in the lung, surrounding blood vessels and bone.

Although the metabolic defect has not been defined, there appears to be an abnormal accumulation of cholestanol, (52) a saturated sterol usually associated with cholesterol and structurally different by the absence of one double bond. This observation suggests defective membrane transport of cholesterol out of cells as being the basic defect, with resulting conversion of cholesterol to cholestanol (52). Elevated tendon xanthoma or plasma cholestanol levels may aid in the diagnosis of the slowly progressive disorder. Although only a few patients have had the diagnosis confirmed, the frequency of affected children of parents who are cousins and the occurrence of affected siblings strongly support an autosomal recessive genetic basis for the disease (53).

Acid Phosphatase Deficiency

A rapidly fatal disease of early infancy, associated with deficient activity of lysosomal acid phosphatase, has been described in one family. Three sibs were affected; their symptoms were vomiting, lethargy, hypotonia, opisthotonos and hepatomegaly. Death occurred before the end of the first year of life. Large vacuoles containing lipid material were found in their hepatocytes and kidney cells. Acid phosphatase activity in the lysosomal fraction of liver, brain, kidney, spleen and cultured fibroblasts was practically absent, and the total acid phosphatase level in cell lysates was decreased. Cultured fibroblasts and lymphocytes from the parents and some of the relatives showed a decrease in acid phosphatae activity consistent with an autosomal recessive mode of inheritance. Prenatal diagnosis of an affected fetus was confirmed after termination of pregnancy. A second family was reported in which two sibs, with symptoms of lethargy, opisthotonos and bleeding, died two days after birth. Total acid phosphatase activity was undetectable in cultured skin fibroblasts from the patients.

Finally a large family with several affected children has been reported. Their symptoms included fever, vomiting, diarrhea and seizures leading to death within the first year of life. Acid phosphatase activity was severely deficient in serum and leukocytes, but not in cultured fibroblasts. Although acid phosphatase has been widely used as a marker enzyme in studies on lysosomes, its natural substrates and metabolic function are not well understood. The significance of the enzyme deficiency and its relationship to the lipid material detected in liver and kidney cells in the patients previously described, and with the clinical features of the disease are not known. Interestingly this is the only disease associated with a lysosomal enzymatic deficiency in which pharmacologic therapy has been successful (54). Cultured fibroblasts from patients with lysosomal acid phosphatase deficiency were treated with prednisolone, following which acid phosphatase activity in the lysosomal fraction of the cells increased to normal levels. Two patients were then successfully treated with prednisolone, the drug bringing about remission of symptoms for extended periods.

Aspartylglycosaminuria

Features of the disease include insidious mental deterioration with onset in early childhood that advances to severe mental retardation, coarse facies with broad nose, cutaneous manifestations, short stature, skeletal and connective tissue abnormalities and recurrent respiratory infections. Large quantities of aspartylglycosamine are excreted in the urine, and the patients usually survive to adulthood. The generalized nature of the lysosomal storage inferred from earlier histologic and ultrastructural studies of biopsy specimens has recently been confirmed by the first postmortem study of one patient (55). Liver and brain cells were most severely affected, with extensive cytoplasmic vacuolation of hepatocytes, Kupffer cells, neurons and glial cells. Large electron-lucent cytoplasmic vacuoles were evident by electron microscopy.

The storage material (about 1% of tissue weight) was identified as aspartylglycosamine, the compound excreted in the urine. The osmotic properties of the storage compound, namely, negative charge and small molecular weight, explain the apparent discrepancy between the relatively modest amounts stored in liver tissues and the remarkable increase in number and size of vacuoles. Deficiency of the lysosomal enzyme, N-aspartyl-β-glucosaminidase, was found in patients' serum, semen, brain, liver tissue, and cultured fibroblasts. This enzyme has a neutral pH optimum in vitro, which is unusual for lysosomal hydrolases; its natural substrate, N-aspartylglycosamine, forms the linkage between carbohydrate and polypeptide moieties of many glycoproteins.

Most likely this enzyme deficiency is the primary defect in the disease, but the deficiency is not complete, and further enzymatic studies are needed to support this contention. From pedigree analysis aspartylglycosaminuria is inherited as an autosomal recessive trait, and most of the patients identified so far were from Finland.

Neuronal Ceroid Lipofuscinosis

The collective term neuronal ceroid lipofuscinosis (NCL) is preferred to various eponyms (Batten-Mayou disease, Batten-Spielmeyer-Vogt syndrome; late infantile amaurotic idiocy of the Jansky-Bielschowsky, Spielmeyer-Sjogren, and Kufs type). It subsumes a clinically heterogeneous group of genetically determined, progressively degenerative disorders of the central nervous system (58). In the three main clinical types of NCL—*late infantile, juvenile* and *adult*—the age of onset may vary from early infancy to adulthood, and the course may be rapid or insidious. Symptoms include blindness, ocular fundic changes with optic atrophy and retinitis dementia. Signs of decortication are evident in the terminal phase of the disease. The order in which the symptoms become manifest and the absence of a particular feature e.g., blindness, are useful in defining the clinical subtype (56).

Electroencephalographic and electroretinographic changes, vacuolated lymphocytes and hypergranulated neutrophils are also diagnostically important. The three clinical types of NCL are transmitted as autosomal recessive traits, and an autosomal dominant form of the adult type is also described. The unifying histopathologic finding, namely, the accumulation of autofluorescent lipopigments (most prominently in CNS tissues), is usually associated with cerebral atrophy, but may also be in peripheral nerves, all visceral organs and skeletal muscle. In addition to "multilamellar cytosomes" or "curvilinear bodies" typical of the disorder, diverse lysosomal storage bodies can be observed by electron microscopy, inclusions ranging from amorphous to granular, "fingerprint-like" and crystalloid (57).

The lipid material accumulated in the tissues consists of lipofuscin and ceroid, complex lipids the structures of which have not been fully elucidated. Lipofuscin is common in the nervous tissue of aged individuals; in contrast, ceroid has only been found in patients with NCL. Both lipids are thought to be formed by oxidation and cross-linking of cytoplasmic and cell organelle lipid constituents, which become enclosed in autophagic vacuoles. The enzymatic block resulting in accumulation of the lipopigments, especially ceroid, is not understood. Since lipofuscin and ceroid appear to be resistant to lysosomal degradation, excess oxidation of unsaturated lipid precursors, rather than deficiency of a lysosomal degradative enzyme, is a possible pathogenetic mechanism.

Increased cellular concentration of peroxides owing to a peroxidase deficiency might be responsible for cellular injury; the lipopigments would be the end product of oxidative injury to subcellular constituents. The morphologically prominent lysosomal storage would thus be a secondary phenomenon. This hypothesis has recently been strengthened by the finding of decreased leukocytic peroxide activity in patients with the late infantile and juvenile types, and the dominant form of the adult type, of NCL. A similar deficiency is seen in brain and visceral tissues of some patients. A variable reduction of leukocytic peroxidase activity is found in parents and sibs of patients. Because of our still imperfect knowledge of the leukocytic peroxidase system, it is not yet clear if this deficiency reflects the primary enzyme defect or a secondary phenomenon (58).

THERAPEUTIC ASPECTS

The fusion processes characterizing the lysosome–vacuole complex make it possible to correct lysosomal storage by exposing the cell to exogenous, normal enzyme, which can be internalized by endocytosis and eventually come in contact with and degrade the stored material. This possibility, borne out by cross-correction experiments with cultured fibroblasts led to several attempts at enzymatic replacement therapy in patients. Administration of homologous and heterologous enzymes, intact leukocytes and plasma did not yield significant therapeutic results. The principles and prerequisites emerging from these early trials, however, have made these attempts more successful. Clearly a rational approach to enzymatic replacement must be based on a firm understanding of the disease; sound knowledge of the biochemical characteristics of the normal enzyme; availability of a convenient source of enzyme for purification; demonstrable enzyme purity, stability and activity toward the natural substrate; a significant uptake of enzyme by cells; and evidence that the enzyme is stable and effective in reducing endocellular storage.

The real possibility of delivering the enzyme to the target organ as well as the lack of complications (especially immunologic) following administration of the enzyme must also be demonstrated before trial in human subjects. In most of the early attempts, only some of these ideal prerequisites were met. The subject has recently been reviewed (59), and only the most significant aspects will be briefly discussed here. In general, extremely short half-lives of the injected enzyme in the recipient's circulation (10-20 minutes) has been observed when enzymes purified from human placenta or urine were used. The enzymes were almost totally sequestered by the liver. Plasma-derived hydrolases, (β-hexosaminidase, β-glucuronidase), however, remained in the patient's circulation longer, with a half-life of a few hours.

Lysosomal enzymes from plasma appear to differ from the organ enzymes by additional sialic acid residues. These residues, linked to galactose, possibly interfere with galactose-dependent cell recognition, which is necessary for hepatic uptake of glycoproteins. Experiments with enzyme-deficient cultured fibroblasts also indicate that selective uptake of a given exogenous enzyme may depend on either the tissue of origin or the net charge of the administered enzyme. At least three different carbohydrate-dependent recognition mechanisms are known, which modulate glycoprotein uptake by cells, and there is evidence for organ-specificity of these mechanisms. Thus carbohydrate-chain heterogeneity in lysosomal enzymes, owing to the tissue of origin or induction by preparative procedures, may play an important role in effective organ targeting of administered enzymes.

Preferential uptake by the liver or reticuloendothelial system can be exploited in diseases in which these are target organs, as exemplified by the encouraging results obtained by enzyme replacement in patients with Gaucher disease type I and III (60). The use of cultured fibroblasts to estimate in vivo efficiency of exogenous enzyme depends on whether a given compound can be detected in a patient's fibroblasts. So far this method seems limited to the mucopolysaccharidoses. A recent report on storage of Gm_2 ganglioside in cultured cerebellum cells derived from a fetus with Tay-Sachs disease (61) suggests that cells other than fibroblasts may be used advantageously in these studies.

Immunologic complications are reported so far only in a few instances, possibly because of the limited number or duration of trials. This well-recognized danger of replacement therapy may be less of a problem than is generally thought. The inactive enzyme (CRM) in the patient, the poor antigenic response to intravenously injected proteins, and, most important, the apparent long-lasting effects of a single administration of the enzyme (60) may all decrease this danger. Thus enzyme entrapment in liposomes, RBCs or leukocytes may not be required; this would result in more immediate access of the enzyme to the stored substrate and avoid technical or biologic complications.

The effectiveness of organ transplantation in storage diseases seems largely limited to the immediate problem of organ failure (e.g., kidney failure in Fabry's disease). The recent results obtained by skin and intact cultured fibroblast implantation in patients with MPS II (62) are promising, but need further, long-term evaluation.

The most important problem in replacement therapy for storage diseases with primary neurologic involvement is the blood-brain barrier, which in all cases studied has effectively prevented the injected enzyme proteins from reaching the brain parenchyma. Although numerous experimental treatments are known that can render the barrier permeable to proteins, none is of immediate therapeutic use. Ethical and practical considerations make it important that experiments aimed at solving the previously outlined problems of human storage diseases be carried out as much as possible in animal models. A number of them have been described.

In conclusion, numerous obstacles impede the use of enzyme replacement therapy. At present the only practical alternative is prenatal diagnosis and selective abortion (excluding adoption or otherwise avoiding natural childbirth). Rigorous basic and clinical research aimed at devising rational and effective therapeutic measures is desirable, since it may ultimately allow affected children to be treated and possibly provide an alternative to selective abortion.

ABNORMALITIES IN CARBOHYDRATE METABOLISM

R. Neil Schimke

A number of inborn errors of glucose metabolism are known, the majority of them being abnormalities in degradation of glycogen with subsequent excessive storage of this substance in various tissues; hence, the use of the collective term, *glycogen storage diseases*. Some of the individual disorders bear eponyms; all have an assigned Roman numeral, although controversy exists concerning the numbering system.

GLUCOSE

All of the disorders described in this section are inherited as autosomal recessive defects except for one form of phosphorylase kinase deficiency. The salient features of these disorders are summarized in Table 6, and the metabolic interrelationship of glucose, fructose and galactose are shown in Figure 3.

Glycogen Storage Diseases

The commonest (type I), accounting for perhaps one-third of all cases, results from a deficiency of *glucose-6-phosphatase* (Step 1, Fig. 3) (1). The diagnostic hallmarks are extreme hepatomegaly and hypoglycemia. The lack of available plasma glucose during fasting leads to increased reliance on lipids, resulting in hyperlipidemia, eruptive xanthomas, ketoacidosis, hyperlactic acidemia and

186

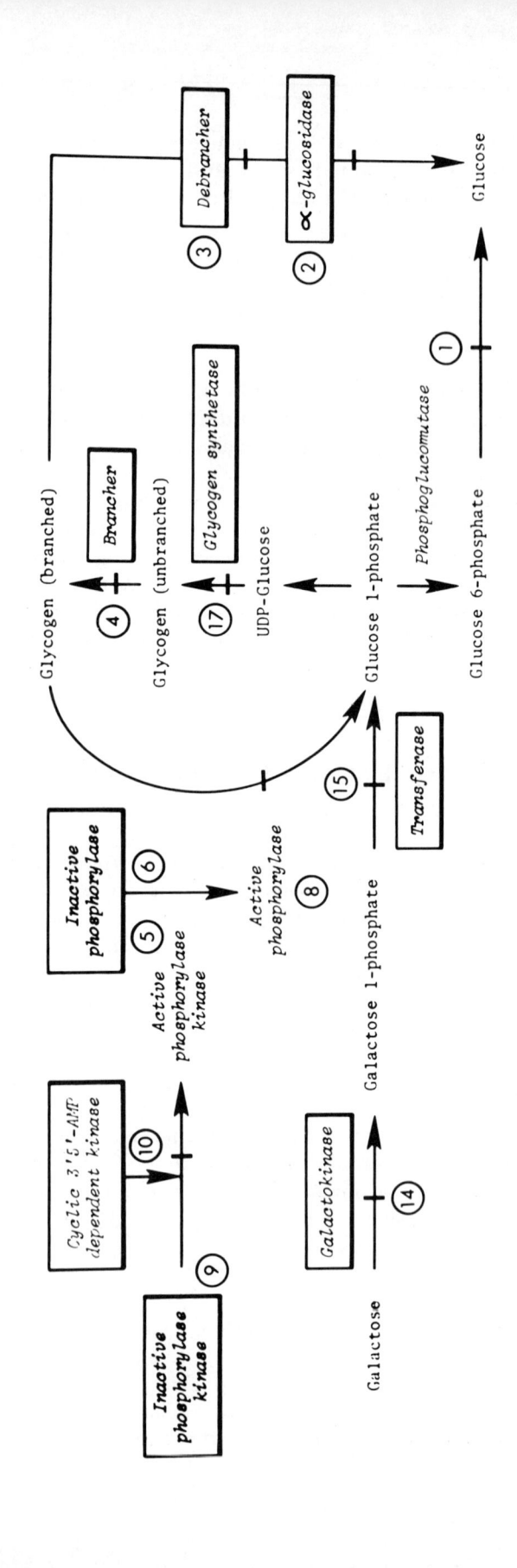

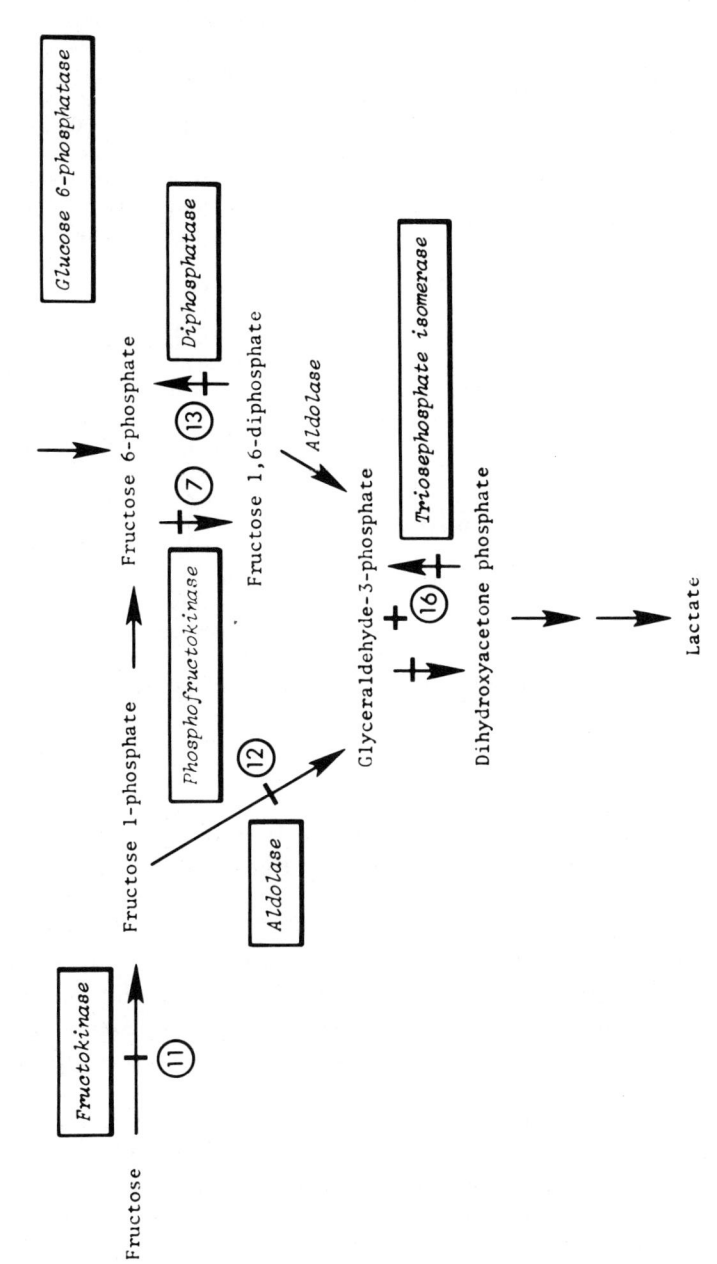

Figure 3. Various pathways of carbohydrate metabolism. Deficiencies of enzymes enclosed in boxes are described in text. Reactions are also numbered to facilitate reference.

Table 6. Glycogen Storage Disease*

Type	Eponym	Enzyme Defect	Clinical Features
I	von Gierke's disease	Glucose-6-phosphatase	Hepatomegaly, hypoglycemia, acidosis, hyperlipidemia, retarded growth; symptoms improve with adolescence
IIa	Pompe's disease	α-glucosidase	Hypotonia, hepatomegaly, cardiac failure; lethal in infancy
IIb	None	α-glucosidase	Muscular weakness beginning in 3rd-5th decade, slowly progressive
III	Cori disease; Forbes disease	Debrancher	Similar to Type I but milder
IV	Anderson's disease	Brancher	Cirrhosis with liver failure; fatal in childhood
V	McArdle's disease	Myophosphorylase	Muscle weakness and cramping with exercise; occasional myoglobinuria
VI	None	Hepatic phosphorylase	Similar to Type III
VII	None	Phosphofructokinase	Myopathic symptoms similar to Type V; hemolysis due to abnormal RBC enzyme
VIII	None	? Abnormal phosphorylase	Hepatomegaly, cerebral degeneration, fatal in childhood; not well characterized
IXa, b	None	Phosphorylase kinase	Autosomal recessive and X-linked recessive, both clinically benign with only hepatomegaly that diminishes with time
X	None	AMP-dependent kinase	Same as Type IX

*All the disorders listed are autosomal recessive ones except for Type IX, which has two genetic forms.

hyperuricemia. Platelet dysfunction may result in prolonged bleeding times. Most affected children grow poorly, although patients are described who tolerated the disease well (2). The enzyme deficiency can be diagnosed by liver biopsy. A regimen of frequent small feedings perhaps supplemented with medium-chain triglycerides is effective in some individuals (3). Portacaval shunts have been performed on selected patients with some success (4), and diazoxide has been used to relieve severe hypoglycemic episodes. In most cases symptomatic improvement occurs with adolescence, but hepatic carcinoma is reported as a late complication (5).

Generalized glycogenosis (type IIa) (Step 2, Fig. 3) is uniformly fatal in early childhood, intractable heart failure being the commonest terminal event. Hypoglycemia and acidosis are uncommon. Muscle function may be so impaired in the early stage of the disease that a diagnosis of amyotonia congenita is entertained. Electron microscopy of various tissues shows two types of accumulated glycogen, one cytoplasmic, the other lysosomal (6). The lysosomal enzyme, β-glucosidase (acid maltase) is deficient in this condition and probably accounts for the metabolic derangement, although the persistent cytoplasmic glycogen remains a puzzle. The enzyme deficiency has been found in liver, heart, leukocytes and cultured fibroblasts. Prenatal diagnosis can be made by enzymatic assay on cultivated amniotic fluid cells. Deficient urinary excretion of this enzyme has also been seen (7). There is no effective treatment. For a general discussion of lysosomal-storage disorders, see earlier sections of this chapter.

A milder form of *acid maltase deficiency (IIb)* is described in which the muscles are primarily involved (8). Affected persons develop muscular weakness in the third to fifth decade. The pattern of involvement may simulate adult-onset muscular dystrophy of the limb-girdle type. The symptoms are insidious, and it is possible that a partial enzymatic deficiency accounts for the milder clinical course. Prenatal diagnosis is also possible with this variant.

Massive hepatomegaly with a liver glycogen content of nearly 20% is seen in *glycogenosis, type III* (Step 3, Fig. 3). In this disorder the defective enzyme, amylo-1, 6-glucosidase (debrancher) does not permit degradation of glycogen past the branch points (limit dextran). Symptoms are similar to those in type I glycogenosis, but milder. Profound hypoglycemia is uncommon and extreme hyperlipidemia is rare. Prognosis is fair to good, although strenuous physical activity and cardiodepressant drugs are probably both better avoided. The diagnosis can be made by enzyme assay in leukocytes. Heterozygotes frequently show reduced enzyme levels, supporting an autosomal recessive inheritance pattern.

The rarest glycogenosis is *brancher disease (type IV)* (Step 4, Fig. 3). Because of the absence of a transglucosylase, the growing glycogen molecules have an abnormal structure, rendering them less soluble. Apparently this molecular form of glycogen incites a fibrotic reaction in the liver, leading to cirrhosis with concomitant ascites and varices. Patients usually survive for only a few years, dying of liver or heart failure. Assay of leukocytic enzyme or cultured skin fibroblasts establishes the diagnosis (9).

Study of Figure 3 reveals that a series of metabolic errors may lead to abnormal phosphorolysis of glycogen. In muscle, active phosphorylase (phosphorylase a) is a tetramer derived from two dimers (phosphorylase b), ATP and a specific phosphorylase kinase. To make the situation more complex, the kinase itself exists in an active and inactive form, and conversion to the active molecule requires still another kinase called cyclic 3'5'-AMP–dependent kinase. Immunologic and biochemical studies show that muscle and liver phosphorylases are different enzymes in that both the active and inactive forms of hepatic phosphorylase have the same molecular weight. Activation of hepatic phosphorylase still requires ATP as an energy source and a kinase enzyme system.

Patients with *muscle phosphorylase deficiency (type V)* (Step 5, Fig. 3), or the McArdle syndrome, suffer muscular pain, stiffness and weakness on exertion. Myoglobinuria may occur with extreme exercise potentially to the point of renal failure. These symptoms result from the subject's inability to degrade stored muscle glycogen and thus occur when utilization of glucose exceeds the rate of glucose entry into the cell from the plasma. Continuous infusion of glucose prevents the symptoms.

A clue to the diagnosis can be obtained by measuring blood lactate levels before and after exercise in an extremity in which arterial flow has been occluded by a sphygmomanometer. Enzymatic assay of muscle obtained at biopsy is confirmatory. Some evidence suggests more than one form of the disease, i.e., one in which the relevant enzyme is antigenically detectable although functionally absent, and the other in which neither attribute can be detected (10). Glycogen deposition is not massive, and the symptoms may not be disabling until later in life when muscular weakness may become more profound.

Liver phosphorylase deficiency (type VI) (Step 6, Fig. 3). glycogenosis results in a clinical profile similar to that in types I and III. Heart and skeletal muscle are not affected. The hepatic-type phosphorylase is also present in leukocytes, so the diagnosis can be suspected without liver biopsy. The enzyme is labile, however, and a firm diagnosis probably requires an increased hepatic glycogen content, normal levels of glucose-6-phosphatase, the debranching enzyme and an intact kinase system.

Glycogenosis types IX and X are caused by abnormalities in the *phosphorylase kinase* (Step 9, Fig. 3) and in the *cyclic AMP-dependent kinase* (Step 10, Fig. 3) respectively. In reported cases neither of these latter enzymes was entirely absent so that except for hepatomegaly, no other symptom is present. Hypoglycemia and muscular weakness are not seen. Interestingly phosphorylase-kinase deficiency is reported to be inherited as both an autosomal and an X-linked recessive trait; they are clinically indistinguishable (11). Phosphorylase kinase appears to be composed of three polypeptide chains (12), one of which is activated (phosphorylated) by the cyclic AMP-dependent kinase. In mice, in which phosphorylase-kinase deficiency of muscle occurs as an X-linked recessive disorder, this particular polypeptide chain is absent. Perhaps a similar situation holds in man, and one of the other polypeptide chains, coded by an autosomal gene, is abnormal in the autosomal recessive variety. However, the muscle enzyme in the human X-linked disorder is normal.

The reactive sequence for phosphorylase activation may be even more complex, as evidenced by a patient with *type VIII glycogenosis* (Step 8, Fig. 3). The affected child had not only hepatomegaly but also ataxia, progressive spasticity and decerebration leading to death (6). No enzyme deficiency was noted. Liver phosphorylase level was normal, but virtually all of the enzyme was in the inactive form despite a normal activating system. Apparently the genetic defect resides in abnormal control over activation.

The remaining glycogen storage disease, type VII, results from a deficiency of *phosphofructokinase* (Step 7, Fig. 3), and shows symptoms similar to those of the McArdle syndrome. Unlike the McArdle syndrome, however, hemolysis also occurs in this entity as a result of deficiency of the erythrocyte enzyme (13). This abnormality severely limits the generation of ATP necessary for a host of metabolic reactions. Liver dysfunction has not been reported so that different enzymes in muscle and liver analogous to the phosphorylases are possible. One might speculate that a severe defect in liver phosphofructokinase would be lethal because of its central role in intermediary metabolism.

Other Enzyme Errors in Glucose Degradation

The erythrocyte, having neither DNA nor mitochondria, relies totally on anaerobic glycolysis for an energy (ATP) source. Erythrocyte glycolytic enzyme deficiency results in hemolysis, often severe. The various red cell enzyme errors are discussed in detail elsewhere. For the most part only the erythrocyte suffers from this type of enzymopathy. However, in patients with *triosephosphate isomerase deficiency* (Step 16, Fig. 3), hemolysis was accompanied by generalized spasticity, not felt to be due to kernicterus or any other recognized neurologic disorder. Deficiency of this enzyme leads to inefficient use of glucose and decreased production of glyceraldehyde 3-phosphate, the glycerol moiety of triglycerides and other lipid substances, including those in the central nervous system. It is possible that expression of this enzyme deficiency in the brain results in abnormal development with subsequent neurologic dysfunction.

Glycogen Synthetase Deficiency

A family has been described in which sibs suffered from fasting hypoglycemia and glucose intolerance (14). The liver showed virtually no stored glycogen and enzyme studies suggested a defect in hepatic glycogen synthetase (Step 17, Fig. 3). Frequent feedings prevent symptoms. Autosomal recessive inheritance of this abnormality seems likely.

FRUCTOSE

There are at least three disorders of fructose metabolism, two of which are symptomatic. The third is benign and results from deficiency of the enzyme *fructokinase*. Because of this defect, fructose cannot enter the cell (Step 11, Fig. 4), and with high intake, fructosuria may occur and be confused with glucosuria unless testing is done with glucose oxidase reagent strip. An erroneous diagnosis of diabetes mellitus may be made if the usual nonspecific reducing reagent (Clinitest) is used. Fructosuria is the result of an autosomal recessive inherited trait.

Fructosemia or *fructose intolerance* results from an abnormality in hepatic fructose-1-phosphate aldolase (Step 12, Fig. 3), and diagnosis requires enzyme assay of liver tissue (15). Symptoms usually develop when fruits or juices are introduced into the infant's diet. Vomiting, hypoglycemia and acidosis develop, and proximal renal tubular dysfunction is usually present as evidenced by glucosuria, aminoaciduria, phosphaturia and renal tubular acidosis secondary to bicarbonate loss (Fanconi syndrome). Hepatomegaly with increased glycogen stores is found on biopsy, causing initial confusion with the hepatic glycogen-storage diseases. The aldolase defect causes fructose-1-phosphate to accumulate, excessive quantities of which substance are toxic in three potential ways: 1) Excessive amounts in the proximal tubule in some fashion interfere with renal transport systems. 2) Fructose-1-phosphate interferes with fructose 1,6-diphosphate aldolase, thereby interrupting the whole of intermediary glycolytic metabolism*, i.e., glucose utilization is retarded and since all three carbon intermediates must pass through this step in synthetic reactions, gluconeogenesis is halted. Hypoglycemia results from a combination of the absence of this protective mechanism and from 3) inhibition by fructose-1-phosphate of phosphorylase, thereby halting glycogenolysis (16). The disorder is an autosomal recessive trait. If fructose (and also sucrose and sorbitol) is eliminated from the

---

* Controversy exists as to whether there is one aldolase with different affinities for the two substrates, fructose-1-phosphate and fructose 1,6-diphosphate, or two discrete molecules.

diet, normal growth and development ensue. Interestingly affected individuals show a strong early aversion to fructose-containing foods.

The other recognized defect in fructose metabolism involves deficiency of the enzyme *fructose 1,6-diphosphatase* (Step 13, Fig. 3). Hepatomegaly, fasting hypoglycemia and metabolic acidosis are the paramount symptoms (17). The enzyme deficiency prevents gluconeogenesis, which action explains the frequent concomitant finding of elevated levels of plasma lactate and alanine. Excessive reliance on lipid metabolism probably accounts for the metabolic ketoacidosis. Apparently fructose 1,6-diphosphate excess, like fructose-1-phosphate, interferes with phosphorylase, thereby also blocking glycogen breakdown. The enzyme deficiency is expressed in liver, kidney and leukocytes; the muscle enzyme is apparently distinct. The condition is an autosomal recessive one. Unlike persons with fructose intolerance secondary to fructose-1-phosphate aldolase deficiency, children with the fructose 1,6-diphosphatase deficiency have no aversion to sweets. However, dietary restriction is mandatory.

Other families are reported in which fructose intolerance occurs with normal aldolase and fructose diphosphatase (18), but the metabolic defect remains obscure.

GALACTOSE

Two enzyme defects are known in the intermediary metabolism of galactose. In the first, *galactokinase deficiency* (Step 14, Fig. 3) interferes with galactose entry into the cell (19). Galactosuria occurs, and as with fructosuria the sugar may be confused with glucose in the urine unless glucose oxidase reagents are used for testing. Excessive galactose accumulates in the lens of the eye, causing cataracts early in life. The children are otherwise asymptomatic. Elimination of dietary galactose in the first month or 6 weeks may effect regression of the lenticular opacities. The condition is an autosomal recessive disorder and occurs in about 1 in 50,000 births (20). Excessive urinary galactose should be sought in any child with infantile cataracts.

Cataracts are also a feature of *galactosemia*, another autosomal recessive condition in which the enzyme galactose-1-phosphate uridyl transferase is deficient (Step 15, Fig. 3) (21). However, additional symptoms include vomiting, diarrhea, hypoglycemia and eventually cirrhosis with jaundice, ascites and mental retardation. Accumulation of galactose-1-phosphate inhibits phosphoglucomutase, thereby interfering with interconversion of glucose-1-phosphate and glucose-6-phosphate—hence the hypoglycemia. Excessive galactose, galactose-1-phosphate or even the sugar alcohol, galactolol, may be individually or collectively responsible for the ocular and hepatic findings. The retardation may be related to aberrant metabolism of galactose-containing cerebral lipids.

The enzyme deficiency can be demonstrated in liver, leukocytes, erythrocytes or cultured fibroblasts, and prenatal diagnosis has been accomplished (22). Elimination of dietary galactose is mandatory in infancy and early childhood. Evidently alternative pathways for galactose disposal become operational with time so that in late childhood dietary restrictions are usually lifted or at least modified. However, the long-term consequences of this dietary liberalization are not known.

Galactosemia is heterogeneous. By direct enzyme assay, as well as electrophoretic mobility and other studies, at least five transferase enzyme abnormalities have been detected, in addition to the "classic" variety (23). They are the *Duarte variant* with 50% normal erythrocyte transferase activity in the homozygote, the *Rennes variant* with 7-10% residual activity, the *Indiana variant* with transferase activity intermediate between these two and the *Los Angeles variant*, in which RBC transferase activity is actually increased to 140% of normal. In another variety, the so-called *Negro variant*, erythrocyte transferase is absent, but liver and intestinal activity may approach 10%. Heterozygotes for the Negro variant have normal erythrocyte transferase activity in contrast to heterozygotes for the classic type, in which activity is reduced to 50% of normal. Obviously the more enzymatic activity, the fewer the symptoms; patients with the Duarte variant are usually asymptomatic. It seems likely that these various genes all result from structural mutations and are thus allelic, although no proof of this contention exists.

A single infant, detected in a galactosemia screening program, lacked *epimerase*, the enzyme necessary for conversation of UDP-galactose to UDP-glucose (24). The patient was asymptomatic. Since both parents had an intermediate level of enzyme activity, the condition was likely an autosomal recessive one. The long-range consequences of this defect are unknown, but may involve early cataract formation analogous to galactokinase deficiency.

PENTOSURIA; ERRORS IN PYRUVATE METABOLISM

Pentosuria is a benign autosomal recessive condition resulting from deficiency of the enzyme L-xylulose dehydrogenase. L-Xylulose cannot be converted to D-xylulose and thus enter the pentose pathway. Again the excessive pentosuria (xylulosuria) may simulate glucosuria and the patient be misdiagnosed as diabetic. The condition occurs almost exclusively in Ashkenazic Jews in whom the frequency may be as high as 1 in 2,500 births (25).

Strictly speaking, pyruvate is not a carbohydrate, but errors in metabolism of this substance may have profound effects on synthesis and degradation of this foodstuff, depending on the tissue in which the inborn error is expressed. *Pyruvate kinase deficiency* is an autosomal recessive defect the expression of which is confined to the erythrocyte. The clinical presentation is hemolysis. Hepatic defects in pyruvate metabolism may be reflected as lactic acidosis, often profound, or hypoglycemia, or both. Moreover, the conditions may be acute and fatal or chronic and intermittent. One route of pyruvate metabolism, e.g., involves *pyruvate dehydrogenase* (formerly called pyruvate decarboxylase), a complex of at least three enzymes, including pyruvate dehydrogenase, lipoate acetyltransferase and lipomide dehydrogenase, all of which catalyze the conversion of pyruvate to acetyl-CoA and CO_2 (26).

A deficiency in the first unit, pyruvate dehydrogenase, has been reported in several patients all of whom show profound congenital lactic acidosis (27). A defect in the second (or possibly the third) unit has been reported in a 3-year-old girl who had acidosis, mental retardation and neurologic dysfunction (28). Patients with deficiency of *pyruvate carboxylase*, the enzyme necessary for conversion of pyruvate to oxaloacetate, exhibited varying features, depending on the degree of deficiency. Some with low enzyme levels have had ataxia, hypotonia, convulsions and gradual mental and motor deterioration. The pathologic findings were those of Leigh's subacute necrotizing encephalopathy (29). Other patients with higher residual enzyme levels showed intermittent ataxia and choreoathetosis, often precipitated by infection (30).

A defect in *phosphoenolpyruvate carboxykinase* interferes with gluconeogenesis and results in profound hypoglycemia (31). The various reactions are complex and some of the enzymes unstable, rendering interpretation of the clinical data difficult. The genetics are similiarly ill delineated since the conditions are rare, although autosomal recessive inheritance is likely.

ABNORMALITIES IN AMINO ACID METABOLISM
R. Neil Schimke

The widespread use of chromatography for screening both plasma and urine has led to the discovery of a host of abnormalities in amino acid metabolism. In many of the disorders the exact enzymatic defect has been established; in others the evidence is merely suggestive; in yet others the biochemical lesion remains obscure. All the conditions are rare, having a population incidence of approximately 1 in 15,000 for classic phenylketonuria (PKU) to conditions, such as hypervalinemia, that have been described only on a single occasion.

Most of the disturbances in amino acid metabolism are best treated by eliminating or sharply reducing dietary intake of the accumulating amino acid (e.g., PKU); some of them require a more generalized protein restriction (e.g., urea cycle defects); and a smaller group seems at least partially to respond to pharmacologic amounts of vitamin cofactors (e.g., pyridoxine-responsive homocystinuria).

DEFECTS IN AROMATIC AMINO ACID METABOLISM

Probably the commonest disorder in amino acid metabolism is PKU, which occurs in 1 in 15,000 to 1 in 20,000 live births (1). Children with classic PKU are normal at birth, but if untreated develop neurologic deterioration, seizures and mental retardation. Pigmentation may be reduced and eczema is common. On a regular diet, plasma phenylalanine accumulates to a level greater than 20 mg/100 ml, and the various ketoacid metabolites spill into the urine. If dietary phenylalanine is restricted early in life, normal development is possible.

Some patients with classic PKU escape severe retardation for reasons still not totally clear, although some degree of intellectual handicap is invariable. Women treated for PKU in childhood and who are intellectually normal may have retarded children, since phenylalanine crosses the placenta and damages the developing fetal brain (2). The metabolic error results from a defect in the hepatic enzyme phenylalanine hydroxylase, and the inheritance pattern is autosomal recessive. Many states have obligatory newborn screening programs using urinary chromatographic or bacterial inhibition tests. Commercially available testing tapes are also used in some areas. All these methods are presumptive, and definitive diagnosis requires more specific plasma phenylalanine assay.

At least four other conditions may exhibit elevated plasma and urine phenylalanine levels (1), one of which is *transient* and likely results from delayed development of the hydroxylating enzyme system. Another is a *milder variant* in which phenylalanine levels rarely exceed 15 mg/100 ml. This latter condition is regarded by some workers as an isozyme abnormality. A disturbance in an enzyme in an alternative disposal pathway, *phenylalanine transaminase*, accounts for elevated phenylalanine levels in a few patients. None of the patients in these three categories is symptomatic and no treatment is necessary. Inappropriate restriction of dietary phenylalanine may retard growth and somatic development. For this reason it is probably prudent to rechallenge the child with a regular diet at age 3-6 months to insure that he or she suffers from the classic form of the disease and not from one of the milder variants, which latter generally require no therapy.

Neurologic deterioration in a child with presumptive diagnosis of PKU despite efficient dietary control of plasma phenylalanine levels should suggest the diagnosis of *dihydropteridine reductase deficiency* (3, 4). This enzyme is necessary for synthesis of tetrahydrobiopterin, a quinoid cofactor essential for the activity of phenylalanine hydroxylase and for other hydroxylating systems active in the synthesis of dopamine and serotonin. Dietary control of phenylalanine salvages just one of several hydroxylating reactions; hence it is not surprising that the neurologic abnormalities develop, since neurotrasmitter function would continue to be grossly disturbed. Autosomal recessive inheritance is probable. Unfortunately no therapy is currently available.

Elevated plasma concentrations of tyrosine in the neonatal period are not uncommon, especially in premature infants on a high protein diet; the event is probably due to immaturity of the aromatic amino acid hydroxylating and transaminating systems (5). One of these enzymes, p-hydroxylphenylpyruvate hydroxylase, has a high requirement for vitamin C, and relative deficiency of this vitamin may result in hypertyrosinemia, tyrosyluria and even hyperphenylalaninemia. Large doses of ascorbic acid are curative. Three other conditions also show tyrosyluria. *Tyrosinemia type I* is a complex disorder involving aberrant metabolism of tyrosine, methionine, porphyrin and carbohydrates. The disease may be fulminating with progressive hepatic failure as the paramount feature, together with Fanconi-type nephropathy, although patients with a more indolent course have also been described. These latter have increased incidence of hepatoma (6).

Both the acute and chronic forms have been found in the same sibship. A deficiency of p-hydroxylphenylpyruvate hydroxylase has been found in some patients (7), although the condition is likely heterogeneous. The clinical and biochemical findings are not specific since cirrhosis from any cause—and even entities such as hereditary fructosemia—can show the same abnormalities. Restriction of dietary tyrosine has been beneficial. Inheritance is autosomal recessive (8).

Patients with *tyrosinemia type II* (Richner-Hanhart syndrome) have mental retardation, corneal opacities, hyperkeratotic skin rash and occasionally self-multilating behavior, without hepatorenal disease (9). Although the disorder is not well characterized, a defect in hepatic tyrosine transaminase has been found in some patients (5). A third defect, *tyrosinosis*, has even a more obscure etiology since only a single patient, a 44-year-old man with symptoms of myasthenia gravis, has been described (10). No enzyme deficiency was established.

Alkaptonuria, ochronosis and homogentisic aciduria are often used synonymously to describe a deficiency of homogentisic acid oxidase, a substance derived from the intermediary metabolism of aromatic amino acids (11). This is one of the earliest reported inborn errors, having been described by Garrod in 1908. The patients have dark scleral pigmentation and arthritis, particularly of the spine. The intervertebral disks prematurely degenerate, leading to vertebral fusion, but the smaller joints show little abnormality. The pigment is derived from aberrant oxidation of homogentisic acid. The relationship of the pigment deposition in cartilage and the subsequent joint degeneration is unknown. The condition is inherited as an autosomal recessive trait.

DISORDERS OF UREA CYCLE ENZYME METABOLISM

All the urea cycle enzyme defects feature hyperammonemia (Table 7). In more severely affected children, vomiting, lethargy, flaccidity, stupor, seizures and coma lead inexorably to death unless a low protein diet is instituted (12). Symptoms usually develop shortly after milk feeding begins. Older patients with milder degrees of enzyme deficiency may have symptomatic episodes especially after high-protein intake or during infection-induced stress. However, very high ammonia levels are frequently well tolerated for long intervals. Hepatomegaly and mental retardation are common. Frequent feedings of high carbohydrate intake commensurate with restriction of protein is the only therapy currently available. All the disorders are inherited as autosomal recessive traits except for ornithine transcarbamylase deficiency, which is X-linked, generally with lethality in the male hemizygote (13).

Carbamylphosphatase synthetase (CPS), the first enzyme in the urea cycle, catalyzes the reaction, trapping free ammonia. Two CPS enzymes exist, one mitochondrial, the other in the soluble fraction of the cell. It is the former that is deficient in this form of hyperammonemia. Few patients with the defect have been reported, and the deficiency from a biochemical point of view has been poorly characterized (14). The symptoms are the same as in the other urea cycle defects. Ketosis has been seen and elevated levels of glycine in plasma and urine have been detected for reasons not entirely clear. No other amino acid is consistently elevated.

Another mitochondrial enzyme, *ornithine transcarbamylase* (OTC), promotes the conversion of carbamylphosphate and ornithine to citrulline. Many affected female patients show the usual neurologic symptoms, although others have difficulties only from stress or high protein intake. Still others have symptoms early, but later develop an aversion to meat and other protein products and become virtually asymptomatic. In view of the fact that this disorder is X-linked, the

Table 7. Differential Diagnosis of Hyperammonemia in Infancy and Childhood

Congenital:
  Carbamylphosphate synthetase deficiency
  Ornithine transcarbamylase deficiency
  Citrullinemia
  Argininosuccinic aciduria
  Argininemia
  Hyperornithinemia, type II
  Propionic acidemia
  Methylmalonic acidemia
  α-Methylacetoacetic aciduria
  Familial protein intolerance
  Nonketotic hyperglycinemia
  Lysine intolerance

Acquired:
  Reye's syndrome
  Liver failure of any cause

---

clinical variability in female patients is not surprising. Special histochemical staining of liver tissue may show two populations of cells, one with, the other devoid of, OTC activity in female heterozygotes.

Generally the severer the earlier symptoms before protein restriction was imposed, the greater the likelihood of intellectual impairment. Affected male babies die in the neonatal period (16), although those with somewhat atypical hepatic enzyme assays have been reported, indicating possible heterogeneity (17). Urinary ornithine levels are usually not increased, but the alternative disposal pathway for carbamylphosphate leads to overproduction of orotic acid, urinary levels of which metabolite in a hyperammonemic female patient should suggest the diagnosis. Since, however, there are no characteristic abnormalities in urine or blood amino acids in patients with either CPS or OTC deficiency, diagnosis must be made by direct enzymatic assay of liver tissue.

Parenthetically it is important to recognize that at least two forms of "true" *hyperornithinemia* exist. In one, ornithine ketoacid aminotransferase, a degradative enzyme outside the urea cycle, was deficient (18). Severely affected individuals showed gyrate atrophy of choroid and retina (19). In hyperornithinemia type II, patients show infantile spasms, intermittent ataxia and mental retardation with hyperammonemia. The metabolic error in type II disease is unknown, but defects in either ornithine decarboxylase (20) or ornithine transport in the mitochondria (21) are suggested.

At least three different variants of *citrullinemia* are described (13). In the neonatal form typical symptoms of hyperammonemia developed during the first few days of life and all affected infants expired. In the second type, symptoms developed later, during the first year of life, and included vomiting, lethargy, mental retardation, ataxia, seizures, hepatomegaly (frequently with abnormal liver function tests) and osteoporosis. Fasting ammonia levels often were normal, but postprandial elevations were invariable. The third variant is apparently benign and was discovered only coincidentally. Although not proven, all three forms likely are allelic defects in the enzyme argininosuccinate synthetase. Hyperammonemia, hypercitrullinemia and citrullinuria are observed in the severe forms, but only excessive plasma and urinary citrulline levels are found in the benign form.

Three forms of *argininosuccinic aciduria* also exist, namely, 1) a neonatal form the symptoms of which are lethargy and seizures leading to death; 2) an infantile form with onset, in the first few weeks of life, of vomiting, hepatomegaly and failure to thrive; and 3) a chronic form with seizures, ataxia and mental retardation, together with a peculiar hair anomaly called trichorrhexis nodosa, becoming apparent at age 1-2 years (22). Normally the degradative enzyme, argininosuccinase, is found in liver, kidney and brain. The liver enzyme in one neonate was absent, but brain and kidney activities were normal. This phenomenon as well as the clinical variability may be explained by isozymic variation or different regulatory mechanisms in different tissues. All three forms show large quantities of argininosuccinic acid in plasma and urine, and prenatal diagnosis has been accomplished.

Another rare cause of hyperammonia is *arginase* deficiency (23). The elevated urinary arginine concentration interferes with the renal reabsorption of other dibasic amino acids and cystine and the urine findings may simulate cystinuria, although no cystine stones have been described. The diagnosis can be made by RBC enzymatic assay.

DISORDERS OF SULFUR AMINO ACID METABOLISM

The discovery of homocystine in the urine is of itself not specific as a number of conditions may give rise to excessive excretion of this substance. These various abnormalities, together with other defects in the sulfur pathway, are shown in Figure 4 with the various steps enumerated for reference. Common usage associates the term *homocystinuria* with a deficiency in the enzyme cystathionine synthase (Step 2, Fig. 4). Clinically the enzyme deficiency results in a condition that superficially may mimic the Marfan syndrome (24). The patients are frequently tall, exhibit kyphosoliosis and pectus excavatum and have dislocated lenses. The joints are not lax, however, and early-onset osteoporosis is common. Mental retardation and seizures are also frequent.

Apparently either homocysteine or the disulfide derivative interfere with collagen cross-linkage, leading to a diffuse connective tissue abnormality. A hallmark of the disease is the propensity to develop spontaneous thromboses of both arteries and veins, probably related to small intimal tears that develop in the abnormal vascular connective tissue and predispose the patient to thrombosis (25). The CNS abnormalities are likely due to recurrent vascular injury. Excess methionine, as well as homocystine, is found in plasma and urine. A positive urinary cyanide–nitroprusside test suggests the diagnosis, but chromatography or high-voltage electrophoresis is required to differentiate homocystine from cystine. Definitive diagnosis requires direct enzymatic assay of liver or cultured skin fibroblasts, although peculiarly the enzyme is not expressed in normal skin.

At least two forms of cystathionine synthase deficiency exist (26). One of them is responsive to pharmacologic doses of pyridoxine (vitamin B_6), administration of which agent causes plasma methionine and homocystine concentrations to normalize and the homocystinuria to disappear. The excessive vitamin cofactor

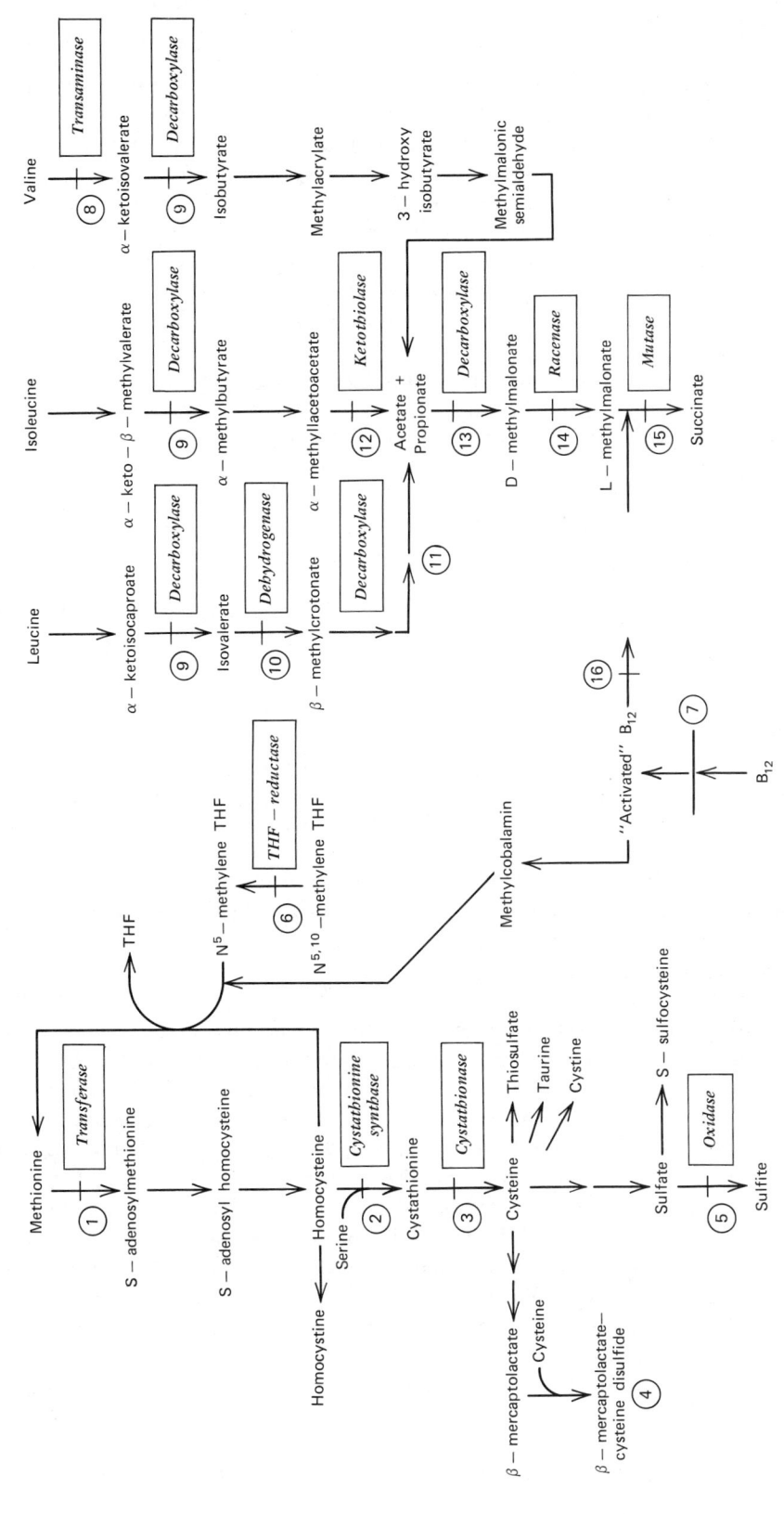

Figure 4. Abnormalities and defects arising from impairment of sulfur metabolic pathway.

may partially stabilize a defective apoenzyme, leading to the development of some (3-5%) residual activity. Although several B6-responsive individuals have been treated, the symptoms are so variable that more patient-time experience will be necessary before the potential success of this therapeutic modality can be assessed. Non vitamin B6-responsive patients should be treated with a low methionine (low animal protein) diet, probably with supplemental cystine. Both forms may require supplemental doses of folic acid because of the increased alternative use of the folate-related methytransferase pathway (27). Inheritance of both forms of homocystinuria is autosomal recessive. Prenatal diagnosis, however, is possible.

One patient was discovered in a mass screening survey to have isolated hypermethioninemia. Study of this asymptomatic infant revealed a defect in *methionine adenoxyltransferase* (Step 1, Fig. 4) (28). At age 1 the child was normal with a normal liver profile although EM study of hepatocytes showed mitrochondrial abnormalities. Remember that hypermethioninemia may occur transiently in newborn infants and in any child with diffuse hepatic dysfunction.

Patients with aberrant vitamin B_{12} absorption or metabolism from whatever cause can excrete excessive amounts of homocystine, since B_{12} metabolites act as a cofactor in the remethylation from homocysteine to methionine (29). A condition has been described in which neither the methylated B_{12} derivative (necessary for conversion of homocysteine to methionine) nor the deoxyadenosyl B_{12} derivative (necessary for methylmalonic acid metabolism) is available (Step 7, Fig. 4) (30). Affected persons excrete both homocystine and methylmalonic acid. Only a few such patients have been described, all of whom showed some mental and growth retardation. One of them had episodic megaloblastic anemia. The defect has been termed N^5-*methyltetrahydrofolate homocysteine—methionine transferase deficiency* (30a), although more than one entity may be represented in this category since the basic biochemical mechanisms of the reactive sequence are unknown. In no patient with deranged B_{12} metabolism have the characteristic clinical features of cystathionine synthase deficiency been reported. The inheritance pattern in unknown.

$N^{5,10}$-*Methylenetetrahydrofolate reductase deficiency* results in reduced quantities of N^5-methyltetrahydrofolate, a methyl donor for the homocysteine–methionine transferase reaction (Step 6, Fig. 4). Inadequate methylation leads to homocysteine accumulation and increased disulfide excretion (31). In contrast to homocystinuria due to cystathionine synthetase deficiency, hypermethioninemia does not occur. The clinical features, based on a few cases, include mental retardation and various neurologic abnormalities. Schizophrenia-like episodes, noted in one patient, appeared to reverse after combined folate and pyridoxine therapy. Interestingly morphologic studies of the arterial vessels of these patients were strikingly similar to those seen in homocystinuria due to cystathionine synthetase deficiency, supporting the theory that homocystine per se is responsible for the pathologic features of these diseases (30a). Inheritance is probably an autosomal recessive one.

Cystathionuria was first discovered on screening studies of the mentally retarded. However, wide experience suggests that retardation is not a consistent feature and some investigators feel that the biochemical lesion is benign, the variable clinical features being independent (32). The defective enzyme, cystathionase (Step 3, Fig. 4) has an absolute requirement for vitamin B_6, and high doses of the vitamin have resulted in markedly reduced cystathionine excretion. In view of the clinical variability and likely inconsequential nature of the enzyme defect, the therapy is questionable. Although prenatal diagnosis has been made, it too is of questionable value. The disorder is inherited as an autosomal recessive trait.

Four patients with *β-mercaptolactate-cysteine disulfiduria* have been described, two of whom were perfectly normal. The other two were retarded and, interestingly, one had dislocated lenses. The peculiar compound is formed by transamination of one molecule of cysteine to β-mercaptolactate. Another molecule of cysteine combines with this substance to form a mixed disulfide. The reasons for the excessive excretion of this peculiar compound are unknown, but may represent overflow into an alternative pathway (Step 4, Fig. 4). Two of the three families were consanguineous. If a single entity is represented by this defect, it is likely a rare autosomal recessive trait.

A 2½-year-old male is the only reported case of *sulfite oxidase deficiency*, although three of his sibs died in the first few days of life and may have been affected (32). Severe retardation, neurologic abnormalities and dislocated lenses were found, together with an unusual urinary amino acid, S-sulfocysteine. The metabolic block is in the conversion of sulfite to sulfate in the sulfur disposal pathway (Step 5, Fig. 4). The defect is likely a rare autosomal recessive one.

Defects of Branched-chain Amino Acid Metabolism

The various abnormalities in the metabolism of the branched chain amino acids leucine, isoleucine and valine are also shown in Figure 4. Except for a defect in the initial transamination in which only the precursor amino acid may be expected to accumulate, all the other degradative errors lead to the overproduction of various organic acids with systemic acidosis (Table 8). These acids can be detected by urinary screening with 2,4-dinitrophenylhydrazine or ferric chloride; however, separation of these various moieties—and thus definitive diagnosis—requires gas chromatographic analysis of plasma and urine. The clinical features of these disorders are remarkable similar and include ketoacidosis (may be episodic); vomiting; flaccidity or spasticity, or both; seizures; coma; leukopenia; thrombocytopenia; occasional hyperammonemia; mental retardation (older patients); osteoporosis; autosomal recessive inheritance; and prenatal diagnosis (theoretically possible in most patients).

Maple syrup urine disease (MSUD) is the prototype of branched-chain amino acid defects and results from defective decarboxylation of all the keto derivatives of the branched-chain amino acids (Step 9, Fig. 4) (33). These substances impart to urine, sweat and cerumen a characteristic odor. Affected children are normal at birth, but begin to deteriorate during the first week of life unless dietary restriction is imposed. Classic MSUD is usually fatal if untreated, and retardation occurs in surviving children in whom treatment is appreciably delayed. Prenatal diagnosis has been accomplished. Both a mild and an episodic variant of MSUD have been described, and yet a fourth type—also with milder symptoms—is thiamine responsive. If these four conditions are allelic is not known.

Hypervalinemia is described in one patient in whom vomiting, failure to thrive

Table 8. Some Heritable Causes of Acidosis in Infancy and Childhood

Lactic acidosis:
  Diabetes mellitus
  Hepatic glycogen storage diseases
  Fructose dyphosphatase deficiency
  Pyruvate decarboxylase deficiency
  Pyruvate dehydrogenase deficiency

Branched-chain amino acid metabolic defects:
  Maple syrup urine disease
  Isovaleric acidemia
  β-Methylcrotonylglycinuria
  α-Methyl-β-hydroxybutyric acidemia
  Propionic acidemia
  Methylmalonic aciduria

Miscellaneous causes:
  Pyroglutamic acidemia
  Lysosomal acid phosphatase deficiency
  Phosphoenolpyruvate carboxykinase deficiency
  Succinyl-CoA transferase deficiency
  Renal tubular acidosis:
  1. proximal—primary or secondary to other inborn errors, i.e., fructosemia, cystinosis
  2. distal

and mental retardation were noted (34). A defect in leukocytic valine transaminase was demonstrated (Step 8, Fig. 4). A separate leucine–isoleucine transaminase apparently exists, but no defect in the latter system has been described.

Children with *isovaleric acidemia* (Step 10, Fig. 5) have a peculiar body odor, described as that of sweaty feet, in addition to the usual ketoacidosis. Some patients are severely affected and die in the neonatal period; others experience a more benign course and may not be retarded, indicating that genetic heterogeneity is likely (35).

β-Methylcrotonylglycinuria is rare, having been noted in only a few patients. One patient responded to high-dose biotin supplementation with reduction of urinary metabolic acids. The defect results from impaired decarboxylation of β-methylcrotonic acid, the glycinated product being derived via an alternative disposal pathway (Step 11, Fig. 4). β-Hydroxyisovaleric acid also appears in the urine. Severe lactic acidosis coupled with hepatic dysfunction may cause nonglycinated β-methylcrotonic aciduria as a potential phenocopy of the genetic defect (36).

α-Methyl-β-hydroxybutyric acidemia, a defect in isoleucine metabolism, results from an inability to cleave α-methylacetoacetate to acetate and propionic acid (Step 12, Fig. 4). It is one cause of so-called ketotic hyperglycinemia, the others being propionic acidemia and methylmalonic acidemia, discussed in the following paragraphs. The reason for the excessive glycine is unknown. Hyperammonemia also may occur and is probably secondary to an acidosis-induced secondary defect in mitochondrial function.

Patients with genetic *propionic acidemia* have striking ketoacidosis, occasioned by even minor infections (37). Elevated blood ammonia levels are noted, simulating the various disorders of urea metabolism. Propionic acid, and probably other organic acids as well, inhibits both fatty-acid oxidation and ureagenesis; hence the common clinical features of these various disorders (37a). Propionic acidemia is presumably due to defective propionic acid decarboxylation (Step 13, Fig. 4). A less severe variant may exist, and biotin-responsiveness has been reported in still another patients. Genetic heterogeneity is likely.

Methylmalonic aciduria may be found in patients with a defect in vitamin B_{12} absorption since this vitamin is essential for methylmalonic acid (MMA) metabolism. Subsequent studies reveal four classes of patients who exhibit elevated urinary and plasma levels of MMA without hematologic abnormalities, but with signs and symptoms of defective ketoacid decarboxylation (38). Hyperammonemia may also be present in conjunction with the systemic acidosis. Types I and II are due to defects in D-MMA racemase (Step 14, Fig. 4) and MMA-mutase (Step 15, Fig. 4), respectively. Treatment requires protein restriction with prompt treatment of acidosis. Type III MMA actually results from an abnormality in the synthesis of a vitamin B_{12} metabolite, 5-deoxyadenosyl cobalamin, the cofactor in the MMA-mutase reaction (Step 16, Fig. 4).

Similarly type IV is a defect in the conversion of vitamin B_{12} to more active metabolites, one of which, methylcobalamin, is a cofactor in the remethylation of homocysteine to methionine (Step 7, Fig. 4). This last metabolic error leads to excessive urinary excretion of both MMA and homocystine. Both types III and IV respond to large doses of parenteral B_{12}, although a low-protein diet may be necessary in times of stress to avoid accumulation of ketosis-producing metabolites. Prenatal therapy of type III disease has been accomplished (38), and a heritable benign form of type II (mutase deficiency) is described in adults (38a).

DEFECTS IN PROLINE, HYDROXYPROLINE AND GLYCINE METABOLISM

Elevated plasma and urinary proline levels result from two possible defects in proline metabolism (39). Abnormalities both in the first degradative step in *proline oxidase* and in the subsequent *dehydrogenation* lead to proline accumulation. Since proline, hydroxyproline and glycine share a common renal tubular transport mechanism, all three amino acids will appear in the urine, thereby simulating the independent renal tubular defects collectively called iminoglycinuria. In none of the renal tubular defects is the plasma proline level elevated. Both the errors in proline metabolism are inherited as autosomal recessives, and it is likely that both are incidental findings of no clinical importance.

In *hydroxyprolinemia*, only hydroxyproline appears in the urine, probably because plasma levels of this substance, although elevated, are not high enough to saturate the common renal transport mechanism. The metabolic defect resides in hydroxyproline oxidase. The condition is clinically benign and likely an autosomal recessive defect.

Hyperglycinemia occurs with and without associated ketosis. Ketotic hyperglycinemia is usually due to defective metabolism of branched-chain amino acids, but the reason for the secondary glycine excess in these conditions is unknown. In the nonketotic form of hyperglycinemia, the mechanism of the abnormality, although not precisely known, evidently involves defective oxidative degradation of this neutral amino acid (40). Affected infants have marked neurologic dysfunction with seizures, myoclonus, spasticity and even decerebrate rigidity. Hyperammonemia has been found in some of them. Excessive plasma and urinary levels of glyceric acid were found in one case (40a). Respiratory embarrassment may occur early, leading to infant death. Older patients are usually severely retarded. There is, unfortunately, no effective treatment. Inheritance is probably autosomal recessive. Another entity in which there was both hyperglycinuria and glucosuria in the absence of elevated plasma levels of these substances was reported in three generations of a Swiss family (41). This asymptomatic dominant disorder is a renal tubular one and is mentioned only for differential diagnosis purposes.

Sarcosine (N-methylglycine) may play a role in one-carbon metabolism. Elevated levels of urinary and plasma sarcosine with associated deficiency of sarcosine dehydrogenase have been found in a few individuals with variable clinical findings, most of which were related to mental and growth retardation (42). There is a question whether the biochemical and clinical findings in these patients are related. The condition is autosomal recessive.

DEFECTS IN LYSINE METABOLISM

Several abnormalities in lysine metabolism are reported. In one, lysine is not converted to saccharopine because of deficiency of the enzyme lysine ketoglutarate reductase, and the patients develop *lysinuria* (Step 1, Fig. 5) (43). The condition may be benign although some patients have shown dislocated lenses. Autosomal recessive inheritance has been established.

Two patients, both of whom were retarded, are reported with *saccharopinuria* secondary to an abnormality in the next degradative step (Step 2, Fig. 5) (44).

An error in the alternative disposal pathway for lysine results in *pipecolic aciduria* (Step 3, Fig. 5). Two children with this defect are described who showed hepatomegaly and mental retardation. Both of them died in infancy of progressive neurologic deterioration (45).

α-*Aminoadipic aciduria* (Step 4, Fig. 5)—a benign condition—has been seen in two brothers (46). Similarly α-*ketoadipic aciduria* has been reported in sibs, only one of whom was retarded (47). An error in the next metabolic step (Step 6, Fig. 5) results in *glutaric aciduria*, a condition also described in sibs (48). In none of these cases has the defect been well characterized and there is question if the symptoms are at all related to the biochemical findings.

Patients with hyperlysinuria of unknown cause are reported, some of whom show hyperammonemia (49). In some patients multiple enzyme defects in the degradative pathway have been found, raising the question of a genetic abnormality in a controller mechanism (49a). The conditions are not understood. Hyperlysinuria may also occur in familial protein intolerance (hyperdibasic aminoaciduria) for reasons equally obscure.

MISCELLANEOUS DEFECTS OF AMINO ACID METABOLISM

Histidinemia is relatively rare and the clinical profile is unclear, since affected individuals showed a spectrum of abnormalities including nonspecific behavior and scholastic problems, mental and growth retardation, congenital anomalies

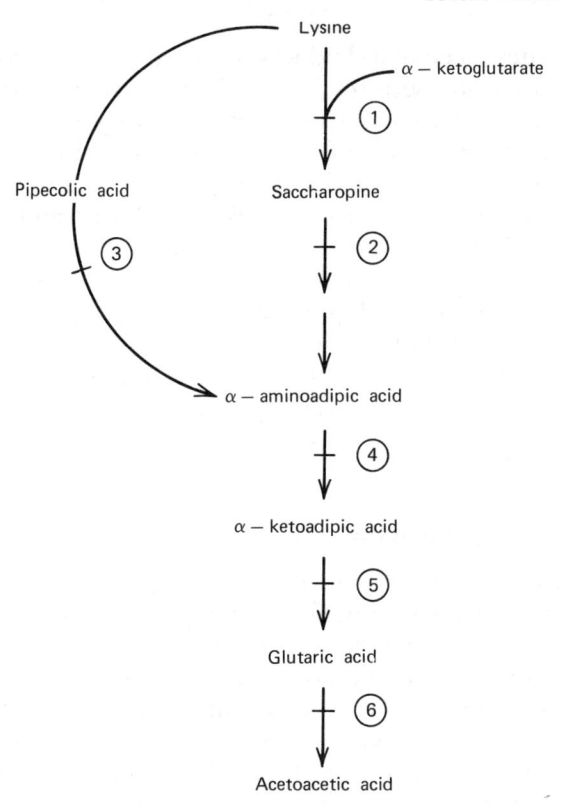

Figure 5. Metabolic pathway of lysine.

and speech difficulties (50). Still others were totally normal and it is conceivable that the metabolic error is relatively harmless, the behavioral and other problems being coincidental, although various speech difficulties seem to be fairly consistent findings. The degradative enzyme histadase, which converts histidine to urocanic acid, is deficient in this condition, and histidine is shunted into alternative pathways yielding imidazolepyruvic, imidazolelactic and imidazoleacetic acids. The enzyme assay can be performed on stratum corneum. The disorder is inherited as an autosomal recessive trait.

Condensation of histidine with β-alanine yields the dipeptide *carnosine*. The substance occurs in skeletal muscle, and normal persons on a high meat diet may excrete appreciable quantities in the urine. Children with a deficiency of carnosinase are described in whom carnosinemia and carnosinuria persisted while they were on a meat-free diet (51). Some of them have mental retardation, seizures and spasticity, although it is not clear if the enzyme defect is at all related to the neurologic abnormalities. Autosomal recessive inheritance is likely.

Pyroglutamic acid (5-oxoproline) was found in the urine and blood of three patients, two of whom were sibs. All of them suffered from episodic acidosis. The single child was retarded and exhibited neurologic dysfunction. The sib pair was neurologically intact and other than acidosis showed only low-grade hemolysis. The unusual amino acid results from an abnormality in glutathione synthetase (52). In view of the pivotal role of glutathione in maintenance of erythrocytic integrity, the hemolysis is not surprising.

A single case is described of a child who suffered from somnolence from birth and subsequent intractable seizures (53). He had *hyper-β-alaninemia*. When excessive quantities of this substance were excreted, other β-amino acids, notably β-aminosobutyric acid (BAIB) and taurine, appeared in the urine, probably because of a common saturable transport mechanism. γ-Aminobutyric aciduria (GABA) was also detected. The proposed block was in β-alanine-α-ketoglutarate transferase. Some secondary interference with GABA content in the CNS was postulated as the cause of the neurologic dysfunction. The usual form of alanine is in the alpha configuration. Hyperalaninemia and hyperalanuria occur as a consequence of impaired lactate metabolism, as in the various forms of lactic acidosis (Table 8). Elevated levels of urinary BAIB are excreted as the expression of an autosomal recessive isolated renal tubular defect.

DEFECTS IN OXATE METABOLISM

Because oxalic acid is not an amino acid, consideration of oxalosis should more properly be considered a part of nephrology since nephrolithiasis represents the prime symptom of metabolic errors in the disposal of oxalate precursors (54). However, at least one form of excessive oxalate production seems to be tied to abnormalities in serine metabolism; hence the rationale for the brief discussion of the subject here. In *type I hyperoxaluria* a defect in the decarboxylation of glyoxylate causes overproduction of glycolate and oxalate. Affected persons develop calcium oxalate nephrolithiasis and renal failure at an early age. *Hyperoxaluria type II* may result from an abnormality in the enzyme O-glyceric acid dehydrogenase. Because of this proposed defect, hydroxypyruvate accumulates (derived from transamination of serine) and is shunted into L-glyceric and oxalic acids.

Only a few patients with the type II disease have been described. Despite excessive urinary oxalate, the condition is not so severe as the type I abnormality. Both are autosomal recessive disorders. It must be remembered that secondary hyperoxaluria may occur at least transiently in several conditions, e.g., excessive dietary intake, small bowel disease, methoxyflurathane anesthesia and pyridoxine deficiency. The anesthetic is structurally similar to oxalate and may be directly converted to this compound. With vitamin B_6 deficiency, transamination of glyoxalate to glycine is presumably impeded, leading to ultimate overproduction of oxalate. The treatment regimen is discussed in Chapter 16, "Heritable Diseases of the Kidney."

RENAL AMINOACIDURIAS

More properly regarded as renal tubular (and frequently intestinal cell) transport defects, they may cause diagnostic confusion when urinary screening alone detects inborn errors of amino-acid metabolism. Plasma levels of the various amino acids are not elevated. *Cystinuria* of any of the three known types usually is manifested by renal colic. *Iminoglycinuria* is an incidental finding of no pathologic significance. Homozygotes excrete proline, hydroxyproline and glycine. Heterozygotes may have excessive glycinuria only.

Excessive urinary glycine is noted in *hereditary glucoglycinuria*, as mentioned previously (41). Patients with *Hartnup disease* have a pellegra-like skin rash and ataxia in addition to excessive urinary neutral amino acids. The symptoms are secondary to defective intestinal absorption of these same amino acids, particularly trypthophan, which is necessary for nicotinamide synthesis. Vitamin therapy eliminates symptoms, but not the aminoaciduria. More profound defects in tryptophan metabolism apparently exist, but they are poorly characterized (55).

Hyperdibasic aminoaciduria (also called familial protein intolerance with excretion of lysine, ornithine and arginine *not* cystine) occurs in two forms. One is an autosomal recessive defect associated with protein intolerance, vomiting, diarrhea, hepatomegaly and hyperammonemia. A defect in glutaminase is suggested. The other form is more benign and shows only urinary abnormalities.

Disorders with generalized aminoaciduria—such as cystinosis, galactosemia, fructosemia and Wilson's disease—should be diagnosed by the associated clinical findings. These diseases are considered in more detail elsewhere.

EARLY SCREENING AND DIAGNOSIS

The question frequently asked is: Who should have amino-acid screening studies? Obviously it would be advantageous to screen every newborn baby after milk or formula is instituted. Although ideal, this procedure is simply not practical. However, the foregoing discussion offers some general guidelines, and it seems sensible to screen any child with:

1. Unexplained failure to thrive, especially if there is familial history of similar problem in sibs;
2. Ketoacidosis;
3. Hyperammonemia;
4. Dislocated lenses (adults also);
5. Episodic acute illness, without obvious cause;
6. Progressive mental or motor retardation, or both.

Most metropolitan hospitals can undertake routine chromatographic studies, and more sophisticated evaluation generally is available at a nearby medical school with an active program in genetics. Early diagnosis cannot be overemphasized since many of the abnormalities in amino-acid metabolism are responsive to vitamin or diet therapy, or both, and prompt therapeutic intervention may prevent developmental handicap.

In giving genetic counseling to these families, the physician should stress the availability of prenatal diagnosis where available. Almost all of the disorders have an autosomal recessive mode of inheritance; this places the family at 25% risk for recurrence of affected offspring because the genes are rare, and the chance of family members marrying a carrier is low. Carrier tests are generally unavailable for these conditions except where specifically noted.

DISORDERS OF PORPHYRIN METABOLISM
Laird G. Jackson

The porphyrias are a group of disorders resulting from defects in the biosynthesis of heme. The first clinical descriptions appeared in the late nineteenth century (1), but chemical studies were first done extensively by Waldenström in Europe (2, 3) and by Watson in the United States (4), thus providing the initial basis for the understanding and classification of these conditions. Other studies include the observation of Sachs (5) that the urine of patients with acute intermittent porphyria contains an abnormal porphyrin metabolite and the subsequent demonstration of the hepatic origin of porphyrin metabolic abnormalities in the disease (6). Granick and Urata (7) then showed that the liver enzyme δ-aminolevulinic acid (ALA) synthetase was inducible, which led to the establishment of the role of overproduction of this enzyme in the disease (8).

PORPHYRIN SYNTHESIS AND HEME PATHWAY

The principal sites of porphyrin synthesis are hematopoietic tissue and liver. The basic steps are the joining of glycine and succinate (Step 1, Fig. 6) to form ALA, two molecules of which are then dehydrated to form the five-sided pyrrole ring porphobilinogen, the building block of heme (Step 2, Fig. 6). Four of these molecules are joined in a tetrapyrrole structure called uroporphyrinogen, coproporphyrinogen or protoporphyrinogen, depending on the side chains on the basic tetrapyrrole ring. (Steps 3 and 4, Fig. 6). Two isomers of uroporphyrinogen are synthesized, the I and III forms, (Fig. 7), but only the III isomer is eventually bound to a central iron molecule to form heme proteins—such as hemoglobin, cytochrome, myoglobin, catalase and peroxidase. When it has served its function, the tetrapyrrole ring is ruptured, forming a linear tetrapyrrole, bilirubin, which is conjugated and excreted. The porphyrins (uro-, copro-, or protoporphyrins) are oxidation products of the basic porphyrinogen molecules.

Various intermediary metabolites in the heme pathway are normally found in urine and feces. Urinary excretion of porphobilinogen is about 1–1.5 mg/day, but the substance is not detected in the usual urinary tests. Excretion may be increased in hepatic malignancy or hepatitis. The principal porphrin in urine is coproporphyrin, excreted at a rate of 100–300 µg/day. Hemolytic anemia, liver disease and lead poisoning are conditions that may raise the urinary excretion of coproporphyrin. Uroporphyrin I is also detectable in normal urine, whereas uroporphyrin III is only in trace amounts (9). Normally protoporphyrin is only present in feces and bile. Fecal porphyrins may arise either from metabolic products excreted with bile or by conversion of heme proteins of food in the intestine. The levels therefore vary. Coproporphyrin is present at 0–40 µg/g feces and protoporphyrin at 0–100 µg/g feces. Free prophyrins are principally in the erythrocytes.

Porphyrins produce pharmacologic effects that may correlate with clinical symptoms. Certain porphyrins are visibly pigmented so that, e.g., the change of

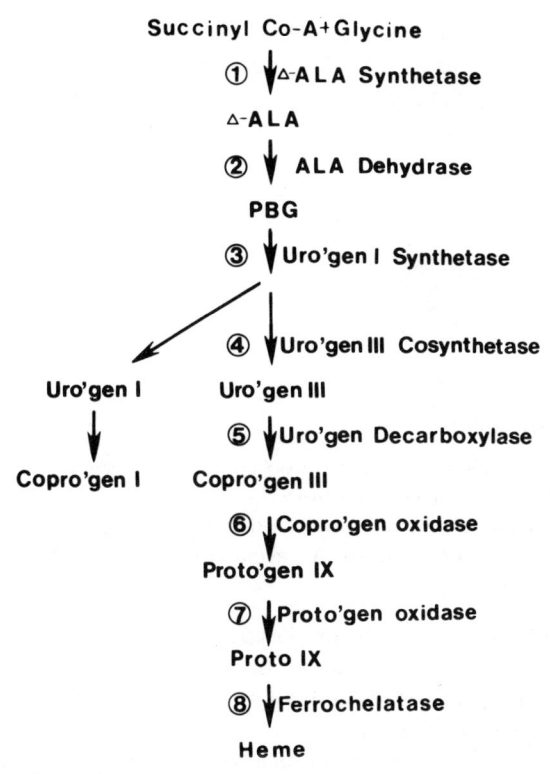

```
Succinyl Co-A+Glycine
    ① ↓ΔALA Synthetase
    Δ-ALA
    ② ↓ ALA Dehydrase
    PBG
    ③ ↓Uro'gen I Synthetase
    ④ ↓Uro'gen III Cosynthetase
Uro'gen I    Uro'gen III
  ↓          ⑤ ↓Uro'gen Decarboxylase
Copro'gen I  Copro'gen III
             ⑥ ↓Copro'gen oxidase
             Proto'gen IX
             ⑦ ↓Proto'gen oxidase
             Proto IX
             ⑧ ↓Ferrochelatase
             Heme
```

Figure 6. Synthesis and Metabolism of the Porphyrins.

Abbreviations: ΔALA—delta-amniolevulinic acid
CoA—coenzyme A
PBG—porphobilinogen
Uro'gen—uroporphyrinogen
Copro'gen—coproporphyrinogen
Proto'gen—protoporphyrinogen
Proto—protoporphyrin

Scheme of the biosynthesis and metabolism of the porphyrins. Enzymatic defects have been demonstrated at steps 3–6 and 8 for inherited forms of human porphyria: acute intermittent porphyria (3), erythropoietic porphyria (4), porphyria cutanea tarda (5), coproporphyria (6) and erythropoietic protoporphyria (8). See text for details.

uroporphyrin to porphyrin by light and air is responsible for the red color of urine in patients excreting these substance in excess. Porphyrins have photosensitivity activity, which is not clearly understood (10), although it is thought to be related to their intense fluorescent response to long-wave ultraviolet light at the 400 nm range. This absorption, characteristic of porphyrins in the skin or superficial vasculature, may produce photochemical damage to adjacent membranes

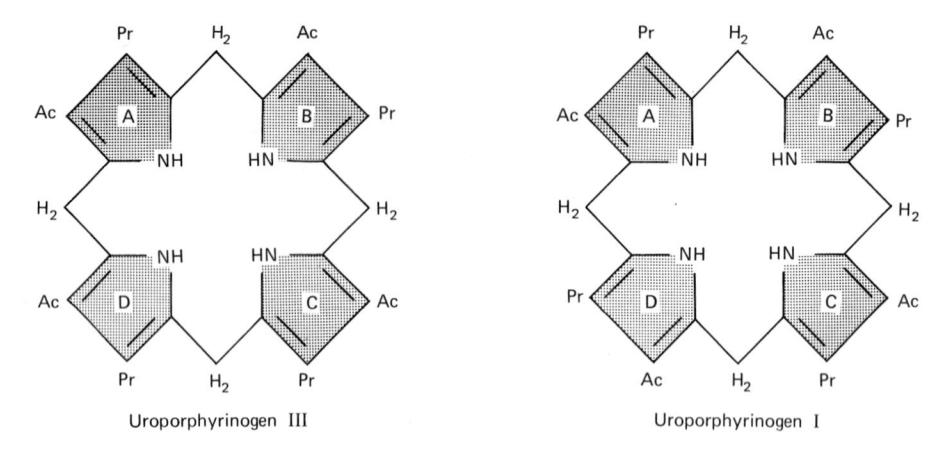

Figure 7. Tetrapyrrole molecule. Composed of four 5-sided pyrrole rings with their nitrogen groups facing the center where the iron of heme will ultimately be bound, I and III forms of any of the compounds differ only by mirror reversal of exposed groups of D pyrrole. Abbreviations: Ac, acetate; Pr, pyruvate.

and therefore cutaneous symptoms. Porphobilinogen and its products may be inhibitory to the acetylcholine that is normally released after neuromuscular end-plate depolarization and thus lead to the neuromuscular toxicity seen in hepatic forms of porphyria (11).

Currently possibly five forms of clinical porphyria are associated with specific enzyme abnormalities (Fig. 6 [11a]). The hereditary forms of human porphyria are usually classified as *erythropoietic* porphyrias and *hepatic* porphyrias (Table 9) (12).

ERYTHROPOIETIC PORPHYRIAS

Heritable erythropoietic porphyrias are characterized by cutaneous photosensitivity with the accumulation of large amounts of porphyrins in peripheral erythrocytes and erythroid bone-marrow elements. The cutaneous symptoms are thought to result from skin or superficial vascular porphyrin absorption of 400 nm light energy and subsequent photochemical injury to adjacent membranes.

Erthropoietic Porphyria

This is a rare clinical disorder with less than 100 cases reported in the literature (13). Prominent photosensitivity accompanied by massive porphyrinuria are hallmarks of the disease.

An early sign is the excretion of red urine in infancy. Photosensitivity is usually absent in the neonatal period, but becomes obvious in the first year as exposure to sunlight increases. Vesicular or bullous eruptions occur on the exposed body parts, and the term *hydroa aestivale* refers to the seasonal recurrence of the phenomenon. The lesions contain serous fluid, may exhibit red fluorescence, heal slowly and tend to become secondarily infected with subsequent scarring. Hypertrichosis is a frequent finding.

Table 9. Classification of Porphyrias

Type	Inheritance*	Clinical Features	Laboratory Findings
Erythropoietic porphyria	AR	Photosensitivity, anemia, massive porphyrinuria	Elevated urinary uroporphyrin I, decreased uroporphyrinogen III cosynthetase
Erythropoietic protoporphyria	AD	Photosensitivity, chronic skin changes	Elevated blood and fecal protoporphyrin, normal urine porphyrins
Acute intermittent porphyria	AD	Abdominal pain, neurologic symptoms	Elevated urine porphobilinogen, decreased uroporphyrinogen I synthetase
Variegate porphyria	AD	Photosensitivity and chronic skin changes, abdominal and neurologic symptoms usually precipitated by drugs	Increased fecal coproporphyrin and protoporphyrin
Coproporphyria	AD	Symptoms similar to variegate form and precipitated by drugs	Elevated fecal coproporphyrin III, decreased coproporphyrinogen oxidase
Porphyria cutanea tarda	Nongenetic	Late onset of photosensitivity and chronic skin changes	Elevated urine uroporphyrin

*AD, autosomal dominant; AR, autosomal recessive.

The teeth may be discolored and are always fluorescent. Splenomegaly is an almost constant feature, and hemolytic activity is usually increased with normochromic anemia and normoblastic hyperplasia of the bone marrow, the latter which shows fluorescent nucleated RBCs and reticulocytes.

Diagnostic signs are the red urine (uroporphyrin excretion increases to 500 mg/day in extreme cases, the majority of which is uroporphyrin I). Urinary coproporphyrin is also increased, but not as markedly. The feces contain large amounts of coproporphyrin and less uroporphyrin.

Biochemically a decrease in the levels of the enzyme uroporporhyrinogen III cosynthetase (CoS) in erythrocytes has been demonstrated (14) (Step 4, Fig. 6). Enzyme studies show a decrease of CoS in cultured skin fibroblasts (15) of homozygotes and in erythrocytes of heterozygotes (16). These findings support the pedigree conclusions for an autosomal recessive pattern of inheritance. However, other workers find conflicting results, suggesting that erythropoietic porphyria cannot be entirely explained by an isolated CoS deficiency (17). Hence more research is necessary before carrier detection and prenatal diagnosis are possible. Management is difficult, but splenectomy may reverse the hemolytic process and either diminish or reverse the photosensitivity.

Erythropoietic Protoporphyria

A commoner disorder than the preceding one, erythropoietic protoporphyria is inherited as an autosomal dominant trait. Signs and symptoms are frequently subtle and intermittent, and absence of excess urinary protoporphyrin makes recognition difficult. Excess protoporphyrin levels are measurable in blood and feces and should be sought in clinically suspicious light-sensitive patients. Excess protoporphyrins are formed in erythroid precursor cells of the bone marrow. Fluorocytes may be seen in bone marrow or blood by examination of thin blood smears under a fluorescent microscope (18).

Clinical signs are frequently present from early childhood and include sensations of "burning" or "stinging" in the skin exposed to sunlight. This is followed in hours by erythema, edema and occasionally petechiae. Less commonly vesiculation, crusting and superficial scarring will occur, giving the skin a prematurely aged appearance over the dorsa of the hands with fine, shallow, pitted scars over the cheeks and nose and around the mouth. Clinical variability is wide and cases of "nonpenetrance" are common in pedigrees (19). Deaths, reported in rare instances, were apparently correlated with severe liver involvement. Although the basic defect is unknown, a defect in the conversion of protoporphyrinogen to heme involving a deficiency of the enzyme ferrochelatase has been suggested (Fig 6, step 8) ([19a]).

HEPATIC PORPHYRIAS

Excessive levels of porphyrins and porphyrin precursors are formed in the liver in three genetically determined diseases: acute intermittent porphyria, variegate porphyria and hereditary coproporphyria. These disorders have in common: 1) an acute phase with enhanced activity of ALA synthetase; 2) exacerbations that may be precipitated by exposure to certain drugs, most of which induce ALA synthetase in hepatocytes (20); 3) clinical exacerbations associated with a neurologic syndrome; and 4) development of symptoms in late puberty.

It is postulated that the primary defect in the hepatic porphyrias is an increased activity of ALA synthetase, perhaps from an operator gene mutation resulting in such overactivity. Alternative theories include a partial block in the heme synthetic pathway or a block in the participation of heme in the feedback control of ALA synthetase. Recent findings in the study of acute intermittent porphyria suggest that the loss of activity of specific enzymes in the heme pathway can lead to decreased levels of heme for negative feedback control of the liver production of ALA synthetase (21). It is likely that similar mechanisms operate in the other forms of hepatic porphyria, which will be elucidated in future studies.

ACUTE INTERMITTENT PORPHYRIA

Abdominal pain of varying degree and frequency is usually the initial symptom in acute intermittant porphyria (AIP). Pain is localized or general, but accompanied by a soft abdomen without marked tenderness. Radiographs show areas of distention and spasm. Constipation is frequent and marked. These symptoms are intermittent, but may lead to weight loss with time and also cause fluid imbalance and anorexia—with vomiting if protracted. Neurologic symptoms are variable and are those referable to peripheral or autonomic nerves, brainstem, cranial nerves or cerebral function. The basic pathophysiologic profile is not clear. Photosensitivity and skin eruption are almost always absent in AIP. Many patients have hypertension, some have fever and leukocytosis. Onset of symptoms is usually at puberty or shortly thereafter.

Acute attacks may be fatal, and the mortality in one study in which 50 patients with AIP were followed for five years was 24% (22). Later surveys suggest that this death rate has fallen as drug risks and latent disease are more expertly recognized (23). Nonfatal attacks are often followed by long latentcy. Various endocrinologic and metabolic changes may occur in an acute attack the most serious of which are water and electrolyte modifications compatible with inappropriate ADH secretion. Exacerbations of AIP are known to be correlated with the menstrual cycle (24). The effect of pregnancy is variable, but in itself pregnancy probably does not affect porphyria (25). Glucose tolerance abnormalities are observed as well as hypercholesterolemia, and they both apparently accompany the attack.

The urine shows increased levels of porphobilinogen and (usually) ALA during acute attacks. The levels of these compounds may not be elevated in the urine during latency, and they are almost always too low for detection by Watson-Schwartz test prior to puberty in persons in whom the disease is latent (26). In contrast to the urinary findings, fecal porphyrin excretion is usually not significantly disturbed. At autopsy the liver contains large amounts of porphobilinogen.

Recently several laboratories have demonstrated a reduction in hepatic and erythrocyte uroporphyrinogen-I-synthetase activity (URO-S) of about 50% in affected individuals (26-28) (Step 3, Fig. 6). This enzyme catalyzes in one step the conversion of porphobilinogen to uroporphyrinogen, and the block may result in accumulation of porphobilinogen and an increase in ALA synthetase activity. The 50% reduction in activity is consistent with a single autosomal dominant mode of genetic transmission as seen in pedigree studies. Activity of URO-S is abnormal in children of affected persons, even in infancy (29). The test therefore appears to be reliable within families at risk 1) to determine the status of "latent" individuals and 2) afford counseling for drug risks and preventive dietary management.

A high carbohydrate diet is frequently effective in aborting or treating an attack, and a daily intake of 450 to 500 g of carbohydrate is now used (1). Phenothiazines, particularly chlorpromazine, are often helpful in controlling abdominal pain, and meperidine may be used if necessary. The most important consideration is the avoidance of symptomatic provocation by drugs, alchohol or chemical and food additives. Some of these agents are summarized in Table 10 (23).

VARIEGATE PORPHYRIA

First delineated by Dean and Barnes in studies of South African persons of Dutch descent (30, 31), variegate porphyria has now been described more widely. The disorder differs from AIP by 1) a familial history of chronic skin involvement with or without abdominal and neurologic changes; 2) photo- and mechanical sensitivity of exposed skin; 3) episodic abdominal pain and neurologic symptoms—usually precipitated by drugs, notably barbiturates; 4) increased fecal excretion of protoporphyrin and coproporphyrin, even in asymptomatic persons.

The skin manifestations are chronic and involve mainly light-exposed skin areas, although photosensitivity varies from negligible to severe. Trivial trauma produces abrasions, erosions, bullae and scarring, and cutaneous symptoms are the only manifestation of the gene in about one-half the affected cases. Acute abdominal and neurologic symptoms similar to those in AIP are found in one-half the cases, and they are usually but not invariably preceded by skin sensitivity. Although mortality during acute attacks is about 25%, Dean's data suggest that lifespan was not appreciably reduced prior to recent times in the affected Afrikaner population (32), but introduction of barbiturates and sulfonamides apparently altered this finding. Acute attacks are frequently associated with water and electrolyte changes similar to those in AIP and are similarly compatible with inappropriate ADH secretion (33).

The characteristic laboratory finding consists of large amounts of coproporphyrin and protoporphyrin in the feces, a feature that is also demonstrable in asymptomatic children of patients with variegate porphyria, who appear therefore to be "latent porphyrics." During acute attacks increased amounts of ALA, porphobilinogen and uroporphyrin may be found in the urine of patients. Although there is no documented enzymatic defect, the inheritance is clearly autosomal dominant with almost complete clinical "penetrance" of the gene.

Table 10. Safe and Unsafe Drugs in Acute Intermittent Porphyria

Safe or Probably Safe	Unsafe or Probably Unsafe
Analgesics:	
Salicylates (Aspirin), propoxyphene (Darvon), meperidine (Demerol) morphine and codeine	Antipyrine, aminopyrine, phenylbutazone (Butazolidin) and pentazocine (Talwin)
Sedatives, tranquilizers, hypnotics:	
Chlorpromazine (Thorazine), chloral hydrate (Noctec) and trifluoperazine (Stelazine)	Barbiturates (all types), meprobamate (Miltown), isopropyl meprobamate (Soma), glutethimide (Doriden), methyprylon (Noludar), ethchlorvynol (Placidyl), ethinamate (Valmid), carbromal, chlordiazepoxide (Librium) and diazepam (Valium)
Anticonvulsants:	
Bromides, acetazolamide (Diamox), magnesium sulfate and paraldehyde	Barbiturates (mephobarbital [mebaral], primidone [mysoline[and secobarbital [Nembutal])
	Hydantoins (ethotoin [Peganone], mephenytoin [Mesantoin] and phenytoin [Dilantin])
	Succinimides (ethosuximide [Zarontin], methsuximide [Celontin] and phensuximide [Milontin])
	Iminostilbenes (carbamazepine [Tegretol])
	Oxazolidinediones (paramethadione [Paradione] and trimethadione [Tridione])
Antihypertensives:	
Propranolol (Inderal), chlorthiazides (Hydrodiuril, Diuril) and guanethidine (Ismelin)	Hydralazine (Apresoline)
	Spironolactone (Aldactone, Aldactazide)
Reserpine (Serpasil, hydropres) and diazoxide (Hyperstat)	-methyldopa (Aldomet, Aldoril, Aldoclor)
Antibiotics:	
Penicillin, streptomycin, tetracycline (Achromycin) rifampin (Rifadin), nitrofurantoin (Furadantin) and resorcinal	Sulfonamides, griseofulvin (Fulvicin, Frifulvin), pyrazinamide and chloramphenicol (Chloromycetin)

Table 10. (Continued)

Safe or Probably Safe	Unsafe or Probably Unsafe
Antihistamines: Prochlorperazine (Compazine), diphenhydramine (Benadryl), chlorpheniramine (Chlortrimeton), tripelennamine (Pyribenzamine) and meclizine (Bonine)	
Hypoglycemics: Insulin and biguanides (Phenformin, metformin)	Tolbutamide (Orinase), chlorpropamide (Diabinese) and tolazamide (Tolinase)
Anesthetics and adjuncts: Ether, nitrous oxide, succinylcholine, droperidol (Inapsine, Innovar) and promethazine (Phenergan)	Barbiturates, halothane and ketamine
Analeptics: Methylphenidate (Ritalin)	Bemegride, nikethamide (Coramine) and pentylenetetrazol (Metrazol)
Alkylating agents:	Busulphan (Myleran), chlorambucil (Leukeran) and cyclophosphamide (Cytoxan)
Psychotropic drugs: Amytriptylin (Elavil)	Imipramine (Tofranil), pargyline (Eutonyl) and tranylcypromine (Parnate)
Miscellaneous: Digitalis preparations, dicoumerol (Coumadin), atropine, anticholinesterases, ganglionic blocking agents, glucocorticoids, EDTA, clofibrate (atromid), vitamin B group and vitamin C	Metyrapone (Metopirone), aminoglutethimide (Elipten), op' DDD, theophyline, ergot preparations, chloroquine (Aralen), probenecid (Benemid), estrogens, progestational agents and oral contraceptives

HEREDITARY COPROPORPHYRIA

Although clinical findings resemble those of variegate porphyria, the abnormality appears to have its own characteristic pattern of porphyrin excretion. Fecal excretion of large amounts of coproporphyrin III is constant, whereas all other porphyrin levels may be minimally elevated to normal in both feces and urine. Liver ALA synthetase activity is increased as in the other hepatic porphyrias, and there appears to be a decrease in coproporphyrinogen oxidase (34) (Step 5, Fig. 6). Symptoms are minimal or absent unless provoked by drugs such as barbiturates, anticonvulsants or tranquilizers. Mild episodic and ill-defined abdominal, neurologic or psychiatric manifestations may be present between attacks. Almost all of the reported cases demonstrate a familial occurrence in successive generations, suggesting an autosomal dominant inheritance.

PORPHYRIA CUTANEA TARDA

Of the hepatic porphyrias, this variety is the only one with almost exclusively cutaneous manifestations. Vesicular and ulcerative eruptions occur on exposed areas, leaving scars on healing. Acute attacks of abdominal or neurologic symptoms do not occur. The onset is usually in middle age and is frequent in patients with unrelated chronic infections or parasitic infestations. Although cases are reported from all over the world, the Bantu population of South Africa is most frequently involved. Urinary excretion of uroporphyrin is the primary chemical defect, fecal porphyrins remaining undisturbed. A recent report documents autosomal dominant transmission and demonstrates a defect in uroporphyrinogen decarboxylase activity in erythrocytes in some families (35).

DISORDERS OF PURINE METABOLISM
Laird G. Jackson

LESCH-NYHAN SYNDROME

The Lesch-Nyhan syndrome is an inherited form of cerebral palsy, characterized by muscle spasticity, choreoathetosis and mental retardation. The disease is of great interest because it associates a specific biochemical defect with a behavioral disorder, self-mutilation. Current information suggests that research in this defect will also increase our understanding of other purine metabolic disorders, especially the X-linked inherited form of gout (1). Lesch and Nyhan in 1964 originally described two sibs, one of whom had hematuria associated with uric acid crystalluria and the characteristic neurologic features (2). Now more than 150 patients have been reported with this condition.

The affected child usually appears normal for several months after birth (3, 4). Between ages 6 and 9 months the persistent uric aciduria begins to cause brownish-orange flecks in the diaper due to precipitation of uric acid (5). Developmental delay is usually evident by 1 year; choreoathetosis may develop subsequently, and most children exhibit dysarthria with speech development. An early symptom observed by many parents is persistent vomiting, perhaps related to dysphagia but also explained by compulsive aggressive behavior. Self-mutilation may begin early, but may also be delayed and is the least constant symptom. Symptomatic severity, onset and progression have broadened with time and experience.

Although the child is retarded by intelligence testing, he appears to be aware of the consequences of self-mutilation and frequently communicates the request for protection from his own compulsive behavior. The application of splints to prevent moving the fingers to the mouth or protective mittens usually calms the child, suggesting an understanding of his problem beyond our ability to measure. Some investigators report an increase in self-mutilation during stress and a decrease when anxiety is minimized. Aggressive behavior characterizes these boys from an early age with biting and pinching of others as well as swearing, kicking and hitting out at attendants and visitors. In contrast they retain an awareness of activities surrounding them and laugh and react appropriately to situations as they occur. Gouty arthritis or renal symptoms do not appear until later, but uremia is usually a contributing factor in mortality. As with other heritable, degenerative, metabolic disorders, longevity depends to a large extent on the quality of supportive care. Respiratory infection usually leads to death in early adolescence or young adult life.

Enzymatic Deficiency

Biochemical studies demonstrate a deficiency of hypoxanthine–guanine phosphoribosyl transferase (HPRT) activity, the enzyme which converts either hypoxanthine or guanine to the ribonucleotide form (6) (Fig. 8). Activity in RBCs is less than 0.01% of normal and in cultured skin fibroblasts is 2–3% of normal. The deficiency of HPRT leads to an apparent overabundance of phosphoribosyl-1-pyrophosphate (PRPP), the purine precursor substance, and this in turn accelerates purine synthesis. The deficiency of HPRT activity can be demonstrated in brain tissue of affected children at autopsy, since the enzyme is normally abundant throughout the brain, especially in the basal ganglia. It is no surprise, therefore, that the major neurologic symptoms are related to basal ganglia dysfunction. It is this defect in enzymatic activity, rather than an accumulation of uric acid or other secondary substance that seems most closely correlated with the neurologic symptoms (1).

All affected children have been males, and cell culture (7) and hair-root assays (8) demonstrate heterozygosity in the mothers of these children, supporting an X-linked mode of inheritance; HPRT activity may also be assayed in cultured amniotic fluid cells (9), making the disorder amenable to prenatal diagnosis by midtrimester amniocentesis (10). The finding of 2-3% residual HPRT activity in cultured skin fibroblasts suggests a structural gene alteration rather than a gene deletion (1).

At present screening for the Lesch-Nyhan syndrome on any more than a small-scale examination of uric acid levels in developmentally delayed spastic children is not available. Enzymatic determination of hair follicles or cloned fibroblasts is indicated for normal female relatives of affected males, preferably before the girls are reproductively active. Prevention through carrier detection within families and prenatal diagnosis is currently the only effective clinical tool for this strange but highly interesting disease. Therapy is limited to allopurinol to diminish renal damage and diazepam to diminish the severity of self-mutilation.

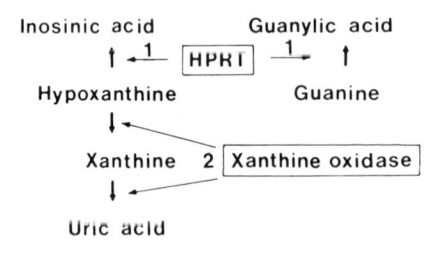

Figure 8. Pathways of interconversion of purines to nucleotide forms or to breakdown products, showing sites of action of hypoxanthine–guanine phosphoribosyl transferase (HPRT) (1), which is blocked in the Lesch-Nyhan syndrome, and of xanthine oxidase (2), which is blocked in xanthinuria.

GOUT

Gout, another disorder of purine metabolism, is frequently familial and more often seen in men than in women. In most cases no single enzymatic defect is responsible for the symptoms; the condition is likely heterogeneous, with both overproduction of uric acid and abnormalities in urinary excretion having been implicated in pathogenesis. Excretion of more than 600 mg uric acid/24 hours in urine collected during days 4, 5 and 6 of a purine-free diet in patients free of renal disease has demonstrated enzymatic defects in a small group (less than 5%) of all gouty patients (11, 12). These exhibit 0.1 to 17% of normal HPRT activity in RBCs, synthesize uric acid in roughly inverse proportion to their enzymatic levels, have a more aggressive progression of gouty arthritis and suffer earlier and more extensive renal calculi and renal damage. A subgroup with less than 1% residual HPRT activity also shows diverse neurologic symptoms. No common presentation has been found, and genetic heterogeneity or clinical variability is certainly present in the overall group of X-linked uric acidurias. Further investigation may yield more information applicable to the broader problem of gouty arthritis.

The purine precursor PRPP alone may be increased intracellularly in response to increased PRPP synthetase (13). This may be due to increased specific activity of the enzyme, high affinity of ribose-5-phosphate for the enzyme, or both (Fig. 9). Clinically the patients produce excessive uric acid with a tendency to gout and kidney stones early in adult life. Maintenance with allopurinol and high fluid intake is beneficial. Apparently the gene for PRPP is located on the X-chromosome next to that for HGPRT (14), making this another X-linked recessive disorder.

XANTHINURIA

Xanthinuria, a rare disorder characterized by replacement of uric acid by xanthine and hypoxanthine in the urine, results from a gross deficiency of tissue xanthine oxidase. Most patients have a relatively benign disorder, one usually suggested by a very low serum uric acid value in a routine laboratory survey (15, 16). In 8 of the 23 subjects xanthine stones were observed, and at least one person developed hydronephrosis and eventual renal failure. Two patients had a myopathy associated with crystalline deposits of xanthine and hypoxanthine inside the muscle cells. Another patient had recurrent polyarthritis, suggesting a crystal-induced synovitis. Diagnosis is made by demonstrating excessive (> 100 mg) excretion of hypoxanthine and xanthine and very little uric acid (usually < 30 mg) in the 24-hour urine sample.

Xanthine oxidase catalyzes both the oxidation of hypoxanthine to xanthine and of xanthine to uric acid (Fig. 8). Found primarily in liver and small intestinal mucosa (17), it acts complexly in the two-step conversion of hypoxanthine to uric acid. Studies of patients demonstrate < 0.1% of normal enzyme activity in jejunal muscosal and liver biopsy material. The increased clearances of hypoxanthine and xanthine appear to be normal responses to elevated serum levels, suggesting overload rather than an abnormal renal mechanism (18). This over-

$$\text{ATP} + \text{Ribose-5-P} \xrightarrow{P_i} \text{PRPP} + \text{AMP}$$

Figure 9. Synthesis of phosphoribosyl-l-pyrophosphate (PRPP), which is apparently under control of PRPP synthetase, an X-linked gene (13). **Key:** P_i, inorganic phosphate; ATP, adenosine triphosphate; AMP, adenosine monophosphate.

load may result in xanthine stones, although they also occur in patients without genetic xanthinuria or in studies insufficient to make this determination.

The genetics of the disorder are consistent with an autosomal recessive mode of inheritance, with at least one sib pair being affected and all parents showing no alteration of urinary xanthine excretion. So far no consistent abnormality has been found in heterozygotes.

Prevention of xanthine stones is the main goal of therapy. Xanthine is poorly soluble in acid solutions, but long-term alkali therapy is potentially hazardous. High fluid intake to maintain a large urinary volume is useful as with other stone-forming conditions, and dietary restriction of purines will probably be helpful. Allopurinol may reverse the xanthine/hypoxanthine ratio (19) and may be useful in selected cases.

Other Purine Disorders

Three children have been described with *adenine phosphoribosyl transferase (APRT) deficiency*, blocking adenine metabolism and leading to kidney stones composed of 2-8 dioxyadenine, a poorly soluble oxidation product of adenine. The disorder is probably autosomal recessive (20). The patients' urine shows excessive adenine, and 2-8 dioxyadenine and allopurinol seem beneficial. A rare form of hemolytic anemia is associated with increased *adenosine deaminase* (ADA) activity. This apparently shunts adenosine from utilization in ATP generation and leads to anemia, reticulocytosis and splenomegaly inherited as a dominant disorder (21).

REFERENCES

Lysosomal Storage Diseases

1. Hers HG: Inborn lysosomal diseases. *Gastroenterology* 48:625, 1965
2. DeDuve C, Wattiaux R: Functions of lysosomes. *Ann Rev Physiol* 28:435, 1966
3. Dingle JT, Fell HB (ed): Lysosomes in Biology and Pathology, Vols. I and II. New York, American Elsevier, 1969
4. Dingle JT (ed): Lysosomes in Biology and Pathology, Vol. III. New York, American Elsevier, 1973
5. Hers HG, VanHoof F (ed): Lysosomes and Storage Diseases. New York, Academic, 1973
6. Bondy PK, Rosenberg LE (ed): Duncan's Diseases of Metabolism, ed. 7. Philadelphia, Saunders, 1974
7. Stanbury JB, Wyngaarden JB, Fredrickson DS (ed): The Metabolic Basis of Inherited Diseases, ed. 4. New York, McGraw-Hill, 1978

8. Volk BW, Aronson SM (ed): Sphingolipids, Sphingolipidoses and Allied Disorders. New York, Plenum, 1972
9. Neufeld EF, Timple WL, Shapiro LJ: Inherited disorders of lysosomal metabolism. *Ann Rev Biochem* 44:357, 1975
10. Hirschhorn R, Weissman G: Genetic disorders of lysosomes. *Progr Med Genet* (new series) 1:49, 1976
11. Brady RO: The chemistry and control of hereditary lipid diseases. *Chem Phys Lipids* 13:271, 1974
12. Malone MJ: The cerebral lipidoses. *Pediatr Clin North Am* 23:303, 1976
13. Neufeld EF: The biochemical basis for mucopolysaccharidoses and mucolipidoses. *Progr Med Genet* 10:81, 1974
14. Dorfman A, Matalon R: The mucopolysaccharidoses (A Review). *Proc Nat Acad Sci USA* 73:630, 1976
15. O'Brien JS: Molecular genetics of Gm$_1$ β-galactosidase. *Clin Genet* 8:303, 1975
15a. Tallman JF, Brady RO, Quirk JM, et al: Isolation and relationships of human hexosaminidases. *J Biol Chem* 249:3489, 1974
16. Srivastava S, Beutler E: Studies on human β-D-N acetyl hexosaminidase. III. Biochemical genetics of Tay-Sachs and Sandhoff's diseases. *J Biol Chem* 249:2054, 1975
17. Geiger B, Arnon R: Chemical characterization and subunit structure of human N-acetyl hexosaminidase A and B. *Biochemistry* 15:3484, 1976
18. Conzelman E, Sandhoff K: AB variant of infantile Gm$_2$ gangliosidosis: deficiency of a factor necessary for stimulation of hexosaminidase A-catalyzed degradation of ganglioside Gm$_2$ and glycolipid GA$_2$. *Proc Natl Acad Sci USA* 75:3979, 1978
18a. Navon R, Padeh, B, Adam, A: Apparent deficiency of hexosaminidase A in healthy members of a family with Tay-Sachs disease. *Am J Hum Genet* 25:287, 1973
19. Romeo G, D'Urso M, Pisacane A, et al: Residual activity of α-galactosidase A in Fabry's disease. *Biochem Genet* 13:615, 1975
20. Wenger DA, Sattler M, Clark C, et al: Lactosyl ceramidosis: Normal activity for two lactosyl ceramide β-galactosidases. *Science* 188:1310, 1975
21. Rouser G, Kritchevsky G, Yamamoto A, et al: Accumulation of a glycerolphospholipid in classic Niemann-Pick Disease. *Lipids* 3:237, 1968
22. Stremmel W, Debuch H: Bis(monoacylglycerin) phosphorsäure-ein Marker-lipid sekundärer lysosomen? *Hoppe-Seyler's Z Physiol Chem* 357:803, 1976
23. Callahan JW, Khalil M, Philippart M: Sphingomyelinases in human tissues. II. Absence of a specific enzyme from liver and brain of Niemann-Pick Disease, Type C. *Pediatr Res* 9:908, 1975
24. Callahan JW, Khalil M: Sphingomyelinases in human tissues. III. Expression of Niemann-Pick Disease in cultured skin fibroblasts. *Pediatr Res* 9:914, 1975
25. Gal AE, Brady RO, Hibbert SR, et al: A practical chromogenic procedure for the detection of homozygotes and heterozygous carriers of Niemann-Pick Disease. *New Engl J Med* 293:632, 1975
26. Shapira E, Nadler L: The nature of the residual arylsulfatase activity in metachromatic leukodystrophy. *J Pediatr* 86:881, 1975
27. Rattazzi MC, Marks JS, Davidson RG: Electrophoresis of arylsulfatase from normal individuals and patients with Metachromatic Leukodystrophy. *Am J Hum Genet* 25:310, 1973
28. Dubois G, Turpin JC, Baumann N: Arylsulfatase isoenzymes in metachromatic leukodystrophy. Detection of a new variant by electrophoresis. *Biomedicine* 23:116, 1975
29. Rattazzi MC, Carmody PJ, Davidson RG: Studies on human lysosomal β-D-N-acetylhexosaminidase and arylsulfatase isozymes, in Markert CL (ed): Isozymes II. Physiological Function. New York, Academic, 1975, p. 439
30. Tanaka H, Suzuki K: Lactosylceramide β-galactosidase in human sphingolipidoses. *J Biol Chem* 250:2324, 1975
31. Dulaney JT, Milunsky A, Sidbury JB, et al: Diagnosis of lipogranulomatosis (Farber disease) by use of cultured fibroblasts. *J Pediatr* 89:59, 1976

32. Fishman PH, Max SR, Tallman JF, et al: Deficient ganglioside biosynthesis: A novel human sphingolipidosis. *Science* 187:68, 1975
33. Tanaka J, Garcia JH, Max SR, et al: Cerebral sponginess and Gm3 gangliosidosis: Ultrastructure and probable pathogenesis. *J Neuropathol Exp Neurol* 34:249, 1975
34. McKusick VA: Heritable Disorders of Connective Tissue, ed 4. St. Louis, Mosby, 1972, p. 521
35. Wappner R, Brandt IK: Hurler syndrome: α-L-iduronidase activity in leukocytes as a method for heterozygote detection. *Pediatr Res* 10:629, 1976
36. Klein U, Kresse H, Von Figura K: Evidence for degradation of heparan sulfate by endoglycosidases: Glucosamine and hexuronic acid are reducing terminals of intracellular heparan sulfate from human skin fibroblasts. *Biochem Biophys Res Comm* 69:158, 1976
37. Liebaers I, Neufeld EF: Iduronate sulfatase activity in serum, lymphocytes and fibroblasts—Simplified diagnosis of the Hunter syndrome. *Pediatr Res* 10:733, 1976
38. Von Figura K, Kresse H: Sanfilippo disease type B: Presence of material cross reacting with antibodies against α-N-acetyl-glucosaminidase. *Eur J Biochem* 61:581, 1976
39. Singh J, Donnelly PV, DiFerrante N, et al: Sanfilippo disease: Differentiation of types A and B by an analytical method. *J Lab Clin Med* 84:438, 1974
40. Couchot J, Pluot M, Schmauch M-A, et al: La mucosulfatidose. Etude de trois cas familiaux. *Arch Franc Ped* 31:775, 1974
41. Chern CJ, Croce CM: Assignment of the structural gene for human β-glucuronidase to chromosome 7 and tetrameric association of subunits in the enzyme molecule. *Am J Hum Genet* 28:232, 1976
42. Franke U: The human gene for β-glucuronidase is on chromosome 7. *Am J Hum Genet* 28:357, 1976
43. Turner BM, Turner VS, Beratis NG, et al: Polymorphism of human α-fucosidase. *Am J Hum Genet* 27:651, 1976
44. Desnick RJ, Sharp HL, Grabowsky GA, et al: Mannosidosis: Clinical, morphologic, immunologic and biochemical studies. *Pediatr Res* 10:985, 1976
44a. Spranger J: Mucolipidosis I. in Bergsma D (ed.): Disorders of connective tissue. *Birth Defects* 11:279, 1975
44b. Rapin I, Goldfischer S, Katzman R, et al: The Cherry-red Spot-Myoclonus Syndrome. *Ann Neurol* 3:234, 1978
45. Hickman S, Neufeld EF: A hypothesis for I-cell disease: defective hydrolases that do not enter lysosomes. *Biochem Biophys Res Comm* 49:992, 1972
45a. Vladutiu GD, Rattazzi MC: The excretion-reuptake route of β-hexosaminidase in normal and I-cell disease cultured fibroblasts. *Amer J Hum Genet* (in press) 1978
46. Neufeld EF, Sando GN, Garvin J, et al: The transport of lysosomal enzymes. *J Supramol Struc* 6:95, 1977
46a. Vladutiu GD, Rattazzi MC: I-cell disease—Desialylation of β-hexosaminidase and its effect on uptake by fibroblasts. *Biochim Biophys Acta* 539:31, 1978
46b. Berman ER, Livni, N, Shapira E, et al: Congenital corneal clouding with abnormal systemic storage bodies: a new variant of mucolipidosis. *J Pediatr* 84:519, 1974
46c. Bach G, Ziegler M, Kohn G, et al: Mucopolysaccharide accumulation in cultured skin fibroblasts derived from patients with Mucolipidosis IV. *Am J Hum Genet* 29:610, 1977
47. Cortner JA, Coates PM, Swoboda E, et al: Genetic variation of lysosomal acid lipase. *Pediatr Res* 10:927, 1976
48. Brown MS, Goldstein JL: Receptor-mediated control of cholesterol metabolism. *Science* 191:150, 1976
49. Brown MS, Sobhani MK, Brunschede GY, et al: Restoration of a regulatory response to low density lipoprotein in acid lipase-deficient human fibroblasts. *J Biol Chem* 251:3277, 1976
50. Van Bogaert H, Schever HJ, Froehlich A, et al: Une deuxième observation de cholestérinose tendinéuse symetrique aux symptomes cérébraux. *Ann Med* 42:69, 1937
51. Van Bogaert L, Scherer HJ, Epstein E: Une Forme Cérébrale de la Cholestérinose Généralisee. Paris, Mason, 1937

52. Menkes JH, Schimstock JR, Swanson PD: Cerebrotendinous xanthomatosis: The storage to cholestanol within the nervous system. *Arch Neurol* 19:47, 1968
53. Phillipart M, Van Bogaert L: Cholestanosis (cerebrotendinous xanthomatosis): A follow-up study on the original family. *Arch Neurol* 21:603, 1969
54. Nadler HL: Treatment of acid phosphatase deficiency disorders. *Birth Defects* 9:195, 1973
55. Haltia M, Palo J, Autio S: Aspartylglycosaminuria: A generalized storage disease. *Acta Neuropathol* 31:243, 1975
56. Zeman W, Donahue S, Dyken P, et al: The neuronal ceroid lipofuscinoses (Batten-Vogt syndrome), in Vinken PJ, Bruyn GW (eds): Handbook of Clinical Neurology, Vol 10. Amsterdam, North Holland, 1970, p. 588
57. Carpenter, S, Karpati G, Andermann F: Specific involvement of muscle, nerve and skin in late infantile and juvenile amaurotic idiocy. *Neurology* 22:120, 1972
58. Pilz H, O'Brien JS, Heipertz R: Human leukocyte peroxidase: Activity of a soluble and membrane-bound enzyme form in normal persons and patients with neuronal ceroid-lipofuscinosis. *Metabolism* 25:561, 1976
58a. Wolfe LS, Ng Ying Kin NMK, Barker RR, et al: Identification of retinoyl complexes as the autofluorescent component of the neuronal storage material in Batten disease. *Science* 195:1360, 1977
59. Desnick RJ, Thorpe SR, Fiddler, MB: Towards enzyme therapy for lysosomal storage diseases. *Physiol Rev* 56:57, 1976
60. Pentchev PG, Brady RO, Gal AE, et al: Replacement therapy for inherited enzyme deficiency. Sustained clearance of accumulated glucocerebroside in Gaucher's disease following infusion of purified glucocerebrosidase. *J Mol Med* 1:73, 1975
61. Hoffman LM, Amsterdam D, Schneck L: Gm_2 ganglioside in fetal Tay-Sachs disease brain cultures: A model system for the disease. *Brain Res* 111:109, 1976
62. Dean MF, Button LR, Boylston A, et al: Enzyme replacement therapy by fibroblast transplantation in a case of Hunter syndrome. *Nature* 261:323, 1976

Abnormalities in Carbohydrate Metabolism

1. Howell RR: The glycogen storage diseases, in Stanbury JB, Wyngaarden JB, Fredrickson DS (eds): Metabolic Basis of Inherited Disease, 4th ed. New York, McGraw-Hill, 1978 p. 149
2. Fine RN, Frasier SD, Donnell GN: Growth in glycogen storage disease type I. *Am J Dis Childh* 117:169, 1969
3. Bondy PK, Felig P: Disorders of carbohydrate metabolism, in Bondy PK, Rosenberg LE (eds): Duncan's Diseases of Metabolism, 7th ed. Philadelphia, Saunders, 1974. p. 315
4. Field RA: The glycogenoses. *Am J Clin Pathol* 50:20, 1968
5. Zangeneh F, Limbeck GA, Brown BI, et al: Hepatorenal glycogenosis (type I glycogenosis) and carcinoma of the liver. *J Pediatr* 74:73, 1969
6. Hug G: Glycogen storage diseases. *Birth Defects* (in press)
7. Salafsky IS, Nadler HL: Deficiency of acid alpha glycosidase in the urine of patients with Pompe's disease. *J Pediatr* 82:294, 1973
8. Askanas V, Engel WK, DiMauro S, et al: Adult-onset acid maltase deficiency. *New Engl J Med* 294:573, 1976
9. Howell RR, Kaback MM, Brown BI: Type IV glycogen storage disease: Branching enzyme deficiency in skin fibroblasts and possible heterozygote detection. *J Pediatr* 78:638, 1971
10. Grunfeld J-P, Ganeval D, Chanard, J, et al: Acute renal failure in McArdle's disease: Report of two cases. *New Engl J Med* 286:1237, 1972
11. Schimke RN, Zakheim RM, Corder RC, et al: Glycogen storage disease type IX: Benign glycogenosis of liver and hepatic phosphorylase kinase deficiency. *J Pediatr* 83:1031, 1973
12. Lederer B, Van Hoof, F Van den Berge G, et al: Glycogen phosphorylase and its convertor enzymes in hemolysates of normal human subjects and of patients with type VI glycogen-storage disease. *Biochem J* 147:23, 1975

13. Layzer RB, Rowland LP, Bank WJ: Physical and kinetic properties of human phosphofructokinase from skeletal muscle and erythrocytes. *J Biol Chem* 244:3823, 1969
14. Ranald J, Dykes W, Spencer-Peet J: Hepatic glycogen synthetase deficiency. *Arch Dis Childh* 47:558, 1972
15. Froesch ER: Essential fructosuria and hereditary fructose intolerance, in Stanbury JB, Wyngaarden JB, Fredrickson DS (eds): Metabolic Basis of Inherited Disease, 4th ed. New York, McGraw-Hill, 1978
16. Koster JR, Sleer RG, Fernandes J: On the biochemical basis of hereditary fructose intolerance. *Biochem Biophys Res Com* 64:289, 1975
17. Melancon SB, Khachaderian AK, Nadler HL, et al: Metabolic and biochemical studies in fructose 1,6-diphosphate deficiency. *J Pediatr* 82:650, 1973
18. Corbeel LM, Eggermont E, Bettens W, et al: Fructose intolerance with normal liver aldolase. *Helv Paediatr Acta* 25:626, 1970
19. Monteleone JA, Bentler E, Monteleone PL, et al: Cataracts, galactosuria and hypergalactosemia due to galactokinase deficiency in a child. *Am J Med* 50:403, 1971
20. Sidbury JB: Some inferences from galactokinase deficiency. *Pediatrics* 53:309, 1974
21. Cohn RM, Segal S: Galactose metabolism and its regulation. *Metabolism* 22:627, 1973
22. Fensom AH, Benson PF, Blunt S: Prenatal diagnosis of galactosemia. *Br Med J* 4:386, 1974
23. Hammersen G, Houghton S, Levy HL: Rennes-like variant of galactosemia: Clinical and biochemical studies. *J Pediatr* 87:50, 1975
24. Ginzelmann R, Steinmann B: Uridine diphosphate galactose 4-epimerase deficiency. II. Clinical follow-up, biochemical studies and family investigation. *Helv Paediatr Acta* 28:497, 1973
25. Hiatt HH: Pentosuria, in Stanbury JB, Wyngaarden JB, Frederickson DS (eds): Metabolic Basis of Inherited Disease, 4th ed. New York, McGraw-Hill, 1978
26. Hayakawa T, Kanzaki T, Kitamura T, et al: Mammalian α-keto acid dehydrogenase complex. *J Biol Chem* 244:3660, 1969
27. Stromme JH, Borud O, Moe PJ: Fatal lactic acidosis in a newborn attributable to a congenital defect of pyruvate dehydrogenase. *Pediatr Res* 10:60, 1976
28. Blass JP, Schulman JD, Young DS, et al: An inherited defect affecting the tricarboxylic acid cycle in a patient with congenital lactic acidosis. *J Clin Invest* 51:1845, 1972

Abnormalities in Amino Acid Metabolism

1. Koch R, Blaskovics M, Wenz E, et al: Phenylalaninemia and phenylketonuria, in Nyhan WL (ed): Heritable Disorders of Amino Acid Metabolism. New York, Wiley, 1974, p. 109
2. Perry TL, Hansen S, Tischler B, et al: Unrecognized adult phenylketonuria. *New Engl J Med* 285:395, 1973
3. Smith I, Clayton BE, Wolff OH: New variant of phenylketonuria. *Lancet* 1:1108, 1975
4. Kaufman S, Holtzman NA, Milstien S, et al: Phenylketonuria due to a deficiency of dihydropteridine reductase. *New Engl J Med* 293:785, 1975
5. Buist NRM, Kennaway NG, Fellman JA: Disorders of tyrosine metabolism, in Nyhan WL (ed): Heritable Disorders of Amino Acid Metabolism. New York, Wiley, 1974, p. 160
6. Weinberg AG, Mise CE, Worthen HG: The occurrence of hepatoma in the chronic form of hereditary tyrosinemia. *J Pediatr* 88:434, 1976
7. Gentz J, Jagenburg R, Zetterstrom R: Tyrosinemia. *J Pediatr* 66:670, 1965
8. Scriver CR, Larochelle J, Silverberg M: Hereditary tyrosinemia and tyrosyluria in a French-Canadian geographic isolate. *Am J Dis Childh* 113:41, 1967
9. Goldsmith LA, Kareg E, Bienfang DC, et al: Tyrosinemia with plantar and palmar keratosis and keratitis. *J Pediatr* 83:798, 1973
10. Medes G: A new error of tyrosine metabolism: Tyrosinosis. *Biochem J* 26:917, 1932
11. LaDu B: Alcaptonuria, in Stanbury JB, Wyngaarden JB, Fredrickson DS, (eds): Metabolic Basis of Inherited Disease, 4th ed. New York, McGraw-Hill, 1978

12. Hsia YE: Inherited hyperammonemia syndromes. *Gastroenterology* 67:347, 1974
13. Bachman C: Urea cycle, in Nyhan WL (ed): Heritable Disorders of Amino Acid Metabolism. New York, Wiley, 1974, p. 361
14. Gelehrter TD, Snodgrass PJ: Lethal neonatal deficiency of carbamyl phosphate synthetase. *New Engl J Med* 290:430, 1974
15. Ricciuti FC, Gelehrter TD, Rosenberg LE: X chromosome inactivation in human liver: Conformation of X-linkage of ornithine transcarbamylase. *Am J Hum Genet* 28:332, 1976
16. Campbell AGM, Rosenberg LE, Snodgrass PJ, et al: Ornithine transcarbamylase deficiency. *New Engl J Med* 288:1, 1973
17. Palmer T, Oberholzer VG, Burgess EA, et al: Hyperammonaemia in 20 families. *Arch Dis Child* 49:443, 1974
18. Shih VE, Schulman JD: Ornithine-keto-acid transaminase activity in human skin and amniotic fluid cell culture. *Clin Chem Acta* 27:73, 1970
19. Simell O, Takki K: Raised plasma ornithine and gyrate atrophy of the choroid and retina. *Lancet* 1:1031, 1973
20. Shih VE, Efron ML, Moser HW: Hyperornithinemia, hyperammonemia and homocitrullinemia: A new disorder of amino acid metabolism associated with myoclonic seizures and mental retardation. *Am J Dis Childh* 117:83, 1969
21. Fell V, Pollett RJ, Sampson GA, et al: Ornithinemia, hyperammonemia and homocitrullinuria. *Am J Dis Childh* 127:752, 1974
22. Glick NR, Snodgrass PJ, Schafer IA: Neonatal argininosuccinic aciduria with normal brain and kidney but absent liver argininosuccinate lyase activity. *Am J Hum Genet* 28:22, 1976
23. Terheggen HG, Lowenthal A, Lavinha F, et al: Familial hyperargininaemia. *Arch Dis Childh* 50:57, 1975
24. Schimke RN, McKusick VA, Huang T, et al: Homocystinuria: Studies of 20 families with 38 affected members. *JAMA* 93:711, 1965
25. Harker LA, Slichter SJ, Scott CR, et al: Homocystinemia. *New Engl J Med* 291:537, 1974
26. Gaull G, Sturman JA, Schaffner F: Homocystinuria due to cystathionine synthase deficiency: Enzymatic and ultrastructural studies. *J Pediatr* 84:381, 1974
27. Morrow G, III, Barner LA: Combined vitamin responsiveness in homocystinuria. *J Pediatr* 81:946, 1972
28. Gaull G, Tallaw HH: Methionine adenosyltransferase deficiency: New enzymatic defect associated with hypermethioninemia. *Science* 186:59, 1974
29. Finkelstein JD: Methionine metabolism in mammals: The biochemical basis for homocystinuria. *Metabolism* 23:387, 1974
30. Mudd SH: Homocystinuria and homocysteine metabolism: Selected aspects, in Nyhan WL (ed): Heritable Disorders of Amino Acid Metabolism. New York, Wiley, 1974, p. 429
30a. Kanwas YS, Manaligod JR, Wong PWK: Morphologic studies in a patient with homocystinuria due to 5,10-methylenetetrahydrofolate reductase deficiency. *Pediatr Res* 10:598, 1976
31. Freeman JM, Finkelstein JD, Mudd SH: Folate-responsive homocystinuria and "schizophrenia." *New Engl J Med* 292:491, 1975
32. Frimpter GW: Cystathioninuria, sulfite oxidase deficiency and beta-mercaptolactate-cysteine disulfiduria, in Stanbury JB, Wyngaarden JB, Fredrickson DS (eds): Metabolic Basis of Inherited Disease, 4th ed. New York, McGraw-Hill, 1978
33. Dancis J, Levitz M: Abnormalities of branched-chain amino acid metabolism, in Stanbury JB, Wyngaarden JB, Frederickson DS (eds): Metabolic Basis of Inherited Disease, 4th ed. New York, McGraw-Hill, 1978
34. Wada Y, Taka K, Minagawa A, et al: Idiopathic hypervalinemia: Probably a new entity of inborn error of value metabolism. *Tohaku J Exp Med* 81:46, 1963
35. Nyhan WL: New disorders of branched-chain amino acid metabolism in Heritable Disorders of Amino Acid Metabolism. New York, Wiley, 1974, p. 98
36. Roth K, Cohn, R, Yandrasitz J, et al: Beta-methylcrotonic aciduria associated with lactic acidosis. *J Pediatr* 88:229, 1976

37. Wadlington WB, Kilroy A, Ando T, et al: Hyperglycinemia and propionyl CoA carboxylase deficiency and episodic severe illness without consistent ketosis. *J Pediatr* 86:707, 1975

37a. Glasgow AM, Chase HP: Effect of propionic acid on fatty acid oxidation and ureagenesis. *Pediatr Res* 10:683, 1976

38. Ampola MG, Mahoney MJ, Nakamura E, et al: Prenatal therapy of a patient with vitamin B$_{12}$-responsive methylmalonic aciduria. *New Engl J Med* 293, 313, 1975

38a. Giorgio AJ, Trowbridge M Bonne AW, et al: Methylmalonic aciduria without vitamin B$_{12}$ deficiency in an adult sibship. *New Engl J Med* 295:310, 1976

39. Scriver CR, Efron ML: Disorders of proline and hydroxy-proline metabolism, in Stanbury JB, Wyngaarden JB, Fredrickson DS (eds): Metabolic Basis of Inherited Disease, 4th ed. New York, McGraw-Hill, 1978

40. Baumgartner R, Ando T, Nyhan WL: Non-ketotic hyperglycinemia. *J Pediatr* 75:1022, 1969

40a. Brandt NJ, Rasmussen K, Brandt S, et al: D-Glyceric-acidemia and non-ketotic hyperglycinaemia. *Acta Paediatr Scand* 65:17, 1976

41. Kaser H, Cottier P, Antener I: Glucoglycinuria, a new familial syndrome. *J Pediatr* 61:386, 1962

42. Gerritsen T: Sarcosine dehydrogenase deficiency, the enzyme defect in hypersarcosinemia. *Helv Paediatr Acta* 27:33, 1972

43. Dancis J, Hutzler J, Cox RP, et al: Familial hyperlysinemia with lysine-ketoglutarate reductase deficiency. *J Clin Invest* 48:1447, 1969

44. Sinell O, Visakorpi JE, Donner M: Saccharopinuria. *Arch Dis Childh* 47:52, 1972

45. Thomas GH, Haslam HA, Batshaw ML, et al: Hyperpipecolic acidemia associated with hepatomegaly, mental retardation, optic nerve dysplasia and progressive neurological disease. *Clin Genet* 8:376, 1975

46. Fisher MH, Gerritsen T, Optiz JM: α-Aminoadipic aciduria, a non-deleterious inborn metabolic defect. *Humangenetik* 24:265, 1974

47. Wilson RW, Wilson CM, Gates SC, et al: α-Ketoadipic aciduria: a description of a new metabolic error in lysine-tryptophane degradation. *Pediatr Res* 9:522, 1975

48. Goodman SI, Mackey SP, Moc PG, et al: Glutaric aciduria; a "new" disorder of amino acid metabolism. *Biochem Med* 12:12, 1975

49. Brown JH, Fabre LF, Jr, Farrell GL, et al: Hyperlysinuria with hyperammonemia. *Am J Dis Childh* 124:127, 1972

49a. Dancis J, Hutzler J, Woody NC, et al: Multiple enzyme defects in familial hyperlysinemia. *Pediatr Res* 10:686, 1976

50. Ghadimi H: Histidinemia: Emerging clinical picture, in Nyhan WL, (ed): Heritable Disorders of Amino Acid Metabolism. New York, Wiley, 1974, p. 265

51. Perry TL, Hansen S, Tischler B, et al: Carnosinemia. *New Engl J Med* 277:1219, 1967

52. Wellner VP, Sekura R, Meister A, et al: Glutathione synthetase deficiency, an inborn error of metabolism involving the α-glutamyl cycle in patients with 5-oxo-prolinuria (pyroglutamic aciduria). *Proc Nat Acad Sci* 71:2505, 1976

53. Scriver CR, Pneschel S, Davies E: Hyper β-alaninuria associated with β-aminoaciduria and α-aminobutyricaciduria, somnolence and seizures. *N Engl J Med* 274:635, 1966

54. Smith LH, Jr, Williams HE: Heritable Disorders of oxalate metabolism, in Nyhan WL, (ed): Heritable Disorders of Amino Acid Metabolism. New York, Wiley, 1974, p. 343

55. Wong PWK, Forman P, Tabahoff B, et al: A defect in tryptophane metabolism. *Pediatr Res* 10:725, 1976

Disorders of Porphyrin Metabolism

1. Tschudy DP, Valsalmis M, and Magnussen CR: Acute Intermittent porphyria: Clinical and selected research aspects. *Ann Int Med* 83:851, 1975
2. Waldenstrom J: Studien über Porphyrie, *Acta Med Scand* (Supp 82) 1937
3. Waldenstrom J: The porphyrias as inborn errors of metabolism, *Am J Med* 22:758, 1957
4. Watson CJ: Porphyria. *Adv Int Med* 6:235, 1954

5. Sachs P: Ein Fall von akuter Porphyrie mit hochgradiger Muskelatrophie. *Klin Wochenschr* 10:1123, 1931
6. Schmid R, Schwartz S, Watson CJ: Porphyria content of bone marrow and liver in the various forms of porphyria. *Arch Int Med* 93:167, 1954
7. Granick S, Urata G: Increase in activity of δ-aminolevulinic acid synthetase in liver mitochondria induced by feeding of 3,5-dicarbethoxy 1,4-dihydrocollidine. *J Biol Chem* 238:821, 1963
8. Tshudy DP, Perlroth MG, Marver HS, et al: Acute intermittent porphyria: The first "overproduction disease" localized to a specific enzyme. *Proc Nat Acad Science USA* 53:841, 1965
9. Lockwood WH, Bloomfield B: Uroporphyrins, III. Crystalline uroporphyrin from normal human urine. *Aust J Exp Biol Med Sci* 32:733, 1954
10. Fischer H: Ueber die Giftigkeit die sensibilisierende Wirkung, das spektroskopische Verhalten der natürlichen Porphyrine. *Z Physiol Chem* 97:109, 1916
11. Feldman DS, Levere RD, Lieberman JS, et al: Presynaptic neuromuscular inhibition by porphobilinogen and porphobilin. *Proc Nat Acad Sci USA* 68:388, 1971
11a. Romeo G: Enzymatic defects of hereditary porphyrias: An explanation of dominance at the molecular level. *Hum Genet* 39:261, 1977
12. Meyer UA, Schmid R: The porphyrias, in Stanbury JG, Wyngaarden JB, Fredrickson DS (eds): The Metabolic Basis of Inherited Disease, 4th ed. New York, McGraw-Hill, 1978
13. Poh-Fitzpatrick MB: Erythropoietic porphyrias: Current mechanisms, diagnostic, and therapeutic considerations. *Sem Hematol* 14:211, 1977
14. Romeo G, Levin EY: Uroporphyrinogen III cosynthetase in human congenital erythropoietic porphyria. *Proc Nat Acad Sci USA* 63:856, 1969
15. Romeo G, Kaback MM, Levin EY: Uroporphyrinogen III consynthetase in fibroblasts from patients with congenital erythopoietic porphyria. *Biochem Gent* 4:417, 1970
16. Romeo G, Glenn BL, Levin EY: Uroporphyrinogen III cosynthetase in asymptomatic carriers of congenital erythropoietic porphyria. *Biochem Genet* 4:659, 1970
17. Miyagi K, Petryka ZJ, Bossenmeir I, et al: The activities of uroporphyrinogen synthetase and cosynthetase in congenital erythropoietic porphyria (CEP). *Am J Hematol* 1:3, 1976
18. Cripps DJ, Peteers HA: Fluorescing erythrocytes and porphyrin screening tests on urine, stool and blood. *Arch Dermatol* 96:719, 1967
19. DeLeo VA, Poh-Fitzpatrick MB, Mathews RM, et al: Erythropoietic protoporphyria: Ten years experience. *Am J Med* 60:8, 1976
19a. Bottomley SS, Tanaka M, Everett MA: Diminished erythroid ferrochelatase activity in protoporphyria. *J Lab Clin Med* 86:126, 1975
20. Goldberg A, Rimington C: Disease of porphyrin metabolism. Springfield, Ill., CC Thomas, 1962
21. Strand LJ, Felsher BW, Redeker AG, et al: Enzymatic abnormalities in heme biosynthesis in intermittent acute porphyria: Decreased hepatic conversion of porphobilinogen to porphyrins and increased δ-aminolevulinic acid synthetase activity. *Proc Nat Acad Sci USA* 67:1315, 1970
22. Goldberg A: Porphyria. *Q J Med* 28:183, 1959
23. Rifkind AB: Drug-induced exacerbations of porphyria. *Primary Care* 3:665, 1976
24. Perlroth MG, Marver HS, Tschudy DP: Oral contraceptive agents and the management of acute intermittent porphyria. *JAMA* 194:1037, 1965
25. Stein JS, Tschudy DP: Acute intermittent porphyria: A clinical and biochemical study of 46 patients. *Medicine* 49:1, 1970
26. Meyer UA: Intermittent acute porphyria. Clinical and biochemical studies of disordered heme biosynthesis. *Enzyme* 16:334, 1973
27. Meyer UA, Strand LJ, Doss M, et al: Intermittent acute porphyria—demonstration of a genetic defect in porphobilinogen metabolism. *N Engl J Med* 286:1277, 1972
28. Strand LJ, Meyer UA, Felsher BF, et al: Decreased red cell uroporphyrinogen I synthetase activity in intermittent acute porphyria. *J Clin Invest* 51:2530, 1972
29. Sassa S, Granick S, Bickers DR, et al: A microassay for uroporphyrinogen I synthetase, one of three abnormal enzyme activities in acute intermittent porphyria, and its application to the study of the genetics of this disease. *Proc Nat Acad Sci USA* 71:732, 1974

30. Dean G, Barnes HD: The inheritance of porphyria. *Br Med J* 2:89, 1955
31. Dean G: Porphyria. *Br Med J* 2:1291, 1953
32. Dean G: The Porphyrias: A Study of Inheritance and Environment. London, Pitman, 1963
33. Eales L, Linder GC: Porphyria—the acute attack *South Afr Med J* 36:284, 1962
34. Brodie MJ, Thompson GG, Moore MR, et al: Hereditary coproporphyria *Q J Med* 46:229, 1977
35. Benedetto, AV, Kushner, JP, Taylor, JS: Porphyria cutanea tarda in three generations of a single family. *N Eng J Med* 298:358, 1978

Disorders of Purine Metabolism

1. Seegmiller JE: Inherited deficiency of hypoxanthine-guanine phosphoribosyltransferase in X-linked uric aciduria (the Lesch-Nyhan syndrome and its variants.) *Adv Hum Gen* 6:75, 1976
2. Lesch M, Nyhan WL: A familial disorder of uric acid metabolism and central nervous system function. *Am J Med* 36:561, 1964
3. Nyhan WL: Summary of clinical features, in Seminars on Lesch-Nyhan syndrome. *Fed Proc* 2:1034, 1968
4. Nyhan WL: Clinical features of the Lesch-Nyhan syndrome. *Arch Intern Med* 130:186, 1972
5. Nyhan WL: The Lesch-Nyhan syndrome. *Ann Rev Med* 24:41, 1973
6. Seegmiller JE, Rosenbloom FM, Kelley WN: Enzyme defect associated with a sex-linked human neurological disorder and excessive purine synthesis. *Science* 155:1682, 1967
7. Migeon BR: X-linked hypoxanthine-guanine phosphoribosyl transferase deficiency: Detection of the heterozygotes by selective medium. *Biochem Genet* 4:377, 1970
8. Francke U, Bakay B, Nyhan WL: Detection of heterozygotes for the Lesch-Nyhan syndrome by electrophoresis of hair root lysates, *J Pediatr* 82:472, 1973
9. Fujimoto WY, Seegmiller JE, Uhlendorf BW, et al: Biochemical diagnosis of an X-linked disease in utero. *Lancet* 2:511, 1968
10. Boyle JA, Raivio KO, Adtrin KH, et al: Lesch-Nyhan syndrome: Preventive control by prenatal diagnosis. *Science* 169:688, 1970
11. Kelley WN, Rosenbloom FM, Henderson JE, et al: A specific enzyme defect in gout associated with over-production of uric acid, *Proc Natl Acad Sci USA* 57:1735, 1967
12. Kelley WN, Greene ML, Rosenbloom FM, et al: Hypoxanthine-guanine phosphoribosyl-transferase deficiency in gout: A review. *Ann Intern Med* 70:155, 1969
13. Becker MA, Seegmiller JE: Recent advances in the identification of enzyme abnormalities underlying excessive purine synthesis in man. *Arthritis Rheum* 18:687, 1975
14. Becker MA, Yen RCK, Goss S, et al: Localization of the structural gene for human phosphoribosylpyrophosphate synthetase on the X-chromosome. *Clin Res* 26:114A, 1977
15. Seegmiller JE: Disorders of purine and pyrimidine metabolism, in Freinkel N (ed): The Year in Metabolism 1975–1976. New York, Plenum, p. 213, 1976
16. Seegmiller JE: Disorders of purine and pyrimidine metabolism, in Freinkel N (ed): The Year in Metabolism 1976–1977. New York, Plenum, 1978
17. Watts RWE, Watts JEM, Seegmiller JE: Xanthine oxidase activity in human tissues and its inhibition by allopurinol (4-hydroxy-pyrazolo (3,3,d) pyrimidine). *J Lab Clin Med* 66:688, 1965
18. Hsieh YF, Hsu TC: Xanthine calculus: A case report. *J Formosa Med Assoc* 62:83, 1963
19. Englemen K, Watts RWE, Klineberg JR, et al: Clinical, physiological and biochemical studies of a patient with xanthinuria and pheochromocytoma. *Am J Med* 37:839, 1964
20. Van Acker KJ, Simmonds HA, Cameron JS: Complete deficiency of adenine phosphoribosyl-transferase: Report of a family, in Muller MM, Kaiser E, Seegmiller JE (eds): Advances in Experimental Medical Biology, Purine Metabolism. Man II: Regulation Pathways of Enzyme Defects, 76A. New York, Plenum, p. 292 1977
21. Valentine WN, Paglia DE, Tartaglia AP, et al: Hereditary hemolytic anemia with increased red cell adenosine deaminase (45-to 70-fold) and decreased adenosine triphosphate. *Science* 195:783, 1977

6
Heritable Connective Tissue Disorders

William A. Horton

A large number of disorders affect predominantly the connective tissues, including skin, bone, cartilage, ligament, tendon, fascia, joint capsule, sclera and elements of the heart and blood vessels. Although defects in the constituents of connective tissue have been suspected for many years, it is only in the last decade that the biologic aspects of normal connective tissue have been sufficiently defined to permit proper investigation. Hence many abnormalities are now identified, and many others postulated. A discussion of this group of disorders therefore requires that certain biologic features of normal connective tissue be summarized here since the subject has been extensively reviewed (1-6).

NORMAL BIOLOGY OF CONNECTIVE TISSUE

Despite marked differences in structure and function, the various bodily connective or supportive tissues share the same basic composition: a matrix of two fibrous elements, collagen and elastic fibers, and so-called "ground substance" in which large amounts of proteoglycan (mucopolysaccharides) are found. This matrix is elaborated by fibroblasts, osteoblasts and chondroblasts.

Collagen, the most abundant protein in the body, is actually a heterogeneous group of fibrous proteins supplying tensile strength to connective tissues, and several genetically and chemically distinct types have been identified (Table 1). The molecule is composed of three alpha (α) chains wound around each other to form a triple helix. The chains are identical except for Type I collagen, which is composed of two types. Barring only the small distinctive segments at the terminal regions, the various α chains of all the collagens share several compositional characteristics, including: 1) glycine residues in every third position, 2) high content of proline and hydroxyproline, 3) small amounts of methionine and tryptophan and 4) no cysteine. This peculiar amino acid makeup is apparently essential to the triple helix configuration of the intact molecule (1, 2).

Collagen synthesis is complex, involving many posttranslational modifications to the basic molecule (Fig. 1). Compared to the final product, the molecule as initially synthesized is larger and contains nonhelical portions at each end; hence

Table 1. Types of Collagen

Type	Chain Composition	Distribution
I	$(\alpha[I])_2 \alpha 2$	Skin, bone, tendon fascia, ligament and dentin
II	$(\alpha 1[II])_3$	Cartilage
III	$(\alpha 1[III])_3$	Blood vessel, fetal skin, similar to, although proportionately less than, Type I except for bone, which has little Type III
IV	$(\alpha 1[IV])_3$	Basement membrane

the terminology pro-α chain and procollagen. Following synthesis of the pro-α chain, many of the lysine and proline residues are hydroxylated by the respective hydroxylases (Step 2, Fig. 1), both of which require cofactors, including α-ketoglutarate, ferrous iron, molecular oxygen and ascorbic acid. Subsequently glucose—and sometimes glucose and galactose—is added to some of the hydroxylysine residues by specific glycosyl transferases (Step 3, Fig. 1). Folding into the triple helix of procollagen then occurs, apparently facilitated by the nonhelical portions of the pro-α chain, the so-called registration peptides at each end, which align the individual chains (Step 4, Fig. 1). Procollagen is then secreted from the cell. In the process the registration peptides are cleaved from both ends of the molecule by the enzyme(s) procollagen peptidase(s) (Step 5, Fig. 1), leaving the helical collagen molecule (tropocollagen).

Once outside the cell, specific lysine and hydroxylysine residues are oxidatively deaminated by the membrane-bound, copper-requiring enzyme lysyl oxidase to form reactive peptidyl-bound aldehydes (Step 6, Fig. 1). Individual molecules then aggregate to form fibrils; the latter then aggregate into collagen fibers, which can be seen microscopically (Step 7, Fig. 1). These structures are stabilized by a series of cross-links, which appear to arise spontaneously from condensation of the reactive aldehydes with each other (intramolecular) and with the amino group of other unoxidized lysine or hydroxylysine residues (intermolecular) (Step 8, Fig. 1) (2-4). Although this scheme has been devised from studying synthesis of Type I collagen by fibroblasts, it presumably applies to synthesis of other types as well.

The biochemical features of elastic fibers are not nearly so well understood as that of collagen. They are found in most connective tissues, in particular blood vessel, skin, ligament, cartilage and lung, in which structures stretchability, or "elasticity," is required (1). It is not clear if there is a single type of elastic fiber or different types, each with a specific function as is true with collagen (5). Elastic fibers consist of two elements: glycoprotein microfibrils—serving as a scaffold during fiber formation—and the amorphous elastin molecule itself. Elastin has a unique amino-acid composition consisting largely of apolar amino acids, i.e., alanine, valine and little or no methionine or histidine.

In addition, repeating units of Lys-Ala-Ala-Lys and Lys-Ala-Ala-Ala-Lys often occur (6). The biosynthesis of elastic fibers is thought to involve 1) synthesis and secretion of both the microfibril protein and soluble elastin subunit (tropoelastin), 2) oxidative deamination by lysyl oxidase of certain lysine residues to form aldehydes (as with collagen) and 3) condensation of three lysine derivatives to

```
                    Specific
  Nuclear mb         mRNA
         ╲     ─ ─ ─ ─ ─ ─ ╱
                    │ 1
                    ▼
               Pro α–chain
                    │        ┌──────────────────────┐
                    │ 2      │ Proline Hydroxylase  │
                    │        └──────────────────────┘
                    ▼        ┌──────────────────────┐
                             │ Lysine Hydroxylase   │
                             └──────────────────────┘
                    │
                    │ 3      ┌──────────────────────┐
                    ▼        │ Glycosyl Transferases│
                             └──────────────────────┘
                    │ 4
                    ▼
               Procollagen
  Cell mb ══════              ══════
                    │ 5      ┌──────────────────────┐
                    ▼        │ Procollagen Peptidase│
                             └──────────────────────┘
             (Tropo)collagen
                    │ 6      ┌──────────────────────┐
                    ▼        │ Lysyl Oxidase        │
                             └──────────────────────┘
                    │ 7
                    ▼
              Collagen Fibril
                    │ 8
                    ▼
              Collagen Fiber
```

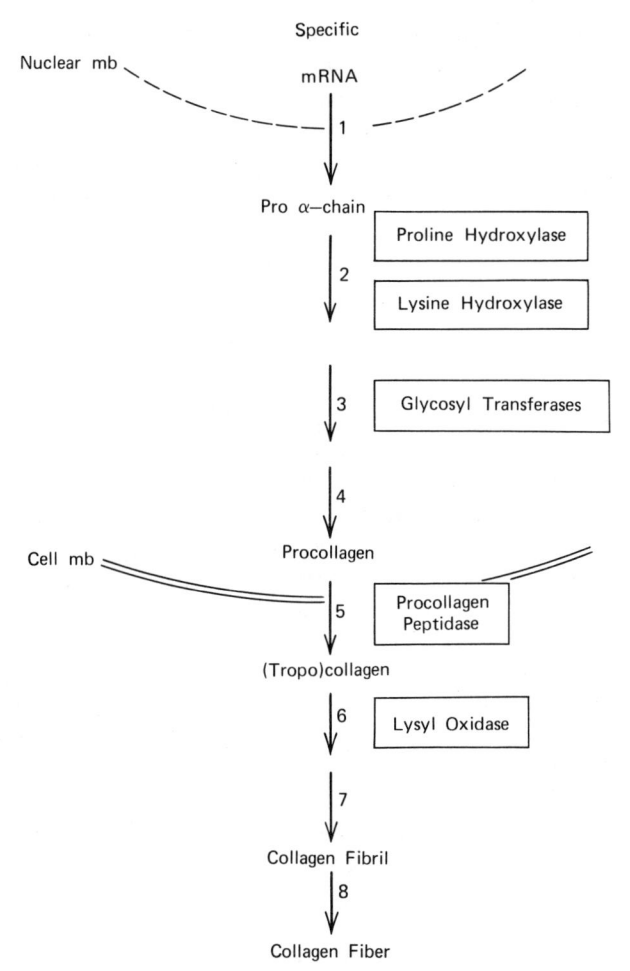

Figure 1. Synthesis of collagen. Key to numbers in the figure: 1) transcription of specific pro-α chain mRNA; 2) hydroxylation of prolyl and lysyl side chains; 3) addition of galactose or glucosylgalactose, or both to certain hydroxylysine residues; 4) triple helix assembly; 5) cleavage of nonhelical portions at both ends; 6) oxidative deamination of lysyl and hydroxylysyl side chains; 7) intra-and intermolecular cross-link formation as molecules aggregate to form fibrils; and 8) further aggregation and crosslinking to form fibers. The abbreviation "mb" in figure stands for "membrane."

form desmosine (or one of its isomers), a cyclic amino acid that can cross-link as many as four different tropoelastin chains (1). The actual mechanism of cross-linking has not been determined; however, the microfibrils seem to be involved in the orientation of the soluble elastin molecules during this process and the repeating sequences appear also to be important. The structure of elastic fibers has not been identified despite many models having been proposed (5, 6).

The nonfibrous element of connective tissue matrix, so-called ground substance consists of many substances including proteoglycan, water, certain extracellular ions and metabolites and diverse other substances whose contribution to connective-tissue integrity is poorly understood (1). Although their actual function is not known, these various molecules probably interact with the fibrous elements and form an integral part of the connective-tissue matrix. Diseases of proteoglycan metabolism (mucopolysaccharidoses) are discussed in detail in Chapter 5, "Genetic Metabolic Disorders."

PRIMARY DEFECTS

Marfan Syndrome

The Marfan syndrome is an autosomal dominant disorder characterized by involvement mainly of skeletal, periarticular, ocular and cardiovascular tissues. Tall stature with disproportionately long, thin limbs, especially distally (i.e., arachnodactyly), is the major skeletal feature; however, dolichocephaly, pectus excavatum, pigeon breast, asymmetry of the thoracic cage, high arched palate and prognathism also occur. The upper:lower segment ratio and metacarpal index are useful ways to document skeletal disproportion (7). "Weakness" of the periarticular tissues—such as joint capsules, ligaments, tendons and fascia—results in generalized joint hypermobility and kyphoscoliosis. The ocular involvement consists of myopia and retinal detachment, both of which defects result from an excessively long globe and ectopia lentis (dislocaion of the lens); the latter is due to redundancy of the suspensory ligaments of the lens. Most patients have some iridodenesis (shimmering of the iris when the head is brought to a sudden halt), although minimal ectopia lentis cannot be excluded except by thorough slit-lamp examination. Other ocular abnormalities include heterochromia irides, keratoconus, megalocornea and glaucoma secondary to ectopia lentis (7).

The greatest morbidity in the Marfan syndrome comes from cardiovascular involvement, which appears to reflect a weakness in the fibrous structure of the heart and great vessels (8). Those areas receiving the greatest hemodynamic pulsatile stress, i.e., aortic ring and ascending aorta, show the greatest damage. Usually aortic ring dilatation occurs first, producing aortic regurgitation, and is soon followed by progressive dilatation of the ascending aorta. Intimal tears may then occur, permitting aortic dissection and often rupture (8). The dissection may interupt bloodflow across the coronary ostia, producing coronary artery insufficiency (7).

Aortic dilatation and its complications lead to death in 80% of patients dying from the syndrome (8). Mitral regurgitation from redundancy of the chordae tendineae also occurs and may produce congestive heart failure; the latter which can often be detected earlier than can aortic regurgitation. Other reported cardiovascular findings include aneurysmal dilatation of the pulmonary artery, aortic coarctation and patent ductus arteriosus (9). Various electrocardiographic abnormalities have been reported, which likely are secondary to the aforementioned anatomic lesions. Bacterial endocarditis, described in several patients, primarily affected the regurgitant mitral valve (10).

Many other clinical manifestations are reported, although less frequently in the Marfan syndrome. They include various hernias, hemivertebrae and spontaneous pneumothorax. Cutaneous striae are common, and a specific lesion called Miescher's elastoma is occasionally observed (7).

The phenotypic expression of the Marfan syndrome varies widely. In fact the variability is sufficient to suggest that genetic heterogeneity exists within the entity. Based on clinical features alone, McKusick recently classified the disorder as either 1) a severe asthenic form characterized by profound skeletal involvement leading to death from cardiovascular lesions, usually in the teens; 2) a milder nonasthenic form in which patients have ectopic lentis and skeletal anomalies, but do not develop cardiovascular lesions until the 4th, 5th or 6th decades; and 3) the Marfanoid form in which joint hypermobility is the most striking feature (11).

Despite the many abnormalities observed in tissues from patients with this syndrome—e.g., degeneration of elastic fibers with an increase in collagenous material in the tunica media of the great vessels (7), increased solubility of skin collagen (12), increased metachromasia in fibroblasts (13) and so on, the basic defect remains unknown. Likewise, no biochemical marker has yet been identified. A primary defect in either collagen or elastin seems the likeliest cause (7).

There are three potential sites at which the natural history of this disorder may be interrupted by treatment. The hormonal induction of premature puberty reduces total height, which may be important, particularly to young girls, and also lessens the kyphoscoliosis, which often worsens markedly during the normal pubertal growth spurt (7). Reduction of left ventricular contractility, which diminishes the pulsatile hemodynamic stress on the aortic ring and ascending aorta, theoretically retards or prevents aortic complications. Although propranolol has been used for this purpose, it is not yet clear if the drug is beneficial (14). Since bacterial endocarditis occurs in the Marfan syndrome, it is recommended that patients with cardiovascular involvement be treated prophylactically with antibiotics prior to procedures like dental extraction (15).

Ehlers-Danlos Syndromes

The Ehlers-Danlos (ED) syndromes are a heterogeneous group of connective-tissue disorders sharing hyperextensibility of joints and skin and increased tissue friability. At least seven discrete entities are recognized on genetic, clinical and biochemical grounds (Table 2) (3, 4, 11, 16-18). The gravis type, *ED-I*, is characterized by generalized severe joint hypermobility. Musculoskeletal deformities, such as pes planus, may occur. Skin hyperextensibility and easy bruising are severe, and the increased fragility leads to skin splitting and subsequent "cigarette-paper" scarring, particularly over forehead, elbows, knees and shins. Varicose veins are common as are mulluscoid pseudotumors and subcutaneous spheroids. Generalized tissue friability may complicate postsurgical or posttraumatic wound healing, as well as premature rupture of fetal membranes.

The mitis type, *ED-II*, resembles, but is milder than, ED-I. Joint laxity is often limited to the hands and feet and cutaneous involvement is minimal. There is a slight tendency to bruising, but little scar formation. Varicose veins are uncommon and tissue friability is rare. Severe hypermobility of all joints, usually without musculoskeletal deformities, characterizes *ED-III*, the benign hypermobility type. Skin changes are minimal. The arterial, ecchymotic or Sachs type, *ED-IV*, is the most malignant due to the tendency to both spontaneous rupture of large

Table 2. Ehlers-Danlos Syndromes

Classification (ED)	Type	Inheritance*	Clinical Features	Basic Defect
I	Gravis	AD	Generalized severe joint hypermobility, skin hyperextensibility, easy bruisability, molluscoid pseudotumors, subcutaneous spheroids, poor wound healing and premature rupture of fetal membranes	Unknown
II	Mitis	AD	Similar to ED-I but milder, joint laxity limited to hands and feet, little cutaneous involvement and tissue friability	Unknown
III	Benign hypermobility type	AD	Severe hypermobility of all joints	Unknown
IV	Arterial, ecchymotic, Sachs	AD/AR	Spontaneous rupture of large arteries, perforation of bowel and thin skin with prominent underlying veins	Reduced synthesis of Type III collagen
V		XL	Marked hyperextensibility of skin, minimal joint hypermobility	Deficiency of lysyl oxidase
VI	Ocular	AR	Severe scoliosis, moderate joint involvement, ocular fragility with scleral rupture, retinal detachment, or both	Deficiency of lysyl hydroxylase
VII	Arthrochalosis multiplex congenita	AR	Short stature, generalized joint hypermobility with multiple subluxations and abnormal facies	Deficiency of pro-collagen peptidase

*AD, autosomal dominant; AR, autosomal recessive; XL, X-linked.

and intermediate arteries and bowel perforation. The skin is very thin and bruises easily; underlying veins are prominent, but stretchability is not a major feature.

The X-linked type, *ED-V*, manifests only minimal joint hypermobility in contrast to marked hyperextensibility of the skin. Cutaneous bruisability and fragility are moderately increased. The ocular type, *ED-VI*, is characterized by severe scoliosis and ocular fragility in addition to moderate joint and skin involvement. Corneoscleral rupture or retinal detachment occur after minor trauma. Short stature and generalized joint hypermobility are characteristic of *ED-VII*. Subluxations of hips, knees, elbows and feet are common, and affected infants are floppy. Skin stretchability and bruisability are moderately increased. Abnormal facies, including hypertelorism, epicanthic folds and scooped out midfacies may be part of the disorder.

Specific abnormalities in collagen biosynthesis are identified in ED-IV, V, VI and VII. In ED-IV, which in some families seems to be inherited as an autosomal dominant trait, but in an autosomal recessive fashion in others (19), there is a defect in Type III collagen synthesis; the skin has reduced amounts and fibroblasts fail to synthesize Type III collagen (20). In Types V, VI and VII, enzymatic deficiencies concerned with collagen cross-linking have been identified. There is reduced activity of lysyl oxidase in ED-V (21), lysyl hydroxylase in ED-VI (22) and procollagen peptidase in ED-VII (23). Although the basic defects have not been identified in the autosomal dominant types (ED-I, II and III), collagen defects seem likely. Thus the general Ehlers-Danlos phenotype, i.e., hyperextensible skin and joints, appears to be the clinical correlate of disturbed collagen integrity.

Treatment varies according to the type; in general, however, patients should avoid trauma and wear protective padding over bony prominences; particular care should be taken during surgical and obstetric procedures (19).

Cutis Laxa

In this disorder, characterized by excessive loose skin over the entire body, the skin is extensible, but in contrast to the Ehler-Danlos syndromes does not return to place upon release, nor is there increased bruisability or friability (24). Joint hypermobility is usually not a feature (25). Both autosomal dominant and recessive forms are reported. Hernias, pulmonary emphysema and diverticuli leading to early death occur in the recessive type, whereas in the dominant form involvement is limited to the skin (25-27). Recently an X-linked form was reported in which deficiency of lysyl oxidase, an enzyme important in cross-linking both collagen and elastin, was identified (28). The basic defect in the autosomal forms has not been determined. Histologic sections of the defective tissues, however, show sparsity and irregularity of elastin fibers (25).

Osteogenesis Imperfecta

Osteogenesis imperfecta (OI) is a generalized disorder of the skeletal, ocular, cutaneous, otologic, dental and vascular tissues. The most characteristic feature is increased susceptibility by the person to fractures, and the resulting deformities may include pseudoarthroses, saber shins and marked bowing of the legs. Short stature can result from these deformities as well as growth retardation from repeated epiphyseal fractures. The latter become less frequent following puberty. Radiologically the bones show severe generalized osteopenia with multiple fractures that heal with excessive callous formation. The vertebral bodies usually have the typical "codfish" appearance, and skull X-rays reveal multiple Wormian bones. The face is usually triangular owing to the broad, domed forehead, temporal bulge and overhanging occiput. Teeth are often yellow, brown or opalescent and break easily (dentogenesis imperfecta). The skin is thin and translucent. Joint laxity is increased and hernias are frequent. Blue sclerae are also common, but not universal. Beyond the 2nd decade, otosclerosis often leads to degrees of deafness. Aortic regurgitation is reported (29-31).

Clinical severity varies considerably. At one end of the spectrum is a severe neonatal form characterized by multiple intrauterine fractures of the limbs and ribs; usually neonatal death occurs from intracranial hemorrhage or respiratory embarrassment. Some patients, however, are so mildly affected that their tendency to fractures, blue sclerae or mild deafness is slight (31). Traditionally the disorder has been classified as OI congenita if it assumes the severe neonatal form, and as OI tarda if it occurs as the milder form of late onset. The striking intrafamilial variability in which both OI congenita and OI tarda are observed casts doubt on the soundness of this division (32).

The validity of other classifications proposed recently remains to be tested (33). In any event most cases appear to be inherited in an autosomal dominant fashion (29). Several reports, however, describe the severe neonatal form in which autosomal recessive inheritance is most likely. Radiologically the bones appear to be broader than those in typical OI congenita, supporting a distinct autosomal recessive form (11). In one such case a defect in the synthesis of Type I collagen was observed (34).

Therapy in osteogenesis imperfecta is limited primarily to orthopedic surgical management of skeletal deformities. Immobilization may aggravate osteopenia and should be eschewed (31).

Pseudoxanthoma Elasticum

Pseudoxanthoma elasticum (PXE), a generalized connective-tissue disorder, appears to reflect defective elastic tissue (11). Based on clinical and genetic data, Pope has tentatively classified PXE into two autosomal dominant and two autosomal recessive types. The dominant Type I is characterized by a subcutaneous yellow, raised rash over flexure sites, particularly the neck, axillae, groin and cubital area; severe chorioretinitis and complications of arterial degeneration (i.e., hypertension, angina pectoris and intermittent claudication). Dominant Type II is milder, showing only a macular rash, stretchable skin, retinal angioid streaks, myopia, high arched palate and blue sclerae. Recessive Type I has skin changes similar to those of the dominant Type I, plus angioid streaks and a predisposition to gastrointestinal hemorrhage. In the rare recessive Type II there are no eye or vascular changes, but rather generalized lax skin infiltrated with degenerative elastic fibers (35, 36). There is no definitive treatment.

SECONDARY DEFECTS

Homocystinuria

Homocystinuria from cystathionine synthetase deficiency* is an autosomal recessive disorder resembling the Marfan syndrome because of skeletal, ocular and cardiovascular involvement (37). (Clinical and biochemical features are discussed in Chapter 5, "Genetic Metabolic Disorders.") The connective-tissue abnormalities are thought to be due to homocysteine, which accumulates proximal to the enzyme block, interfering with normal collagen cross-linking (38, 39).

Alcaptonuria

Alcaptonuria is an autosomal recessive inborn metabolic error, which results from deficient activity of the enzyme homogentisic acid oxidase. The accumulation of this compound, an intermediate in phenylalanine and tyrosine metabolism, is responsible for the manifestations (40). (see also Chapter 5, "Genetic Metabolic Diseases"). Blue-black ochronotic pigment (ochronosis), which presumably is a polymer of homogentisic acid, is deposited in all connective tissues, including tendons, ligaments, endocardium, heart valves, intima of large blood vessels, sclerae and particularly cartilage. This is most evident clinically in the cornea, nasal cartilage and pinna, which may calcify (40, 41). Deposition within both articular cartilage and the annulus fibrosis and nucleus pulposis of intervertebral disks leads to progressive degenerative arthritis, predominantly of the spine, knees, hips and shoulders (42). These manifestations are thought to result from the binding of collagen by the polymerized homogentisic acid (43).

Menke's Kinky Hair Syndrome

Menke's kinky hair syndrome, a rare X-linked recessive disorder, is characterized, as the name states, by sparse, kinky hair that microscopically shows pili torti, progressive neurologic degeneration, metaphyseal irregularities of the long bones, hypothermia and generalized arterial occlusive disease resulting from degeneration of the internal elastic lamina. Defective intestinal absorption of copper is the basic defect (44). Many of the manifestations likely result from copper deficiency alone. The connective-tissue abnormalities probably reflect dysfunction of the copper-requiring enzyme, lysyl oxidase, which is involved in cross-linking both elastin and collagen (18).

OTHER DISORDERS

The Winchester Syndrome

The *Winchester syndrome* is an autosomal recessive disorder characterized by coarse facial features, dwarfism, joint contractures, corneal opacities, os-

---

*Homocystinuria per se occurs in three other genetic disorders, namely, 1) deficiency of $N^{5,10}$-methylene tetrahydrofolate reductase, 2) defect in vitamin B_{12} metabolism and 3) defect in vitamin B_{12} intestinal absorption. Connective tissue abnormalities, however, are not prominent in these disorders (11).

teoporosis and carpal tunnel osteolysis. The destructive joint changes resemble rheumatoid arthritis. Although the basic defect is not known, there is pathologic replacement of bone and cartilage by dense fibrous tissue, and structurally abnormal fibroblasts have been seen (45).

Fibrodysplasia Ossificans Progressiva

Clinical features of fibrodysplasia ossificans progressiva, an autosomal dominant disorder, include progressive ossification of fascia, tendons, ligaments and aponeuroses. The process usually begins in childhood and leads to severe disability. Microdactyly, particularly of the first digits, is frequent. The basic defect, unfortunately, remains unknown (46).

Weill-Marchesani Syndrome

The combination of short stature, brachydactyly, limited joint mobility, myopia and small spherical lenses, which often dislocate, characterize the *Weill-Marchesani syndrome*. It is inherited as an autosomal recessive trait although heterozygotes may have short stature (47).

Genetic Disorders with Secondary Arthropathy

Numerous other inherited disorders show secondary connective tissue manifestations in the form of arthropathy (Table 3) (42, 48), the responsible mechanisms for which vary widely and are discussed in other chapters.

Table 3. Genetic Disorders with Secondary Arthropathy

Disorder	*Inheritance**	*Arthropathy*
Hemochromatosis	AR	Degenerative arthritis, chondrocalcinosis
Hemophilia (and von Willebrand's disease)	XL, (AD)	Repeated hemarthroses with degenerative arthritis
Sickle cell anemia and related hemoglobinopathies†	AR	Repeated synovial infections, vascular necrosis of femoral heads
Gaucher's disease	AR	Avascular necrosis of femoral heads
Fabry's disease	XL	Avascular necrosis of femoral heads
Familial Mediterranean fever	AR	Recurrent inflammatory arthritis
Lesch-Nyhan syndrome and variants	XL	Recurrent inflammation, chronic destructive arthritis (tophaceous gouty arthritis)
Wilson's disease	AR	Chondrocalcinosis
Hyperparathyroidism‡	AD	Chondrocalcinosis
Hyperlipoproteinemia	Varied§	Xanthomatous infiltration in periarticular tissues, occasional inflammation
Chondrodystrophies with epiphyseal dysplasia	Varied§	Premature degenerative arthritis

*See footnote, Table 2.
†SS, SA, SC and sickle-thalassemia syndromes.
‡Includes the multiple endocrine adenoma syndromes.
§Includes disorders with different inheritance patterns.

JUVENILE OSTEOCHONDROSES

The juvenile osteochondroses are a group of disorders in which localized noninflammatory arthropathies result from regional disturbances of skeletal growth (Table 4), and there is ischemic necrosis of either primary or secondary endochondral ossification centers (49). Most of the abnormalities occur sporadically, but familial forms have also been described. Legg-Calvé-Perthes disease is osteonecrosis of the capital femoral epiphysis. Autosomal dominant inheritance is reported (50). Bilateral cases may be easily confused with the mild form of multiple epiphyseal dysplasia (Ribbing type, Chapter 18, "Heritable Skeletal Dysplasias"), which is also an autosomal dominant trait (51).

Osteochondritis dissecans—involving multiple sites, particularly the knees, hips, elbows and ankles—is also described as an autosomal dominant trait (42, 52). Blount's disease, a growth disturbance of the medial aspect of the proximal tibial growth plate, occurs in both infancy and adolescence. The former has been reported as an autosomal dominant, the latter as an autosomal recessive disorder (53, 54). Some overlap is found among the different osteochondroses, e.g., osteochondritis dissecans and Blount's disease within the same family (53). This suggests that the current classification may not be entirely adequate. The incidence of asymptomatic lesions is not known and familial occurrence in all these disorders may be commoner than is generally believed. Further, certain of the lesions may occur as a component of other recognized syndromes, such as Scheuerman's disease with multiple epiphyseal dysplasia (51).

GENETIC FACTORS AND "ACQUIRED" DISORDERS

Several common connective-tissue disorders, generally considered to be acquired, are marked by familial aggregation (Table 5). Although various modes of Mendelian and non-Mendelian inheritance are postulated, in general the role of genetic factors is poorly understood. Familial aggregation may arise by several mechanisms, both genetic and nongenetic. On a statistical basis alone, e.g., common disorders would be expected occasionally in more than one family member. Exposure to a common intrafamilial environmental agent may simulate genetic

Table 4. Juvenile Osteochondroses

Region Affected	Disease (Eponym)	Suspected Inheritance*
Capital femoral epiphysis	Legg-Calvé-Perthes	AD
Tibial tubercle	Osgood-Schlatter	—
Os calcis	Sever's	—
Tarsal of navicular bone	Kohler's	—
Head of second metatarsal	Freiberg's	—
Vertebral bodies	Scheuerman's	—
Medial aspect of proximal tibial epiphysis	Blount's, tibia vara	AD, AR
Subchondral areas of diarthroidal joints (particularly knee, hip, elbow and ankle)	Osteochondritis dissecans	AD
Capitellum of humerus	Panner's	—

*AD, autosomal dominant; AR, autosomal recessive; ——, not relevant, or unknown.

Table 5. "Acquired" Connective Tissue Disorders in which Familial Aggregation is Observed

Disorder	Reference
Ankylosing spondylitis	55–57
Lupus erythematosus*	58
Polymyalgia rheumatica	59
Psoriatic arthritis	60
Rheumatoid arthritis	61
Scleroderma (CRST)†	62, 63
Sjögren's syndrome	64

*Does not include patients with Clr deficiency of complement and lupus-like syndrome (65).
†Individuals with scleroderma and other with the CRST syndrome have occurred within same family (62).

disease, and in some cases genetic phenocopies of the acquired disorder may also exist. The inheritance of certain susceptibility factors, such as histocompatibility genes, may contribute to familial aggregation as well (57). Possibly all these factors as well as others yet to be determined act individually or in concert to produce familial aggregation.

The human leukocyte antigen (HLA) system has been implicated in the pathogenesis of many diseases, including several connective-tissue disorders. Briefly the system consists of a series of histocompatibility antigens that are coded for by genes at four closely linked, but distinct, loci designated as HLA-A,B,C and D, within a genetic region known as the Major Histocompatibility Complex (MHC), which is located on the short arm of chromosome 6 (66).

The antigens derived from the A, B and C loci are on all nucleated cells and are demonstrated by specific antibodies (67). The fourth, or D, locus codes for a gene product that provokes a cell-mediated (lymphocyte) response and is demonstrated by the mixed lymphocyte culture technique (66). All four loci are highly polymorphic, with numerous alleles at each locus, and the particular set of alleles found together on one chromosome is called a *haplotype*. The frequency of specific alleles differs among racial and ethnic groups. Moreover certain combinations of alleles (haplotypes), are commoner than would be expected from their individual frequencies. This phenomenon, termed linkage disequilibrium, may arise by several mechanisms, including 1) recent origin of an allele, such that equilibrium from recombination over many generations has not yet been reached; 2) inbreeding among persons with a particular haplotype; 3) migration of a subpopulation with a particular haplotype into the general population; and 4) natural selection in favor of specific haplotypes (68).

In addition to the four loci just listed, the MHC probably contains immune response (Ir) genes regulating the immune response to certain antigens. This locus (or loci) is thought to be close to the HLA-D locus, although the Ir genes, their products and their functional relationship to other HLA loci has not been defined in man (66, 69).

A relationship between the HLA system and disease has now been shown for several disorders, and in a few cases actual genetic linkage has been detected. Hemachromatosis e.g., appears to be linked to the HLA-A locus (70). In most cases, however, associations between a disorder and a particular allele have been observed and eight such associations are shown in Table 6. Several explanations

Table 6. Human Leukocyte Antigen (HLA) Disease Association*

Disease	HLA Allele	Relative Risk†
Ankylosing spondylitis	B27	87.8
Reiter's syndrome	B27	35.9
Yersinia arthritis	B27	24.3
Psoritic arthritis	B13	4.8
Salmonella arthritis	B27	17.6
Juvenile rheumatoid arthritis	B27	4.7
Rheumatoid arthritis	Dw4	3.0
	Cw3	2.7
Sjögren's syndrome	B8	3.2
	Dw3	5.2

*Adapted from (69).
†ratio of risk of developing disease in persons with particular allele divided by risk for persons lacking allele.

for this phenomenon are proposed. An association may reflect sampling of a subgroup in which a particular allele and disease occur more frequently than in the general population (67). Increased viral or toxic susceptibility may be conferred by a particular allele, i.e., the antigen serves as a receptor for, or resembles, the agent so that the virus or toxin goes undetected and escapes immune attack (69).

Alternatively linkage disequilibrium may exist between the HLA allele and another gene at a nearby locus, such as an Ir gene, that actually confers increased susceptibility. In this latter case the HLA allele would simply represent a marker unrelated to the pathogenesis (67). Thus, although HLA-disease associations exist, they may arise from diverse mechanisms and their true meaning is far from clear. The HLA alleles, however, are inherited in a dominant fashion and if related to pathogenesis, could certainly contribute to familial aggregation (57).

REFERENCES

1. McKusick VA: The biology of normal connective tissue, in Heritable Disorders of Connective Tissue. St. Louis, Mosby, 1972, p. 32
2. Nimmi ME: Collagen, its structure and function in normal and pathological connective tissues. *Semin Arthritis Rheum* 4:95, 1974
3. Kivirikko KI, Risteli L: Biosynthesis of collagen and its alterations in pathological states. *Med Biol* 54:159, 1976
4. Uitto J, Lichtenstein JR: Defects in the biochemistry of collagen in diseases of connective tissue. *J Invest Dermatol* 66:59, 1976
5. Sandberg LB: Elastin structure in health and disease. *Int Rev Connect Tissue Res* 7:159, 1976
6. Rucker RB, Tinker D: Structure and metabolism of arterial elastin. *Int Rev Exp Pathol* 17:1, 1977
7. McKusick VA: The Marfan syndrome, in Heritable Disorders of Connective Tissue. St. Louis, Mosby, 1972, p. 61
8. Murdoch JL, Walker BA, Halpern BL, Kuzma JW, et al: Life expectancy and causes of death in the Marfan syndrome. *N Engl J Med* 286:804, 1972

9. Hirst AE, Gore I: Marfan's syndrome, a review. *Prog Cardiovasc Dis* 16:187, 1973
10. Wunsch CM, Steinmetz EF, Fisch C: Marfan's syndrome and subacute bacterial endocarditis. *Am J Cardiol* 15:102, 1965
11. McKusick VA: The classification of heritable disorders of connective tissue. *Birth Defects* 11:1, 1975
12. Macek M, Hurych J, Chvapil M, Kadlecava, V: Study of fibroblasts in Marfan's syndrome. *Humangenetik* 3:87, 1966
13. Cartwright E, Danks DM, Jack I: Metachromatic fibroblasts in pseudoxanthoma elasticum and Marfan's syndrome. *Lancet* 1:583, 1969
14. Halpern BL, Char F, Murdoch JL, Horton WA, et al: A prospectus on the prevention of aortic rupture in the Marfan syndrome with data on surviorship without treatment. *Johns Hopkins Med J* 129:123, 1971
15. Dowling JN, Ho M: Endocarditis in the Marfan syndrome (ed. reply). *Ann Int Med* 82:432, 1974
16. McKusick VA: The Ehlers-Danlos syndrome, in Heritable Disorders of Connective Tissue. St. Louis, Mosby, 1972, p. 292
17. Beighton P, Price A, Lord J, Dickson E: Variants of the Ehlers-Danlos syndrome, clinical, biochemical, hematological and chromosomal features of 100 patients. *Ann Rheum Dis* 28:223, 1969
18. McKusick VA, Martin GR, Lichtenstein JR, Penttinen RPK, et al: Acquired and heritable defects in collagen synthesis and fibrogeneses. *Trans Am Clin Climatol Assoc* 85:130, 1973
19. Horton WA: The Ehlers-Danlos syndrome, in Birth Defects Compendium. National Foundation–March of Dimes (in press)
20. Pope FM, Martin GR, Lichtenstein JR, Penttinen R, et al: Patients with Ehlers-Danlos syndrome type IV lack type III collagen. *Proc Natl Acad Sci* 72:1314, 1975
21. DiFerrante N, Leachman RD, Angelini P, Donnelly PV, et al: Lysyl oxidase deficiency in Ehlers-Danlos syndrome type V. *Conn Tissue Res* 3:49, 1975
22. Krane SM, Pinnell SR, Erbe RW: Lysyl-protocollagen hydroxylase deficiency in fibroblasts from siblings with hydroxylysine-deficient collagen. *Proc Natl Acad Sci* 69:2899, 1972
23. Lichtenstein J, Kohn L, Byers P, Martin GR, et al: Procollagen peptidase deficiency in a form of the Ehlers-Danlos syndrome. *Trans Am Assoc Phys* 86:333, 1973
24. McKusick VA: Cutis laxa, in Heritable Disorders of Connective Tissue. St. Louis, Mosby, 1972, p. 372
25. Goltz RW, Hult AM, Goldfarb M, Gorlin RJ: Cutis laxa, a manifestation of generalized elastolysis. *Arch Dermatol* 92:373, 1965
26. Beighton PH: The dominant and recessive forms of cutis laxa. *J Med Genet* 9:216, 1972
27. Schreiber MM, Tilley JC: Cutis laxa. *Arch Dermatol* 84:266, 1961
28. Byers PH, Narayanon AS, Bornstein P, Hall JG: An X-linked form of cutis laxa due to deficiency of lysyl oxidase. *Birth Defects* 12:293, 1976
29. McKusick VA: Osteogenesis imperfecta, in Heritable Disorders of Connective Tissue. St. Louis, Mosby, 1972, p. 390-454
30. Bauze BJ: A new look at osteogenesis imperfecta, a clinical radiological and biochemical study of 42 patients. *J Bone Joint Surg* 57B: 1975
31. Horton WA, Uitto J, Lichtenstein JR: Osteogenesis imperfecta, in Birth Defects Compendium. National Foundation–March of Dimes (in press)
32. Rosenbaum S: Osteogenesis imperfecta and osteopsathyrosis, a contribution of their identity and their pathogenesis. *J. Pediatr* 25:161, 1944
33. Francis MJO, Bauze RJ, Smith R: Osteogenesis imperfecta, a new classification. *Birth Defects* 11:99, 1975
34. Penttinen RP, Lichtenstein JR, Martin GR, McKusick VA: Abnormal collagen metabolism in cultured cells in osteogenesis imperfecta. *Proc Natl Acad Sci* 72:586, 1975

35. Pope FM: Two types of autosomal recessive pseudoxanthoma elasticum. *Arch Dermatol* 110:209, 1974
36. Pope FM: Autosomal dominant pseudoxanthoma elasticum. *J Med Genet* 11:152, 1974
37. McKusick VA: Homocystinuria, in Heritable Disorders of Connective Tissue. St. Louis, Mosby, 1972, p. 224,
38. Kang AH, Trelstad RL: A collagen defect in homocystinuria. *J Clin Invest* 52:2571, 1973
39. Siegel RC: The connective tissue defect in homocystinuria (HS). *Clin Res* 23:263A, 1975
40. La Du BN: Alcaptonuria, in Stanbury JB, Wyngaarden JB, Fredrickson DS (eds): The Metabolic Basis of Inherited Disease. New York, McGraw-Hill, 1972, p. 308
41. McKusick VA: Alkaptonuria, in Heritable Disorders of Connective Tissue. St. Louis, Mosby, 1972, p. 455
42. Rodnan BP, McEwen C, Wallace SL (eds): Primer on the rheumatic diseases. *JAMA* 224:1, 1973
43. Mitch RA: Studies of alcaptonuria, binding of homogentisic acid solutions to hide powder collagen. *Proc Soc Exp Biol Med* 106:68, 1961
44. Danks DM, Campbell PE, Walker-Smith J, Stevens BJ, et al: Menkes' kinky hair syndrome. *Lancet* 1:1100, 1972
45. Hollister DW, Rimoin DL, Lachman RS, Cohen AH, et al: The Winchester syndrome, a non-lysosomal connective tissue disease. *J Pediatr* 84:701, 1974
46. McKusick VA: Other heritable and generalized disorders of connective tissue, in Heritable Disorders of Connective Tissue. St. Louis, Mosby, 1972, p. 687
47. McKusick VA: The Weill-Marchesani syndrome, in Heritable Disorders of Connective T sue. St. Louis, Mosby, 1972, p. 282
48. Bennett JC (ed): Twenty second rheumatism review, review of American and English literature for the years 1973 and 1974. *Arthritis Rheum* 19:973, 1976
49. Pappas AM: The osteochondroses. *Pediatr Clin N Am* 14:549, 1967
50. Wamoscher Z, Farhi A: Hereditary Legg-Calvé-Perthes disease. *Am J Dis Childh* 106:97, 1963
51. Rimoin DL: The chondrodystrophies. *Adv Hum Genet* 5:1, 1975
52. Stougaard J: The hereditary factor in osteochondritis dissecans. *J Bone Joint Surg* 43B:256, 1961
53. Tobin WJ: Familial osteochondritis dissecans with associated tibia vara. *J Bone Joint Surg* 39A:1091, 1957
54. Sevastikaglou JA, Erikson I: Familial infantile osteochondrosis deformans tibial idiopathic tibia vara. *Acta Orthop* Scand 38:81, 1967
55. Karten I, DiTata D, McEwen C, Tamer M: A familial study of rheumatoid (ankylosing) spondylitis. *Arthritis Rheum* 5:131, 1962
56. Emery AEH, Lawrence JS: Genetics of ankylosing spondylitis. *J Med Genet* 4:239, 1967
57. Brewerton DA, Hart FD, Nicholls A, Caffrey M, et al: Ankylosing spondylitis and HLA-27. *Lancet* 1:904, 1973.
58. Arnett FC, Shulman LE: Studies in familial lupus erythematosus. *Medicine* 55:313, 1976
59. Liang GG, Simkin PA, Hunder GG, Wilske KR, et al: Familial agregation of polymyalgia rheumatica and giant cell arteritis. *Arthritis Rheum* 17:19, 1974
60. Moll JMH, Wright V: Familial occurrence of psoriatic arthritis. *Ann Rheum Dis* 32:181, 1973
61. Lawrence JS, Ball J: Genetic studies on rheumatoid arthritis. *Ann Rheum Dis* 17:160, 1958
62. Burge KM, Perry HO, Stickler GB: "Familial" scleroderma. *Arch Dermatol* 99:681, 1969
63. Greger RE: Familial progressive systemic scleroderma. *Arch Dermatol* 111:81, 1975
64. Lichenfeld JL, Kirshner RH, Wiernik PH: Familial Sjogren's syndrome with associated primary salivary gland lymphoma. *Am J Med* 60:286, 1976
65. Moncada B, Day NKB, Good RA, Windhorst DB: Lupus-erythematosus-like syndrome with a familial defect of complement. *N Engl J Med* 286:689, 1972

66. Bach FH, van Rood JJ: The major histocompatibility complex-genetics and biology. Parts I, II and III. *N Engl J Med* 295:806, 872, 927, 1976
67. Rosenberg LE, Kidd KK (eds): HLA and disease susceptibility, a primer (ed). *N Engl J Med* 297:1060, 1977
68. Schaller JG, Omenn GS: The histocompatibility system and human disease. *J Pediatr* 88:913, 1976
69. McMichael A, McDevitt H: The association between the HLA system and disease. *Prog Med Genet* 2:39, 1977
70. Simon M, Bourel M, Genetet B, Fanchet R: Idiopathic hemachromatosis, demonstration of recessive transmission and early detection by family HLA typing. *N Engl J Med* 297:1017, 1977

7
Pharmacogenetics

Elliot S. Vesell

Epidemiologic studies reveal that about one in five patients enters a hospital in the United States for treatment of an adverse drug reaction. Further, 15 to 30% of all hospitalized patients have at least one such reaction (1). Although the wide disparity in patients' responses to drugs is only one of the many causes of adverse reactions, it nevertheless constitutes a significant contribution to this major medical problem because it demands individualization of dosage. Multiple factors have been systematically investigated and identified as contributing to wide interindividual variations in drug disposition and response. They include age, sex, time of day or season of drug administration, painful stimuli, disease, hormonal and nutritional status and exposure to inducers or inhibitors of the hepatic microsomal drug-metabolizing enzymes, including chronic administration of any one of several hundred drugs (2-4). Also in the last 20 years multiple genetic factors altering drug disposition and response in man have been discovered (5, 6).

SIMPLE INHERITED CONDITIONS AND DRUG RESPONSE

Pharmacogenetics as a specialty deals with clinically significant hereditary variations in response to drugs. These entities include traditionally recognized, hereditary conditions producing a clinically significant abnormal response to drugs (Table 1). They include disorders causing defects in the metabolism of drugs by the body and disorders in which the drug produces an apparent effect on the body because of altered receptor sites on which it acts (5).

Abnormal Breakdown of Drugs by Body

In 1946, the Japanese otorhinolaryngologist Takahara discovered acatalasia in an 11-year-old Japanese girl. In a series of classic studies he demonstrated that the defect was transmitted as an autosomal recessive trait (7-9). His original patient lacked catalase activity in her oral mucosa and erythrocytes, as did three of her five siblings. The patient's parents were second cousins.

Acatalasia
Mild, moderate and severe expressions of acatalasia have been described (10). The mild form is characterized by ulcers of the dental alveoli; in the moderate

Table 1. Twelve Pharmacogenetics Conditions

Condition	Aberrant Enzyme and Location	Mode of Inheritance*	Agent Provoking Response
Genetic conditions probably transmitted as single factors altering the way the body acts on drugs:			
Acatalasia	Catalase in erythrocytes	AR	Hydrogen peroxide
Slow inactivation of isoniazid	Isoniazid acetylase in liver	AR	Isoniazid, sulfamethazine, sulfamaprine, procainamide, phenelzine, dapsone and hydralazine
Suxamethonium sensitivity or atypical pseudocholinesterase	Pseudocholinesterase in plasma	AR	Suxamethonium or succinylcholine
Diphenylhydantoin toxicity due to deficient parahydroxylation	? Mixed function oxidase in liver microsomes that parahydroxylates diphenylhydantoin	AD or XLD	Diphenylhydantoin
Bishydroxycoumarin sensitivity	? Mixed function oxidase in liver microsomes that hydroxylates bishydroxycoumarin	Unknown	Bishydroxycoumarin
Acetophenetidin-induced methemoglobinemia	? Mixed function oxidase in liver microsomes that deethylates acetophenetidin	AR	Acetophenetidin
Genetic conditions probably transmitted as single factors altering the way drugs act on body:			
Warfarin resistance	? Altered receptor or enzyme in liver with increased affinity for vitamin K	AD	Warfarin
Glucose-6-phosphate dehydrogenase deficiency, favism or drug-induced hemolytic anemia	Glucose-6-phosphate dehydrogenase	XL incomplete codominant	Various analgesics [acetanilide, acetylsalicylic acid, acetophenetidin (phenacetin); antipyrine, aminopyrine (Pyramidon)], sulfonamides and sulfones [sulfanilamide, sulfapyridine, N₂-acetylsulfanilamide, sulfacetamide sulfisoxazole (Gantrisin), thiazolsulfone

			salicylazosulfapyridine (Azulfadine), sulfoxone, sulfamethoxypyridazine (Kynex)], antimalarials [primaquine, pamaquine, pentaquine, quinacrine (Atabrine)], nonsulfonamide antibacterial agents [furazolidone, nitrofurantoin (Furadantin), chloramphenicol, p-aminosalicylic acid], and miscellaneous drugs [naphthalene, vitamin K, probenecid, trinitrotoluene, methylene blue, dimercaprol, (BAL), phenylhydrazine, quinine and quinidine]
Drug-sensitive hemoglobins:			
1) Hemoglobin Zürich	Arginine substitution for histidine at 63rd position of β-chain of hemoglobin	AD	Sulfonamides
2) Hemoglobin H	Hemoglobin composed of 4 β-chains	AR	Same drugs as listed for G6PD deficiency
Inability to taste phenylthiourea or phenylthiocarbamide	U†	AR	Drugs containing N-C⁻S group such as phenylthiourea methyl and propylthiouracil
Glaucoma due to abnormal response of intraocular pressure to steroids	U†	AR	Corticosteroids
Malignant hyperthermia with muscular rigidity	U†	AD	Such anesthetics as halothane, succinylcholine, methoxyfluorene, ether and cyclopropane

*AR, autosomal recessive; AD, autosomal dominant; XLD, X-linked dominant; XL, X-linked.

†U, unknown.

type alveolar gangrene and atrophy occur; and in the severe form recession of alveolar bone develops with exposure of the necks—and eventual loss—of teeth. The enzyme is deficient in tissues such as mucous membrane, skin, liver, muscle and bone marrow. Trace levels of catalase activity occur in some patients, and the term "severe hypocatalasia" seems more appropriate than does acatalasia (11). Heterozygotes who usually have values of catalase activity between those of affected and normal persons would be classified as having "intermediate hypocatalasia." In certain Japanese kindreds, some heterozygotes do not exhibit intermediate levels of catalase activity, but rather have values that overlap the normal range, suggesting heterogeneity (12).

In 1959, Yata reported a Korean patient with acatalasia, the first non-Japanese subject to be described (13). Two years later Aebi and associates found three affected individuals by screening 73,661 blood samples from Swiss Army recruits (14). All three were healthy and showed none of the dental defects typical of the Japanese cases. The Swiss "acatalasics," unlike the Japanese ones, exhibited residual catalase activity, possibly protecting them against the hydrogen peroxide formed by certain microorganisms thought to be responsible for the oral lesions. The catalase from Swiss patients also differed from that of normal persons in electrophoretic differences, pH and heat stabilities and sensitivity to certain inhibitors. These facts suggest that in Swiss families acatalasia is a structural gene mutation (15). Other variants likely have a similar derivation, although more complex regulatory mutations cannot be excluded (Table 2).

Slow Inactivation of Isoniazid

Although isoniazid (INH) was synthesized in 1921, its bacteriostatic effect was not discovered until 1952. Soon great differences were reported in the metabolism of INH in man, and each patient maintained an unchanged pattern of excretion during long-term therapy (16, 17). Slow inactivators show reduced

Table 2. Acatalasia and Related Anomalies

Type	Origin	Residual catalase activity percentage (normal = 100)		Remarks
I	Japan	Hom:	0-3.2	Incomplete recessive inheritance; oral gangrene (Takahara's disease) in ~50% of homozygotes; activity: trimodal distribution curve (no overlap)
	Korea	Het:	37-56	
II	Japan	Hom:	3.2	Complete recessive inheritance
		Het:	~100	
IIIa	Japan	Hom:	0 (?)	Overlap between heterozygous carrier and normal subjects
		Het:	>56	
IIIb	Switzerland	Hom:	0.1-1.3	Synthesis of two different types of catalase in heterozygotes (normal catalase + unstable variant), all homozygotes healthy
		Het:	15-85	
IV	Israel	Hom:	8	Combination with deficiency of G-6-PD; intolerance to fungicide
		Het:	49-67	
V	United States	All:	~100	Allocatalasia: synthesis of a variant catalase; activity and stability as normal catalase

Modified from (15).

activity of acetyl transferase, the liver supernatant enzyme responsible for the metabolism of INH and of sulfamethazine (18), as well as other monosubstituted hydrazines, such as phenelzine and hydralazine (19). Toxic effects of these drugs occur chiefly in slow acetylators. Acetylation of procainamide is polymorphic, and thus the effect of this popular antiarrhythmic agent varies appreciably according to the genotype of the patient to whom it is administered.

The sedative nitrazepam also shows a similar variation in response. Acetylation of other drugs, such as para-aminosalicylic acid and sulfanilamide, is accomplished by a different acetylase. Interestingly neither the slow nor the rapid acetylase genotype is more liable to resistance to tubercle bacilli or reversion, although INH-induced polyneuritis occurs more frequently in slow than in rapid inactivators and is the primary clinical problem related to this polymorphism. The half-life of INH ranges from 45 to 80 min in the plasma of rapid inactivators, whereas the half-life extends from 140 to 200 min in slow inactivators (21). Although slow acetylators may excrete unchanged 30% of a dose, rapid acetylators may excrete unchanged only 3% (22).

Slow inactivation of INH is inherited as an autosomal recessive trait (23). The best evidence suggests that the different phenotypes result from a structural gene mutation. Diverse geographic and racial genetic distributions are reported. Most uncommon in Eskimos, slow inactivation is only slightly more frequent in Far Eastern populations (24). Slow inactivation is also common in blacks and European populations, in nearly 80% of whom the individuals possess the aberrant gene either in the homozygous or heterozygous state (24-27).

Succinylcholine Sensitivity
Shortly after the muscle relaxant succinylcholine was introduced in 1952 and its use became widespread, patients occasionally were found to be extraordinarily sensitive to it; indeed, several deaths associated with its use were reported (28). Normally action of the drug is short (2-3 minutes), and this brevity in normal subjects is due to the exceedingly rapid hydrolysis of succinylcholine by plasma pseudocholinesterase, which catalyzes the sequential removal of choline radicals. Serum pseudocholinesterase activity was reduced in the initially published reports of prolonged apnea. The difficulty can be reversed by transfusion of either normal plasma or a highly purified preparation of human enzyme. The abnormality is the result of a structurally altered enzyme with kinetic properties decidedly different from those of the usual enzyme (29). The abnormal enzyme, e.g., exerts no measurable effect on succinylcholine at concentrations of the drug usually present during anesthesia, whereas the normal enzyme shows marked hydrolytic activity (30).

The atypical enzyme is more resistant than the normal one to many pseudocholinesterase inhibitors, i.e., both fluoride and organophosphorus compounds inhibited the normal and atypical enzyme differentially (31, 32). Dibucaine, also a differential inhibitor of normal and atypical pseudocholinesterases, can distinguish three phenotypes: homozygous normals, heterozygotes and affected individuals who could not be satisfactorily separated simply by measuring the pseudocholinesterase activity of their plasma (29). The percentage inhibition of pseudocholinesterase activity produced by $10^{-5}M$ dibucaine was designated the "dibucaine number," or "DN." Whereas atypical pseudocholinesterase is inhibited only 20%, the normal enzyme is inhibited about 80% and heterozygotes exhibit 50–70% inhibition. Tetracaine, unlike other previously studied compounds, is hydrolyzed faster by atypical than by normal pseudocholinesterase, and an even larger separation of phenotypes apparently can be achieved with the procaine-tetracaine ratio than with the DN. The discovery of additional genetic variants resulted from using sodium fluoride as an inhibitor (33).

In some familial members the DNs do not follow the typical pattern of inheritance. These persons are thought to be heterozygous for a rare, so-called silent gene. Heterozygotes for this gene exhibit two-thirds of the normal serum cholinesterase activity; they widely overlap normal values. A few rare individuals are presumably homozygous for the silent allele, reflecting complete absence of serum and liver pseudocholinesterase activity (34). Apparently normal otherwise, these persons lack all four of the usual isozymes of serum pseudocholinesterase; the absence of antigenically cross-reacting material was revealed by immunodiffusion and immunoelectrophoretic studies.

This silent mutation may affect the controlling element of the gene, thereby completely disrupting protein production. Alternatively a single structural mutation may affect both the active site and the antigenic determinants. Another "silent" allele is described in which there is some (about 2%) residual enzymatic activity, indicating further heterogeneity (35).

Family studies suggest that inheritance of various types of atypical pseudocholinesterase occurs through allelic codominant genes at a single locus (36, 37). Symptoms may occur after treatment with succinylcholine in persons homozygous for any of the variant alleles and in some mixed heterozygotes (11). At least four alleles have been definitely identified with the 10 resultant genotypes: $E_1{}^u E_1{}^u$, $E_1{}^u E_1{}^a$, $E_1{}^a E_1{}^a$, $E_1{}^s E_1{}^u$, $E_1{}^s E_1{}^s$, $E_1{}^s E_1{}^a$, $E_1{}^f E_1{}^u$, $E_1{}^f E_1{}^f$, $E_1{}^f E_1{}^a$ and $E_1\ E_1{}^s$, where E_1 signifies the pseudocholinesterase genetic locus and u, a, s and f indicate the "usual," "atypical," "silent" and "fluoride"-sensitive alleles, respectively. A new allele ($E_1{}^j$) has just been described, which apparently causes reduction of the usual ($E_1{}^u$) molecules by about 60% (38).

The incidence of atypical pseudocholinesterase remains comparatively constant in different geographic areas. The homozygous recessive persons for the atypical allele number about 1 in 2,500 (39). However, Gutsche et al discovered an exceptionally high incidence of the silent mutation in a population of southern Eskimos (40). Prior to this survey in Alaska, only 10 individuals homozygous for the silent gene had been described. The gene frequency of 0.12 in this locality, extending from Hooper Bay to Unalakleet and centered on the lower Yukon River, suggested that 1.5% of this Alaskan population was sensitive to succinylcholine. The isolation and consequent inbreeding of these natives, the authors contended, resulted in the high frequency of the rare silent gene in this region of Alaska, although only two of the 11 Eskimo families are known to be related.

Similarity of gene frequencies of atypical pseudocholinesterase in most populations suggests that either little selective advantage is conferred by the various genotypes or the contributing environmental factors are common to widely differing countries. In several abnormalities—such as thyrotoxicosis, schizophrenia, hypertension, acute emotional disorders and after concussion—plasma pseudocholinesterase activities may be elevated. Increases are also observed as a genetically transmitted condition without apparent clinical consequences, but associated with an electrophoretically slower migrating C_4 isozyme (the Cynthiana variant) (41). The person with this variant had plasma pseudocholinesterase activity more than three times higher than normal. Further investigation of his family revealed a sister and daughter, also with high values. The exceptionally high pseudocholinesterase activity was associated with resistance to the pharmacologic effects of succinylcholine (42).

The Cynthiana variant may result from either a defect of a regulator gene controlling pseudocholinesterase activity or a duplication of a structural gene. Slightly higher than normal pseudocholinesterase activity associated with a retarded electrophoretic mobility of the main isozyme was found in roughly 10% of a random sample of the British population (43). This slower-moving band was designated C_5. The greatly elevated total plasma pseudocholinesterase activity of the U.S. variants distinguished them from the variants described in England.

Deficient Parahydroxylation of Diphenylhydantoin

Since its introduction, diphenylhydantoin (DPH) has become one of the most popular anticonvulsants. However, it can cause multiple toxic reactions, including nystagmus, ataxia, dysarthria and drowsiness, reactions that are clearly dose related. The drug is metabolized in man mainly by parahydroxylation of one of the phenyl groups to yield 5-phenyl-5'-parahydroxyphenylhydantoin (PPHP), which is conjugated with glucuronic acid and then eliminated in the urine (44). Many lipid-soluble drugs, such as DPH, are rendered more water soluble, and hence more excretable, through metabolism by oxidative enzyme systems in liver microsomes. The earliest published example of a genetic defect of mixed function oxidases in human beings is deficient hydroxylation of DPH (45), although only one affected family is described.

A study of two generations of this family revealed two affected and three unaffected members, suggesting that low activity of DPH hydroxylase exhibits dominant transmission. Toxic symptoms developed in the propositus on a commonly used dosage of 4.0 mg/kg, but not on a dose of 1.4 mg/kg. Abnormally low urine levels of the metabolite PPHP occurred in combination with prolonged high blood levels of unchanged DPH. Apparently drugs like phenobarbital and phenylalanine are parahydroxylated by enzymes different from those hydroxylating DPH, since the proband's capacity to parahydroxylate these compounds was normal.

Recently slow inactivation of INH has been identified as a more important cause of DPH intoxication than heritable deficiency of parahydroxylase activity (46). In 29 individuals receiving DPH and INH, all five patients who developed symptoms of DPH toxicity were slow INH inactivators. Both INH and para-aminosalicylic acid interfered with DPH parahydroxylation in rat liver microsomes.

Bishydroxycoumarin Sensitivity

Solomon reported bishydroxycoumarin sensitivity in a patient receiving the drug for an acute myocardial infarction (47). On a dose of 150 mg the patient's plasma bishydroxycoumarin half-life was 82 hr compared to normal values of 27 ± 5 hr.

The patient's mother suffered a spinal cord hematoma, causing permanent paraplegia, while she was receiving a small weekly dose of 2.5–5 mg warfarin. Although familial studies were not performed because of lack of cooperation, this unfortunate event in the treatment of the patient's mother suggests the possibility of hereditary transmission of bishydroxycoumarin sensitivity.

Warfarin and bishydroxycoumarin are extensively hydroxylated in the rat (48). Genetic factors influence responsiveness to anticoagulants in rabbits, as they do in rats, in which resistance to warfarin as a rodenticide is transmitted as an autosomal dominant trait (49). The metabolites in man are not fully characterized, but the patient with bishydroxycoumarin sensitivity just described, and his mother, may have a metabolic defect involving deficiency of a hepatic microsomal hydroxylase.

Increased sensitivity to coumarin anticoagulants also can result from acquired conditions, including vitamin K deficiency, increased turnover of plasma proteins and numerous forms of liver disease that impair the subject's capacity to produce vitamin K-dependent clotting factors. Various drugs can increase the prothrombinopenic response to coumarin anticoagulants. Cinchophen may damage liver cells; phenothiazine may produce cholestasis, thereby diminishing absorption of vitamin K; phenylbutazone increases sensitivity by displacing warfarin from plasma albumin; and phenyramidol inhibits the hepatic microsomal enzymes responsible for metabolism of coumarin drugs. Resistance to warfarin derivatives also has been reported.

Acetophenetidin-Induced Methemoglobinemia
Shahidi reported severe methemoglobinemia and hemolysis in a 17-year-old girl after she had taken phenacetin (acetophenetidin) (50). Multiple studies excluded heritable erythrocytic disorders, including hemoglobinopathies, and extracorpuscular compounds seemed to be causing hemolysis. As much as one-half the patient's hemoglobin was occasionally in the form of methemoglobin. After administration of phenacetin, large amounts of 2-hydroxyphenetidin and 2-hydroxyphenacetin derivatives were discovered in her urine. In normal persons more than 70% of a dose of 2 g phenacetin appears in the urine as N-acetyl-para-aminophenol with only minute amounts of the hydroxylated products, which were so prevalent in the patient's urine. One sister, a brother and both parents of the patient had a normal response to phenacetin, but another sister likewise responded abnormally.

These observations suggest an autosomal recessive inheritance of a defect in which the patient's hepatic microsomal mixed function oxidases were deficient in deethylating capacity. Instead of being deethylated as in normal persons, phenacetin, in the patient and her 38-year-old sister, was hydroxylated.

The toxicity observed after phenacetin administration was probably produced by these hydroxylated products, since induction of the hepatic microsomal phenacetin hydroxylating enzymes prior to administration of phenacetin by phenobarbital exacerbated the condition, i.e., severe neurologic symptoms, including bilateral positive Babinski responses, and profound methemoglobinemia developed. After the same pretreatment a normal volunteer developed neither methemoglobinemia nor neurologic changes.

Genetic Conditions and Drugs

Warfarin Resistance

Genetically controlled resistance to warfarin was found in a patient, age 71 years, receiving anticoagulants for a myocardial infarction (51). Physical and laboratory examination showed no abnormalities other than a reproducible reduction in his one-state prothrombin concentration to about 60% of normal. Anticoagulants were initially withheld because of the patient's low prothrombin time. They were administered after one month, at which time he proved to be resistant, rather than sensitive, to dicoumarol. A daily dose of 145 mg was required to reduce the prothrombin concentration to therapeutic levels, i.e., nearly 50 standard deviations above the mean.

Detailed studies showed that the drug was absorbed normally from the gastrointestinal tract; kinetic and binding studies were also normal. Even after administration of very high doses, warfarin was not excreted unchanged in the urine or stools, and amounts of a metabolite of warfarin similar to those recovered from the urine of normal subjects who were given equivalent amounts of the drug were recovered from the patient's urine. The patient also showed resistance to bishydroxycoumarin and the indanedione anticoagulant, phenindione, but not to heparin.

An enzyme or receptor site with altered affinity for vitamin K or for anticoagulant drugs was postulated by O'Reilly et al as the mechanism responsible for resistance to warfarin in this patient (52).

Five other members of both sexes of the patient's family over three generations were also resistant to warfarin, suggesting autosomal dominant transmission of the trait. A second large kindred of 18 patients with warfarin resistance in two generations is also reported (53).

Various environmental conditions lead to resistance to coumarin anticoagulants as phenocopies of the genetic defect. Most commonly the resistance is related to the simultaneous administration of inducing agents that reduce the blood concentration of anticoagulant drugs by stimulating their metabolism, e.g., barbiturates, glutethimide, chloral hydrate and griseofulvin.

Glucose 6-Phosphate Dehydrogenase (G6PD)

Deficiency of G-6-PD—formerly called primaquine sensitivity, or favism—is the commonest hereditary enzymatic abnormality in man and is transmitted as an X-linked recessive disorder. More than 80 physicochemically discrete molecular variants are described, each being associated with slightly different clinical features (54). Ordinarily only the male hemizygote shows significant drug-related hemolysis. (Table 1). Female subjects may be affected mildly, as would be predicted from the Lyon hypothesis, or more severely, as in populations wherein the gene frequency is high enough that homozygosity is appreciable. A mild self-limited anemia, e.g., is associated with the common variant of G-6-PD found in blacks, in which the drugs listed in Table 1 can be given repeatedly without danger, since only the susceptible, older RBCs are removed from the circulation by hemolysis; they are rapidly replaced by resistant younger cells.

In various Mediterranean G-6-PD variants, hemolysis affects a larger proportion of the total erythrocytic population and occurs more rapidly after administration of smaller doses of drugs. Table 3 relates the severity of hemolysis to the amount of erythrocytic G-6-PD activity. Several properties in addition to symptomatic severity can characterize the variants, including the total erythrocytic G-6-PD activity enzymatic electrophoretic mobility and various kinetic measurements. The specific amino acid substitution in G-6-PD A$^+$ and in G-6-PD Hektoen has been elucidated by microfingerprinting techniques (55, 56).

The exact biochemical mechanisms by which a given drug or its metabolites cause hemolysis remains unknown. The metabolism of the erythrocyte is unusual in that it must function without benefit of a nucleus. It still needs sources of

Table 3. Severity of Hemolysis in Erythrocytic Glucose-6-Phosphate Dehydrogenase (G-6-PD) Deficiencies*

Variant	Population Origin	Frequency
Variants with no or very mild G-6-PD deficiency:		
Inhambane	African Bantu	Rare
Steilacoom	Negro	——†
A+	Negro	Common
Levadia	Greek	——
Lourenzo Marques	African Bantu	Rare
King County	Negro	Rare
Thessaly	Greek	——
Karditsa	Greek	——
Western	Greek	——
Manjacaze	African Bantu	Rare
Baltimore-Austin	Negro	Rare
Ijebu-Ode	Negro	Rare
Minas Gerais	Brazilian	Rare
Tacoma	Negro	——
Madrona	Negro	Rare
Ibadan-Austin	Negro	Rare
Ita-Bale	Negro	Rare
Variants with mild-to-moderate G-6-PD deficiency:		
Barbieri	Italian	Rare
Puerto Rico	Puerto Rican	Rare
A−	Negro	Common
Constantine	Arab	Common
Taipei-Hakka	Chinese	——
Kabyle	Algerian	——
Chibuto	Negro-Bantu	Rare
Melissa	Greek	——
Canton	South Chinese	——
Columbus	Negro	Rare
Athens	Greek	Common
Washington	Negro	——
Benevento	Italian	——
West Bengal	Asiatic Indian	Rare
Mexico	Mexican	——
Seattle	Welsh-Scottish	Rare
Kerala	Asiatic S.E. Indian	Rare
Tel Hashomer	Tunisian Jew	Rare

energy to maintain concentration gradients of sodium and potassium and for continual reduction of methemoglobin. This energy source is supplied by the glycolytic and oxidative pathways of glucose metabolism. Certain enzymes, including G-6-PD, lose activity as the normal cell ages, and G-6-PD activity declines with cell age faster than normal in G-6-PD–deficient cells. Therefore older cells of persons with mutations of their G-6-PD are more susceptible to hemolysis than are younger cells.

Reactions catalyzed by G-6-PD lead to the production of NADPH, a substance necessary for maintenance of sulfhydryl substances, such as glutathione, in the reduced (GSH) state. Sufficient quantities of reduced glutathione appear to be

Table 3. (Continued)

Variant	Origin (Population)	Frequency
Variants with mild-to-moderate G-6-PD deficiency:		
Capetown	Cape Colored/Norwegian	—
Variants with severe G-6-PD deficiency:		
Hualien-Chi	Taiwan	—
San Juan	Puerto Rican	Rare
Markham	New Guinea	Common
Union	Filipino	Common
Teheran	Iran	—
Hualien	Taiwan	—
Indonesia	Indonesia	—
Camplellpur	Pakistani	Common
Mediterranean	Greek, Sardinian, Sephardic Jew, Asiatic	Common
Corinth	Greeks, S.E. Asian	May be common
Panay	Filipino	May be common
Orchomenos	Greek	Common
Lifta	Iraqui Jew	Rare
Carswell	Irish	Rare
Variants with severe G-6-PD deficiency and chronic nonspherocytic hemolytic anemia:		
Ohio	Italian	Rare
Torrance	U.S.	—
Bat-Yam	Iraqui Jew	Rare
Albuquerque	U.S. White	Rare
Bangkok	Thai	—
Oklahoma	West Europe	Rare
Duarte	U.S. White	Rare
Hong Kong	Chinese	—
Chicago	West Europe	Rare
Tripler	U.S. White	—
Alhambra	Finnish/Swedish	—
Milwaukee	Puerto Rican White	Rare
Ramat-Gan	Iraqui Jew	Rare
Ashdod	North African Jew	Rare
Freiburg	German	Rare
Worcester	U.S. White	—

*Modified from (54) with permission.

†No data available.

essential for erythrocytic-membrane integrity. The sequence of events suggested in drug-induced hemolysis related to G-6-PD deficiency (57, 58) includes, first, the metabolism of the drug to a product more amenable to further oxidation. The erythrocyte converts this metabolite to an oxidant intermediate. The latter then damages the erythrocyte membrane (particularly in old cells), perhaps by oxidation of smaller sulfhydryl groups. The younger cells with their higher G-6-PD activities resist the osmotic and oxidant effects of various drugs and their metabolites, whereas the older, more sensitive cells with their greater relative deficiency are eliminated.

Determining in vitro the hemolytic potential of new drugs continues to be of prime importance, especially in geographic areas where the incidence of the disorder is high. Numerous tests have been devised, but none is suitable, in availability and cost, for routine screening of large populations.

Hemolysis apparently may occur spontaneously or during infection in certain G-6-PD variants. Obviously, enough stress can be placed on the metabolism of G-6-PD–deficient erythrocytes to cause hemolysis by several environmental alterations in addition to those produced by drug administration.

Drug-Sensitive Hemoglobins
A life-threatening hemolytic anemia developed in a 2-year-old girl and her father after they received sulfa drugs (59). Both subjects registered an abnormal hemoglobin content, electrophoretic mobility being between that of hemoglobins A and S (60). Further studies showed an abnormality in the beta chain, with arginine taking the place of the usual histidine residue at the 63rd position, where the heme group is attached (61). Fifteen of the 65 relatives examined showed the abnormal hemoglobin feature, designated hemoglobin Zürich, a defect transmitted as an autosomal dominant trait. In another family, discovered in Maryland, with the same substitution the severity of the hemolytic episodes was less than in the Swiss cases (62).

Another drug-sensitive hemoglobin, hemoglobin H, is a special form of α-thalassemia. Composed of four beta chains, hemoglobin H is sensitive to the oxidant drugs described under G-6-PD deficiency. In certain regions, such as Thailand, the frequency of homozygous hemoglobin H is high, i.e., one in 300 individuals.

Phenylthiocarbamide Tasting Ability
The ability to taste phenylthiocarbamide (PTC) is transmitted as an autosomal dominant trait, and tasters may be either heterozygous or homozygous (63). This polymorphism was discovered in 1932 when Fox, who synthesized the compound, noted that he could not detect a bitter taste from dust of the compound arising as it was poured into a container, whereas a colleague in the same room complained of the bitter taste (64).

Although this polymorphism seems to be benign, some workers have related the ability to taste PTC to thyroid disease. Administration of PTC can, e.g., produce goiter in the rat. Compounds related to PTC by possessing the $N-C=S$ group, such as the antithyroid drugs methyl- and propylthiouracil, also have the same bimodality in taste perception exhibited by subjects to PTC. Harris et al found that 41% of 134 patients with nodular goiter were nontasters (65), an observation confirmed by Kitchin et al in 447 patients who underwent thyroidectomy for various reasons (66). In male patients with multiple thyroid adenomas, a marked increase in nontasting frequency also was noted. Nontasters seem to be more susceptible to athyreotic cretinism and also to adenomatous goiter. These data suggest to some investigators that nontasters may be more susceptible than tasters to environmental goitrogens. The physiocochemical basis for the difference in taste perception in affected individuals is unknown.

The frequency of tasting capacity shows geographic variation in that 31.5% of Europeans, 10.6% of Chinese and 2.7% of Africans are nontasters (67, 68). As with the physicochemical findings the reasons for these variations in the gene for PTC tasting are equally obscure.

Intraocular Pressure, Steroids and Glaucoma

A polymorphism exists in the response of ocular pressure of normal subjects to steroids applied topically. Elevations in intraocular pressure in 80 normal persons after local administration of 0.1% ophthalmic solution of dexamethasone 21-phosphate for four weeks exhibited a trimodal distribution. Familial studies confirmed the existence of three genotypes, namely, P_LP_L for low elevations of 5 mm Hg or less, P_LP_H for intermediate increases from 6 to 15 mm Hg and P_HP_H for high increment in pressure of 16 or more mm Hg.

In 1968, Armaly cited an association between certain types of response and glaucoma (69). In a sample of patients with both open-angle hypertensive, and low-tension, glaucomas, the distribution of responses differed from that in the random sample of normal subjects (Table 4). A marked reduction in P_LP_L genotypes and a corresponding increase in P_LP_H and P_HP_H genotypes occurred in both conditions and surprisingly in the uninvolved eye of patients with unilateral post-traumatic glaucoma. Familial studies indicated that the response of high elevations of intraocular pressure after administration of dexamethasone was inherited as an autosomal recessive trait. Although glaucoma can occur with

Table 4. Genotype Classification of Dexamethasone Hypertension*

Category	Subjects Tested	Low (P_LP_L) ΔP < 6 mm Hg	Intermediate (P_LP_H) ΔP 6–15 mm Hg	High (P_HP_H) ΔP > 15 mm Hg
Limits of pressure rise (mm Hg)		5 or less	6–15	16 or more
Mean pressure rise (mm Hg)		1.96	10.0	19.5
Standard deviation (mm Hg)		±2.00	±2.5	**
Genotype		P_LP_L	P_LP_H	P_HP_H
Random sample	80	66%	29%	5%
Open-angle hypertensive glaucoma	33	6%	48%	44%
Low-tension glaucoma	15	7%	53%	40%
Normal eye in recessed-angle glaucoma	15	——†	53%	47%
Normal eye in angle recession without glaucoma	4	75%	25%	——†

*From (69) with permission.

**Range in sample 18–22 mm Hg.

†No data available.

genotypes other than P$_H$P$_H$ and P$_H$P$_L$, Armaly concluded that the P$_H$ gene is closely associated with these types of glaucoma.

Malignant Hyperthermia and Muscular Rigidity

Denborough et al reported that hyperthermia led to death in 10 of 38 family members who had received anesthesia for various surgical procedures (70). This was the first indication that the rare, hitherto seemingly sporadic, malignant hyperthermia afflicting persons exposed to various anesthetic agents might be genetically transmitted. Almost 200 cases of malignant hyperthermia have been identified and shown to have a hereditary basis (71). The condition is associated with muscular rigidity and appears to be transmitted as an autosomal dominant trait. It develops during anesthesia with nitrous oxide, methoxyflurane, halothane, ether, cyclopropane or combinations thereof and is commoner in association with the use of succinylcholine as a preanesthetic agent. During anesthesia body temperature rises rapidly, occasionally reaching 112°F!

The incidence of malignant hyperthermia is in the range of 1 in 20,000 cases of general anesthesia and exhibits no sex preference, but occurs more in younger than in older anesthetized patients. Approximately two-thirds of the patients die usually from cardiac arrest. The degree of rigidity is variable, differing from patient to patient and sometimes being absent. This variability may indicate that the term malignant hyperthermia refers to several discrete diseases.

Occasionally rigidity is so marked that the body literally becomes as stiff as a board, progressing without interruption into rigor mortis. Intravenous administration of procaine or procainamide is reported to alleviate the rigidity and fever in certain cases. Curare is ineffective. Interestingly a limb under tourniquet does not become rigid, suggesting a peripheral rather than a central lesion. Animal models have been produced in dogs treated with halothane and dinitrophenol and in Landrace pigs.

GENETIC FACTORS AND DRUG REACTIONS

Ethanol Metabolism

Atypical alcohol dehydrogenase (ADH), a variant of the enzyme metabolizing ethanol, has been described in man (72). The enzyme occurs in sufficiently high frequencies in Swiss and English populations to be designated a polymorphism. Exceptionally active, the variant occurred in 20% of 59 liver specimens from a Swiss population and in 4% of 50 livers from an English population.

After IV infusion of ethanol, attempts were made to correlate rates of degradation of the drug with liver ADH types from biopsies obtained during surgical procedures (73). Of 23 subjects, two had atypical ADH. Interestingly the capacity to metabolize alcohol was no different in the male subject with atypical ADH than in male subjects with the typical enzyme, whereas the capacity to degrade ethanol was greater in the female subject with atypical ADH than in a small group of female subjects with typical ADH. Liver ADH-specific activity and isoenzyme pattern in biopsy specimens for seven American indians and six whites revealed no racial differences.

The question whether individuals with atypical ADH have increased capacity to degrade ethanol, and possibly to resist alcoholic cirrhosis of the liver, remains unresolved. The subject of ethanol toxicity is difficult to attack experimentally, because of the considerable variation in ability of ethnic groups to metabolize ethanol. Natives of Far Eastern countries have less ethanol-metabolizing capacity than have those of Western countries. Possibly, too, major genetic differences are obscured by elevated rates of ethanol metabolism since induction thereof occurs with chronic administration. There also appear to be racial differences in ethanol sensitivity since Japanese, Taiwanese and Koreans exhibit marked facial flushing and mild-to-moderate symptoms of intoxication after drinking amounts of ethanol that produce no detectable effect on whites (74). These differences in ethanol responsiveness, present since birth, have been attributed to variations in autonomic reactivity.

Blood Groups

To determine possible correlations between adverse reactions to drugs and genetic factors, a survey was made of many hospitalized patients (75, 76). In young women developing venous thromboembolism while taking oral contraceptives, a significant deficit of blood group O individuals relative to those possessing groups A and AB combined was discovered.

A correlation was found between ABO blood groups and the development of arrhythmias after administration of digoxin, with decreased risk in O patients relative to non-O patients.

DRUG RESPONSE AND GENETIC HETEROGENEITY

Depression and Antidepressants

It is suggested that the symptoms of depression are produced by at least two genetically distinct entities and that "endogenous" depressions more frequently benefit from treatment with imipramine, whereas "reactive" depressions improve after administration of monoamine oxidase (MAO) inhibitors (77).

Data supporting this hypothesis are based on similarities in drug response between probands and relatives who also suffered from depression and who also received imipramine, MAO inhibitors or lithium carbonate. The concordance in drug response among depressed relatives and depressed probands was reported to be statistically greater than expected by chance alone. Studies in different patients tended to confirm these initial impressions (78).

Striking differences may exist in the genetic background of lithium responders compared to nonresponders. Genetic determination of large interindividual variations in lithium ion distribution was observed in a study of monozygotic and dizygotic twins (79).

Vitamin-Dependent Genetic Disease

A group of inborn metabolic errors has recently been shown to respond not to physiologic replacement therapy but to pharmacologic doses of various vitamins (80). Table 5 summarizes nine of these disorders, all of which are discussed in detail elsewhere in this work. Several mechanisms have been proposed whereby pharmacologic doses of the vitamin relieve symptoms. If a defect existed in transporting the vitamin across a cell membrane to its target, e.g., this defective transport might exhibit improvement if very high, "saturating" concentrations of vitamin were furnished. If the lesion involved an enzyme responsible for converting a vitamin to its active form (as, e.g., the metabolically inert vitamin B_6, pyridoxal or pyridoxine, is converted by a kinase in the presence of ATP to the biologically active coenzyme pyridoxal phosphate), then also a partial defect in the converting enzyme might be remedied by saturating the converting enzyme with very high concentrations of the vitamin.

If the disorder arose from alteration in the enzyme to which the vitamin in the form of a coenzyme is bound, the avidity of binding might be increased by supplying the enzyme with higher concentrations of coenzyme. Although in some of the disorders shown in Table 5 the precise mechanisms whereby pharmacologic doses of various vitamins are clinically effective remains to be established, some progress has been made. In vitamin B_{12}-responsive methylmalonic aciduria, e.g., the primary defect is a partial defect in the biosynthesis of 5'-deoxyadenosylcobalamin, one of the three natural forms of cobalamin in mammalian tissue. When devising new therapeutic maneuvers, it is important that the investigator recognize that a very slight increase in activity of a defective enzyme can produce substantial clinical improvement in many inborn metabolic errors. Since an enzyme functions as a catalyst, a small change in its catalytic efficiency alone can produce a large alteration either in elevating the concentration of a necessary product of the reaction or in reducing the amount of a toxic metabolite.

Genetic Control and Drug Disposition

Drug accumulation to toxic levels in healthy persons who cannot eliminate the usual doses of a therapeutic agent as fast as other healthy persons is a major cause of adverse reactions. Studies disclose large variations in the rates at which healthy, nonmedicated volunteers clear antipyrine, bishydroxycoumarin and phenylbutazone from their plasma. After a standard dose of the drug, plasma concentrations of chlorpromazine, propranolol, nortriptyline, diphenylhydantoin or procainamide vary widely in normal subjects. The twin method is used to estimate the relative contribution of environmental and genetic factors to the observed differences in drug metabolism that do not follow a strict Mendelian pattern.

Intratwin differences should be of similar magnitude in identical and fraternal twins for traits controlled primarily by environmental factors. There should be a difference in the magnitude of intratwin differences for traits controlled primarily by genetic factors, however, since identical twins share all their genes, whereas fraternal twins, on the average, have in common only one-half of their genes. For traits appreciably influenced by genetic factors intratwin differences should be less in identical than in fraternal twins. The following relationship,

$$\frac{\text{variance within pairs of fraternal twins} - \text{variance within pairs of identical twins}}{\text{variance within pairs of identical twins}}$$

Table 5. Vitamin-Dependent Inborn Errors of Metabolism

Disorder	Clinical Manifestations	Vitamin Required in Pharmacologic Dose	Biochemical Basis
Thiamine-responsive megaloblastic anemia	Megaloblastic anemia	Thiamine (B$_1$)	Unknown
Hartnup's disease	Intermittent cerebellar ataxia, mental retardation	Nicotinamide	Defective intestinal absorption of tryptophan
Vitamin B$_6$-dependent infantile convulsions	Clonic and tonic seizures	Pyridoxine (B$_6$)	Defective glutamic acid decarboxylase (?)
Vitamin B$_6$-responsive anemia	Microcytic, hypochromic anemia	Pyridoxine (B$_6$)	Defective δ-aminolevulinic acid synthetase (?)
Cystathioninuria	Probably none	Pyridoxine (B$_6$)	Defective cystathionase
Xanthurenic aciduria	Mental retardation (?)	Pyridoxine (B$_6$)	Defective kynureninase
Homocystinuria (one type)	Ectopia lentis, arterial and venous thromboses, mental retardation	Pyridoxine (B$_6$)	Defective cystathionine synthetase
Methylmalonic aciduria (one type)	Infantile ketoacidosis, developmental retardation	Cobalamin (B$_{12}$)	Defective biosynthesis of vitamin B$_{12}$ coenzyme
Familial hypophosphatemic rickets	Rickets, short stature	Calciferol (D)	Unknown

Modified from (80) with permission.

provides a range of values from 0, indicating negligible hereditary and complete environmental control over variations observed for any given trait, to 1, indicating virtually complete hereditary influence over interindividual variations. In healthy, nonmedicated twins, almost complete genetic and negligible environmental controls of large interindividual variations in metabolism of phenylbutazone (Fig. 1) and antipyrine (Fig. 2) have been observed (81, 82). In the formula just given, the value for hereditary control was 0.99 and 0.98 for phenylbutazone and antipyrine, respectively. Predominantly genetic control over large interindividual differences in the metabolism of bishydroxycoumarin, halothane, ethanol and nortriptyline has also been found.

The distribution curves for antipyrine and nortriptyline are closer to the

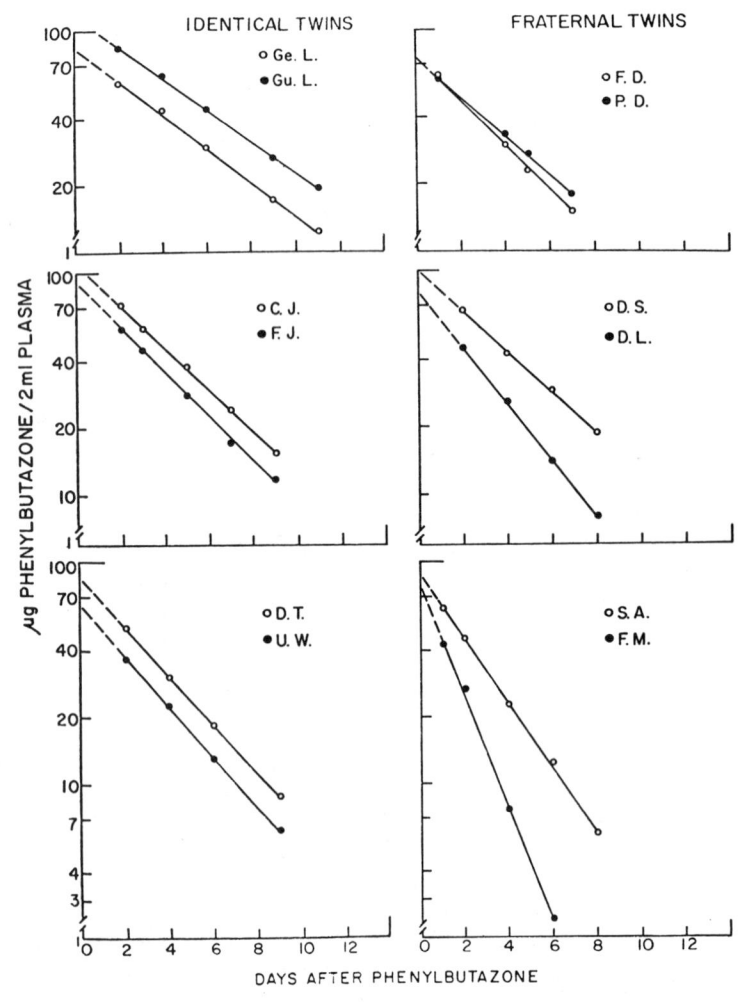

Figure 1. Decline of phenylbutazone in plasma of three sets of identical twins (left) and of three sets of fraternal twins (right) after single oral dose of 6 mg/kg.

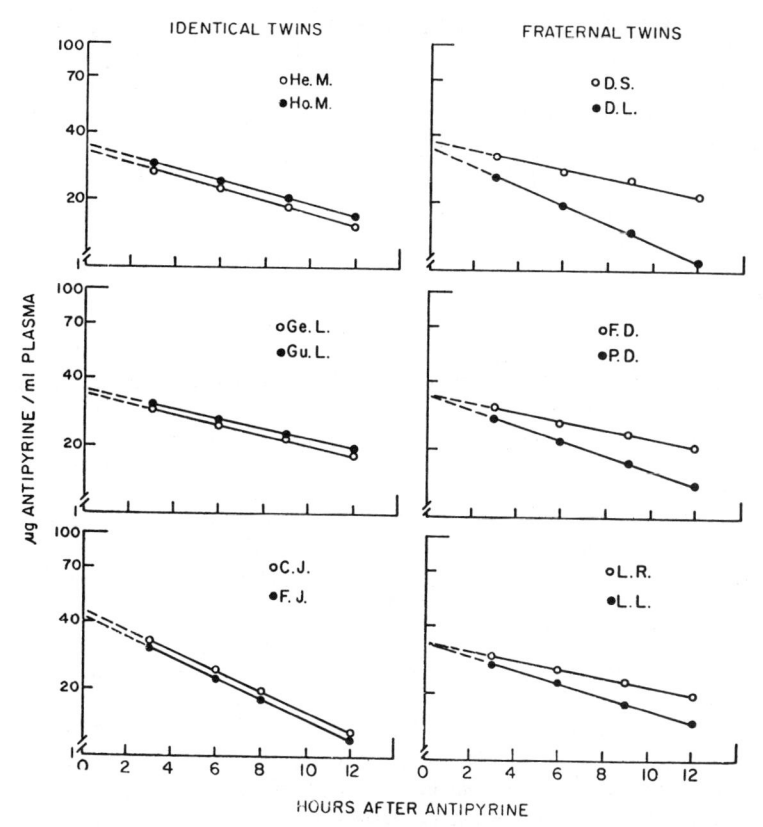

Figure 2. Decline of antipyrine in plasma of three sets of identical twins (left) and of three sets of fraternal twins (right) after single oral dose of 18 mg/kg.

unimodal, continuous curves observed in polygenically controlled traits, in contrast to the polymodal curve for isoniazid excretion. A significant regression of mean offspring value on midparent value also was revealed by the familial studies, a result consistent with polygenic control with metabolism of these two drugs. A recent study in two extensive Swedish pedigrees with high steady-state plasma concentrations of nortriptyline, e.g., suggested that the appreciable individual differences in the concentrations were polygenically controlled (83). So far conclusions for both the twin and the family studies agree that large differences among healthy, nonmedicated volunteers in rates of drug metabolism are primarily controlled by genetic factors.

REFERENCES

1. Cluff LE, Thornton GL, Smith J: Epidemiological study of adverse drug reaction. *Trans Assoc Am Phys* 78:255, 1965
2. Gillette JR: Factors affecting drug metabolism. Drug metabolism in man. *Ann NY Acad Sci* 179:43, 1971

3. Conney AH, Welch R, Kuntzman R, Chang R, et al: Effects of environmental chemicals on the metabolism of drugs, carcinogens and normal body constituents in man. *Ann NY Acad Sci* 179:155, 1971
4. Vesell ES: Factors altering the response of mice to hexobarbital. *Pharmacology* 1:81, 1968
5. Vesell ES: Recent progress in pharmacogenetics. *Adv Pharmacol Chemother* 7:1, 1969
6. Vesell ES (ed): Drug metabolism in man. *Ann NY Acad Sci* 179:1, 1971
7. Takahara S: Progressive oral gangrene probably due to lack of catalase in the blood (acatalasemia). *Lancet* 263:1101, 1952
8. Takahara S, Sato H, Doi M, Mihara S: Acatalasemia. III. On the heredity of acatalasemia. *Proc Japan Acad* 28:585, 1952
9. Takahara S, Doi K: Statistical study of acatalasemia (a review of thirty-eight cases appearing in the literature). *Acta Med Okayama* 13: , 1959
10. Takahara S, Hamilton HB, Neel JV, Kobara TY, et al: Hypocatalasemia: A new genetic carrier state. *J Clin Invest* 39:610, 1960
11. Aebi, H. and Suter, H.: Acatalasemia, ch. 73 in Stanbury JB, Wyngaarden JB, Frederickson DS (eds): The Metabolic Basis of Inherited Disease. New York, McGraw-Hill, p. 1710, 3rd ed., 1972.
12. Hamilton HB, Neel JV: Genetic heterogeneity in human acatalasia. *Am J Hum Genet* 15:408, 1963
13. Yata H: A case of acatalasemia. *Nihou Shika Hyoron* 204:7, 1959
14. Aebi H, Heiniger JP, Butler R, Hässig A: Two cases of acatalasia in Switzerland. *Experientia* 17:466, 1961
15. Aebi H, Suter H: Acatalasemia, in Harris H, Hirschhorn K (eds): Advances in Human Genetics, Vol 2. New York, Plenum, 1971
16. Hughes HB: On the metabolic fate of isoniazid. *J Pharmacol Exp Ther* 109:444, 1953
17. Hughes HB, Biehl JP, Jones AP, Schmidt LH: Metabolism of isoniazid in man as related to the occurrence of peripheral neuritis. *Am Rev Tuberc* 70:226, 1954
18. Price Evans DAP, White TA: Human acetylation polymorphism. *J Lab Clin Med* 63:394, 1964
19. Price Evans DAP: Individual variations of drug metabolism as a factor in drug toxicity. *Ann NY Acad Sci* 123:178, 1965
20. Reidenberg MM, Drayer DE, Levy M, Warner H: Polymorphic acetylation of procainamide in man. *Clin Pharmacol Ther* 17:722, 1975
21. Kalow W: Pharmacogenetics: Heredity and the Response to Drugs. Philadelphia, Saunders, 1962
22. Peters JH: Relationship Between Plasma Concentration and Urinary Excretion of Isoniazid. *Transactions of Conference on Chemotherapy on Tuberculosis,* 18th Conference, 1959
23. Price Evans DAP, Manley K, McKusick VA: Genetic control of isoniazid metabolism in man. *Br Med J* 2:485, 1960
24. Motulsky A: Pharmacogenetics. *Prog Med Genet* 3:49, 1964
25. Sunahara S: Genetical, geographical and clinical studies of isoniazid metabolism. *Proc 16th Intern Tuberc Conf* 2:513, 1961
26. Harris HW: Isoniazid metabolism in humans: Genetic control, variation among races, and influence on the chemotherapy of tuberculosis. *Proc 16th Intern Tuberc Conf* 2:503, 1961
27. Price Evans DAP: Pharmacogenetique. *Med Hyg* 20:905, 1962
28. Evans FT, Gray PWS, Lehmann H, Silk E: Sensitivity to succinylcholine in relation to serum cholinesterase. *Lancet* 262:1229, 1952
29. Kalow W, Genest K: A method for the detection of atypical forms of human serum cholinesterase. Determination of dibucaine numbers. *Can J Biochem* 35:339, 1957
30. Davies RO, Marton AV, Kalow W: The action of normal and atypical cholinesterase of human serum upon a series of esters of choline. *Can J Biochem* 38:545, 1960
31. Kalow W, Davies RO: The activity of various esterase inhibitors toward atypical human serum cholinesterase. *Biochem Pharmacol* 1:183, 1959

32. Harris H, Whittaker M: Differential inhibition of human serum cholinesterase with fluoride; recognition of two new phenotypes. *Nature* 191:496, 1961
33. Harris H, Whittaker M: The serum cholinesterase variants. A study of twenty-two families selected via the "intermediate" phenotype. *Ann Hum Genet* 26:59, 1962
34. Hodgkin WE, Giblett ER, Levine H, Bauer W, et al: Complete pseudocholinesterase deficiency: Genetic and immunologic characterization. *J Clin Invest* 44:486, 1965
35. Goedde HW, Altland K: Evidence for different "silent genes" in the human serum pseudocholinesterase polymorphism. *Ann NY Acad Sci* 151:540, 1968
36. Harris H, Whittaker M, Lehmann H, Silk E: The pseudocholinesterase variants. Esterase levels and dibucaine numbers in families selected through suxamethonium-sensitive individuals. *Acta Genet* 10:1, 1960
37. Lehmann H, Liddell J: Genetical variants of human serum pseudocholinesterase. *Prog Med Genet* 3:75, 1964
38. Garry PJ, Oretz AA, Lubraw T, Ford PC: New allele at cholinesterase locus 1. *J Med Genet* 13:38, 1976
39. LaDu BN: The isoniazid and pseudocholinesterase polymorphisms. *Fed Proc* 31:1276, 1972
40. Gutsche BB, Scott EM, Wright RC: Hereditary deficiency of pseudocholinesterase in eskimos. *Nature* 215:322, 1967
41. Neitlich HW: Increased plasma cholinesterase activity and succinylcholine resistance; a genetic variant. *J Clin Invest* 45:380, 1966
42. Yoshida A, Motulsky AG: A pseudocholinesterase variant (E Cynthiana) with elevated plasma enzyme activity. *Am J Hum Genet* 21:486, 1969
43. Harris H, Hopkinson DA, Robson EB, Whittaker M: Genetical studies on a new variant of serum cholinesterase detected by electrophoresis. *Ann Hum Genet* 26:359, 1963
44. Maynert EW: The metabolic fate of diphenylhydantoin in man. *J Pharmacol Exp Ther* 130:275, 1960
45. Kutt H, Wolk M, Scherman R, McDowell F: Insufficient parahydroxylation as a cause of diphenylhydantoin toxicity. *Neurology* 14:542, 1964
46. Brennan RW, Dehejia H, Kutt H, McDowell F: Diphenylhydantoin intoxication attendant to slow inactivation of isoniazid. *Neurology* 18:283, 1968
47. Solomon HM: Variations in metabolism of coumarin anticoagulant drugs. *Ann NY Acad Sci* 151:932, 1968
48. Ikeda M, Sezesny B, Barnes M: Enhanced metabolism and decreased toxicity of warfarin in rats pretreated with phenobarbital, DDT or chlordane. *Fed Proc* 25:417, 1966
49. Greaves JH, Ayres P: Heritable resistance to warfarin in rats. *Nature* 215:877, 1967
50. Shahidi NT: Acetophenetidin sensitivity. *Am J Dis Child* 113:81, 1967
51. O'Reilly RA, Aggeler PM, Hoag MS, Leong LS, et al: Hereditary transmission of exceptional resistance to coumarin anticoagulant drugs. The first reported kindred. *New Engl J Med* 271:809, 1964
52. O'Reilly RA, Pool JG, Aggeler PM: Hereditary resistance to coumarin anticoagulant drugs in man and rat. *Ann NY Acad Sci* 151:913, 1968
53. O'Reilly RA: The second reported kindred with hereditary resistance to oral anticoagulant drugs. *New Engl J Med* 282:1448, 1970
54. Motulsky AG, Yoshida A, Stamatoyannopoulos G: Variants of glucose-6-phosphate dehydrogenase. *Ann NY Acad Sci* 179:636, 1971
55. Yoshida A: Amino acid substitution (histidine to tyrosine) in a glucose-6-phosphate dehydrogenase variant (G6PD Hekteon) associated with overproduction. *J Mol Biol* 52:483, 1970
56. Yoshida A: A single amino acid substitution (asparagine to aspartic acid) between normal (B+) and the common negro variant (A+) of human glucose-6-phosphate dehydrogenase. *Proc Natl Acad Sci* 57:835, 1967
57. Fraser IM, Tilton BE, Vesell ES: Effects of some metabolites of hemolytic drugs on young and old, normal and G6PD-deficient human erythrocytes. *Ann NY Acad Sci* 71:644, 1971

58. Fraser IM, Tilton BE, Vesell ES: Alterations in normal and G6PD-deficient human erythrocytes of various ages after exposure to metabolites of hemolytic drugs. *Pharmacology* 5:173, 1971
59. Hitzig WH, Frick PG, Betke K, Huisman TH: Hemoglobin Zürich: A new hemoglobin anomaly with sulfonamide-induced inclusion body anemia. *Helv Paediatr Acta* 15:499, 1960
60. Frick PG, Hitzig WH, Betke K: Hemoglobin Zürich, I. A new hemoglobin anomaly associated with acute hemolytic episodes with inclusion bodies after sulfonamide therapy. *Blood* 20:261, 1962
61. Muller CJ, Kingma S: Hemoglobin Zürich: $\alpha 2A\beta 2$-63 Arg. *Biochem Biophys Acta* 50:595, 1961
62. Rieder RF, Zinkham WH, Holtzman NA: Hemoglobin Zürich; Clinical chemical and kinetic studies. *Am J Med* 39:4, 1965
63. Blakeslee AF: Genetics of sensory thresholds: Taste for phenyl thio carbamide. *Proc Natl Acad Sci* 18:120, 1932
64. Fox AL: The relationship between chemical constitution and taste. *Proc Natl Acad Sci* 18:115, 1932
65. Harris H, Kalmus H, Trotter WH: Taste sensitivity to PTC in goitre and diabetes. *Lancet* 257:1038, 1949
66. Kitchin FD, Howel-Evans W, Clarke CA, McConnell RB, et al: PTC Taste Response and Thyroid Disease. *Br Med J* 1:1069, 1959
67. Saldanha PH, Becak W: Taste thresholds for phenylthiourea among Ashkenazic Jews. *Science* 129:150, 1959
68. Barnicot NA: Taste deficiency for phenylthiourea in African Negroes and Chinese. *Ann Eugen* 15:248, 1950
69. Armaly, MF: Genetic factors related to glaucoma. *Ann NY Acad Sci* 151:861, 1968
70. Denborough MA, Forster JFA, Lovell RH, Maplestone PA, et al: Anaesthetic deaths in a family. *Br J Anaesth* 34:395, 1962
71. Kalow W: Topics in pharmacogenetics. *Ann NY Acad Sci* 179:654, 1971
72. Von Wartburg JP, Schürch PM: Atypical human liver alcohol dehydrogenase. *Ann NY Acad Sci* 151:936, 1968
73. Edwards JA, Evans DAP: Ethanol metabolism in subjects possessing typical and atypical liver alcohol dehydrogenase. *Clin Pharmacol Ther* 8:824, 1967
74. Wolff PH: Ethnic differences in alcohol sensitivity. *Science* 175:449, 1972
75. Jick H, Slone D, Westerholm B, Inman WHW, et al: Venous thromboembolic disease and ABO blood type. A cooperative study. *Lancet* 1:539, 1969
76. Lewis GP, Jick H, Slone D, Shapiro S: The role of genetic factors and serum protein binding in determining drug response as revealed by comprehensive drug surveillance. *Ann NY Acad Sci* 179:729, 1971
77. Pare CMB: Differentiation of two genetically specific types of depression by the response to antidepressive drugs. *Humangenetik* 9:199, 1970
78. Pare CMB, Mack JW: Differentiation of two genetically specific types of depression by the response to antidepressant drugs. *J Med Genet* 8:306, 1971
79. Dorus E, Pandey GN, Frazer A, Mendels J: Genetic determinant of lithium ion distribution. I. An in vivo monozygotic-dizygotic twin study. *Arch Gen Psychiatry* 31:463, 1974
80. Rosenberg LE: Vitamin-dependent genetic disease. *Hosp Practice* 5:59, 1970
81. Vesell ES, Page JG: Genetic control of drug levels in man: Phenylbutazone. *Science* 159:1479, 1968
82. Vesell ES, Page JG: Genetic control of drug levels in man: Antipyrine. *Science* 161:72, 1968
83. Åsberg M, Price Evans DA, Sjöqvist F: Genetic control of nortriptyline kinetics in man: A study of relatives of propositi with high plasma concentrations. *J Med Genet* 8:129, 1971

Part 3

ORGAN SYSTEM DISEASES

8
Genetics of Cardiovascular Diseases

James J. Nora
Audrey H. Nora

It is more appropriate to talk about genetic predisposition to, rather than the formal genetics of, cardiovascular disease, because it appears that only a very small percentage of affected persons show monogenic transmission. Either the great majority conform to the multifactorial inheritance mode, or the genetics are not yet clearly defined. The term *multifactorial inheritance* is used in the most limited sense, i.e., a polygenic predisposition usually interacting with an environmental trigger. The term *multifactorial* (not followed by the word inheritance) is a hedge, meaning that many factors are involved, more of which may be genetic although it is likely that those labeled multifactorial will eventually gain the more specific imprimatur of multifactorial inheritance.

An overview of the interaction of heredity and environment in expressing the major cardiovascular diseases is shown in Figure 1. All, except rheumatic fever, cluster midway between predominant genetic and predominant environmental influences. For rheumatic fever the overriding consideration is the environmental trigger, the group A β-hemolytic streptoccal infection.

Rheumatic Fever

In 1889, Cheadle (1) appreciated that rheumatic fever frequently occurred in more than one member of a family. In 1937, Wilson and Schweitzer (2) proposed

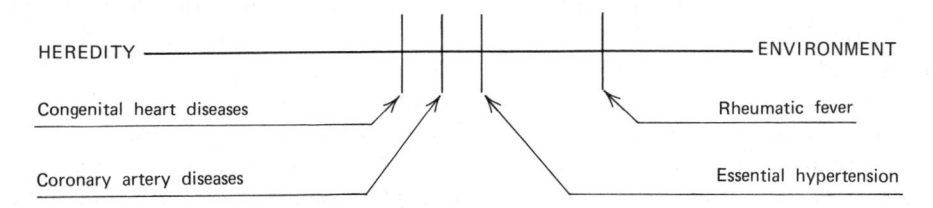

Figure 1. Congenital heart disease as it appears with other cardiovascular disease in continuum of genetic and environmental influences (22).

that susceptibility to rheumatic fever was best explained by inheritance of a single autosomal recessive gene. They stated that the "role of environment and contagion in the acquisition of the disease . . . cannot be determined by these studies." Coburn (3) appeared to be well ahead of his time in 1931, when he proposed a streptococcal etiology. Many workers rejected this possibility during the 1930s (4), and some, even into the 1950s (5) were not prepared fully to embrace this concept.

However, there must be an explanation for the familial nature of rheumatic fever. Rheumatic fever has been all but eliminated, e.g., in Scandanavia and among the Scandanavian populations in the United States through improved standards of living and medical care. Yet 50 years ago this population showed exquisite susceptibility. The Hispanic populations in the United States and Mexico remain extremely susceptible. Untreated streptococcal infections will not be followed by rheumatic fever in one family, but will in another, and once a person has had rheumatic fever the risk for subsequent attacks increases from 0.3 to 50% for the next untreated group A streptococcal infection.

Kaplan has shown that there is a shared antigen between the human myocardium and the group A β-hemolytic streptococcus (6). The person susceptible to rheumatic fever raises streptococcal antibodies, which attack his own myocardium. There likely are differences in cellular or humoral immunity, or both, between those persons susceptible to rheumatic fever and those not. Although these differences have not been identified, they are likely polygenic, although there could easily be, as Wilson originally proposed, a major gene effect.

Essential Hypertension

There are a number of clearly defined causes for systemic arterial hypertension. This discussion, however, will confine itself to the disease that becomes manifest in adult life and which does not have a basis in arterial anomalies, renal disease or other identifiable disorders. Although the definition of hypertension in an adult may range from a systolic pressure (mm Hg)/diastolic pressure (mm Hg) above 120/80 to above 180/110, a commonly accepted level for the normal upper limits in the North American adult population is 140/90 mm Hg. Normal upper limits for children have also been defined in various age groups and populations and are generally lower, the lower the age group. These thresholds are arbitrary and cannot be directly equated with disease.

The familial nature of hypertension (or *presumed* effects of hypertension) was probably first described by Morgagni in 1769, when he noted apparent cerebral vascular episodes in two generations of a family. There are populations (blacks, e.g.) and familial aggregates of people with blood pressures above the accepted normal upper limits. Pickering (7, 8) represents a school of thought that finds blood pressure to be a quantitative trait transmitted by multifactorial inheritance, and independent studies are in agreement with his conclusion (9). Platt (10) and Morrison and Morris (11), among others, have favored a major gene effect, although in later reviews Platt (12) acknowledged that multifactorial inheritance is an alternative not eliminated by his data. It is possible that the category "essential hypertension" is not entirely homogeneous, and that some major gene effects may be identified. It also is unlikely that these major genes account for a large proportion of the cases.

Congenital Heart Disease

As an etiologic overview congenital heart diseases (CHD) appear to fit in the middle of a continuum from primarily genetic to primarily environmental causation (Fig. 1). Most studies attempting to define the causes of CHD have been done in the past two decades. (13-18). No inflexible dogma has emerged regarding etiology, but newly acquired data are generally evaluated through frames of reference like that shown in Table 1. Current experience suggests that in roughly 8% of patients with CHD the cause can be attributed solely to genetic factors, with little contribution from the environment. Another 2% have an environmental cause, with little or no genetic contribution. In the remaining 90% CHD is best explained by a genetic–environmental interaction (multifactorial inheritance) (19), in which the genetic and environmental contributions are of comparable importance.

Chromosomal Aberrations

Gross chromosomal anomalies now account for about 5% of the patients with congenital cardiovascular maldevelopment. To date studies have failed to show chromosomal anomalies in patients with congenital heart defects (both familial and nonfamilial) that were not part of a syndrome or malformation complex. (19, 20).

Newer methods of chromosomal study may reveal that some familial cases of CHD can be attributed to minor chromosomal aberrations, although, even a minor anomaly of a chromosome affects many gene loci and would likely cause abnormalities of structures in addition to the heart. Table 2 lists chromosomal syndromes associated with congenital heart defects, the frequency of the association and the three commonest cardiovascular malformations found with each chromosomal anomaly. Several anomalies have been described in C group chromosomes 7, 8, 9, 10 and 11, including trisomies, partial trisomies, deletions and mosaicism. Since the heart lesions that have been specified are similar, all these chromosomal disorders are grouped together.

Excepting the XO Turner and the XXXXY syndromes, the commonest defect in the general population is also the commonest in the various chromosomal syndromes (actually ventricular septal defect [VSD] and atrioventricular [A-V] canal are about as common in 21 trisomy). The XO Turner and XXXXY syndromes are the only sex chromosomal syndromes represented in this table, since there seems to be very little increase in frequency of association of cardiovascular

Table 1. Etiologic Basis of Congenital Heart Disease

Cause	*Percentage*
Primarily genetic factors:	
Chromosomal	5
Single mutant gene	3
Primarily environmental factors:	
Rubella	≈ 1
Other	≈ 1
Genetic–environmental interaction:	
(Multifactorial inheritance)	≈ 90

Table 2. Congenital Heart Diseases (CHD) in Selected Chromosomal Aberrations

Population Studied	Incidence of CHD (%)	Commonest Lesions 1	2	3
General population	1	VSD*	PDA†	ASD‡
4p-	40	VSD	ASD	PDA
5p- (cri-du-chat)	25	VSD	PDA	ASD
C group anomalies	25–50	VSD	PDA	
13 trisomy	90	VSD	PDA	Dex‖
13q-	50	VSD		
18 trisomy	99+	VSD	PDA	PS¶
18q-	50	VSD		
21 trisomy	50	VSD	A-V canal	ASD
X0 Turner	35	Coarc§	AS**	ASD
XXXXY	14	PDA	ASD	

*VSD, ventricular septal defect.
†PDA, patent ductus arteriosus.
‡ASD, atrial septal defect.
§Coarc, coarctation of aorta.
‖Dex, dextroversion.
¶PS, pulmonic stenosis.
**AS, aortic stenosis

disorders with XXY, XYY and XXX syndromes. The cases of XXXXX are few, but patent ductus arteriosus (PDA) has been reported and some workers have tried to define a role for the X-chromosome in PDA production (21). Certainly the frequency of PDA in female patients is twice as high as in males. Patients with XO Turner syndrome are "deficient" in X-chromosomes and most often have the "male-associated" coarctation of the aorta, whereas the XXXXY patients with three extra X-chromosomes have the "female-associated" disorder, PDA. How this relates precisely to morphogenesis of the cardiovascular system remains to be demonstrated.

Single Mutant Gene

As with the chromosomal causes of cardiovascular diseases, the image that one should have of single mutant gene causes is also that of a syndrome. However, there are exceptions. Idiopathic hypertrophic subaortic stenosis (IHSS) is a dominantly inherited abiotrophy, the earliest diagnostic manifestations of which can be detected by echocardiography. It is not cardiac maldevelopment as such, but a progressive disease of the myocardium. Familial forms of complete heart block also exist, which are inherited as autosomal dominant and recessive disorders.

Since many genes are required in cardiac organogenesis, it is likely that only a gene of large effect will produce cardiac maldevelopment. In general genes of large effect influence the development of more than one structure—thus the expectation rather of a syndrome than of a discrete cardiac anomaly. Families are reported with three generations of discrete congenital heart lesions, most often atrial septal defect. On closer scrutiny, some members of these families had associated minor anomalies of the hands, consistent with the dominantly inherited Holt-Oram syndrome. In other families no skeletal anomalies are apparent, nor is a prolonged P-R interval evident. From the point of view of genetic counseling, the practical consideration in families so infrequently encountered, demonstrating direct inheritance through three generations, is that the risk of recurrence is high—whether one categorizes the family as an autosomal dominant or as a type C multifactorial inheritance one (see section, "Environmental Causation").

It is essential to recognize families in which the congenital heart defects are transmitted by Mendelian inheritance, because their risks are much higher than those of the usual families demonstrating multifactorial inheritance. It is typical that a congenital heart anomaly caused by a single mutant gene will be found as part of a syndrome and have a 25% risk of recurrence *of the syndrome* if the gene is recessive; and a 50% risk of recurrence if the gene is dominant. The CHD may not be present in everyone affected with the syndrome. In Tables 3–5 are summarized some of the single mutant gene syndromes with which are associated cardiovascular disease in a variable percentage of cases. The frequency of cardiovascular involvement in these syndromes may be 5% to almost 100%. Syn-

Table 3. Autosomal Recessive Syndromes with Associated Cardiovascular Abnormalities

Syndrome	Abnormality
Adrenogenital syndrome	Hyperkalemia, broad QRS, arrhythmias
Alkaptonuria	Atherosclerosis, valve disease
Carpenter	Patent ductus arteriosus (PDA)
Conradi	Ventricular septal defect (VSD), PDA
Cockayne	Accelerated atherosclerosis
Cutis laxa	Pulmonary hypertension, peripheral pulmonary artery stenosis (PPAS)
Cystic fibrosis	Cor pulmonale
Ellis-van Creveld	Atrial septal defect (ASD), most commonly single atrium, other congenital heart lesions
Friedreich ataxia	Myocardiopathy
Glycogenosis	Myocardiopathy
Homocystinuria	Coronary and other vascular thromboses
Jervell-Lange-Nielson	Prolonged QT, sudden death
Laurence-Moon-Biedl	VSD and other congenital heart diseases
Mucolipidosis III	Aortic valve disease
Mucopolysaccharidosis (MPS) IH (Hurler)	Coronary artery disease, aortic and mitral valve insufficiency
MPS IS (Scheie), MPS IV (Morquio), MPS VI (Maroteaux-Lamy)	Aortic valve disease, coronary artery disease
Osteogenesis imperfecta	Aortic valve disease
Saldino-Noonan	Transposition of great arteries
Progeria	Accelerated atherosclerosis
Pseudoxanthoma elasticum	Coronary insufficiency, mitral insufficiency (MI), hypertension
Riley-Day	Episodic hypertension, postural hypotension
Refsum	Atrioventricular (A-V) conduction defects
Seckel	VSD, PDA
Sickle cell disease	Myocardiopathy, MI
Smith-Lemli-Opitz	VSD, PDA, and other congenital heart diseases
Thrombocytopenia absent radius (TAR)	ASD, tetralogy, dextrocardia
Thalassemia major	Myocardiopathy
Weill-Marchesani	PDA
Werner	Vascular sclerosis

Table 4. Autosomal Dominant Syndromes with Associated Cardiovascular Abnormalities

Syndrome	Abnormality
Apert	VSD,* tetralogy
Crouzon	PDA,* coarctation of aorta
Ehlers-Danlos	Rupture of large blood vessels, e.g., carotids; dissecting aneurysms of aorta
Familial periodic paralysis	Hypokalemia, supraventricular tachycardia
Forney	MI
Holt-Oram	ASD,* VSD
Idiopathic hypertrophic subaortic stenosis (IHSS)	Subaortic muscular hypertrophy
Leopard	Pulmonic stenosis (PS), prolonged P-R interval, abnormal p waves
Lymphedema (Milroy and Miege)	Lymphedema
Marfan	Great artery aneurysms; aortic insufficiency (AI), MI
Myotonic dystrophy (Steinert)	Myocardiopathy
Neurofibromatosis	PS,* pheochromocytoma with liver hypertension; coarctation of aorta
Osler-Weber-Rendu	Multiple telangiectasias; pulmonary arteriovenous fistulas
Osteogenesis imperfecta	AI
Romano-Ward	Prolonged Q-T, sudden death
Treacher Collins	VSD, PDA, ASD
Tuberous sclerosis	Myocardial rhabdomyoma, aortic aneurysm
Ullrich-Noonan	PS, ASD, IHSS
von Hippel-Lindau disease	Hemangiomas; pheochromocytoma with hypertension

*See footnotes, table 2.

Table 5. X-Linked Recessive XR and Dominant XD Syndromes with Associated Cardiovascular Abnormalities

Syndrome	Abnormality
MPS II (Hunter), X-R	Coronary artery disease, valve disease
Muscular dystrophy (Duchenne), X-R	Myocardiopathy
Muscular dystrophy (Dreifuss), X-D	Myocardiopathy
Incontinentia pigmenti, X-D	PDA,* pulmonary hypertension
Goltz, X-D	Occasional congenital heart defects, telangiectasia

*See footnotes, table 2.

dromes have been selected with a high enough risk of recurrence of congenital heart defects within the syndrome to warrant counseling for that specific problem.

Environmental Causation

Formerly patients with environmentally induced CHD were lumped with the multifactorial inheritance group, even though the environmental considerably outweighed the genetic contribution.

There is a positive family history in some cases of VSD and a rare patient with PDA in the *rubella* syndrome, supporting a genetic contribution. However, the great majority of cases of PDA and virtually all the cases of peripheral pulmonary artery stenosis (PPAS) are without a positive family history, indicating a negligible genetic contribution (22). The percentage of patients with rubella at a pediatric cardiology center varies with time and place. During and immediately after the 1964–65, rubella pandemic 2-3% of pediatric cardiovascular patients in Houston had the rubella syndrome. In 1975, in Denver, far fewer than 1% of pediatric cardiovascular patients had the syndrome. Probably not more than ≈ 1% of the current crop of patients with CHD have rubella as a cause of the cardiovascular defect.

A recent cause of frequent and significant cardiovascular disease that is essentially environmental (but postnatal environment) is the *aggressive perinatal care* of premature infants in nurseries, leading to a frequency of PDA in high-risk premature babies of 20 to 60%. Other prenatal environmental triggers may not require a significant genetic predisposition. Thalidomide is a case in point. Perhaps cytomegalovirus and lithium chloride may be teratogenic without requiring an important polygenic predisposition in the host. At this time an estimated 1% of CHD fall in this category.

Genetic–Environmental Interaction (Multifactorial Inheritance)

This is the major category in the etiology of CHD, since those situations in which the causes are primarily genetic or primarily environmental are uncommon. In this category there is a heriditary predisposition to cardiovascular maldevelopment, which interacts with an environmental trigger (e.g., virus or drug) at the vulnerable period of cardiogenesis. The essential ingredients are that 1) the individual must be genetically predisposed to cardiovascular maldevelopment, 2) there must also be a genetic predisposition to react adversely to the environmental teratogen and 3) the environmental insult must occur at the vulnerable period of cardiac development (i.e., very early in pregnancy). Consider, e.g., that it requires the products of several different genes for the development of the ventricular septum—genes specifying structural proteins, others specifying enzymes—and a precise timetable must be followed (23). The correct building blocks must be laid in the correct sequence and at the correct speed.

The contribution from the endocardial cushions, conus and ventricular septum must arrive at precisely the right time for the ventricular septum to close. If the contribution from the endocardial cushions is late, or the building blocks are flawed, and the septum will not close completely, the threshold from normal to abnormal development has been crossed. This is probably what happens in cardiac maldevelopment of the multifactorial inheritance mode. There is a precarious balance of many genes (which individually may be normal). This balance may be influenced by environmental triggers producing accelerations and delays of primary gene products, thus setting the schedule of orderly development askew and resulting in a malformed structure.

Risk of Recurrence to Sibs
In the typical family with CHD as it presents itself in the counseling situation, the Type B family is one in which one first-degree relative has a cardiovascular anomaly (e.g., a previously affected child). This family is established as having a genetic predisposition. The addition of an environmental trigger upsets the precarious balance, and maldevelopment may result. The risks of recurrence when there is only one affected first-degree relative are given in Table 6. The subcolumn, Exp $\sqrt{p}$, is the risk predicted by one of the simpler models of mul-

Table 6. Observed and Expected Risk of Recurrence in Siblings of Probands with Congenital Heart Lesions*

Anomaly	Probands	Affected Siblings No.	%	Exp. ($\sqrt{p}$)
VSD†	212	24/543	4.4	5.0
PDA†	204	17/505	3.4	3.5
Tetralogy of Fallot	157	9/338	2.7	3.2
ASD†	152	11/342	3.2	3.2
PS†	146	10/345	2.9	2.9
AS†	135	7/317	2.2	2.1
Coarc†	128	5/272	1.8	2.4
Transposition of great vessels	103	4/209	1.9	2.2
Atrioventricular canal	73	4/151	2.6	2.0
Tricuspid atresia	51	1/96	1.0	1.4
Ebstein's anomaly	42	1/96	1.1	0.7
Truncus arteriosus	41	1/86	1.2	0.7
Pulmonic atresia	34	1/77	1.3	1.0
Total	1,478	95/3,377		

*1,478 probands were studied.
†See footnotes, Table 2.

tifactorial inheritance. The "ballpark figure" to remember for the range of risk of recurrence is 1-5%; the commoner the heart defect, the likelier it will recur. Hence VSD is likelier to recur in a family than is tricuspid atresia.

The next point of counseling information is that the risk in multifactorial inheritance increases precipitously with the number of affected first-degree relatives. If there are two affected first-degree relatives, the risk is almost tripled. To illustrate, if there has been only one child with VSD, the risk to the next child is of the order of 5%, but if there have been two children so affected the risk to the next child is of the order of 15%.

This fact leads to the rare, unfortunate family that we have called Type C. The genetic balance is so precarious that most of the first-degree relatives will have cardiac maldevelopment—doubtless without much insult from the environment. Indeed the risk may become even higher than it is in Mendelian inheritance. In autosomal recessive inheritance, the risk is fixed at 25% and in dominant inheritance at 50%, but in multifactorial inheritance the risk may approach 100%. If, e.g., there are three affected first-degree relatives, the family should be classified as Type C, and the risk of recurrence will be of the order of 60-100%.

In summary, the risk of CHD in subsequent pregnancies after the birth of an affected child (and no affected parent) is of the order of 1-5%, depending on how common is the heart defect. If there are two affected first-degree relatives, the risk is tripled. If there are three, the family is probably Type C, with a prohibitive risk of recurrence (one that could approach 100%). These factors apply to most counseling situations, (namely, those in which multifactorial inheritance is the etiologic mode).

Risks to Offspring of Affected Parent

The inevitable question of patients with CHD as they grow to maturity relates to the chance that their children will also have CHD. In accumulating risk figures it is important that the investigator be precise in the categorization of the etiologic modes and the specific lesions. To say that X% of the children of parents with "congenital heart disease" also have "congenital heart disease" is meaningless if not misleading. An asymptomatic adult with pulmonic stenosis (PS) who has no affected children has a risk of 3% recurrence in the next child. If the adult has the Noonan syndrome, the risk is 50% that the next child will have the syndrome, and 25% that there will be heart disease (most often PS).

Table 7 lists the recurrence of congenital heart defects in 1,120 children, each with only one parent having a specific cardiac anomaly that was diagnosed in childhood and (with few exceptions) repaired (24). The special anomalies are the seven commonest malformations compatible with survival to a reproductive age. Most of the patients were personally interviewed by the authors in Madison (Wisconsin), Montreal, Houston and Denver, and this series was combined with two comparable studies from the literature (21, 25, 26). Again the risk of recurrence approximates the exception for recurrence of the anomaly in the multifactorial inheritance mode, given one affected first-degree relative. It is apparent that risks of recurrence are similar no matter which first-degree relative is affected, i.e., sibling or parent, since the expectations in Tables 6 and 7 are comparable.

Table 7. Risk of Recurrence in Offspring of Parents With Common Congenital Heart Lesions*

Anomaly	Affected Offspring No.	%	Exp. ($\sqrt{p}$)
Coarctation†	7/253	2.7	2.4
ASD‡	5/199	2.5	3.2
VSD‡	7/174	4.0	5.0
Tetralogy§	6/141	4.2	3.2
PDA‡	6/139	4.3	3.5
PS‡	4/111	3.6	2.9
AS‡	4/103	3.9	2.1
Total	39/1120	(3.4)	

*1,120 offspring were surveyed.
†Includes cases of Zetterqvist.
‡See footnotes, Table 2.
§Includes cases of Taussig HB (25) and McNamara DG (26).

Environmental Contribution and Genetic–Environmental Interaction

The importance of the environmental contribution to a congenital heart defect is becoming more clearly established as evidence rapidly accumulates. We are able to identify potentially teratogenic exposures to mothers of over one-half of our patients with CHD. Proving that these "potential" teratogens are the true environmental triggers is another story. A respiratory infection or "flu" in the first weeks of pregnancy may have played a role. The probability that the infection was teratogenic may be strengthened by demonstrating antiheart IgM—but definitive proof is still lacking.

Table 8 lists agents that are clearly established as, or highly suspected of being, triggers of cardiovascular maldevelopment. As stated earlier, rubella and

Table 8. Cardiovascular Teratogens

Conclusion	Drugs	Viruses
Proved	Thalidomide	Rubella
Highly suspected	Dextroamphetamine Anticonvulsants Lithium chloride Alcohol Progestogen/estrogen	Cytomegalovirus Herpesvirus hominis B Coxsackie virus B

Table 9. Presumed Vulnerable Period of Teratogenic Influence on Cardiovascular Development

Abnormality	Embryonic Event Completed (days)	Extreme Limit of Vulnerable Period	Most Sensitive Vulnerable Period (days)
Truncoconal septation	34	14–34	18–22
Endocardial cushions	38	14–38	18–33
Ventricular septum	38–44	14–?	18–39
Atrial septum secundum	55	14–?	18–50
Semilunar valves	55	14–?	18–50
Ductus arteriosus	——*	14–?	18–60
Coarctation of aorta	——	14–?	18–60

*——, not applicable.

thalidomide appear to require little interaction with a genetic predisposition. Better models for genetic–environmental interaction may be coxsackie virus B and dextroamphetamine, which appear to produce a low frequency of cardiovascular malformations and—at least for dextroamphetamine—clearly work with a genetic predisposition (22).

The dangerous period for exposure to cardiac teratogens is early in pregnancy—often just at the time when the woman may wonder if she is indeed pregnant. Table 9 shows what we consider to be the extreme limits and most sensitive periods of vulnerability of the developing heart to teratogens. From both the thalidomide data of Lenz (27) and our own dextroamphetamine data (18), it appears that an effective teratogenic insult (at least with these two drugs) occurs about two weeks before the completion of the embryologic event. Therefore, when trying to determine the role that a teratogen may have played in causing, say, transposition of the great vessels, it cannot be implicated if the exposure occurred after the 34th day and is likely not responsible if exposure was after the 27th day.

Certain drugs have come under strong suspicion of playing a teratogenic role in cardiovascular development and should be considered guilty until proven innocent. Thalidomide (27) and folic acid antagonists (28) have been convicted. Those agents that are highly suspected include dextroamphetamine (29), anticonvulsants (30), lithium (31), alcohol (32) and progestogen/estrogen (33). Clearly some drugs, including those strongly suspected of teratogenicity, cannot be eliminated from the regimens of pregnant women who need them, whereas other drugs, such as dextroamphetamine and pregnancy-test progestogen can be completely removed. As a general principle, when treating a gravida in a family that has a first-degree relative with CHD, discourage indiscriminate use of prescription and nonprescription drugs.

Atherosclerosis–Coronary Heart Disease

This entity is responsible for more deaths in the United States than *all* other diseases combined. The familial aspects of the coronary heart disease (with xanthomatosis) were first described in the English literature by Fagge, in 1873 (34). In 1897, Osler discussed possible genetic features of coronary heart disease by calling attention to the distinguished Arnold family of England in which was recorded, in consecutive generations, sudden deaths following relatively short courses of chest pain. In all fairness, it would be appropriate to credit a member of this family, the poet and critic Matthew Arnold, with one of the earliest written observations of the familial nature of coronary heart disease. He experienced his first attack of angina in 1887 and wrote to a friend: "I began to think that my time was really coming to an end. I had so much pain in my chest, the sign of a malady which had suddenly struck down in middle life, long before they came to my present age, both my father and my grandfather."

Matthew Arnold lived with angina pectoris for less than one year and died on April 15, 1888, with what was curiously described by one biographer as "heart failure ... sudden and quite unexpected." Although Matthew survived to age 65, his father, Thomas, lived only to age 46, and his grandfather, William, died when Thomas was only 6 years old.

The preceding anecdote is meant to provide perspective for the study of genetics of coronary heart disease, with all its complexities, rather than scrutinize only one risk factor that is the subject of major genetic interest, namely, hyperlipemia. However, the importance of hyperlipemia in this complex scheme must not be minimized. It is probably the most important of the inherited pathogenetic components of atherosclerosis.

Hyperlipidemia and Hyperlipoproteinemia

The speculation over a century ago was that xanthomatosis might develop from hyperlipidemia (35). The association of elevated levels of serum cholesterol with familial coronary heart disease (and xanthomatosis) has been appreciated for almost four decades (36). Considerable impetus to investigations in this area came from the Fredrickson group (37), which proposed a phenotypic classification of hyperlipoproteinemias as a means of reaching a genetic definition of lipid disorders. More extensive examination, using ultracentrifugation and electrophoresis, furnishes the definitive phenotype of those who are suspected of having a lipoprotein abnormality (Table 10).

Clearly the state of the art requires more than identifying individuals and families at risk because of hyperlipidemia merely on the basis of serum cholesterol levels. There are, e.g., epidemiologic data that relate directly to the Fredrickson phenotypes (39). However, it is apparent that these phenotypes are several steps removed from primary gene products and subject to considerable variability. Within the same family, e.g., three or four hyperlipoproteinemia phenotypes may be present, and even the same individual may be phenotyped differently under varying circumstances. Table 11 shows a classification of the hyperlipop-

Table 10. Characteristics of Hyperlipidemias*

Type	Electrophoresis	Ultracentrifugation
I (hyperchylomicronemia)	Chylomicrons	Chylomicrons
II (IIa) (hyperbetalipoproteinemia)	Beta†↑	LDL‡↑
(IIb) (combined hyperlipidemia)	Broad beta	Intermediates
III (broad beta disease)	Prebeta ↑	VLDL§ ↑
IV (hypertriglyceridemia)	Chylomicrons, prebeta ↑	Chylomicrons, VLDL ↑
V (mixed hyperlipidemia)	Beta ↑, prebeta ↑	LDL ↑, VLDL ↑

*From (37, 38, 40).
†Beta = beta lipoprotein.
‡LDL = low-density lipoproteins.
§VLDL = very low-density lipoproteins.

Table 11. Known Hyperlipoproteinemias Presented in Modified Fredrickson Format

Type	Inheritance	Defect
Ia$_1$	Autosomal recessive	Lipoprotein lipase deficiency
Ib	Multifactorial	Diabetes, dysglobulinemia
IIa$_1$, IIa$_2$	Autosomal dominant	Defects in HMG CoA reductase* Regulation, heterogeneous
IIb	Multifactorial	Most individuals with hypercholesterolemia
IIIa	? Autosomal dominant	Rare
IIIb	Multifactorial	Most cases
IVa	Autosomal dominant	Hypertriglyceridemia
IVb	Multifactorial	Hypertriglyceridemia
Va	? Monogenic	Most often in Type IV families
Vb	Multifactorial	Diabetes, nephrosis, lupus erythematosus
VIa	Autosomal dominant	Combined hyperlipemia
VIb	Multifactorial	Combined hyperlipemia, most cases

*See text.

roteinemias modified to accommodate advances in knowledge of specific entities within the present highly heterogeneous phenotypes. The letter "a" (as in IIa) specifies single gene disorders; the letter "b" serves to define abnormalities more consistent with a multifactorial etiology. Subscripts 1, 2, . . . further define entities in which the basic defect is confidently identified. Type IIa, e.g., could serve as a generic for presumed single gene familial hypercholesterolemia and type IIa$_1$ for one of the defects in 3-hydroxy-3 methylglutaryl coenzyme A (HMG CoA) reductase regulation.

Type Ia$_1$ is a rare autosomal recessive defect in removal of chylomicrons, presumably secondary to a deficiency in activity of lipoprotein lipase (41). The disease is characterized by striking chylomicronemia (producing a "creamy" plasma after the blood cells have settled), extremely high plasma triglyceride levels (2,000–3,000/ml is common) and normal or high plasma cholesterol levels. More definitive biochemical evaluation reveals that the elevated levels of triglycerides and cholesterol are *not* accompanied by abnormally high beta, low-density, lipoproteins (LDL) or high prebeta, very low-density lipoproteins (VLDL). The excess lipids are carried in the chylomicrons. Clinically the disorder is recognized in childhood because of episodic abdominal pain (with occasionally fatal pancreatitis), xanthomatosis and hepatosplenomegaly. Type Ib₁, an uncommon subgroup of an already uncommon disorder, includes cases associated with dysglobulinemia, diabetes, lupus erythematosus, hypothyroidism and administration of oral contraceptives. The relative contributions of polygenic predisposition and environmental triggers in these conditions have not yet been assessed. It is essential to look for these associated diseases before assuming that a patient has the monogenic form of the disease.

Type II (hyperbetalipoproteinemia-hypercholesterolemia) is the group of greatest interest and concern. In 1970, a WHO committee split this type into IIa, in which patients had high beta (LD) lipoproteins alone; and IIb, in which patients had high prebeta (BDL) lipoproteins in addition to the high beta lipoproteins. Type IIa₁, or monogenic hyperbetalipoproteinemia, is related to a defect in the regulation of HMG CoA reductase, secondary to a defective cell–surface receptor for the lipoprotein–cholesterol complex. Evidence for heterogeneity exists, i.e., both a receptor-negative and a receptor-defective mutation have been identified (42, 43). If one takes the conventional approach to genetic diseases, namely, that with a frequency greater than 1/1,000 a disorder is classified as common, the Type IIa heterozygote is at the threshold of being regarded by some estimates as common (0.1-0.2% population frequency).

The autosomal dominant form is of serious clinical consequence, producing coronary heart disease as early as the third and fourth decades. It is the homozygous manifestation of the disease, however, that is disastrous. Tuberous xanthomas are present in the homozygotes in infancy and childhood and in the heterozygotes in the third and fourth decades. Angina pectoris, congestive heart failure, progressive aortic stenosis and frank myocardial infarction ensue. Children in the second and even first decades of life die of coronary heart disease, and it is the exceptional homozygote who survives into adult life.

A significant breakthrough in the exploration and management of this problem was provided by Starzl and co-workers (44). They demonstrated a dramatic decrease in plasma cholesterol levels (from 1,000 mg/100 ml to 300 mg/100 ml) and striking regression of xanthomas and the gradient of aortic stenosis with improvement in the coronary arteriograms in a 12-year-old homozygous type IIa₁ patient following portacaval shunt. More than 30 additional patients have benefited from surgical intervention, which finding must represent an early example of documented regression of a genetic disease following metabolic surgical management.

Most patients with hypercholesterolemia do not have the monogenic familial form, but rather a multifactorial disease (*Type IIb*). Depending on where one sets the threshold to define hypercholesterolemia, as high as 10–75% of American adults have elevated cholesterol levels. The relationship of diet, exercise and stress to the level of plasma cholesterol requires precise evaluation. Medical conditions known to be associated with hypercholesterolemia include hepatic disease, porphyria, diabetes, nephrosis, hypothyroidism and dysglobulinemia.

Type III-(broad-beta disease) (41) is characterized biochemically by an abnormal lipoprotein with a high content of both triglycerides and cholesterol, and it appears as a broad-beta band on electrophoresis. The orange-yellow lipid deposits in the creases of the hands are highly characteristic. As in patients with Types I and IIa, large tuberoeruptive and planar xanthomas are also seen.

Premature coronary—and especially peripheral vascular—disease is a feature of this disorder, as are abnormal glucose tolerance and hyperuricemia. Some extensively studied pedigrees suggest an autosomal dominant inheritance (*Type IIIa*). However, patients with Type IV disease appear frequently enough in such pedigrees to raise the question of the same basic abnormality's being responsible for both Types III and IV phenotypes. Most familial cases do not fit comfortably into an autosomal dominant mode (Type IIIb). As in other hyperlipoproteinemias, abnormalities that may play an etiologic role (or are at least associated with the disorder) are diabetes, dysglobulinemia and hypothyroidism.

Type IV is a very common disorder characterized biochemically by abnormally high levels of triglycerides, VLDL and prebeta lipoproteins. Early-onset coronary heart and peripheral vascular diseases are found in both familial and nonfamilial hypertriglyceridemia. The triglyceride levels of most patients are raised by dietary intake of carbohydrates and lowered by dietary restriction thereof. Abnormal glucose tolerance is common. Xanthomas are not a feature, but hyperuricemia and diabetes are frequent.

Clearly pedigrees exist in which autosomal dominant inheritance appears (*Type IVa*). There are within certain families individuals who conform to different phenotypes. In so-called Type IV families, there may be patients with Types II, III, IV, V or VI (mixed hyperlipidemia). Some familial members do not fit a dominant mode, and there are many "sporadic cases" (Type IVb). The interaction of obesity and carbohydrate indiscretion with a familial predisposition is well recognized.

In Type V (mixed hyperlipidemia) exogenous chylomicrons and increased beta (LD) and prebeta (VLD) lipoproteins appear. Eruptive xanthomas, abdominal pain, pancreatitis, hyperuricemia, abnormal glucose tolerance and (possibly) some prematurity of vascular disease also occur. The phenotype appears to be highly heterogeneous, patients with the Type V phenotype not infrequently appearing in families in which the predominating phenotype is Type IV and the frequent mode of inheritance autosomal dominant (Va). The proposal for autosomal recessive inheritance is based on a pedigree in which inbreeding was identified and on the tenuous findings of affected siblings without apparent parental phenotypic expression.

The Type Vb phenotype—even more than some of the other hyperlipoproteinemias—is associated with other diseases, such as lupus, diabetes, nephrosis and alcoholism. Some Type IV individuals may readily convert to Type V (or Type III) after sizable ingestion of alcohol the day before their blood is drawn—even following the traditional 12–14-hour fast.

Type VI (combined hyperlipemia) is proposed as a discrete entity caused by a single gene, which can produce within different individuals elevations in both cholesterol and triglyceride levels, cholesterol level alone or triglyceride level alone. In some families the Type III or V phenotypes may also appear. Hence it is possible for members of the *same* family to fulfill the biochemical criteria for five of the six phenotypes.

Possibly a single gene abnormality produces multiple phenotypes, inasmuch as the lipoprotein phenotypes are not primary gene products. However, when dealing with common disorders, the possibility of frequent simulation of Mendelism cannot be completely dismissed. The preponderance of familial studies leads us to believe that most patients with Type VI as well as Types II, III, IV and V do not have single gene disorders.

If we admit that our understanding of the genetics of hyperlipoproteinemias is deficient, our understanding of the ultimate question—the genetics of coronary heart disease is negligible. There certainly are multiple, coagulation, immunologic, hormonal and other factors, many of which have genetic implications. Most factors are probably continuously distributed, but there is reason to believe that major genes are operating in a small number of cases. It will be important to distinguish the monogenic from polygenic causes to provide both optimal assessment of individual and family risk and specific therapeutic approaches.

REFERENCES

1. Cheadle WB: Harveian lectures on the various manifestions of the rheumatic state as exemplified in childhood and early life. *Lancet* 1:821, 871, 921, 1889
2. Wilson MG, Schweitzer MD: Rheumatic fever as a familial disease. Environment, communicability and heredity in their relation to observed familial incidence of disease. *J Clin Invest* 16:555, 1937
3. Coburn AF: The Factor of Infection in the Rheumatic State. Baltimore, Williams & Wilkins Co., 1931
4. Wilson MG, Ingerman E, DuBois RO, et al: The relation of upper respiratory infections to rheumatic fever in children. I. The significance of hemolytic streptococci in the pharyngeal flora during respiratory infection. *J Clin Invest* 14:325, 1935
5. Levine SA: Clinical Heart Disease, ed. 4. Philadelphia, Saunders, 1958
6. Kaplan MH: Immunologic relation of streptococcal and tissue antigens. I. Properties of an antigen in certain strains of group A streptococci exhibiting an immunologic cross-reaction with human heart tissue. *J Immun* 90:595, 1963
7. Pickering GW: *High Blood Pressure*, ed. 2. London, Churchill, 1968
8. Hamilton M, Pickering GW, Roberts JAF, et al: Etiology of essential hypertension; arterial pressure in general population. *Clin Sci* 13:11, 1954
9. Ostfeld AM, Paul O: The inheritance of hypertension. *Lancet* 1:575, 1963
10. Platt R: Heredity in hypertension. *Lancet* 1:899, 1963
11. Morrison S, Morris J: Nature of essential hypertension. *Lancet* 2:829, 1960
12. Platt R: The natural history and epidemiology of essential hypertension. *Practitioner* 193:5, 1964
13. McKeown T, MacMahon B, Parsons CG: The familial incidence of congenital malformations of the heart. *Br Heart J* 15:273, 1953
14. Polani PE, Campbell M: An aetiological study of congenital heart disease. *Ann Hum Genet* 19:209, 1955
15. McKusick VA: A genetical view of cardiovascular disease. *Circulation* 30:326, 1964
16. Nora JJ, Meyer TC: Familial nature of congenital heart diseases. *Pediatrics* 37:329, 1966
17. Lamy M, DeGrouchy J, Schweisguth O: Genetic and non-genetic factors in the etiology of congenital heart disease: A study of 1188 cases. *Am J Hum Genet* 9:17, 1957
18. Nora JJ, Nora AH: The evolution of specific genetic and environmental counseling in congenital heart disease. *Circulation* 57:205, 1978
19. Nora JJ: Multifactorial inheritance hypothesis for the etiology of congenital heart disease: The genetic-environmental interaction. *Circulation* 38:604, 1968
20. Anders JM, Moores EC, Emanuel R: Chromosome studies in 156 patients with congenital heart disease. *Br Heart J* 27:756, 1965

21. Zetterqvist P: A clinical and genetic study of congenital heart defects. M.D. Thesis. Institute for Medical Genetics. University of Uppsala, Sweden, 1972
22. Nora JJ, Fraser FC: Medical Genetics: Principles and Practice. Philadelphia, Lea & Febiger, 1974
23. Nora JJ, Wolfe RR, Miles VN: Etiologic aspects of cardiovascular disease and predisposition detectable in the infant and child, in Friedman WF, Lesch M, Sonnenblick EH (eds): Neonatal Heart Disease. New York, Grune & Stratton, 1973, p. 279
24. Nora JJ, Nora AH: Recurrence risks in children having a parent with congenital heart disease. *Circulation* 53:701, 1976
25. Taussig HB, Crocetti A, Eshagh E: Follow-up of Blalock-Taussig operation. I. Results of first operation. *Hopkins Med J* 129:243, 1971
26. McNamara DG: Unpublished data,
27. Lenz W, Knapp K: Die thalidomid—embryopathie. *Dtsch Med Wochenschr* 87:1232, 1962
28. Wilson JG: Present status of drugs as teratogens in man. *Teratology* 7:3, 1973
29. Nora JJ, Vargo TA, Nora AH: Dexamphetamine a possible environmental trigger in cardiovascular malformations. *Lancet* 1:1290, 1970
30. Meadow SR: Anticonvulsant drugs and congenital abnormalities. *Lancet* 2:1296, 1968
31. Nora JJ, Nora AH, Toews WH: Lithium, Ebstein's anomaly and other congenital heart defects. *Lancet* 2:594, 1974
32. Jones KL, Smith DW, Ulleland CN, et al: Pattern of malformations in offspring of chronic alcoholic mothers. *Lancet* 1:1267, 1973
33. Nora AH, Nora JJ: A syndrome of multiple congenital anomalies associated with teratogenic exposure. *Arch Environ Health* 30:17, 1975
34. Fagge CH: General xanthalasma or vitiligoidea. *Trans Pathol Soc* 24:242, 1873
35. Quinquaud M: Recherches hematochimiques et dermatochimiques. *Bull Soc Chim*, Paris, 1878, p. 259
36. Muller C: Angina pectoris in hereditary xanthomatosis. *Arch Intern Med* 64:675, 1939
37. Fredrickson DS, Lees RS: System for phenotyping hyperliporoteinemia. *Circulation* 31:321, 1965
38. WHO Memorandum: Classification of hyperlipidemias and hyperlipoproteinemias. *Circulation* 45:501, 1972
39. Slack J: Risks of ischaemic heart disease in familial hyperlipoproteinemic states. *Lancet* 2:1380, 1969
40. Goldstein JL, Hazzard WR, Schrott HG, et al: Hyperlipidemia in coronary heart disease. II. Genetic analysis of lipid levels in 176 families and delineation of a new inherited disorder, combined hyperlipidemia. *J Clin Invest* 52:1544, 1973
41. Fredrickson DS, Levy AI: Familial hyperlipoproteinemia, in Stanbury, JB, Wyngaarden JB, Fredrickson DS (eds): The Metabolic Basis of Inherited Disease, ed. 3. New York, McGraw-Hill, 1972
42. Brown MS, Goldstein JL: Familial hypercholesterolemia: Defective binding of lipoproteins to cultured fibroblasts associated with impaired regulation of HMG CoA reductase activity. *Proc Nat Acad Sci* 71:788, 1974
43. Fogelman AM, Edmond J, Seager J, et al: Abnormal induction of 3-hydroxy-3-methylglutaryl coenzyme reductase in leukocytes from subjects with heterozygous familial hypercholesterolemia. *J Biol Chem* 250:2045, 1975
44. Starzl TE, Chase HP, Putnam CW, et al: Follow-up of patient with portacaval shunt for treatment of hyperlipidemia. *Lancet* 2:714, 1974

9
Genetics of Endocrine Diseases

Angelo M. DiGeorge

Many endocrine disorders are the direct result of single mutant genes, whereas in others there is a potent background substrate of genetic factors leading to the endocrinopathy. This latter group, exemplified by the autoimmune endocrine disorders, is only now in the process of being unraveled. The pathogenetic mechanisms that may lead to genetic endocrine disorders are extremely varied and have been reviewed by Rimoin and Schimke (12). In some instances new insights into genetic mechanisms for disease were first established in the endocrine system. Examples include disorders as divergent as autoimmune mechanisms (lymphocytic thyroiditis), deficiency of receptor protein (testicular feminization) and the production of antireceptor antibodies (Graves' disease). The endocrine system also serves as an important model for the delineation of interrelationships between genetic and oncogenic mechanisms as manifested in the multiple endocrine neoplasia syndromes.

With the increasing ability to measure hormones in pico- and femtogram amounts, the number and variants of genetic endocrine disorders are increasing rapidly, and this chapter makes no pretense at an exhaustive review of the subject. It does attempt, however, to survey the commoner endocrine disorders, particularly those that have genetic implications for the counselor. As is always the case, precise diagnosis is critical to any genetic counseling since different pathogenetic mechanisms with distinct modes of inheritance can result in the same endocrine deficiency. Although treatment is not considered herein, it is worth noting that in no other medical discipline has the treatment of genetic disorders been more satisfactory than in those of the endocrine system.

ANTERIOR PITUITARY

Congenital Defects

Aplasia, or hypoplasia of the pituitary, is usually associated with developmental abnormalities of the skull and brain no matter what the underlying cause, and it

Supported in part by General Clinical Research Center Grant RR-75 of the U.S. Public Health Service.

occurs in such defects as anencephaly, holoprosencephaly (cyclopia, cebocephaly, oribital hypotelorism) and septo-optic dysplasia. Most information about these conditions is deduced from postmortem observations, and limited data are available concerning pituitary function during life. Recent observations suggest that pituitary hypoplasia in anencephaly is secondary to the hypothalamic defect (1). Many of these conditions are fatal early in life, but partial defects have been recognized on rare occasions. A child is reported with isolated deficiency of growth hormone and mild hypotelorism, who had two siblings with holoprosencephaly and hypopituitarism (2). Moreover, a study of 200 children with isolated cleft defect demonstrated that 12% of them were less than the third percentile in height, and eight of them had total or partial growth hormone deficiency (3).

A *solitary maxillary central incisor* in association with short stature should also alert the clinician to the probability of growth hormonal deficiency (4). This anomaly probably represents another example of the association between midline facial and dental defects and growth hormonal deficiency.

Septo-optic dysplasia should be suspected in blind children who are short. The optic nerves are hypoplastic and the fundus exhibits hypoplastic disks with typical double rims and sparse retinal vessels. Air encephalographic or computerized tomographic studies usually reveal absence of the septum pellucidum and dilatation of the chiasmatic cistern. The defect is believed to reside in the hypothalamus and is usually sporadic. The hormonal deficiency may involve only growth hormone or may result in panhypopituitarism, including diabetes insipidus (5).

Aplasia of the pituitary without abnormalities of the brain or skull is rare, but should be considered in male neonates with microphallus (6, 7). A number of affected infants have also had the neonatal hepatitis syndrome, but the relationship to the hypopituitarism is not known (8). The condition is reported in siblings of both sexes and consanguinity has been noted in two families, pointing to an autosomal recessive mode of inheritance (9). Some pediatric endocrine studies place the defect in the hypothalamus, suggesting that this may be a heterogenous group of disorders (6). Since these infants have hypoadrenalism as well as other pituitary deficiencies, it may be possible to monitor pregnancies for this condition once it has occurred in a family. Deficiency of maternal urinary estriol levels strongly suggests fetal hypoadrenalism (10).

Idiopathic hypopituitarism is a heterogeneous condition in that many affected patients have a defect in the hypothalamus rather than in the pituitary. In most instances the disorder is sporadic, and on the one hand several instances are reported in which only one monozygotic twin had the condition, reflecting a nongenetic mechanism (11). On the other hand, several affected kindreds are described, and genetic forms of the defect are well documented (12). There is, e.g., convincing evidence for autosomal recessive forms of the disorder. It is described in inbred genetic isolates, such as the Hutterites in the northern United States and Canada (12), on the Yugoslavian island of Krk (13) and in Oriental Jews in Israel (14). These recessive forms may be subdivided on the following basis: In isolated growth hormone deficiency the patients are short at birth, and hypoglycemia is conspicuous; puberty may be markedly delayed, but occurs spontaneously ("sexual ateliosis"); pregnancy is followed by normal lactation; the defect may be in the pituitary or hypothalamus, and even this subgroup may be heterogenous (12).

A distinct type of hypopituitarism was originally described by Laron in 1966 in 22 patients from 14 families in Israel, all of Oriental Jewish origin (14). Affected children have all the clinical findings of isolated deficiency of growth hormone, but plasma levels of growth hormone are elevated (15). Levels of somatomedin are low and the response in metabolic parameters to administration of growth hormone is minimal or absent. Neuroendocrine mechanisms mediating growth hormone secretion, however, are normal (16), as is the growth hormone molecule produced by these patients; current evidence suggests that the primary defect is a deficiency of growth hormonal receptors (17). Multiple affected sibs of both sexes and a high incidence of consanguinity clearly establish this disorder as an autosomal recessive one. Several non-Jewish cases have now been reported (15).

In congenital panhypopituitarism several pituitary hormonal deficiencies may be seen, especially that of gonadotropin ("asexual ateliosis"). The disorder is observed among the Hutterites and inbred populations on the island of Krk. Many other pedigrees with affected siblings are assumed to have this autosomal recessive form of panhypopiutitarism. However, Rona and Tanner (see discussion that follows) now have evidence suggesting a multifactorial mode of inheritance to explain the disorder in some of these nonisolates (18).

A small number of kindreds are reported in whom there is evidence for an *X-linked recessive* mode of inheritance. In two families, a mother had two affected boys by two marriages (19), and in another family first cousins and their maternal uncle had the disorder (20). In another family, the endocrine deficiency was isolated to growth hormone, but in the other two there were multiple pituitary deficiencies (21).

Whether isolated growth hormone deficiency is ever inherited as an *autosomal dominant* is not known for certain, although the condition is encountered in successive generations, including father and son, and dominant inheritance seems possible.

An extensive analysis of 140 cases of idiopathic growth hormone deficiency in England and Wales led Rona and Tanner to propose a *multifactorial* etiology for the disorder (18). They postulated an underlying liability to the hypopituitarism, depending on polygenic and environmental factors. Of particular interest was the observation—in pregnancies of patients with idiopathic growth hormone deficiency—of a significantly greater incidence of breech births (13.7%), forceps deliveries (5.6%) and early vaginal bleeding (8.1%).

Panhypopituitarism with an abnormally small sella turcica is reported in two sisters (22). Conversely a new syndrome of *enlargement of the sella turcica* is reported in three siblings of short stature who were deficient in growth hormone and thyroid-stimulating hormone (TSH) (23). According to the authors, familial pituitary tumor or excessive somatostatin production could not be excluded.

A relative *unresponsiveness to growth hormone* seems to be present in African pygmies who have normal levels of growth hormone and somatomedin (24). The failure of growth hormone to elicit any metabolic changes leads to the conclusion that these recipients have peripheral unresponsiveness. The mode of inheritance is not known.

Hypopituitarism occurs as a manifestation of other syndromes in which it may readily be overlooked. There are now, e.g., four well-documented cases of growth hormone deficiency in children with the *Fanconi syndrome (constitutional infantile panmyelopathy)* (25). A pedigree is reported in which isolated deficiency of growth hormone has occurred as an inconstant component of *Rieger syndrome* (iris-dental dysplasia) (26). This autosomal dominant disorder is characterized by malformation of the iris, pupillary anomalies, hypodontia and hypoplasia of the teeth, and maxillary hypoplasia. A common embryologic developmental defect of the neural crest is postulated to explain the association.

Pituitary insufficiency may develop in patients with *hemochromatosis* secondary to iron deposition or with *hemoglobinopathies* secondary to infarction. About one-half of patients with *histocytosis* X have deficiency of growth hormone (27). Growth hormonal deficiency is also reported in a child with deletion of the short arm of chromosome 18 (28). Although the association suggests that a locus-controlling growth hormone synthesis is on the deleted segment, coincidence cannot be ruled out.

Hyperpituitarism

Hyperfunction of the pituitary is a normal finding in conditions in which deficiency of a target organ gives decreased hormonal feedback. Primary hypersecretion of pituitary hormones is usual with suspected or confirmed pituitary neoplasia, although occasionally, as in Cushing's syndrome, it is not settled if the pituitary disorder is secondary to a primary hypothalamic defect. Most of these conditions are sporadic, but familial forms of pituitary hyperfunction occur in the Type I multiple endocrine neoplasia (MEN) syndrome (see Chapter 3, "Genetics and Cancer"). It is not clear if the rare reports of familial acromegaly or of Chiari-Frommel and Forbes-Albright syndromes represent partial forms of the MEN syndrome.

Several disorders associated with accelerated growth are suspected of having an element of hyperpituitarism. *Cerebral gigantism* (Sotos syndrome) is characterized by large size at birth, accelerated growth in early childhood with accompanying advanced bone age, early dental eruption, normal sexual development and relatively normal final adult stature (29). Physical features include macrocrania, together with dolichocephaly, prognathism, large hands and feet, an increased arm span and variable mental retardation. Extensive endocrine studies have failed to define the pathogenesis of the accelerated growth. Patients appear to be at increased risk for development of malignancies, for a hepatocarcinoma and a Wilms' tumor are reported in two children with the syndrome (30, 31).

Familial occurrence has been described on a number of occasions and includes affected first cousins, monozygous twins and siblings, suggesting a recessive mode of inheritance. However, families with affected mother and son, or affected father and daughter, suggest a dominant, probably autosomal, mode of inheritance (31).

The salient features of the *Beckwith-Wiedemann syndrome* consist of exomphalos, macroglossia and gigantism. Other frequent manifestations are nephromegaly, hepatomegaly, facial flame nevus, earlobe anomalies, microcephaly, hemihypertrophy, umbilical hernia and neonatal hypoglycemia. Widespread and bizarre pathologic changes occur, and patients have a marked propensity for developing malignant tumors, including nephroblastoma, adrenal carcinoma, hepatoblastoma and rhabdymosarcoma (33). In the author's experience roughly 20% of patients with the syndrome have developed tumors, including two with adrenal cortical tumors. Most cases are sporadic, but the typical syndrome has occurred in a mother and son, and autosomal dominant inheritance with incomplete penetrance and variable expressivity is likely (33, 34). More than a dozen pedigrees show irregular inheritance through both male and female carriers. A mechanism of delayed mutation is suggested to explain the transmission through unaffected carriers of the condition (35); a family in which three normal sisters gave birth to eight affected infants is an especially striking example of this phenomenon (34).

POSTERIOR PITUITARY

Diabetes Insipidus

Diabetes insipidus is characterized by polyuria and polydipsia and results from lack of antidiuretic hormone. Destruction of the supraoptic and paraventricular nuclei or division of the supraoptic–hypophyseal tract above the median eminence results in permanent diabetes insipidus. Transection of the tract below the median eminence or removal of just the posterior lobe may result in transitory polyuria, but release of hormone into the median eminence prevents diabetes insipidus. Any lesion of the neurohypophyseal unit may cause diabetes insipidus, inducing various tumors, encephalitis, sarcoidosis, tuberculosis, actinomycosis, basal skull fractures and operative procedures in the pituitary or hypothalamus. Vasopressin deficiency may also result from diverse complex syndromes of the posterior pituitary or hypothalamus, or both, and it may be seen with the congenital defects of the pituitary mentioned earlier.

Diabetes insipidus is hereditary in a small number of instances. Autosomal dominant and X-linked recessive modes of transmission are known (12), and affected males with either type are indistinguishable. These genetic forms of the disorder show marked reduction of the neurosecretory cells of the supraoptic and paraventricular nuclei. Oxytocin secretion is normal as demonstrated by women with the autosomal dominant form of the disorder who undergo normal pregnancy and delivery and successfully nurse their children.

Diabetes Insipidus, Diabetes Mellitus, Optic Atrophy and Deafness (Wolfram Syndrome)

Diabetes insipidus also occurs as part of a rare syndrome in association with diabetes mellitus, optic atrophy and sensory nerve deafness (36). Various incomplete combinations of these clinical manifestations are reported in patients as well as in their siblings, and the order of appearance of these components varies with the patient. Male and female subjects are equally affected, and diabetes mellitus appears to be the predominant manifestation. Evidence suggests that the disorder is caused by an autosomal recessive mutant gene that, in the homozygote, causes juvenile diabetes mellitus as well as one or more of the symptoms mentioned earlier. Heterozygotes appear to have an increased probability of developing juvenile diabetes mellitus (37).

Nephrogenic Diabetes Insipidus

This disorder closely mimics vasopressin deficiency, but hormonal levels in plasma and urine are normal, and patients show no antidiuresis even with large doses of vasopressin. Symptoms appear shortly after birth, being those of diabetes insipidus. In addition to polyuria and polydipsia, fever, irritability, vomiting, constipation and failure to thrive are common. The basic genetic defect appears to be renal resistance to the effects of antiduretic hormone (ADH) (38).

Vasopressin resistance occurs primarily in males as an X-linked dominant trait (39). Heterozygous female subjects are usually asymptomatic, but may exhibit a variable defect in concentration, explicable by the Lyon hypothesis. A large number of kindreds have been traced to Ulster Scotsmen who arrived in Halifax in 1761 and who settled in Nova Scotia (40). The disorder is also reported in others, groups including blacks (41).

Precocious Puberty

Strictly speaking, isosexual precocious puberty is not a pituitary disorder. However, in view of the potential difficulties which the diagnosis engenders, it is presented in this section for completeness. In the United States about 95% of girls have at least one sign of puberty by age 13 years (42). In boys the average age of onset of puberty is about 6 months later than in girls. Variations, however, are wide in the sequence of changes of growth, breast, pubic hair and genital development. Precocious puberty is difficult to define because of the marked variation in the age at which puberty begins normally. Onset of puberty before 8½ years of age in girls and 10 years in boys may be precocious, but these are arbitrary guidelines (43).

The causes of precocious puberty are given in Table 1. Very few of these conditions are genetically determined. Congenital virilizing adrenal hyperplasia, a recessively inherited disease, is one of the commonest causes of precocious pseudopuberty in boys and of virilization in girls (see section, "Adrenal Glands").

Idiopathic true precocious puberty is the commonest cause of true puberty in girls but is rarely genetically determined. On the other hand, in boys the condition may be familial. The usual form of inheritance is probably sex-limited autosomal dominant; however, families have been reported in which there is only male-to-male transmission and still other families in which the trait is transmitted to affected males only through unaffected females (12). These findings support genetic heterogeneity.

THYROID GLAND

Congenital Hypothyroidism

The commonest cause of congenital hypothyroidism is a developmental defect (aplasia or arrested development and descent); the condition is often referred to as athyrotic cretinism or thyroid dysgenesis. The recent institution of neonatal screening programs for this condition reveals that it occurs in about 1 in 6,000

Table 1. Causes of Precocious Puberty

Precocious pseudopuberty in males:
  Congenital adrenal hyperplasia
  Virilizing adrenal tumors
  Virilizing testicular tumors (Leydig cell)
Precocious pseudopuberty in females:
  Feminizing adrenal tumors
  Ovarian follicular cysts
  Feminizing ovarian tumors (granulosa-theca cell)
Precocious True Puberty:
  Tumors producing gonadotropins (pinealoma, teratomas, hepatoblastoma and choriocarcinoma of any site)
  CNS lesions (hypothalmic tumors like Gliomas, Astrocytomas and Hamartomas; destructive pineal tumors; tumors of other areas of the brain; lesions of the third ventricle, including cysts and hydrocephalus)
  CNS manifestations of systemic diseases (McCune-Albright syndrome, neurofibromatosis and postinflammatory diseases, such as tuberculous meningitis and encephalitis)
  Endocrine deficiency syndromes (hypothyroidism, hypopituitism and primary adrenal hypofunction)
  Idiopathic

births in Canada, North America and the Netherlands (44, 45); it occurs twice as often among female as among male subjects. The incidence of the disorder in blacks is less than in whites (46). On rare occasions the disorder has been noted in two or more siblings; both male and female patients are affected, suggesting recessive inheritance in rare instances or a hereditary tendency to the condition (47). Both discordance and concordance for hypothyroidism in monozygotic twins is reported, and it is important to know that the onset of hypothyroidism in the second twin may be delayed. In one pair of monozygotic twins, e.g., hypothyroidism was diagnosed at age 4 months, whereas in the second twin the ectopic thyroid did not lose adequate function until 4-6 years (48).

Congenital hypothyroidism is reported in a mother and daughter. Lingual thyroid is the most extreme form of failure of migration of the thyroid, occurring sporadically. In a report of a rare familial occurrence of lingual thyroid, two siblings were affected and a third had hypoplasia of one lobe of a normally placed thyroid (50). A remarkable family is reported in which a mother with lymphocytic thyroiditis gave birth to six children with congenital hypothyroidism, suggesting the operation of immune factors in thyroid maldevelopment (51). A significant excess of nontasters for phenylthiocarbamide (PTC), a recessively inherited trait, is found among congenitally athyrotic individuals (52). The compound, PTC, is structurally related to thiourea, propylthiouracil and goitrin, compounds interfering with synthesis of thyroid hormones. It is suggested, without sound evidence, that perhaps mothers of athyrotic infants have an abnormal intake of similar goitrogens. The incidence of PTC nontasting is not increased in patients with lingual and sublingual aberrant thyroid, which finding suggests that these conditions may be etiologically unrelated to athyrotic hypothyroidism.

Iodides, propylthiouracil, tapazole and thyrotropin-releasing hormone (4) readily cross the placenta, and the administration of radioactive iodine during pregnancy for treatment of cancer of the thyroid or of hyperthyroidism has caused congenital hypothyroidism (53). Administration of radioactive iodine to lactating women is also contraindicated since the substance is readily excreted in milk.

Rarely the hypothyroid child may exhibit generalized muscular hypertrophy, particularly in the calf muscles *(Kocher-Debre-Semelaigne syndrome)* (54). The disorder seems to represent the effect of-long standing hypothyroidism on the muscles, and it is observed in athyrotic children as well as in those with enzymatic defects of thyroid hormone synthesis. Children with congenital hypothyroidism have been reported with a constellation of associated abnormalities, including low birth weight, aplastic alae nasi, absent permanent tooth buds, midline ectodermal scalp defects, microcephaly, deafness, mental retardation and rectourogenital abnormalities (Johnson-Blizzard Syndrome) (55). Growth retardation is severe and only partially corrected by thyroid therapy. The cause is unknown and all cases have been sporadic.

Rarely isolated deficiency of TSH in otherwise normal individuals can cause congenital hypothyroidism. Few well-documented reports are available and the condition is usually sporadic. In one instance isolated deficiency of thyrotropin was established in two Japanese children, the offspring of a consanguineous mating, suggesting that the disorder may result from a rare autosomal recessive mutant gene (56). Congenital hypothyroidism resulting from panhypopituitarism occurs only one-tenth as frequently as does congenital aplasia or hypoplasia of the thyroid. Some of these panhypopituitary conditions may have a genetic basis (see previous discussion).

Thyroid Hormone Unresponsiveness

Three of six siblings of a consanguineous marriage are reported as having goiter, deaf-mutism, stippled epiphyses and clinical euthyroidism. Their levels of circulating thyroid hormones and radioactive iodine uptake were normal, as was conversion of T_4 to T_3 (57). However, thyrotropin (TSH) levels were normal or slightly elevated, a finding that reflects tissue resistance to the effect of thyroid hormone at a cellular level. The severity of the syndrome decreases with time. Partial target organ resistance to thyroid hormone is reported in several patients with goiter; levels of TSH were normal or only slightly elevated in most patients but markedly elevated in one patient (58). Dominant inheritance is suggested as the cause, but the evidence is not convincing.

A congenitally nongoitrous hypothyroid boy of a consanguineous mating is reported with an elevated level of biologically active TSH and normal uptake of radioactive iodine (59). Absence of response to thyrotropin was shown in vivo and in metabolism of thyroid tissue in vitro, indicating an impaired ability of the thyroid to respond to TSH. The family history revealed no similar defect.

Defective Synthesis of Thyroxine (Goitrous hypothyroidism)

Defects in the biosynthesis of thyroid hormone may cause congenital hypothyroidism (60, 61). When the defect is incomplete or mild, manifestations of hypothyroidism may be delayed, not appearing until childhood or later. The hallmark of these defects is a goiter due to compensatory hyperplasia. Various defects in thyroid biosynthesis are identified, all of which appear to be transmitted in an autosomal recessive manner. They include:

1. iodide trapping defect;
2. iodide organification defect (absent or reduced peroxidase, peroxidase

apoprotein–prosthetic group binding defect, other abnormality in peroxidase structure or function and the Pendred syndrome);
3. diminished H_2O_2 production;
4. iodotyrosine coupling defect;
5. iodotyrosine deiodination defect; and
6. defective thyroglobulin synthesis.

The phenotype and biochemical findings exhibit intrafamilial homogeneity; thus a mild defect in one familial member heralds a similar degree of deficiency in other affected members. However, a rare exception to this rule is recently reported in a pair of identical twins with goiter who were grandnieces of a sibship of four children with severe hypothyroidism, mental retardation and a complete defect in organification of iodide (62). Surprisingly the twins exhibited a much milder degree of hypothyroidism, having only slight retardation of bone age, no mental retardation and normal levels of T_4, T_3 and TSH. They also had a defect in iodide organification. The explanation for this anomalous situation is not clear, but it was suggested that the twins have allelic defects at a gene locus related to iodide organification. One of these alleles, when present in double dose, would result in a complete block in organification, whereas the other might not cause clinical thyroid disease, but when paired with the first it would be responsible for defective iodide organification, resulting in goiter.

Defective organification of iodide, when accompanied by deafness, is known as the Pendred syndrome, an autosomal recessive disorder (63). This syndrome is not to be confused with the deaf-mutism that may occur in severely hypothyroid persons. Hearing loss can vary, but is usually severe, present at birth and most pronounced in high frequencies. The condition may be the cause of 5-10% of congenital deafness. The goiter generally appears at puberty or later, but may be present during childhood. It may be barely detectable or pronounced and it tends to become nodular in adult life. Affected individuals are usually euthyroid, but hypothyroidism may develop even during childhood.

The syndrome has wide geographic distribution; the incidence in Europe is 1 to 8 in 100,000. Although administration of perchlorate causes significant discharge of iodide from the thyroid gland—indicating a defect in organification—none of the usual biochemical defects in the organification process has been identified (64). The goiter or hypothyroidism, or both and the deafness are thought to be independent effects of the gene.

In the autosomal recessive neuronal storage disease known as *Batten-Spielmeyer-Vogt disease* there is a deficiency of peroxidase in WBCs. Examination of the thyroid in two patients revealed decreased levels of peroxidase although thyroid function was grossly normal (65). It is not known if this is the same peroxidase which is deficient in patients with goitrous hypothyroidism. It is proposed that thyroid biopsy is more advantageous than brain biopsy for the diagnosis since it provides tissue for both enzymatic analysis and histologic demonstration of the characteristic intracytoplasmic structures.

Prenatal Diagnosis of Congenital Thyroid Disorders

Although pituitary–thyroid automony is complete in the fetus by 12 weeks, it is functioning at a very low level and concentrations of T_4, T_3 and TSH are low before midgestation in both fetal serum and amniotic fluid. There is virtually no placental transfer of maternal T$_4$, T$_3$ or TSH. Nor is there correlation between maternal and serum concentrations of T$_4$, free T$_4$, T$_3$, free T$_3$ and TSH at any time during gestation. On the other hand the level of reverse T$_3$ (3,3',5'—triiodothyronine) is relatively high in amniotic fluid, appears to be of fetal origin and may prove useful in evaluating intrauterine hypothyroidism (66). Direct injection of thyroid hormone into amniotic fluid or into the fetus are possible modes of therapeutic intervention in which congenital hypothyroidism is detected (67).

Lymphocytic or Hashimoto's thyroiditis is one of the commonest disorders of the thyroid. It is almost certainly due to a primary immunologic disturbance, and considerable evidence for an autoimmune pathogenesis has been accumulated (68). The genetic predisposition to the development of autoimmunity is strong, but the mode of inheritance has not been delineated. A close relationship exists between Graves' disease, idiopathic juvenile hypothyroidism, idiopathic myxedema of adults and lymphocytic thyroiditis. The unity of these disorders is reflected by their coexistence within the same thyroid gland, their concurrence within the same families and circulating autoantibodies in a large proportion of affected patients and members of their families. Thyrotoxicosis and lymphocytic thyroiditis have been observed separately in a pair of monozygotic twins (69).

Familial clusters of lymphocytic thyroiditis are common; the incidence in siblings or parents of affected children, or both may be as high as 25%. The disorder has been described on a number of occasions in both of identical twins providing further evidence for a genetic factor (70). The frequency of thyroid antibodies is increased in relatives of probands (71). Hashimoto's thyroiditis is noted in association with other autoimmune disorders more often than would be expected by chance alone (72). These disorders include, among others, idiopathic adrenal atrophy, pernicious anemia, insulin-dependent diabetes mellitus, myasthenia gravis, alopecia areata or totalis and idiopathic thrombocytopenis purpura. Moreover the association of lymphocytic thyroiditis with Turner syndrome is well established (73).

Most patients are euthyroid and may exhibit only goiter or antithyroid antibodies, or both, but frank hypothyroidism may occur. Since growth failure is a cardinal manifestation of Turner syndrome, diminishing thyroid function is readily overlooked. A similar association between Down syndrome and autoimmune thyroid disease is documented (74). The mothers of children with Turner and Down syndromes have a higher incidence of thyroid antibodies. Hence it is suggested (without sufficient evidence) that the antibodies or related factors may be predisposing agents in the development of chromosomal aberrations.

Current evidence suggests that multifactorial inheritance underlies the organ-specific autoimmune diseases in man. This applies not only to lymphocytic thyroiditis but also to Graves' disease, Addison's disease, idiopathic hypoparathyroidism and other autoimmune endocrine disorders.

Graves' disease is also thought to have an autoimmune basis. It is unique in that it is the only known disease caused by a stimulatory antibody (68) capable of binding to the receptor for TSH and triggering autonomous thyroxine biosynthesis and thus hyperthyroidism. A statistical correlation between histocompatibility antigen HLA-B8 and thyrotoxicosis is found in whites (75). Persons with this HLA antigen appear to be more susceptible to Graves' disease than those without it. Other disorders with an increased frequency of HLA-B8 antigens occurring in association with Graves' disease include myasthenia gravis, lupoid hepatitis and systemic lupus erythematosus. The most attractive hypothesis for this association is that it represents a link between HLA and an immune response gene controlling the response to thyroid antigens.

Congenital hyperthyroidism is usually caused by transplacental passage of thyroid stimulating immunoglobulins in infants of mothers who have a history of recently active Graves' disease (76). In some infants the condition does not abate, but persists for several years or longer. Because of this persistence the role of transplacental passage of maternal factors in the pathogenesis of the disorder is questioned; it is suggested that neonatal Graves' disease is an autosomal dominant trait with a predilection for the female subject (76).

It seems likely that neonatal Graves' disease is a heterogenous disorder; in most instances it is transitory, resulting from placental transfer of thyroid stimulating immunoglobulin; in a few instances the infant has true Graves' disease with persistence of the thyrotoxic manifestations.

Other Genetic Disorders Affecting the Thyroid

Acquired primary hypothyroidism is being increasingly recognized in children with *nephropathic cystinosis*, resulting from the progressive destruction of the thyroid gland by deposition of cystine crystals (77). The complication appears to be age related, and the signs and symptoms of hypothyroidism are easily overlooked. Periodic measurement of serum concentration of TSH is indicated in all children with cystinosis, particularly after the first few years of life. *Hereditary thyroid tumors* occur and are discussed in Chapter 3. At least three genetic defects of *thyroxine-binding globulin* (TBG) are known (78). Usually they are not associated with symptoms and do not require treatment; however, since they are associated with decreased levels of thyroxine, they may be a source of diagnostic confusion; *TBG-deficiency* is detected in screening programs for congenital hypothyroidism in about 1 in 14,000 newborn infants (79), whereas in a large thyroid clinic the prevalence of TBG-deficiency is about 0.6 in 1,000 (80). Most often TBG deficiency is inherited as an X-linked dominant trait, although rare instances of an autosomal dominant form of the disorder also occur (78). Linkage studies have so far failed to reveal proximity of this gene to that for Xg or color blindness (81). In affected patients TBG is absent or low and serum thyroxine levels are low, but TSH levels are normal and the patient is euthyroid. Heterozygous female patients have intermediate levels of TBG and levels of T_4 are low normal.

Elevation of TBG also occurs as a genetic disorder and is inherited as an X-linked dominant trait (78). Levels of TBG and T_4 are elevated, but affected patients are euthyroid. In one four-generation pedigree, goiter was present in 13 of 15 family members with high levels of TBG (82). The data in this family suggest that both the high TBG and goiter were transmitted as X-linked characters, but closeness of linkage could not be determined.

PARATHYROID GLANDS

Hypoparathyroidism

The *congenital* form may result from aplasia or hypoplasia of the parathyroid glands. Frequently there are other developmental defects, particularly right-sided aortic arch with or without other cardiovascular anomalies and other defects arising from the III and IV pharygeal pouches, such as aplasia or hypoplasia of the thymus *(DiGeorge syndrome)* (83). The disorder is usually sporadic although in one instance a son and daughter were affected and the mother had mild manifestations (84). Congenital hypoparathyroidism is reported in an infant with ring chromosome 18 (85) and in another with ring chromosome 16 (86). Isolated congenital hypoparathyroidism is reported in two brothers and a maternal male cousin, suggesting that the disorder may have been transmitted as a sex-linked recessive trait (87). In other instances autosomal dominant inheritance seemed possible (88).

Transient hypoparathyroidism may occur in infants born to mothers with hyperparathyroidism caused by suppression of fetal parathyroids by elevated maternal levels of calcium. In some instances two or three newborn siblings were affected before the maternal hyperparathyroidism was suspected. Thus an environmental cause for congenital hypoparathyroidism may mimic a genetic etiology, and maternal hyperparathyroidism should be excluded in all cases of the congenital form.

Idiopathic Hypoparathyroidism

This disorder may be acquired at any age. It is believed to be caused by an autoimmune mechanism, since over one-third of such patients have parathyroid antibodies (90). This belief is supported by the frequent association of hypoparathyroidism with other disorders that are thought to have a similar origin, as mentioned previously. Mucocutaneous candidiasis also occurs frequently in patients with idiopathic hypoparathyroidism or Addison's disease, or both (91). The evidence is abundant that the fungal infection is not the cause of the endocrinopathy but that both are the result of aberrant immunologic function. Siblings are recently reported who manifested nephrosis and nerve deafness in association with idiopathic hypoparathyroidism (92); an autoimmune cause of this syndrome also appears likely.

Idiopathic hypoparathyroidism is often observed in siblings, either alone or in association with other autoimmune disorders. In some families one sibling may manifest hypoparathyroidism, another Addison's disease and yet another both conditions. Although in some families the condition often appears to be recessively inherited, the genetic mechanism is much more complex and, as for the other autoimmune disorders, a multifactorial mode of inheritance is likelier.

Pseudoidiopathic Hypoparathyroidism

A young adult is reported with laboratory findings of hypoparathyroidism from 8 years of age. His serum contained normal-to-high levels of immunoreactive parathyroid hormone by several assay systems (93). He exhibited a normal response to exogenous parathyroid hormone. Possible defects include an abnormality in conversion of proparathyroid hormone to parathyroid hormone in the gland or to aberrant peripheral conversion of a secreted precursor of parathormone to an active form. Since this case is unique, nothing is know about its possible genetic origin.

Familial Hypomagnesemia

A small number of infants have been reported who developed tetany and seizures in the early weeks of life and who exhibited both hypocalcemia and hypomagnesemia (94). Administration of calcium proved ineffective, but administration of magnesium promptly corrected both calcium and magnesium levels. The cause of the hypomagnesemia is believed to be an inborn defect, leading to impaired intestinal absorption. Since the disease only affects boys, X-linked recessive inheritance appears likely.

Pseudohypoparathyroidism (Albright's hereditary osteodystrophy)
Patients have a short, stocky build, round face, brachydactyly and other skeletal abnormalities, such as short and wide phalanges, bowing, exostoses, thickening of the calvaria and general demineralization of the bones. They frequently have calcium deposits and metaplastic bone formation subcutaneously. Mental retardation is common as are calcifications of the basal ganglia and lenticular cataracts (95).

The disorder is caused by a genetic defect in receptors to parathyroid hormone (PTH), particularly in the kidney and skeleton. The parathyroids are hyperplastic, and both endogenously and exogenously administered PTH fails to raise the serum level of calcium or to lower the serum level of phosphorus. Ordinarily administration of PTH evokes a marked increase in intracellular cyclic AMP and in excretion of urinary cyclic AMP, whereas in most patients with pseudohypoparathyroidism urinary levels of cyclic AMP do not rise after administration of PTH. The condition is likely heterogeneous in that a few patients are recognized in whom PTH activates intracellular cyclic AMP, and its urinary excretion is elevated both in the basal state and after stimulation; however, serum calcium level does not increase (96). It is suggested that the defect in these latter patients is an inability of the target cells to respond to the intracellular cyclic AMP signal; hence the disorder is classified as pseudohypoparathyroidism, Type II.

The issue is confused by the observation that in several patients there is no relationship between the effect of parathyroid extract on urinary excretion of cyclic AMP and the phosphaturia, since normalization of the serum calcium concentration with vitamin D results in disappearance of resistance to parathyroid extract in regard to phosphaturia, but not for cyclic AMP (97). In some patients with pseudohypoparathyroidism, resistance to PTH appears to be limited to the kidneys, the bones normally being responsive to the elevated levels of circulating hormone. These patients exhibit subperiostal bone resorption, ostitis fibrosa—and in children widening and irregularity of the epiphyseal plates—and the condition is called *pseudohypoparathyroidism with osteitis fibrosa* (98). There are also patients with the usual anatomic stigmata of pseudohypoparathyroidism in whom, however, the levels of serum calcium and phosphorus are normal. The term pseudopseudohypoparathyroidism describes these patients. The term should be abandoned, however, since transition from the normo- to the hypocalcemic form is observed, and there are pedigrees with normocalcemic and hypocalcemic forms in different members.

The disorder is regarded as X-linked dominant; however, homozygous males are not more severely affected than heterozygous females, as might be expected. A father and son with the incomplete syndrome are reported; the father's sister had the complete syndrome (99). Autosomal dominant inheritance with enhanced expression in the female line was suggested. In another family autosomal recessive inheritance has been suggested (100). In view of the previously mentioned biochemical heterogeneity, different forms or degrees of PTH, or both probably exist under the rubric of pseudohypoparathyroidism.

The confusing terminology for this entire group of disorders will not in all likelihood be alleviated until more precise etiologic and genetic determinants are delineated.

The association of hypothyroidism with pseudohypoparathyroidism is reported frequently. Eight of 10 euthyroid patients with pseudohypoparathyroidism stimulated with thyrotropin-releasing hormone exhibited an excessive thyrotropic response in contrast to eight patients with idiopathic hypoparathyroidism, in all of whom response was normal (101). Prolactin deficiency has also been identified (102).

Hyperparathyroidism
Primary hyperparathyroidism may be caused by single adenoma, multiple adenomas or hyperplasia of the parathyroid glands; most cases are sporadic. Increasing evidence reflects an association between exposure of the head and neck to irradiation and the subsequent development of hyperparathyroidism (103). The disorder may be familial, most often occurring as part of the multiple endocrine neoplasia syndromes (104).

Neonatal Primary Hyperparathyroidism
When hyperparathyroidism has its onset in the early weeks of life, it may lead rapidly to death if diagnosis is delayed (105). The condition is usually caused by *clear-cell hyperplasia* of the parathyroids; subtotal parathyroidectomy is indicated. Affected siblings are recorded in two families, and there was parental consanguinity in two other instances, findings that suggest an autosomal recessive mode of inheritance (105). A newborn infant with hyperparathyroidism caused by *chief-cell hyperplasia* is recently reported (106). This patient was a member of a large kindred in which the disorder was transmitted in an autosomal dominant fashion and involved adults and children. Thus whether there is a distinct recessive form of early-onset hyperparathyroidism is moot.

Other Causes of Hypercalcemia
Increased production of parathyroid hormone often occurs as a compensatory phenomenon, usually aimed at correcting hypocalcemic states of diverse origins (secondary hyperparathyroidism). Ten other causes of hypercalcemia are unrelated to any known disorder of the parathyroid glands. They include:

- idiopathic hypercalcemia of infancy
- famial benign hypercalcemia
- vitamin D excess
- hypervitaminosis A
- thyrotoxicosis
- hypophosphatasia
- prolonged immobilization
- subcutaneous fat necrosis
- leukemia
- metaphyseal chondrodysplasia (Jansen type)

Since hypercalcemia of any origin results in a similar clinical pattern, differentiation from hyperparathyroidism is necessary.

Idiopathic hypercalcemia of infancy (Williams syndrome) is a disorder of unknown origin. Usually sporadic, the abnormality is reported in both monozygotic and dizygotic twins, siblings and second cousins (107, 108). The genetic nature of the condition remains to be established. Serum phosphorus level is usually elevated, and roentgenographically the increased bone density contrasts sharply with the rarefaction of primary hyperthyroidism. Affected persons often have characteristic elfin facies and supravalvular aortic stenosis. The hypercalcemia is usually self limiting and disappears by one year of age.

Familial benign hypercalcemia has been observed in four generations of a family (109). Extensive studies, including surgical exploration of the proband, failed to reveal any abnormality of parathyroid hormone secretion or of the parathyroids. The pedigree suggests autosomal dominant inheritance although instances of male-to-male transmission were absent.

ADRENAL GLANDS

Adrenal Cortical Hypofunction

Numerous conditions can cause adrenal hypofunction (Table 2), and it is important to note that some of the known errors of steroid biosynthesis feature both hypofunction (from decreased cortisol synthesis) and hyperfunction (from increased androgen production).

Corticotropin Deficiency
Congenital hypoplasia or aplasia of the pituitary is almost always associated with secondary hypoplasia of the adrenals, as well as with other hormonal deficiencies (see section on "Hypopituitarism"). Isolated deficiency of corticotropin is a rare lesion in all ages. More often it occurs in association with deficiency of growth hormone in patients with idiopathic hypopituitarism; indirect evidence suggests that the deficiency is secondary to deficient corticotropin releasing factor (CRF). Destructive lesions of the pituitary are the commonest cause of the deficiency, but rarely, autoimmune hypophysitis may be responsible.

Adrenal Aplasia and Hypoplasia
Aplasia or hypoplasia, or both have been noted in siblings, and the disorder seems to be a defect of organogenesis without demonstrable disturbance of pituitary function (110). Both X-linked recessive and autosomal recessive modes of inheritance are suggested by analysis of the reported cases. Most of the sporadic cases are male patients, suggesting that the X-linked form of the disorder is the commonest defect (111). Much less frequently, male and female siblings are reported, pointing to autosomal recessive inheritance. Histologic examination of the hypoplastic adrenal cortex in patients with the X-linked form of the disorder reveals disorganization and cytomegaly, findings suggesting that the gene-mediated defect may be one of disturbed cellular differentiation of adrenal cortical cells. In the recessive form of the disorder, the adrenals are miniature, but with normal architecture, resembling those of patients with corticotropin deficiency.

Onset of symptoms usually occurs during the first week of life, and until recent years relatively few patients were diagnosed and successfully treated with

Table 2. Etiologic Classification of Adrenocortical Hypofunction

Corticotropin-releasing factor deficiency:
  Hypothalmic defects (e.g., anencephaly and holoprosencephaly)
  Destructive lesions (e.g., tumor and trauma)
  Idiopathic (e.g., idiopathic hypopituitarism)
Corticotropin deficiency:
  Pituitary hypoplasia or aplasia
  Destructive lesions of pituitary (e.g., craniopharyngioma and trauma)
  Autoimmune hypophysitis
Primary adrenal hypoplasia or aplasia:
  X-linked
  Autosomal recessive
  Sporadic
Familial glucocorticoid deficiency
Inborn Defects of Steroidogenesis:
  Congenital adrenal hyperplasia (lipoid adrenal hyperplasia [desmolase defect], 3-β-hydroxysteroid dehydrogenase deficiency, 21-hydroxylase deficiency [complete defect, i.e., [salt losers) and partial defect, i.e., (nonsalt-losers])
  Isolated defects of aldosterone synthesis (18-hydroxylation deficiency, 18-oxidase deficiency [corticosterone methyl oxidase deficiency])
Unresponsiveness to mineralocorticoids:
  Pseudohypoaldosteronism
Destructive lesions of adrenal cortex:
  Granulomatous lesions
    Autoimmune adrenalitis (idiopathic Addison's disease) (isolated, associated with other autoimmune endocrinopathies or mucocutaneous candidiasis [e.g., Schmidt syndrome], adrenoleukodystrophy, X-linked, neonatal hemorrhage, acute infection [Waterhouse-Friedrichsen syndrome] and Wolman syndrome [lysosomal acid lipase deficiency])
Iatrogenic:
  Abrupt cessation of exogenous corticosteroids or corticotropin; removal of functioning adrenal tumor; adrenalectomy for Cushing syndrome; drugs (aminoglutethimide, mitotane [o, p'-DDD] and metyrapone)
Fetal adrenal suppression:
  Maternal Cushing syndrome?

adrenocortical replacement therapy. Of the six or seven surviving patients, most have now reached the usual age of puberty, but have failed to develop normally and exhibit deficiency of gonadotropins (112). Although relationship between the adrenal hypoplasia and gonadotropin deficiency is not settled, it appears that adrenal cortical steroids may be necessary in triggering the hypothalamo-pituitary-gonadal axis.

Utilization of maternal urine and serum estriol measurements may be useful in managing pregnancies where fetal adrenal hypoplasia is suspected (113). Approximately 90% of the estriol in maternal serum or urine is derived from placental conversion of fetally produced steroid precursors. In addition to fetal adrenal hypoplasia, other causes of low maternal estriol levels in the third trimester of pregnancy are fetal death, high-dose corticosteroid therapy and anencephaly, a condition complicated by adrenal hypoplasia.

More recently *placental steroid sulfatase deficiency,* an X-linked inborn error of sulfated steroid metabolism of the placenta is also found to be characteristic of very low levels of maternal estriol and must be considered in the differential diagnosis. In fetal adrenal hypoplasia, amniotic fluid concentrations of cortisol and dehydroepiandrosterone sulfate (DHEA-S) are also low, whereas in placental steroid sulfatase deficiency, levels of these steroids are normal (114).

Placental steroid sulfatase deficiency is established in a dozen patients and appears to occur in less than 1 in 5,000 pregnancies. In primiparas the defect is associated with prolonged gestation and occasional inability to induce labor, whereas the complications have not occurred in multiparas. All affected pregnancies have resulted in male infants who appeared to be clinically normal at birth. However, cultured fibroblasts from these male infants reveal that the steroid sulfatase deficiency is not only not limited to the placenta but also affects other somatic tissue and persists throughout life. After three months the infants develop ichthyosis (X-linked), which is life long (115). The defect does not affect the activities of lysosomal sulfatase, however. Estriol levels to detect the condition are useful only in the third trimester, but it should be possible to diagnosis the condition early in pregnancy by enzymatic assay in cultured amniotic fluid cells.

Familial Glucocorticoid Deficiency

This form of chronic adrenal insufficiency is characterized by isolated deficiency of glucocortiocoids and elevated levels of corticotropin, together with normal aldosterone production (116). Histologically there is marked adrenocortical atrophy with relative sparing of the zona glomerulosa. Consequently salt-losing manifestations do not occur. Instead presenting symptoms of patients include mainly hypoglycemia, seizures and pigmentation. All zones of the adrenal cortex are incapable of responding to ACTH.

The disorder affects both sexes equally and appears to be inherited in an autosomal recessive manner. The defect is not known, but it is suggested that the unresponsiveness of the adrenal cortex may be at the adrenal ACTH receptor or postreceptor site (116). If such be the case, one would expect the unresponsiveness to be demonstrable early in life. However, the prospective study of siblings of affected patients reveals normal responsiveness to ACTH early in life, with progressive deterioration later (117). These findings suggest that a progressive inherited degeneration of the adrenals may be the pathogenetic basis of the syndrome, at least for some cases. The syndrome likely consists of a heterogeneous group of disorders.

Isolated Deficiency of Aldosterone

In patients with this disorder the aldosterone secretion rate is decreased, especially given the rate of sodium depletion, whereas 17-ketosteroids, cortisol and pregnanetriol levels are normal (118). Pigmentation is also normal, and clinical manifestations are primarily those of salt loss. The error is caused by a defect either in the 18-hydroxylation of corticosterone or in the dehydrogenation of 18-hydroxycorticosterone. Since amelioration of the salt-losing manifestations is evident with increasing age, some adaptation or compensation is likely. Nevertheless the biosynthetic defect persists and can be demonstrated in adults.

The disorder has been described in siblings in three families and in 14 Iranian Jews in nine families (119). Genetic analysis of these families reveals a high coefficient of inbreeding and strongly suggests an autosomal recessive mode of transmission.

Selective hypoaldosteronism, which first develops in adults, is sporadic and usually the result of decreased renin formation together with mild-to-moderate renal failure (120).

Adrenoleukodystrophy

In this X-linked recessive disorder adrenal insufficiency is accompanied by cerebral demyelinization. The error is described as diffuse sclerosis, melanodermic leukodystrophy, sudanophilic leukodystrophy and Siemerling-Creutzfeldt disease (121-122). Symptoms usually begin between age 3 and 12, but may also begin in adulthood. The CNS clinical manifestations dominate, consisting of behavorial changes, disturbance of gait, dysarthria, dysphagia and loss of vision. Eventually seizures, spastic quadraparesis and decorticate posturing occur. About one-third of patients exhibit signs and symptoms of adrenal insufficiency, the latter which are insidious and may antedate or appear concomitantly with the neurologic manifestations. Reduced adrenal reserve may be demonstrable, even in children, without clinical manifestations of adrenal insufficiency. Cytoplasmic striations are evident in the cells of the fasiculata and reticularis zones of the adrenal, but the metabolic defect is not known.

Attempts to identify heterozygotes based on endocrinologic investigations have been unsuccessful. However, it is reported that cultured fibroblasts from affected male subjects develop characteristic morphologic anomalies, and two populations of cells have been cloned from a heterozygote (123). These early observations remain to be confirmed.

Idiopathic Addison's Disease

Although tuberculosis was once the commonest cause of Addison's disease, currently "idiopathic atrophy" is most often noted. The adrenal glands may not be visible at autopsy, and remnants of remaining tissue reveal lymphocytic infiltration. About one-half of affected persons have circulating antibodies against adrenal tissue (124), and antibodies to thyroid, ovary, testis and stomach may be detected even if there is no functional or clinical defect in the other glands. The disorder is now believed to have an autoimmune origin similar to that of lymphocytic thyroiditis. In fact the association of idiopathic adrenal insufficiency occurs frequently enough with lymphocytic thyroiditis or hypothyroidism that it has acquired the eponymic designation of Schmidt syndrome and often is complicated by diabetes mellitus (125).

Addison's disease often occurs in siblings and is reported in three pairs of identical twins, but the genetic complexities are unclear. Heterogeneity within families is common, e.g., one sibling may have adrenal insufficiency and another hypoparathyroidism (126). As for other autoimmune disorders, multifactorial inheritance appears likely. A few families are reported with Addison's disease affecting only males in more than one generation, suggesting X-linked inheritance (127). Affected persons probably do not have autoimmune adrenalitis because only males are affected, there are no other autoimmune disorders and organ specific antibodies are absent. These patients may have a milder form of congenital adrenal hypoplasia of infancy (128).

Adrenal insufficiency accompanied by calcifications within the adrenal glands occur in a variety of situations, some serious, others of no consequence. Of particular interest is the fatal infantile form of a rare autosomal recessive lipid-storage disorder characterized by extensive bilateral calcifications of the adrenal glands known as *Wolman disease* (129). Deposition of cholesterol esters and triglycerides is especially heavy in the adrenal, but the cause of the calcifications is not known. The disorder is caused by deficiency of lysosomal acid lipase. Other clinical manifestations include hepatosplenomegaly, gastrointestinal symptoms and failure to thrive with rapid deterioration, leading to death by age 3 to 4 months.

A mild form of the disorder, *cholesterol ester storage disease,* is characterized by a benign clinical course and hepatomegaly (130). These two conditions are thought to be allelic (131). Prenatal diagnosis of Wolman disease can be established by identification of the missing lysosomal acid lipase in cultured fibroblasts (132).

Adrenocortical Hyperfunction

Infants with genetic defects in cortisol synthesis and resulting virilizing adrenal hyperplasia often exhibit adrenocortical insufficiency. About one-half the infants with 21-hydroxlase defect, all the infants with lipoid adrenal hyperplasia and most infants with 3-β-hydroxysteroid dehydrogenase defect manifest salt-losing symptoms in the newborn period. Both cortisol and aldosterone are deficient.

Congenital Adrenal Hyperplasia
This autosomal recessive disorder is caused by an inborn defect in the biosynthesis of adrenal corticoids (133). Five different enzymatic defects in the pathway are known; three are characterized clinically by virilization, the others do not permit excessive androgen secretion. In many infants salt loss may also be manifested. The incidence of the condition varies with the population, but is probably of the order of 1 in 15,000 births (134). The enzymatic defects can be distinguished not only by biochemical studies but often by phenotype.

Deficiency of 21-hydroxylase accounts for 95% of affected patients. Two clinical variants occur, a salt-losing form in which the enzymatic defect is complete and a nonsalt-losing form in which the enzymatic defect is incomplete, and there is sufficient production of cortisol and aldosterone to avert salt loss. Each defect is genetically specific; if one form occurs in a family, subsequently affected infants will always have the same form—often with the same severity. In both variants excessive production of androgen results in pseudohermaphroditism in the female and in precocious pseudopuberty in the male.

Many approaches to the detection of the heterozygous carrier for congenital virilizing adrenal hyperplasia have been made, but none has been completely successful. It has recently been established that the gene for 21-hydroxylase is closely linked to the HLA-B locus. This makes it possible to detect the heteroxygous carriers and opens the way for prenatal diagnosis in families with an affected child (134a).

Deficiency of 11-β-hydroxylase is the second commonest enzymatic defect causing this syndrome. Of the 50 or so cases reported, most have been adults or children over 2 years of age. Excessive production of desoxycorticosterone occurs and results in hypertension, a characteristic of this defect. Salt-losing manifestations do not occur. In a young infant we have studied and in another previously reported infant with this defect, hypertension was absent (135). As in the deficiency of 21-hydroxylase, females have ambiguous genitalia and males are virilized.

Deficiency of 3-β-hydroxysteroid dehydrogenase is reported in less than a score of patients. Deficiency of both cortisol and aldosterone occurs. Salt wasting is usual, but incomplete defects without salt-losing manifestations are also reported (136). Girls are only slightly virilized at birth, and the enlargement of the clitoris may be mild and escape detection. Labial fusion is usual, but a female with normal genitalia has been observed. Male babies are usually incompletely virilized and manifest varying degrees of hypospadias with or without bifid scrotum and cryptorchidism. The enzyme is required for the biosynthesis of testicular hormones; its absence in fetal testes explains the incomplete virilization of males during embryonic life.

Lipoid adrenal hyperplasia is reported in about 18 patients and results from a defect very early in the biosynthetic pathway. There is failure of conversion of cholesterol into pregnenolone, causing marked accumulation of lipids and cholesterol in the adrenal cortex with total failure of synthesis of any adrenal steroids (137). The enzymatic defect in the adrenal is also in the testis, thus preventing synthesis of testicular hormones. Consequently boys are highly feminized, and girls exhibit no genital abnormality. Salt-losing manifestations are usual, and most infants die in early infancy.

Because urinary 17-ketosteroids are not elevated in this form of adrenal hyperplasia, affected infants are liable to be confused with those with adrenal hypoplasia, and all affected male babies are assumed to be females at birth. In one instance a phenotypic female infant with salt-losing manifestations at age 2 months, who was thought to have "Addison's disease," developed inguinal testes at 6 years of age and was found to have an XY karyotype; these findings led to the diagnosis of lipoid adrenal hyperplasia (138). In another instance an incomplete enzymatic defect led to partial masculinization of a male infant who did not exhibit hypoadrenalism until age 17 months (139).

17-Hydroxylase deficiency is described in only 14 adult patients (140). When this enzymatic defect occurs in the genotypic male, the fetal testis is also involved and genitalia may be ambiguous with cryptorchidism and a rudimentary vagina, or the patients may be completely feminized with inguinal testes. Female patients exhibit no secondary sexual characteristics, and amenorrhea and absence of pubic hair are common.

Virilizing Adrenocortical Tumors

Usually sporadic, however, these tumors are apt to occur in association with certain congenital defects, such as hemihypertrophy (141) and the Beckwith syndrome (142). These tumors also tend to be associated with genitourinary tract and CNS abnormalities and hamartomatous defects. Rarely the tumors are bilateral, and in at least eight instances the contralateral adrenal was absent (143). Other tumors and perhaps an increased incidence of tumors in first degree relatives has been noted.

Cushing Syndrome

Cushing syndrome, whether caused by adrenal tumor or bilateral adrenal hyperplasia, is usually a sporadic condition with little evidence of genetic factors. It is seen with Albright's syndrome and may be secondary to ectopic production of ACTH by a variety of tumors. When it occurs in families, it is almost invariably as a facet of the multiple endocrine neoplasia syndrome.

Familial Glucocorticoid Suppressible Aldosteronism

The cause of this condition is unknown, but it mimics primary hyperaldosteronism. As in the latter condition patients have hypertension, mild alkalosis, hypokalemia and increased levels of aldosterone, which are not altered by restriction or excess of sodium. Plasma renin activity is low. Administration of dexamethasone results in marked suppression of aldosterone and in the remission of the hypertension. The condition is described in siblings in two families (144-145). Its occurrence in two brothers and their mother suggests an autosomal dominant mode of transmission, but this remains to be proved.

Adrenal Medullary Hyperfunction

Catecholamine-secreting neural tumors, such as pheochromocytomas and neuroblastomas, occasionally are familial. Familial catecholamine-secreting neural tumors include Pheochromocytomas (alone, with neurofibromatosis, with hemangioblastoma of the retina and cerebellum [Lindau-vonHippel syndrome], with medullary thyroid carcinoma and parathyroid disease [MEN II] and with medullary carcinoma and mucosal neuroma [MEN III]) and neuroblastomas.

GONADS

Gonadal hypofunction may be caused primarily by a defect of testis or ovary (primary hypogonadism) or may be secondary to deficiency of pituitary gonadotropic hormones (secondary hypogonadism). Accordingly hypogonadism is classified as hyper- or hypogonadotropic. Except in children, the two are usually readily differentiated by measurement of levels of gonadotropins in blood or urine. When the levels are low, the defect may be in the pituitary or in the hypothalamus, and it is necessary to differentiate the site of the defect by stimulation tests with gonadotropin-releasing hormone.

Primary Male Hypogonadism

This disorder is caused by decreased testicular androgen production. Defects in which testosterone production involves the fetal testes result in male pseudohermaphroditism and are discussed later. The commonest cause of androgen deficiency is Klinefelter syndrome and all its variants; patients invariably have small testes.

Congenital *anorchia* is found in a small percentage of boys with bilateral cryptorchidism who are otherwise normal. In this condition it is presumed that a noxious factor injured the fetal testis sometime after sexual differentiation had taken place. The condition has been described in one of monozygotic twins and in boys who had a sib with unilateral anorchia (144, 145).

Bilateral rudimentary testes is a syndrome consisting of minute testes and a small penis and occurs in XY males with normal male ductal development. Affected siblings are reported (146).

Neither of these conditions is to be confused with the *XY gonadal agenesis syndrome* in which the externalia show ambiguity and are more nearly female.

Neither female nor male ducts are present or when present, are rudimentary. Most patients have been reared as females. Of the 12 cases reported, there was one pair of siblings. Testicular tissue was presumably present in the fetus long enough to inhibit müllerian duct development, but its Leydig cell function was minimal (147).

XY Turner Phenotype (Noonan syndrome)
This disorder is identified by various designations, including male Turner syndrome, familial Turner phenotype and Ullrich syndrome. It also occurs in female patients (XX Turner phenotype). Affected persons have normal chromosomes together with many of the somatic anomalies that occur in 45X Turner syndrome, including short stature, webbing of the neck, pectus carinatum or pectus excavatum, cubitus valgus, congenital heart disease and a characteristic facies (hypertelorism, epicanthus, antimongoloid palpebral slant, ptosis, micrognathia and ear abnormalities). Other abnormalities, such as clinodactyly, hernias and vertebral anomalies, occur less frequently. Lymphedema is less common than in 45X Turner syndrome, although mental retardation is common. The cardiac defect is most often pulmonary valvular stenosis or atrial septal defect.

In a small percentage of patients, congenital hypoplastic anemia or congenital stem cell dysfunction is reported. There is a wide spectrum of gonadal defects, varying from severe deficiency to apparently normal sexual development. Male babies frequently have cryptorchidism (over 50%) and small testes. Late puberty may be followed by completely normal development.

Although the disorder is often sporadic, partial expression of the syndrome is often present in first-degree relatives when they are examined closely, and it is considered to be caused by an autosomal dominant gene with variable expressivity (148-150). It is not understood how an autosomal genetic mutation results in a phenotype so closely resembling that which is produced by a chromosomal deficiency syndrome (45X). It is postulated that these patients may have an undetectable deletion of a homologous portion of either the X- or Y-chromosome; however, banding studies in one affected person were normal (151).

Marinesco-Sjögren Syndrome
Two kindreds are reported in which this syndrome has occurred, together with hypergonadotropic hypogonadism in nine of 10 individuals. Evidence was presented that this was the result of close linkage of the two disorders (152). The nature of the gonadal lesion was not well delineated other than to be described as germinal aplasia.

Secondary Hypogonadism

In hypogonadotropic hypogonadism there is deficiency of follicle-stimulating hormone (FSH) or luteinizing hormone (LH), or both. The primary defect may be in the anterior pituitary, or there may be a deficiency of gonadotropin-releasing hormone (LRH) in the hypothalamus. The testes are normal, but remain in the prepubertal state owing to lack of stimulation by gonadotropins. Since LRH has only recently become available, the classification of these disorders is still being evolved.

Gonadotropin deficiency is most often secondary to injury to the pituitary by tumors, hemorrhage, anoxia, and surgical maneuvers. In these instances the deficiency of gonadotropins is usually associated with other deficiencies of pituitary hormones. Deficiency of gonadotropins may also be associated with familial deficiency of growth hormone. Most often, genetic defects of the pituitary or hypothalamic gonadotropin system are isolated disorders.

Kallman Syndrome (Anosmia—hypogonadism syndrome)
Patients either fail to develop sexually or exhibit only minimal development at puberty. Another major component of the syndrome is anosmia (153). Inability to smell is present from early childhood, but it is usually not discovered except on direct questioning. Agenesis of the olfactory lobes accounts for the defect. Both LH and FSH remain at prepubertal levels in adult life. Administration of LRH to affected patients produces an increase in FSH and LH, suggesting that the deficiency is in the hypothalamus (154).

Other somatic defects have been observed in some patients with Kallman syndrome, particularly cryptorchidism, congenital deafness, harelip or cleft palate, and renal abnormalities. The disorder occurs in both males and females, and familial occurrence is reported. Although X-linked inheritance is suggested by many authors, autosomal dominance with decreased penetrance may be likelier. The expression is variable; in some kindreds there are anosmic individuals without, as well as with, hypogonadism; in other kindreds there are individuals with only harelip or cleft palate, or with only hypogonadism or anosmia. More male than female patients have been recognized with the syndrome. The author has studied one adult male with anosmia who eventually went into normal puberty at 20 years of age.

Fertile Eunuch Syndrome (Isolated LH deficiency)
In this condition the Leydig cells fail to mature at puberty (155). The size of the testes may be normal, however, and spermatogenesis may occur. Administration of chorionic gonadotropin reveals a response indicating normal Leydig-cell precursors (156). Fertility has occasionally been noted, but evidence suggests that testicular androgen is necessary for completely normal spermatogenesis. Isolated deficiency of LH is observed, but administration of LRF results in an increase in LH and in testosterone establishing the defect in the hypothalamus (156). There has been no documentation of an isolated deficiency of FSH in the male subject. This rare syndrome has been noted in brothers, but the mode of inheritance is still unclear.

A number of families have been reported in which *ataxia* was associated with *hypogonadism*. The ataxia usually is difficult to classify and the mode of inheritance usually appeared to be autosomal dominant with male sex limitation or X-linked recessive. In one well-documented family, however, two brothers and two sisters were affected out of 15 children born to second cousins, supporting an autosomal recessive mode of inheritance (157).

Several pedigrees are described in which hypogonadotropic *hypogonadism* is associated with *congenital ichthyosis*. In two families only males were affected through several generations, and the pattern of inheritance appeared to be X-linked recessive. Affected males are eunuchoid, with small genitalia and atrophic testes, which may also be undescended. In one family anosmia and mental retardation was present in all affected individuals who were examined (158). There appeared to be no linkage between the syndrome and the Xg locus.

The *Prader-Willi syndrome* is characterized by short stature, obesity, acromicria, mental retardation and cryptorchidism. During the first years of life, patients exhibit severe hypotonia with failure to thrive. As the hypotonia improves, an insatiable appetite develops, leading to generalized obesity. Many patients develop diabetes in adolescence. The face is characterized by almond-shaped eyes and a fish-like mouth. Hypogonadotropic hypogonadism occurs but is variable (159). Of the 130 patients reported in the literature, there are three instances of affected siblings. The disorder shows all gradations of severity and has a predilection for males and a low but significant risk of recurrence in siblings. It is suggested that the disorder represents the consequences of a single early localized defect in brain development (160).

Deficiency of gonadotropins and secondary hypogonadism are reported in the *mutiple lentigenes syndrome* (161), and the author has observed it in adolescents with the *Carpenter* and *Lowe syndromes* (162). In the *Laurence-Moon-Biedl* syndrome primary hypogonadism may occur on occasion, but deficiency of gonadotropic hormone is observed in a brother and sister (162). Hypogonadism is recorded as a manifestation of many other syndromes, but since the affected patients were prepubertal or not investigated by modern endocrinologic studies the cause of the hypogonadism is unknown. Neither has it been settled for most of the disorders with low gonadotropin levels if the defect is in the pituitary or in the hypothalamus.

Primary Hypogonadism in Females

As with males, primary female hypogonadism is caused most often by chromosomal abnormalities, particularly Turner syndrome (45,X), mosaics (46,XX/45,X), isochromosomes, deletion and rings of the X-chromosome. It is now established that two normal X-chromosomes are required for normal development of the ovaries.

XX Pure Gonadal Dysgenesis
Normal female subjects have been found with hypogonadism and gonadal lesions identical to those in 45,X patients (163). Since the former have none of the somatic features of Turner syndrome and since they have normal chromosomes (46,XX), their condition is termed "Pure gonadal dysgenesis." An analogous condition with similar clinical findings but with 46,XY chromosomes is discussed subsequently with male pseudohermaphroditism. the two conditions are distinct entities; in no instance have XX and XY gonadal dysgenesis been reported in the same family.

The disorder is rarely recognized in children because the externalia are normal and no other abnormalities are visible. At puberty sexual maturation fails to occur and gonadotropin levels are elevated. Affected siblings, parental consanguinity and failure to uncover mosaicism (even in the streak gonads) all point to autosomal recessive inheritance. Two families are reported in which all affected persons were deaf, suggesting that there may be two distinct genetic forms of the disorder. Gonadal tumors have not been reported in these patients.

Other Causes of Female Hypogonadism

Patients with the *Noonan syndrome* have a phenotype resembling that of Turner syndrome, but their sex chromosomes are normal. The disorder is described in the section on XY Turner phenotype.

An increasing number of young women are being seen with "streak" gonads, which contain no or only occasional germ cells. No chromosomal abnormality is found and gonadotropin levels are elevated. The cause in many instances is not known, but it is clear that *ovarian failure* may result from an autoimmune disorder of the ovary. *Autoimmune ovarian disease* occurs particularly in association with Addison's disease; in a large group of patients with the disease, 25% had amenorrhea (164).

Affected girls may not develop sexually (primary amenorrhea), or the menses may cease during the reproductive years (secondary amenorrhea). The ovaries may have lymphocytic infiltration or simply appear as streaks. Most patients have demonstrable circulating steroid cell antibodies. The risk of recurrence is presumably the same as for other autoimmune disorders, such as Addison's disease.

Girls with *ataxia-telangiectasia*, a recessively inherited disorder, often have streak gonads (ovarian hypoplasia), and there are two reports of bilateral dysgerminoma in such gonads (165, 166).

Most of the causes of *secondary hypogonadism* already discussed in the section of hypogonadism in the male also result in hypogonadism in the female. In addition isolated deficiency of FSH is described in female subjects, and it is documented that the defect resides in the hypothalamus (167).

GYNECOMASTIA

Gynecomastia, or the occurrence of mammary tissue in the male, is a common condition. Approximately two-thirds of boys develop varying degrees of subalveolar hyperplasia of the breast during their pubertal development. The condition is considered to be physiologic. In the postpubertal years Klinefelter syndrome is one of the commoner causes of gynecomastia. Other genetic causes of gynecomastia are rare (Table 3).

DISORDERS OF SEXUAL DIFFERENTIATION

A discrepancy between the morphology of the gonads and of the external genitalia is termed *hermaphroditism*. It is now well established that certain chromosomal aberrations can result in ambiguity of external genitalia; these conditions are discussed in Chapter 2. Here are discussed only those conditions of disordered sexual differentiation which are imposed on the XX or XY genotype (Table 4). An increasing number of such conditions can now be explained owing to advances in the understanding of normal sexual differentiation.

Embryonic Sexual Differentiation

In normal differentiation the final form of all sexual structures is consistent with a normal complement of the sex chromosomes (XX or XY). A complete 46,XX

Table 3. Causes of Gynecomastia

Gynecomastia of puberty
Endocrine tumors (Leydig cell, adrenocortical and chromophobe)
Chromosomal abnormalities (Klinefelter syndrome and XXYY)
Defects in steroidogenesis (17α-hydroxylase, 17, 20-desmolase, and 17-ketoreductase)
Defects in androgen action (testicular feminization syndrome, Reifenstein syndrome and Rosewater syndrome)
Familial gynecomastia with normal sexual development
delCastillo syndrome
Drug-induced (estrogens, testosterone and anabolic steroids, spironolactone, digitalis and chlorpromazine, phenothiazine and related drugs)

Table 4. Etiologic Classification of Hermaphroditism

Female pseudohermaphroditism:
1. Androgen-exposure
    Fetal source:
        Congenital adrenal hyperplasia
            (21-hydroxylase deficiency, 11-β-hydroxylase deficiency and 3-β-hydroxysteroid dehydrogenase deficiency)
        Adrenal tumor?
    Maternal source
        Virilizing tumor (ovary, adrenal)
        Androgenic drugs
        Progestational drugs
2. Undetermined origin (usually associated with other defects like skeletal, urinary and gastrointestinal tracts)

Male pseudohermaphroditism:
1. Defect in testicular differentiation (deletion short arm of Y-chromosome, XY pure gonadal dysgenesis [Swyer syndrome] and XY gonadal agenesis syndrome)
2. Defect in testicular hormones (Leydig-cell aplasia, inborn errors of testosterone synthesis [20, 22-desmolase deficiency, 3β-hyrodxysteroid dehydrogenase deficiency, 17-hydroxylase deficiency and 17-β-hydroxysteroid oxireductase deficiency])
    Defect in müllerian-inhibiting factor (defective synthesis and defective response)
3. Defect in androgen action
    Defect in conversion of testosterone to dihydrotestosterone
        5-α reductase deficiency (pseudovaginal perineoscrotal hypospadias)
    Testicular feminization syndrome (absent cytoplasmic receptor, defect in translocation of steroid–receptor complex to nucleus?)
    Incomplete testicular feminization
    Reifenstein syndrome (Lubs; Gilbert-Dreyfus; Rosewater syndromes)
        Decreased cytoplasmic receptor
        Normal cytoplasmic receptor
4. Undetermined
    Syndrome of male pseudohermaphroditism, Wilms' tumor, aniridia and nephrosis
    Other genetic syndromes

True hermaphroditism:
  XX
  XY
  XX/XY chimeras

chromosomal complement is necessary for the development of normal ovaries. Deletion of any portion of an X-chromosome results in streak gonads.

The Y-chromosome determines the male sex by causing the initially indifferent gonad to differentiate as a testis in the fifth to sixth week of intrauterine life (168). It is now clear that there are genes on the short arm of the Y chromosome that code for H-Y antigen, a cell-surface component associated with testicular differentiation (169). Those H-Y genes and the primary male-determining genes are probably identical. A serologic assay for H-Y antigen is now available to detect the Y-chromosome genes even in the absence of any evidence of the Y-chromosome in the karyotype. Thus both XX males and XX true hermaphrodites have H-Y antigenic material and therefore may harbor a Y/X or Y/autosome translocation.

Deletion of the short arm of the Y results in female phenotype with streak gonads, but even extreme deletions of the long arm of the Y are compatable with normal male development (170). In the XX fetus, the female phenotype develops simply because of the absence of the Y determining genes and absence of H-Y antigen. The same bipotential gonad in the H-Y negative fetus develops into an ovary, but this does not occur until about the 12th week. Once the indifferent gonad has differentiated into a testis, it begins to produce hormones and masculinization of the fetus begins at about 8 weeks (171). Absence of these hormones during the critical period results in failure of masculinization. The fetal testes secrete two substances during this period of masculinization, one of which is testosterone.

Secretion of testosterone probably occurs in response to stimulation by placental chorionic gonadotropin (HCG). Testosterone apparently acts directly to initiate virilization of the wolffian ducts into the epididymis, vas deferens and seminal vesicle. Testosterone is also converted by 5α-reductase to another active metabolite, dihydrotestosterone, the androgen causing virilization of the urogenital sinus and the external genitalia (172). A functional androgen receptor controlled by an X-linked gene is required for testosterone to give a masculine phenotype to XY individuals. Even in the presence of H-Y antigen of testes and of normal androgen receptors, normal masculinization may not occur if there is a defect in the synthesis of testosterone. The pathway for testosterone biosynthesis is given in Figure 1, which also indicates the various biosynthetic defects.

The second substance produced by the fetal testis is the müllerian-inhibitory factor (MIF), a high molecular-weight protein produced by the Sertoli cells (173); MIF causes the müllerian ducts to regress; in its absence, they persist. Essentially it can be said that maleness is imposed on a basic female fetus by these substances secreted by the fetal testis. Defects are now known at a number of developmental stages, and an appreciation of these basic principles greatly enhances an understanding of various types of hermaphroditic conditions.

Female Pseudohermaphroditism

In the femal pseudohermaphrodite the genotype is XX, the gonads are ovaries and the internal genitalia are female, but the external genitalia are virilized or ambiguous. The mechanisms in normal female differentiation are considerably less complex than those required for male differentiation and the varieties and causes of female pseudohermaphroditism are fewer.

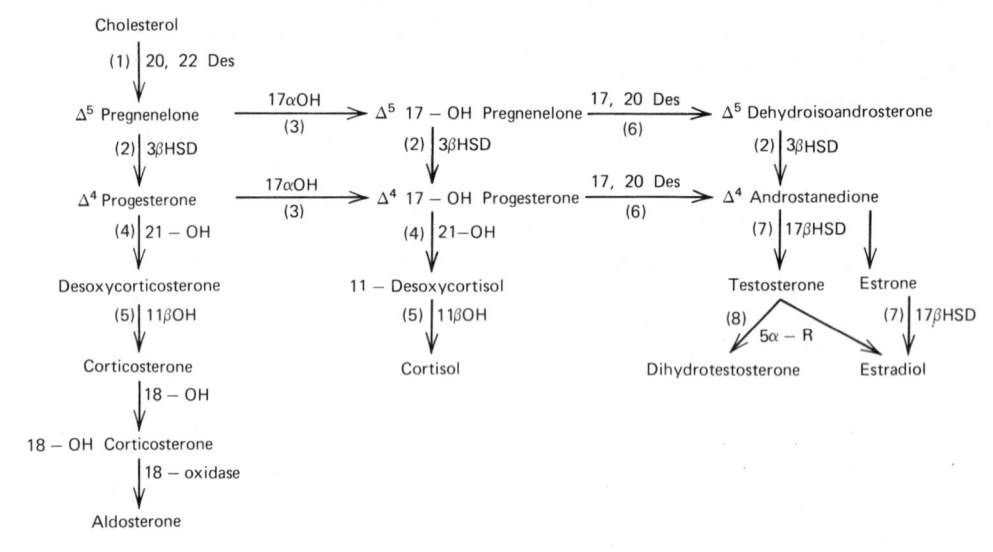

```
Cholesterol
   │ (1) 20, 22 Des
   ▼
Δ⁵ Pregnenelone ──17αOH(3)──▶ Δ⁵ 17-OH Pregnenelone ──17,20 Des(6)──▶ Δ⁵ Dehydroisoandrosterone
   │ (2) 3βHSD                    │ (2) 3βHSD                              │ (2) 3βHSD
   ▼                              ▼                                        ▼
Δ⁴ Progesterone ──17αOH(3)──▶ Δ⁴ 17-OH Progesterone ──17,20 Des(6)──▶ Δ⁴ Androstanedione
   │ (4) 21-OH                    │ (4) 21-OH                              │ (7) 17βHSD
   ▼                              ▼                                        ▼
Desoxycorticosterone           11-Desoxycortisol                       Testosterone        Estrone
   │ (5) 11βOH                    │ (5) 11βOH                    (8) ╱ 5α-R ╲ (7) 17βHSD
   ▼                              ▼                          Dihydrotestosterone   Estradiol
Corticosterone                 Cortisol
   │ 18-OH
   ▼
18-OH Corticosterone
   │ 18-oxidase
   ▼
Aldosterone
```

Figure 1. Enzymatic defects in synthesis of cortisol (congenital adrenal hyperplasia) and in synthesis of testosterone (male pseudohermaphroditism). **Key:**

1. 20,22 desmolase
2. 3β-hydroxysteroid dehydrogenase
3. 17α-hydroxylase
4. 21-hydroxylase
5. 11β-hydroxylase
6. 17,20 desmolase
7. 17β-hydroxysteroid oxireductase
8. 5α-reductase.

Congenital Adrenal Hyperplasia

This is by far the commonest cause of the condition. Females with the 21-hydroxylase and 11-hydroxylase defects are the most highly virilized, although minimal virilization also occurs with the 3β-hydroxysteroid dehydrogenase defect. Salt-losers tend to have greater degrees of virilization than have nonsalt-losers. The masculinization may be so intense as to result in a completely penile urethra and may mimic a male with cryptorchidism (see section on "Adrenal Glands").

Masculinizing Maternal Tumors

In several instances the female fetus has been virilized by a maternal androgen-producing tumor (174, 175). The lesion may be an adrenal adenoma or an ovarian tumor, particularly, an arrhenoblastoma, luteoma or Krukenberg tumor. The newborn infant shows both enlargement of the clitoris of varying degrees and often labial fusion.

Administration of androgenic drugs to women during pregnancy can result in female pseudohermaphroditism. In some instances testosterone and 17-methyltestosterone are reported to be the masculinizing agents, but the most cases have resulted from the certain progestational compounds for the treatment of threatened abortion (176). Within the past decade most of these progestins have been replaced by nonvirilizing ones; accordingly this form of female pseudohermaphroditism has become less frequent.

Infants with female pseudohermaphroditism are reported for whom no masculinizing agent can be identified. In such instances the disorder is usually associated with other congenital defects, particularly of the urinary and gastrointestinal tracts. No etiologic factors are known.

Male Pseudohermaphroditism

In male pseudohermaphroditism the genotype is XY, the internal gentalia are usually but not always male and the external genitalia are incompletely virilized, ambiguous or completely female. When gonads can be found, they are invariably testes although their development may be either rudimentary or normal. Because of the complexity of normal virilization in the fetus, it is not surprising that many varieties of male hermaphroditism have been delineated; the topic is extensively covered in a recent review (177).

XY Pure Gonadal Dysgenesis (Swyer syndrome)

This disorder is known as "pure" gonadal dysgenesis to distinguish it from other forms of chromosomal origin associated with somatic anomalies. Patients have a completely female phenotype, including vagina, uterus and fallopian tubes, but the gonads consist of almost totally undifferentiated streaks despite a Y-chromosome. At puberty there is primary amenorrhea and failure of breast development. There may be hilar cells in the undifferentiated gonad capable of producing some androgens; accordingly some postpubertal virilization, such as clitoral enlargement, may occur. Growth is normal and there are no other associated defects.

It is believed that the gonads in these patients have failed to differentiate and cannot accomplish even the most primative testicular functions, such as regression of müllerian ducts. Several pedigrees are identified; the disorder is probably determined by an X-linked locus, suggesting that a factor on the X-chromosome is necessary for development of the indifferent gonad into a normal testis. The streak gonads undergo neoplastic changes, specifically gonadoblastomas and dysgerminomas, apparently earlier than in the testicular feminization syndrome. The gonads should therefore be removed shortly after ascertainment, irrespective of age (178).

Pure gonadal dysgenesis also occurs in XX individuals (see section on "XX Pure Gonadal Dysgenesis").

XY Gondal Agenesis Syndrome

In this rare syndrome the externalia are slightly ambiguous, but are more nearly female than male (179, 180), with hypoplasia of the labia, some degree of labioscrotal fusion, a small clitoris-like phallus, a perineal urethral opening and (usually) no vagina. There is neither a uterus nor gonadal tissue. At puberty there is no sexual development, and gonadotropin levels are elevated. Most patients have been reared as females. It is presumed that testicular tissue was in the fetus long enough to inhibit mullerian-duct development, but that its Leydig cell function was minimal.

In at least one young child with XY gonadal agenesis in whom no testes could be found, the rise in testosterone levels after stimulation with human chorionic gonadotropin was significant, indicating that some functional Leydig cells were present. Two siblings with the disorder are reported, and the syndrome may have a genetic basis, although the mode of inheritance is uncertain.

The disorder differs from bilateral *anorchia*, a condition in which the testes are absent but there is a complete male phenotype. In the latter condition it is presumed that tissue with fetal testicular function was present at the critical period of genital differentiation but that sometime later it was damaged. Bilateral anorchia is reported in identical twins, and unilateral anorchia is reported in both identical twins and siblings, suggesting that it too may have a genetic predisposition (144, 145).

Defects in Testicular Hormone Synthesis

The fetal testis produces testosterone, and five genetic defects in its enzymatic synthesis have been delineated. More recently a defect in Leydig cell differentiation has been described. All these defects result in male pseudohermaphroditism from inadequate masculinization of the XY fetus.

The one patient with *Leydig cell aplasia* is a phenotypic female adult with a shallow vagina, absent uterus and bilateral inguinal testes (181). Unlike patients with testicular feminization, she had no breast development; pubic hair distribution was that of a normal female, and levels of testosterone were low and did not rise after stimulation with chorionic gonadotropin. Plasma levels of LH were elevated, but FSH levels were normal; no Leydig cells were in the testes. All these findings can be explained on the basis of selective agenesis of the Leydig cells.

In *20, 22 desmolase deficiency*, the enzyme 20, 22 desmolase is required early in the biosynthetic pathway for both hydrocortisone and testosterone. Deficiency is also known as lipoid adrenal hyperplasia (see section on "Adrenal Glands"). The fetal testis cannot synthesize testosterone, and affected males are regarded as normal females until salt-losing symptoms intervene. In at least one instance a partial defect resulted in a partially masculinized male with ambigious genitalia who had delayed onset of salt-losing symptoms (139).

Males with *3-β-hydroxysteroid dehydrogenase deficiency*, a form of congenital adrenal hyperplasia (see section on "Adrenal Glands") have varying degrees of hypospadias, with or without bifid scrotum and cryptorchidism. Affected infants usually develop salt-losing manifestations shortly after birth; incomplete defects are reported, and in a pubertal boy the defect seemed to be complete in the adrenal, but only partial in the testis (182).

Deficiency of 17-α-hydroxylase has thus far been detected in only a few patients (183). The genitalia are ambiguous, with hypospadias, cryptorchidism and a rudimentary vagina. The phallus may be so small as to suggest a female phenotype, and several patients were reared to adult life as females. Because of the overproduction of deoxycorticosterone (DOC) and corticosterone, hypertension and hypokalemic alkalosis are characteristic, although in less severly affected males the blood pressure may be normal in early life. With failure of adrenal and testicular synthesis of androgens, puberty does not occur and the patient remains eunuchoid. Absence of müllerian duct remnants indicates that fetal production of MIF is normal.

The diagnosis can be suspected after puberty on the basis of low levels of 17-ketosteroids, 21-hydroxycorticoids and plasma androgens. To establish the diagnosis before puberty, it is necessary to determine secretion rates of DOC, corticosterone, cortisol and compound S. The defect is inherited in autosomal recessive fashion.

A kindred is reported in which two first cousins and a maternal "aunt" had ambiguous external genitalia and XY constitution, with a defect in testosterone secretion *(deficiency of steroid 17, 20-desmolase)* (133). Biochemical studies established a deficiency of the enzyme that cleaves to the side-chain of 17-αhydroxypregnenolone and 17-α-hydroxyprogesterone (Fig. 1). Consequently there was deficiency of testosterone and of dehydroepiandrosterone (DHA). The enzymatic deficiency also involved the adrenal, since ACTH administration failed to increase DHA excretion. The inguinal testes in these patients revealed no specific abnormalities; there were no mullerian structures in the two cousins. The defect was probably incomplete, since one would anticipate complete feminization with total absence of testosterone. The diagnosis can be suspected in male pseudohermaphrodites with histologically normal testes who fail to exhibit a rise in plasma level of testosterone following stimulation with HCG. The mode of transmission is not clear, both X-linked and autosomal recessive inheritance being possible.

A *defect in testicular 17-ketosteroid reductase* (17-β-hydroxysteroid oxireductase) is identified as a cause of male pseudohermaphroditism (184). Affected persons were completely feminized and reared as girls until virilization, primary amenorrhea and, in some patients, gynecomastia occurred at puberty. A shallow vagina is present, but no cervix or uterus. The defect results in low plasma levels of testosterone and in marked accumulation of its precursor, androstanedione (Fig. 1). The defect has been detected only in adults.

In prepubertal children the disorder can be easily confused with the testicular feminization syndrome to be discussed subsequently. In one family the occurrence in siblings with first-cousin parents suggests autosomal recessive transmission. Removal of the defective testes prevents or halts virilization. Replacement therapy with estrogens is indicated.

In the *uterine–hernia syndrome* fetal testosterone production is normal and affected males are completely virilized (185). However, deficiency of MIF results in persistence of müllerian ducts, which are usually detected when surgical correction of an inguinal hernia in an otherwise normal male discloses uterus and fallopian tubes. The degree of müllerian development is variable and may be asymmetrical. Testicular function, including spermatogenesis, may be normal. At least six different sibships, each with multiple affected males, are reported. This suggests that the disorder is inherited as a recessive trait, either X-linked or autosomal. Treatment consists of removal of as much of the müllerian structures as possible without damaging testis, epididymis or vas deferens.

Defective Androgen Action
In the following group of disorders it is now established that fetal synthesis of testosterone is normal and that the defective virilization results from an inherited abnormality in androgen action.

Most patients with pseudovaginal perineoscrotal hypospadias (PPSH)—a descriptive term used before enzymatic defects in testosterone metabolism were recognized—have a deficiency in *steroid 5-α-reductase* (186). Affected boys have a small phallus, bidfid scrotum, urogenital sinus with perineal hypospadias and blind vaginal pouch. Testes are in the inguinal canals or labioscrotal folds and are histologically normal. There are no mullerian structures, and the vas deferens, epididymus and seminal vesicles are present. At puberty masculinization occurs normally, the phallus enlarges, the testes decend and grow normally and spermatogenesis occurs. Beard growth is scanty, acne is absent, the prostate is small and temporal hairline recession does not occur.

Plasma levels of testosterone are normal, but 5-α-dihydrotestosterone is markedly reduced. Virilization of the wolffian duct is due to the action of testosterone itself, whereas masculinization of the urogenital sinus and external genitalia depends on the availability of dihydrotestosterone during the critical period of fetal masculinization. Growth of facial hair and of the prostate also appears to depend on dihydrotestosterone. The disorder is inherited as an autosomal recessive trait and is limited to male subjects; homozygous girls are normal with normal fertility, indicating that there is no role in females for dihydrotestosterone in sexual differentiation or in ovarian function later in life. An extended pedigree of 38 affected males in 24 related families from the Dominican Republic has been extensively studied (186).

One of the commoner and extreme examples of failure of virilization is the *testicular feminization syndrome* (177). These XY patients appear to be female at birth and are invariably reared accordingly. The external genitalia are female, the vagina ends blindly in a pouch and the uterus is absent. The gonads are testes, which consist largely of seminiferous tubules. They are usually intra-abdominal, but may descend into the inguinal canal. At puberty there is normal development of breasts and the habitus is female, but menstruation does not occur and sexual hair often is absent. The psychosexual orientation of these persons is entirely female.

The testes of affected adults produce normal male levels of testosterone. It is firmly established that the absence of androgenic effects is due to a striking resistance to the action of endogenous or exogenous testosterone at the peripheral cellular level. Affected persons can convert testosterone to $5\text{-}\alpha\text{-}$dihydrotestosterone, the biologically active androgen at the cellular level. Current evidence suggests that of the two distinct genetic variants (187), the one is characterized by a deficiency of the cytosol recepetor for dihydrotestosterone. In the second type the cytosol receptor is present and the precise biochemical defect remains unknown. These two variants are identical clinically. Failure of normal male differentiation in fetal life reflects the defective response to testicular androgens.

Prepubertal affected children are often recognized when inguinal masses prove to be testes, or when a testis is unexpectedly found during herniorrhaphy in an apparent female. Examination of the chromosomes is indicated for any girl with an inguinal hernia, since 1 to 2% of them have this syndrome.

The disorder is inherited as an X-linked recessive trait. There is an analogous condition of androgen insensitivity in the mouse (tfm), which is also X-linked. Affected persons should always be reared as female. The testes should be removed since there is about a 4% incidence of tumors by age 25, and 22% by age 50 (188). Some investigators recommend not removing the testes until after puberty and completion of secondary sexual development, but a testicular tumor has been found in an affected 18-month-old infant. To relieve parental anxiety and avoid adverse effects on the child's psychosexual orientation, it is recommended that the testes be removed as soon as the condition is recognized.

In the disorder known as *incomplete testicular feminization,* patients exhibit some degree of masculinization and at birth may have phallic enlargement and labioscrotal fusion. The vagina ends blindly and the uterus is absent. Testes are present in the inguinal canal or in the labioscrotal folds. At puberty breast development occurs as well as axillary and pubic hair. The hereditary pattern is not clear, and the "complete" and "incomplete" forms have not been reported in the same family. Presumably these patients have a lower degree of insensitivity to androgens than those with the complete syndrome, but the precise biochemical defect is not known.

Another type of male pseudohermaphroditism is the *Reifenstein syndrome*. Caused by decreased end-organ responsiveness rather than by decreased androgen production, the disorder is best described as *partial androgen insensitivity* (189). In childhood the disorder is characterized by severe perineal hypospadias and by small testes, which may be in the scrotal sac or may be cryptorchid. Phallic size is usually normal, and affected persons are regarded as male. After puberty there is inadequate masculinization and lack of facial hair and voice change. Female escutcheon, azospermia and infertility are usual findings.

Study of affected families reveals considerable variability, ranging from a mild defect (microphallus and bifid scrotum) to the severe abnormality already described (189). It is now believed that three other syndromes of defective virilization described by Lubs, Gilbert-Dreyfus, and Rosewater and co-workers represent variable manifestations of the same defect as in the Reifenstein syndrome. In adults, plasma levels of testosterone and of dihydrotestosterone are normal or elevated. Levels of LH, and often of FSH, are also elevated. Androgen receptor studies in skin fibroblasts reveal low or normal androgen binding capacity. Thus there appear to be two variants of the syndrome, as in the testicular feminization syndrome. The cause of the androgen insensitivity in patients with normal dihydrotestosterone binding is not known. The disorder is believed to be inherited as an X-linked recessive trait.

There are other XY male pseudohermaphrodites, with much variability in external and internal genitalia and with varying degrees of phallic and mullerian development (177). Testes may be histologically normal or rudimentary, or there may be only one. Some of the reported cases may belong to one of the previously discussed categories, but have not been adequately studied by the newer techniques. Since genital ambiguity is associated with a wide range of chromosomal aberrations, they must always be considered in the differential. The commonest condition in this category is the XO/XY syndrome. It may be necessary to examine various tissues to establish the mosaic condition. A large number of complex genetic syndromes also exist, many of them resulting from single gene mutation, which are associated with degrees of ambiguity of the genitalia, particularly in the male. These entities must be identified on the basis of the associated extragenital malformations.

About a dozen XY male pseudohermaphrodites are reported to have developed *Wilms' tumors*, usually in the first 2 years of life (190). Some patients also had glomerulonephritis, the nephrotic syndrome, aniridia and/or gonadoblastomas (191). The reason for these associations is not understood, but they are consistent with the increasingly better recognized relationship between oncogenesis and certain types of congenital malformations, particularly those of the genitourinary tract.

True Hermaphroditism

In true hermaphroditism both ovarian and testicular tissue are in either the same or opposite gonads (177). The phenotype may be male or female, but ambiguity of the externalia is most frequent. The commonest mosaic conditions are XO/XY, XX/XXY and XX/XY.

Chromosomal examination of 119 true hermaphrodites discloses that 50% of them are XX, 20% are XY and only 30% are mosaics (192). Mosaicism may, on the one hand, be difficult to establish, requiring study of many different tissues and its possibility can never be eliminated. On the other hand, some instances of XX true hermaphroditism have been very intensively investigated, with no evidence being found of a Y-chromosome. It was proposed that in such patients the portion of the Y-chromosome containing the male-determining genes had been translocated to the X-chromosome or to one of the autosomes (193). In a few cases cytologic evidence for such translocations has been presented. More recently a number of XX true hermaphrodites with no cytologic evidence of translocation have been found with H-Y antigen, indicating Y-linked male-determining genes and Y-X or Y-autosomal translocation (193). The situation is comparable to that which occurs in XX males.

Patients with XX/XY mosaicism are of special interest and the best understood of the true hermaphrodites. Of the 12 reported cases, nine were whole-body chimeras, i.e., they were derived from more than one zygote, an observation that is usually established by blood group studies (194). The presence of both paternal alleles for some blood groups and of both maternal alleles for other blood groups is clear evidence for chimerism. Diverse mechanisms are possible, including fusion of early zygotes or double fertilization of a double-nucleated ovum.

Endocrine Pancreas

A variety of disease processes may cause pancreatic endocrine hypofunction; most are not heritable. The most common pancreatic disease generally classified as being genetic in origin is diabetes mellitus. Current evidence supports the contention that the clinical condition can no longer be considered a single entity, but a syndrome with a multiplicity of causes, many of which have at least a partial genetic component (195). There would appear to be at least five different general types of diabetes mellitus and within each type additional heterogeneity has either already been established or is suggestively present (Table 5).

Table 5. Various Types of Diabetes Mellitus

Juvenile-onset

Associated with HLA-B8/D locus
Associated with HLA-B15/C locus

Maturity-onset
Maturity-onset in the young
Various syndromes with diabetes

    Diabetes as primary facet of defective gene action
    Diabetes as a secondary event

Diabetes secondary to acquired disease

Juvenile-Onset Diabetes (JOD)

Insulin-dependent diabetes was classically considered to be juvenile-onset disease. Recent immunologic studies have suggested that the basic problem may be similar in older, insulinopenic diabetics (196). JOD has been found to be positively associated with certain HLA antigens, particularly HLA-B8 and BW15 of the serologically defined portion of the major histocompatibility complex (MHC) and DW3 as determined by mixed lymphocyte reactivity (197). Many workers feel that the HLA region is linked to certain immune response (Ir) genes and that it is actually abnormalities in these genes that make an individual unduly susceptible to infectious, diabetogenic factors. Thus, the primary genetic error might be in the immune system with resultant inability to respond to certain viral agents, damage to the islet cells, and subsequent autoantibody formation. A number of epidemiologic studies have implicated a variety of viral agents in the etiology of JOD including coxsackie B, rubella, mumps, cytomegalovirus, EB virus and infectious hepatitis virus and it is possible that any of these viruses might be responsible for the initiation of symptoms in any given patient or set of patients. A classification of JOD which includes autoimmune, viral and intermediate types has been developed although supportive information is hardly firm at the present time (198).

It has been postulated that JOD is a recessively inherited disorder, with 50% penetrance, the gene determining the condition being closely linked to the HLA-D locus (199). However, the data are equally compatible with other interpretations including multifactorial or additive inheritance or even overdominance (200). Moreover, there is good evidence to support the contention that JOD associated with the B8 haplotype is clinically different from that seen in conjunction with BW15 (201). For example, it is known that there is only 50% concordance for monozygotic twins with JOD, whereas for maturity-onset disease concordance is nearly 100%. The low figure has been used to support a stronger environmental component in the etiology of JOD. However, a closer look at the data reveals that only in concordant twins is diabetes associated with B8, whereas BW15 is associated with both the concordant and discordant pairs. The relative risks for both HLA antigens are additive; i.e. the risk for individuals (and their sibs) who possess both B8 and BW15 is greater than if either is present alone. It also appears that the association of JOD with HLA-B8 occurs because of linkage disequilibrium of B8 with DW3; i.e., the increase in B8 is secondary. The increase in BW15 also may be secondary to linkage disequilibrium with the C locus. Since the D and C loci are on opposite sides of B in the MHC, the implication is that there are at least two distinct diabetogenic genes. Clinically, this supposition is supported by the finding that the diabetes found in conjunction with B8/DW3 tends to develop earlier, is more commonly associated with islet-cell antibodies and cell mediated immunity to islet-cells, shows an increased susceptibility to microangiopathy, and occurs more commonly with other autoimmune endocrine disease (202). Patients with this form of autoimmune diabetes also would appear to be at significant risk for developing other endocrine gland failure. Thus, while the chance that a sib of a patient with JOD (who bears the same B8-DW3 haplotype) will himself develop JOD may be relatively small; i.e. from 5-10% by actual family studies, the risk that he will develop any one or more of the endocrine conditions in the autoimmune spectrum may be substantially higher. Families have now been reported with multiple members affected with autoimmune endocrinopathy, including insulin-dependent diabetes, in which the predisposition to disease of this type is strongly correlated with B8/DW3 and behaves for practical purposes like an autosomal dominant trait (203,204). Possession of this genetic predisposition may well confer on its recipient a greatly increased risk of developing pluriglandular deficiency.

Maturity-Onset Diabetes (MOD)

This type of diabetes characteristically appears in the fifth decade or later, shows no HLA association, is frequently seen in conjunction with obesity and may actually be associated with hyperinsulinism. As mentioned previously, monozygotic twin concordance in MOD approaches 100% when the index patient is 45 years of age or older. Despite this high twin concordance, no clear pattern of inheritance has emerged. This may be due to the relatively late onset of clinical disease and the necessity to wait until offspring are 40 or over before considering an individual as being affected. Attempts have been made to define the prediabetic state in younger children of such diabetic patients, generally without much success. Then too, it is possible that multiple genotypes are contained within the population of patients with MOD, as has been shown by studies of various different ethnic groups who, despite similarities in diet and activity, have widely different incidences of MOD, and diabetic complications (195).

Maturity-Onset Diabetes in the Young (MODY)

A subgroup of patients has been identified who have mild, noninsulin-requiring diabetes frequently with childhood or adolescent onset, a prolonged clinical course generally free of microangiopathic complications and a family history compatible with autosomal dominant inheritance (205). The existence of such patients again stresses the need for more careful attention to both the phenotype and to the pattern of transmission of the diabetes in a given family since inappropriate "lumping" obscures etiologic heterogeneity.

Syndromes With Diabetes

A number of distinct genetic syndromes have been identified in which glucose intolerance has been seen consistently enough to qualify as an integral feature of the condition (Table 6). In certain of these disorders, (panhypopituitarism, isolated growth hormone deficiency, pheochromocytoma) the glucose intolerance is simply a chemical phenomenon. In others, like the muscular dystrophies and some of the neurodegenerative disorders, the glucose intolerance may be due to a combination of factors including delay in disposal of glucose occasioned by a decreased muscle mass, inactivity and even poor nutrition. However, in other neuromuscular degenerative disorders such as Huntington's chorea and myotonic dystrophy, hyperinsulinemia has been found in conjunction with hyperglycemia. The carbohydrate intolerance generally has not been demonstrated prior to the diagnosis of the neuromuscular degeneration so a cause and effect relationship cannot be excluded. By the same token, patients with ataxia-

Table 6. A Partial List of Genetic Syndromes Associated with Glucose Intolerance

Hereditary pancreatitis
Cystic fibrosis
Hemochromatosis
Isolated growth hormone deficiency
Panhypopituitarism
Pheochromocytoma
Glycogen storage disease type I
Acute intermittent porphyria
Hyperlipidemias, some types
Ataxia telangiectasia
Myotonic dystrophy
Lipoatrophic diabetes
Muscular dystrophies
Huntington's chorea
Optic atrophy-diabetes mellitus syndrome
Friedreich ataxia
Alstrom syndrome
Laurence-Moon syndrome
Cockayne syndrome
Werner syndrome
Prader-Willi syndrome
Infantile-onset diabetes with epiphyseal dysplasia
Trisomy 21
Klinefelter syndrome
Turner syndrome

telangiectasia who also have insulin-resistant diabetes, have a variety of other primary immunologic abnormalities that may be more important than either neuromuscular degeneration or poor nutrition in the pathogenesis of their glucose intolerance. Some conditions, notably the Alstrom, Laurence-Moon and Prader-Willi syndromes, feature pronounced obesity and it is quite possible the glucose intolerance in affected individuals is a secondary phenomenon. The various syndromes with partial or complete lipoatrophy are interesting, since all feature nonketotic insulin-resistant diabetes, and some have been shown to have insulin receptor abnormalities (206). In older individuals without a positive family history, antireceptor antibodies have been found, along with other autoimmune phenomena. In adolescent-onset disease, which is probably a dominant disorder, a primary receptor defect has been postulated. The status of the insulin receptor in the congenital form of the condition (the Seip-Bernardinelli syndrome), an autosomal recessive disease, is unknown.

All of the foregoing supports the contention that diabetes mellitus is a phenotype due to any number of genetic errors. Although accurate genetic counseling will have to await more precise delineation of the molecular lesions involved in the production of hyperglycemia, some strides have been made such that reasonable genetic advice can be offered.

REFERENCES

1. Rimoin DL: Hereditary forms of growth hormone deficiency and resistance. *Birth Defects* 12:15, 1976

2. Hintz HL, Menking M, Sotos JF: Familial holoprosencephaly and endocrine dysgenesis. *J Pediatr* 72:81, 1968
3. Rudman D, Patterson JB, Heymsfield SB, et al: Prevalance of growth hormone deficienty in children with cleft lip or palate. *Clin Res* 26:72A, 1978
4. Rapport EB, Ulstrom RA, Gorlin RJ, et al: Solitary maxillary central incisor and short stature. *J Pediatr* 91:924, 1977
5. Brook CGD, Sanders MD, Hoare R: Septo-optic dysplasia. *Br Med J* 2:811, 1973
6. Steiner MM, Boggs JD: Absence of pituitary gland, hypothyroidism, hypoadrenalism and hypogonadism in a 17-year-old dwarf. *J Clin Endocrinol Metab* 25:1591, 1965
7. Lovinger RD, Kaplan SL, Grumbach MM: Congenital hypopituitarism associated with neonatal hypoglycemia and microphallus. *J Pediatr* 87:1171, 1975
8. Human SP: Hepatic dysfunction associated with neonatal hypopituitarism. *J Pediatr* 89:336, 1976
9. Sadeghi-Nejed A, Senior B: A familial syndrome of isolated "aplasia" of the anterior pituitary; diagnostic studies and treatment in the neonatal period. *J Pediatr* 84:79, 1974
10. Birkeland SA: The urinary excretion of oestriol in a case of pregnancy with normocephaly and bilateral adrenal aplasia. *Dan Med Bull* 16:249, 1969
11. Rosenfield RL, Root AW, Bongiovanni AM, et al: Idopathic anterior hypopituitarism in one of monozygous twins. *J Pediatr* 70:114, 1967
12. Rimoin DL, Schimke RN: Genetic Disorders of the Endocrine Glands. St. Louis, Mosby, 1971
13. Zergollern L: A follow-up on Hanhart's Dwarts of Krk. *Birth Defects* 10:28, 1971
14. Laron Z, Pertzelan A, Mannheimer S: Genetic pituitary dwarfism with high serum concentrations of growth hormone. *Isr J Med Sci* 2:152, 1966
15. Laron Z: The syndrome of familial dwarfism and high plasma immuno reactive growth hormone. *Birth Defects* 10:23, 1974
16. Lemons RD, Costin G, Kogut MD: Laron dwarfism: Growth and immunoreactive insulin following treatment with human growth hormone. *J Pediatr* 88:247, 1976
17. Jacobs LS, Sneid DS, Garland JU, et al: Receptor-active growth hormone in Laron Dwarfism. *J Clin Endocrinol Metab* 42:403, 1976
18. Rona RJ, Tanner JM: Aetiology of idiopathic growth hormone deficiency in England and Wales. *Arch Dis Childh* 52:197, 1977
19. Schimke RN, Spaulding JJ, Hallowell JG: X-linked congenital panhypopituitarism. *Birth Defects* 10:21, 1971
20. Phelan PD, Connelly J, Martin FR, et al: X-linked recessive hypopituitarism. *Birth Defects* 10:24, 1971
21. Zipf WB, Kelch RP, Bacon GE: Variable X-linked recessive hypopituitarism with evidence of gonadotrophin deficiency in two pre-pubertal males. *Clin Genet* 11:249, 1977
22. Perrier PE, Stone EF Jr: Familial pituitary dwarfism associated with an abnormal sella turcica. *Pediatrics* 43:858, 1969
23. Parks JS, Tenore A, Kirkland RT, et al: Familial hypopituitarism with large sella turcica. *Clin Res* 25:68A, 1977
24. Merimee TJ, Rimoin DL, Cavalli-Sforza LL: Metabolic studies in the African pygmy. *J Clin Invest* 51:395, 1972
25. Clark W, Weldon V: Idopathic growth hormone deficiency and Fanconi's anemia. *J Pediatr* 86:814, 1975
26. Sadeghi-Nejad A, Senior B: Autosomal dominant transmission of isolated growth hormone deficiency in iris-dental dysplasia (Rieger's syndrome). *J Pediatr* 85:644, 1974
27. Latorre H, Kenney FM, Lahey ME, et al: Short stature and growth hormone deficiency in histocytosis X. *J Pediatr* 85:813, 1974
28. Lesti J, Lesti S, Perheentupa J, et al: Absence of IgA and growth hormone deficiency associated with short arm deletion of chromosome 18. *Arch Dis Childh* 48:320, 1973
29. Sotos JF, Cutler SA: Cerebral gigantism. *Am J Dis Childh* 131:625, 1977

30. Sugarman GI, Heuser ET, Reed WB: A case of cerebral gigantism and hepatocarcinoma. *Am J Dis Childh* 131:631, 1977
31. Zonana J, Sotos JF, Romshe CA, et al: Dominant inheritance of cerebral gigantism. *J Pediatr* 91:251, 1977
32. Satelo-Avila C, Gooch WM III: Neoplasms associated with Beckwith-Wiedemann syndrome. *Perspect Pediatr Pathol* 3:155, 1976
33. Ben-Galim E, Gross-Kieselstein GE, Abrahamov A: Beckwith-Wiedemann syndrome in a mother and her son. *Am J Dis Childh* 131:801, 1977
34. Sommer A, Cutler EA, Cohen BL: Familial occurrence of the Wiedemann-Beckwith syndrome and persistent fontanel. *Am J Med Genet* 1:59, 1977
35. Herrmann J: Clinical aspects of gene expression. *Birth Defects* 13:25, 1977
36. Cremers CWFJ, Wijdereld PGAB, Pinckers AFLG: Juvenile diabetes mellitus, optic atrophy, hearing loss, diabetes insipidus, atonia of the urinary tract and bladder and other abnormalities (Wolfram syndrome). *Acta Paediatr Scand* [Supp] 264, 1977
37. Frazer FC, Gunn T: Diabetes mellitus, diabetes insipidus, and optic atrophy. An autosomal recessive syndrome. *J Med Genet* 14:190, 1977
38. Fishman MP, Brooker G: Deficient renal cyclic adenosine 3',5'-monophosphate production in nephrogenic diabetes insipidus. *J Clin Endocrinol Metab* 35:35, 1972
39. Bode HH, Miettinen OS: Nephrogenic diabetes insipidus: Absence of close linkage with Xg. *Am J Hum Genet* 22:221, 1970
40. Bode HH, Crawford JD: Nephrogenic diabetes insipidus in North America—the Hopewell hypothesis. *N Engl J Med* 280:750, 1969
41. Feigin RD, Rimoin DL, Kaufmann RL: Nephrogenic diabetes insipidus in a Negro kindred. *Am J Dis Childh* 120:64, 1970
42. Root AW: Endocrinology of puberty I. Normal sexual maturation. *J Pediatr* 83:1, 1973
43. Root AW: Endocrinology of puberty II. Aberrations of sexual maturation. *J Pediatr* 83:187, 1973
44. Mitchell ML, Larsen PR, Levy HL, Bennett AJE, Madoff MA: Screening for congenital hypothyroidism. *JAMA* 239:2348, 1973
45. DeJonge GA: Congenital hypothyroidism in the Netherlands. *Lancet* 2:143, 1976
46. Scott RR, Jenkins ME: Hypothyroidism in Negro children. *J Pediatr* 44:307, 1954
47. Neel JV, Carr EA, Bierwalters WH, et al: Genetic studies on the congenitally hypothyroid. *Pediatrics* 27:269, 1967
48. Greig WR, Henderson AS, Boyle JA, et al: Thyroid dysgenesis in two pairs of monozygotic twins and in a mother and child. *J Clin Endocrinol Metab* 26:1309, 1966
49. Neinas FW, Groman CA, Devine KD, et al: Lingual thyroid. Clinical characteristics of 15 cases. *Ann Intern Med* 79:205, 1973
50. Orti E, Castells S, Quazi QH, et al: Familial thyroid disease. Lingual thyroid in two siblings and hypoplasia of a thyroid lobe in a third. *J Pediatr* 78:675, 1971
51. Goldsmith RE, McAdams AJ, Larsen PR, et al: Familial autoimmune thyroiditis: Maternal-fetal relationship and the role of generalized autoimmunity. *J Clin Endocrinol Metab* 27:265, 1973
52. Shepard TH, Anderson HJ: Phenylthiocarbamide non-tasting among different types of cretinism and thyroid disorders. *Acta Endocrinol* [Suppl] 89:32, 1964
53. Staffer SS, Hamburger JI: Inadvertant ^{131}I therapy for hypothyroidism in the first trimester of pregnancy. *J Nucl Med* 17:146, 1976
54. Najjar SS: Muscular hypertrophy in hypothyroid children. *J Pediatr* 85:236, 1974
55. Johanson A, Blizzard R: A syndrome of congenital aplasia of the alae nasi, deafness, hypothyroidism, dwarfism, absent permanent teeth and malabsorption. *J Pediatr* 79:182, 1971
56. Miyai K, Azukizawa M, Kumahara Y: Familial isolated thyrotropin deficiency with cretinism. *N Engl J Med* 285:1043, 1971
57. Refetoff S, DeGroot LJ, Bernard B, et al: Studies of a sibship with apparent hereditary

resistance to the intracellular action of a thyroid hormone. *Metabolism* 21:723, 1972
58. Elewant A, Mussche M, Vermeulen A: Familial partial target organ resistance to thyroid hormones. *J Clin Endocrinol Metab* 43:575, 1976
59. Stanbury JB, Rocmans P, Buhler UK, et al: Congenital hypothyroidism with impaired thyroid responsiveness to thyrotropin. *N Engl J Med* 279:1132, 1968
60. Standbury JB: Familial goiter, in Stanbury JB, Wyngaarden JB, Fredrickson DS (eds): The Metabolic Basis of Inherited Disease, 3d ed. New York, McGraw-Hill, 1972
61. DeGroot LJ, Niepomniszcze H: Biosynthesis of thyroid hormone: Basic and clinical aspects. *Metabolism* 26:665, 1977
62. Perez-Cuvit E, Crigler JR Jr, Stanbury JB: Partial and total organification defect in different sibships of a kindred. *Am J Human Genet* 29:142, 1977
63. Fraser GR: Association of congenital deafness with goiter (Pendred's syndrome). *Ann Hum Genet* 28:201, 1965
64. Cave WT Jr, Dunn JT: Studies on the thyroidal defect in an atypical form of Pendred's syndrome. *J Clin Endocrinol Metab* 41:590, 1975
65. Amstrong D, VanWormer DE, Neville H, et al: Thyroid peroxidase deficiency in Batten-Spielmeyer-Vogt Disease. *Arch Pathol* 100:430, 1975
66. Fisher DA: Reverse tri-iodothyronine and fetal thyroid status. *N Engl J Med* 293:770, 1975
67. VanHerle AJ, Young RT, Fisher DA, et al: Intrauterine treatment of a hypothyroid fetus. *J Clin Endocrinol Metab* 40:474, 1975
68. Volpe R: The role of autoimmunity in hypoendocrine and hyperendocrine function. With special emphasis on autoimmune thyroid disease. *Ann Int Med* 87:86, 19-7
69. Jayson M, IV, Doniach D, Benhamov-Glynn N, et al: Thyrotoxicosis and Hashimoto goitre in a pair of monozygotic twins with serum long-acting thyroid stimulator. *Lancet* 2:15, 1967
70. Foley TP Jr, Schubert WK, Marnell RT, et al: Chronic lymphocytic thyroiditis and juvenile myxedema in uniovular twins. *J Pediatr* 72:201, 1968
71. Hall R, Dingle PR, Roberts DF: Thyroid antibodies: A study of first degree relatives. *Clin Genet* 3:319, 1972
72. Irvine WJ (ed): Autoimmunity in endocrine disease. *Clin Endocrinol Metab* 4:227, 1975
73. Pai GS, Leach DL, Weiss L, et al: Thyroid abnormalities in 20 children with Turner syndrome. *J Pediatr* 91:267, 1977
74. Murduch JC, Ratcliff WA, McLarty DG, et al: Thyroid dysfunction in adults with Down's syndrome. *J Clin Endocrinol Metab* 44:453, 1977
75. Grumet FC, Payne RO, Konishi J, et al: HL-A antigens as markers for disease susceptibility and autoimmunity in Graves' disease. *J Clin Endocrinol Metab* 39:115, 1974
76. Hollingsworth DR, Mabry CC: Congenital Graves' disease. *Am J Dis Childh* 130:148, 1976
77. Rezvani I, DiGeorge AM, Cote ML: Primary hypothyroidism in cystinosis. *J Pediatr* 91:340, 1977
78. Rivas ML, Merritt AD, Oliver L: Genetic varients of thyroxine-binding globulin (TBG). *Birth Defects* 10:34, 1971
79. Sussault JH, Letarte J, Guyda H, et al: Serum thyroid hormone and TSH concentrations in newborn infants with congenital absence of thyroxine-binding globulin. *J Pediatr* 90:264, 1977
80. Horowitz DL, Refetoff J: Graves' disease associated with familial deficiency of thyroxine-binding globulin. *J Clin Endocrinol Metab* 44:242, 1977
81. Bode HH, Rothman JK, Danon M: Linkage of thyroxine-binding globulin deficiency to other X-chromosome loci. *J Clin Endocrinol Metab* 37:25, 1973
82. Shane SR, Seal US, Jones JE: X-chromosome linked inheritance of elevated thyroxine-binding globulin in association with goiter. *J Clin Endocrinol Metab* 32:587, 1971
83. DiGeorge AM: Congenital absence of the thymus and its immunologic consequences, concurrence with congenital hypoparathyroidism. *Birth Defects.* 4:116, 1968
84. Steele RW, Limus C, Thurman GB, et al: Familial thymic aplasia. Attempted reconstitution with fetal thymus in a millipore diffusion chamber. *N Engl J Med* 287:787, 1972
85. Olambiwonnu NO, Ebbin AJ, Frasier SD: Primary hypoparathyroidism associated with ring

chromosome 18. *J Pediatr* 80:833, 1972
86. Pergament E, Pietra CG, Kadotani T, et al: A ring chromosome No 16 in an infant with primary hypoparathyroidism. *J Pediatr* 76:745, 1970
87. Peden VH: True idiopathic hypoparathyroidism as a sex-linked recessive trait. *Am J Hum Genet* 12:323, 1960
88. Gorodischer R, Aceto T Jr, Terplan K: Congenital familial hypoparathyroidism. Management of an infant, genetics, pathogensis of hypoparathyroidism and fetal undermineralization. *Am J Dis Childh* 119:74, 1970
89. Hartenstein J, Gardner LI: Tetany of the newborn associated with maternal parathyroid adenoma. *N Engl J Med* 274:266, 1966
90. Blizzard RM, Chee D, Davis W: The incidence of parathyroid and other antibodies in the sera of patients with idiopathic hypoparathyroidism. *Clin Exp Immunol* 1:119, 1966
91. Irvine WJ, Barness EW: Addison's disease, ovarian failure and hypoparathyroidism. *Clin Endocrinol Metab* 4:379, 1975
92. Barakat AY, D'Albora JB, Martin MM, et al: Familial nephrosis, nerve deafness and hypoparathyroidism. *J Pediatr* 91:61, 1977
93. Nusynowitz ML, Klein MH: Pseudoidiopathic hypoparathyroidism. Hypoparathyroidism with ineffective parathyroid hormone. *Am J Med* 55:677, 1973
94. Paunier L, Radde IC, Occh SW, et al: Primary hypomagnesemia with secondary hypocalcemia in an infant. *Pediatrics* 41:385, 1968
95. Chase LR, Melson GL, Auerbach GD: Pseudohypoparathyroidism: Defective excretion of 3′-5′-AMP in response to parathyroid hormone. *J Clin Invest* 48:1832, 1969
96. Drezner M, Neelson FA, Lebovitz HE: Pseudohypoparathyroidism Type II. A possible defect in the reception of the cyclic AMP signal. *N Engl J Med* 289:1056, 1973
97. Stogmann W, Fischer JA: Pseudohypoparathyroidism. *Am J Med* 59:140, 1975
98. Frame B, Harison CA, Frost HM, et al: Renal resistance to parathyroid hormone with osteitis fibrosa. "Pseudohypohyperparathyroidism." *Am J Med* 52:311, 1972
99. Weinberg AG, Stone RT: Autosomal dominant inheritance in Albright's hereditary osteodystrophy. *J Pediatr* 79:997, 1971
100. Cederbaum SD, Lippe BM: Probable autosomal recessive inheritance in a family with Albright's hereditary osteodystrophy and evaluation of the genetics of the disorder. *Am J Hum Genet* 25:638, 1973
101. Werder EA, Illig R, Pernasconi S: Excessive thyrotropin response to thyrotropin-releasing hormone in pseudohypoparathyroidism. *Pediatr Res* 9:12, 1975
102. Carlson AE, Brickman AS, Boltazzo GF: Prolactin deficiency in pseudohypoparathyroidism. *N Engl J Med* 296:140, 1977
103. Paloyan E, Laurence AW, Prinz RA, et al: Radiation-associated hyperparathyroidism. *Lancet* 1:949, 1977
104. Jackson CE, Frame B: Relationship of hyperparathyroidism to multiple endocrine adenomatosis. *Birth Defects* 10:66, 1971
105. Rhone DP: Primary neonatal hyerparathyroidism: Report of a case and review of the literature. *Am J Clin Pathol* 64:488, 1975
106. Spiegel AM, Harrison HE, Marx SJ, et al: Neonatal primary hypoparathyroidism with autosomal dominant inheritance. *J Pediatr* 90:269, 1977
107. Jones KL, Smith DW: The Williams elfin facies syndrome. *J Pediatr* 86:718, 1975
108. White RA, Preus M, Watters GV, et al: Familial occurrence of the Williams syndrome. *J Pediatr* 91:615, 1977
109. Foley TP Jr, Harrison HC, Arnaud CD: Benign familial hypercalcemia. *J Pediatr* 81:1060, 1972
110. Sperling MA, Wolfsen AR, Fischer DA: Congenital adrenal hypoplasia. An isolated defect of organogenesis. *J Pediatr* 82:444, 1973
111. Weiss L, Mellinger RC: Congenital adrenal hypoplasia—an X-linked disease. *J Med Genet* 7:27, 1970

112. Kelly WF, Joplin GF, Norfolk E: Gonadotropin deficiency and adrenocortical hypoplasia. *Lancet* 2:1035, 1977
113. Beischer NA, Bhargava VL, Brown JB, et al: The incidence and significance of low oestriol excretion in an obstetric population. *J Obstet Gynaecol Br Commonwth* 75:1024, 1968
114. Shipiro LJ, Cousins L, Fluharty AL, et al: Steroid sulfatase deficiency. *Pediatr Res* 11:894, 1977
115. Shipiro LJ, Weiss R, Webster D, et al: X-linked ichthyosis due to steroid-sulfatase deficiency. *Lancet* 1:70, 1978
116. Spark RF, Etzkorn JR: Absent aldosterone response to ACTH in familial glucocorticoid deficiency. *N Engl J Med* 297:917, 1977
117. Moshang T, Rosenfield RL, Bongiovanni AM, et al: Familial glucocorticoid insufficiency. *J Pediatr* 82:821, 1973
118. Rosler A, Rabinowitz D, Theodore R, et al: The nature of the defect in a salt-wasting disorder in Jews of Iran. *J Clin Endocrinol Metab* 44:279, 1977
119. Cohen T, Theodore E, Rosler A: Selective hypoaldosteronism in Iranian Jews: An autosomal recessive trait. *Clin Genet* 11:25, 1977
120. Michelis MF, Murdaugh HV: Selective hypoaldosteronism. *Am J Med* 59:1, 1975
121. Schaumburg HH, Powers JM, Raine CS, et al: Adrenal leukodystrophy. A clinical pathological study of 17 cases. *Arch Neurol* 33:557, 1975
122. Ropers H, Barmeister P, von Petrykowski W, et al: Leukodystrophy, skin hyperpigmentation and adrenal atrophy: Siemerling-Creutzfeldt disease. Transmission through several generations in two families. *Am J Hum Genet* 27:547, 1975
123. Ropers HH, Zimmermann J, Wienker T: Adrenoleukodystrophy (Siemerling-Creutzfeldt disease): Heterozygote with two clonal fibroblast population. *Clin Genet* II:114, 1977
124. Irvine WJ, Barnes EW: Adrenocortical insufficiency. *Clin Endocrinol Metab* 1:1549, 1972
125. Carpenter CC, Solomon J, Silverberg I, et al: Schmidt's syndrome (thyroid and adrenal insufficiency): A review of the literature and a report of fifteen new cases including ten instances of coexistent diabetes mellitus. *Medicine* 43:153, 1964
126. Hooper MJ, Carter JN, Stel JN: Idiopathic hypoparathyroidism and idiopathic hypoadrenalism occuring separately in two siblings. *Med J Aust* 1:990, 1972
127. Tanae A, Aoyama M, Egi S, et al: Familial Addison's disease. *Acta Paediatr Jap* 13:1, 1971
128. Martin MM: Familial Addison's disease. *Birth Defects* 8:98, 1971
129. Crocker AC, Vawter GF, Neuhauser EBO, et al: Wolman's disease: Three new patients with recently described lipidosis. *Pediatrics* 35:627, 1965
130. Beaudet AZ, Ferry GD, Nichols BL, et al: Cholesterol ester storage disease: Clinical, biochemical and pathological studies. *J Pediatr* 90:910, 1977
131. Cortner JA, Coates PM, Swoboda E, et al: Genetic variation of lysosomal acid lipase. *Pediatr Res* 10:927, 1976
132. Cortner JA, Swoboda E: Wolman's disease. Prenatal diagnosis: Identification of the missing lysosomal acid lipase. *Am J Hum Genet* 26:23A, 1974
133. Lee PL, Plotnick LP, Kowarski A, et al (eds): Treatment of Congenital Adrenal Hyperplasia: A Quarter of a Century Later. Baltimore, University Park Press, 1977
134. Qazi QH, Thompson MW: Incidence of salt-losing form of congenital virilizing adrenal hyperplasia. *Arch Dis Childh* 47:302, 1972
134a. Levine LS, Zachman M, New MI, et al: Genetic mapping of the 21-hydroxylase-deficiency gene within the HLA linkage group. *N Engl J Med* 299:911, 1978
135. Zachmann M, Vollmin JA, New MJ, et al: Congenital adrenal hyperplasia due to deficiency of 11-β-hydroxylation of a 17-α-hydroxylated steroids. *J Clin Endocrinol Metab* 35:501, 1971
136. Kenny FM, Reynolds JW, Green O: Partial 3β-hydroxysteroid dehydrogenase (3B-HSD) deficiency in a family with congenital adrenal hyperplasia. Evidence for increasing 3β-HSD activity with age. *Pediatrics* 48:256, 1971
137. Degenhart HG, Visser HKA, Boon H, et al: Evidence for deficient 20α-cholesterol-hydroxylase activity in adrenal tissue of a patient with lipoid adrenal hyperplasia. *Acta Endrocrinol* 71:512, 1972

138. Kirkland RT, Kirkland JL, Johnson CM, et al: Congenital lipoid adrenal hyperplasia in an eight-year-old phenotypic female. *J Clin Endocrinol Metab* 36:488, 1973
139. Comacho AM, Kowarski A, Migeon J, et al: Congenital adrenal hyperplasia due to a deficiency of one of the enzymes in the biosynthesis of pregnenolone. *J Clin Endocrinol* 28:873, 1968
140. Kershnar AK, Borut D, Kogut MD, et al: Studies in a phenotypic female with 17-α-hydroxylase deficiency. *J Pediatr* 89:395, 1976
141. Hicker BN, Schulman NH, Schneider KM: Adrenocortical carcinoma and congenital hemihypertrophy. *J Pediatr Pathol* 3:255, 1976
142. Sotelo-Avila C, Gooch WM III: Neoplasms associated with the Bechwith-Wiedemann syndrome. *Perspect Pediatr Pathol* 3:255, 1976
143. Fraumeni JF Jr, Miller RW: Adrenocortical neoplasia with hemihypertrophy, brain tumors and other disorders. *J Pediatr* 70:129, 1967
144. Giebink GS, Gotlin RW, Biglieri EG, et al: A kindred with familial glucocortical-suppressible aldrosteronism. *J Clin Endocrinol Metab* 36:715, 1973
145. New MI, Siegal EJ, Peterson RE: Dexamethasone-suppressible hyperaldosteronism. *J Clin Endocrinol Metab* 37:93, 1973
146. Najjar SS, Takla RJ, Nassar VH: The syndrome of rudimentary testes: Occurrence in five siblings. *J Pediatr* 84:119, 1974
147. Wu RH, Goyar RM, Knight R, et al: Endocrine studies in a phenotypic girl with XY gonadal agenesis. *J Clin Endocrinol Metab* 43:506, 1976
148. Collins E, Turner G: The Noonan syndrome: A review of the clinical and genetic features of 27 cases. *J Pediatr* 83:941, 1973
149. Bolton MR, Pugh DM, Mattioli LF, et al: The Noonan syndrome: A family study. *Ann Intern Med* 80:626, 1974
150. Quazi QH, Arnon RG, Paydar MH, et al: Familial occurrence of Noonan syndrome. *Am J Dis Childh* 127:696, 1974
151. Barlow MJ, Neu RL, Gardner LI: X-chromosome banding in Noonan syndrome. *Am J Dis Childh* 126:656, 1973
152. Skre H, Berg K: Linkage studies on the Marinesco-Sjogren syndrome and hypergonadotropic hypogonadism. *Clin Genet* 11:57, 1977
153. Santen RJ, Paulsen CA: Hypogonadotropic eunuchoidism. I. Clinical study of the mode of inheritance. *J Clin Endocrinol Metab* 36:47, 1973
154. Zarak A, Kastin AJ, Soria J, et al: Effect of synthetic luteinizing hormone-releasing hormone (LH-RH) in two brothers with hypogonadotropic hypogonadism and anosmia. *J Clin Endocrinol Metab* 36:612, 1973
155. Faiman C, Hoffman DL, Ryan RJ, et al: The "fertile eunuch" syndrome: Demonstration of isolated luteinizing hormone deficiency by radio-immunoassay technique. *May Clin Proc* 43:661, 1968
156. Williams C, Wieland RG, Zorn EM, et al: Effect of synthetic gonadotropin-releasing hormone (GnRH) in a patient with the "fertile eunuch" syndrome.
157. Neuhauser G, Opitz JM: Autosomal recessive syndrome of cerebellar ataxia and hypogonadotropic hypogonadism. *Clin Genet* 7:426, 1975
158. Perrin JCS, Ideomoto JY, Sotos JF, et al: X-linked syndrome of congenital ichthyosis, hypogonadism, mental retardation and anosmia
159. Tolis G, Lewis W, Verdy M, et al: Anterior pituitary function in the Prader-Willi syndrome. *J Clin Endocrinol Metab* 39:1061, 1974
160. Clarren Sk, Smith DW: Prader-Willi syndrome. *Am J Dis Childh* 131:78, 1977
161. Swanson Sl, Santen RJ, Smith DW: Multiple lentigenes syndrome. *J Pediatr* 78:1037, 1971
162. Dekaban AS, Parks JS, Ross GT: Laurence-Moon syndrome: Evaluation of endocrinological function and phenotypic concordance and report of cases. *Med Ann DC* 41:687, 1972
163. Simpson JL, Christakos AC, Horwith M, et al: Gonadal dysgenesis in individuals with apparently normal chromosomal complements: Tabulation of cases and the compilation of genetic data. *Birth Defects* 10:215, 1971

164. Irvine WJ, Barnes EW: Addison's disease, ovarian failure and hypoparathyroidism. *Clin Endocrinol Metab* 4:379, 1975
165. Miller ME, Chatten J: Ovarian changes in ataxia telangiectasia. *Acta Paediatr Scand* 56:559, 1967
166. McFarlin DE, Stuber W, Waldmann TA: Ataxia telangiectasia. *Medicine* 51:281, 1972
167. Spitz IM, et al: Isolated gonadotropin deficiency. *N Engl J Med* 290:10, 1974
168. Jost A: A new look at the mechanisms controlling sex differentiation in mammals. *Johns Hopkins Med J* 130:38, 1972
169. Silvers WK, Wachtel SS: H-Y antigen behavior and function. *Science* 195:956, 1977
170. Book JA, Eilon B, Hallirecht I, et al: Isochromosome Y (46, Xi, [Yq]) and female phenotype. *Clin Genet* 4:410, 1973
171. Winter JSD, Faiman C, Reyes FL: Sex steroid production by the human fetus: Its role in morphogenesis and control by gonadotropins. *Birth Defects* 13:41, 1977
172. Imperato-McGinley J, Peterson RE: Male pseudohermaphroditism: The complexities of male phenotypic development. *Am J Med* 61:251, 1976
173. Josso N: Mullerian-inhibiting activity of human fetal testicular tissue deprived of germ cells by in vitro irradiation. *Pediatr Res* 8:755, 1974
174. Haymond MW, Weldon, VV: Female pseudohermaphroditism secondary to a maternal virilizing tumor. *J Pediatr* 82:682, 1977
175. Verhoeven ATM, Mastboom HA, Van Leusden IM, et al: Virilizing in pregnancy coexisting with an ovarian cysadenoma, case report and review of virilizing ovarian tumors of pregnancy. *Obstet Gynecol Surg* 28:597, 1973
176. Grumbach MM, Ducharme JR: The effects of androgens on fetal sexual development. *Fertil Steril* 11:157, 1960
177. Simpson JL: Disorders of Sexual Differentiation. New York, Academic, 1976
178. Isurigi K, Aso Y, Ishida H, et al: Prepubertal XY gonadal dysgenesis. *Pediatrics* 59:569, 1977
179. Sarto GE, Opitz JM: The XY gonadal agenesis syndrome. *J Med Genet* 10:288, 1975
180. Wu RH, Boyer RM, Knight R, et al: Endocrine studies in a phenotypic girl with XY gonadal agenesis. *J Clin Endocrinol Metab* 43:506, 1976
181. Berthezene F, Forest MG, Grimaud JA, et al: Leydig cell agenesis. A cause of male pseudohermaphroditism. *N Engl J Med* 295:969, 1976
182. Schneider G, Genel M, Bongiovanni AM, et al: Persistent testicular Δ^5 isometase-3-hydroxylsteroid dehydrogenase (Δ^5-3β-HSD) deficiency in the Δ^5-3β-HSD from of congenital adrenal hyperplasia. *J Clin Invest* 55:681, 1975
183. Hershnar AK, Borurt D, Kogut MD, et al: Studies in a phenotypic female with 17α-hydroxylase deficiency. *J Pediatr* 89:395, 1976
184. Pittaway DE, Anderson RN, Givens JR: Deficient 17β-hydroxysteroid oxireductase activity in testes from male pseudohermaphrodite. *J Clin Endocrinol Metab* 43:457, 1976
185. Brook CGB, Wagner H, et al: Familial occurrence of persistent mullerian structures in otherwise normal males. *Br Med J* 1:771, 1973
186. Peterson RE, Imperato-McGinley Jr, Gautier T, et al: Male pseudohermaphrodite due to steroid 5α-reductase deficiency. *Am J Med* 62:170, 1977
187. Meyer WJ III, Migeon BR, Migeon CJ: Locus on human X chromosome for dihydrotestosterone receptor and androgen insensitivity. *Proc Natl Acad Sci* 72:1469, 1975
188. Manuel M, Katayama KP, Jones HW Jr: Age of occurrence of gonadal tumors in intersex patients with a Y chromosome. *Am J Obstet Gynecol* 124:293, 1976
189. Arnhein JA, Walsh PC et al: Partial androgen insensitivity. *N Engl J Med* 297:350, 1977
190. Bond JV: Wilms's tumor, hypospadias, and cryptorchidism in twins, *Arch Dis Childh* 52:243, 1977
191. Spear GS, Hyde TR, Gruppo RA, et al: Pseudohermaphroditism, glomerulonephritis with the nephrotic syndrome and Wilms tumor in infancy. *J Pediatr* 79:677, 1971
192. Benirscke K, Naftolen G, Gittes R, et al: True hermaphroditism and chimerism. *Am J Obstet Gynecol* 113:449, 1972

193. Wachtel SS, Koo GC, Greg WR, et al: Serologic detection of a Y-linked gene in XX males and XX true hermaphrodites. *N Engl J Med* 295:750, 1976
194. Corey MJ, Miller JR, MacLean JR, et al: A case of XX/XY mosaicism. *Am J Hum Genet* 19:378, 1967
195. Rimoin DL, Schimke RN: Genetic Disorders of the Endocrine Glands. St. Louis, Mosby, 1971, p. 150
196. Irvine WJ: Classification of idiopathic diabetes. *Lancet* 2:638, 1977
196a. McMichael A, McDevitt H: The association between the HLA system and disease. *Prog Med Genet* 2(NS):39, 1977
197. Bottazzo GF, Coniach D: Pancreatic immunity and HLA antigens. *Lancet* 2:800, 1976
198. Rubenstein P, Suciu-Foca N, Nicholson JF: Genetics of juvenile diabetes mellitus. *N Engl J Med* 297:1036, 1977
199. Neel JV: Genetics of juvenile-onset-type diabetes mellitus. *N Engl J Med* 297:1062, 1977
200. Rotter JI, Rimoin DL: Heterogeneity in diabetes mellitus—update, 1978. *Diabetes* 27:599, 1978
201. Van Thiel DH, Smith WI, Rabin BS, Fisher SE, Lester R: A syndrome of immunoglobulin A deficiency, diabetes mellitus, malabsorption and a common HL-A haplotype. *Ann Int Med* 86:10, 1977
202. Eisenbarth G, Wilson P, Ward F, Lebovitz HE: HLA type and occurrence of disease in familial polyglandular failure. *N Engl J Med* 298:92, 1978
203. Tattersall RB, Fajans SS: A difference between the inheritance of classical juvenile-onset and maturity-onset diabetes in young people. *Diabetes* 24:44, 1975
204. Kahn CR, Flier JS, Bar RS, Archer JA, Gorden P, Martin MM, Roth J: The syndromes of insulin-resistance and acanthosis nigricans. *N Engl J Med* 294:739, 1976

10
Genetics of the Gastrointestinal Tract

Eberhard Passarge

The gastrointestinal (GI) tract is involved in diverse hereditary disorders and congenital developmental defects. It may be affected at one or multiple sites and in conjunction with disorders of other organ systems. This chapter is keyed to the main clinical manifestations rather than to specific genetic categories, e.g., Mendelian disorders, chromosomal abnormalities and so forth. Tables 1 and 2 list rare autosomal dominant and recessive conditions involving the GI tract (1). No definite X-linked disorders have been described, however. The genetic complexities of the GI tract have been the subject of numerous reviews, which the interested reader is advised to consult for more explicit clinical information (2-6).

Table 1. Mendelian Disorders of Gastrointestinal Tract*: Autosomal Dominant

Anal stenosis
Cancer of colon
Cancer, hepatocellular
Carcinoid, intestinal
Celiac sprue
Cirrhosis, familial
Ehlers-Danlos syndrome, Type IV
Endocrine adenomatosis, Type I*
Hemangiomas of small intestine
Hernia, hiatus*
Hyperbilirubinemia, Type I (Gilbert)*
Hyperbilinubinemia, Type III (Rotor)
Hyperbilirubinemia, Type IV (Arias)*
Keratosis palmaris and plantaris (tylosis) with esophageal cancer*
Lymphangiectasia, intestinal
Lymphedema, early onset, Type Nonne-Milroy*
Neurofibromatosis*
Neuromas, mucosal, with endocrine tumors (endocrine adenomatosis, Type III)*
Pancreatitis, hereditary*
Phenylthiocarbamide (PTC) tasting*

Research for the material discussed in this chapter was supported by grants from the Deutsche Forschungsgemeinschaft.

Table 1. (Continued)

Polydactyly, imperforate anus, vertebral anomalies
Polyposis coli, juvenile type*
Polyposis, familial, of entire GI tract
Polyposis, intestinal, Type I (colon)*
Polyposis, intestinal, Type II (Peutz-Jeghers syndrome)*
Polyposis, intestinal, Type III (Gardner)*
Polyposis, intestinal, Type IV (scattered discrete polyps)
Polyposis, skin pigmentation, alopecia and fingernail changes (Cronkhite-Canada)
Schinzel syndrome (ulnar defects, small penis, delayed puberty, obesity, immune defects and anal atresia)
Telangiectasia, hereditary hemorrhagic (Rendu-Osler-Weber syndrome)*
Tracheoesophangeal atresia
VATER association
Volvulus of midgut*

*Indicates that mode of inheritance of specific entity appears to be certain. According to McKusick (1), nonasterisked entities are those in which familial aggregation is recorded consistent with the proposed mode of inheritance, but firm data are lacking. This list implies not that all cases of, e.g., liver cancer are heritable, but only that affected families are described.

Table 2. Mendelian Disorders of Gastrointestinal Tract: Autosomal Recessive

Abetalipoproteinemia*
Achalasia, familial, esophageal
Acrodermatitis enteropathica*
Alpers disease*
Analphalipoproteinemia (Tangier)*
Anal-sacral anomalies
Antitrypsin deficiency
Anus, imperforate
Arteriohepatic dysplasia
Ascites, chylous
Asplenia with cardiovascular anomalies*
β-Sitosterolemia*
Biliary atresia, extrahepatic
Biliary malformation with renal tubular insufficiency
Blue diaper syndrome
Byler disease*
Celiac artery stenosis
Chloride diarrhea, familial*
Cholestasis and lymphedema*
Cholesterol ester storage disease of liver
Cirrhosis, familial
Cornelia de Lange syndrome
Crigler-Najjar syndrome*
Cystic fibrosis*
Deafness, nerve type, with mesenteric diverticula of small bowel and progressive neuropathy
Cystinuria*
Dibasicaminoaciduria, Type II*
Disaccharide intolerance, Type I (congenital sucrose intolerance)*
Disaccharide intolerance, Type II (congenital lactose intolerance)*
Disaccharide intolerance, Type III (adult lactose intolerance)*
Diverticulosis of bowel, hernia, retinal detachment
Duodenal atresia*
EMG (*e*xomphalos-*m*acroglossia-*g*igantism) syndrome*
Enterokinase deficiency*

Table 2. (Continued)

Enteropathy, protein-losing*
Fatty metamorphosis of viscera
Folic acid, intestinal transport defect*
Glycogen storage disease, Type IV (brancher deficiency)*
Glycogen storage disease, Type VI (liver phosphorylase deficiency)*
Hartnup disease*
Hyperbilirubinemia, Type II (Dubin-Johnson syndrome)*
Hyperbilirubinemia, transient familial neonatal
Hyperpipecolatemia
Hypoglycemia with absent pancreatic alpha cells
Intestinal atresia, multiple*
Intrahepatic cholestasis
Jejunal atresia ("apple peel")*
Lipase, congenital absence of pancreatic*
Lipid transport, defect of intestine
Liver cancer
Mediterranean fever, familial*
Megacolon, aganglionic (Hirschsprung disease)
Microcolon
Mulibrey nanism*
Pancreatic insufficiency and bone-marrow dysfunction (Shwachman)*
Pancreatic insufficiency, combined exocrine
Pernicious anemia, juvenile, due to selective intestinal malabsorption of vitamin B$_{12}$*
Pseudoxanthoma elasticum*
Pyloric atresia*
Pyloric stenosis, infantile
Regional enteritis
Situs inversus viscerum
Skeletal dysplasia with visceral anomalies (Saldino-Noonan)*
Smith-Lemli-Opitz syndrome*
Spleen, absent (asplenia syndrome)
Splenoportal vascular anomalies
Trypsinogen deficiency*
Trypsinemia*
Wilson disease*
Wolman disease*

*See footnote, Table 1.

FACIAL CLEFTS

Facial clefts and cleft lip or cleft palate, or both are relatively common birth defects with a diverse etiology. The basic fault is generally assumed to be delayed proximate growth of the primary palatal shelves and subsequent fusion of these processes. The clefts are frequently part of a syndrome of cytogenetic or single gene origin. Trisomy 13, e.g., and deletion of the short arms of chromosome 4 are commonly associated with severe facial clefts. Cleft lip with lip pits, cleft lip and pterygia are examples of autosomal dominant genetic syndromes with clefts.

Etiologically experimental evidence suggests that the rate of closure of the primary and secondary palates are quasi-continuous variables controlled by multiple factors, some of which are genetic. These factors establish a "threshold of risk" beyond which the person is affected. Clefts are distributed as roughly 25% cleft lip, 50% cleft lip and palate and 25% cleft palate alone. Cleft lip may be

Table 3. Risk of Recurrence of Cleft Lip (CL) (with or without Cleft Palate) or Cleft Palate (CP)*

Pedigree	CL (P) %	CP %
Random—no family history	0.1	0.04
Affected siblings	4.0	2.0
Affected siblings and relative	4.0	7.0
2 affected siblings	9.0	10.0
Affected parent	4.0	6.0
Affected parent and siblings	17.0	15.0
Affected parent and 2 siblings	25.0	24.0
Both parents affected	35.0	25.0
Both parents and 1 sibling	45.0	40.0
Both parents and 2 siblings	50.0	45.0

*Adapted from (7) and (47).

unilateral or bilateral, 80% of cases being unilateral. Of that 80%, 70% extend to include the palate. Bilateral cases make up 20% of the total, and 80% of them include the palate. Recurrence rates are given in Table 3 (7).

ULCERATION AND BLEEDING

Gastric and duodenal peptic ulcers may show familial aggregation as independent entities, but they are not inherited in a simple Mendelian manner. Both of them occur three times as often among close relatives as in the general population. The families of children with duodenal ulcer show a particularly high incidence of about 10% in first-degree, 7% in second-degree and 3.4% in third-degree relatives, compared to about 2% in controls. The risk of recurrence for sibs has not been accurately determined, but appears to be greater for male than for female relatives (8.9% vs 4%, compared to 2% and 1%, respectively, in control families) (8). The condition is so frequent in the population that an occasional pedigree may resemble a Mendelian pattern of inheritance. Monozygous twins are concordant for duodenal ulcer three times more often than are dizygotic twins.

Certain blood groups and other genetic markers can be regarded as predisposing factors as described later in this chapter. On the whole, environmental and behavioral factors appear to contribute more to the etiology of the usual form of peptic ulcers than do genetic mechanisms.

Peptic Ulcer and Genetic Disease

About three-fourths of patients with autosomal dominant multiple endocrine adenomatosis, Type I, develop duodenal ulcers, some of which occur in the distal portion of the duodenum or proximal portion of the jejunum. Hypertrophic gastric folds (sometimes with ulcerations), duodenojejunal dilatation and mucosal edema, together with gastric acid hypersecretion and diarrhea, suggest the diagnosis of the Zollinger-Ellison (Z-E) syndrome. This conclusion implies the simultaneous presence of a gastrin-producing islet cell adenoma or adenocarcinoma. Probably most patients under age 50 with this syndrome represent partial manifestation of the multiple endocrine adenomatosis (MEA), Type I gene. For genetic counseling it is important to separate the sporadic nongenetic Z-E syndrome from the genetic form, a distinction that at present is beset with uncertainties in any case in which other endocrine tumors are absent. Treatment for the classic Z-E syndrome is total gastrectomy. Other aspects of the endocrine system should also be evaluated in every patient.

Peptic ulcers may occur in several other genetic disorders, such as hemophilia, pseudoxanthoma elasticum and the self-healing squamous epithelioma of Ferguson-Smith. Esophageal ulceration may occur in epidermolysis bullosa dystrophica.

Duodenal ulcer may be the presenting sign in an autosomal dominant syndrome consisting of congenital nystagmus, essential tremor, narcolepsy-like sleep disturbance and ulcer disease (9).

Gastrointestinal Bleeding

Gastrointestinal hemorrhage, reflected by hematemesis, melena or hematochezia—frank or occult—is the manifestation of an underlying genetic disease in about 5-10% of cases (10). The vascularity of the GI tract makes it vulnerable to any disturbance of normal hemostasis—in particular to platelet disorders—although other bleeding sites predominate in genetic defects of blood clotting.

Genetic disorders that must be considered in the evaluation of GI hemorrhage include pseudoxanthoma elasticum; hereditary hemorrhagic telangiectasis (Rendu-Osler-Weber syndrome); neurofibromatosis Peutz-Jeghers syndrome; Ehlers-Danlos syndrome, Type IV; Turner syndrome; genetic forms of thrombocytopenia; and heritable disorders leading to hypoprothrombinemia.

OBSTRUCTION

Developmental causes of intestinal obstruction may result from developmental defects, partial or total, of the gross anatomic structures or from functional impairment of intestinal mobility.

Atresia

Atresia of the GI tract usually involves the esophagus, with or without fistula of the trachea, or the small intestine, or both. Atresia and stenosis of the esophagus account for about one-third of the congenital intestinal obstructions. As a rule they are not genetically determined unless part of genetic syndromes. Familial occurrence of esophageal atresia is reported in first-degree relatives (but in less than 5% of all cases); hence the risk of recurrence is low. About 10% of intestinal atresia involves the duodenum or the colon; familial recurrence risk at these sites is also low.

In contrast, the syndrome comprising duodenal atresia, multiple intestinal atresias with intraluminal calcifications and "apple peel" jejunal atresia with obliterated superior mesenteric artery has occurred in sibs—most likely as the result of an autosomal recessive gene defect (11). Duodenal atresia in trisomy 21 and esophageal or intestinal atresia in Cornelia de Lange syndrome are characteristic complications of these disorders. Prenatal vascular accidents and malrotation presumably play a role in the pathogenesis of atresia, and it is important to anticipate additional atretic sites at surgery.

Anal atresia has shown familial aggregation suggestive in some instances, of a Mendelian disorder, including possible X-linked inheritance. It occurs in a number of multiple developmental defects (12), including the nongenetic VATER association: *V*ertebral defects, *A*nal atresia, *T*racheoesophageal fistula with *E*sophageal atresia, *R*adial dysplasia, *C*ongenital heart defects and *R*enal anomalies.

Recently sirenomelia has been added to the list of this wide spectrum, nonrandom pattern of malformations (13). Other conditions featuring anal atresia in a high proportion of patients include the cat eye syndrome (partial trisomy 22), a familial disorder of endocrine dysfunction (obesity and delayed puberty), immune defects, imperforate anus (observed in four males over three generations [13a]), and trisomy for the short arm of chromosome 4 (14). Imperforate anus is seen in patients with combined proteolytic and lipolytic pancreas deficiency disease.

A familial syndrome involving imperforate anus affecting male subjects has recently been described in patients with mental deficiency, ventricular septum defect (VSD), frontal upsweep of hair, anomalies of the hands and cleft scrotum (16, 17). In the absence of these associated anomalies or a strong familial history of anal atresia, the recurrence risk to later sibs is low, namely, 5%.

About one-third of these patients have symptoms of duodenal obstruction at birth. The duodenal obstruction may not be the result of an *annular pancreas*, but rather a concurrent defect of the duodenum enveloped in the pancreatic ring. The condition is generally not considered to be heritable, although one family with autosomal dominant inheritance is described. The condition also complicates the Down syndrome.

No definite genetic factors are recognized in intestinal duplications and malrotations. A comprehensive review, however, is given by Warkany (15).

Diverticula

No heritable factors have been established in the etiology of common colon diverticula. Persistence of the omphalomesenteric duct results in the Meckel diverticulum, which is present after birth in about 2-3% of the population (with a 3:1 M:F ratio of male to female). It is a frequent concomitant of generalized malformation syndromes affecting the GI tract.

Multiple intestinal diverticula, absent gastric motility and progressive sensory neuropathy was observed in three sisters (18). Sensory nerve deafness was present in the affected as well as in other family members. Diverticula of the large and small bowel and of the urinary bladder, recurrent femoral and inguinal hernias, myopia and retinal detachment were reported in sibs. Since the parents were first cousins, this may be a rare autosomal recessive trait.

Impaired Intestinal Motility

Two groups of disorders with a polygenic etiology fall into this category, including 1) congenital hypertrophic pyloric stenosis and 2) congenital intestinal aganglionosis (Hirschsprung disease). The genetic risk of recurrence has been determined by empirical figures for both. In addition both may be part of other disorders with genetic causes.

An elongated, thickened, almost cartilaginous pylorus, leading to obstruction in varying degrees, occurs in about 1 in 300 live births (about 1 in 150 males and 1 in 750 females). The genetic analysis of the familial aggregation (19, 20) reveals the following pattern: affected males will, on the average, have 5% affected sons and 2.5% affected daughters; affected females will have 20% affected sons and 7% affected daughters. The same risk exists for normal parents of one affected child who plan to have more children. The disorder was the first to illustrate a genetic influence related to the sex of the affected individual, i.e., the familial incidence is higher in first-degree relatives of the sex that is less commonly affected in the population. Infantile pyloric stenosis must be heterogeneous because it occurs in several disorders with different genetic causes, e.g., in chromosomal disorders (trisomy 21, trisomy 18), single gene disorders (Smith-Lemli-Opitz syndrome, cerebrohepatorenal syndrome), secondary to prenatal exogenous conditions (rubella and thalidomide embryopathies) and in disorders of unknown etiology (Cornelia de Lange syndrome). Geographic and seasonal differences in the incidence suggest other causative factors.

The congenital or early infantile megacolon occurring in Hirschsprung disease results from absent mucosal and submucosal ganglion cells. It may involve the colon and the small intestine to varying degrees, although the aganglionic segment typically extends from the lower sigmoid to about the splenic flexure (Type I or short segment involvement in about 90% of patients). In Type II, or long segment involvement, the aganglionic segment may reach beyond the colon into the entire ileum and even into the jejunum. Most likely the ganglion cells fail to migrate from the neural crest into the mucosal and submucosal myenteric plexus. Total neonatal or early infantile obstruction may be the presenting sign of the long aganglionic segment disease rather than the usual obstipation.

The genetic risk of recurrence differs according to the length of the aganglionic segment (21). In short segment disease (Type I) the genetic risk is low, especially for sibs of males. Since males are about five times more often affected in the population, the risk of recurrence is about 1% for sisters and 6% for brothers. For the long segment disease (Type II), the recurrence risk for sibs of males is about 10% for sisters and 7% for brothers. The risk of recurrence to sibs of affected females is 3% for sisters and 8% for brothers in the short segment disease, whereas it is 9% and 18%, respectively, for long segment disease. These figures, together with an almost even sex ratio for the long segment disease (1.86:1 vs 5.2:1 for Type I), indicate a relatively high proportion of genetic factors in the long segment disease. The possibility exists that total intestinal aganglionosis (colon and small intestine) can occur in some families as an autosomal recessive trait.

The failure of the mucosal and submucosal ganglion cells to develop normally is etiologically heterogeneous, a view supported by the diversity of disorders associated with aganglionosis, including trisomy 21 syndrome (in about 2% of cases), occasionally with others of the rarer chromosomal aberrations, cartilage-hair hypoplasia, congenital deafness, neuroblastoma, pheochromocytoma, rubella embryopathy and colonic atresia (21, 22). It is important to remember that functional megacolon may occur in various genetic and nongenetic diseases (23).

Meconium ileus is usually a presenting sign in the neonate with cystic fibrosis. However, it has also been observed in two brothers who did not have this disease; hence the possibility exists of an isolated, genetically determined form.

Intestinal signs, presenting as acute megacolon, intestinal pseudoobstruction, malabsorption, ileus, dysphagia and uncoordinated esophageogastric mobility, may occur in some muscular dystrophies, particularly in myotonic dystrophy. Impaired intestinal motility may be a prominent feature of the Riley-Day syndrome (familial dysautonomia).

INFLAMMATORY BOWEL DISEASE

Both ulcerative colitis and regional enteritis have a tendency to appear in familial aggregations, probably as a result of genetic influences in their complex etiology. Approximately 2-10% of affected persons with inflammatory bowel disease (IBD) have a relative who is also affected. The relative of a proband with regional enteritis is likelier to be affected with either the same disorder or ulcerative colitis, whereas an affected relative of a proband with ulcerative colitis usually has the same disorder. However, intermingling occurs. Individual pedigrees of multiple occurrences in several generations have been noted. Monozygotic twins with ulcerative colitis are about twice as often concordant as are dizygotic twins. In general, however, no major genetic influences are recognized in the etiology of IBD.

Inflammatory bowel disease is associated with ankylosing spondylitis in either the same patient or in other members in about 3-5% of families, although even higher rates of association are reported (24). Other associated conditions include allergic disorders, rheumatoid arthritis and chronic liver disease. Rarely IBD occurs in the Turner syndrome (26). No definite association pattern exists with blood groups or other genetic markers.

MALABSORPTION

Genetic causes of impaired intestinal absorption are implicated in a wide variety of disorders. Table 4 summarizes the pertinent features of genetic malabsorption syndromes. Genetically determined malabsorption may be classified into primary and secondary malabsorption states and according to the major absorptive defect, namely, whether carbohydrate, protein, amino acid, lipid, mineral, vitamins or mixed forms.

Often malabsorption is a secondary feature of other genetic disorders, such as cystic fibrosis or the various immunodeficiency diseases. Genetic factors are involved in the etiology of gluten intolerance (*celiac disease*). The gliadin fraction of gluten in wheat and rye causes functional and morphologic injury to the small

Table 4. Partial List of Genetic Malabsorption Syndromes

Substance Malabsorbed	Disorder	Major Clinical Problem	Inheritance*
Sucrose/isomaltose	Disaccharide malabsorption	Neonatal watery diarrhea, sucrose intolerance	AR
Lactose	Congenital lactose intolerance (different from adult form)	Neonatal watery diarrhea, vomiting, lactose intolerance	AR
Glucose and galactose	Glucose-galactose malabsorption	Neonatal watery diarrhea, pH of stool is 4 or 5	AR
Lysine, cystine, arginine and ornithine	Cystinuria, some types	Renal calculi, no known consequences of gut defect	AR
Lysine only	Hyperdibasic amino-aciduria, Type I	Mild malabsorption	AD
Neutral amino acids	Hartnup disease	Pellagra-like rash, cerebellar ataxia, aminoaciduria	AR
Tryptophan	Blue diaper syndrome	Hypercalcemia, nephrocalcinosis, indicanuria	U
Lipids	Abetalipoproteinemia	Steatorrhea, hypocholesterolemia, acanthocytosis, pigmentary retinopathy, ataxia	AR
Chloride ion	Congenital chloridorrhea	Polyhydramnios, watery diarrhea, ileus, hypoelectrolytemia	AR
Zinc	Acrodermatitis enteropathica	Diarrhea, dermatitis, alopecia, failure to thrive	AR
Magnesium	Hypomagnesemia	Muscle weakness, seizures, diarrhea	AR
Calcium	Vitamin D-dependent rickets	Hypocalcemia, increased vitamin D requirements	AR
Copper	Menkes disease	Mental retardation, pili torti	XR
Folic acid	Megaloblastic anemia	Infantile megaloblastic anemia, ataxia, athetosis, convulsions	AR
Vitamin B_{12}	Juvenile pernicious anemia	Megaloblastic anemia	AR

*AD, autosomal dominant; AR, autosomal recessive; XLR, X-linked recessive; U, uncertain.

intestinal mucosa, followed by various secondary effects related to malabsorption. About 8-10% of first-degree relatives of an affected patient show signs of the disease, either clinically or by biopsy. In a recent study the familial incidence was 18% (27) compared to a population incidence ranging from 1 in 300 to 1 in 8,000.

About one-half of dizygotic and most monozygotic twins are concordant. Approximately 75-80% of patients carry the histocompatibility antigen HLA-B8 compared to 25-30% of controls. The HLA-B1 antigen is in about 70% compared to 33% of controls. Both antigens are found together in 66% of patients, but in only 20% of controls. Their presence although nonspecific, may be of additional diagnostic value in patients with diarrhea. Immunologic factors may be involved in the pathogenesis of this disorder, but simple Mendelian inheritance appears unlikely.

NEOPLASIA

Intestinal neoplasia occurs usually as an isolated event, not obviously related to genetic factors. However, familial aggregation of some tumors is described, and several genetic disorders with a distinct risk of tumor development have been observed (28-31). From a genetic standpoint it is noteworthy that most of these disorders are inherited as autosomal dominant traits. The most significant of them are the familial polyposis coli syndromes. In these families periodic surveillance of at-risk members by radiography and surgical prophylaxis of affected bowel appear to be integral parts of a cancer preventive program. The general topic of genetics and neoplasia is discussed elsewhere. In some kindreds familial carcinoma of the colon may also occur without polyposis (so-called family cancer syndrome) as an autosomal dominant trait with variable penetrance ([31] See Chapter 3).

PANCREATIC INSUFFICIENCY

Juvenile/early adulthood pancreatitis may occur as an autosomal dominant trait (*hereditary pancreatitis*). The age of onset varies from 4 to 16 years, the mean being 12-14 years. Other than an earlier age of onset and the lack of recognizable precipitating factors, the hereditary form differs in no way from acquired pancreatitis.

A second type of hereditary pancreatitis may occur in adults, often precipitated by chronic alcoholism and malnutrition, but its clinical and genetic features are not yet fully delineated.

Pancreatitis may occur secondarily in hereditary hyperparathyroidism and in any hereditary disorder of which cholelithiasis is a complication, e.g., hemolytic anemia.

Pancreatic Deficiency

Five distinct genetic disorders with features of exocrine pancreatic deficiency are known (32), all of which are inherited as autosomal recessive traits. The main features are summarized in Table 5. Clinical features of these disorders include failure to thrive, vomiting, malabsorption of proteins or fats, anemia and a relatively strong response to parenterally or orally administered hydrolyzed protein.

Cystic Fibrosis

Cystic fibrosis of the pancreas (mucoviscidosis) is not actually a primary pancreatic insufficiency disease, but it is included here because pancreatic exocrine deficiency is a major clinical feature.

Its clinical and genetic features have been well summarized in recent years (33-36). Gastrointestinal involvement occurs in about 85-90% of patients and can be related basically to progressive loss of pancreatic function. Pancreatic enzyme replacement, medium-chain triglyceride supplement and administration of water-miscible forms of the liposoluble vitamins are therefore important in

Table 5. Genetic Disorders of Exocrine Pancreatic Insufficiency

Signs and Symptoms	Shwachman Syndrome	Lipase Deficiency	Trypsinogen Deficiency	Proteolytic/ Lipolytic Deficiency	Entero kinase Deficiency
Age of onset	3 weeks-6 months			1/2-1 yr	
Failure to thrive	+	(+)	+	+	+
Diarrhea	+	−	+	−	+
Steatorrhea	+	+	−	+	−
Hypoproteinemia	−	−	−	+	+
Neutropenia	+	−	−	−	−
Metaphyseal dysostosis	+	−	−	−	−
Dwarfism	+	−	−	−	−
Immune dysfunction	+	−	−	−	−
Deficiency of:					
Chymotrypsinogen	−	−	−	−	−
Chymotrypsin	−	−	+	+	+
Trypsinogen	−	−	+	−	−
Trypsin	−	−	+	+	+
Carboxypeptidase	−	−	+	+	+
Lipase	−	+	−	+	−
Intestinal enterokinase	−	−	−	−	+
Response to hydrolyzed protein	−	−	+	+	+/−

therapy (33). Cirrhosis of the liver from biliary obstruction occurs in about 1-2% of cases. The disorder is inherited as an autosomal recessive trait.

About 1 in 22 persons of Caucasian background is an asymptomatic heterozygous carrier. At present heterozygous carriers of the cystic fibrosis gene cannot be reliably recognized, and no prenatal detection of an affected fetus is possible. About 1 in 2,000 liveborn infants is affected, making this disease the most prevalent recessive one in this population. In contrast, the incidence is only one-sixth as high in American black children, and only 1/25 as high in Asian populations. Current evidence suggests that the condition is genetically heterogeneous.

HEPATOBILIARY DISEASE

Genetic disorders of the liver and bile ducts are conveniently classified as primary developmental defects, problems of bilirubin metabolism and abnormalities in bile excretion or other specific hepatic functions, which may eventually lead to secondary anatomic lesions.

Atresias of the intra- as well as the extrahepatic bile ducts (*biliary atresia*) are developmental defects that may show a familial aggregation, although most cases are isolated. The overall genetic risk of recurrence in sibs of both types can be regarded as low, i.e., less than 5%. Probably not all intrahepatic biliary atresias are developmental defects in the strict sense, because some of them may be the result of prenatal hepatitis and thus be solely environmental.

Familial occurrence is frequently observed in *neonatal hepatitis* and *cirrhosis*, but the genetic principles are far from clear. Some familial reports may have been due to unrecognized metabolic genetic defects, such as tyrosinemia; Wilson disease; galactosemia; glycogen storage disease, Type IV; α-1-antitrypsin deficiency disease; or other well-known disorders associated with neonatal cirrhosis or hepatitis. Nevertheless a small group of patients with pure familial neonatal cirrhosis exists. However, if any of the disorders just cited can be ruled out, the genetic risk is low, probably not exceeding 5-8%.

A fairly high proportion of affected sibs (about 17%) and parental consanguinity support the existence of an autosomal recessive form of *neonatal giant cell hepatitis*. At present the proportion of affected sibs deviates from the expected 25% for unknown reasons, although incomplete manifestation of the genotype, etiologic heterogeneity and difficulties in formal genetic analysis are possible considerations. It is likely that some familial intrahepatic biliary atresias actually belong in this group. Multiple cysts occur in the liver, (*cystic liver disease*) together with polycystic renal disease.

Hereditary hemorrhagic telangiectasia (Osler-Rendu-Weber Syndrome) is an autosomal dominant vascular disease that may affect the liver. Hepatic telangiectasias and large hemangiomas may lead to profound hepatic dysfunction.

Metabolic Disorders of the Liver: Bilirubin

Blood to the bile transfer of bilirubin is governed by three principal steps, including hepatic uptake, bilirubin conjugation and excretion. Genetic disorders are known in each of these areas, and they have contributed to the understanding of the physiology of bilirubin metabolism. In particular the discovery of specific receptor proteins in liver cells (Y and Z proteins) (37) has advanced our understanding of some hepatic disorders and their physiologic basis. The Y protein, a basic protein of molecular weight of about 32,000 dalton, is the major binding protein of the liver. Almost absent at birth, it reaches mature levels in the second or third week of life. Relative deficiency of this protein may be the basis for "physiologic" neonatal jaundice.

The Z protein is present in adult levels at birth, but has less organic anion binding capacity with a molecular weight of about 10,000 daltons. Defective uptake and binding to these proteins may be at least as important as bilirubin conjugation to glucuronic acid by glucuronyl transferase in the pathogenesis of neonatal jaundice. Detailed discussions of this topic are available in the literature (37-40).

Defective Bilirubin Uptake and Conjugation

The commonest example is Gilbert disease (also named Meulengracht disease in some European countries). This is most likely an etiologically heterogeneous group of disorders characterized by chronic intermittent jaundice (bilirubin, 2-5 mg/100 ml). The liver remains normal. Familial occurrence suggests autosomal dominant inheritance.

Two disorders, the Crigler-Najjar syndrome (or hyperbilirubinemia, Type I of Arias) and hyperbilirubinemia, Type II, can be distinguished on clinical and genetic grounds. The autosomal recessive Crigler-Najjar syndrome is due to a total defect of bilirubin conjugation, which leads to persistently high levels of unconjugated bilirubin (20-40 mg/100 ml and more) and a poor prognosis. No treatment is available. The other disorder, hyperbilirubinemia Type II, is inherited as an autosomal dominant trait and shows an incomplete defect in bilirubin conjugation. Bilirubin levels reach 4-20 mg/100 ml and kernicterus is rare.

This disorder shows a dramatic response to phenobarbital, which may reduce the bilirubin concentration to about 10% of its original level. Two other forms of glucuronyl transferase deficiency are identified. One is a defect in steroid metabolism leading to the excretion of an isomer of pregnanediol into breast milk, which inhibits glucuronyl transferase activity. Infants of such mothers develop benign jaundice as soon as they are breast fed. In the other disorder, not yet delineated, severe jaundice and often kernicterus occur in babies of healthy mothers whose serum contains an unidentified inhibitor of glucuronyl transferase activity. The genetic nature of these two abnormalities has not been established.

Defective Bilirubin and Bile Excretion

Genetic disorders in this function include the Dubin-Johnson syndrome (mild, persistent conjugated hyperbilirubinemia with characteristic hepatic melanin pigmentation, abdominal pain and increased coproporphyrin excretion, possibly due to a coproporphyrin III cosynthetase defect) as an autosomal recessive trait, and the so-called Rotor syndrome as a conjugated hyperbilirubinemia of indistinct genetic etiology (possibly autosomal dominant). These are functionally harmless disorders, but the jaundice poses cosmetic problems and occasionally leads to an erroneous diagnosis of chronic liver disease. Women with the Dubin-Johnson syndrome may develop pronounced jaundice while taking oral contraceptives or during pregnancy.

Several genetic disorders feature total bile excretory failure with the usual attendant symptoms. The severest form is progressive intrahepatic cholestasis (Byler disease), an autosomal recessive trait beginning in early childhood; it is usually fatal during the first decade of life. It is accompanied by malabsorption and retarded development. The bile salts show a characteristic pattern with predominance of dihydroxy bile salts. Bromsulphthalein (BSP) excretion is reduced in heterozygous carriers of the gene. Cholestyramine usually reduces pruritus and jaundice.

Benign familial recurrent cholestasis causing cholestasis without other problems may be an autosomal recessive trait. A familial tendency to develop cholestasis during pregnancy (third trimester) may well be shown to have a genetic basis related to bile-salt excretion and pregnanediol metabolism. Autosomal dominant inheritance is suggested. Neonatal hepatitis without persistent cholestasis occurs as an autosomal recessive trait. Neonatal cholestasis with lymphedema due to hypoplastic vessels and associated with recurrent jaundice is also reported as an autosomal recessive disorder.

Among other metabolic disorders with major involvement of the liver is Wilson's disease. The features leading to the definite diagnosis of this autosomal recessive disorder include early cirrhosis, jaundice, neurologic deficits, psychic disturbances, Kayser-Fleischer corneal rings, hemolytic anemia, decreased levels of ceruloplasmin (under 20 mg/100 ml) and increased urinary copper excretion.

α_1-Antitrypsin Deficiency

There are more than 20 genetic variants of the protease inhibitor (Pi) system, α_1-antitrypsin. Two allelic forms, PiZ and PiS, result in low-serum antitrypsin activity levels of 15% and 65%, respectively. Individuals homozygous for the Z allele are at risk to develop either neonatal liver disease or early adult obstructive pulmonary disease. Individuals heterozygous for the Z allele also have an increased risk for liver disease. Recent data (40) indicate a risk of about 7% for infants with PiZ phenotype to develop clinically overt liver disease, and an additional 3% to have subclinical jaundice. In addition about 50% of (apparently) healthy PiZ infants show laboratory evidence of disturbed liver function.

The clinical picture may include low birth weight, failure to thrive, late umbilical bleeding, hepatosplenomegaly, prolonged obstructive jaundice, increased levels of conjugated bilirubin and abnormal laboratory liver function tests. The endoplasmic reticulum of hepatocytes contains PAS-positive inclusion bodies consisting of a qualitatively altered α_1-antitrypsin deficient in carbohydrate components (41). Paucity of bile ducts is another morphologic feature (42).

About two-thirds of children with cirrhosis of the liver and a history of neonatal hepatitis have the PiZ phenotype (most of them are homozygous ZZ except for a few with a silent allele Pi$^-$ or Pi null) and are deficient in α_1-antitrypsin activity. In some children without overt clinical liver disease, insidious cirrhosis may develop, sometimes complicated by malignant hepatoma.

Primary Hemochromatosis

Increased serum iron levels due to increased intestinal iron absorption occur as a familial disorder. Neither the etiology nor the genetic mechanisms are clear, although autosomal dominant inheritance is the best explanation for many of the families with multiple affected members. Female members are usually less severely affected than male members due to menstrual blood loss. Several genetic and nongenetic forms probably exist, but they cannot be reliably distinguished at present. Liver cirrhosis, diabetes mellitus and melanotic skin pigmentation are other clinical features of the hemochromatosis disorders. Prophylactic venesection is recommended.

The liver is involved in a number of metabolic disorders, generally related to excessive storage of some substance secondary to an inborn error of metabolism (39, 43). These conditions are listed in Table 6. The details of these conditions including specific hepatic involvement is considered elsewhere in this work.

Progressive encephalopathy and fatty infiltration of the viscera and liver dysfunction with ammonia intoxication *(Reye syndrome)* have been observed with increasing frequency in recent years. Wide clinical variability seems to be possible in this usually fatal disorder. Affected sibs are described, but no firm evidence for genetic factors in the etiology has been found.

GASTROINTESTINAL TRACT AND CHROMOSOMAL DISORDERS

Most major GI tract malformations occur in the three autosomal trisomy syndromes, trisomy for chromosome 13, 18 and 21.

For *trisomy 13* an analysis of 32 necropsies (44) revealed malformations of the

Table 6. Liver Involvement in Metabolic Lysosomal Disorders

Carbohydrate disorders:
Galactosemia
Glycogen storage diseases
Fructose intolerance

Aminoacid disorders:
Tyrosinosis
Argininosuccinic aciduria
Dibasic aminoaciduria-Type II
de Toni-Fanconi syndrome

Mucopolysaccharidoses:
Type I   (Hurler syndrome)
Type II   (Hunter syndrome)
Type III   (Sanfilippo syndrome)
Type VI   (Maroteaux-Lamy syndrome)
Type VII   (Sly syndrome)

Sphingolipidoses:
GM_1 gangliosidosis (Sandhoff)
GM_2 gangliosidosis (Tay-Sachs disease)
Gaucher disease
Krabbe Disease

Mucolipidoses:
Type II (I cell disease)

GI tract in 21. The various anomalies included malrotation of the colon (8), unattached mesentery (5), mesenteric cyst (1), Meckel diverticulum (3), accessory spleen (9), ectopic spleen tissue within the pancreas (2), hypoplastic bile ducts (1), abnormal lobulation of the liver (1), true diverticula of the appendix (1), omphalocele and other umbilical defects (13).

In *trisomy 18* among 84 necropsies (44), malformations of the GI tract were present in 64 (85%). They included Meckel diverticulum (19), malrotation (13), accessory spleen (6), ectopic pancreas (4), tracheoesophageal fistula (3), biliary atresia or stenosis (4), imperforate anus (2) and pyloric stenosis (2).

Characteristic complications involving the GI tract can be expected in a small percentage of patients with trisomy 21 (Down syndrome) according to different data reviewed by Warkany et al (44). They include esophageal atresia (two mongoloids among 233 infants with this defect), duodenal atresia (10 of 32), annular pancreas (12 of 56), congenital intestinal agangliosis in about 2% and anorectal malformations (10 patients with Down syndrome among 697 infants with such anomalies).

Rectal hemorrhage resulting from multiple GI telangiectasias or IBD may be presenting signs of the *Turner syndrome*.

GASTROINTESTINAL DISORDERS AND GENETIC MARKERS

Certain genotypes, such as blood groups, serum proteins or the HLA system, show a statistical association with some GI disorders. Although the significance of these associations is unknown, it may ultimately be related to etiologic factors of the disease. The subject is reviewed in more detail by Langman (45).

Carcinoma of the stomach is 20% more frequent among persons with blood group A than among members of the other groups of the ABO system. A similar association exists for pernicious anemia. In two studies, blood group O was observed with increased frequency in persons with nasopharyngeal and hypopharyngeal cancer.

Individuals with blood group O appear to be at 30-35% greater risk than persons with blood groups A or B for *peptic ulcer*. Moreover hemorrhage and perforation can be expected more often in ulcer patients with blood group O than with A or B.

Nonsecretors of ABO blood groups are found about 50% more often among duodenal ulcer patients and 40% among gastric ulcer patients. Gastric cancer does not appear to be associated with the secretor status.

None of the disorders of the GI tract with a defined mode of inheritance shows an association with antigens determined by the major histocompatibility gene complex *HLA* (46). In contrast, about 75% of patients with celiac disease carry the HLA-B8 antigen compared to about 24% of unaffected persons. Chronic autoimmune hepatitis is associated with HLA-B27 (88%). Again the significance of these associations from a diagnostic, prognostic or etiologic standpoint at present remains unclear, and they are noted solely for their heuristic value.

REFERENCES

1. McKusick VA: Mendelian Inheritance in Man, 4th ed. Baltimore, Johns Hopkins University Press, 1975
2. McConnell RB: The Genetics of the Gastrointestinal Disorders. London, Oxford University, 1966
3. McConnell RB: Genetics and the gastrointestinal system. *Prog Med Genet* 6:63, 1969
4. McConnell RB (ed): Genetics of gastrointestinal disorders, in Clinics in Gastroenterology, Vol. 2. Philadelphia, Saunders, 1973
5. Lehmann W: Anomalien, Fehlbildungen und Krankheiten der Verdauungsorgane, in Becker PE (ed): Handbuch der Humangenetik, Band III/2, Stuttgart, Thieme, 1972
6. Gryboski I: Gastrointestinal problems in the infant, in Major Problems in Clinical Pediatrics, Vol. 13, 1975
7. Witkop CJ: in Bergsma D (ed): Birth Defect Atlas and Compendium. Baltimore, Williams & Wilkins, p. 260, 1973
8. Cowan WK: Genetics of duodenal and gastric ulcer. *Clin Gastroenter*, 2:539, 1973
9. Neuhauser B, Daley RF, Magnelli NC, et al: Essential tremor, nystagmus and duodenal ulceration. *Clin Genet* 9:81, 1976
10. Woodrow IC: Inherited disorders causing gastrointestinal bleeding. *Clin Gastroenter* 2:703, 1973
11. Martin CE, Leonidas FC, Amoury RA: Multiple gastrointestinal atresias, with intraluminal calcifications and cystic dilatation of bile ducts: A newly recognized entity resembling "a string of pearls." *Pediatrics* 57:268, 1976
12. Bergsma D (ed): Birth Defects. Atlas and Compendium. Baltimore, Williams & Wilkins, 1973
13. Smith DW: Recognizable Patterns of Human Malformations, 2d ed. Philadelphia, Saunders, 1976
13a. Schinzel A: Personal communication, 1976
14. Owen L, Martin B, Blank CE, et al: Multiple congenital defects associated with trisomy for the

short arm of chromosome 4. *J Med Genet* 11:291, 1974
15. Warkany J: Congenital Malformations. p. 690, Chicago, Year Book, 1971
16. Naveh Y, Friedman A: Familial imperforate anus. *Am J Dis Childh* 130:441, 1976
17. Keller MA, Jones KL, Nyhan WL, et al: New syndrome of mental deficiency with craniofacial, limb, and anal abnormalities. *J Pediat* 88:589, 1976
18. Hirschowitz BI, Groll A, Ceballos R: Hereditary nerve deafness in three sisters with absent gastric motility, small bowel diverticulitis and ulceration and progressive sensory neuropathy. *Birth Defects* 8:27, 1972
19. Carter CO, Evans KA: Inheritance of congenital pyloric stenosis. *J Med Genet* 6:233, 1969
20. Dodge J: Genetics of hypertrophic pyloric stenosis. *Clin Gastroenterol* 2:523, 1973
21. Passarge E: The genetics of Hirschsprung's disease. *New Engl J Med* 276:138, 1967
22. Passarge E: Genetics of Hirschsprung's disease. *Clin Gastroenterol* 2:507, 1973
23. Simpson AJ, Khilnani MT: Gastrointestinal manifestations of the muscular dystrophies. *Am J Roentgenol* 125:948, 1976
24. Kirsner FB: Genetic aspects of inflammatory bowel disease. *Clin Gastroenterol* 2:557, 1973
25. Weinrieb IJ, Fineman RM, Spiro HM: Turner syndrome and inflammatory bowel disease. *New Engl J Med* 294:1221, 1976
26. Scriver CR, Rosenberg LE: Amino Acid Metabolism and Its Disorders. Philadelphia, Saunders, 1973
27. Stokes PL, Asquith P, Cook WT: Genetics of celiac disease. *Clin Gastroenterol* 2:547, 1973
28. Harper PS: Hereditary and gastrointestinal tumors. *Clin Gastroenterol* 2:675, 1973
29. Erbe RW: Current concepts in genetics: Inherited gastrointestinal-polyposis syndromes. *New Engl J Med* 294:1101, 1976
30. Lovett E: Family studies in cancer of the colon and rectum. *Br J Surg* 63:13, 1976
31. Lovett E: Familial cancer of the gastrointestinal tract. *Br J Surg* 63:19, 1976
32. Townes PL: Trypsinogen deficiency and other proteolytic deficiency diseases. *Birth Defects* 8:95, 1972
33. Shwachman H: The heterogeneity of cystic fibrosis. *Birth Defects* 8:102, 1972
34. Bearn AG: Genetics of cystic fibrosis. *Clin Gastroenterol* 2:515, 1973
35. Bowman BH, Mangos JA: Current concepts in genetics: Cystic fibrosis. *New Engl J Med* 294:937, 1976
36. Forstner G, Crozier DN, Sturgess JM: Cystic fibrosis: Present status and future prospects in detection of patients and carriers. *Can Med Ass J* 113:550, 1975
37. Arias IW: Genetic disorders in bilirubin metabolism, in McKusick VA, Clairborne R, (eds): Medical Genetics. H.P. Inc., New York, 1973
38. Brunt PW: Genetics of liver disease. *Clin Gastroenter* 2:615, 1973
39. Berk PD, Howe RB, Berlin NI: Disorders of bilirubin metabolism, in Bondy PK, Rosenberg LE, (eds): Duncan's Diseases of Metabolism, 7th ed. Philadelphia, Saunders, 1974
40. Sveger T: Liver disease in alpha$_1$-antitrypsin deficiency detected by screening of 200,000 infants. *New Engl J Med* 294:1316, 1976
41. Jeppsson JO, Larrson C, Erikson S: Characterization of alpha$_1$-antitrypsin inclusion bodies from the liver in alpha$_1$-antitrypsin deficiency. *New Engl J Med* 293:576, 1975
42. Hadchouel M, Gautier M: Histopathologic study of the liver in early cholestatic phase of alpha$_1$-antitrypsin deficiency. *J Pediat* 89:211, 1976
43. Hug G: Nonbilirubin genetic disorders of the liver, in The Liver. International Academy of Pathology Monograph No. 13, Baltimore, Williams & Wilkins, 1972
44. Warkany J, Passarge E, Smith LB: Congenital malformations in autosomal trisomy syndromes. *Am J Dis Childh* 112:502, 1966
45. Langman MTS: Blood groups and alimentary disorders. *Clin Gastroenterol* 2:497, 1973
46. Svejgaard A, Hauge M, Jersild C, et al: The HLA System. Monographs in Human Genetics, No. 7. Basel, Karger, 1975

11
Genetics
and Hematology

Peter Hathaway
R. Neil Schimke

Broadly defined, hematology includes the study of all things circulating. The genetic principles of such a subject would usurp a large section of this textbook; hence this discussion will be limited to disorders of the formed elements and the coagulation factors. The various proliferative syndromes and the immunodeficiency disorders are considered elsewhere. Mixed syndromes with hematologic features may not be found in this chapter, and the index should be consulted for reference to a specific condition.

HEMOGLOBIN VARIANTS

Hemoglobin has been and continues to be one of the most extensively studied of human proteins. More than 200 structural hemoglobin variants have been discovered to date, mostly single amino-acid substitutions in the alpha or beta chain. Most of them have been detected by screening surveys using various types of electrophoresis, and they are usually not associated with disease. Except for hemoglobin S, C and E, which are polymorphic in some populations, these mutant hemoglobins are rare. A full understanding of the clinical implications of these altered molecules requires a brief review of the genetics of human hemoglobin (1, 2).

Table 1 delineates the various types of hemoglobin tetramers and the structural formulas as found in normal man from embryonic to adult life. Adults have hemoglobin (Hb) A and A$_2$, perhaps with traces of F. In the fetus Hb F predominates and is preceded in the embryo by Hb Portland, Gower 1 and Gower 2 respectively. The zeta chain in Hb Portland and Gower 1 is felt to be a primitive alpha chain and the epsilon chain a primitive gamma chain. Gower 1 was formerly thought to be a tetramer of epsilon chains, but currently is held to have a zeta$_2$, epsilon$_2$ structure. The primitive hemoglobins have been found only in stillborn infants with hydrops secondary to severe types of hemoglobinopathy or—as in Gower 2—in trisomy 13.

Beta chain and thus HbA synthesis begins about 8-10 weeks of intrauterine

Table 1. Hemoglobin Tetramers in Man

Hemoglobin	Structure
A	$\alpha_2 \beta_2$
A$_2$	$\alpha_2 \delta_2$
F	$\alpha_2 \gamma_2$
Gower 2	$\alpha_2 \epsilon_2$
Gower 1	$\zeta_2 \epsilon_2$
Portland	$\zeta_2 \epsilon_2$

life, but persists as only a small proportion (about 10%) of total hemoglobin content until about 34 weeks' gestation when synthesis rapidly increases. That normal and mutant beta chain syntheses begin so early potentially allows for intrauterine detection of deleterious variants by direct sampling of placental vessels (2a). Exactly what causes the switch from Hb F to A and A$_2$ synthesis in the third trimester is unknown.

Structural homology among the various hemoglobin chains has led to the suggestion that they were sequentially derived from more primitive ancestral genes by unequal crossing over, with resultant gene duplication and later independent mutation. The delta and the beta chains differ by only 10 amino acids, and this fact coupled with the observation that HbA$_2$ has been found only in primates suggests that the duplication leading to the delta chain is a fairly recent evolutionary event. Linkage studies and detection of delta-beta fusion genes (Hb Lepore) have established that the two loci are adjacent. Moreover, in another hemoglobin variant, Hb Kenya, the nonalpha chain has a structure consistent with being derived from a gamma-beta fusion gene. This observation—together with the discovery that all human beings possess two types of gamma chain genes that do not segregate and whose products differ from one another only by a glycine-alanine interchange at position 136 (designated Gγ and Aγ)—has lead to tentative mapping of the nonalpha chains as Gγ- Aγ-δ-β, i.e., the gene loci are all contiguous. The alpha locus is unlinked and may be on another chromosome. The location of the more primitive globin chain genes is unknown.

Study of alpha chain variants has led to the contention that the alpha chain locus is also duplicated in some populations, i.e., four alpha chain genes per diploid set of chromosomes. Individuals with two alpha chain mutations, such as HbJ-Buda and HbG-Pest, were found, e.g., to synthesize considerable amounts of HbA as well. Although the duplicated alpha locus is in many populations, it is not in everyone, and its absence confers no apparent selective disadvantage unless an alpha chain mutation or deletion supervenes.

Point Mutations

For the most part the mutant hemoglobin chains differ from the normal ones by a single amino acid, and the substitutions are almost always consistent with a single nucleotide change in the DNA, a *point mutation* (Table 2). There are some variants, such as HbC-Harlem, with two mutations. This hemoglobin causes sickling, because in addition to having the same amino-acid change in the sixth position from the amino terminal end of the beta chain (valine for glutamic acid) as in HbS, it also possesses an aspargine for aspartic acid exchange at beta 73, as

Table 2. Hemoglobin Varients Seen in Man

Molecular Event	Typical Examples
Point mutation	Hbs, S, C, E (see [3] for list and citation to original description)
Intragene crossovers	HbC-Harlem (Between β_s and β Korle Bu)
	HbC-Ziquincher (between β_s and β Dhofar)
	Hb Arlington Park (between β_c and β_n-Baltimore)
Chain termination variants (elongated)	Hb Constant Spring, Seal Rock, Icaria, Koya Dora (all α)
Chain termination variants (shortened)	Hb McKees-Rock (β chain)
Triplet (or multiples of 3) deletions	Hb Leiden (β^6)*, Lyon (β^{17-18}), Freiburg (β^{23}), Tochigi (β^{56-59}), St. Antoine (β^{74-75}), Tours (β^{87})
Insertions	Hb Grady ($\alpha^{116-118}$ duplicated)
Frame shift	Hb Wayne, Tak, Cranston (extended C-terminal sequences of β chain)

*Number in parenthesis indicates location and number of amino acids lost in deletion.

found in Hb-Korle Bu. Either an *intragenic* crossover occurred in an individual who was doubly heterozygous for two different beta chain mutations—his offspring receiving the doubly mutant gene—or a new mutation occurred at beta 73 in the germ cell of a person who already had the HbS mutation. Both mechanisms are possible and only family studies can differentiate between them. This hemoglobin is a "C" rather than an "S" hemoglobin because it migrates electrophoretically like classic HbC, the extra positive charge being conferred by the amino group of the aspargine molecule.

Point mutations of particular interest are those occurring around the histidine residues, since they may interfere with the iron moiety of heme and produce *methemoglobinemia*. There are both alpha and beta HbM variants, and most of them involve tyrosine for histidine substitutions at alpha 53, alpha 87, beta 63 or beta 92. The carboxy group of the tyrosine molecule irreversibly binds the ferric ion, eliminating the ferric–ferrous electron shift and therefore oxyhemoglobin formation. An interesting exception to these mutations is seen in HbM-Milwaukee in which glutamic acid has replaced valine at beta 67, four positions removed from the critical histidine group. These four positions represent one turn of the peptide alpha helix, such that the polar COOH group of glutamic acid in the mutant hemoglobin is placed in proximity to the iron molecule and, as with tyrosine, binds ionically. It is important to remember that methemoglobinemia has many causes, among which are two autosomal recessive gene defects, deficiency of NADA-dependent and NADPH-dependent methemoglobin reductase, only the former producing the syndrome.

Other point mutations may cause alterations in the contact points between the alpha and beta chains; they are the *altered affinity hemoglobin* variants, so named because most of them (e.g., Hb Chesapeake, Capetown, Rainier) have a high affinity for oxygen and are associated with polycythemia. One exception is Hb Kansas in which the oxygen affinity is very low. Carriers of this variant have cyanosis with low, rather than high, hemoglobin levels. Heterozygotes for neither high nor low affinity hemoglobins have any major disability.

Unstable hemoglobin variants commonly cause significant disease in heterozygotes, as is seen with Hb Zürich and Hb Köln. Other variants may replace a nonpolar group (Hb Shepard's Bush) with a polar one or insert a proline residue (Hb Santa Ana). The polar substitutions are felt to be energetically unstable; the proline substitution modifies the alpha helical structure of the polypeptide with resultant instability. Amino-acid substitutions at a heme contact point (Hb Hammersmith) also may give rise to instability and hemolysis. Most of the patients with those unstable variants have Heinz bodies in their red blood cells (RBCs), and the degree of congenital hemolytic disease depends on the severity of the molecular lesions.

Point mutations in a chain terminating codon lead to *chain termination variants* that are longer than the usual alpha or beta chain. The most extensively studied is Hb Constant Spring, which has an elongated alpha chain of 172 rather than 141 amino-acid residues. In individuals with only two, rather than four, alpha chain genes hetero- and homozygosity for Hb Constant Spring produce, respectively, thalassemia minor and major, probably because of altered tertiary and quarternary structure engendered by the extra amino acids. A mutation resulting in a *premature termination codon* can give rise to a shortened gene product, e.g., the beta chain variant Hb McKees-Rock, which lacks the two carboxy-terminal amino acids. *Deletion* mutations have been described as have internal duplications (called *insertions*).

Genetic Deletions

Deletions occur when a codon of three nucleotides or a multiple thereof is lost. A loss of one or two nucleotides leads to a *frameshift mutation*, only a few of which have been found—for obvious reasons. Equally obvious is that those discovered have had the mutation near the C-terminal end of the molecule. A frameshift mutation near the N-terminal end either would probably not be translated (nonsense mutation) or would not code for a functional protein. In either case the measured result would be a diminution of synthesis of that particular chain, with thalassemia-like features.

Deletion of whole genes theoretically could be responsible for some of the thalassemia syndromes. Thus deletion of one beta gene could give rise to thalassemia minor in which the amounts of Hb A_2 and Hb F are relatively increased. Deletion of a beta and a delta gene could yield so-called δ-β thalassemia, a condition in which the homozygote has severe anemia and can synthesize only HbF. Some patients with *hereditary persistence of fetal hemoglobin* may have deletions of this sort; however, most of them have normal or near-normal hemoglobin levels and are asymptomatic. These latter apparently have a heritable abnormality in the "switch" mechanism from gamma to beta and delta chain synthesis.

Deletions of the alpha loci produce intrauterine problems. Thus deletion of two alpha loci (of 4) results in alpha thalassemia minor that can be distinguished from the beta chain type by the production of Hb Barts (γ4) and HbH (β4). Homozygosity for alpha deletions (all 4 loci lost) is incompatible with life and results in hydrops fetalis. Some of the thalassemia variants, with findings in homozygote and heterozygote, together with the possible molecular pathology, are shown in Table 3.

Table 3. Partial List of Thalassemia Syndromes

Type of Thalassemia	Homozygote	Heterozygote	Possible Defect
α-Thal 1	Hydrops fetalis	Thal. minor with <5% Hb Barts	Deletions of 2 (heterozygote or 4 (homozygote) α genes
α-Thal 2	Thal. minor	Prob. silent	Deletion of 1 (of 2) α genes
Hb Constant Spring	Thal. minor (+HbCS)	Silent (? some HbCS)	Chain termination mutation
β⁺-Thal	Thal. major HbA, A₂, F	Thal. minor HbA, A₂, F	Reduced β chain mRNA, reasons ?
β⁺-Thal (Negro)	Thal. intermedia HbA, A₂, F	Thal. minor HbA₂	?
β⁰-Thal	Thal. major HbA₂, F	Thal. minor HbA₂	β genes present, β mRNA?
(δ-β)⁰-Thal	Thal. major HbF only	Thal. minor HbA₂ normal HbF elevated	Deletion of δ and β genes
(δ-β)⁺-Lepore	Thal. major	Thal. minor	δ-β fusion gene, ? unstable mRNA

Fusion Genes

Fusion genes were mentioned previously as Hb Lepore and Hb Kenya. They result from a crossover in an area of "slippage" or mispairing. Once gene duplication has occurred by unequal crossover, the resultant genes resemble one another in nucleotide sequence for a long evolutionary interval; hence mispairing may occasionally occur and if a crossover takes place within one of these misaligned areas, a fusion gene will result. The delta–beta fusion gene products are the Lepore hemoglobins, of which three are well characterized (Holland, Baltimore and Boston), and they are differentiated by where the crossover occurred. The antithesis of Lepore is the beta-delta gene (called anti-Lepore) of which two are known (Hb Miyada, P-Congo). Hb Kenya is due to a crossover within an area of even greater misalignment so that the delta chain gene has been deleted.

Inheritance of Hemoglobin Abnormalities

The various hemoglobin abnormalities are inherited in an intermediate fashion. Thus in those variants in which symptoms occur, such as HbS, the abnormal hemoglobin is inherited as a dominant (or codominant) factor since it is regularly detectable by electrophoretic techniques in the heterozygote. However, only homozygous patients, i.e., those with sickle cell disease rather than trait, are generally symptomatic. The *disease* therefore behaves as if it were recessively inherited—hence the designation "intermediate inheritance." Even a disease the symptoms of which are as well known as those of sickle cell anemia may be clinically heterogeneous. A relatively benign form of sickle cell disease has been found in a group of Shi'ite Arabs from Saudia Arabia (3a). It appears that these individuals possess a genetically determined ability to produce large amounts of fetal hemoglobin, which partially compensates for the HbS mutation.

Obviously then a detailed exposition of the symptoms associated with either hetero- or homozygosity for the numerous hemoglobin variants is beyond the scope of this text, and the interested reader is advised to consult standard hematology texts or the original source (3) for greater detail.

HEREDITARY ANEMIA

Anemia obviously has a multiplicity of causes. Basically they may be divided into decreased production or increased destruction of erythrocytes, but the two processes are not mutually exclusive. There is no accurate way to characterize the anemias from an overall nosologic point of vew, since they do not fit into a tight hemolytic or nonhemolytic, microcytic or macrocytic system. The various defects will be presented in terms of the specific biochemical features when they are known, and in terms of morphologic or clinical symptoms, or both, when the exact genetic lesion is obscure.

Inborn Errors of Red Cell Metabolism

The mature erythrocyte lacks a nucleus and subcellular organelles, and its entire energy supply is derived from conversion of one mole of glucose to two moles of lactate with the generation of adenosine triphosphate (ATP) via the anaerobic glycolytic (Embden-Meyerhof) pathway (Fig. 1). Important in maintaining ion gradients across the cell membrane, ATP is thus responsible for the integrity of the membrane. Whereas the red cell lacks a complete tricarboxylic acid (Krebs) cycle, remnants remain as do elements of the hexose monophosphate shunt, as exemplified by the enzyme glucose-6-phosphate dehydrogenase (G-6-PD). This enzyme helps to control the amount of reduced nicotinamide-adenine dinucleotide phosphate (NADPH) and reduced glutathione in the cell. The latter compound protects the cell from destruction by peroxides. Deficiency of any of the requisite enzymes of red cell glucose metabolism or for glutathione synthesis results in hemolytic anemia.

These various disorders are usually grouped as *congenital nonspherocytic hemolytic anemias* (4). All, except G-6-PD deficiency, are both rare and recessively inherited. The degree of anemia—and hence symptoms—depends greatly on the amount of residual enzyme remaining. However, defects in the metabolic pathway prior to the generation 2,3 diphosphoglycerate (2,3-DPG) are symptomatically severer than defects after this step (which lead to 2,3-DPG excess). The phenomenon is related to the fact that 2,3-DPG avidly binds to deoxygenated hemoglobin, and this facilitates tissue "unloading" of oxygen. Thus with the same reduced hematocrit a patient with hexokinase deficiency is more symptomatic than one with pyruvate kinase (PK) deficiency.

The enzymes deficient in the red blood cells (RBCs) are frequently but not always deficient in other tissues as well, the exceptions relating to those enzymes, such as PK with isozymic forms, in which the mutation may affect only the RBC isozyme (5). That the enzymatic deficiency in other tissues is generally of no significance may be related to two factors. First the tissue may have sufficient

INBORN ERRORS OF ERYTHROCYTE METABOLISM

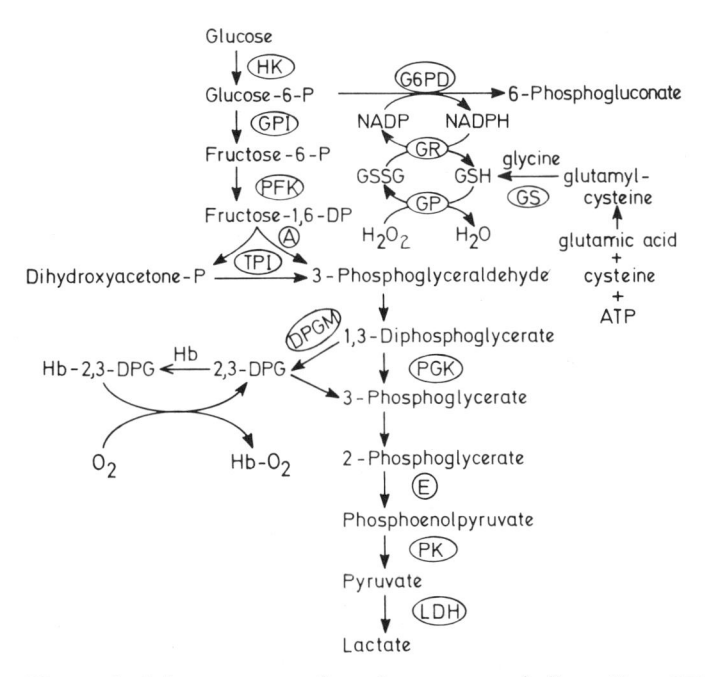

Figure 1. Inborn errors of erythrocyte metabolism. **Key:** HK, hexokinases; GPI, glucose phosphophate isomerase; PFK, phosphofructokinase; A, aldolase; TPI, triose phosphate isomerase; DPGM, diphosphoglucomutase; G6PD, glucose-6-phosphate dehydrogenase; GR, glutathione reductase; GP, glutathione peroxidase; GS, glutathione synthetase; PGK, phosphoglycerokinase; E, enolase; PK, pyruvate kinase; and LDH, lactate dehydrogenase.

alternative ways of obtaining ATP, as seems to be the case in one form of PK deficiency in which the isozyme common to both RBC and liver is defective. The liver, with an intact mitochondrial system, presumably generates enough ATP that the PK deficiency is of no consequence. Second some cells have a half-life shorter than that of the mutant enzyme. In some forms of G-6-PD deficiency, e.g., the enzyme half-life is longer than that of the white blood cell (WBC) and no leukocytic enzyme deficiency ensues.

In other forms of G-6-PD deficiency the enzyme half-life is a matter of hours and hence both red and white cells express the enzyme deficiency. Some patients have a form of chronic granulomatous disease. Patients with combined RBCs and muscle phosphofructokinase deficiency are described who had not only hemolysis but also a myopathic syndrome. In triosephosphate isomerase and phosphoglycerokinase deficiencies, the patients have neurologic abnormalities; these lesions possibly result from CNS deficiencies of those enzymes that are pivotal in brain lipid metabolism. Alternatively any severe hemolysis has the potential to produce early hyperbilirubinemia and the symptoms of kernicterus, so that it is difficult to sort out primary and secondary effects of the mutant

Table 4. Glycolytic Enzyme Deficiencies*

Enzyme Deficiency	Hemolysis	Inheritance	Other Symptoms
Hexokinase	Mild	AR	Rarely may be associated with Fanconi anemia features (6)
Glucosephosphate isomerase	Variable	AR	Anemia exacerbated during infection and stress; RBCs may be spherocytic
Phosphofructo kinase	Moderate	AR, ?XR	Some forms have only hemolysis; others have anemia and myopathy
Phosphoglycero kinase	Severe	AR	Neurologic abnormalities and retardation
Pyruvate kinase	Variable	AR	Commonest error in glycolytic pathway
Triosephosphate isomerase	Severe	AR	Rare; neurologic symptoms even in absence of hemolysis; early death common

*The six enzymes listed all have variant forms.
AR, autosomal recessive; XR, X-linked recessive.

enzymes. Heterozygotes for any of these related enzymopathies are asymptomatic. Table 4 lists the various properties of the congenital nonspherocytic hemolytic anemias from deficiencies of RBC glycolytic enzymes. In addition to them, abnormalities in other glycolytic enzymes are reported, namely, *glyceraldehyde-3-phosphate dehydrogenase* (7), *2,3 diphosphoglycerate phosphatase* (8a), *lactate dehydrogenase* (9), *enolase* (10) and *aldolase A* (11). All of them are poorly characterized and at present not established entities.

Other red cell enzymopathies cause hemolytic anemia, but they do not, strictly speaking, involve glycolysis. Chief among them is *G-6-PD deficiency*. The enzyme G-6-PD catalyzes the first steps in the hexose monophosphate shunt (Fig. 1), is composed of a number of subunits and tightly bound to NADP (12). The NADPH formed from the G-6-PD catalyzed reaction is cofactor in the reduction of glutathione and helps to protect the cell from oxidative damage. The deficiency was discovered 25 years ago in a study of drug-related hemolysis, a topic discussed more fully in Chapter 7, "Pharmacogenetics". In the intervening years the enzyme has been found to be definitely polymorphic; nearly 100 variants have been described, only one-fourth of which are associated with hemolysis, either spontaneous or drug related (13, 14).

Most of the G-6-PD variants are detected by electrophoretic techniques, but studies of thermolability and various kinetic measurements are also used (15). Two of the commonest electrophoretic types not associated with enzymatic deficiency are designated G-6-PD-B+, as found in most whites, and G-6-PD-A+, seen in American and West African blacks. These alleles differ each from the other by a single amino-acid substitution (aspargine vs aspartic acid) (16). One of the commoner deficiency types is found in blacks and is designated as A−. Presumably this enzyme, which is present in 10-15% of Negro males in the United States, harbors an additional nonelectrophoretically detectable mutatant that reduces enzymatic activity.

It is suggested that the A− and other deficiency variants found in high frequency in various parts of the world, particularly in the Mediterranean, are somehow related to selective pressure exerted by malaria in these areas, similar to that postulated for the sickle cell hemoglobin mutation. The variants probably differ from one another by point mutations, analogous to the abnormal hemoglobins, as are found for A+ and B+ types. However, little peptide mapping has been done to substantiate this contention. One interesting variant, G-6-PD–Hektoen, consists of a histidine-to-tyrosine exchange that confers increased G-6-PD activity to the red blood cell (17). Most of the other mutations cause some diminution of enzymatic activity although, as indicated earlier, they are not always clinically significant.

The high frequency of an X-linked deficiency state, like A− in black males, indicates that it will be found in 2-3% of homozygous black females, since this particular mutation is not known to interfere appreciably with fertility. Heterozygous (carrier) female subjects harbor two populations of cells by cloning techniques, an observation that is applied to the study of single vs multicell origin of malignancy. Cells other than erythrocytes may show deficiency, as discussed previously. The relatively common Mediterranean G-6-PD variant is associated with *favism*, a condition in which hemolysis occurs after ingestion of the fava bean, a dietary staple in that area. The bean contains substances with redox potential and thus acts as a pharmacologic hemolysin.

Interestingly the next enzyme in the shunt pathway, which is also a dehydrogenase, *phosphogluconate dehydrogenase,* is not X-linked; in fact the structural locus of the enzyme polypeptide is on chromosome 1. Severe deficiency is reported in sibs and partial deficiency over four generations. In neither situation was hemolysis recorded (18). Apparently the NADPH derived from the G-6-PD reaction suffices for the metabolic needs of the red blood cell.

Still other inborn errors occur in enzymes critical for generation or maintenance of cellular glutathione. As mentioned, glutathione is important in reducing organic and inorganic peroxides (Fig. 1). Any defect in its synthesis or in conversion of the oxidized disulfide to reduced glutathione would be expected to result in accumulation of peroxides and eventual hemolysis. The contribution of G-6-PD to this system has already been discussed. In addition at least three autosomal recessive disorders are associated with abnormalities of glutathione metabolism and hemolytic anemia.

In *glutathione synthetase deficiency* (19), total glutathione concentration is reduced, whereas in *glutathione reductase deficiency* (20) only the reduced form is diminished. Deficiency of the reductase is heterogeneous, since a variant is described in which decreased affinity for nucleotide cofactor was found (21). The anemia was cured by the administration of vitamin B_2. *Glutathione peroxidase deficiency* of mild degree is normal in the newborn infant and seen in various acquired conditions (22). Homozygotes have about one-third normal enzyme levels and show a compensated hemolytic anemia. Heterozygosity for the deficiency may be benign, but large doses of oxidant drugs possibly produce hemolysis. All these deficiencies are rare, and the anemia seems to be relatively mild.

Purine and Pyrimidine Metabolic Defects

There are two known defects in purine and one in pyrimidine metabolism that give rise to hemolytic anemia. Deficiency of *adenylate kinase,* the enzyme catalyzing the conversion of ADP to ATP was found in two sibs, one of whom also had

G-6-PD deficiency (23). Presumably this enzyme functions in a salvage pathway for generation of high-energy phosphate. The deficiency is autosomal recessive. Electrophoretic variants of the enzyme have been found in normal persons and, as expected, are inherited as codominants. The structural gene locus is on chromosome 1.

Deficiency of *adenosine triphosphatase* (ATP-ase) leads to hemolytic anemia in some patients for obscure reasons (24). Although it can be postulated that excessive ATP is inherently toxic or that its accumulation interferes with one or more of the glycolytic kinase reactions, dominantly inherited high levels of erythrocytic ATP are found in some families who show no anemia (25). Not all patients with ATP-ase deficiency have recognizable hemolysis. The deficiency is inherited as an irregular dominant trait, an observation of some interest since most enzymatic deficiencies are only symptomatic in the homozygote (however, see "porphyrin metabolism" in Chapter 5, "Genetic Metabolic Disorders"). *Pyrimidine-5-nucleosidase deficiency* in the RBC is associated with Heinz bodies and hemolytic anemia in homozygotes (26). Again the reason for the hemolysis is unclear, but may be related to nucleoside inhibition of the various kinase reactions. The Heinz bodies probably represent incompletely degraded ribosomal RNA due to this enzyme deficiency. The genetic mechanisms of this disorder are unknown.

The other abnormalities in RBC pyrimidine metabolism result in hypochromic anemia with a megaloblastic marrow and are distinguished by *orotic aciduria* (27). Affected children grow poorly and show neutropenia. The symptoms respond to replacement therapy with pyrimidine nucleotides or with uridine. Two forms of orotic aciduria are known; Type I involves a dual enzymatic defect in both orotidylic pyrophosphorylase and orotidylic decarboxylase, sequential enzymes necessary for conversion of orotic acid to uridine-5-phosphate. In Type II there is an isolated deficiency of the decarboxylase. Both types are rare, and probably both are autosomal recessive defects.

Heterozygotes for Type I disease have partial enzymatic deficiencies, but are asymptomatic. Since two enzymes are deficient in Type I disease, a defect in a regulator gene is postulated as the basic lesion, but this has not been proved. A mild variant of Type I disease is also reported. Remember that orotic aciduria can occur as a result of therapy in man with 6-azauridine or allopurinol and in the genetic defect associated with hepatic ornithine transcarboxylase deficiency.

MACROCYTIC ANEMIA

The effects of disordered vitamin B_{12} and amino-acid metabolism are discussed elsewhere in this work; this section will confine itself to a discussion of abnormalities in vitamin B_{12} absorption and transport affecting the erythrocyte.

Defective Vitamin B_{12} Absorption

Vitamin B_{12} is necessary for normal erythroid maturation, and either defective absorption or activation leads to megaloblastic anemia. The vitamin requires both a gastric glycoprotein (intrinsic factor) and a specific ileal receptor for absorption. Once in the circulation, B_{12} is bound to transport proteins, transcobalamin I (TCI) and II (TCII), the former an alpha, the latter a beta, globulin. The most important transport protein for newly absorbed B_{12} is TCII; it also facilitates cellular uptake. The role of TCI is not known (28).

Intrinsic factor (IF) deficiency with normal gastric morphology and acidity results in a form of juvenile pernicious anemia (29). Another results from an *altered IF* that has reduced affinity for the ileal receptor (30), and a third disorder is caused by a lesion in the *ileal transport* system (31). Each disorder is associated with early-onset megaloblastic anemia with low serum B_{12} levels, and all of them are reported in male and female sibs, indicating probable autosomal recessive inheritance. All these disorders respond to parenteral, but not oral, administration of vitamin B_{12}. Curiously proteinuria—a feature of the syndrome—persists even after B_{12} therapy.

Transcobalamin II deficiency is reported in two children with infantile megaloblastic anemia who also had thrombocytopenia and leukopenia (32). Serum B_{12} levels were normal, but hematologic remission occurred only after large doses of B_{12} were parenterally administered every other day. Apparently only by maintaining high free B_{12} levels in the serum can tissue requirements be met, perhaps by passive diffusion. In contrast partial *transcobalamin I deficiency* is benign, although total deficiency may be expected to cause anemia if TCI and TCII-bound B_{12} are in equilibrium (33). The transcobalamin deficiencies are probably inherited as autosomal recessives traits, but so few cases are recorded that no firm statement can be made (28).

As previously mentioned, defective tissue utilization of B_{12} (the methylmalonic acidurias) are discussed elsewhere. Although some of these conditions may show associated megaloblastic changes in the bone marrow, the metabolic features are more striking. Moreover, although homocystine and methylmalonic acid may occasionally be found in the urine of any patient with a B_{12} absorption defect from whatever cause, their presence is not usually associated with symptoms of acidosis or connective tissue problems, as is seen in the more fundamental metabolic vitamin B_{12} defects.

Remember that *pernicious anemia* is reported in multiple members of some families, usually in concert with autoimmune endocrine disease, such as thyroiditis, adrenalitis and diabetes mellitus (Schmidt's or the polyglandular deficiency syndrome). The disease may appear in childhood, but is much commoner in adult life and is generally associated with antiparietal cell or anti-IF antibodies, together with other autoimmune phenomenon. The childhood forms of pernicious anemia already described do not feature immunologic abnormalities. The inheritance of the syndrome in adults is unclear, but may be an autosomal dominant one with reduced penetrance and variable expressivity. There seems to be an association of this type of immunologic disease with HLA-B8 antigen.

Abnormal Folate Metabolism

Like vitamin B_{12}, folic acid also has a complex (and poorly understood) intestinal transport system and may undergo, numerous conversions to different coenzymes (20). A defect in *folate absorption* is described in children with megaloblastic anemia, mental retardation, ataxia and athetosis (34). Parenteral administration of folate caused hematologic, but not neurologic, remission. Although the basic abnormality remains obscure, autosomal recessive inheritance is likely. Autosomal recessive inheritance is probable for megaloblastic anemia responsive to N^5-formyltetrahydrofolate (N^5-THF), but not to folic acid, a condition described in children whose livers showed *dihydrofolate reductase deficiency* (35, 36).

Other errors of folate metabolism are related to tissue utilization of folate and do not result in anemia, but are associated with major neurologic dysfunction. These enzyme deficiencies *(formiminotransferase, cyclohydrolase* and N^5, N^{10}-*methylenetetrahydrofolate reductase)* are not well characterized (28), although abnormalities in the last-mentioned enzyme are associated with homocystinuria.

Other Causes of Macrocytic Anemia

A single case of *thiamine-responsive megaloblastic anemia* is reported in an 11-year-old girl (37). Serum thiamine levels were normal as were the activities of several thiamine-dependent enzymes. The biochemical basis of the condition, however, is unknown.

Three rather unusual forms of macrocytic anemia are described in which the bone marrow shows erythroblastic multinuclearity; they are designated in the aggregate as *congenital dyserythropoietic anemia*. Type II is the commonest and distinguished from the others by showing a positive acidified serum (HAM) test. It is designated as HEMPAS (*H*ereditary *E*rythroblastic *M*ultinuclearity with *P*ositive *A*cidified *S*erum test) (38). Increased susceptibility to hemolysis in anti-I or anti-i antiserum is also found. On electron microscopic examination double cytoplasmic membranes are usual. Type II is an autosomal recessive trait and although heterozygotes may show serologic abnormalities, usually they are otherwise normal.

Type I shows a megaloblastoid marrow (39) and Type III features multinucleated erythroblasts, so-called "gigantoblasts," with as many as 12 nuclei (40). Type III is an autosomal dominant, Type I, a recessive, trait. Type III probably also includes the entity described in Sweden as *hereditary benign erythroreticulosis*. A further possible subtype (IV) is suggested on the basis of patients whose red blood cell morphology is similar to Type II, but without the characteristic EM changes (41). The specificity of these morphologic alterations is moot.

MISCELLANEOUS ANEMIAS

Hypochromic anemia is most commonly acquired and related to deficient nutritional intake of iron or to gastrointestinal (GI) bleeding. The hereditary form of hypochromic anemia is differentiated by the fact that it is often detected in childhood and associated with hyperferricemia, early death from hemosiderosis disproportionate to transfusions, and sideroblasts in the blood after splenectomy; hence the alternative names, primary siderocytic, sideroblastic or sideroachrestic anemia. It is usually inherited as an X-linked recessive disorder (42). Heterozygotes may show two populations of cells, in keeping with the Lyon hypothesis. The condition is likely heterogeneous since 1) a pyridoxine responsive form is described (43), 2) some families show severely affected females—suggesting an autosomal recessive form of hypochromic anemia (44) and 3) an association of the hematologic finding is described in brothers who also had myopathy with mitochondrial inclusion bodies and chronic lactic acidosis (45). The *pyridoxine-responsive* form may be heterogeneous since it is seen in families, also as an autosomal recessive form of hypochromic anemia, but without iron overload (46).

Increased body iron—whether nutritional, related to transfusions (hemosiderosis) or heritable (hemochromatosis)—has the same eventual consequences, i.e., hepatic, cardiac and endocrine failure from deposition of excess iron in the various bodily organs with ultimate destruction. Genetic *hemochromatosis* is the subject of dispute, some investigators favoring a dominant, others a recessive, hypothesis; the genetic interpretation is rendered more difficult by menstrual blood loss. Most workers now favor an intermediate form of inheritance with increased iron stores in both hetero- and homozygotes (dominant inheritance), but with true clinical manifestations only in homozygotes (recessive inheritance), a situation analogous to that seen in sickle cell trait and sickle cell anemia. Patients with *atransferrinemia* show hypochromic microcytic anemia with tissue hemosiderosis (46a). The condition is inherited as an autosomal recessive one, and heterozygotes have roughly half-normal values of the plasma protein. In view of these various hematologic syndromes involving iron metabolism either directly or indirectly, it is not surprising that pure genetic hemochromatosis is a difficult diagnosis to establish.

Pure red cell aplasia is called the *Blackfan Diamond Syndrome,* and it has a multiplicity of causes (47). In adults, e.g., it may be an acquired accompaniment of thymoma. It is also seen in children with congenital hypogammaglobulinemia (48). In some cases it is the result of lymphocyte-mediated suppression of erythropoiesis (49), in still other cases the cause is unknown. It is occasionally encountered in the newborn baby, and an abnormality of tryptophan metabolism has been discovered in some individuals (47). Sibs are only rarely affected, but two-generation transmission is noted (50). Whether a heritable form of the pure syndrome exists as such is not clear; if it does, current evidence probably best supports polygenic inheritance. A related disorder involving *congenital anemia and triphalangeal thumbs* is described in brothers (51); whether it is distinct is moot, since some of Diamond's original cases had thumb abnormalities. The association of hematologic with upper extremity abnormalities is interesting because it includes several conditions in addition to that described by Blackfan and Diamond and the anemia–triphalangeal thumb syndrome, i.e., Poland anomaly (absent pectoralis muscle with leukemia), thrombocytopenia–absent radius (TAR) syndrome and Fanconi pancytopenia (radial anomalies with aplastic anemia–leukemia).

Of these conditions, *Fanconi pancytopenia* has received the most attention. This autosomal recessive disorder comprises childhood onset aplastic anemia; pigmentary anomalies of the skin; congenital malformations of kidney, heart and skeleton; endocrine disturbances; and growth failure and frequently terminates in acute leukemia (52). Chromosomal instability is a common accompanying feature, probably due to defective excision-repair of DNA after gamma irradiation (53). Not all affected persons have the full constellation of defects; it is interesting, however, that heterozygotes for the condition seem to have a three- to four-fold increased risk of nonleukemic malignancy.

Fanconi's anemia is reported in a patient whose hematologically normal father had a radial anomaly, a finding perhaps indicative of the heterozygous state (54). The condition may be heterogeneous, i.e., two brothers are described who had pancytopenia, recurrent infections and low IgA levels (55). One of them developed multiple cutaneous malignancies and died from metastatic disease. In the other the bone marrow responded to methyltestosterone. This may represent an immunologic form of the Fanconi syndrome akin to that described in the anemia of Blackfan and Diamond. Aplastic anemia may be a feature of other disorders with unknown etiology, such as the Shwachman syndrome or result from bone overgrowth and crowding, as is seen in juvenile osteopetrosis.

Autoimmune hemolytic anemia is a common facet of diverse immunologic disorders. Rarely it may occur as an isolated familial entity, but most cases are sporadic or follow the inheritance pattern of the primary disease (56). Hemolytic anemia secondary to Rhesus or other blood group incompatibility is dealt with in standard textbooks and will not be discussed here.

ALTERATIONS IN RED CELL MEMBRANES

The red blood cell membrane is composed of three basic structural substances, including glycophorin, a myosin-like protein called spectrin and a microfilamentous protein similar to actin. These substances are important in maintaining RBC plasticity and viability. A structural abnormality in them may be expected to lead to decreased deformability, splenic sequestration with failure of glucose delivery, loss of the energy-dependent cation pump and eventual hemolysis. It is logical to presume that the various hereditary disorders of red cell shape involve abnormal membrane proteins, but so far this contention has not been rigorously established (57).

The commonest intrinsic RBC defect is *spherocytosis*, with a population prevalence of about 1 in 5,000 (58). It occurs in all races, but is particularly common in northern Europeans. The clinical spectrum is well known and described in standard texts. Briefly, affected individuals suffer repeated, occasionally severe, hemolytic episodes from early childhood, splenomegaly and (commonly) gallstones. Splenectomy is clinically curative although spherocytes do not disappear from the peripheral blood. Studies of membrane proteins in this autosomal dominant disorder are conflicting as are investigations into the cation-dependent ATP system (57-59). Variants of the condition are also described, as characterized by normal osmotic fragility, disappearance of spherocytes after splenectomy and atypical autohemolysis, but they are nonspecific findings (60). Spherocytes are the likely result of a number of processes, and genetic heterogeneity would not be surprising.

Such heterogeneity is found in *elliptocytosis* wherein one form of the disease is linked, the other nonlinked to the Rhesus blood group locus. The "unlinked" type reputedly shows more hemolysis than the linked variety, but phenotypic variability even within a given family is common, rendering this observation of little differential value. Some workers suggest that the nonlinked form with hemolysis should be termed *ovalocytosis* (61). Another form shows elliptocytosis with transverse slit-like areas of decreased density, accompanied by no hemolysis, normal survival and a *permeability defect* (62). All the various subtypes are inherited as autosomal dominant traits.

Other dominant pedigrees are reported in which membrane phosphatides or ultrastructural abnormalities are seen (63, 64). The RBCs may assume diverse shapes in these disorders, none of which is well characterized. Extreme microcytosis with marked aniso- and poikilocytosis and hemolysis is reported in children who showed *calcium leak and abnormal thermal sensitivity* of the RBC membrane (65). This disorder, which may be a recessive one, was previously classified as a variant form of elliptocytosis, again emphasizing the lack of specificity of the red cell morphology in the differential diagnosis (66).

Stomatocytosis describes a swollen, cup-shaped erythrocyte that, like the other morphologic variants, shows decreased deformability and abnormal cation-pump function. Both dominant and recessive forms may exist, and hemolysis may be mild. Of interest is the apparently effective treatment of patients with dimethyl adipimidate, an agent known to cause amide formation of free NH$_2$ groups and to enhance molecular cross-linking (67). This observation opens new vistas to the study of hereditary disorders of RBC membranes.

Lipid Metabolic Defects

Abnormalities in lipid metabolism may primarily or secondarily effect the red cell membrane (68). In *abetalipoproteinemia* (Bassen-Kornzweig syndrome), e.g., plasma cholesterol and triglyceride concentrations are low, as are low-density lipoproteins and chylomicrons. Symptoms are mainly those of malabsorption, retinitis pigmentosa and neurologic dysfunction. The red cells have a thorny appearance termed acanthocytosis. Anemia may be present. The condition is an autosomal recessive one. In addition, there are dominant forms of *hypobetalipoproteinemia* in which acanthocytes are only variably present, and other symptoms may be minimal or absent. It is unclear whether the red cell defect is primary or secondary. Patients with high-density lipoprotein deficiency (Tangier disease) have a defective apolipoprotein, store cholesterol esters in many parts of the body and have variable neurologic signs, but not RBC abnormalities or anemia.

In contrast, patients with *lecithin–cholesterol acyltransferase deficiency* (Norum disease) have—in addition to corneal opacities, renal disease and hypercholesterolemia—a moderate normochromic anemia due to a combination of decreased production and increased destruction of RBCs (69). In this autosomal recessive disease (virtually confined to Norway) the unesterified cholesterol presumably interferes with some element in red cell membrane synthesis, although the exact mechanism is unknown.

Anemia may be a feature of some storage diseases, such as Gaucher disease, in some cases perhaps because of simple hypersplenism, but in others it possibly is related to defective RBC membrane constituents (70).

LEUKOCYTIC ABNORMALITIES

Neutropenia may be generalized or selectively involve one or more of the WBC series. The purely lymphopenic conditions are associated with an immunodeficiency and are considered in this context elsewhere (see Chapter 4, "Genetic Disorder of the Immune System"). Severe failure of all white cell elements occurs in *reticular dysgenesis.* Affected persons have no lymphocytes as well, and the disease is generally classified with the immunologic deficiency diseases. It is probably a rare autosomal recessive disorder.

Other recessively inherited neutropenic syndromes include *fatal infantile agranulocytosis* (Kostmann syndrome) (71), *congenital neutropenia with eosinophilia* (72), lazy leukocyte syndrome, Chédiak-Hegashi syndrome, Fanconi pancytopenia and the Shwachman syndrome. Neutropenia may be seen in those conditions associated with abnormalities in glycine and branched-chain amino acid metabolism for reasons not entirely clear. In addition two dominantly inherited, more benign pure forms are described as *chronic familial neutropenia*—in which infections are usually of no major consequence (73)—and *cyclic neutropenia*—in which neutropenia and infections occur every 14-21 days with spontaneous recovery of marrow elements (74). Study of an animal model of the human cyclic disease occurs in the gray collie and suggests a possible stem cell defect (75). Other workers feel that the basic lesion in man may be related to a quantitatively decreased entry of stem cells or granulocytic precursors into granulopoiesis, perhaps owing to an abnormality in production or activity of colony-stimulating factor (76).

Apparently benign *neutrophilia* is reported in a mother and three children who also had hepatosplenomegaly, Gaucher-type histocytes and thickened calvarias (77). Patients with *familial eosinophilia* suffer from chronic recurrent pulmonary infiltrates, bronchitis, and skin infiltration with eosinophils and histocytes without signs of skin allergy (78). Both conditions are inherited as autosomal dominant disorders.

Hereditary morphologic leukocytic variants have been described on a number of occasions (79). Azurophilic cytoplasmic inclusions in the polymorphonuclear leukocytes, the *Alder anomaly*, is inherited as an autosomal dominant disorder. It probably cannot be distinguished from the Reilly bodies seen in the mucopolysaccharide storage diseases. In the May-Hegglin anomaly, RNA leukocytic inclusions (Döhle bodies) are associated with giant platelets, also as a dominant trait. Döhle bodies may be seen transiently in normal persons during acute infection, and they have also been noted in affected sibs who had concomitant leukemia (80). With the autosomal dominant *Pelger-Huët anomaly*, the leukocytes have a rod or dumb-bell shape, rather than the usual segmented nucleus. Interestingly, if patients with this anomaly develop B_{12} deficiency, their leukocytes become normally segmental and revert to the rod shape on therapy. The Pelger-Huët gene is linked to a dominant form of proximal muscular dystrophy (81). Hypersegmentation of the neutrophils, another autosomal dominant disorder, is known as the *Undritz anomaly*. All these conditions are medical curiosities and are not known to interfere with function in any way.

Disorders of leukocyte function are discussed in Chapter 4, "Genetic Disorders of the Immune System," and genetic aspects of neoplastic disease of the bone marrow is considered in Chapter 3, "Genetics and Cancer."

PLATELET DISORDERS

Platelet deficiency or malformation, or both, although usually associated with autoimmune disease or neoplasia, may be familial. Isolated *thrombocytopenia* may be inherited as a dominant, recessive or X-linked disorder (82). It may also be part of more complex conditions, such as Fanconi pancytopenia, the recessively inherited (TAR) syndrome, the Wiscott-Aldrich syndrome and an apparently X-linked disorder comprising thrombocytopenia, elevated IgA and glomerulonephritis (83). Thrombocytopenia is reported as being caused by a deficiency of a plasma platelet-production factor termed *thrombopoietin*, analogous to erythropoietin (84). Patients with a number of metabolic disorders, notably hyperglycinemia and defects of branched-chain amino acid metabolism have shown thrombocytopenia as well as neutropenia. These are likely toxic effects of the aberrant metabolism. Giant hemangiomas, which are rarely familial, may trap platelets and cause a form of consumption thrombocytopenia (Kassbach-Merritt syndrome).

The remaining platelet disorders are more properly termed thrombopathies and include defects in intrinsic platelet factors, failure to release or to aggregate with adenosine diphophastate (ADP) and compound defects. In *Glanzman-Naegli thrombasthenia* there are abnormal platelets, deficient clot retraction and a bleeding diathesis. The condition is genetically heterogeneous, encompassing both dominant and recessive variants, and biochemically variable as well since deficient platelet factor 3 and various platelet enzyme defects are described in different families, especially glutathione reductase and peroxidase (85). Abnormal platelets with thrombocytopenia (*thrombocytopenic thrombopathy*) may be dominant or recessive. Presumably the bleeding problems result from removal of the structurally altered (usually giant) platelets from the circulation. In *storage pool disease*, an autosomal dominant disorder, the platelets are small, but the bleeding diathesis is related to failure of aggregated platelets to release ADP. Platelet serotonin and prostaglandin are also reduced in this disorder (86). Patients with *von Willebrand disease* may have prolonged bleeding time because of a deficiency of the von Willebrand factor, a substance important for normal platelet adhesiveness.

Familial platelet deficiency may not be hereditary, as exemplified by neonatal thrombocytopenia secondary to maternal antiplatelet antibodies. *Thrombotic thrombocytopenia* usually is secondary to collagen disease or hypergammaglobulinemia. Rarely such conditions may be familial, as was observed in a family in which four sibs developed thrombotic thrombocytopenic purpura (87).

POLYCYTHEMIA

The vast majority of patients with polycythemia have an increased RBC mass secondary to chronic hypoxia for whatever reason. The mutant hemoglobins with increased oxygen affinity that result in polycythemia are hereditary and transmitted as autosomal dominant disorders as discussed earlier. More than 20 of them are known, most of which are beta chain point mutations. The stimulus to increased red cell production is defective oxygen release (high O_2 affinity), usually due to the mutation-induced abnormality in alpha and beta chain interaction (88).

Other examples of benign familial polycythemia or primary erythrocytosis exclusive of hemoglobinopathies are reported. In one form, apparently inherited as a recessive trait, defective regulation of erythopoietin production is postulated (89). In another, an expanded red cell precursor pool responsive to erythropoietin has been uncovered, also as an autosomal recessive defect (90). Dominant polycythemia is described, but for the most part the patients were incompletely studied and abnormal hemoglobins were not excluded. By the same token polycythemia vera is reported in familial aggregates, but the diagnosis in relatives is not always completely documented (91). At best the condition may be polygenic.

COAGULATION ABNORMALITIES

A detailed discussion of the various mechanisms underlying normal coagulation is beyond the scope of this discussion. Suffice to say that genetic abnormalities at every stage have been detected (Table 5). The commonest coagulation disorder and the one most intensively studied is *hemophilia A* (Hm A), or factor VIII deficiency. This X-linked recessive disease is about five times commoner than *hemophilia B*, another X-linked, nonallelic (in fact the two loci are not within measurable distance of one another) clotting disorder. It was formerly thought that Hm A was heterogeneous, one form showing an immunologically identifiable but nonfunctional factor VIII, the other having no immunologically identifiable protein. However, with heterologous antibody all patients have detectable antigen.

Factor VIII is apparently a complex of at least two proteins: a high molecular weight carrier protein and a small active procoagulant fragment (92). Only the former is present in HmA (and absent or at least markedly diminished in von Willebrand disease). This finding does not preclude the possibility that the condition is heterogeneous although if it is, there is no clear clinical distinction between the classic and variant forms. Affected individuals have repeated severe bleeding episodes—often with deforming hemarthroses unless they are treated repeatedly with factor VIII. Because of the debilitating and frequently fatal clinical course, vigorous attempts have been made to diagnose the female carrier of the gene, who is usually asymptomatic.

Current techniques utilize a dual approach, namely, testing for an altered

Table 5. Coagulation Deficiency States

Factor	Common Name	Inheritance*
I	Afibrinogenemia	AR; structural variants are AD
II	Hypoprothrombinemia, dysprothrombinemia	AR, AD; structural variant analogous to I
V	Labile factor deficiency	AR
VII	Hypoproconvertinemia	AR
VIII	Hemophilia A, von Willebrand disease	XR; ? heterogeneous AD, ? AR form
IX	Hemophilia B	XR; probably heterogeneous
X	Stuart-Prower deficiency	AR
XI	PTA deficiency	AR; heterozygotes show some bleeding
XII	Hageman factor deficiency	AR
XIII?	Fibrinase deficiency	AR, XR

*AR, autosomal recessive; AD, autosomal dominant; XR, X-linked, recessive.

ratio of factor VIII coagulant activity (VIII:C) to factor VIII-related antigen (VIIIR:AG) in the plasma of the suspected carrier as compared to similar ratios obtained from reference groups (93). Calculation of a probability for the carrier state can then be made utilizing the laboratory information plus the prior (Bayesian) probability obtained from pedigree information (94). The accuracy of this approach in classifying carriers may be as high as 90%, depending on the laboratory.

In severe *von Willebrand disease* all three functional components of the factor VIII system are proportionally decreased, i.e., the clot-promoting activity, the antigenic response and the endothelial cell-synthesized von Willebrand factor, which corrects the platelet-function defect in von Willebrand disease (95). Several variant forms are recognized, suggesting that there is a range of qualitatively deranged molecular forms (96, 97). Some of these variants may be recessively inherited in contrast to the classic form of the disease, which is an autosomal dominant trait.

The other coagulation factor deficiencies, except perhaps for one form of fibrin-stabilizing factor (fibrinase) deficiency, are inherited as autosomal recessive defects (98). They are all rare. It is important to differentiate between congenital *afibrinogenemia*, a recessive disorder, and the various types of structural defects in the fibrinogen protein that, analogous to the hemoglobin variants, are inherited as codominants. However, the term afibrinogenemia may be a misnomer since few of these patients have been studied by sophisticated immunologic means.

Persons with afibrinogenemia may be remarkably asymptomatic for reasons not completely clear. A similar situation is evident in *hypoprothrombinemia* wherein no or little prothrombin is identified, whereas in dysprothrombinemia, the protein is in normal concentration by immunoassay, but the biologic activity of prothrombin is measurably reduced in the two-stage procedure (99). Heterozygotes for either the recognized structural variants or the functional deficiencies (which may be largely structural) may not show any significant clotting abnormalities. The possible exception is in factor XI, or plasma thromboplastin antecedent deficiency, in which heterozygotes have a mild bleeding tendency (100).

Fibrin-stabilizing factor (fibrinase, factor XIII) is a plasma constituent that cross-links adjacent molecules of fibrin by peptide bonds to form a stable clot. Some evidence points to both autosomal and X-linked forms of the deficiency, indicating perhaps that either the reaction is complex and involves two or more steps or that a compound protein promotes the reaction and that one constituent polypeptide chain is under the control of an autosomal and the other under an X-linked locus (101). A third possibility is a regulatory locus on either an X-chromosome or an autosome.

In addition to these recognized factor-deficiency states, there seem to be several "private" coagulation defects, such as Fletcher factor and Flood factor deficiency, confined to one or at best a few families. The exact lesion in these families is in general poorly characterized.

REFERENCES

1. Weatherall DJ, Clegg JB: Molecular genetics of human hemoglobin. *Ann Rev Genet* 10:157, 1976

2. Lang A, Lorkin PA: Genetics of human haemoglobins. *Br Med Bull* 32:239, 1976
2a. Alter BP, Modell CB, Fairweather D, et al: Prenatal diagnosis of the hemoglobinopathies. *New Engl J Med* 295:1437, 1976
3. McKusick VA: Mendelian Inheritance in Man, 4th ed. Johns Hopkins Press, 1975, p. 124
3a. Perrine RP, Brown MJ, Clegg JB, et al: Benign sickle cell anaemia. *Lancet* 2:1163, 1972
4. Piomelli S, Corash N: Hereditary hemolytic anemia due to enzyme defects of glycolysis. *Adv Hum Genet* 6:165, 1976
5. Bigley RH, Stenzel P, Jones RT, et al: Tissue distribution of human pyruvate kinase isozymes. *Enzym Biol Clin* 9:10, 1968
6. Löhr GW, Waller HD, Auschutz F, et al: Biochemische Defecte in den Blutzellen bei familiärer Panmyelopathie (Typ Fanconi). *Humangenetik* 1:383, 1965
7. Harkness DR: A new erythrocytic enzyme defect with hemolytic anemia. *J Lab Clin Med* 68:879, 1966
8. Cartier P, Labie P, Leroux JP, et al: Déficit familial en diphosphoglycerate mutase: Etude hematologique et biochemique. *Nouv Rev Fr Hematol* 12:269, 1972
8a. Gilman PA: Hemolysis in the newborn infant resulting from deficiencies of red blood cell enzymes: Diagnosis and management. *J Pediatr* 84:625, 1974
9. Miura S, Nishina T, Kakehashi Y, et al: Studies on erythrocyte metabolism in a case with hereditary deficiency of H subunit of lactate dehydrogenase. *Acta Haematol Jap* 34:228, 197
10. Stefanini M: Chronic hemolytic anemia associated with erythrocyte enolase deficiency exacerbated by ingestion of nitrofurantoin. *Am J Clin Pathol* 58:408, 1972
11. Beutler E, Scott S, Bishop A, et al: Red cell aldolase deficiency and hemolytic anemia: A new syndrome. *Trans Assoc Am Physiol* 86:154, 1973
12. Beutler E: Glucose-6-phosphate dehydrogenase deficiency, in Stanbury JC, Wyngaarden JB, Fredrickson DS (eds): The Metabolic Basis of Inherited Disease, 3d ed. New York, McGraw-Hill, 1972, p. 1358
13. Yoshida A, Beutler E, Motulsky AG: Human glucose-6-phosphate dehydrogenase variants. *Bull WHO* 45:243, 1971
14. Luzzatto L: Inherited hemolytic states: Glucose-6-phosphate dehydrogenase deficiency. *Acta Haematol* 4:83, 1975
15. Luzzatto L: Genetic heterogeneity and pathophysiology of G6PD deficiency. *Br J Haematol* 28:151, 1974
16. Yoshida A: A single amino acid substitution (aspargine to aspartic acid) between normal (B+) and the common Negro variant (A+) of human glucose-6-phosphate dehydrogenase. *Proc Nat Acad Sci* 57:835, 1967
17. Yoshida A: Amino acid substitution (histidine to tyrosine) in a glucose-6-phosphate dehydrogenase variant (G6PD Hektoen) associated with overproduction. *J Mol Biol* 52:483, 1970
18. Parr CW: Erythrocyte phosphogluconate dehydrogenase polymorphism. *Nature* 210:487, 1966
19. Mohler DH, Majerus PW, Minnick V, et al: Glutathione synthetase deficiency as a cause of hereditary hemolytic disease. *New Engl J Med* 283:1253, 1970
20. Lohr GW, Waller HD: Eine neue enzymopenische haemolytische Anaemie mit Glutationreductase—Mangel. *Med Klin* 57:1521, 1962
21. Staal GEJ, Helleman PW, DeWae J, et al: Purification and properties of an abnormal glutathione reductase from human erythrocytes. *Biochem Biophys Acta* 815:63, 1969
22. Necheles TF, Steinberg MJ, Cameron D: Erythrocyte glutathione-peroxidase deficiency. *Br J Haematol* 19:605, 1970
23. Szeinberg A, Kahana D, Gavendo S, et al: Hereditary deficiency of adenylate kinase in red blood cells. *Acta Haematol* 42:111, 1968
24. Hanel HK, Cohn J, Harvald B: Adenosine-triphosphatase deficiency in a family with non-spherocytic haemolytic anemia. *Hum Hered* 21:313, 1971
25. Brewer GJ: A new inherited abnormality of human erythrocyte-elevated erythrocyte

adenosine triphosphate. *Biochem Biophys Rev Comm* 18:430, 1965
26. Valentine WN, Fink K, Paglia DE, et al: Hereditary hemolytic anemia with human erythrocyte pyrimidine 5-nucleosidase deficiency. *J Clin Invest* 54:866, 1974
27. Smith LH, Huguley CM Jr, Bain JA: Hereditary orotic aciduria, in Stanbury JB, Wyngaarden JB, Fredrickson DS (eds): Metabolic Basis of Inherited Disease, 3d ed. New York, McGraw-Hill, 1972, p. 1003
28. Rosenberg LE: Vitamin-responsive metabolic disorders. *Adv Hum Genet* 6:1, 1976
29. Mohamed SD, McKay E, Galloway WH: Juvenile familial megaloblastic anemia due to selective malabsorption of vitamin B_{12}. *Q J Med* 35:433, 1966
30. Katz M, Lee SK, Cooper BA: Vitamin B_{12} malabsorption due to biologically inert intrinsic factor. *New Engl J Med* 287:425, 1972
31. Grasbeck R: Familial selective vitamin B_{12} malabsorption. *New Engl J Med* 287:358, 1972
32. Hakami N, Neiman PE, Canellos GP, et al: Neonatal megalobastic anemia due to inherited transcobalamin II deficiency in two siblings. *New Engl J Med* 285:1163, 1971
33. Carmel R, Herbert V: Deficiency of vitamin B_{12} binding alpha-globulin in two brothers. *Blood* 33:1, 1969
34. Santiago-Borrero PJ, Santini R Jr, Perez-Santiago E, et al: Congenital isolated defect of folic acid absorption. *J Pediatr* 82:450, 1973
35. Walters TR: Congenital megaloblastic anemia responsive to N^5-formyl tetrahydrofolic acid administration. *J Pediatr* 70:686, 1967
36. Tauro GP, Danks DM, Rowe PB, et al: Dihydrofolate reductase deficiency causing megaloblastic anemia in two families. *N Engl J Med* 294:466, 1976
37. Rogers LE, Porter FS, Sidbury JB Jr: Thiamine-responsive megaloblastic anemia. *J Pediatr* 74:494, 1969
38. Verwilghen RL, Lewis SM, Dacie JV, et al: Hempas: Congenital dyserythropoietic anaemia (type II). *Q J Med* 42:257, 1973
39. Wendt F, Heimpel H: Kongenitale dyserythopoietische Anämie bei einem zweieiigen Zwillingspaar. *Med Klin* 62:172, 1967
40. Wolff JA, von Hofe M: Familial erythroid multinuclearity. *Blood* 6:1274, 1951
41. Benjamin JT, Rosse W, Dalldorf FG, et al: Congenital dyserythropoietic anemia—type IV. *J Pediatr* 87:210, 1975
42. Rundles RW, Falls HF: Hereditary (sex-linked) anemia. *Am J Med Sci* 211:641, 1946
43. Elves MW, Bourne MS, Israels MCG: Pyridoxine-responsive anemia determined by an X-linked gene. *J Med Genet* 3:1, 1966
44. Stavern P, Saltvedt E, Elgjo K, et al: Congenital hypochromic microcytic anaemia with iron overload of the liver and hyperferraemia. *Scand J Haematol* 10:153, 1973
45. Rawles JM, Weller RO: Familial association of metabolic myopathy, lactic acidosis and sideroblastic anemia. *Am J Med* 56:891, 1974
46. Cotton HB, Harris JW: Familial pyridoxine-responsive anemia. *J Clin Invest* 41:1352, 1962
46a. Goya N, Miyazaki S, Kodata S, et al: A family of congenital atransferrinemia. *Blood* 40:239, 1972
47. Diamond LK, Allen DW, Magill FB: Congenital (erythroid) hypoplastic anemia: A 25-year study. *Am J Dis Childh* 013:403, 1961
48. Brookfield EG, Singh P: Congenital hypoplastic anemia associated with hypogammaglobulinemia. *J Pediatr* 85:529, 1974
49. Hoffman R: Diamond-Blackfan syndrome: Lymphocyte-mediated suppression of erythropoiesis. *Science* 193:899, 1976
50. Falter ML, Robinson MG: Autosomal dominant inheritance and aminoaciduria in Blackfan-Diamond anemia. *J Med Genet* 9:64, 1972
51. Case JM, Smith DW: Congenital anemia and triphalangeal thumbs: A new syndrome. *J Pediatr* 74:471, 1969

52. Schroeder TM, Tilgen D, Kruger J, et al: Formal genetics of Fanconi's anemia. *Hum Genet* 32:257, 1976
53. Remsen JF, Cerutti PA: Deficiency of gamma-ray excision repair in skin fibroblasts from patients with Fanconi's anemia. *Proc Nat Acad Sci* 73:2419, 1976
54. Altay C, Sevgi Y, Pirnam T: Fanconi's anemia in offspring of patient with congenital radial and carpal hypoplasia. *N Engl J Med* 293:151, 1975
55. Abels O, Reed WB: Fanconi's-like syndrome. *Arch Dermatol* 107:419, 1973
56. Dobbs CE: Familial auto-immune hemolytic anemia. *Arch Int Med* 116:273, 1965
57. Jacob HS: Tightening red-cell membranes. *N Engl J Med* 294:1234, 1976
58. Jandl JH, Cooper RA: Hereditory spherocytosis, in Stanbury JB, Wyngaarden JB, Fredrickson DS (eds): The Metabolic Basis of Inherited Disease. 3d ed. New York, McGraw-Hill, 1972, p. 1323
59. Kirkpatrick FH, Woods GM, LaCelle PL: Absence of one component of spectrin adenosine triphosphatase in hereditary spherocytosis. *Blood* 46:945, 1975
60. Wiley JS, Firkin BG: An unusual variant of hereditary spherocytosis. *Am J Med* 48:63, 1970
61. Cutting HO, McHugh WJ, Conrad FG, et al: Autosomal dominant hemolytic anemia characterized by ovalocytosis. *Am J Med* 39:21, 1965
62. Honig GR, Lacson PS, Mauer HS: A new familial disorder with abnormal erythrocyte morphology and increased permeability of the erythrocytes. *Ped Res* 5:159, 1971
63. Shohet SB, Nathan DG, Livermore BM, et al: Hereditary hemolytic anemia associated with abnormal membrane lipid. *Blood* 42:1, 1973
64. Danon D, DeVries A, Djaldetti M, et al: Episodes of acute haemolytic anaemia in a patient with familial ultrastructural abnormality of the red-cell membrane. *Br J Haematol* 8:724, 1962
65. Wiley JS, Gill FM: Red cell calcium leak in congenital hemolytic anemia with extreme microcytosis. *Blood* 47:197, 1976
66. Zarowsky HS, Mohandas N, Speaker OB, et al: A congenital hemolytic anemia with thermal sensitivity of the erythrocyte membrane. *Br J Haematol* 29:537, 1975
67. Mentzer C, Lubin BH, Emmons S: Correction of permeability defect in hereditary stomatocytosis by dimethyl adipimidate. *N Engl J Med* 294:1200, 1976
68. Fredrickson DS, Gotto AM, Levy RI: Familial lipoprotein deficiency, in Stanbury JB, Wyngaarden JB, Fredrickson DS (eds): The Metabolic Basis of Inherited Disease, 3d ed. New York, McGraw-Hill, 1972, p. 493
69. Norum KR, Glomset JA, Gjone E: Familial lecithin: cholesterol acyl transferese deficiency, in Stanbury JB, Wyngaarden JB, Fredrickson DS (eds): The Metabolic Basis of Inherited Disease 3d ed. New York, McGraw-Hill, 1972, p. 531
70. Jackson LG: Genetic counselling in haematological disorders and pregnancy. *Clin Haematol* 2:587, 1973
71. Hedenberg F: Infantile agranulocytosis of probably congenital origin. *Acta Paediat* 48:77, 1959
72. Andrews JP, McClellan JT, Scott CH: Lethal congenital neutropenia with eosinophilia occurring in two siblings. *Am J Med* 29:358, 1960
73. Cutting HO, Lang JE Jr: Familial benign chronic neutropenia. *Ann Int Med* 61:876, 1964
74. Hahneman BM, Alt HL: Cyclic neutropenia in a father and daughter. *JAMA* 163:270, 1958
75. Weiden PL, Robinett B, Graham TC, et al: Canine cyclic neutropenia. *J Clin Invest* 53:950, 1974
76. Greenberg PL, Bax I, Levin J, et al: Alteration of colony-stimulating factor output, endotoxemia, and granulopoiesis in cyclic neutropenia. *Am J Hematol* :375, 1976
77. Herring WB, Smith LG, Walker RI, et al: Neutrophilia. *Am J Med* 56:729, 1974
78. Naiman JL, Oski FA, Allen FH, et al: Hereditary eosinophilia. *Am J Hum Genet* 16:195, 1964
79. Davidson WM: Inherited variations in leukocytes. *Br Med Bull* 17:190, 1961
80. Goudsmit R, Leeuwen AM, James J: Döhle bodies and acute myeloblastic leukemia in one family: A new familial disorder. *Br J Haematol* 20:557, 1971

81. Schneiderman LJ, Sampson WI, Schoene WC, et al: Genetic studies of a family with two unusual autosomal dominant conditions: Muscular dystrophy and Pelger-Huet anomaly. *Am J Med* 46:380, 1969
82. Hathaway WE: Bleeding disorders due to platelet dysfunction. *Am J Dis Childh* 121:127, 1972
83. Gutenberger J, Trygstad CW, Stiehm ER, et al: Familial thrombocytopenia elevated serum IgA levels and renal disease. *Am J Med* 49:729, 1970
84. Shulam I, Pierce M, Lukens A, et al: Studies of thrombopoiesis. *Blood* 16:943, 1960
85. Karpatkin S, Weiss HJ: Deficiency of glutathione peroxidase associated with high levels of reduced glutathione in Glanzman's thrombasthenia. *N Engl J Med* 287:1062, 1972
86. Willis AL, Weiss HJ: A congenital defect in platelet prostaglandin production associated with impaired hemostasis in storage pool disease. *Prostaglandins* 4:793, 1973
87. Wallace DC, Lovric A, Clubb JS, Carseldine DB: Thrombotic thrombocytopenic purpura in four siblings. *Am J Med* 58:724, 1975
88. Adamson J: Familial polycythemia. *Sem Hematol* 12:383, 1975
89. Adamson JW, Stamatoyannopoulos G, Kontras S, et al: Recessive familial erythrocytosis: Aspects of marrow regulation in two families. *Blood* 41:641, 1973
90. Greenberg BR, Golde DW: Erythropoiesis in familial erythrocytosis. *N Engl J Med* 296:1080, 1977
91. Modan B: Polycythemia: A review of epidemiological and clinical aspects. *J Chron Dis* 18:605, 1965
92. Cooper HA, Wagner RH: The defect in hemophilic and von Willebrand's disease plasmas studied by recombination technique. *J Clin Invest* 54:1093, 1974
93. Klein HG, Aledort LM, Bouman BM, et al: A cooperative study for the detection of the carrier state of classic hemophilia. *N Engl J Med* 296:959, 1977
94. Graham JB: Genetic counselling in classic hemophilia A. *N Engl J Med* 296:996, 1977
95. Nachman RL: von Willebrand's disease and the molecular pathology of hemostasis. *New Engl J Med* 296:1059, 1977
96. Hoyer LW: von Willebrand's disease. *Prog Hemostasis Thromb* 3:231, 1976
97. Italian Working Group: Spectrum of von Willebrand's disease. *Br J Haematol* 35:101, 1977
98. Ratnoff OD, Bennett B: The genetics of hereditary disorders of blood coagulation. *Science* 179:1291, 1973
99. Shapiro SS, Martinez J, Holburn RR: Congenital dysprothrombinemia: An inherited structural disorder of human prothrombin. *J Clin Invest* 48:2251, 1969
100. Rapaport SI, Proctor RR, Patch MJ, et al: The mode of inheritance of PTA deficiency: Evidence for the existence of major PTA deficiency and minor PTA deficiency. *Blood* 18:149, 1961
101. Ratnoff OD, Steinberg AG: Inheritance of fibrin-stabilizing factor deficiency. *Lancet* 1:25, 1968

12
Genetic Diseases of Skeletal Muscle

Salvatore DiMauro

"Myopathies" are best defined in a negative way as those disorders of muscle that, by clinical, pathologic and electrophysiologic criteria, do not appear to be secondary to dysfunction of the central or peripheral nervous system. This definition includes disorders limited to skeletal muscle, as well as those involving multiple systems, and does not require that the primary abnormality actually be localized within muscle. A large number of myopathies thus defined are genetically determined. Unfortunately only for very few is the primary biochemical error known, and for fewer still is the pathogenetic mechanism understood. As a consequence, the classification of hereditary myopathies—based on genetic, clinical and pathologic criteria—is still unsettled and controversial.

For most of these disorders, prenatal diagnosis is unavailable and detection of heterozygotes uncertain, making genetic counseling difficult. A brief description of the clinical features and morphologic abnormalities of muscle in each disorder follows. Additional details can be found in specialized textbooks (1, 2) and recent proceedings (3, 4).

DYSTROPHIES

Under this generic heading are grouped five hereditary, progressive and degenerative muscle diseases (5) (Table 1). Common features include progressive muscle weakness and wasting, increased serum enzymes, variant electromyographic pattern (small potentials of short duration in relatively normal number) and morphologic changes (variation of fiber size, degeneration and necrosis of fibers, connective and fat tissue infiltration). However, neither clinical nor laboratory characteristics are specific, and a truly rational definition and classification of dystrophies will be possible only when the primary biochemical errors are known.

Research for the material discussed in this chapter was supported by Center Grants NS-11766-05 from the National Institute of Neurological and Communicative Disorders and Stroke, and from the Muscular Dystrophy Association.

Table 1. Muscular Dystrophies

Type (Syndrome)	Onset	Clinical Features	Inheritance*
Duchenne	Early childhood	Proximal muscle weakness and wasting, calf pseudohypertrophy; rapidly progressive	XR
Becker	Rarely before 5 years	Same as Duchenne except slower progression	XR
Emery-Dreifuss	4-5 years	No pseudohypertrophy; early contractures, cardiac features prominent	XR
Congenital	Birth	Muscular weakness with contractures; slowly progressive; may be static; CPK elevated early	AR
Facioscapulohumeral	Adolescence	Facial and shoulder muscle weakness; slower progression than XR forms; CPK may be normal	AD
Scapuloperoneal	Adolescence	Weakness and wasting of proximal arm muscles and distal leg muscles; heterogeneous	AD
Limb-girdle	Childhood-adolescence	Progression is slow; heterogeneous CPK may be normal late	AR
Oculopharyngeal	Late (4-5th decade)	Dysphagia and ptosis; CPK usually normal	AD

*XR, X-linked recessive; AR, autosomal recessive; AD, autosomal dominant.

Three major hypotheses are proposed to explain the pathogenesis of muscular dystrophies, including abnormal blood supply with microinfarcts, abnormal neuronal influence and a primary abnormality of muscle cell membrane. Several experimental observations contradict the first two postulated mechanisms, whereas there is increasing evidence of structural and functional abnormalities of the sarcolemma (4, 6). The primary genetic defects, however, remain elusive.

Duchenne Muscular Dystrophy

The most severe of these disorders, onset of Duchenne muscular dystrophy is in early childhood, with evidence of proximal muscle weakness, difficulty running and rising from the ground and a waddling gait. There is wasting of proximal muscles and, frequently, pseudohypertrophy of the calves. Because of the increasing weakness and subsequent contractures, patients are generally confined to a wheelchair by early adolescence. Involvement of respiratory muscles increases vulnerability to pulmonary infections, and death generally intervenes before the third decade. Heart involvement is reflected by characteristic electrocardiographic (ECG) abnormalities (7), although cardiac symptoms are rare. Central nervous system involvement is suggested by the high incidence of mental retardation (30%), although most studies of brain morphology are negative (8).

Serum levels of several enzymes, presumably of muscle origin (adolase, lactate dehydrogenase, glutamic-oxaloacetic transaminase and creatine phosphokinase) are greatly increased in the initial stages, but decrease as the disease continues.

No specific therapy is available. Physical therapy is aimed at preventing contractures and delaying immobilization.

Transmission is by an X-linked recessive mechanism, and prevalence rate is approximately 30 per 100,000 males (Survey to age 5 years) (5). Duchenne dystrophy is therefore the commonest genetic disease of muscle, with about 10,000 affected boys in the United States. Since most patients die without transmitting the mutant gene, the rate of spontaneous mutation is considered to be very high (one-third of all cases). More recent data suggest that the figure is much lower, since more complex biochemical studies may identify many more carriers than expected in the group of mothers with only one affected son (9).

Because of the crippling nature of the disease and the lack of specific therapy, detection of female carriers is of crucial importance in the prevention of new cases. Mothers of Duchenne patients are divided into two groups, including definite carriers who have an affected son and an affected brother or uncle, and those who have two or more affected sons and who should be counseled as carriers; possible carriers have only one affected son. Detection of carriers is made difficult by our ignorance of the primary biochemical defect. However, by a combination of serum creatine phosphokinase (CPK) measurement and electromyogram (EMG) examination, about 70% of carriers can be identified.

Prenatal diagnosis is not available, but prenatal sex determination can identify the male fetus at risk if the mother is a definite carrier.

Becker Dystrophy

Clinical and laboratory findings are virtually identical to those in Duchenne dystrophy, but onset is rarely before age 5 years and often not until adolescence. The course is slower, and death generally does not intervene before the fourth decade (10). In greater contrast to Duchenne dystrophy, no consistent ECG abnormalities occur. The incidence of Becker dystrophy is approximately one-tenth that of the Duchenne form, and the disorder is also an X-linked recessive one. That Becker and Duchenne dystrophy are not alleles is suggested by different linkage relationship with the locus for deutan color blindness.

Emery-Dreifuss Dystrophy

Another X-linked form of muscular dystrophy features an early onset of weakness (4-5), but differs from the other two X-linked disorders in that pseudohypertrophy is rare, contractures occur early and cardiac involvement with dysrhythmia is common. If heart involvement is not extensive, survival into the sixth decade is possible (11).

Another relatively benign form of X-linked dystrophy (Mabry type) is suggested, but at present evidence is insufficient to support this type as a discrete entity. Remember, however, that with the benign forms of X-linked dystrophy, fertility is unimpaired in those surviving to adult life and therefore *all* daughters of affected males will be carriers of the trait. Rarely such females may be clinically affected by the operation of the Lyon hypothesis. Other ways in which female subjects can be affected is if they have 45X gonadal dysgenesis with the single X bearing the mutant gene; or, for the benign forms, if they are homozygous (i.e., have an affected father and a carrier mother).

Congenital Muscular Dystrophy

In congenital muscular dystrophy weakness is evident at birth and is often accompanied by contractures and joint immobilization.*Biopsy findings resemble those of other dystrophies and do not show the peculiar structural abnormalities or fiber-type disproportion that are characteristic of other congenital myopathies (see the following discussion). Weakness is diffuse, more proximal than distal, with frequent involvement of facial but not extraocular muscles. There is no clinical evidence of heart or CNS involvement in most patients, and the brain was normal in those cases studied at autopsy (12, 13). Although this disorder is considered to be nonprogressive or only slowly progressive, the clinical course varied greatly in different patients in a large series from Finland (14).

Serum enzymes are generally moderately elevated, but may be very high in the first or second year. Autosomal recessive transmission is suggested by the observation that both sexes are affected, the disorder is often present in siblings and there is parental consanguinity in some families. The condition is obviously heterogeneous. A distinct form of congenital muscular dystrophy, prevalent in Japan, is also a recessive disorder. It features associated severe mental retardation with CNS abnormalities, including cerebral and cerebellar micropolygyria (15). An autosomal recessive form with congenital muscular dystrophy and patent ductus arteriosus was observed in an Arab kindred (16). Another congenital, nonprogressive or slowly progressive myopathy was described in seven members of an isolated Norwegian family. In two siblings, a man and a woman, the muscle disorder was associated with cataract and hypogonadism (17).

Facioscapulohumeral (FSH) Dystrophy

An autosomal dominant disorder characterized by early involvement of facial (but not ocular) and shoulder girdle muscles, the onset of this dystrophy is generally in adolescence and the course is more benign than in the Duchenne type. However, as in other diseases with autosomal dominant inheritance, the expression varies greatly in different patients, from a virtually asymptomatic condition (often considered to be nothing more than a curious family trait, like sleeping with the eyes open or not being able to whistle) to severe disability. The common initial symptoms are difficulty in raising the arms overhead and lifting objects. Later, leg muscles are also affected, the patient having difficulty in walking. Serum enzyme levels are normal or only slightly elevated. In some families the muscle biopsy shows striking inflammatory changes (18).

Scapuloperoneal Syndrome

A relatively benign neuromuscular disease with onset in adolescence and generally slow progression, the disease is characterized by weakness and wasting, predominantly of the proximal muscles in the arms and distal muscles in the legs. Lack of prominent facial weakness is the main clinical distinction from the FSH dystrophy. The simultaneous presence of myogenic and neurogenic features in

---

*Congenital contractures are often called *arthrogryposis multiplex congenita*, no matter what the cause. The condition is heterogenous, with both neuropathic and myopathic causes.

both EMG and muscle biopsy in many cases makes differentiation from juvenile spinal muscular atrophy (Kugelberg-Welander disease) difficult, and the syndrome is likely heterogeneous. Many affected persons have autosomal dominant inheritance. In a few families transmission is by an X-linked recessive mechanism, and the neuromuscular disease is accompanied by severe cardiac abnormalities with atrial arrythmias, including bradycardia and atrial paralysis (19-21). Many workers feel that at least some patients with this variety of X-linked scapuloperoneal syndrome with cardiac involvement actually have the Emery-Dreifuss form of muscular dystrophy.

Limb-Girdle Dystrophy

This is a catchall classification for genetic myopathies that do not fall into other, better-defined categories. Transmission is usually autosomal recessive, and the onset is usually in adolescence or early adult life. There is slowly progressive proximal weakness, generally affecting one limb-girdle first, then spreading to the other. Serum enzyme levels are normal or slightly increased.

Distal Muscular Dystrophy

First recognized in Sweden, the disorder appears to be largely confined to Scandinavia (22). However, a few families are described in other countries, including the United States (23). In a few cases symptoms began in infancy (24), but in most patients the disease starts in the fourth or fifth decade and proceeds very slowly. There is weakness and wasting of distal muscles, generally involving the arms first, whereas facial and neck muscles are spared. Serum enzyme levels are normal. Postmortem examination of spinal cord, spinal ganglia, nerve roots and peripheral nerve was normal in several patients. The disorder is an autosomal dominant trait, with men being more commonly affected than women.

Ocular Muscular Dystrophy

The classification of progressive ophthalmoplegias is still uncertain because the peculiar features of ocular muscles make electromyographic and morphologic criteria of diagnosis of little value in distinguishing neurogenic from myopathic disorders. In many cases the association of ophthalmoplegia with clearly neurogenic disorders suggests a neural origin for the ocular manifestations. In other cases, however, the disorder is limited to muscle, and both EMG and muscle biopsy suggest a myopathic process. Histochemical and ultrastructural studies of both ocular and limb muscles in these patients frequently show mitochondrial abnormalities, but the pathogenic significance of these findings is uncertain. Onset may be at any age, generally with ptosis, followed by weakness of ocular, proximal limb, trunk and, sometimes pharyngeal, muscles. Although autosomal dominant inheritance was demonstrated in some families, most cases appear to be sporadic.

A better-defined clinical and genetic entity is represented by *oculopharyngeal muscular dystrophy*. Onset is late (fourth or fifth decade), and the course is slowly progressive. Usually dysphagia is the initial symptom, followed by ptosis. Extraocular, facial and limb-girdle muscles are less frequently involved. Autosomal dominant inheritance with equal distribution in both sexes is well documented in several families (25), and in a large French-Canadian kindred the disease was traced back several generations to a common ancestor (26). However, autosomal recessive transmission is reported in one family with earlier onset of ptosis, suggesting genetic heterogeneity (27).

External ophthalmoplegia was one element of a familial multisystem disorder the other features of which were oligophrenia, short stature, cataract, pyramidal signs, ataxia, skeletal changes (funnel chest, scoliosis, pes planovalgus) and hypergonadotropic hypogonadism (28). The myopathy, documented by EMG and muscle biopsy, started around puberty and progressed slowly, affecting mainly proximal limb and facial muscles. Inheritance was probably autosomal recessive.

Other Dystrophies

A few characteristic clinical and genetic entities are described that escape classification into any of the nosologic groups already cited. A benign myopathy, with onset in childhood and slow course, involving limb girdle and facial muscles but sparing ocular muscles, was described in 11 members (nine males, two females) of several Hutterite colonies of southern Manitoba (29). Serum CPK levels were elevated in all patients. Inheritance appeared to be autosomal recessive.

Another benign myopathy, with autosomal dominant transmission, was reported in 28 patients of three unrelated Dutch families (30). Onset was in childhood, and progression was extremely slow, allowing normal lifespan and relatively normal activity in most patients. There was involvement of proximal more than distal, and extensor more than flexor, muscles. In 22 patients involvement of the extensor digitorum communis muscle caused a characteristic flexion contracture of the interphalangeal joints of the last four fingers. Four patients had congenital torticollis.

Another slowly progressive myopathy affecting proximal muscles and starting in the fifth decade was described in nine male members of a family in three successive generations (31). Autosomal dominant transmission was also apparent in another benign myopathy with adult onset and predominant involvement of the proximal muscles, particularly the quadriceps (32).

MYOTONIAS

Myotonia is a defect in muscle relaxation characterized electromyographically by repetitive high-frequency discharges waxing and waning in frequency and amplitude and persisting after curarization and peripheral nerve block. It appears to be due to electrical instability of the muscle surface membrane, the biochemical basis of which is not known. Several hereditary muscle diseases are characterized by myotonia (Table 2).

Myotonia congenita (Thomsen disease) is an autosomal dominant disease in which symptoms are limited to myotonia. Strength is fully preserved and many patients excel in athletics. This form is distinguished from the more frequent

Table 2. Myotonias

Type	Inheritance*	Clinical Features
Myotonia congenita	AD	Myotonia alone
	AR	Myotonia with pronounced muscle hypertrophy; frequent muscle weakness
Myotonic dystrophy	AD	Slowly progressive; variable severity with muscle weakness; wasting; myotonia; cardiopathy; cataract; male baldness; hypercatabolism of IgG; endocrine dysfunction
Chondrodystrophic	AR	Multisystem disease; myotonia is abolished by curare
Paramyotonia congenita	AD	Myotonia provoked by cold and aggravated by repeated effort
Hyperkalemic periodic paralysis	AD	Symptoms similar to paramyotonia; serum K^+ increased by attacks, and attacks precipitated by administration of K^+

*See footnote, Table 1.

autosomal recessive myotonia, in which muscle hypertrophy is more pronounced than in the dominant form, and muscle weakness is often seen (33). Although no specific biochemical defect is known in either of these forms, abnormal fatty-acid composition of muscle phosphatides was found in the autosomal dominant but not in the recessive form (34).

Myotonic dystrophy is a multisystem disorder characterized by muscle weakness, wasting and myotonia, cardiopathy, cataract, baldness, hypogonadism and other endocrine dysfunction, such as diabetes and hypothyroidism. Onset is generally in adolescence and the course is slowly progressive. Severity varies greatly—some patients are asymptomatic, others have crippling weakness. Weakness of facial muscles, with wasting of temporal muscles and ptosis, causes a characteristic elongated and droopy facial appearance. The sternocleidomastoid muscles are small, and limb weakness is more pronounced distally.

Involvement of smooth muscle may cause impaired esophageal or intestinal motility, or uterine dystocia. Abnormalities of cardiac conduction are seen in most patients. Biochemical studies of both erythrocytes and skeletal muscle show decreased endogenous cell-membrane phosphorylation and protein kinase activity (35), but these changes may be due to some more basic and possibly more diffuse membrane abnormality (36). However, the primary biochemical error in myotonic dystrophy is still unknown. This pleiotropic disorder is transmitted in an autosomal dominant fashion. In a genetic–epidemiologic study in Switzerland, the incidence of myotonic dystrophy was estimated to be 13.5 per 100,000 people, and the mutation rate 1.1 per 100,000 gametes (37).

Affected infants frequently have very different clinical features characterized by severe weakness, facial diplegia, difficulty in sucking and swallowing and respiratory distress (38). Arthrogryposis multiplex congenita is not uncommon, and mental retardation is often seen in later life. The disproportionate prevalence of congenital myotonic dystrophy in children of affected mothers cannot be explained simply by decreased fertility of affected men and is attributed to an intrauterine environmental factor. The nature of this "maternal factor," however, remains to be determined. Although the specific biochemical defect in myotonic dystrophy is unknown, the proximity of the myotonic dystrophy and secretor genes (responsible for secretion of ABH blood group substance in body fluids, including amniotic fluid) allows prenatal prediction of the disease by linkage analysis (39). Unfortunately this diagnostic approach is possible only in a percentage of families (10-20%) that are informative for linkage. An additional frustrating element of this type of prenatal prediction is the difficulty of proving morphologically or biochemically if a fetus, aborted because of predicted high risk, had the disease.

Fibroblast cultures from four patients with myotonic dystrophy grew to abnormally high cell densities and contained large amounts of material with the staining characteristics of acid muchopolysaccharide (40). These changes, however, are not specific, and no studies are reported on the characteristics of fibroblast cultures obtained by amniocentesis from fetuses at risk.

Myotonic phenomena in all these disorders are reduced by "membrane stabilizers," such as quinine, procainamide and diphenylhydantoin. Diphenylhydantoin (0.3-0.5 g daily) is the drug of choice.

Chondrodystrophic myotonia (Schwartz-Jampel syndrome) is also a multisystem disorder affecting mainly the skeletal system and muscle. It is characterized by dwarfism, hip dysplasia, pigeon breast, micrognathia and clinical myotonia (41-42). In contrast to true myotonia, however, the continuous repetitive discharges of the Schwartz-Jampel syndrome are abolished by curare. Abnormal sensitivity of the postsynaptic end-plate membrane is suggested by the persistence of spontaneous electrical activity after peripheral nerve injury. The combination of micrognathia, pursed mouth and blepharospasm compensated for by overactivity of frontal muscles, results in a characteristically rigid facial expression, sometimes described as "frozen smile."

The limited range of motions in many joints contributes to the stiff appearance. There is moderate weakness of proximal muscles, contrasting in some patients with hypertrophy of the same muscle groups. Myopathic features are confirmed by EMG and biopsy. Usually serum enzyme levels are slightly increased. Onset is in infancy and the course is slowly progressive. Intelligence is normal. The disorder is described in at least three pairs of siblings of both sexes, whereas parents were normal in all cases, indicating autosomal recessive transmission.

In families with paramyotonia congenita myotonic symptoms occur only in cold weather and, in contrast to other forms of myotonia, tend to worsen with repeated effort ("myotonia paradoxica"). In many families attacks of hyperkalemic periodic paralysis may also occur, and it is often difficult to distinguish between "pure paramyotonia congenita" and "familial hyperkalemic periodic paralysis with myotonia" (43). Transmission is autosomal dominant.

PERIODIC PARALYSES

These muscle disorders of unknown pathogenesis are characterized by recurrent attacks of weakness variable in severity, extent and duration (43).

In *familial hypokalemic periodic paralysis*, attacks of flaccid paralysis involve limb and trunk muscles and last from a few hours to several days; respiratory and facial muscles are not usually affected. Large carbohydrate meals and rest after vigorous exercise are predisposing factors. During attacks, EMG shows electrical silence, and serum enzyme levels are not increased. Serum potassium concentration is characteristically decreased, apparently because of a shift of potassium into the muscle. However, there is no close correlation between serum potassium levels and degree of weakness.

Light microscopy of muscle biopsies during attacks shows numerous apparently empty vacuoles, which may become a permanent morphologic change in later stages of the disease. With electron microscopy the vacuoles appear to be derived from the sarcoplasmic reticulum. Various pathogenetic hypotheses have been considered, including a disorder of carbohydrate metabolism, intermittent adrenocortical hypersecretion and sarcoplasmic reticulum dysfunction. Electrophysiologic studies of single fibers in vitro suggest that the basic abnormality may affect the sarcolemma and result in increased Na^+ permeability (44). Attacks can be induced by administration of glucose and insulin, a procedure often used as a diagnostic aid.

During attacks, treatment consists in oral administration of potassium salts. Acetazolamide is used as a prophylactic drug. A few sporadic cases are reported, but in the vast majority of patients the disorder is inherited as an autosomal dominant trait, with markedly reduced penetrance in women, as indicated by the striking prevalance of affected men in all series (with sex ratios as high as 12:1).

In *hyperkalemic periodic paralysis*, attacks are generally shorter than in the hypokalemic form (rarely lasting more than a few hours) and are often precipitated by rest after strenuous exercise and exposure to cold. Clinical or electromyographic myotonia is present in most affected families (45). The serum potassium level is moderately increased during attacks, and serum enzyme levels may be slightly elevated. Morphologic changes of muscle are similar to, though usually less severe than, those in the hypokalemic form. Attacks may be induced by administration of potassium and treated or prevented by acetazolamide. Involvement of the sarcolemma is suggested by electrophysiologic studies and by the association with myotonia. There are few sporadic cases, and transmission is autosomal dominant, with equal involvement of both sexes.

Normokalemic periodic paralysis is also dominantly inherited and has features in common with both the hypokalemic and hyperkalemic forms, from which it is distinguished by the normal levels of serum potassium during attacks.

METABOLIC MYOPATHIES

Five hereditary storage diseases of glycogen metabolism (*glycogen storage diseases*) affect muscle to a greater or less extent. They are Types II, III, IV, V and VII (46, 47, 50) and are discussed in Chapter 5, "Genetic Metabolic Disorders."

Disorders of Lipid Metabolism

A defect of the mitochondrial enzyme *carnitine palmityltransferase* was found in two brothers with recurrent myoglobinuria precipitated by prolonged exercise and fasting (48, 49). There was hypertriglyceridemia, and utilization of long-chain but not medium-chain, fatty acids was impaired. Muscle morphology was normal. The enzymatic defect was also expressed in leukocytes, and intermediate values were found in WBCs from the asymptomatic mother (50). Deficiency of muscle *carnitine* is reported in several patients, most of whom had normal serum carnitine levels (50, 51); the biochemical defect may involve carnitine transport from blood into muscle. In one case, however, carnitine content was also decreased in liver and serum (52), suggesting a defect in hepatic synthesis of carnitine. In all these cases there was progressive muscle weakness and increased serum enzyme levels, but no myoglobinuria. Excessive accumulation of lipid droplets within muscle fibers is the morphologic hallmark of the disorder. Evidence of genetic transmission was found in one family, in whom muscle carnitine content was decreased in both asymptomatic parents and a maternal aunt of the propositus (50), suggesting autosomal recessive inheritance.

Mitochondrial Myopathies

Besides carnitine and carnitine palmityltransferase deficiencies, a few other hereditary disorders of muscle involve mitochondrial metabolism. Different defects of the *pyruvate dehydrogenase* complex have been described in children with various neurologic abnormalities and lactic acidosis (53). Although there was no detectable weakness, muscle biopsy in one 9-year-old child showed accumulation of lipid droplets, and activity of pyruvate decarboxylase, the first enzyme of the pyruvate dehydrogenase complex, was less than 30% of normal in muscle, leukocytes and cultured fibroblasts (54). Leukocytic and fibroblastic enzyme activity was decreased in the father and borderline in the mother, both of whom were asymptomatic. Kinetic studies in cultured fibroblasts from members of the affected family and controls suggest the presence of multiple forms of the enzyme (55).

In *Menkes' disease* (trichopoliodystrophy; steely hair disease), the primary disorder is a defect of copper absorption in the intestine, resulting in low blood levels of copper and ceruloplasmin. The disease starts in infancy and progresses rapidly with developmental regression, seizures, hypotonia, sometimes spasticity, hair abnormalities, temperature instability, scorbutic bone changes and arterial abnormalities. Muscle biopsy in one patient showed abnormal mitochondria, excessive accumulation of glycogen, vacuolation and hyaline changes (56, 57). Mitochondria isolated from muscle and brain in the same patient had very low concentration of cytochrome-oxidase, one of the copper-dependent enzymes. Menkes' disease is transmitted as an X-linked recessive character. The copper concentration was consistently elevated in skin fibroblasts cultured from patients with Menkes' disease (58). If expressed by amniotic cells in culture, this genetic marker will be a valuable tool in the antenatal diagnosis of this untreatable disease.

Reduced content of cytochrome b, abnormal insensitivity to antimycin A and "loose coupling" of respiration and phosphorylation were reported in isolated muscle mitochondria from a 46-year-old man and his 16-year-old son with progressive ataxia, proximal muscle weakness, areflexia, dementia and bilateral Babinski sign (59). The son also had chorioretinitis, intracranial calcification and partial external ophthalmoplegia. Three other children (two girls and a boy) were normal.

In four sporadic patients with findings of achondroplasia, isolated muscle mitochondria showed a defect of oxidative phosphorylation apparently involving the first of the three coupling sites (60). The significance of this finding is unknown.

Other metabolic myopathies

Larsson et al (61) described 14 patients in five Swedish families with a chronic myopathy starting in childhood and dominated by exercise intolerance. Even moderate physical activity caused pain and tenderness of exercising muscles, accompanied by dyspnea, tachycardia, nausea and vomiting. Episodes of pigmenturia occurred in nine patients, and myoglobinuria was documented in three. Blood lactate concentration was normal at rest, but increased disproportionately during exercise. Muscle showed nonspecific histologic changes. The metabolic defect is unknown, but an abnormality of glycolysis is postulated. Transmission appears to be autosomal recessive.

A similar disorder was reported in two brothers, with the added feature of sideroblastic anemia (62). Ultrastructural examination of muscle showed abnormal mitochondria with paracrystalline inclusions, whereas no alteration of glycolysis was detected in biochemical studies of muscle homogenate. The disorder was therefore considered to be secondary to a defect rather of mitochondrial function than of glycolysis. The asymptomatic father had chronic lactic acidosis, suggesting autosomal recessive inheritance.

Myoclonus epilepsy is an autosomal recessive disorder characterized clinically by the triad of epilepsy, myoclonus and dementia and pathologically by characteristic intraneuronal accumulations of storage material (Lafora bodies). Muscular involvement is shown by ultrastructural and biochemical studies, and an abnormality of carbohydrate metabolism is suggested by the polysaccharidic nature of the storage material (63, 64).

MALIGNANT HYPERTHERMIA

This hereditary, often lethal abnormality develops following anesthesia with halothane or succinylcholine as a symptom complex characterized by rapidly rising temperature, tachycardia, tachypnea, cyanosis, respiratory and metabolic acidosis and usually (but not invariably) rigidity and myoglobinuria (65). A few cases with rigidity but without hyperthermia are reported, particularly in children (66). In most cases susceptibility is transmitted as an autosomal dominant character. Although patients are generally asymptomatic prior to an attack, nonspecific myopathic changes have been found in muscle biopsies from several members of affected families. Several lines of evidence indicate that the primary biochemical defect may involve the sarcoplasmic reticulum and result in defective calcium uptake and increased cytoplasmic concentration of calcium. This in turn may cause contracture and uncoupling of mitochondrial oxidative phosphorylation, hence hyperthermia (67).

The catastrophic course of attacks makes detection of persons at risk vital. Serum CPK activity is often increased in susceptible individuals, but not uniformly (70). Morphologic changes in muscle biopsies and functional abnormalities of muscle fibers in vitro (e.g., increased sensitivity to caffeine-induced contracture) appear to be more reliable criteria in the identification of susceptible members within affected families, but cannot be used to screen the general population.

Malignant hyperthermia is also reported in association with other autosomal dominant myopathies, myotonia congenita and central core disease (68, 69).

MORPHOLOGICALLY DISTINCT "CONGENITAL" MYOPATHIES

This is a group of usually benign, relatively nonprogressive myopathies, with proximal or diffuse muscle weakness and no or only slight increase of serum enzyme levels. Skeletal anomalies are often present, such as kyphoscoliosis, pes cavus and dislocated hips. Specific entities are defined by distinct (but not absolutely specific) structural features.

In *central core disease* central or paracentral areas (cores), generally extending along the entire length of muscle fibers, show decreased stain with histochemical reactions for oxidative enzymes and phosphorylase. The lesions are typically limited to Type 1 fibers, but the number of affected fibers varies greatly from case to case. In some patients marked preponderance of Type 1 fibers was also observed. In the electron microscopic examination the cores are composed of closely packed myofibrils, with scanty intermyofibrillar space; sarcomeres are shorter than in nonaffected areas, and Z disks are wider and often irregular. Mitochondria, SR profiles and glycogen are decreased with the cores. Although a few sporadic cases are reported, the disease appears to be transmitted in most instances as an autosomal dominant character (71).

Multicore disease was originally described in two children with nonprogressive proximal muscle weakness present since birth or early infancy (72). The structural abnormalities of the "cores" were similar to those of central core disease: decrease of the mitochondrial population, disintegration of the Z disks and loss of sarcomere integrity. In contrast to central core disease, multiple cores were present in each affected fiber, the cores did not extend along the entire length of the fiber and both fiber types were involved. Clinical signs of neuromuscular disorder were present in a maternal uncle and in the maternal grandmother of one of the patients, suggesting autosomal dominant inheritance.

Nemaline myopathy was first described in 1965 and is characterized by "nemaline" or "rod" structures in muscle fibers, which by light microscopy are best seen in paraffin sections stained with phosphotungstic acid–hematoxyline (PTAH) or in fresh-frozen sections stained with the modified Gomori trichrome stain (73). Although both fiber types may be affected, rods are generally more abundant in Type I fibers. Type I fiber preponderance was also observed in some cases. The ultrastructural characteristics of the rods are virtually identical to those of Z disks, from which they appear to originate. Presence of α-actinin within the rods has been suggested by immunocytochemical studies (74). Although weakness may not be apparent until adulthood, in most cases onset is in early infancy, with floppiness, weak cry, feeding difficulties and delayed motor development (75). Weakness is generalized, but paraspinal and proximal limb muscles are more severely affected. There is involvement of facial, but only rarely of ocular, muscles.

Weakness is often static or slowly progressive, but death from respiratory muscle weakness is reported in both infantile and "late-onset" cases. Most patients with nemaline myopathy have long, thin, expressionless faces. Other dysmorphic features include kyphoscoliosis, pigeon chest and pes cavus. A few apparently sporadic cases are reported, but hereditary transmission is well documented in a number of families with several affected members in the same or subsequent generations. The disease is about twice as common in female as in male members. The most probable mode of transmission is autosomal dominant, with reduced penetrance (76).

Myotubular or centronuclear myopathy is clinically and genetically heterogeneous (76a). The characteristic structural abnormalities are the rows of central nuclei in both fiber types, but predominantly in Type I fibers. In some cases Type I fiber preponderance and hypotrophy and Type II fiber hypertrophy have been described. The central nuclei are generally surrounded by an area of cytoplasm devoid of myofibrils and containing variably increased oxidative enzymes, phosphorylase and glycogen, a picture reminiscent of myotubes. Because of the similarity of centronucleated fibers to myotubes, it was suggested that the disorder might be due to an arrest of normal muscle development. A disorder of innervation of Type I fibers was also considered.

A distinctive clinical feature is a combination of ophthalmoplegia and facial diplegia. In many cases fetal movements are diminished and severe weakness is present at birth, but in others onset is later in childhood or even in the second or third decade. The clinical course also varies considerably in different patients from benign, static or slowly progressive myopathy to profound, generalized weakness with respiratory failure (77). Unlike most other "congenital" myopathies, serum enzyme levels are elevated in some patients. "Myotubular" myopathy is not a single genetic entity. Autosomal dominant transmission was suggested by the involvement of several individuals of both sexes in subsequent generations in many families (78). In some families, however, sex-linked recessive inheritance was clearly demonstrated (79). Severe respiratory failure at birth was present in children from these families, and scattered fibers with central nuclei were found in muscle biopsies from asymptomatic mothers.

Sarcotubular myopathy, a congenital nonprogressive myopathy with abnormal accumulations of sarcotubular vesicles, mainly in Type II fibers, was described in two brothers of a Hutterite family with parental consanguinity (80). Neither parents nor other members of previous generations had muscle weakness, and transmission was probably autosomal recessive.

Accumulation of finely granular material in peripheral areas of Type I fibers was reported in a brother and sister with another form of benign congenital myopathy, attributed to breakdown of myofibrils (81). No other member of the family was clinically affected. In two other forms of "structural" congenital myopathies, *fingerprint body myopathy* (82) and *reducing body myopathy* (83), there was no evidence of hereditary transmission.

REFERENCES

1. Walton JN (ed): Disorders of Voluntary Muscle. Edinburgh and London, Churchill-Livingstone, 1974
2. Rowland LP, Layzer RB: Muscular dystrophies, atrophies, and related diseases, in Baker AB (ed): Clinical Neurology. Vol. 3 New York, Harper & Row, 1975, p. 37
3. Bradley WG, Gardner-Medwin D, Walton JN (eds): Recent Advances in Myology. New York, American Elsevier, 1975
4. Rowland LP: Pathogenesis of Muscular Dystrophies. *Arch Neurol* 33:315, 1976
5. Walton JN, Gardner-Medwin D: Progressive muscular dystrophy and the myotonic disorders, in Walton JN (ed): Disorders of Voluntary Muscle. Edinburgh and London, Churchill-Livingstone, 1974, p. 561
6. Mokri B, Engel AG: Duchenne dystrophy: Electron microscopic findings pointing to a basic or early abnormality in the plasma membrane of the muscle fiber. *Neurology* 25:111, 1975
7. Perloff JK, Roberts WC, DeLeon AC et al: The distinctive electrocardiogram of Duchenne's progressive muscular dystrophy. *Am J Med* 42:179, 1967
8. Dubowitz V, Crome L: The central nervous system in Duchenne muscular dystrophy. *Brain* 92:805, 1969
9. Roses AD, Roses MJ, Miller SE, et al: Carrier detection in Duchenne muscular dystrophy. *New Engl J Med* 294:193, 1976
10. Emery AEH, Skinner R: Clinical studies in benign (Becker type) X-linked muscular dystrophy. *Clin Genet* 10:189, 1976
11. Emery AEH, Dreifuss FE: Unusual type of benign X-linked muscular dystrophy. *J Neurol Neurosurg Psychiatr* 29:338, 1966
12. Short JK: Congenital muscular dystrophy. *Neurology* 13:526, 1963
13. Wharton BS: An unusual variety of muscular dystrophy. *Lancet* 1:248, 1975
14. Donner M, Rapola J, Somer H: Congenital muscular dystrophy: A clinicopathological and follow-up study of 15 patients. *Neuropadiätrie* 6:239, 1975
15. Kamoshita S, Konishi Y, Segawa M, et al: Congenital muscular dystrophy as a disease of the central nervous system. *Arch Neurol* 33:513, 1976
16. Lebenthal E, Scochet SR, Adam A, et al: Arthrogryposis multiplex congenita—23 cases in an Arab kindred. *Pediatrics* 46:891, 1970
17. Bassöe HH: Familial congenital muscular dystrophy with gonadal dysgenesis. *J Clin Endocrinol* 16:1614, 1956
18. Munsat TL, Piper D, Cancilla P, et al: Inflammatory myopathy with facioscapulohumeral distribution. *Neurology* 22:335, 1972
19. Rotthause HW, Mortier W, Beyer H: Neuer Typ einer recessiv x-chromosomal vererbten Muskeldystrophie: Scapulo-humero-distale Muskeldystrophie mit frühzeitigen Kontrakturen und Herzrythmusstörungen. *Humangenetik* 16:181, 1972
20. Waters DD, Nutter DO, Hopkins LC, et al: Cardiac features of an unusual X-linked humeroperoneal neuromuscular disease. *New Engl J Med* 293:1017, 1975
21. Thomas PK, Calne DB, Elliott CF: X-linked scapuloperoneal syndrome. *J Neurol Neurosurg Psychiatr* 35:208, 1972
22. Welander L: Myopathia distalis tarda hereditaria. *Acta Med Scand* 141 (Suppl 265):1, 1951
23. Markesbery WR, Griggs RC, Leach RP, et al: Late-onset distal myopathy. *Neurology* 24:127, 1974
24. Van Der Does De Villebois AEM, Bethlem J, Meyer AEFH: Distal myopathy with onset in early infancy. *Neurology* 18:383, 1968
25. Murphy SF, Drachman DB: The oculopharyngeal syndrome. *J Am Med Assoc* 203:1003, 1968
26. Barbeau A: The syndrome of hereditary late-onset ptosis and dysphagia in French Canada, in Kuhn E (ed): Symposium über progressive Muskeldystrophie, Myotonie, Myasthenie. Berlin, Springer-Verlag, 1966, p. 102

27. Fried K, Arlozorov A, Spira R: Autosomal recessive oculopharyngeal muscular dystrophy. *J Med Genet* 12:416, 1975
28. Lundberg PO: Hereditary myopathy, oligophrenia, cataract, skeletal abnormalities and hypergonadotropic hypogonadism. *Europ Neurol* 10:261, 1973
29. Shokeir MHK, Kobrinsky NL: Autosomal recessive muscular dystrophy in Manitoba Hutterites. *Clin Genet* 9:197, 1976
30. Bethlem J, Van Wijngaarden GK: Benign myopathy, with autosomal dominant inheritance. *Brain* 99:91, 1976
31. DeCoster W, De Reuck J, Thiery E: A late-autosomal dominant form of limb-girdle muscular dystrophy. *Europ Neurol* 12, 159, 1974
32. Bacon PA, Smith B: Familial muscular dystrophy of late-onset. *J Neurol Neurosurg Psychiatr* 34:93, 1971
33. Becker PE: Genetic approaches to the nosology of muscle disease: Myotonias and similar diseases. *Birth Defects* 7:52, 1971
34. Kuhn E: Biochemische Besonderheiten und Unterschiede der autosomal dominant und autosomal recessiv vererbten Myotonia congenita. *Klin Wochenschr* 48, 1134, 1970
35. Roses AD, Appel SH: Phosphorylation of component of the human erythrocyte membrane in myotonic muscular dystrophy. *J Membr Biol* 20:51, 1975
36. Butterfield DA, Chesnut DB, Roses AD, et al: Electron spin resonance studies of erythrocytes from patients with myotonic muscular dystrophy. *Proc Nat Acad Sci* 71:905, 1974
37. Todorov A, Jequier M, Klein D, et al: Analyse de la ségrégation dans la dystrophie myotonique. *J Genet Hum* 18:387, 1970
38. Sarnat HB, O'Connor T, Byrne PA: Clinical effects of myotonic dystrophy on pregnancy and the neonate. *Arch Neurol* 33:459, 1976
39. Schrott HG, Omenn GS: Myotonic dystrophy: Opportunities for prenatal diagnosis. *Neurology* 25:789, 1975
40. Swift MR, Finegold MJ: Myotonic muscular dystrophy: Abnormalities in fibroblast culture. *Science* 165:294, 1969
41. Schwartz O, Jampel RS: Congenital blepharophimosis associated with a unique generalized myopathy. *Arch Ophthalmol* 68:52, 1962
42. Fowler WM, Layzer RB, Taylor RG, et al: The Schwartz-Jampel syndrome. Its clinical, physiological and histological expressions. *J Neurol Sci* 22:127, 1974
43. Pearson CM, Kalyanaraman K: The periodic paralyses, in Stanbury JB, Wyngaarden JB, Fredrickson DS (eds): The Metabolic Basis of Inherited Disease (3d ed). New York, McGraw-Hill, 1972, p. 1181
44. Hoffman WW, Smith RA: Hypokalaemic periodic paralysis studied *in vitro*. *Brain* 93:445, 1970
45. Layzer RB, Lovelace RE, Rowland LP: Hyperkalemic periodic paralysis. *Arch Neurol* 16:455, 1967
46. Rowland LP, DiMauro S, Bank WJ: Glycogen storage diseases of muscle. *Birth Defects* 7:43, 1971
47. Huijing F: Glycogen metabolism and glycogen-storage diseases. *Physiol Rev* 55:609, 1975
48. DiMauro S, Melis-DiMauro PM: Muscle carnitine palmityltransferase deficiency and myglobinuria. *Science* 182, 929, 1973
49. Bank WJ, DiMauro S, Bonilla E, et al: A disorder of muscle lipid metabolism and myoglobinuria. *New Engl J Med* 282:697, 1975
50. DiMauro S, Eastwood AB, Disorder of glycogen and lipid metabolism, in Griggs RC, Moxley T (eds): Treatment of Neuromuscular Diseases. New York, Raven Press, 1977, p. 123
51. Engel AG, Angelini C: Carnitine deficiency of human skeletal muscle with associated lipid storage myopathy: A New Syndrome. *Science* 179:899, 1973
52. Karpati G, Carpenter S, Engel AG, et al: The syndrome of systemic carnitine deficiency. *Neurology* 25:16, 1975
53. Blass JP, Cederbaum SD, Gibson GE: Clinical and metabolic abnormalities accompanying

deficiencies in pyruvate oxidation, in Hommes FA, Van den Berg CJ (eds): Normal and Pathological Development of Energy Metabolism. London, Wiley, 1975, p. 193
54. Blass JP, Kark P, Engel WK: Clinical studies of patient with pyruvate decarboxylase deficiency. *Arch Neurol* 25:449, 1971
55. Blass JP, Avigan J, Uhlendorf W: A defect of pyruvate decarboxylase in a child with intermittent movement disorder. *J Clin Invest* 49:423, 1970
56. Ghatak NR, Hirano A, Poon TP, et al: Trichopoliodystrophy. Pathological changes in skeletal muscle and nervous system. *Arch Neurol* 26:60, 1972
57. French JH, Sherard ES, Lubell H, et al: Trichopoliodystrophy: Report of a case and biochemical studies. *Arch Neurol* 26:229, 1972
58. Goka TJ, Stevenson RE, Hefferan PM, et al: Menkes disease: A biochemical abnormality in cultured human fibroblasts. *Proc Nat Acad Sci* 73:604, 1976
59. Spiro AJ, Moore CL, Prineas JW, et al: A cytochrome-related inherited disorder of the nervous system and muscle. *Arch Neurol* 23:103, 1970
60. Mackler B, Haynes B, Inamdar AR, et al: Oxidative energy deficiency. II. Human achondroplasia. *Arch Biochem Biopys* 159:885, 1973
61. Larsson LE, Linderholm H, Müller R, et al: Hereditary metabolic myopathy with paroxysmal myoglobinuria due to abnormal glycolysis. *J Neurol Neurosurg Psychiatr* 27:361, 1964
62. Rawles JM, Weller RO: Familial association of metabolic myopathy, lactic acidosis and sideroblastic anemia. *Am J Med* 56:891, 1974
63. Carpenter S, Karpati G, Andermann F, et al: Lafora's disease: Peroxisomal storage in skeletal muscle, *Neurology* 24:531, 1974
64. Coleman DL, Gambetti PL, DiMauro S, et al: Muscle in Lafora disease. *Arch Neurol* 31:396, 1974
65. Britt BA, Kalow W, Gordon A, et al: Malignant hyperthermia: An investigation of five patients. *Can Anaesth Soc J* 20:431, 1973
66. Schmitt HP, Simmendinger HJ, Wagner H, et al: Severe morphological changes in skeletal muscles of a five-month-old infant dying from an anesthetic complication with general muscle rigidity. *Neuropäd* 6:102, 1975
67. Britt BA: Malignant hyperthermia: A pharmacogenetic disease of skeletal and cardiac muscle. *New Engl J Med* 290:1140, 1974
68. King JO, Denborough MA, Zapf PW: Inheritance of malignant hyperpyrexia. *Lancet* 1:365, 1972
69. Denborough MA, Dennett X, Anderson RM: Central-core disease and malignant hyperpyrexia. *Br Med J* 1:272, 1973
70. Moulds RFW, Denborough MA: Identification of susceptibility to malignant hyperpyrexia. *Br Med J* 1:245, 1974
71. Dubowitz V, Roy S: Central core disease of muscle: Clinical, histochemical and electron microscopic studies of an affected mother and child. *Brain* 93:133, 1970
72. Engel AG, Gomez MR, Groover RV: Multicore disease: A recently recognized congenital myopathy associated with multifocal degeneration of muscle fibers. *Mayo Clinic Proc* 46:666, 1971
73. Shy GM, Engel WK, Somers JE, et al: Nemaline myopathy: New congenital myopathy. *Brain* 86:793, 1963
74. Sugita H, Masaki T, Ebashi S, et al: Staining of the nemaline rod by fluorescent antibody against 10S actinin. *Proc Japan Acad* 50:237, 1974
75. Heffernan LP, Rewcastle NB, Humphrey JG: The spectrum of rod myopathies. *Arch Neurol* 18:529, 1968
76. Neustein HB: Nemaline myopathy. A family study with three autopsied cases. *Arch Pathol* 96:192, 1973
77. Schochet SS, Zellweger H, Ionasescu V, et al: Centronuclear myopathy: Disease entity or a syndrome? *J Neurol Sci* 16:215, 1972

78. Mortier W, Michaelis E, Becker J, et al: Centronucleäre Myopathie mit autosomal dominantem Erbgang. *Humangenetik* 27, 199, 1975
79. Barth PG, Van Wijngaarded GK, Bethlem J: X-linked myotubular myopathy with fatal neonatal asphyxia. *Neurology* 25:531, 1975
80. Jerusalem F, Engel AG, Gomez MR: Sarcotubular myopathy. *Neurology* 23:897, 1973
81. Cancilla PA, Kalyanaraman K, Verity MA, et al: Familial myopathy with probable lysis of myofibrils in type I fibers. *Neurology* 21:579, 1971
82. Engel AG, Angelini C, Gomez MR: Fingerprint body myopathy. *Mayo Clin Proc* 47:377, 1972
83. Brooke MH, Neville HE: Reducing body myopathy. *Neurology* 22:829, 1972

13
Heritable Neurologic Diseases

Harry H. White

A significant number of patients with progressive neurologic disease are affected with disorders of unknown cause. The term "degenerative" has traditionally been used as a classification for these cases in which the standard methods of clinicopathologic examination have failed to determine a specific cause (i.e., neoplastic, infectious, vascular, toxic, traumatic or metabolic). Recent refinements in laboratory techniques have uncovered specific causes in some diseases previously classified as "degenerative" (subacute sclerosing panencephalitis and Creutzfeldt-Jakob disease). It is reasonable to expect that other degenerative diseases will eventually be explained by damage to the nervous system caused by infectious agents, altered immune responses or intoxications.

In many cases of degenerative disease the familial occurrence of similar illness in the same or successive generations has led to its designation as a "heredodegenerative" or "heredofamilial" disease of the nervous system. The recognition of the heritable nature of a particular disease is important in the delineation of the specific abnormality at the molecular level. Information thus derived will provide the basis for informed genetic counseling and a rational approach to treatment. The nosology of the heritable diseases of the nervous system has traditionally been based on similarities and differences in both Clinical and Pathologic studies. Phenotypic overlap, commonly encountered in neurologic disease, has resulted in the use of terms such as "forme fruste," "variant" and "abortive" to link together—in some cases—basically dissimilar genetic diseases. Further nosologic confusion has resulted from the frequent inter- and intrafamilial variability of phenotypic expression in the same disease. This has lead to a failure to recognize cases having dissimilar clinical manifestations as examples of genetic pleiotropism.

Classification of the genetically determined diseases of the nervous system has depended heavily on analysis of phenotypes and assessment of the mode of inheritance. Although these strictly clinical methods have served well in the delineation of certain disorders (Huntington chorea), the lack of the easy availability of living tissue for study has hampered the biochemical classification of nervous system diseases in comparison to those of other body organs. Many of the disorders affecting the nervous system are discussed in other chapters of this book. The present chapter is devoted to a presentation of the commoner disorders in which the nervous system is or appears to be the major site of functional disability.

CEREBRAL HEMISPHERES

Congenital malformations of the central nervous system (CNS) constitute an etiologically heterogeneous group of disorders. Environmental teratogens, chromosomal abnormalities and multifactorial inheritance represent the major causes of such defects. Some of the localized malformations are due to the effects of single mutant genes (1).

The neural-tube defects constituting the *anencephaly—meningomyelocele—encephalocele complex* may be caused by teratogens, chromosomal abnormalities and multifactorial inheritance (2). Determination of α-fetoprotein levels in amniotic fluid may be useful in antenatal diagnosis of these lesions in at-risk families (2a).

The *Meckel syndrome* is a clearly defined combination of malformations consisting of occipital encephalocele, microcephaly, abnormal facies including cleft lip and palate, polydactyly and polycystic kidneys. The syndrome is caused by the effects of an autosomal recessive gene (3).

Hydrocephalus secondary to *aqueductal stenosis* is established in some families as caused by X-linked recessive inheritance. The severity of the hydrocephalus shows both intra- and interfamily phenotypic variability, but the condition differs in no other respects from aqueductal stenosis of acquired or nongenetic causes (4).

Agenesis of the corpus callosum has resulted from X-linked recessive inheritance and was manifest by seizures from birth and severe developmental retardation (5). Two sisters are described with agenesis of the corpus callosum who were thought to represent examples of autosomal recessive inheritance (6).

The anomaly *holoprosencephaly* is produced by a failure of the forebrain (prosencephalon) to cleave properly into cerebral hemispheres during embryogenesis. The telencephalon and its ventricle remain as a single structure and cavity. Median facial defects plus orbital hypotelorism are present and may be used to correctly predict the severe, underlying anomaly of the brain (7). The anomaly likely has several different etiologies (8), including chromosomal abnormalities and the effect of a single mutant gene (9, 10). Holoprosencephaly, which precludes any possibility for normal psychomotor development, must be distinguished from the median cleft face syndrome in which similar facial anomalies are associated with orbital *hypertelorism* and a more favorable prognosis for psychomotor development (11).

Microcephaly is usually defined as being present if the circumference of the skull falls two standard deviations below normal for the patient's sex and age. Mental subnormality correlates strongly with microcephaly (12), and severe mental deficiency is most often present. Microcephaly is a secondary or associated feature of some genetically determined diseases and chromosomal abnormalities (13) and may also be caused by exogenous agents (14). "Pure" or "true" microcephaly designates cases in which there are no other associated abnormalities (15). Most cases of microcephaly from autosomal recessive inheritance are placed in this category (16).

Premature closure of one or more cranial sutures without other associated abnormalities most often occurs on a sporadic basis. Autosomal recessive and autosomal dominant inheritance is established in some pedigrees (16).

Acrocephalosyndactyly (Apert syndrome) is a well-defined syndrome consisting of multiple craniostenoses producing acrocephaly, flat facies, depressed nasal bridge, proptosis, downward slanting palpebral fissures, low-set ears and syndactyly. Mental retardation, frequently mild, is usually present. Although most cases seem to occur on a sporadic basis, an autosomal dominant mode of inheritance is strongly supported in some families (16).

Patients with features similar to those of the Apert syndrome in addition to polydactyly, *acrocephalopolysyndactyly,* have been described as a result of autosomal recessive (Carpenter syndrome) and autosomal dominant (Noack syndrome) inheritance (16).

Craniofacial dysostosis (Crouzon syndrome) phenotypically resembles the Apert syndrome, but syndactyly is absent. The disease is transmitted by an autosomal dominant gene (17).

The syndrome of mental retardation, pigmentary retinopathy, polydactyly, obesity and hypogenitalism has usually been associated with the combined eponyms of *Laurence, Moon, Bardet and Biedl*. McKusick pointed out that the patients described by Bardet and Biedl showed this constellation of signs, whereas those of Laurence and Moon represented a distinctly different entity characterized by mental retardation, pigmentary retinopathy, hypogenitalism and spastic paraplegia (16). Both the Bardet-Biedl and the Laurence-Moon syndromes are due to the expression of an autosomal recessive gene. Additional varieties and atypical forms are described (18).

The triad of oligophrenia, congenital ichthyosis and spasticity comprise the clinical symptoms of the autosomal recessive disorder called the *Sjögren-Larsson syndrome* (19). The clinical course may vary from a mild, stationary deficit to one of a progressive deterioration of intellect and movement (20).

Familial dysautonomia (*Riley-Day syndrome*) is a disease that occurs primarily in Ashkenazic Jews and is inherited in an autosomal recessive manner. The clinical signs and symptoms include autonomic nervous system (ANS) manifestations (defective lacrimation, vasomotor instability, abnormal temperature control and episodic hyperhidrosis); sensory disturbances (corneal anesthesia, insensitivity to pain and impaired taste sensation); and neuromuscular dysfunction (poor coordination and depressed tendon reflexes) (21, 22). Absent or reduced levels of plasma dopamine-β-hydroxylase activity are reported in familial dysautonomia (23).

Progressive cerebral degeneration in infancy, known eponymously as Alpers disease, is applied to a progressive infantile encephalopathy associated with myoclonic and generalized seizures and spasticity. The pathologic changes in the brain involve primarily the gray matter, in contrast to the predominant involvement of white matter in the leukodystrophies and diffuse sclerosis (24). Most authors agree that these cases do not represent a single disease entity (25).

Anoxic, genetic and metabolic factors are proposed as possible etiologic mechanisms (26). A small number of familial cases are recorded, however, in which autosomal recessive inheritance seems well established (25).

Huntington chorea is an autosomal dominant disorder characterized by choreiform movements and progressive mental deterioration. The age of onset of symptoms varies from the first to the sixth decades. The average duration of life after the onset of symptoms is 15 years. Intercurrent infections are the usual cause of death.

The *Alzheimer disease* is a pathologic entity characterized by atrophy of the brain, loss of neurons, neurofibrillary tangles, granulovascular changes and neuritic or senile plaques. Identical changes are found in the brains of patients with senile dementia, and the identity of the two diseases is now widely accepted (27). The symptoms consist of a progressive organic mental syndrome, frequently accompanied by disturbances in the higher functions of speech. A number of pedigrees supporting an autosomal dominant mode of inheritance are described (16).

The symptoms of the *Pick presenile dementia* are clinically indistinguishable from those of Alzheimer disease. The pathologic changes in the brain are distinct from those in the Alzheimer dementia and include a predominately frontotemporal atrophy, swollen neurons and argentophilic cytoplasmic inclusions (Pick bodies) in cortical neurons. Although most cases, are sporadic, autosomal dominant inheritance is occasionally established (28).

Tuberous sclerosis is inherited as an autosomal dominant condition. The clinical manifestations vary, but in the fully developed form the patient is severely retarded and subject to seizures, and the characteristic adenomas sebaceum are present on the face. Other associated lesions include ocular phakoma; tumors or hamartomas of the lungs, heart, kidney or brain; shagreen patches and depigmented nevi of the skin; and evidence of endocrine and metabolic dysfunction (29).

Mild forms of the disease with normal mentality are usually found in affected parents who have produced an affected offspring. Most cases of tuberous sclerosis arise as a result of a new mutation (30).

Neurofibromatosis (*Von Recklinghausen disease*) is a genetically determined autosomal dominant disorder. The major features of the disease include cutaneous pigmentation, fibromatous tumors of the skin and multiple tumors of the spinal or cranial nerves (16, 31). Diverse abnormalities indicating involvement of the bone, endocrine and other visceral organs are reported in the voluminous literature on this disease.

The two characteristics of the autosomal dominant disorder called *von-Hippel-Lindau disease* are angioma of the retina and cerebellar hemangioblastoma. Angiomatous cysts of various visceral organs, pheochromocytoma and renal cell tumors are present in some patients.

"Epilepsy" like dyspnea, does not constitute a disease, but rather a clinical symptom reflecting diverse etiologies. Patients whose seizures can be ascribed to a specific cause (trauma, tumor, metabolic disease and so on) are said to have symptomatic seizures, i.e., the seizures are secondary to or symptomatic of some defined disease. Patients in whom presently available methods of investigation fail to uncover a cause for the seizures are often designated as having "idiopathic" seizures. Ultimately, when brain function is better understood, a cause will be established for all patients with idiopathic seizures. There seems to be little doubt that heredity plays a significant role in some patients currently classified in one large, heterogeneous group of idiopathic epilepsy. At present, however, the specific role of heredity in seizure disorders is unknown. A recent review nicely summarizes the current state of uncertainty on the subject of epilepsy and heredity (32).

The frequent familial occurrence of *migraine* has been noted by many authors on the subject. Until the cause(s) and pathogenesis are known, the question of whether migraine is a genetic entity will remain unsolved. Incomplete dominance and multifactorial inheritance are suggested as probable mechanisms of genetic transmission (33).

Lafora and Glueck described a single, sporadic case showing myoclonus, major seizures and evidence of a generalized neuronal degeneration progressing to dementia and terminating in death (*Lafora-body disease*) (34). The nosologic entity and genetic basis of this disorder were later established by the report of subsequent familial cases showing the peculiar neuronal inclusion bodies first described by Lafora and Glueck. From the numerous pathologically confirmed cases in the literature (35-39), it appears that the disease is transmitted by an autosomal recessive mechanism and typically becomes symptomatic during the second decade of life. As mentioned, it is characterized by myoclonic and generalized seizures, progressive neurologic deterioration and dementia leading to death within 3-10 years following the onset of symptoms. May and White (40) have reviewed the literature on "progressive myoclonic epilepsy" and "Unverricht-Lundborg's myoclonic epilepsy" and find no justification for a nosologic association between the disease described by Lafora and Glueck, the family reported by Unverricht and the heterogenous cases described by Lundborg.

BASAL GANGLIA

Parkinson disease does not have a single etiology. Convincing evidence of at least two different causes of the disease—those existing prior to the epidemic of encephalitis lethargica as described by James Parkinson, and the postencephalitic variety occurring in and after the epidemic of 1917–1926—have been reviewed by Duvoisin and his associates (41). It seems likely that the nonencephalitic cases themselves are heterogeneous, containing cases of both acquired and genetically determined disease. Familial cases are postulated as occurring from autosomal dominant (42) and multifactorial inheritance (43). Most cases are sporadic.

Hallervorden and Spatz in 1922 described a progressive disorder of the nervous system that occurred in five children of a 12-member sibship (*Hallervorden-Spatz syndrome*). The onset of the disease is usually in the first or second decade. The appearance of involuntary movements and muscular rigidity indicate involvement of the "extrapyramidal" motor system, whereas spasticity, hyperactive reflexes and extensor plantar responses indicate dysfunction of the corticospinal system. The disease is progressive, accompanied by dementia and seizures, leading to death in early adulthood. The neuropathologic features include destructive lesions of the globus pallidus, pars reticulata and substantia nigra; widely disseminated axonal swellings (spheroids); and the accumulation of iron-containing pigment in the affected areas of the brain (44). The pattern of familial occurrence is consistent with autosomal recessive inheritance.

The word tremor describes involuntary alternating movements of the body, which are more or less rhythmical. "Attitude" tremor refers to one that is most pronounced during sustained posture. It may also be present to some degree with resting posture and phasic movements of the limb, but disappears with muscle relaxation (45). This type of tremor is distinguished from that seen in parkinsonism, which is principally a resting posture tremor and the action (phasic movement) tremor of cerebellar disease. Attitude tremor may be caused by anxiety, thyrotoxicosis or the administration of epinephrine (46). When none of the factors known to produce tremor is present, it is called *essential tremor*. Essential tremor frequently occurs on a familial basis and follows an autosomal dominant mode of inheritance (47).

Bilateral, symmetrical calcification of the basal ganglia, frequently with concomitant calcific deposits in the dentate nuclei of the cerebellum and occasionally other regions of the brain, is clearly a heterogeneous disorder (*familial calcification of the basal ganglia*). The commonest cause is disordered calcium-phosphorus metabolism; hypoparathyroidism and Albright's hereditary osteodystrophy account for two-thirds of the reported cases (48). However, familial cases are described in which no evidence of parathyroid disease was present. A syndrome consisting of calcified basal ganglia, onset of progressive mental deterioration in the first few months of life and seizures leading to early death is reported in several families and follows an autosomal recessive mode of inheritance (49). Four children in a sibship of 16, all showing basal ganglia calcifications, slow psychomotor development with mental retardation, steatorrhea and episodes of pseudointestinal obstruction are described by Cockel and associates (50). The pedigree pattern of this family is consistent with autosomal recessive inheritance.

Moskowitz and associates studied five patients in two generations of one family who had calcifications of the basal ganglia (51). One patient developed choreoathetosis, cerebellar ataxia, dysarthria and memory loss in the fourth decade. Another developed a parkinsonian syndrome in the fifth decade, whereas the remaining three were asymptomatic at ages 30, 15 and 13 years, respectively. The pedigree suggests an autosomal dominant disorder. Boller and associates studied a family in which four members showed calcification of the basal ganglia (52). Two were symptomatic with palilalia (compulsive repetition of words), dementia and chorea; the other two cases were asymptomatic. The pattern of inheritance is consistent with autosomal dominant transmission. The family of Moskowitz and that of Boller share similar clinical characteristics and may represent interfamilial variability of the same disease.

Hepatolenticular degeneration (*Wilson disease*) is a hereditary disease of copper metabolism, affecting principally the brain and liver. The disease results from the homozygous expression of an autosomal recessive gene. Although the precise nature of the biochemical defect is unknown, a rational form of treatment is available and the treatment of presymptomatic cases can prevent the development of symptoms (53).

Dystonia musculorum deformans refers to involuntary movements of basal-ganglia origin characterized by slow, sustained, writhing or twisting motions of the musculature. Dystonic movements may involve any somatic muscle, but predominantly affect those of the head, neck, trunk and proximal limb girdles.

Dystonic movements may occur as a result of kernicterus, perinatal anoxia, Wilson disease, Hallervorden-Spatz syndrome or administration of phenothiazines (54). Familial cases of torsion dystonia, in which none of the diseases listed is present, have been extensively studied by Eldridge (55). Autosomal dominant inheritance is established in some families and a form of dystonia transmitted by an autosomal recessive gene is more prevalent among Ashkenazic Jews. An excellent review of the clinical manifestations, prognosis and pathology in idiopathic torsion dystonia has recently been presented by Marsden and Harrison (54).

Familial paroxysmal choreoathetosis is a relatively benign movement disorder first described by Mount and Reback (56). The onset occurs in the first two decades of life and is characterized by the sudden appearance of brief, intermittent choreoathetoid movements of the face, neck and extremities. The episodes may occur several times daily, resulting in bizarre posturing, but there is no loss of consciousness. The involuntary movements are commonly precipitated by sudden movements after a period of rest or immobility. The attacks are usually terminated by the administration of diphenylhydantoin. Autosomal dominant inheritance with incomplete penetrance is the usual mode of genetic transmission. Jung and associates have provided an excellent review of the subject and suggest the possibility of autosomal recessive inheritance in one of their families (57).

A genetically determined form of chorea commencing in infancy and childhood was first described by Pincus and Chutorian (58-59). The disease *hereditary nonprogressive chorea*, begins in the first two decades of life and is not associated with seizures, neurologic or intellectual deterioration. The mode of inheritance appears to be autosomal dominant (60). Chun and associates suggest that an indistinguishable clinical picture may also be caused by the homozygous expression of an autosomal recessive gene (61).

Rosenberg and associates (62) have described an illness (*striatonigral degeneration*) in more than 50 patients in eight generations of a Portuguese family characterized by onset in the 2nd to 4th decade of parkinsonian rigidity, progressive spasticity, dysarthria, nystagmus, facial fasciculations, myokymia and lingual fasciculations. Neuropathologic studies in one case disclosed neuronal loss in the substantia nigra, corpus striatum, dentate nucleus of the cerebellum and nucleus ruber of the midbrain. The disease follows an autosomal dominant mode of inheritance.

CEREBELLUM

The term *spinocerebellar ataxia*(s) (degenerations) refers to a heterogeneous group of progressive diseases characterized clinically by disturbances of the coordination of movement and pathologically by degeneration of the afferent and efferent neuronal systems on which smooth and coordinated movements depend. Hereditary forms of ataxia contribute significantly to the overall number of these cases. Table 1 tentatively separates these diseases on the basis of

Table 1. Spinocerebellar Ataxias

Type	Age Onset	Age Death	Transmission*	Clinical Findings	Pathologic Findings
Predominantly spinal forms: Friedreich's ataxia	1-2 decade	3-4 decade	AR	Limb and gait ataxia, dysarthria, pes cavus, kyphoscoliosis, nystagmus, cerebellar and sensory ataxia; absent reflexes, loss of position and vibratory sense; extensor plantar responses	Degeneration of posterior funiculi, pyramidal and spinocerebellar tracts of spinal cord
Spastic ataxia of Sanger-Brown	11-45 yrs	2-27 yrs from onset; average 20 yrs	AD	Ataxia, dysarthria, dysphagia, oculomotor palsies and failing vision; ataxia of limbs, hyperactive reflexes and optic atrophy; no skeletal deformity	Degeneration of posterior funiculi, dorsal spinocerebellar tract and Clark's column; olives, brainstem and cerebellum essentially normal
Dominant cerebellar ataxia of Becker (67)	3rd-6th decade	6th-7th decade	AD	Nystagmus, ataxia of gait and limbs, and dysarthria; pyramidal tract signs and mild dementia	Cortical cerebellar atrophy, loss of pontocerebellar fibers, degeneration of vestibular nuclei and degeneration of spinocerebellar tracts and Clarke's column
Spinal and Cerebellar Forms: Olivopontocerebellar atrophies (OPCA):					
1. OPCA I, Menzel type	13-59 yrs	40-60 yrs	AD	Ataxia of gait, dysarthria, dysphagia, tremor of head; increased reflexes, normal position sense	Cerebellar atrophy, olivary and pontine atrophy and pyramidal and spinocerebellar tract atrophy in spinal cord
2. OPCA II, Fiekler-Winkler type	7-20 yrs	33-59 yrs	AR	Ataxia of limbs, head tremor, dysarthria; sensation and reflexes normal	Atrophy of cerebellum, transverse pontine fibers and olives; spinal cord normal

3. OPCA III, with retinal degeneration	1-34 yrs	1½-75 yrs	AD	Progressive visual loss, ataxia and tremor; cerebellar ataxia of limbs, spasticity, retinal pigmentary degeneration and occasionally ophthalmoplegia and nystagmus	Atrophy of cerebellum, basis pontis, olives, substantia nigra and retina
4. OPCA IV, Schut-Haymaker type	17-35 yrs	26-48 yrs	AD	Limb and gait ataxia, dysarthria, dysphagia, variable reflexes and plantar responses and variable sensory loss; involvement of lower cranial nerves	Atrophy of cerebellum, basis pontis, olives; variable changes in brainstem and spinal cord
5. OPCA V, with dementia ophthalmoplegia and extrapyramidal signs	7-45 yrs	11-63 yrs	AD	Ataxia of gait and limbs, dysarthria, extrapyramidal signs, ophthalmoparesis and dementia	Olivopontocerebellar atrophy; scattered cellular loss in oculomotor nucleus, substantia nigra, cerebral cortex, lenticular and caudate nuclei
Spinopontine atrophy	4th-6th decade	6th-7th decade	AD	Ataxia of gait and hands, nystagmus and dysarthria, hypotonia, increased reflexes, extensor plantar responses, impaired vibration and tactile sensation	Atrophy of basis pontis; spino-cerebellar tract and posterior funiculi degeneration in cord; minimal cell loss in cerebellum, red and subthalamic nuclei
Cerebellar forms:					
Cerebelloparenchymal disorders (CPD)					
1. CPD I, late onset dominant type	5th-6th decade	6th-8th decade	AD	Ataxia of gait, dysarthria and dementia	Marked loss of Purkinje cells; some degeneration in olives; pons not affected
2. CPD II, late onset recessive type	4th-5th decade	6th-7th decade	AR	Ataxia, dysarthria, mental disorder or deterioration	Loss of Purkinje cells; olives and pons normal
3. CPD III, congenital granular cell hypoplasia and mental retardation	Birth	—	AR	Congenital cerebellar ataxia and mental retardation	Small cerebellum with hypoplasia of granular cell layer; heterotopic Purkinje cells

Table 1. (Continued)

Type	Age Onset	Age Death	Transmission*	Clinical Findings	Pathologic Findings
4. CPD IV, cerebellar vermis agenesis	Birth	2½-3 mos. alive at 8 yrs.	AR ?	Abnormal breathing noted at birth; psychomotor retardation, episodic hyperpnea, abnormal eye movements and ataxia	Partial or complete agenesis of cerebellar vermis; neuronal heterotopia of cerebellar cortex
5. CPD V, dyssynergia cerebellaris myoclonica of Hunt	Early adulthood	37 yrs	AR ?	Ataxia, myoclonic and generalized seizures; dysarthria	Atrophy of posterior columns, spinocerebellar tracts, dentate nuclei and superior cerebellar peduncles
6. CPD VI, cerebellar granular cell hypertrophy, myelinated molecular layer axons and megalencephaly	2 yrs	25-51 yrs	AD ?	"Mental dullness"; large head noted early in life	Megalencephaly; enlarged neuron somas in granular cell layer of cerebellum; myelinated axons in molecular layer

*AR, autosomal recessive; AD, autosomal dominant.

the predominant involvement as an adaptation of the work of Greenfield (63) and Konigsmark and associates (64-66). In addition there are other syndromes in which ataxia is a primary feature. Farmer and Mustian (68), e.g., described 16 members in one family who suffered *vestibulocerebellar ataxia* with periodic attacks of vertigo, diplopia and ataxia commencing in adult life. A progressive cerebellar ataxia with nystagmus developed in most patients. The pedigree suggests that the disease is due to an autosomal dominant gene. No pathologic data were available.

The *Marinesco-Sjögren syndrome* is an autosomal recessive disorder characterized by cataracts, cerebellar ataxia and impaired mental and physical development. The onset of symptoms is usually during the first two years of life. Pathologic studies show severe cerebellar atrophy and nerve cell loss in the inferior olives and pontine nuclei (69).

The *Bassen-Kornzweig syndrome* is a rare disease, inherited by an autosomal recessive gene. It is of some historic interest because it is the first disease among the "hereditary ataxias" in which a biochemical defect has been found (70). The disease is characterized by steatorrhea in the first months of life, acanthocytosis of the RBCs, absence of β-lipoprotein in blood, retinitis pigmentosa and a slowly progressive ataxia leading to severe physical disability in early adulthood (71). Neurologic findings include cerebellar ataxia of gait, trunk and extremities; proprioceptive sensory loss; areflexia; and in some cases muscular weakness, kyphoscoliosis, ophthalmoparesis, extensor plantar responses and cutaneous sensory loss. Pathologic findings in one case included demyelinating lesions of the anterior columns, spinocerebellar tracts and cerebellum with loss of anterior horn and cerebellar nuclei. Sections of peripheral nerve showed focal demyelination (72).

BRAIN STEM

Subacute necrotizing encephalomyelopathy (Leigh disease) is a genetically determined disease that is transmitted by an autosomal recessive gene. The onset of symptoms usually begins in infancy or early childhood and commonly include loss of appetite, vomiting, regurgitation, dysphagia, psychomotor retardation or regression, hypotonia and weakness (73). The abnormal findings on neurologic examination implicate the common involvement of the brainstem and include nystagmus; dysconjugate eye movements; ptosis; strabismus; pupillary abnormalities; and pyramidal, extrapyramidal and cerebellar signs (74). The disease is progressive, with the development of seizures and respiratory arrhythmias, leading to death in most cases during the first five years of life. The neuropathologic lesions are characteristic and include spongiform degeneration, scarring and increased vascularity in the basal ganglia and brainstem.

Lesions are also found in the white matter of the cerebral hemispheres, cerebellum and spinal cord (75). The brain contains reduced amounts of thiamine triphosphate, and the urine of affected persons contains a nondialyzable inhibitor of an enzyme from the brain that catalyzes the formation of thiamine triphosphate from its precursor, thiamine pyrophosphate (76-77). At least one form may be due to an error in pyruvate metabolism.

The designation *Möbius syndrome* is traditionally applied to a nonprogressive, congenital facial diplegia associated with paralysis of lateral gaze. Recent reviews of the subject clearly indicate the pathologic heterogeneity of the syndrome (78-79). The syndrome may be produced by lesions of diverse location (supranuclear; nuclear; peripheral nerve; fibrosis, absence, and myopathic disorders of muscle) and cause, including the myotonic and facioscapulohumeral forms of muscular dystrophy (79). The latter two diseases are determined by an autosomal dominant gene, and it is likely that other, but not all, cases of Möbius syndrome are expressions of a genetic defect (80-83).

Bulbar paralysis of childhood (Fazio-Londe disease) is a rare disease of neural amyotrophy affecting the brainstem nuclei of infancy; it has been reported in sibs of two separate families (84). In one family the patients were the product of a consanguineous mating. The mother of the original case reported by Fazio (84) also had a progressive bulbar palsy. Alexander and associates (85) believe that the Fazio-Londe disease is a clinical variant of the spinal muscular atrophies, rather than a separate entity.

SPINAL CORD

Three major types of *amyotrophic lateral sclerosis (ALS)* are now generally recognized, including the sporadic (commonest) form, the Western Pacific (Guam, Kii Penisula and New Guinea) form and the familial form (86).

Familial ALS accounts for 5-10% of cases in most clinical series. The clinicopathologic features of familial ALS are usually the same as those of the sporadic form, although lesions outside the motor system are found in some patients with the hereditary form of the disease (87-88). The majority of the familial cases are transmitted by an autosomal dominant gene. Horton and associates (89) propose a tentative classification of the dominant variety into three separate groups based on the duration of the illness and the presence of pathologic findings beyond the confines of the motor system. An adult-onset variety of short duration is reported in two sisters who were the product of a consanguineous mating. The pedigree in this family suggests autosomal recessive inheritance (90).

The *spinal muscular atrophies* (SMA) constitute a group of diseases in which degeneration of the anterior horn cells of the spinal cord results in muscular atrophy and weakness. The syndrome of SMA may be produced by a wide variety of causes, including trauma, infections and neoplasia (91), as well as lead intoxication (92). Familial cases, however, contribute significantly to the total number of patients with SMA. The genetically determined cases themselves represent a heterogeneous group of patients with variations in age of onset, clinical features and mode of inheritance (93). Marked intrafamilial variability is also noted (94).

Table 2 contains a tentative classification of hereditary SMA as adopted from Emery (91) and Marsden (93), who have provided excellent reviews of the subject. The association of SMA with chronic progressive external ophthalmoplegia in a few recently reported cases suggests the possibility of yet another form of hereditary SMA (95-97).

Table 2. Hereditary Spinal Muscular Atrophy

Proximal spinal muscular atrophy:
   Type I, autosomal recessive (common)
     Infantile form (Type I-I, Werdnig-Hoffman syndrome)
     Juvenile form (Type I-J, Wohlfart-Kugelberg-Welander syndrome)
     Adult form (Type I-A)
   Type II, autosomal dominant (rare)
   Type III, X-linked recessive (extremely rare)

Distal spinal muscular atrophy:
   Type I, autosomal recessive
   Type II, autosomal dominant

Scapuloperoneal spinal muscular atrophy:
   Type I, autosomal dominant
   Type II, autosomal recessive

Facioscapulohumeral spinal muscular atrophy:
   Type I, ? autosomal recessive
   Type II, autosomal dominant

Familial spastic paraplegia (FSP) is a rare disorder, which is heterogeneous in regards to both the mode of inheritance and its clinicopathologic features (Table 3). The onset of symptoms may be present in early childhood and manifested by a delay in learning to walk, or it may occur as late as the 6th or 7th decade with stiffness and weakness in the legs (98). The disease may remain stationary or show variable rates of progression with the development of pes cavus or talipes equinovarus. The main pathologic lesion consists of a degeneration of the corticospinal tracts, which is most pronounced in the lower thoracolumber segments of the spinal cord. Lesions have also been found in other parts of the neuraxis. The corticospinal tract lesions are responsible for the spastic weakness of the legs, increased tendon reflexes and extensor plantar responses.

The disease may be transmitted by autosomal dominant, autosomal recessive and X-linked recessive genes (99-100). Autosomal recessive inheritance appears to be the commonest variety of FSP, whereas the X-linked recessive type is reported only rarely (101). The report of Thurmon and Walker indicates at least two distinct autosomal dominant varieties of FSP (102). One form starts in the teenage years and is slowly progressive, with later involvement of urinary and rectal sphincters. The other is apparent in infancy and shows no evidence of significant progression. Silver has described two families with dominant FSP in which amyotrophy of the hands constituted the chief disability (103).

There are, at the least, several different forms of FSP transmitted by autosomal recessive genes (104). Families with recessive FSP and retinal degeneration have been reported by Mahloudji and Chuke (99) and by Stiefel and Todorov (105). Kjellin has recorded the clinical and genetic data on two sets of siblings from different families with recessive FSP, amyotrophy of the hands, oligophrenia and central retinal degeneration (106). Cross and McKusick have described two distinct forms of recessively inherited FSP in an Amish isolate in Ohio (107-108).

X-linked recessive forms of FSP have been reported by Johnson and McKusick (109) and by Thurmon and colleagues (110).

Table 3. Familial Spastic Paraplegias

Disorder	Characteristic
Autosomal dominant, Type I	Onset late teens to early 20s, very slowly progressive
Autosomal dominant, Type II	Onset with walking, very little progression, normal lifespan
Autosomal recessive	"Pure" spastic paraplegia with this inheritance pattern is rare and is usually accompanied by mental retardation, optic signs or other associated symptoms
X-linked spastic paraplegia	Early onset, slow progression, long survival with eventual involvement of cerebellum, cerebral cortex and optic nerves
Familial spastic paraplegia with amyotrophy of hands	Amyotrophy of hands is predominant feature; autosomal dominant
Spastic paraplegia with retinal degeneration	Autosomal recessive forms with associated retinal degeneration
Mast syndrome	Onset of spastic paraplegia in late teens to early 20s with slow progression, basal ganglia symptoms and presenile dementia; autosomal recessive
Troyer syndrome	Early childhood onset of spastic paraplegia with distal muscle wasting; progression with contractures, spasticity and inability to walk by 3-4th decade; autosomal recessive

PERIPHERAL NERVES

Insensitivity and Indifference to Pain

Various disorders, both acquired and inherited, with differing lesions of the central and peripheral nervous system may result in a patient's failure to perceive or react to painful stimuli normally. Although any lesion of the nervous system that interrupts the transmission of painful stimuli from the periphery to the brain will alter pain perception, the group of disorders under discussion is noteworthy in that the lack of a normal appreciation of pain constitutes the sole or major feature of the illness and is responsible for the production of the frequently disabling effects of the disease.

Ogden and his associates (111) have emphasized the important concept of distinguishing between *indifference* and *insensitivity* to pain. Patients with *indifference to pain* have no objective neurologic abnormality; they recognize all types of stimuli at normal thresholds, but fail to evince the usual clinical manifestations associated with pain-producing stimulation. Adequate evidence for a lack of any anatomic abnormality of the sensory neurons should include studies employing quantitative sensory stimuli, the in-vitro study of an excised nerve and histologic studies of number and sizes of unmyelinated fibers as outlined by Dyck and Ohta (112). In contrast, patients with *insensitivity to pain* manifest objective clinical or laboratory evidence of impairment of sensory neuron function either within or distal to the dorsal root ganglion.

A classification of these inherited sensory diseases is modified from the excellent review of Comings and Amromin (Table 4) (113). Excluded from discussion are other diseases of the nervous system known to be accompanied by loss of the appreciation of painful stimuli (i.e., parietal lobe lesions, syringomyelia, mental deficiency, leprosy and hysteria).

Indifference to pain is most common on a sporadic basis or following a pattern of autosomal recessive inheritance. Only two examples of a dominant form of

Table 4. Disorders of Pain Insensitivity and Pain Indifference

Hereditary indifference to pain:
   Autosomal recessive (commonest inherited form)
   Autosomal dominant (rare)
   Sporadic cases (?AR)*

Hereditary insensitivity to pain (hereditary sensory neuropathy):
   Hereditary sensory radicular neuropathy, (AD)
   Hereditary sensory neuropathy, (AR)
   Hyperplastic myelinopathy, (AD)
   Hereditary sensory neuropathy with anhidrosis, (AR)
   Biemond syndrome, (AR)
   Familial dysautonomia, (AR)
   Sporadic cases

*See footnote, Table 1.

the disorder are cited in the literature (113). Although these patients can distinguish between sharp and dull and hot and cold, they do not recognize a painful component to the extremes of sensory stimulation. Corneal anesthesia, repeated fractures and neurotrophic joints are common. They voice no complaints of common forms of pain, such as headache, toothache, abdominal pain or thermal burns. The peripheral axon reflex arc is intact as evidenced by a normal flare response to the intradermal injection of histamine, and microscopic examination of the peripheral nerves is normal. An unspecified "central lesion" of the cerebral cortex is presumed to account for the indifference to pain.

The various forms of insensitivity to pain have been thoroughly reviewed by Dyck and Ohta (112). The salient features of *hereditary sensory radicular neuropathy* (HSN, Type I of Dyck and Ohta) include autosomal dominant inheritance with onset in the 2nd decade of life or later of indolent, perforating ulcers on the plantar aspect of the feet, resulting in bone and soft tissue deformities of the distal portions of the lower extremities. Distal sensory loss and depressed or absent tendon reflexes are found in the lower extremities. Sweating is reduced or lost in the zones of sensory loss. Dyck and Ohta emphasize the preservation of sphincter and sexual functions in these patients as a major point of distinction from dominantly inherited amyloidosis (112).

The autosomal recessive form of hereditary sensory neuropathy (HSN, Type II of Dyck and Ohta) is characterized by onset in infancy or early childhood of mutilating acropathy of the hands and feet due to paronychia and whitlows with resulting ulceration. Unrecognized fractures are common in the hands and feet. All types of cutaneous sensation are lost in the distal portions of the extremities, and tendon reflexes are either depressed or absent in all four limbs.

Comings and Amromin (113) have reported a unique type of congenital insensitivity to pain, which was inherited in an autosomal dominant fashion by five members in three generations of one family. Microscopy of sural nerve biopsies showed thickening and irregular wrinkling of the myelin sheath and a paucity of myelinated nerve fibers. The unmyelinated fibers were unaffected. The authors propose the designation "hyperplastic myelinopathy" to describe this disorder. The two cases reported by Appenzeller and Kornfeld (114) are probably examples of this disorder.

Hereditary sensory neuropathy with anhidrosis (HSN, Type IV of Dyck and Ohta) is

Table 5. Hereditary Polyneuropathies

Disorder	Inheritance*	Age of Onset	Clinical Features	Laboratory Findings	Reference
Peroneal muscular atrophy (Charcot-Marie-Tooth syndrome)	AD AR XR	1st-2nd decade	Weakness and wasting in legs, clubfeet, paresthesias, muscle cramps, later involvement of hands, loss of tendon reflexes in legs, distal diminution of superficial and deep sensation, slowly progressive course	Low conduction velocities of peripheral nerves; axonal atrophy, segmental loss of myelin and onion bulb formations of peripheral nerves; degenerative changes in posterior columns and anterior horn cells of spinal cord	121-124
Roussy-Lévy syndrome	AD	1-2 yrs	Distal weakness in extremities, loss of all deep reflexes, impairment of deep sensation, attitude tremor of hands, very slowly progressive with late appearance of atrophy	Diminished nerve condition velocities; reduction of myelinated axons, loss of myelin and onion bulb formation by Schwann cell proliferation	125-127
Hypertrophic interstitial neuritis (Dejerine-Sottas disease)	AR	Infancy to 3rd decade	Motor development delayed in cases with onset in infancy, progressive weakness and atrophy of muscles of all extremities, loss of reflexes, decreased cutaneous and deep sensation in distal limbs and enlarged nerves	Low conduction velocities of peripheral nerves; loss of myelin in peripheral nerves, Schwann cell hyperplasia with onion bulb formation	128-129
Refsum disease	AR	Early childhood to 3rd decade	Night blindness (pigmentary retinal degeneration), progressive (neural) deafness, distal sensorimotor neuropathy, cerebellar ataxia, cardiomypathy, ichthyosis and skeletal changes	Increased amounts of phytanic acid in body tissues; defect in alpha oxidation of betamethyl fatty acids; elevated protein in cerebrospinal fluid. Schwann cell proliferation with onion bulb formation; lipid granules in Schwann cell cytoplasm	130
Fabry disease	XR	First two decades of life	Tender feet and burning sensations in lower legs, angiokeratoma in umbilical, scrotal, inguinal, and gluteal regions, progressive renal failure, hypertension and cerebrovascular disease	Storage of glycolipid, ceramidetrihexoside in kidney, liver, spleen and blood vessels of skin and nervous system, due to reduced ceramidetrihexosidase activity in tissues	131

Disease	Inheritance	Onset	Clinical features	Pathology	Ref
Tangier disease	AR	Early childhood	Tonsillar hypertrophy with yellow-orange discoloration; hepatosplenomegaly with hypersplenism; 50% of patients have had sensorimotor polyneuropathy involving fine terminal branches of cranial and spinal nerves; resembling mononeuritis multiplex	Severe deficiency of high-density lipoproteins in plasma; storage of cholesterol esters in many tissues; cholesterol ester containing histiocytes within and around terminal nerves; low-serum cholesterol and normal or elevated triglyceride levels	132-133
Hereditary hepatic porphyria:					
Acute intermittent	AD	1-3 decades	Recurrent abdominal pain; organic and psychiatric mental symptoms; seizures; skin photosensitivity in variegate type; cranial and peripheral neuropathy (rare in coproporphyria)	Increased levels of urinary and/or fecal prophyrins or their precursors (varies with type); axonal degeneration of involved nerves	134-138
Variegate	AD	1-3 decades			
Coproporphyria	AD	1-3 decades			
Hereditary amyloidosis:					
Portuguese (Andrade) type	AD	25-35 yrs	Pain and paresthesias of lower limbs; sensory loss; trophic ulcers; impotence; GI symptoms; and areflexia; nephropathy rare. Progressive course with death 5-10 years from onset	Amyloid deposition in blood vessels in virtually all tissues; deposits of amyloid in vessels of peripheral nerves, in perineurium and between nerve fibers; abnormalities found in Schwann cell and myelin sheaths in asymptomatic patients prior to amyloid deposition	139-140
Japanese type	AD	30-47 yrs	Similar to Portuguese type	Similar and perhaps identical to Portuguese type	141
Swedish type	AD	29-63 yrs	Similar to Portuguese type except for later onset; course more benign in one family	—†	142
Indiana (Rukavina) type	AD	5th decade	Carpal tunnel syndrome at onset; progression to generalized polyneuropathy; GI complaints, sphincter disturbances, impotence and foot ulcers rare; duration of disease variable	Amyloid deposits in flexor retinaculum, sheaths of sural nerves and within walls of blood vessels	143

Table 5. (Continued)

Disorder	Inheritance*	Age of Onset	Clinical Features	Laboratory Findings	Reference
Iowa (Van Allen) type	AD	3rd and 4th decades	Chronic progressive sensorimotor neuropathy with greater involvement of lower extremities; nephropathy a constant feature; peptic ulcer and hearing loss common; death from renal failure 10-15 years from onset	Vascular and mesenchymal deposition of amyloid in many organs, including spinal and sympathetic ganglia	144
Finnish (Meretoja) type	AD	3rd decade	Corneal lattice dystrophy; cranial nerve palsies; mild peripheral neuropathy; variable systemic manifestations	Vascular and parenchymal deposition of amyloid in many organs, including cranial and peripheral nerves	145
Hereditary recurrent brachial neuropathy	AD	Most often in 1st decade	Repeated attacks of pain in upper limb girdle followed by paresthesia, weakness and reflex loss; gradual recovery usual, but residual weakness may persist; attacks provoked by pregnancy and infections; hypotelorism, epicanthic folds, facial asymmetry and cleft palate noted in some patients	Nerve conduction times generally normal in distal nerves; EMG evidence of denervation; pathology in two cases showed focal thickenings of myelin sheaths and segmental demyelination	146-149
Hereditary neuropathy with liability to pressure palsies	AD	1st-5th decades	Mononeuropathies or multiple mononeuropathies of peripheral nerves develop following mild traction or compression; frequent, recurring episodes may lead to residual neurologic deficits; asymptomatic relatives may show prolonged conduction times of peripheral nerves	Prolonged nerve conduction times; pathologic findings include focal thickening of myelin sheaths and demyelination similar to that observed in hereditary recurrent brachial neuropathy	150-151
Hereditary polyneuropathy	AD AR	—	Miscellaneous pedigrees of patients with sensorimotor polyneuropathies	—	152-154
Neuropathy in leukodystrophies	AR	—	Discussed in Chapter 5 Genetic Metabolic Diseases	—	155

*AD, autosomal dominate; AR, autosomal recessive; XL, X-linked recessive. †Data unavailable or inapplicable.

characterized by autosomal recessive inheritance, loss of pain sensibility, mental retardation, repeated traumatic and thermal injuries, autonomic dysfunction (Horner's syndrome, poor temperature control, anhidrosis, and vasomotor instability), and occasionally self-mutilating behavior. Examples are reported by Swanson and colleagues (115, 116), Pinsky and DiGeorge (117), Brown and Podosin (118) and Chatrian and associates (119).

Patients with *familial dysautonomia* characteristically show insensitivity to pain. The demonstration of a marked reduction in unmyelinated nerve fibers in this disease qualifies its inclusion as a form of hereditary sensory neuropathy (120).

The two cases of the *Biemond syndrome* have been reviewed by Comings and Amromin (113). Loss of pain sensation is associated with pathologic findings of deficient development of posterior root ganglia, posterior roots, posterior horns of spinal gray matter and posterior columns.

Hereditary Polyneuropathies (Sensorimotor, Mixed Neuropathies)

There are a number of inherited disorders in which involvement of the peripheral nervous system is responsible for virtually all the signs and symptoms of the disease (i.e., peroneal muscular atrophy, hypertrophic neuropathy of Dejerine-Sottas and Roussy-Lévy syndrome). In other diseases dysfunction of the peripheral nerves constitutes only a part of a more widespread disease process (i.e., Refsum's disease, Fabry's disease and porphyria).

Table 5 contains a summary of the inherited disease that involve the peripheral nervous system. Detailed discussions of each disease may be found in the references cited.

REFERENCES

1. Holmes LB: Inborn errors of morphogenesis. *New Engl J Med* 291:763, 1974
2. Holmes LB, Driscoll SG, Atkins L: Etiologic heterogeneity of neural-tube defects. *New Engl J Med* 294:365, 1976
2a. Cowchock FS: Use of alpha-fetoprotein in prenatal diagnosis. *Clin Obstet Gynecol* 19:871, 1976
3. Hsia YE, Bratu M, Herbordt A: Genetics of the Meckel syndrome (Dysencephalia splanchnocystica). *Pediatrics* 48:237, 1971
4. Holmes LB, Nash A, Zurhein GM, et al: X-linked aqueductal stenosis: Clinical and neuropathological findings in two families. *Pediatrics* 51:697, 1973
5. Menkes JH, Phillippart M, Clark DB: Hereditary partial agenesis of corpus callosum. *Arch Neurol* 11:198, 1964
6. Shapira Y, Cohen T: Agenesis of the corpus callosum in two sisters. *J Med Genet* 10:266, 1973
7. DeMyer W, Zeman W, Palmer CG: The face predicts the brain: Diagnostic significance of median facial anomalies for holoprosencephaly (arhinencephaly). *Pediatrics* 34:256, 1964
8. DeMyer W, Zeman W: Alobar holoprosencephaly (arhinencephaly) with median cleft lip and palate: Clinical, electroencephalographic and nosologic considerations. *Confin Neurol* 23:1, 1963
9. Hintz RL, Menking M, Sotos JF: Familial holoprosencephaly with endocrine dysgenesis. *J Pediatr* 72:81, 1968

10. Holmes LB, Driscoll S, Atkins L: Genetic heterogeneity of cebocephaly. *J Med Genet* 11:35, 1974
11. DeMyer W: The median cleft face syndrome. *Neurology* 17:961, 1967
12. O'Connell EJ, Feldt RH, Stickler GB: Head circumference, mental retardation, and growth failure. *Pediatrics* 36:62, 1965
13. Smith DW: Recognizable Patterns of Human Malformation, ed. 2. Philadelphia, Saunders, 1976, p. 504
14. Bergsma D (ed): *Birth Defects. Atlas and Compendium.* Baltimore, Williams & Wilkins, 1972, p. 1006
15. Holmes LB, Moser HW, Halldorsson S, et al: *Mental Retardation.* New York, Macmillan, 1972, p. 430
16. McKusick VA: Mendelian Inheritance in Man, ed. 4. Baltimore, Johns Hopkins University Press, 1975, p. 837
17. Carter CO, Fairbank TJ: The Genetics of Locomotor Disorders. London, Oxford University Press, 1974, p. 170
18. Klein D, Ammann F: The syndrome of Laurence-Moon-Bardet-Biedl and allied diseases in Switzerland. *J Neurol Sci* 9:479, 1969
19. Theile U: Sjögren-Larsson syndrome. *Humangenetik* 22:91, 1974
20. McLennan JE, Gilles FH, Robb RM: Neuropathological correlation in Sjögren-Larsson syndrome. *Brain* 97:693, 1974
21. McKusick VA, et al: The Riley-Day syndrome—observations on genetics and survivorship. *Isr J Med Sci* 3:372, 1967
22. Pearson J, Budzilovich G, Finegold MJ: Sensory, motor, and autonomic dysfunction: The nervous system in familial dysautonomia. *Neurology* 21:486, 1971
23. Weinshilboum RM, Axelrod J: Reduced plasma dopamine-β-hydroxylase activity in familial dysautonomia. *New Engl J Med* 285:938, 1971
24. Alpers BJ: Progressive cerebral degeneration of infancy. *J Nerv Ment Dis* 130:442, 1960
25. Skullerud K, Torvik A, Skaare-Botner L: Progressive degeneration of the cerebral cortex in infancy. *Acta Neuropathol* 24:153, 1973
26. Blackwood W, et al: Diffuse cerebral degeneration in infancy (Alpers' disease). *Arch Dis Childh* 38:193, 1963
27. Katzman R: The prevalence and malignancy of Alzheimer disease. *Arch Neurol* 33:217, 1976
28. Wisniewski HM, Coblentz JM, Terry RD: Pick's disease. *Arch Neurol* 26:97, 1972
29. Sareen CK, Ruvalcaba RHA, Scotvold MJ, et al: Tuberous sclerosis. *Amer J Dis Child* 123:34, 1972
30. Bundey S, Evans K: Tuberous sclerosis: A genetic study. *J Neurol Neurosurg Psychiatr* 32:591, 1969
31. Merritt HH: A Textbook of Neurology, ed. 5. Philadelphia, Lea & Febiger, 1973, p. 309
32. Basic Statistics on the Epilepsies, prepared by Epilepsy Foundation of America. Philadelphia, Davis, 1975, p. 24
33. Refsum S: Genetic aspects of migraine, in Vinken PJ, Bruyn GW (eds): Handbook of Clinical Neurology, Vol. 5. Amsterdam, North Holland, 1968, p. 258
34. Lafora GR, Glueck B: Contributions to the histopathology and pathogenesis of myoclonic-epilepsy. *Bull Gov Hosp Insane* 3:96, 1911
35. Janeway R, et al: Progressive myoclonus epilepsy with Lafora inclusion bodies. *Arch Neurol* 16:565, 1967
36. Schwarz GA, Yanoff, M: Lafora's Disease. *Arch Neurol* 12:172, 1965
37. Seitelberger F, et al: Die Myoklonuskörperkrankheit. *Fortschr Neurol Psychiatr* 32:305, 1964
38. Yanoff M, Schwarz GA: Lafora's disease: A distinct genetically determined form of Unverricht's syndrome. *J Genet Hum* 14:235, 1965
39. Namba M: The microscopic, submicroscopic structure and histochemistry of the inclusion body seen in myoclonus epilepsy. *Bull Yamaguchi Med Schl* 11:103, 1964

40. May DL, White HH: Familial myoclonus, cerebellar ataxia and deafness. *Arch Neurol* 19:331, 1968
41. Duvoisin RC, Yahr MD, Schweitzer MD, et al: Parkinsonism before and since the epidemic of encephalitis lethargica. *Arch Neurol* 9:232, 1963
42. Martin WE, Resch JA, Baker AB: Juvenile parkinsonism. *Arch Neurol* 25:494, 1971
43. Kondo K, Kurland LT, Schull WJ: Parkinson's disease: Genetic analysis and evidence of a multifactorial etiology. *Mayo Clinic Proc* 48:465, 1973
44. Dooling EC, Schoene WC, Richardson EP Jr: Hallervorden-Spatz syndrome. *Arch Neurol* 30:70, 1974
45. Poirier LJ: Recent views on tremors and their treatment, in Williams D (ed): Modern Trends in Neurology, 5th series. London, Butterworths, 1970, p. 80
46. Winkler GF, Young RR: Efficacy of chronic propranolol therapy in action tremors of the familial, senile or essential varieties. *N Engl Med* 290:984, 1974
47. Larsson T, Sjögren T: Essential tremor: A clinical and genetic population study. *Acta Psychiatr Neurol Scand* 36:1, 1960, (Suppl. 144)
48. Bennett JC, Maffly RH, Steinbach HL: The significance of bilateral basal ganglia calcification. *Radiology* 72:368, 1959
49. Babbitt DP, Tang T, Dobbs J, et al: Idiopathic familial cerebrovascular ferrocalcinosis (Fahr's disease) and review of differential diagnosis of intracranial calcification in children. *Am J Roentgen* 105:352, 1969
50. Cockel R, Hill EE, Rushton DI, et al: Familial steatorrhoea with calcification of the basal ganglia and mental retardation. *Q J Med* 42:771, 1973
51. Moskowitz MA, Winickoff RN, Heinz ER: Familial calcification of the basal ganglions. *N Engl J Med* 285:72, 1971
52. Boller F, Boller M, Denes G: Familial palilalia. *Neurology* 23:1117, 1973
53. Bearn AG: Wilson's disease, in Stanbury JB, Wyngaarden JB, Fredrickson DS (eds): The Metabolic Basis of Inherited Disease, 3d ed. New York, McGraw-Hill 1972, p. 1033
54. Marsden CD, Harrison MJG: Idiopathic torsion dystonia. *Brain* 97:793, 1974
55. Eldridge R: The torsion dystonias: Literature review and genetic and clinical studies. *Neurology* 20 (part 2):1, 1970
56. Mount LA, Reback S: Familial paroxysmal choreoathetosis: Preliminary report on a hitherto undescribed clinical syndrome. *Arch Neurol Psychiatr* 44:841, 1940
57. Jung S, Chen K, Brody JA: Paroxysmal choreoathetosis. *Neurology* 23:749, 1973
58. Pincus JH, Chutorian AM: Familial benign chorea. *Trans Am Neurol Assoc* 91:319, 1966
59. Pincus JH, Chutorian AM: Familial benign chorea with intention tremor: A clinical entity. *J Pediatr* 70:724, 1967
60. Haerer AF, Currier RD, Jackson JF: Hereditary nonprogressive chorea of early onset. *N Engl J Med* 276:1220, 1967
61. Chun RWM, Daly RF, Mansheim BJ: Benign familial chorea with onset in childhood. *JAMA* 225:1603, 1973
62. Rosenberg RN, Nyhan WL, Bay C, et al: Autosomal dominant striatonigral degeneration. *Neurology* 26:703, 1976
63. Greenfield JG: The Spino-Cerebellar Degenerations. Springfield, Ill., Thomas, 1954, p. 112
64. Konigsmark BW, Weiner LP: The olivopontocerebellar atrophies: A review. *Medicine* 49:227, 1970
65. Weiner LP, Konigsmark BW: Hereditary disease of the cerebellar parenchyma, *Birth Defects* 6:192, 1971
66. Taniguchi R, Konigsmark BW: Dominant spino-pontine atrophy. *Brain* 94:349, 1971
67. Becker PE, Sabuncu N, Hopf HC: Dominant erblicher Typ von "Cerebellar Ataxie." *Z Neurol* 199:116, 1971
68. Farmer TW, Mustian VM: Vestibulocerebellar ataxia. *Arch Neurol* 8:471, 1963
69. Mahloudji M, Amirhakimi GH, Haghighi P, et al: Marinesco-Sjögren syndrome. *Brain*

95:675, 1972
70. Schwartz JF, Rowland LP, Eder, et al: Bassen-Kornzweig syndrome: Deficiency of serum betalipoprotein. *Arch Neurol* 8:438, 1963
71. Kornzweig AL: Bassen-Kornzweig syndrome. *J Med Genet* 7:271, 1970
72. Sobrevilla LA, Goodman ML, Kane CA: Demyelinating central nervous system disease, macular atrophy and acanthocytosis (Bassen-Kornzweig syndrome). *Am J Med* 37:821, 1964
73. Montpetit VJA, Anderman F, Carpenter S: Subacute necrotizing encephalomyelopathy. *Brain* 94:1, 1971
74. Gordon N, Marsden HB, Lewis DM: Subacute necrotizing encephalopathy in three siblings. *Devel Med Childh Neurol* 16:64, 1974
75. Dayan AD, Ockenden BG, Crome L: Necrotizing encephalomyelopathy of Leigh. *Arch Dis Childh* 45:39, 1970
76. Pincus JH, Cooper JR, Piros K, et al: Specificity of the urine inhibitor test for Leigh's disease. *Neurology* 24:885, 1974
77. Murphy JV, Craig, LJ, Glew, RH: Leigh disease. *Arch Neurol* 31:220, 1974
78. Pitner SE, Edwards, JE, McCormick, WF: Observations on the pathology of the Moebius syndrome. *J Neurol Neurosurg Psychiatr* 28:362, 1965
79. Hanson PA, Rowland LP: Möbius syndrome and facioscapulohumeral dystrophy. *Arch Neurol* 24:31, 1971
80. Hanissian AS, Fuste F, Hayes WT, et al: Möbius syndrome in twins. *Am J Dis Childh* 120:472, 1970
81. Sugarman GI, Stark HH: Möbius syndrome with Poland's anomaly. *J Med Genet* 10:192, 1973
82. Rubinstein AE, Lovelace RE, Behrens MM, et al: Möbius syndrome in association with peripheral neuropathy and Kallmann syndrome. *Arch Neurol* 32:480, 1975
83. Becker-Christensen F, Lund HT: A family with Möbius syndrome. *J Pediatr* 84:115, 1974
84. Gomez MR, Clermont V, Bernstein J: Progressive bulbar paralysis in childhood (Fazio-Londe's disease). *Arch Neurol* 6:317, 1962
85. Alexander MP, Emery ES, Koerner FC: Progressive bulbar paresis in childhood. *Arch Neurol* 33:66, 1976
86. Bobowick AR, Brody JA: Epidemiology of motor-neuron diseases. *N Engl J Med* 288:1047, 1973
87. Finlayson MH, Guberman A, Martin JB: Cerebral lesions in familial amyotrophic lateral sclerosis and dementia. *Acta Neuropathol* 26:237, 1973
88. Takahashi K, Nakamura H, Okada E: Hereditary amyotrophic lateral sclerosis. *Arch Neurol* 27:292, 1972
89. Horton WA, Eldridge R, Bordy JA: Familial motor neuron disease. *Neurology* 26:460, 1976
90. Orthner H, Becker PE, Müller D: Recessiv erbliche Amyotrophische Lateralsklerose mit "Lafora-Körpern." *Arch Psychiatr Nervenkr* 217:387, 1973
91. Emery AEH: The nosology of the spinal muscular atrophies. *J Med Genet* 8:481, 1971
92. Boothby JA, deJesus PV, Rowland LP: Reversible forms of motor neuron disease. *Arch Neurol* 31:18, 1974
93. Marsden CD: Inherited neuronal atrophy and degeneration predominantly of lower motor neurons, in Dyck PJ, Thomas PK, Lambert EH (eds): Peripheral Neuropathy, Vol. 2. Philadelphia, Saunders, 1975, p. 771
94. Paunier L, Pagnamenta F, Monnard E, et al: Spinal muscular atrophy with various clinical manifestations in a family. *Helv Paediatr Acta* 28:19, 1973
95. Matsunaga M, Inokuchi T, Ohnishi A, et al: Oculopharyngeal involvement in familial neurogenic muscular atrophy. *J Neurol Neurosurg Psychiatr* 36:104, 1973
96. Bundey S, Lovelace RE: A clinical and genetic study of chronic proximal spinal muscular atrophy. *Brain* 98:455, 1975
97. Danta G, Hilton RC, Lynch PG: Chronic progressive external ophthalmoplegia. *Brain* 98:473, 1975
98. Roe PF: Hereditary spastic paraplegia. *J Neurol Neurosurg Psychiatr* 26:516, 1963

99. Mahloudji M, Chuke PO: Familial spastic paraplegia with retinal degeneration. *Johns Hopkins Med J* 123:142, 1968
100. Cross HE, McKusick VA: A survey of neurological disorders in a genetic isolate. *Neurology* 17:743, 1967
101. Aagenaes Ø: Hereditary spastic paraplegia. *Acta Psychiatr Neurol Scand* 34:489, 1959
102. Thurmon TF, Walker VA: Two distinct types of autosomal dominant spastic paraplegia. *Birth Defects* 6:216, 1971
103. Silver JR: Familial spastic paraplegia with amyotrophy of the hands. *J Neurol Neurosurg Psychiatr* 29:135, 1966
104. Bell J, Carmichael EA: On the heredity of ataxia and spastic paraplegia, in Treasury of Human Inheritance. London, Cambridge University Press, 1939, 4 (part 3), p. 169
105. Stiefel JW, Todorov AB: Recessive spastic paraplegia with retinal degeneration. *Birth Defects* 16:343, 1974
106. Kjellin K: Familial spastic paraplegia with amyotrophy, oligophrenia, and central retinal degeneration. *Arch Neurol* 1:133, 1959
107. Cross HE, McKusick VA: The Mast syndrome. *Arch Neurol* 16:1, 1967
108. Cross HE, McKusick VA: The Troyer syndrome. *Arch Neurol* 16:473, 1967
109. Johnston AW, McKusick VA: A sex-linked recessive inheritance of spastic paraplegia. *AM J Hum Genet* 14:83, 1962
110. Thurmon TF, Walker BA, Scott CI, et al: Two kindreds with a sex-linked recessive form of spastic paraplegia. *Birth Defects* 6:219, 1971
111. Ogden TE, Robert F, Carmichael EA: Some sensory syndromes in children: Indifference to pain and sensory neuropathy. *J Neurol Neurosurg Psychiatr* 22:267, 1959
112. Dyck PJ, Ohta M: Neuronal atrophy and degeneration predominantly affecting peripheral sensory neurons, in Dyck PJ, Thomas PK, Lambert EH (eds): Peripheral Neuropathy, Vol. 2. Philadelphia, Saunders, 1975, p. 791
113. Comings DE, Amromin GD: Autosomal dominant insensitivity to pain with hyperplastic myelinopathy and autosomal dominant indifference to pain. *Neurology* 24:838, 1974
114. Appenzeller O, Kornfeld M: Indifference to pain. *Arch Neurol* 27:322, 1972
115. Swanson AG: Congenital insensitivity to pain with anhidrosis. *Arch Neurol* 8:299, 1963
116. Swanson AG, Buchan GC, Alvord EC Jr: Anatomic changes in congenital insensitivity to pain. *Arch Neurol* 12:12, 1965
117. Pinsky L, DiGeorge AM: Congenital familial sensory neuropathy with anhidrosis. *J Pediatr* 68:1, 1966
118. Brown JW, Podosin R: A syndrome of the neural crest. *Arch Neurol* 15:294, 1966
119. Chatrian GE, Farrell DF, Canfield RC, et al: Congenital insensitivity to noxious stimuli. *Arch Neurol* 32:141, 1975
120. Aguayo AJ, Nair CPV, Bray GM: Peripheral nerve abnormalities in the Riley-Day syndrome. *Arch Neurol* 24:106, 1971
121. Merritt HH: A Textbook of Neurology, ed. 5. Philadelphia, Lea and Febiger, 1973, p. 514
122. Dyck PJ: Inherited Neuronal Degeneration and Atrophy Affecting Peripheral Motor, Sensory, and Autonomic Neurons, in Dyck PJ, Thomas PK, Lambert EH (eds): Peripheral Neuropathy, Vol. 2. Philadelphia, WB Saunders Co., 1975, p. 825
123. Hughes JT, Brownell, B: Pathology of peroneal muscular atrophy (Charcot-Marie-Tooth disease). *J Neurol Neurosurg Psychiat* 35:648-657, 1972
124. Pratt RTC: *The Genetics of Neurological Disorders*. London, Oxford University Press, 1967, p. 82
125. Oelschlager R, White HH, Schimke RN: Roussy-Lévy syndrome: Report of a kindred and discussion of the nosology. *Acta Neurol Scand* 47:80-90, 1971
126. Lapresle J, Salisachs P: Onion bulbs in a nerve biopsy specimen from an original case of Roussy-Lévy disease. *Arch Neurol* 29:346-348, 1973
127. Kriel RL, Cliffer KD, Berry J, Sung JH, Bland CS: Investigation of a family with hypertophic neuropathy resembling Roussy-Lévy syndrome. *Neurology* 24:801-809, 1974
128. Merritt HH: *A Textbook of Neurology*, ed. 5. Philadelphia, Lea and Febiger, 1973 p. 517

129. Dyck PJ: Inherited Neuronal Degeneration and Atrophy Affecting Peripheral Motor, Sensory, and Autonomic Neurons, in Dyck PJ, Thomas PK, Lambert EH (eds): Peripheral Neuropathy, Vol. 2. Philadelphia, WB Saunders Co., 1975, p. 855

130. Refsum S, Stokke O, Eldjarn L, Fardeau M: Heredopathia Atactica Polyneuritiformis (Refsum's Disease), in Dyck PJ, Thomas PK, Lambert EH (eds): Peripheral Neuropathy, Vol. 2. Philadelphia, WB Saunders Co., 1975, p. 868

131. Brady RO, King FM: Fabry's Disease, in Dyck PJ, Thomas PK, Lambert EH (eds): Peripheral Neuropathy, Vol. 2. Philadelphia, WB Saunders Co., 1975, p. 914

132. Fredrickson DS, Gotto AM, Levy RI: Familial lipoprotein deficiency, in Stanbury JB, Wyngaarden JB, Fredrickson DS (eds): The Metabolic Basis of Inherited Disease, ed. 3. New York, McGraw-Hill, 1972, p. 493

133. Pleasure DE: Abetalipoproteinemia and Tangier disease, in Dyck PJ, Thomas PK, Lambert EH (eds): Peripheral Neuropathy, Vol. 2. Philadelphia, Saunders, 1975, p. 928

134. White HH: Disorders of porphyrin metabolism, in Merritt HH (ed): A Textbook of Neurology, ed. 5. Philadelphia, Lea and Febiger, 1973, p. 675

135. Elder GH, Gray CH, Nicholson DC: The porphyrias: A review. *J Clin Pathol* 25:1013, 1972

136. Ridley A: Porphyric neuropathy, in Dyck PJ, Thomas PK, Lambert EH (eds): Peripheral Neuropathy, Vol. 2. Philadelphia, Saunders, 1975, p. 942

137. Case Records of the Massachusetts General Hospital (Case 41-1975). *N Engl J Med* 293:817, 1975

138. Watson CJ: Hematin and porphyria. *N Engl J Med* 293:605, 1975

139. Cohen AS, Benson MD: Amyloid neuropathy, in Dyck PJ, Thomas PK, Lambert EH (eds): Peripheral Neuropathy, Vol. 2. Philadelphia, Saunders, 1975, p. 1067

140. Carvalho J, Coimbra A, Andrade C: Peripheral nerve fibre changes in asymptomatic children of patients with familial amyloid polyneuropathy. *Brain* 99:1, 1976

141. Araki S, Mawatari S, Ohta M, et al: Polyneuritic amyloidosis in a Japanese family. *Arch Neurol* 18:593, 1968

142. Andersson R: Hereditary amyloidosis with polyneuropathy. *Acta Med Scand* 188:85, 1970

143. Mahloudji M, Teasdall RD, Adamkiewicz JJ, et al: The genetic amyloidoses. *Medicine* 48:1, 1969

144. Van Allen MW, Frohlich JA, Davis JR: Inherited predisposition to generalized amyloidosis. *Neurology* 19:10, 1969

145. Meretoja J: Genetic aspects of familial amyloidosis with corneal lattice dystrophy and cranial neuropathy. *Clin Genet* 4:173, 1973

146. Guillozet N, Mercer RD: Hereditary recurrent brachial neuropathy. *Am J Dis Childh* 125:884, 1973

147. Erikson A: Hereditary syndrome consisting in recurrent attacks resembling brachial plexus neuritis, special facial features, and cleft palate. *Acta Paediatr Scand* 63:885, 1974

148. Geiger LR, Mancall EL, Penn AS, et al: Familial neuralgic amyotrophy—report of three families with review of the literature. *Brain* 97:87, 1974

149. Bradley WG, Madrid R, Thrush DC, et al: Recurrent brachial plexus neuropathy. *Brain* 98:381, 1975

150. Behse F, Buchthal F, Carlsen F, et al: Hereditary neuropathy with liability to pressure palsies. *Brain* 95:777, 1972

151. Görke W: Clinical and electoneuromyographical findings in hereditary neuropathy with liability to pressure palsies. *Neuropaediatrie* 5:358, 1974

152. Mahloudji M: A recessively inherited mixed polyneuropathy of early onset. *J Med Genet* 6:411, 1969

153. Bradley WG, Aguayo A: Hereditary chronic polyneuropathy. *J Neurol Sci* 9:131, 1969

154. Lundberg PO: Hereditary polyneuropathy, oligophrenia, premature menopause and acromicria. *Eur Neurol* 5:84, 1971

155. Bischoff A: Neuropathy in leukodystrophies, in Dyck PJ, Thomas PK, Lambert EH (eds): Peripheral Neuropathy, Vol. 2. Philadelphia, Saunders, 1975, p. 891

14
Heritable Eye Diseases

Irene H. Maumenee
Laird G. Jackson

Generally speaking, there are two groups of patients with hereditary diseases of the eye: the first includes those persons whose genetic problem is limited to the eye, part of the eye or the ocular adnexa. The second includes persons who have a heritable condition involving multiple body systems of which the eye is simply one. Persons in the latter category can be subdivided into those in whom the ocular manifestation is a pleiotropic effect of the mutant gene, e.g., ptosis in the Noonan syndrome, and another group in which eye lesions are a secondary effect, as exemplified by the corneal clouding observed in the various mucopolysaccharide storage diseases. Complex syndromes of which eye disease may be only a part are dealt with elsewhere in this text and the reader should consult the index for these accompanying ocular problems. An exception can be usefully made for the inborn metabolic errors since a large proportion of these diseases have related eye abnormalities.

Table 1 summarizes the ocular findings in the major categories of metabolic disease. More detailed information about other facets of these conditions can be found in the relevant chapters. Tables 2 through 7 describe the ocular findings in some of the more important syndromal conditions. These diseases, too, are discussed elsewhere in this text. The remainder of this discussion then is devoted to the group of patients mentioned initially, namely, those persons whose genetic defect is either limited to or expressed significantly in the eye and ocular adnexae or in whom the eye findings are of major diagnostic importance (1-6).

The cardinal symptoms of a problem in the eye or the visual pathways relate to reduced acuity or visual field loss, or both. Severely reduced visual acuity, if bilateral, is usually detected early, but if the defect is unilateral, it may go undetected for many years. Strabismus or childhood esotropia, even if intermittent, provides a major clinical clue indicative of possible ocular disease. Similarly nystagmus may herald a marked bilateral reduction in visual acuity, and a more

Research for the material in this chapter was supported in part by Grant number 1 RO EY 01773.

Table 1. Ocular Findings in Metabolic Disorders

Disorder	Cornea	Intraocular Pressure	Lens	Retina	Optic Nerve	Comments
Mucopolysaccharidoses:						
Hurler syndrome (MPS 1-H)	Diffuse clouding	Increased	Clear	Retinal degeneration	Frequent papilledema, late optic atrophy	Blindness occurs in small proportion of patients
Scheie syndrome (MPS 1-S)	Diffuse clouding	Frequently open-angle glaucoma fourth to fifth decade	Clear	Retinal degeneration slowly progressive with onset in teens	Papilledema not observed	
Hurler-Scheie compound (MPS 1-HS)	Diffuse clouding	Normal	Clear	Slowly progressive retinal pigmentary degeneration	Papilledema or optic atrophy seen in several patients	Corneal transplantation may be useful
Hunter syndrome (MPS II, severe and mild form)	Clear, even at advanced age	Probably normal	Clear	Retinal degenerative changes	Chronic papilledema and secondary optic atrophy often seen	In severe type, complete blindness may be seen;—only XR type of MPS
Sanfilippo syndromes (MPS III, Types A and B)	Clear	Normal	Clear	Degenerative process	Pallor, probably secondary to retinal degeneration	Blindness usually in teens
Morquio syndromes (MPS IV)	Clouded	Normal	Clear	Normal	Normal	Corneal clouding compatible with normal visual acuity
Maroteaux-Lamy syndromes (Types A and B)	Cloudy, in mild phenotype there is often more prominent peripheral than central clouding	Normal	Clear	Normal	Acute and chronic papilledema observed	
MPS VII (β-glucuronidase deficiency)	Cloudy	Unknown	Clear	Unknown	Unknown	

Mucolipidoses:

Mucolipidosis I	Cloudy	Normal	Cherry-red spot	Late atrophy
Mucolipidosis II	Minimal cloudiness	Normal	Normal	Normal
Mucolipidosis III	Mild ground glass cloudiness from early childhood	Normal	Normal	Frequent pseudo-papilledema or true papilledema	Mild-to-moderate hyperopic astigmatism very common
Mucolipidosis IV	Corneal clouding, primarily central	Normal	ERG changes in childhood; atrophic retinal degenerative changes in teens	Pallor at a later age	Ocular examination may be diagnostic

Sphingolipidoses and related disorders:

GM1, gangliosidosis, Type 1	Occasional mild corneal clouding	Normal	Cherry-red spot in 50% of patients; retinal hemorrhages	Optic atrophy	Decreased visual acuity is common
Sandhoff disease, GM2 gangliosidosis, Type II	Clear	Normal	Cherry-red spot present in 100% of patients after age 6 months	Delayed optic atrophy	Early decreased visual acuity
Fucosidosis (severe and mild forms)	Clear	Normal	Retinal vascular tortuosity	Normal	In mild form conjunctival vessels are dilated with fusiform and saccular microaneurysms; in severe disease no ocular features have been described
Mannosidosis	? Slight opacities	Spoke-shaped cataract in posterior cortex	Normal	Probably normal	
Ceroid lipofuschinosis (infantile late-infantile) and juvenile types)	Clear	Normal	Hypopigmented; macular area hyperpigmented	Atrophic waxy disks	

Table 1. (Continued)

Disorder	Cornea	Intraocular Pressure	Lens	Retina	Optic Nerve	Comments
Farber lipogranulomatosis	Clear	Normal	Normal	Red fovea with gray retina early; later 'peppery' retinopathy	Pallor and optic atrophy	
Lipoid proteinosis (Urbach-Wiethe disease)	Clear	Normal	Normal	Drusen	Normal	Lids with bead-like nodules of hyaline infiltrate
Metachromatic leukodystrophy	Clear	Normal	Normal	Grayness in macula with mild cherry-red spot	Optic atrophy later	
Sea-blue histiocyte syndrome	Clear	Normal	Normal	Doughnut-shaped areas of yellow-white granules surrounding fovea	Normal	
Fabry disease	Cornea verticillata	Normal	Cataract as in mannosidosis	Retinal vascular tortuosity	Normal	Conjunctival vascular dilatations and aneurysms are common
Gaucher disease (Types I, II, III)	Clear	Normal	Normal	Normal	Normal	
Krabbe	Clear	Normal	Clear	No gross changes	Optic atrophy	
Niemann-Pick disease (Types A,B,C)	Normal	Normal	Normal	Cherry-red spot in Type A only	Normal except probably late in Type A	
Ophthalmoplegic lipidosis	Normal	Normal	Normal	Normal	Normal	
Tay-Sachs disease	Clear	Normal	Clear	Macular cherry-red spot present by age 3-4 months	Optic atrophy usually present by age 2 years	Progressive loss of extraocular motor function

Amino acid disorders

Albinism	Normal	Normal	Normal	Striking hypopigmentation with absent foveal differentiation	Normal	Much heterogeneity exists but in tyrosinase positive variety an improvement in visual acuity is seen in some patients; in dominant variety visual acuity is only mildly affected
Alkaptonuria	Corneal pigmentation near limbus	Normal	Normal	Normal	Normal	
Cystinosis	Normal	Normal	Normal	Multiple depigmented and hyperpigmented retinal areas peripherally and ventrally; usually accompanied by sharp visual acuity	Normal	Retinal and corneal lesions are diagnostic
Homocystinuria	Normal	Glaucoma is seen in at least 25% of patients	Progressive dislocation first noticed by age 3 in youngest patient; present in 100% of patients by adolescence	Frequent retinal detachments regardless of age	Normal	
Hyperornithinemia gyrate atrophy	Normal	Normal	Early posterior subcapsular cataracts	Retinal degeneration	Waxy late in course	Deficient ornithine ketoacid transaminase
Sulfite oxidase deficiency	Normal	Normal	Dislocated	Normal	Normal	
Tyrosinemia	Stellate herpes simplex-like corneal lesions and ulcers	Normal	Normal	Normal	Normal	

419

Table 1. (Continued)

Disorder	Cornea	Intraocular Pressure	Lens	Retina	Optic Nerve	Comments
Carbohydrate disorders						
Diabetes mellitus, diabetes insipidus and optic atrophy	Normal	Normal	Normal	Normal	Slowly progressive atrophy	
Galactokinase deficiency	Normal	Normal	Cataracts	Normal	Normal	Cataracts regress on milk-free diet
Galactosemia	Normal	Normal	Cataracts	Normal	Normal	Partial-to-complete reversal of cataracts through galactose-free diet
Lipoprotein disorders:						
Abetalipoproteinemia	Clear	Normal	Normal	Pigment atrophy	Disk pallor late	Retinal degeneration
Lecithin–cholesterol acyltransferase (LCAT) deficiency	Stromal opacities	Normal	Normal	Normal	Normal	
Refsum syndrome	Clear	Normal	Normal	Retinitis pigmentosa	Normal	
Tangier diesease	Corneal clouding	Normal	Normal	Normal	Normal	Occasionally yellow streaking in conjunctiva
Metal or mineral disorders						
Menke disease	Clear	Normal	Normal	Normal	Slightly pale	Seizures with tonic deviation of eyes; brows white, twisted and broken
Pseudohypoparathyriodism pseudopseudohypoparathyroidism	Normal	Normal	Cataracts	Normal	Normal	? Blue sclerae
Wilson disease	Kayser-Fleischer rings	Normal	Sunflower cataract rare finding	Normal	Normal	

Table 2. Ocular Findings in Heritable Connective-Tissue Disease

Syndrome	Disease	Inheritance*
Pseudoxanthoma elasticum	Heterogeneous disorder; angioid streaks in 85% develop in 2d or 3d decade; retinal pigment epithelium often proliferates in or near macula; salmon spots	AR
Ehlers-Danlos	Heterogeneous disorder; type VI with retinal detachment	AR, XR
Marfan	Ectopia lentis	AD
Weill-Marchesani	Acute glaucoma due to lens obstructing outflow tract	AR

*AR, autosomal recessive; AD, autosomal dominant; XR, X-linked recessive.

Table 3. Ocular Findings in Hamartoses

Disorder or Syndrome	Abnormalities	Inheritance*
Nevoid basal-cell carcinoma syndrome	Congenital blindness due to corneal opacity, cataract, glaucoma and/or coloboma of optic nerve; strabismus	AD
Focal dermal hypoplasia (Goltz)	Chorioretinal and iris coloboma most frequent; strabismus, nystagmus, obstruction of tear duct and microphthalmia may occur	XD?
Hereditary hemorrhagic telangiectasia (Osler-Weber-Rendu)	Palpebral conjunctival telangiectasia; rarely retinal involvement	AD
Incontinentia pigmenti	Strabismus, cataract, optic atrophy, retinal detachment and changes similar to those in retrolental fibroplasia are all seen (25%-35% of cases)	XD?
Multiple lentigenes (Leopard)	Ocular hypertelorism and ptosis	AD
Multiple neuroma syndrome (MEN III)	Eyelid margins thickened with pedunculated nodules in 60% of patients; cornea with white medullated nerve fibers visible with slit lamp	AD
Sturge-Weber (S-W)	Choroidal hemangiomas (40%) may cause retinal detachment; glaucoma 30%, of which 60% present at birth; these changes may also be seen in association with Klippel-Trenaunay-Weber (K-T-W) syndrome	Neither S-W nor K-T-W syndrome appears to be genetic
Tuberous sclerosis	About 50% of patients with unilateral or bilateral retinal tumor (phakoma)	AD

*See footnote, Table 2.

Table 4. Ocular Changes in Skeletal Dysplasias

Disorder or Syndrome	Abnormalities	Inheritance*
Arthro-ophthalmopathy (Stickler syndrome)	High myopia, retinal detachments, cataract	AD
Bloom syndrome (dwarfism, telangiectasia)	Conjunctivitis, conjunctival telangiectasia, drusen at posterior pole	AR
Chondrodysplasia punctata	Cataracts (20% in AD, 70% in AR type); poor retinal ganglion cell development in severe, AR type	AD, AR
Cleidocranial dysplasia	Orbital height > width, mild exophthalmos	AD
Cockayne syndrome	Enophthalmos; poor pupil dilatation to drugs; cataracts; salt and pepper retinopathy with extinguished ERG; hyperopia usual; corneal dystrophy occasionally	AR
Cornelia de Lange syndrome	Long, curly eyelashes; synophrys of eyebrows	Sporadic
Coffin-Lowry syndrome	Prominent supraorbital ridges; hypertelorism; antimongoloid slant, thickened upper lids	XD
Craniometaphyseal dysplasia	Hypertelorism; narrowing of optic nerve canals from hyperostoses	AD
Hallerman-Streiff syndrome	Microphthalmos and congenital cataracts; spontaneous rupture of lens is frequent with late iridocyclitis, glaucoma and retinal detachment following	Sporadic
Kniest syndrome	Myopia and retinal detachment	AD (primarily new mutation)
Noonan syndrome	Ptosis; epicanthus, antimongoloid slant	AD
Oculodentodigital syndrome	Microcornea; shallow orbit; small palpebral fissures; juvenile glaucoma	AD
Osteogenesis imperfecta	Blue sclerae; prominent Schwalbe's line; posterior embryotoxon	AD
Rubenstein-Taybi syndrome	Antimongoloid slant; epicanthus; strabismus	Sporadic
Seckel dwarfism	Antimongoloid slant; epicanthus; glaucoma; chorioretinopathy; strabismus	AR
Spondyloepiphyseal dysplasia congenita	Myopia in 50% of patients	AD
Weill-Marchesani syndrome	Ectopia lentis; glaucoma 2° to obstruction	AR
Werner syndrome	Juvenile cataracts	AR

*See footnote, Table 2.

Table 5. Ocular Findings in Craniofacial Dysplasias

Disorder or Syndrome	Abnormalities	Inheritance*
Acrocephalopolysyndactyly (Carpenter)	Dystopia canthorum; microcornea; corneal opacity; minimal optic atrophy	AR
Blepharophimosis syndrome (Marden-Walker)	Strabismus; blepharophimosis	AR
Blepharochalasis and double lip (Ascher syndrome)	Blepharachalasis; atrophy and drooping of lid follow edema	AD
Crouzon syndrome	Proptosis 2° to shallow orbits; divergent strabismus; nystagmus, 80% may have optic nerve involvement	AD
Cryptophthalmos	Microphthalmia with skin-covered globes (cryptophthalmos); absent brows	AR
Hemifacial microsomia (heterogeneous)	Epibulbar dermoid; coloboma of superior lid; Microphthalmia occasionally	Sporadic
Pfeiffer syndrome	Hypertelorism	AD
Cerebrohepatorenal syndrome	Ocular hypertelorism; cataracts; epicanthal folds	AR
Lenz microphthalmia syndrome	Microphthalmia; iris and choroidal coloboma	XR?

*See footnote, Table 2.

Table 6. Ocular Findings in Skin Disorders

Disorder or Syndrome	Abnormalities	Inheritance*
Acrodermatitis enteropathica	Blepharitis; conjunctivitis; photophobia some with corneal opacities	AR
Chédiak-Higashi	Pink eye; photophobia; nystagmus as in ocular albinism	AR
Dyskeratosis congenita	Keratinization of conjuctiva; obstruction of lacrimal puncta	?AR, XR
Epidermolysis bullosa (EB)	Only recessive dysplastic type (EB dystrophica) has conjunctival scarring; corneal involvement	AR
Ichthyosis	Ichthyosis of lid; corneal opacities; peripheral granular retinal pigmentation	AD, AR, XR
Keratosis follicularis spinulosa	Corneal opacities	XD
Keratosis palmaris et plantaris with corneal dystrophy (Richner-Hanhart)	Corneal haze or herpetiform epithelial corneal lesions	AR
Congenital Poikiloderma (Rothmund-Thomson)	Cataract beginning at first few weeks of life; occasional corneal dystrophy	AR
Sjögren-Larsson syndrome	Retinitis pigmentosa; Normal ERG	AR
Xeroderma pigmentosum	Photophobia; lacrimation; hyperemia of conjunctiva	AR

*See footnote, Table 2.

Table 7. Ocular Manifestations in Chromosomal Disorders

Ocular Defect	Disorder*
Palpebral fissures slanted:	
upward/outward	del 9p; tri 21; 48, XXXX; 49, XXXXX; 49, XXXXY
downward/outward	tri 10p; r (21); 45, X
Hypertelorism	del 4p; del 5p; tri 12p; del 13; 49, XXXXY
Hypotelorism	tri 5p; tri 6p; tri 14, r (22)
Epicanthus	del 4p; del 5p; tri 12p; del 13; tri 21; r (22) 45, X; 48XXXX; del 10p
Exophthalmos	del 4p; del 9p
Deep-set eyes	tri 9p; tri 9q; tri 9; tri 15; del 18q
Ptosis	del 4p; del 18p; r (22)
Blepharophymosis	tri 6p
Brushfield spots	tri 21
Small eyes	tri 10q
Minor ocular anomalies	del 4p; tri 4q; del 13; del 18q; 45,X
Severe microphthalmia/ anophthalmia	tri 13
Retinoblastoma	del 13
Coloboma	tri 13; tri 22; tri 13q; tri 22q; del 4p

*Del, deletion; tri, trisomy.

extensive evaluation is indicated. Any child with a white reflex from one or both eyes should be immediately referred to an opthalmologist, since the finding may signify diseases as varied as cataracts, persistent primary hyperplastic vitreous, retrolental fibroplasia, toxocara canis cyst, other intraocular inflammation, retained foreign body, retinal detachment or even retinoblastoma.

Photophobia in a child may be due to simple conjunctivitis, but may also be caused by a host of abnormalities in the refracting media, such as the corneal crystalline deposits in cystinosis, increased intraocular pressure and distortion of the normal collagen structure in congenital or acquired glaucoma, by cataracts or by retinal degenerative processes, such as albinism or the cone dystrophies. A painful eye may be related to infection or trauma, but it may also be a sign of congenital or acquired glaucoma or associated with certain forms of hereditary corneal dystrophy. A small eye may be seen with abnormal brain development. All of these findings are general clues to primary ocular lesions. In the following sections more specific defects will be considered, using an—admittedly arbitrary—anatomic approach.

DISORDERS OF ADNEXAE

Isolated *colobomas* of the lids are rarely heritable, but coloboma of the lower lid is a feature of mandibulofacial dysostosis (7). *Epicanthus* is frequently seen in neonates and infants, but disappears later in childhood. Persistent epicanthus may be transmitted as a dominant trait, but it is commoner as a component of several syndromes, especially those associated with chromosomal anomalies. *Blepharophimosis* (short palpebral fissure) may be transmitted as an isolated dominant trait, accompany more complex eye malformations, such as anophthalmia or microphthalmia, or be a part of a condition like the Waardenburg syndrome (see Chapter 19, "Genetic Disorders of the Skin"). *Distichiasis*, or double rows of eyelashes, occurs as a dominant trait alone (8) or in combination with pubertal lymphedema (9) also transmitted as a dominant. *Tristichiasis*, or three rows of lashes, is also seen as a rare dominant (10).

Several varieties of *congenital ptosis* occur, most of which may occur in hereditary patterns. Simple ptosis (blepharoptosis) is fairly common and may occur unilaterally or bilaterally with variable severity, but usually begins in childhood. The defect is thought to be a localized form of muscle dystrophy caused by an incompletely penetrant autosomal dominant gene. It occurs in both unilateral and bilateral forms within the same family, and it is usually nonprogressive (11). There is a later-onset variety of ptosis in which progression occurs, and it is also inherited as an autosomal dominant disorder.

Of course ptosis may accompany any form of muscular dystrophy, particularly the oculopharyngeal type and may also be due to neuropathic ophthalmoplegia. Ptosis with jaw-winking (Marcus-Gunn phenomenon) is due to interconnecting neural fibers between the nucleus of the levator muscle and the trigeminal motor branch (12). It usually occurs as a unilateral phenomenon and is transmitted as an autosomal dominant trait. Ptosis may also occur as part of a host of complex syndromes and as such is usually a minor consideration.

The origin of *congenital ophthalmoplegias* is probably myopathic rather than neuropathic. The disorders are poorly understood, and the pedigree patterns are not always clear. Total involvement of cranial nerves III, IV and VI has not been recorded although partial involvement has been seen in a family, suggesting recessive transmission (13). Internal ophthalmoplegia is rare, and X-linked inheritance is suggested. The commonest variety is external ophthalmoplegia, which is usually incomplete with considerable intrafamilial variation although the levator palpebrae muscles are regularly involved, and ptosis is a cardinal sign. Autosomal dominance is the usual inheritance pattern although one X-linked family is reported. Paresis of the external rectus muscle alone can be transmitted as an autosomal dominant disorder. When it occurs with retraction of the bulb, it is known as Duane's phenomenon, also an autosomal dominant trait, but with considerable variability. Sporadic cases are common.

So-called adult-onset ophthalmoplegia actually may commence from late childhood to middle-age and usually occurs as a dominant trait. It is progressive and ends in complete extraocular muscle paralysis. Ptosis is commonly associated, and the disorder may be due to a heritable abnormality in the motor end plate.

Congenital squint (strabismus) may be secondary to several congenital structural abnormalities, but it also occurs as a primary disorder. In the latter cases most of the evidence favors a multifactorial inheritance pattern, whether the squint is convergent or divergent. Occasional families show what borders on a dominant pattern and obviously must be counseled on this basis.

Congenital nystagmus is usually a secondary phenomenon. It may also occur, however, as a distinct affliction, and X-linked recessive, autosomal dominant and (rarely) autosomal recessive pedigrees are recorded (14). Unusual latent or myoclonic forms are also reported; they are rare and probably neurogenic, and no genetic implications are established.

Facial cleft syndromes, of which there are several, may extend to the *lacrimal apparatus* and lead to dyskeratosis or corneal ulceration. Heavy desquamation of skin in the area of the lacrimal puncta may result in obstruction and degeneration of the tear ducts, as is seen in some ichthyosis syndromes. By a similar mechanism carcinomas in the same area may injure the ductular apparatus, and such trauma may be a facet of xeroderma pigmentosum or the basal-cell nevus syndrome. One multiple congenital anomaly syndrome involving lacrimal ducts, ears, teeth and hands is reported (15).

DEVELOPMENTAL EYE DISORDERS

Colobomas result from failure of closure of the optic cleft at various times during embryogenesis in any layer of the eye. Typical colobomas of the iris and uvea are usually found in the lower part where the optic cleft actually occurs and are usually dominant (2). Atypical colobomas are those occurring elsewhere in the iris, and they are considered to be partial manifestations of aniridia. (Various forms of aniridia are listed in Table 10.) Colobomas of the iris or of the optic nerve are both associated with choroidal coloboma or microphthalmia as dominant traits, although occasional recessive or X-linked pedigrees are reported.

Coloboma frequently leads to microphthalmia or even anophthalmia at the end of embryonic development and is heritable (16). Pure *microphthalmia,* in contrast, is due to reduction in eyeball volume without a developmental segmentation fault. Microphthalmus is generally recognized rather by simple clinical evaluation than by accurate measurement and is frequently part of a more complex condition (Table 8) (17-22).

REFRACTIVE ERRORS AND CORNEAL ABNORMALITIES

Study of the genetic mechanisms of the refractive components of vision shows that they generally follow a continuous distribution curve compatible with mul-

Table 8. Microphthalmia

Condition	Clinical Features	Inheritance*
Colobomatus microphthalmia or anophthalmia	Colobomatous defect may be obscured either by degenerative changes or by extreme microphthalmos simulating anophthalmos	AD, AR
Pure microphthalmia or anophthalmia	Severe defects, ranging from rudimentary eye to anophthalmos	AD, AR
Complicated microphthalmia	Several forms associated with other malformations	
with Corneal opacities, cataract; vitreous abnormality and myopia	Mental retardation; microcephaly; kyphoscoliosis; dental, urogenital and cardiovascular anomalies	XR
with Congenital cataract	Various forms in different families	AD
with retinitis pigmentosa and glaucoma	Full syndrome not always present	AD
with high hypermetropia	Poor vision from defective foveal development	AR
Cryptophthalmos	Lid closure; unusual hair pattern; deafness; skeletal and urogenital malformations	AR

*See footnote, Table 2.

tifactorial inheritance. Abnormal axial length produces refractive error at either end of the normal scale with resultant hypermetropia or myopia. In most instances there is no clear evidence for a single gene determinant, although there are prominent exceptions, especially when there are associated ocular problems, e.g., myopia with night-blindness occurs in both an X-linked and autosomal recessive form and myopia with late retinal detachment may be inherited as an autosomal dominant trait.

The normal cornea is an avascular, transparent structure shaped like a watch glass. It has a refractive power between 35 and 50 diopters, which is heritably controlled by what appears to be a multifactorial system. Primary microcornea may have no effect on refraction although glaucoma is said to be a complication in 20% of cases. In some instances microcornea may segregate as an autosomal dominant trait, alone or in association with cataract (23). Megalocornea usually occurs as an X-linked trait, but may occasionally be dominantly inherited (24). It must be distinguished from buphthalmos. Keratoconus occurs in syndromes like trisomy 21, but may also be seen alone as an autosomal dominant or recessive trait with onset usually delayed until after puberty (25, 26). In cornea plana, the cornea is small and flat and has poor refractive power. The condition is most often recessively inherited (27).

Corneal opacification is usually dystrophic. There are several primary corneal dystrophies that are generally hereditary, bilateral and either stationary or slowly progressive. Although they may be evident at birth, it is commoner for them to appear later in life with a similar age of onset within a given family. Symptoms of corneal dystrophy include haziness, decreased visual acuity, blurring, dazzling or halo formation due to opacification and crystalline deposition or discoloration of the cornea itself. The cornea is divided into five layers from the anterior to the posterior surface, including epithelium, Bowman's membrane, stroma, Descemet's membrane and endothelium. Most classification schemes use this anatomic separation of pathologic involvement, and Table 9 is modified from (26) and (28) using this topographic approach.

DISORDERS OF IRIS, ANGLE AND ANTERIOR CHAMBER

There are a large number of iris and pupillary abnormalities most of which have little or no pathologic significance. Some of them, however, are inherited and most of them with significance are forms of aniridia, the anterior chamber cleavage syndrome or are pigmentary anomalies. (Tables 10 and 11)

Aniridia, or absence of the iris, is usually an incomplete phenomenon occurring bilaterally, and remnants of the iris may be found on careful examination. It is thought to be due to an embryonic error at the 11-12 week stage when ectodermal layers of the iris and fovea are laid down. Because of this developmental timing there is a frequent association with lens opacities, macular hypoplasia, ciliary process hypoplasia, nystagmus and photophobia. Many cases are apparently due to an autosomal dominant gene with considerable variation within families in clinical expression (29). The well-known association with Wilms' tumor occurs only in the sporadic cases.

Pigmentary anomalies are perhaps better characterized as depigmented conditions. Absence of iris color occurs in both ocular and oculocutaneous albinism

Table 9. Corneal Dystrophies

Condition (Syndrome)	Clinical Features	Inheritance*
Epithelium and Bowman's membrane:		
Dystrophic recurrent erosion	Early onset with frequent relapses and subsidence after age 50; diminished corneal sensation	AD
Juvenile epithelial dystrophy (Meesman)	May be slightly progressive but rarely leads to ocular symptoms	AD
Ring-shaped dystrophy (Reis-Bucklers)	Childhood onset with acute "attacks" simulating neuroparalytic keratitis; ring-like epithelial and subepithelial lesions become confluent later with gross effect on vision; reduced corneal sensation	AD
Band-shaped keratopathy	Frequency secondary to other primary ocular disease; white or gray horizontal band of calcific material is seen on cornea; inherited form infrequent	AR
Dermochondrocorneal dystrophy	Central white irregular corneal patches associated with abnormal ossification of hands and feet and xanthomata, very rare	?AR
Anterior membrane dystrophy	May be same as Reis-Bucklers	AD
Honeycomb dystrophy	Early onset nonprogressive course of recurrent erosion and honeycomb-like opacities of subepithelial layer	AD
Anterior crocodile-shagreen dystrophy	Rare; Probably related to band keratopathy	AD
Stromal forms:		
Dominant granular dystrophy (Groenouw I)	Onset age 5 with small white central dots which increase in size and stud central cornea until visible to naked eye at age 50; may radiate as lines; usually asymptomatic	AD
Recessive macular dystrophy (Groenouw II)	Onset first decade with superficial veil, progressive haziness and increasing confluent opacities; progressive opacification with decreased corneal sensitivity and attacks of photophobia and lacrimation; severe and relatively early blindness	AR
Dominant lattice dystrophy	Onset second decade or earlier with cobweb lines in lattice pattern which increase in number and thickness; acute attacks frequent; visual effect less severe than in macular type but legal blindness occurs by age 40-60	AD
Congenital hereditary dystrophy	Both dominant and recessive pedigrees with no difference between types; fairly stationary disorder	AD, AR
Disciform crystalline dystrophy (Schnyder)	Early onset with slow progression of axial needle-like crystals in superficial stromal layers; steady decline in vision	AD
Marginal crystalline dystrophy	One family with fine dot opacities in margin of cornea; vision unaffected	AR?
Speckled dystrophy	Early onset with slow, asymptomatic progress of small dots in stroma	AD

Table 9. (Continued)

Condition (Syndrome)	Clinical Features	Inheritance*
Central cloudy dystrophy	Defect confined to stroma in posterior aspect, one family reported	AR?
Deep parenchymatous dystrophy	Unusual punctate dystrophy seen in one family	AR?
Progressive stromal dystrophy	Early onset of irritative episodes with snowflake type opacities which increase in number with age	AD
Descemet's membrane and endothelial forms:		
Cornea guttata	Early onset, sometimes congenital with late involvement of parenchyma; may be sporadic	?AD
Polymorphous posterior endothelial dystrophy	Vesicles and polymorphic opacities at posterior limiting membrane	AD
Congenital endothelial dystrophy (Maumenee)	Both dominant and recessive pedigrees reported; vision usually affected early in recessive	AD, AR
Late endothelial dystrophy (Fuchs)	Onset after age 50 with endothelial changes first then progressive epithelial changes, many sporadic cases	AD
Annular endothelial dystrophy	Benign, nonprogressive dystrophy of peripheral posterior cornea; vision unaffected	AD

*See footnote, Table 2.

Table 10. Hereditary Iris Anomalies

Condition (Syndrome)	Clinical Features	Inheritance*
Structural:		
Hyperplasia	Mesodermal (appears as doubling) or ectodermal (flocculi iridis)	AD
Hypoplasia	Ectodermal hypoplasia probably part of aniridia complex; mesodermal also occurs	AD
Aniridia:		
Type I	Aniridia, nystagmus; 2° glaucoma; hypoplastic fovea	AD
Type II	Aniridia with preserved visual acuity	AD
Type III	Aniridia with mental retardation	AR
Type IV	Aniridia and Wilms' tumor	Sporadic
Persistent pupillary membrane	Part of anterior cleavage complex or isolated	AD
Pigmentary:		
Piebald pattern (partial heterochromia)	Usually sporadic but may be heritable	AD
Heterochromia	Isolated or as part of Waardenburg syndrome	AD
Heterochromia with Horner syndrome	Several families; combination is reported as sporadic event with neuroblastoma	AD
Pupillary:		
Ectopic pupils	Usually seen with ectopia lentis (AR), but may be isolated	AD
Anisocoria	Several families	AD

*See footnote, Table 2.

Table 11. Anterior Chamber Cleavage Malformations

Condition (Syndrome)	Clinical Features	Inheritance*
Posterior embryotoxon	Prominent Schwalbe ring; no other symptoms	Sporadic or occasional AD
Axenfeld syndrome	Prominent Schwalbe ring; iris strands between Schwalbe ring and trabeculum, few patients have glaucoma	AD
Rieger syndrome	Prominent Schwalbe ring; iris strands between Schwalbe ring and trabeculum with hypoplastic anterior iris stroma; midfacial hypoplasia, hypodontia omphalolcoele	AD

*See footnote, Table 2.

(see Chapter 19, "Genetic Disorders of the Skin"). The former is an X-linked trait, whereas the latter is an autosomal recessive one. Heterochromia irides may occur as a simple dominant condition and is usually unilateral. When associated with cataract and corneal opacities, it may occur as a dominant trait (see section on "Fuchs' syndrome") although the evidence for this finding is unconvincing. The well-known association of partial albinism, heterochromia irides and deafness constitutes the Waardenburg syndrome (see Chapter 19).

Pupillary abnormalities as isolated findings are uncommon. Congenital anisocoria is reputed to occur as an "irregular dominant condition," but is more likely multifactorial. Ectopic pupils alone are reported in dominant pedigrees, but usually this is associated with ectopic lenses as a recessive disorder (30, 31).

The *anterior chamber cleavage anomalies* were defined as such by Reese and Ellsworth (32) and subsequently classified as in Table 11 (33). Although they are referred to as separate disorders, they share many features, overlap in their clinical presentation and even occur together in a single family (34). These characteristics suggest that they are variable manifestations of a single gene.

GLAUCOMA

Primary glaucoma can be divided into three types, including open-angle, angle-closure and congenital. Each of them may be, to some extent, inherited. In addition, glaucoma may be part of certain syndromes (see Tables 1, 2-4 and 8).

Open-angle glaucoma is characterized by typical visual field defects and anterior chamber angles that are open by gonioscopy. It is estimated that between 13 and 47% of open-angle glaucoma is inherited or familial (35, 36). Since glaucoma usually develops with age, the age-related manifestations compound the difficulties of monitoring families for possible inherited disease. Most case figures ascertained at a single point in time tend to underestimate the prevalence of the disease. The heredity of open-angle glaucoma is unclear and disputed by two sets of studies.

Becker believes that it is an autosomal recessive trait that correlates with the gene for intraocular pressure response to chronic topical corticosteroid application (37). Armaly suggests multifactorial causation due to the combined effect of several quantitatively inherited parameters, such as intraocular pressure, outflow facility and horizontal cup:disk ratio (38). Both of these investigators have apparently demonstrated a monogenic control of the intraocular pressure response to steroids with approximately 5% high responders, 35% intermediate and 60% low responders in the population, representing the two homozygotes and heterozygotes for a single gene pair (39, 40). However, recent twin studies have found no difference in frequency of steriod response in mono- and dizygotic twins, so perhaps the system is more complex than is readily apparent (41). Finally open-angle glaucoma is frequently associated with trabecular pigmentary changes and also with myopia.

For screening purposes, several factors should increase the suspicion of primary open-angle glaucoma, including familial history; high intraocular pressure or low outflow facility; high intraocular pressure response to topical corticosteroids; large, unequal or vertically elongated optic nerve cupping; overt or latent diabetes mellitus; high myopia; Krukenberg's spindles; or central vein occlusion. Within families periodic screening for elevated intraocular pressure is suggested.

Primary angle-closure glaucoma occurs when the peripheral iris bows forward against the trabecular meshwork, blocking the outflow from the anterior chamber (42). Diagnosis rests on demonstrations of a closed anterior-chamber angle in the absence of precipitating cause. Intermittent closure usually occurs, giving rise to episodic pain, congestion, blurred vision and halos. Hypermetropia is frequent as are small eyes, small anterior chambers or small corneas. A common denominator may be a disproportion between size of lens and eyeball (43). As these factors are anatomic and quantitatively determined, it is likely that inheritance of angle-closure glaucoma is multifactorial. Racial differences occur, with angle-closure glaucoma being unusually chronic and insidious in blacks (44) and common in orientals (45).

Primary congenital glaucoma (buphthalmos) is a rare disorder with only the outflow blockade being present at birth, and photophobia, corneal enlargement, edema and optic atrophy following in several weeks to months. Inheritance is probably autosomal recessive for most cases (46).

LENS DEFECTS

Congenital subluxation of the lens is reported in families with a dominant pedigree pattern, but these reports came before careful studies for subtle manifestations of the Marfan syndrome were done. It is likely that most of these families represent persons with lens manifestations of a larger syndrome. The dominantly inherited Marfan syndrome and the recessively inherited Weill-Marchesani syndrome both include ectopic lens (47). Two disorders of sulfur amino acid metabolism, homocystinuria and sulfite oxidase deficiency, both inherited as autosomal recessives, also feature dislocated lens (see Chapter 5, "Genetic Metabolic Disorders").

Both dislocated lenses and spherophakia are seen in the rare inborn error of metabolism, hyperlysinemia (see Chapter 5), a finding not surprising in view of the importance of lysine and its metabolities in collagen cross-linkage. Ectopic lens with ectopic pupils occurs as a recessive disorder.

Cataracts assume a confusing array of shapes and forms (Table 12). Total (diffuse) congenital cataracts are observed in dominant, recessive and X-linked recessive patterns (48). Total nuclear cataracts affect only the embryonic nucleus and are seen in dominant patterns with less than complete penetrance as well as in an X-linked family pattern (49). Anterior polar cataracts occur as capsular cataracts of small opacities attached to the anterior lens pole, usually not interfering seriously with vision. Both autosomal recessive and dominant families are reported (50).

Posterior polar cataracts are associated with persistence of part or all of the hyaloid artery and consist of circular opacities limited to the posterior capsule (51). Vision is usually not affected, and transmission is as an autosomal dominant pattern. Suture cataracts are usually static deposits along the lines of the anterior or posterior Y sutures of the fetal nucleus (52, 53). They occur in the Down syndrome and may also occur as a dominant trait. Lamellar (zonular) cataracts consist of fine white dots in a single layer of the lens and probably represent the most frequent form of congenital cataract. They may occur as a dominant trait or be caused by environmental factors.

Cataract may of course be associated with diverse syndromes, some of which are noted in Table 13.

Table 12. Isolated Cataract

Type	Clinical Features*	Inheritance†
Total congenital	Apparently rare	AD, AR
Zonular, (lamellar)	May be due to poor maternal nutrition (hypocalcemia)	AD
Nuclear	Common severe form with dense or patterned opacity, AR, XR families may occur	AD
Anterior polar	Wide variability in visual effect; unilateral and some sporadic bilateral cases	AD
Posterior polar	Wide variability	AD
Suture (stellate, coralliform)	Clinically minimal	AD
Crystalline, coralliform	Fine crystals in axial region of lens	AD
Crystalline, acuneiform (frosted)	Similar to coralliform type	AD
Floriform	Rare form; reduced vision	AD
Pulvurulent zonular (Coppock; diskoid)	Form with dust-like or granular opacities	AD
Nuclear, diffuse nonprogressive	Opacity limited to fetal nucleus; nonprogressive	AD
Congenital; total with microcornea	Carriers show posterior Y-suture opacities and reduced corneal diameter	XR
Coronary cataract	First noticed at puberty; confined to periphery	Sporadic or multifactorial
Presenile cataract of adult nucleus	Nuclear opacities	AD
Presenile posterior cortical cataract		Uncertain

*NB—There tends to be overlap between clinical manifestations of types.
†See footnote, Table 2.

Table 13. Congenital Cataract in Systemic Disorders and Syndromes

Condition or Disorder	Abnormalities	Inheritance*
Metabolic	Galactosemia	AR
	Lowe syndrome	XR
	Alport syndrome	AD
Skin	Rothmund-Thomson syndrome	AR
	Werner syndrome	AR
	Congenital icthyosis	AR
	Incontinentia pigmenti	XD
Neurologic	Sjörgren syndrome	AR
	Marinesco-Sjögren syndrome	AR
	Sjögren-Larsson syndrome	AR
	Myotonic dystrophy	AD
	Cockayne syndrome	AR
Skeletal	Chondrodysplasia punctata	AR, AD
	Mandibulo-facial dysotosis	AD
	Stickler (hereditary arthro-ophthalmopathy)	AD

*See footnote, Table 2.

VITREOUS ABNORMALITIES

Inherited disorders of the vitreous commonly involve the retina as well. Frequently the central retina is most directly involved, but detachment of the entire retina is not infrequent. Retinal detachment alone is usually a secondary phenomenon seen most frequently in association with myopia, sickle cell disease and the Ehlers-Danlos syndrome. There are five—and recently perhaps six—syndromes in which retinal detachment is associated with hereditary vitreoretinal dysplasia.

Congenital retinoschisis is an X-linked recessive disorder in which developmental splitting of the retina into an inner and outer layer occurs congenitally (54, 55). Although not usually noticed until after age two months, it is probably present at birth. Vitreous hemorrhage and the frequent appearance of a translucent membrane in the vitreous cavity are characteristic (vitreous veil). Diagnosis is often delayed until the child reaches school age and is reflected as poor visual acuity in class or strabismus, or both. There is a bilateral defect with ballooning of the breaks, which develops most commonly in the inner retinal layer. The macula is usually involved, and the retina may become fully detached. Progression is usually slow after childhood, and regression with collapse of the bullous areas and improvement of the visual acuity may actually occur.

A similar disorder, *familial foveal retinoschisis,* is reported in three young women who are the offspring of nonconsanguinous normal parents (56). The girls manifested mild visual loss associated with bilateral foveal dystrophy, which resembled the macular involvement in X-linked retinoschisis. Electro- and psychophysiologic tests showed less severe involvement than the latter. The disorder is assumed to be an autosomal recessive trait.

Wagner's disease is a vitreoretinal degeneration following an autosomal dominant inheritance pattern (57, 58). Impairment of vision occurs by the third decade with lens opacities and choroidal and retinal atrophy. The vitreous is described as empty optically and devoid of normal structure. Cataracts, retinal detachment and end-stage absolute glaucoma finally occur.

Favre syndrome is an extensive hyaloideoretinal (vitreous) degeneration associated with peripheral retinal degeneration (59). It is transmitted as an autosomal recessive trait. A characteristic finding is abolishment of the b-wave in the electroretinograph, together with early hemeralopia.

Familial exudative vitreoretinopathy (falciform detachment) mimics retrolental fibroplasia, but a history of oxygen exposure in the neonatal period is not present (60). There are vitreous detachments, membranes, retinal exudates, breaks and degeneration with slow progressive changes. From the few reports it appears to follow an autosomal recessive pattern.

Facial clefting syndrome is an autosomal dominant condition of midfacial clefting with myopia, radial perivascular lattice degeneration of the retina and retinal detachment (61). Similar changes are described in the variable autosomal dominant condition called the Stickler syndrome (62) as well as other syndromes with bone dysplasia. Ophthalmologists refer to the vitreoretinal changes in all as the Wagner syndrome.

Pseudoglioma is, properly speaking, a lesion resembling the retinal tumor, retinoblastoma (see Chapter 3, "Genetics and Cancer"). In 1961, Warburg called attention to Norrie's disease, which occurs as the result of a highly pleiotropic X-linked recessive gene (63). The disorder, present at birth, is frequently not recognized immediately. The first sign is usually a retrolental mass in the male child. The ciliary processes are elongated, and the lens becomes opaque early in childhood, the eye then beginning to shrink and corneal opacities appearing. Retinal detachment is frequent. About 25% of afflicted boys have mental retardation and psychosis and seizures or electroencephalograph changes. Hearing impairment also occurs in about one-fourth of them. Unfortunately no methods exist for either carrier discrimination or prenatal diagnosis so that genetic counseling is based on the X-linked family pedigree pattern alone.

RETINA AND CHOROID

Congenital anomalies occur in the retinal and choroidal areas and may be either stationary or progressive. Among the former are the various colobomas, opaque nerve fibers, asymptomatic macular defect and the various forms of congenital stationary night blindness. In the progressive category are the forms of retinal aplasia, macular cysts and the sex-linked disorders of pseudoglioma (Norrie disease) and retinoschisis. Table 14 lists these disorders, together with their mode of inheritance and clinical features.

Congenital retinal aplasia or congenital amaurosis as described by Leber (64) probably accounts for 10-20% of congenitally blind children (2). The fundus is apparently normal at birth, but later one sees pigmentary changes with narrowing of the vessels and optic atrophy. Recessive inheritance is assumed to be due to high consanguinity, but reports of normal offspring from the mating of two affected parents suggest heterogeneity in the disorder (65).

Abiotropic abnormalities of the fundus are the common disorders of retinitis pigmentosa and the various macular dystrophies (66-88). They are apparently

Table 14. Retinal Abnormalities

Condition (Syndrome)	Clinical Features	Inheritance*
Stationary defects:		
Macular coloboma (MC)	Predominantly inflammatory; moderately pigmented type; probably often heritable	AD
MC with brachydactyly	One family reported with hypoplasia of distal two phalanges ("B" type brachydactyly)	AD
MC with brachydactyly and cleft palate	Milder skeletal changes	AR
Opaque nerve fibers	Bilateral type probably heritable; unilateral type is sporadic	AD
Asymptomatic macular defect	Resembles macular dystrophy but is congenital without loss of central vision	AD
Night blindness (NB); congenital stationary (hemeralopia)	French family Nougaret reported with large number affected in dominant pedigree; fundi normal	AD
NB with myopia	X-linked type most frequent; recessive type with more severe myopia	XR, AR
NB with abnormal fundus color (Oguchi disease)	Grey reflex on fundal exam which returns to normal after one hour in dark (Mizuo sign). Almost confined to Japanese	AR
Progressive defects:		
macular degeneration, juvenile (Stargardt)	Bilateral and usually symmetrical; defect in pigment epithelium or neuroepithelium and seen in fovea with loss of central vision	AR
Macular degeneration, senile	Uncertain entity as fundus change occurs very late (70), making study of families difficult	AD
Macular degeneration, vitelliform (Best)	Bilateral round elevated abnormality of macula resembles yolk of egg, giving name to entity; visual function does not necessarily correlate with ophthalmoscopic picture; EOG may be diagnostic and should be used to investigate other "at risk" family members	AD
Macular dystrophy with aminoaciduria	One family described; relationship to aminoaciduria unclear	AD
Macular dystrophy: butterfly fovea	Changes begin in childhood with metamorphopsia and photophobia common; visual acuity preserved	AD
Macular dystrophy	X-linked variety apparently distinct from others	XR
Fundus flavimaculatus	Recessive form same as Stargardt's; Dominant families have, uniform distribution of many yellow spots over fundus; visual acuity preserved	AD, AR
Cone dystrophy	Color vision affected early; ERG and EOG affected and distinctive; Bull's eye macular lesion common	
Congenital amaurosis of Leber	Probably much heterogeneity; British use term "retinal aplasia" and have described dominant family	AD, AR
Retinal aplasia with renal dysplasia	Congenital amaurosis, vasopressin-resistant diabetes insipidus with progressive azotemia	AR

Table 14. (Continued)

Condition (Syndrome)	Clinical Features	Inheritance*
Retinitis pigmentosa	Occurs in all three simply inherited forms; 3-4% dominant, 5% X-linked and remainder recessive (figures approximate)	AD, AR, XR
Usher syndrome	Retinitis pigmentosa and congenital deafness may be two heterogeneous forms, one less severe than the other	AR
Doyne honeycomb degeneration of retina	Small round white spots in posterior pole become mosaic or "honeycomb" in pattern as they progress; drusen occur and condition reported as drusen of Bruch's membrane is identical	AD
Retinal degeneration with spastic paraplegia	Onset in 4th decade; rare entity; must be distinguished from olivopontinecerebellar atrophy with retinal degeneration, a dominant; similar clinical picture to OPCA with retinal dystrophy is seen in recessive pedigree, suggesting heterogeneity	AR
Lattice degeneration of retina with detachment	Observed in familial pattern in nonmyopic persons	AD
Retinal detachment	Isolated phenomenon occurring in several reported families	AD
Retinal detachment and occipital encephalocele	One family with 5 of 10 sibs with vitreoretinal degeneration, retinal detachment and encephalocele	AR
Retinal dystrophy; reticular pigmentary of posterior pole	Network of black pigmented lines in posterior pole of retina; drusen late	AR
Choroidal sclerosis	Many reports of sibs, several from consanguineous families; choriocapillaris atrophy seen throughout posterior eyeground	AR, XR
Choroideremia	X-linked disorder with progressive loss of vision (central); night blindness; constricted fields beginning at early age; choroid and retina atrophy; female heterozygotes asymptomatic but often show irregular retinal pigmentation	XR
Choroidoretinal degeneration with retinal reflex in heterozygous female	Resembles choroideremia, but female has tapetal-like retinal reflex	XR
Choroidoretinal dystrophy	Resembles retinitis pigmentosa more than choroideremia, one family reported	XR
Van den Bosch syndrome	Choroideremia with mental deficiency, acrokeratosis verruciformis, anhidrosis and skeletal deformity	XR

*See footnote, Table 2.

the result of progressive wear-and-tear insults to congenitally and gene-determined hypersusceptibility of the target tissue. Both of the types just mentioned are characterized by pigmentary disturbance, with the macular lesions showing mottling and atrophy instead of the bone corpuscle reaction seen in classic retinitis pigmentosa. Other forms are gyrate atrophy with predominate atrophy and exposure of white sclera; choroideremia with extensive atrophy and occasional bone corpuscle pigment; and Doyne's choroiditis with pigment proliferation, whitish exudates and possibly large colloid bodies. Especially in the macular dystrophies, there is a bewildering array of categories, and some confusion results. Table 14 lists the various abiotrophic disorders of retina and choroid.

Color Vision

Almost 100% of women and more than 92% of men have trichromatic vision wherein color is matched by the three primary colors. Dichromatic vision must use two primary lights to match any other light. Of color-vision deficient persons, about 8% of males have trichromatic vision with different coefficient curves and therefore difference in detail of trichromatic vision (anomalous trichromatism). About 2.5% of males are dichromats. Monochromats are rare and include both the male and female subjects (3).

Monochromatism apparently occurs as a rare autosomal recessive disorder in which photophobia, low visual acuity and nystagmus occur. Cone vision is apparently lacking, and the range of severity is considerable.

Dichromatism and anomalous trichromatism occur as X-linked disorders with occasional manifesting female heterozygotes. Protanopia and deuteranopia are the common forms of dichromatism and represent the effect of separate genes. Tritanopia is an unusual disorder not revealed by the usual charts and inherited as an autosomal dominant trait. Protanomaly and deuteranomaly are the trichromatic versions of the dichromatic disorders just mentioned and are inherited in X-linked fashion also. In families in which both disorders have occurred and a female patient has received two mutant X-chromosomes, the "anomalous" version is dominant to the "-opia" form. Deuteranomaly and deuteranopia are thus alleles as are protanomaly and protanopia, with the normal allele being dominant to the less severe anomaly, which in turn is dominant to the severer anomaly. Tritanomaly also occurs, but is insignificant and little studied.

OPTIC NERVE PROBLEMS

Optic colobomas are frequently found in association with other ocular defects, but may occur as isolated phenomena and do so in either AD or AR patterns. Elevated colloid or hyaline bodies in the optic disk area, termed *drusen*, are also reported as an autosomal dominant disorder. Juvenile optic atrophy occurs as an autosomal dominant disorder. It is usually noticed later in childhood and produces severe visual impairment with nystagmus, pale optic disks and attenuation of the vasculature. The electroretinographic results are normal, and this finding distinguishes coloboma from tapetoretinal degeneration.

Infantile optic atrophy in association with mild mental deficiency, spasticity, hypertonia and ataxia is termed Behr's syndrome and is inherited in an autosomal recessive manner (89). Onset is in childhood with variable progression and then stabilization. Disk pallor usually is temporal, and nystagmus and strabismus are frequently present. Infantile optic atrophy also is associated with diabetes mellitus, diabetes insipidus and deafness in a unique autosomal recessive disorder (see Chapter 9, "Genetics of Endocrine Diseases") (90).

Leber's optic atrophy was originally described in the nineteenth century (91) and probably represents a heterogeneous group of disorders. It occurs as a sudden loss of central vision in adult life in the 2nd or 3rd decade with a hereditary transmission that is apparently X-linked. Males predominate, and there is no male-to-male transmission, but heterozygous female subjects are frequently affected and may be severely involved. Following the initial severe loss, there may be stabilization of visual impairment.

REFERENCES

1. Waardenburg PJ, Franceschetti A, Klein D (eds): Genetics and Ophthalmology. Springfield, Ill., Thomas, 1961
2. Francois J: Heredity in Ophthalmology. St. Louis, Mosby, 1971
3. Sorsby A: Ophthalmic Genetics, 2d ed. New York, Appleton-Century-Crofts, 1970
4. Goldberg MF (ed): Genetic and Metabolic Eye Disease. Boston, Little-Brown, 1974
5. Murphree AL, Maumenee IH: The Eye in Genetic Diseases: An Atlas. St. Louis, Mosby, 1979
6. Duke-Elder S: System of Ophthalmology. St. Louis, C. V. Mosby, 1963
7. Franceschetti A, Klein D: Mandibulo-facial dysostosis: New hereditary syndrome. *Acta Ophthalmol* (Kbh) 27:143, 1949
8. Fox SA: Distichiasis. *Am J Ophthalmol* 53:14, 1962
9. Jester HG: Lymphedema-distichiasis. *Hum Genet* 39:173, 1977
10. Danforth CH: Studies on hair, with special reference to hypertrichosis. *Arch Dermatol Syph* 11:494, 1925
11. Briggs HH: Hereditary congenital ptosis with report of 64 cases conforming to the Mendelian rule of dominance. *Am J Ophthalmol* 2:408, 1919
12. Falls HF, Kruse WT, Cotterman CW: Three cases of Marcus Gunn phenomenon in two generations. *Am J Ophthalmol* 32:53, 1949
13. Holt WF, Nachtigaller H: Anomalies of ocular motor nerves: Neuroanatomic correlates of paradoxial innervation in Duane's syndrome and related congenital ocular motor disorders. *Am J Ophthalmol* 60:443, 1965
14. Harcourt B: Hereditary nystagmus in early childhood. *J Med Genet* 7:253, 1970
15. Hollister DW, Klein SH, DeJaeger NJ, et al: The lacrimo-auriculo-dento-digital syndrome *J Pediatr* 83:438, 1973
16. Hoefnagel D, Keenan ME, Allen FH Jr: Heredofamilial bilateral anophthalmia, *Arch Ophthalmol* 69:760, 1963
17. Francois J, Haustrate-Gosset MF: Le conseil genetique dans la micropthalmie et l'anophtalmie. *J Genet Hum* 24 (supp):35, 1976
18. Goldberg MF, McKusick VA: X-linked colobomatous microphthalmos and congenital anomalies, a disorder resembling Lenz's dysmorphogenetic syndrome. *Am J Ophthalmol* 71:1128, 1971

19. Capella JA, Kaufman HE, Lill FJ, et al: Hereditary cataracts and microphthalmia. *Am J Ophthalmol* 56:454, 1963
20. Hermann P: Le syndrome: Microphtalmie-retinite pigmentaire-glaucome. *Arch Ophthalmol* 18:17, 1958
21. Franceschetti A, Gernet H: Diagnostic ultrasonique d'une microphtalmie sans microcornee, avec macrophakie, haute hypermetropie associee d'une degenerescene tapeto-retinienne, une disposition glaucomateuse et des anomalies dentaires (nouveau syndrome familiare). *Arch Ophtalmol* (Paris) 25:105, 1965
22. Friedman MW, Weight ES: Hereditary microcornea and cataract in five generations, *Am J Ophthalmol* 35:1017, 1952
23. Gronhol V: Ueber die Vererbung der Megalokornea nebst einem Beitrag zur Frage des genetischen Zusammenhanges zwischen Megalokornea und Hydrophthalmus. *Klin Mbl Augenheilk* 67:1, 1921
24. Riddell WJB: Uncomplicated hereditary megalocornea. *Ann Eugen* 11:102, 1941
25. Falls HF, Allen AW: Dominantly inherited keratoconus. Report of a family. *J Genet Hum* 17:317, 1969
26. Goldberg MF: A review of selected inherited corneal dystrophies associated with systemic diseases. *Birth Defects* 7:13, 1971
27. Forsius H: Studien über Cornea plana congenita bei 19 Kranken in 9 Familien. *Acta Ophthalmol* (Kbh) 39:203, 1961
28. Maumenee AE: An introduction to corneal dystrophies. *Birth Defects* 7:3, 1971
29. Shaw MW, Falls HF, Neel JV: Congenital aniridia. *Am J Hum Genet* 12:389, 1960
29a. Elsas, TJ, Maumenee JH, Kenyon KR, Yoder T: Familial aniridia with preserved ocular function. *Am J Ophthalmol* 83:718, 1977.
30. Diethelm W: Über Ectopia Lentis ohne Arachnodaktylie und ihre Beziehungen zur Ectopia Lentis et Pupillae. *Ophtalmologia* 114:16, 1947
31. Franceschetti A: Ectopia lentis et pupillae congenita als rezessives Erbleiden und ihre Manifestierung durch Konsanguinität. *Klin Mbl Augenheilk* 78:351, 1927
32. Reese A, Ellsworth R: The anterior chamber cleavage syndrome. *Arch Ophthalmol* 75:307, 1966
33. Waring G, Rodrigues M, Laibson P: Anterior chamber cleavage syndrome. A stepladder classification. *Surv Ophthalmol* 20:3, 1975
34. Fitch N, Kabak M: The Axenfeld syndrome and the Rieger syndrome. *J Med Genet* 15:30, 1978
35. Francois J: Genetics and primary open-angle glaucoma. *Am J Ophthalmol* 61:652, 1966
36. Dodinval P, Prijot E, Weekers R: L'heredite du glaucome a angle ouvert. *J Med Genet* 7:244, 1970
37. Becker B: The genetic problem of chronic simple glaucoma, in Solanes MP (ed): International Congress of Ophthalmology. Proceedings of the XXI Congress, Mexio, 1970. Amsterdam: Excerpta Medica, 1971, p. 286
38. Armaly MF: The genetic problem of chronic simple glaucoma, in Solanes MP (ed): International Congress of Ophthalmology Proceedings of the XXI Congress, Mexico, 1970. Amsterdam: Excerpta Medica, 1971, p. 278
39. Becker B: Intraocular pressure response to topical corticosteroids. *Invest Opthalmol* 4:198, 1965
40. Armaly MF: Topical dexamethasone and intraocular pressure, in Leydhecker W (ed): International Symposium on Glaucoma. Basel, Karger 1967, p. 73.
41. Schwartz JT, Reuling FH, Feinleib M, et al: Twin heritability study of the corticosteroid response. *Trans Am Acad Ophthalmol Otolaryngol* 77:126, 1973
42. Miller SJH: Genetics of closed angle glaucoma. *J Med Genet* 7:250, 1970
43. Lowe RF: Aetiology of the anatomical basis for primary angle-closure glaucoma: Biometrical comparisons between normal eyes and eyes with primary angle-closure glaucoma. *Br J Ophthalmol* 54:161, 1972
44. Alper MG, Laubach JL: Primary angle-closure glaucoma in the American Negro. *Arch Ophthalmol* 79:663, 1968

45. Mann I: Culture, Race, Climate and Eye Disease: An Introduction to the Study of Geographical Ophthalmology. Springfield, Ill., Thomas, 1966
46. Delmarcelle Y: Considerations sur l'heredite du glaucome infantile, *J Genet Hum* 6:33, 1956
47. McKusick VA: Heritable Disorders of Connective Tissue, 4th ed. St. Louis, Mosby, 1972
48. Merin S, Crawford JS: Etiology of congenital cataracts: A survey of 386 cases. *Can J Ophthalmol* 6:178, 1971
49. Fraccaro M, Morone GM, Manfredini U, et al: X-linked cataract. *Ann Hum Genet* 31:45, 1967
50. Fraser GR, Friedmann AI: The Causes of Blindness in Childhood: A Study of 776 Children with Severe Visual Handicaps. Baltimore, Johns Hopkins Press, 1967
51. Tulloh CG: Heredity of posterior polar cataract with report of a pedigree. *Br J Ophthalmol* 39:37, 1955
52. Jordan M: Stammbaumuntersuchungen bei cataracta stellata coralliformis. *Klin Mbl Augenheilk* 126:469, 1955
53. Gifford SR, Puntenney I: Coralliform cataract and a new form of congenital cataract with crystals in the lens. *Arch Ophthalmol* 17:885, 1937
54. Kraushar MF, Schepens CL, Kaplan JA, et al: Congenital retinoschisis, in Bellows JG (ed): Contemporary Ophthalmology, Honoring Sir Steward Duke-Elder. Baltimore, Williams & Wilkins, 1972
55. Manschot WA: Pathology of hereditary juvenile retinoschisis. *Arch Ophthalmol* 88:131, 1972
56. Lewis RA, Lee GB, Martonyi CL, et al: Familial foveal retinoschisis. *Arch Ophthalmol* 95:1190, 1977
57. Jansen LM: Degeneratio hyaloideo-retinalis hereditaria. *Ophthalmologica* 144:458, 1962
58. Wagner, H.: Einhisher unbekhanntis Erbleiden des Auges (Degeneratio hyaloideoretinalis hereditaria), beobachtet im Kanton Zurich. *Klin Mbl Augerheilk* 100:840, 1938.
59. Francois J: The role of heredity in retinal detachment, in McPherson A (ed): New and Controversial Aspects of Retinal Detachment. New York, Harper and Row, 1968
60. Criswick VG, Schepens CL: Familial exudative vitreoretinopathy. *Am J Ophthalmol* 68:578, 1969
61. Knobloch WH, Layer JML: Clefting syndromes associated with retinal detachment. *Am J Ophthalmol* 73:517, 1972
62. Stickler GN, Belau B, Farrel PG: Hereditary progressive arthro-ophthalmopathy. *Mayo Clin Proc* 40:433, 1965
63. Warburg M: Norrie's disease, a congenital progressive oculo-acustico-cerebral degeneration. *Acta Ophthalmol* (Kbh) 89:1, 1961
64. Leber T: Über Retinitis pigmentosa und angeborene Amaurose. *Arch Ophthalmol* 15:1, 1869
65. Franceschetti A: Heredity of hydrophthalmia. *J Pediatr Ophthalmol* 2:35, 1965
66. Deutman AF: The Hereditary Dystrophies of the Posterior Pole of the Eye. Assen, The Netherlands, Royal Van Gorcum, 1971
67. Klien BA: Some aspects of classification and differential diagnosis of senile macular degeneration. *Am J Ophthalmol* 58:927, 1964
68. Krill AE, Morse PA, Potts AM, et al: Hereditary vitelliruptive macular degeneration. *Am J Ophthalmol* 61:405, 1966
69. Lefler WH, Wadsworth JAC, Sidbury JB Jr: Hereditary macular dystrophy and aminoaciduria. *Am J Ophthalmol* 71:224, 197
70. Deutman AF, Van Blommestein A, Henkes HE, et al: Butterfly-shaped pigment dystrophy of the fovea. *Arch Ophthalmol* 83:558, 1970
71. Braley AE: Dystrophy of the macula. *Am J Ophthalmol* 61:1, 1966
72. Klien BA, Krill AE: Fundus flavimaculatus: Clinical, functional and histopathologic observations. *Am J Ophthalmol* 64:3, 1967
73. Krill AE, Deutman AF, Fishman M: The cone degenerations. *Doc Ophthalmol* 35:1, 1973
74. Gillespie FD: Congenital amaurosis of Leber. *Am J Ophthalmol* 61:874, 1966
75. Schimke RN: Hereditary renal-retinal dysplasia. *Ann Int Med* 70:735, 1969

76. Reese AB, Straatsma BR: Retinal dysplasia. *Am J Ophthalmol* 45:199, 1958
77. Klein, D, Franceschetti, A., Hussels I, Race RR and Sander R: X-linked retinitis pigmentosa and linkage studies with the Xg blood group. *Lancet* 1:974, 1967
78. Heck AF: Presumptive X-linked intermediate transmission of retinal degenerations: Variations and coincidental occurrence with ataxia in a large family. *Arch Ophthalmol* 70:143, 1968
79. Usher CH: Bowman Lecture on a few hereditary eye affections. *Trans Ophthalmol Soc* UK 55:164, 1935
80. Deutman AF, Jansen LM: Dominantly inherited drusen of Bruch's membrane. *Br J Ophthalmol* 54:373, 1970
81. Mahloudji M, Chuke PO: Familial spastic paraplegia with retinal degeneration. *Johns Hopkins Med J* 123:142, 1968
82. Everett WG: Study of a family with lattice degeneration and retinal detachment. *Am J Ophthalmol* 65:229, 1968
83. Knobloch WH, Layer JM: Retinal detachment and encephalocele. *J Pediatr Ophthalmol* 8:181, 1971
84. Deutman AF, Rumke AM: Reticular dystrophy of the retinal pigment epithelium: Dystrophia reticularis laminae pigmentosae retinae of H. Sjögren. *Arch Ophthalmol* 82:4, 1969
85. Francois J: Sex-linked chorioretinal heredodegenerations. *Birth Defects* 7:99, 1971
86. Krill AE, Archer D: Classification of the choroidal atrophies. *Am J Ophthalmol* 72:562, 1971
87. Hoare GW: Choroido-retinal dystrophy. *Br J Ophthalmol* 49:449, 1965
88. Van den Bosch J: A new syndrome in three generations of a Dutch family. *Ophthalmologica* 137:422, 1959
89. Behr C: Die komplizierte, hereditär-familiäre Optikusatrophie des Kindesalters. *Klin Monatsbl Augenheilkd* 47:138, 1909
90. Rorsman G, Söderström N: Optic atrophy and juvenile diabetes mellitus with familial occurrence. *Acta Med Scand* 182:419, 1967
91. Leber T: Ueber hereditäre und congenitalangelegte Sehnervenleiden. Albrecht von Graefes. *Arch Ophthalmol* 17:259, 1971

15
Hereditary Hearing Loss

Frederick R. Bieber
Walter E. Nance

Hearing impairment is currently one of the most widespread chronic disabilities in the United States. Over 14 million persons suffer sufficient hearing impairment to interfere with their ability to understand conversational speech and to affect their capacity to function in both social and vocational settings. About 5½ million of these persons are over 65 years of age (1). Hearing problems in children not only interfere with their ability to communicate with others, but can have profound and often irrevocable effects on their linguistic and psychological development, causing serious academic problems if the hearing loss is not identified early and managed appropriately.

Deafness has many causes. It may be genetic, either congenital or postnatal, or it may be acquired as a result of trauma or environmental effects in the pre-, peri- or postnatal periods. Toxic drugs, viral infections, otitis media, meningitis, prematurity, erythroblastosis fetalis and maternal rubella are among the major recognized environmental causes of hearing loss (2).

More than 70 types of inherited hearing loss are described, which differ in their pattern of inheritance, audiologic characteristics, age of onset, clinical course and associated abnormalities (3-6). This heterogeneity should not be surprising when one considers the complexity of the hearing organ. The interaction of hundreds of genes must be involved in its normal development, and consequently defects in any one of many genes can give rise to genetically distinct forms of hearing loss that, when viewed superficially, may appear to be homogeneous.

This is paper #25 from the Department of Human Genetics at the Medical College of Virginia, and was supported in part by USPHS Grant PO1 HD-10291-01 and by an A. D. Williams Fellowship awarded to F.R.B.

THE EAR: STRUCTURAL DEVELOPMENT

In higher animals hearing provides both sound perception and spatial orientation. Whereas one ear alone permits the reception of sound, two enable the hearer to localize its direction and facilitate the discrimination of meaningful signals in a noisy background. In the adult the ear forms one anatomic unit functioning as an organ of both hearing and balance. In the embryo it develops from three distinct parts. At about 22 days' gestation the developing ear primordium can first be seen as thickenings of the surface ectoderm, the otic placodes. These placodes invaginate to form otic vesicles, which later divide into a ventral portion—forming the saccule and cochlear duct—and a dorsal part—forming the utricle and semicircular canals. The inner ear reaches its full adult size and form by the end of the fourth fetal month. Since the cochlear end organ is the last of the labyrinthine structures to develop, it is more liable to developmental anomalies than is the vestibular system.

The middle ear, or tympanic cavity, and the eustachian tube are derived from the first pharyngeal pouch, an outpocketing of the pharynx. This pouch, of endodermal origin, appears in the embryo at about four weeks' gestation. The malleus and the incus are derived from cartilage of the first pharyngeal arch, and the crus of the stapes from the second arch.

The auricle develops from the fusion of mesenchymal swellings or hillocks surrounding the first pharyngeal cleft. The external auditory canal arises from inward growth of the first pharyngeal cleft. The tympanic membrane consists of an ectodermal epithelium at the base of the auditory meatus, an endodermal lining in the tympanic cavity and intermediate connective tissue.

INCIDENCE OF HEARING LOSS

In general, there are two types of hearing loss. Conductive deafness is a result of a block in sound transmission up to and including the stapedovestibular joint. Sensorineural deafness can result from a cochlear lesion (sensory) or from a lesion affecting the peripheral pathway or central projection of cranial nerve VIII (neural). In many cases lesions of both types contribute to the hearing loss. From both diagnostic and therapeutic standpoints, it is important to determine whether the patient suffers from conductive or sensorineural deafness, or both, and to ascertain the degree and pattern of the hearing loss. Standard texts should be consulted for various testing procedures (7).

Brown (8) has summarized the data on the prevalence of deafness in different populations, which ranges from a low of 45 per 100,000 in Northern Ireland and Denmark to a high of 160 per 100,000 among Chicago elementary school children. However, the incidence of deafness in the United States has fluctuated markedly with time, corresponding to rubella epidemics (2), and this factor may account for much of the reported variation in prevalence.

It is generally agreed that genetic factors account for about 40-60% of childhood deafness. Fraser (9) concluded that 49.9% of 2,330 deaf school children in the British Isles had genetic deafness. Two-thirds of the genetic cases were thought to be autosomal recessive inheritance, 31% autosomal dominant and about 3% X-linked. In a survey of deaf-mutism in Northern Ireland, Stevenson and Cheeseman concluded that the 70% of probands who were born deaf had inherited deafness (10). Chung, Robison and Morton (11) reanalyzed Stevenson and Cheeseman's data, using maximum likelihood methods and estimated that only 9% of these congenitally deaf probands were actually sporadic, and that 74% of the genetic cases were autosomal recessive, 24% autosomal dominant and 2% X-linked recessive types.

Rose (12) analyzed family data from three deaf populations in the United States and derived estimates of the proportion of sporadic deafness and of autosomal dominant and recessive deafness. Among deaf sibships recorded by E. A. Fay (13), Rose estimated that 45.1% of cases were sporadic, 6.6% dominant and 48.3% recessive. Additional analysis showed that at least 10 loci contributed to recessive deafness in this population.

In a National Survey of 35,285 deaf children enrolled in special educational programs in the United States during 1969–70, Rose found 49.3% of deafness to be sporadic, 7.6% dominant and 43.1% recessive. Finally in a survey of 1,021 deaf students enrolled at Gallaudet College during 1973–74, she found a lower proportion of sporadic cases (23.5%), reflecting a smaller environmental component and a higher percentage (17.1%) of dominant cases, reflecting a larger proportion of students with deaf parents; 59.6% were autosomal recessive.

INHERITED DEAFNESS

On the following pages we will review the characteristics and patterns of inheritance of the commoner forms of severe genetic deafness and syndromes in which profound hearing loss is a major component. Although new forms of hereditary deafness continue to be reported, it becomes increasingly difficult to differentiate between new entities and phenotypic variants of recognized syndromes. Konigsmark and Gorlin's recent compilation of the recognized forms of hereditary deafness and genetic entities involving hearing loss (5) should be consulted for an encyclopedic tabulation.

Hearing Loss and External Ear Malformations

Several pedigrees are reported in which external ear malformations are transmitted as an inherited trait in association with hearing loss (14-19). These abnormalities can extend from total anotia to large prominent auricles. Frequently there are normal auricles with branchial fistulas or preauricular pits and tags, or both, and the malformations may be unilateral, bilateral or asymmetric. Minor ear malformations occur in about 35 per 10,000 live births and are often sporadic. However, pinna anomalies often occur in conjunction with congenital conductive hearing loss.

Whenever an individual shows a family history of ear malformations, one should be particularly alert to the possibility of an associated clinically significant hearing loss. Rowley (14) reported a family with mandibular hypoplasia, earpits and branchial fistulas in two of seven affected family members. Mixed hearing loss was present in four family members, whereas a fifth had a conductive deficit.

Surgical exploration in one subject revealed a malformed incus and an immobile and presumably congenitally fixed stapes. Melnik et al (20) found renal anomalies and aplasia of the lacrimal ducts in patients with this symptom complex.

Whether families with associated renal abnormalities represent allelic or nonallelic variants is at present uncertain. Studies of the temporal bone of a single infant with this syndrome have recently been published. Fitch et al (21) found gross bilateral abnormalities in the form and relationship of middle ear spaces, the middle cranial fossa and the inner ear. The ossicular chain and joint surfaces were normal. The cochlear cavity was small, the modiolus underdeveloped and there was a decrease in the number of cochlear neurons in the basal and apical regions. The stria vascularis was partly deformed and atrophic in the apical coil.

The occurrence of inner, middle and external ear anomalies in this syndrome may indicate a disruption of a common morphogenic pathway, since both middle and external ears originate from the branchial apparatus, whereas the inner ear arises from the otic vesicle. Altman noted that in one-third of patients with moderate middle ear malformations, malformations of the inner ear occur as well (22).

A dominant syndrome characterized by congenital conductive hearing loss as a result of absence of the incudostapedial joint, together with hypertrophic earlobes, has been described in two families (23, 24). Tympanotomy on three patients revealed a normal malleus, absent stapedial head and a curved hook-like crus of the incus. Insertion of a prosthesis resulted in hearing improvement.

An autosomal recessive syndrome of malformed low-set ears and conductive hearing loss was discovered in six children of two related Mennonite sibships (25). Four of the six affected children had a 70–80 dB conductive hearing loss in one or both ears. The pinnae were small and low-set, with thick helices. Of four affected children who were administered psychological tests, three were retarded. Systolic heart murmurs were noted in five of the six affected children. Hypogonadism was present in three of four affected males, two of whom also had cryptorchidism. Tympanotomy on one patient revealed a malformed malleus and absent incus and stapes. The common embryologic origin of the malleus and the short process of the incus, as well as of the tragus and portions of the helix, may explain the association of middle and external ear anomalies in this syndrome.

A disorder characterized by congenital conductive deafness and microtia with atresia of the external auditory meatus is reported in three sibships (26-28). Two sibs in one family had absent auricles and a small dimple in place of the external meatus. In a second family two sibs had unilateral microtia, their left ears represented by small ridges of cartilage and downward folding of the helices of the right pinnae. Each had unilateral atresia of the external meatus. One of these children had 70 dB air conduction loss and a 30–50 dB bone conduction loss. Although most cases of microtia are thought to be sporadic, microtia with conductive deafness is probably transmitted as an autosomal recessive trait. In a family documented by Dar and Winter (28), the normal parents of affected children were consanguineous.

DOMINANTLY INHERITED HEARING LOSS

Dominantly inherited hearing loss, first involving the *low frequencies* (225–1000 Hz) followed by gradually increasing loss with age in the higher frequencies as well, is described in several large kindreds (29, 30). Pure-tone loss varied from about 20–60 dB, whereas vestibular tests showed no consistent abnormality. Audiologic studies suggest that a cochlear defect probably involving the apical segment initially is responsible for the hearing loss in these individuals.

A dominantly inherited form of progressive sensorineural hearing loss involving the *middle frequency* range (1000–4000 Hz) in childhood, and all frequencies at older ages, is reported in several families (31-34). The degree of hearing deficit varies greatly among and within families, and vestibular function is usually normal. Temporal bone studies in one case showed atrophic stria vascularis and deficiency of ganglion cells in the basal turn of the cochlea.

Nance and McConnell (2) studied four family members affected with abrupt dominant *high-frequency* hearing loss. The hearing loss increased with age to include the lower frequencies also. Crowe et al (35) found atrophy in the basal turn of the organ of Corti, with atrophy of a portion of the auditory nerve supplying the basal turn, and called attention to the distinction between abrupt high-tone loss and high-tone loss with a sloping audiogram.

Otosclerosis

Otosclerosis is a relatively common progressive disease of the middle ear. It typically commences in middle age and leads to a conductive hearing loss of varying severity in about 1% of whites. The disease results from a bony overgrowth and sclerosis of structures of the middle and inner ear. If the process involves the region of the oval window, it can lead to ankylosis of the stapes footplate (36).

Larsson (37) reviewed 202 cases of otosclerosis and found a positive family history in 90% of patients with the disease. He concluded that the disorder was transmitted as an autosomal dominant trait with 25-40% penetrance. Morrison and Bundy (38) found that 70% of 150 patients with otosclerosis had a positive family history. They agreed that the inheritance was by autosomal dominant transmission of the trait and estimated the gene's penetrance to be about 40%.

Morrison estimated the incidence of clinically manifested otosclerosis in an adult English population to be about 3 per 1,000 persons (39), a figure implying a gene frequency of about 3.5×10^{-3} (2). Such a frequency seems low in light of the marked incidence of histologic otosclerosis found in autopsy studies. Genetically determined otosclerosis possibly includes all cases of stapedial fixation, and those patients with histologic disease but no deafness include nonpenetrant genetic cases and nongenetic, environmental phenocopies. Fowler's observations in 10 pairs of monozygotic twins with otosclerosis tend to support this view, since all 10 pairs were concordant for otosclerosis, but four of the co-twins were affected only unilaterally (40).

Progressive Deafness and Nephritis—Alport Syndrome

Alport first described a syndrome of chronic nephritis and progressive sensorineural deafness, occurring predominantly in males (41), and since then many reports have appeared in the literature. The nonauditory features of this condition are discussed in Chapter 16, "Heritable Diseases of the Kidney". Alport syndrome is now known to account for more than 1% of hereditary deafness, although there is considerable variation in the severity of deafness in this group of patients. One study of seven families with Alport syndrome revealed a hearing loss of at least 20 dB at 4,000 Hz in 55% of affected male and in 39% of affected female members (42); the impairment was usually progressive. Audiometric findings indicate that a cochlear defect is responsible for the hearing loss (43).

Histologic studies of the temporal bone in Alport syndrome are reported by several investigators, but without consistent findings. Gregg and Becker (44) and Babai and Bettez (45) saw evidence of degeneration of the stria vascularis and hair cells, particularly in the basal turn of the cochlea, with absence of the tectorial membrane. The ganglion cells showed no evidence of degeneration. Winter et al (46) reported a 50% loss of spiral ganglion cells in the basal turn, with all other structures being intact. Several other groups (47-50) have failed to find any characteristic temporal bone changes.

Pooled data from several studies indicate that heterozygotic female subjects transmit the gene for Alport syndrome to about 65% of their children. Among offspring of affected males, about 75% of daughters and only 45% of sons have Alport syndrome. It is proposed (51) that the syndrome is due to an autosomal gene that shows nonrandom segregation in female subjects and preferential segregation with the X-chromosome during spermatogenesis. Preus and Fraser (52) suggested that an unfavorable intrauterine environment in an affected mother may lead to increased penetrance of the gene in her sons, in contrast to a reduced penetrance in males whose mothers are normal. Mayo (53) noted that there is probably considerable genetic heterogeneity in the reported pedigree data.

Although the etiology of Alport syndrome is unknown, the kidney and cochlea are similar with regard to fluid and electrolyte balance and common ototoxic and nephrotoxic drugs, and they share common antigens (54, 55).

Deafness and Skeletal Malformations

The *Mohr syndrome* (orofacial digital II syndrome) has been recently reviewed by Gorlin (56). The congenital hearing loss is conductive. Other features include autosomal recessive inheritance, cleft lip, an unusual lobulated nodular tongue, mandibular hypoplasia and polysyndactyly and syndactyly.

Craniometaphyseal dysplasia occurs in two heritable forms, resulting in hearing loss in addition to severe skeletal anomalies. The dominant form typically includes cranial sclerosis, a characteristic facies, broad nasal bridge, enlarged paranasal area and metaphyseal dysplasia or splaying of the long bones. The recessive form is more severe with marked mandibular prognathism, complete nasal obstruction and grotesque leonine facies. There is metaphyseal splaying of the long bones as in the dominant form, and cranial nerve paralysis is common. Hearing loss is largely conductive. This condition is a result of abnormal bone resorption.

Symphalangism, the absence of the proximal interphalangeal joints, often accompanied by conductive hearing loss, is described in several families, segregating as an autosomal dominant condition (57, 58). Flexion of the fingers is impossible, whereas fusion of the carpal and tarsal bones may cause complaints of pain in wrist or ankle. Stapes fixation may lead to a conductive hearing loss in infancy or early childhood.

In *craniofacial dysostosis* (Crouzon syndrome) craniosynostosis, ocular hypertelorism, proptosis and malar hypoplasia are found in combination with atresia of the auditory canals, abnormalities of the auditory ossicles and mixed hearing loss (56). Inheritance is consistent with an autosomal dominant gene with complete penetrance.

Mandibulofacial dysostosis (Treacher Collins syndrome) is an autosomal dominant condition, resulting in a characteristic facial appearance and conductive deafness (56). The palpebral fissures have an antimongoloid slant as a result of zygomatic hypoplasia. Other findings include coloboma of the lower eyelids with absence of cilia medial to the coloboma; malocclusion of the jaws; and anomalies of the pinnae, middle ear and external auditory canals. The gene responsible has variable expressivity and is incompletely penetrant. It is estimated that about one-half of cases are sporadic and represent new mutations (59). Nance (2) estimated the incidence to be 2.2 affected individuals per 100,000 births. The deformities of the auditory ossicles observed at tympanotomy can be variable.

Deafness and Pigmentary Anomalies

Waardenburg syndrome, a dominantly inherited condition, is thought to account for about 2% of the congenitally deaf. The most striking feature is the white forelock, present in about one-fifth of reported cases. Other often-associated characteristics include dystopia canthorum, broad-based nose, hyperplasia of the medial eyebrows, heterochromia irides and sensorineural deafness. Inner ear histologic studies of the temporal bones from a 3-year-old girl revealed complete absence of the organ of Corti, and the spiral ganglion was almost totally devoid of neurons (60). Similar associations between pigmentary abnormalities and deafness in other mammalian species (61) suggest that a defect in neural crest cells is responsible, since both melanocytes and ganglionic precursors are derived from the neural crest. There is extreme variability in expression of the many features in the syndrome, including unilateral deafness in many affected persons. Hearing aids may be useful for those with substantial residual hearing. Hageman and Dellman (61a) have carefully reviewed over 1,200 reported cases of the Waardenburg syndrome from the literature and presented family data supporting the delineation of this syndrome into two genetically distinct types: type I, Waardenburg syndrome with dystopia canthorum: and type II, Waardenburg syndrome without dystopia canthorum. Family data indicate that both types have an autosomal dominant mode of inheritance. Their observation that bilateral deafness occurs in about one fourth of the patients with type I and in about one-half of the patients with type II is of obvious importance for genetic counseling. The probability of bilateral hearing loss in any child of a type I patient is approximately one-eighth, whereas the chance of deafness in any child of a type II patient would be about one-fourth.

The *Leopard syndrome*—a variably expressed dominant syndrome characterized by multiple lentigines, ECG abnormalities, ocular hypertelorism, pulmonary stenosis, genital abnormalities, growth retardation and sensorineural deafness—is reported in several families, and the literature has been reviewed by Seuanez et al (62). There is considerable variation in the degree of cardiac abnormality, although at autopsy the pulmonary valve is frequently found to be involved (63). Growth retardation is common, as is genital hypoplasia in male and delayed puberty and late menarche in female subjects. Hearing loss, although not a frequent component of the syndrome, is probably congenital and of the sensorineural type.

Hereditary Acoustic Neuroma

Young et al (64) carried out a 40-year follow-up study on a family with bilateral acoustic neuromas, which had been described initially in 1930. They found 55 affected members (25 males, 30 females) in eight generations with an additional 42 possibly affected. The pedigree was entirely consistent with autosomal dominant inheritance. The average age of onset of symptoms was 20 years, with hearing loss the first symptom in about one-half the patients and tinnitus in another one-third. Other early symptoms were imbalance, vertigo and facial spasms. Visual loss as a result of increased intracranial pressure was common; more than one-half those affected were blind at death. Vestibular involvement was often noted on caloric testing even before a hearing loss was recognized. Fatal accidents were common in this family, three persons having drowned in their teens.

Although surgical removal of the tumors is the therapy of choice, that only three of 12 persons from this kindred who underwent operative treatment survived more than five years discouraged other affected family members from the procedure.

Hereditary acoustic neuromas are bilateral and should be distinguished from both sporadic unilateral neuromas of cranial nerve VIII and neurofibromatosis, which is associated with café-au-lait spots and cutaneous neurofibromatosis, not found in patients with hereditary bilateral neuromas of nerve VIII.

Sensory Radicular Neuropathy

Hicks (65) described a family with a syndrome of progressive peripheral sensory loss, severe ulcers of the feet, shooting leg pains and progressive sensorineural deafness. The pedigree for three generations of this kindred implies that the syndrome is a dominantly transmitted trait. Onset of sensory loss in the distal portion of legs and arms with concomitant foot ulcers began between 15 and 36 years of age. Over several years bilateral hearing loss increased to total deafness. Cranial nerve function was normal except for the auditory nerve. An autopsy of a 53-year-old woman in this family revealed a small brain and a marked loss of ganglion cells in the sacral and lumbar dorsal root ganglia (66). Temporal bone histologic examination (67) showed atrophic stria vascularis and loss of hair cells in the organ of Corti. Fitzpatrick et al (68) recently found this syndrome in multiple members of four generations of a single family. Audiologic evaluation indicated a cochlear lesion, and the propositus manifested reduced vestibular response.

Hearing Loss and Renal Amyloidosis

Muckle and Wells (69) described a family in which eight persons in four generations had recurrent urticaria. Five of them developed nephritis and deafness, one had nephropathy and another developed nephropathy, hyperglycemia and hearing loss. Black (70) reported another family with five affected persons in three generations. In both families episodic urticarial rash, fever, limb pain and concurrent progressive sensorineural hearing loss began in adolescence. In time a uremic condition developed, and the affected individuals succumbed in middle age. Histopathologic studies of three patients (69) revealed amyloid deposits throughout the renal parenchyma, glomeruli, tubules and vessel walls. Spleen, liver, testes and nerve tissue also contained amyloid. Examination of the temporal bone demonstrated absence of the organ of Corti, cochlear nerve atrophy and ossification of the basilar membrane. The syndrome is inherited as an autosomal dominant condition, with variable expression. Although overall prognosis is poor, the hearing of some affected persons may be improved by hearing aids.

Recessive Hearing Loss

Deafness with goiter (Pendred syndrome) consists of severe sensorineural deafness and early developing goiter. Fraser (71) studied 207 families with 334 cases and estimated the population incidence in the British Isles to be about 7.5 affected individuals per 100,000. The goiter, which usually becomes evident before puberty, is due to deficient incorporation of iodine into thyroid hormone despite normal trapping of inorganic iodine by the thyroid gland. Overt hypothyroidism is infrequent. Hearing loss is usually severe and congenital, with audiograms showing a sloping bilateral 40–100-dB loss, which is more profound in the higher frequencies.

Fraser (71) demonstrated depressed vestibular function on caloric testing. Temporal bone findings (72) include absence of hair cells in the organ of Corti, few spiral ganglion cells and lack of fibers in the lamina spiralis. Bilateral malformations of the inner ear may also be present. The Pendred syndrome is inherited as an autosomal recessive trait. Patients have a normal lifespan, and hearing aids may be of value to those with less severe hearing loss.

Pendred syndrome must be distinguished from a form of goitrous cretinism with a similar biochemical lesion, which does not appear to be associated with a primary defect in the organ of hearing.

Although rare in the general population, *deafness with retinitis pigmentosa* (Usher syndrome) which consists of congenital hearing loss, and less frequently mental illness and cataracts, is a relatively common autosomal recessive form of hereditary deafness. It accounts for 3–6% of profound childhood deafness (73). It appears to be commoner among Jews than non-Jews and is the leading cause of deafness and blindness in adults seeking rehabilitation services in the United States. In addition to loss of vision and hearing, some patients lack olfactory sensitivity (74). Vestibular disturbances are common, as are psychiatric disorders (75). Data on hearing loss are variable. Genetic heterogeneity may explain some of the clinical variation observed in patients with the Usher syndrome.

The retinitis pigmentosa associated with the Usher syndrome typically is not noticed until the condition is so advanced that visual complaints occur (75).

DeHaas et al (76) studied three families with the Usher syndrome and reported abnormal electro-oculograms in heterozygotes. Holland et al (77) found no distinctive abnormality in heterozygotes, although the occurrence of gyrate atrophy in several carriers was considered to be suggestive.

Other recognized syndromes in which both hearing loss and retinitis pigmentosa are found include 1) *Alstrom syndrome,* a recessive disease characterized by transient early obesity, adult diabetes mellitus, retinal degeneration and progressive sensorineural hearing loss (78); 2) *Bardet-Biedl syndrome*, which includes obesity, mental deficiency, hypogonadism and polydactyly; and 3) *Laurence-Moon syndrome* consisting of spastic paraplegia, mental retardation and hypogenitalism. Although both the Laurence-Moon and the Bardet-Biedl syndromes include retinitis pigmentosa, deafness is rare in both (79).

Edwards and co-workers (80) described three affected black sibs with retinitis pigmentosa, sensorineural deafness, hypogonadism, glucose intolerance and mental retardation. Although similar in most respects to Alstrom syndrome, the mental retardation may indicate additional genetic heterogeneity in this group of disorders.

Cockayne syndrome, although characterized by retinal degeneration and sensorineural hearing loss, can be easily distinguished by the cachectic dwarfism with senile, bird-like facies and severe mental retardation (81).

Refsum syndrome includes retinitis pigmentosa and progressive sensorineural hearing loss in addition to peripheral neuropathy, mild cerebellar ataxia and nystagmus (82). An enzymatic deficiency leads to elevated plasma levels of phytanic acid, which is thought to cause demyelination of peripheral nerves. Dietary restriction of phytanic acid and phytol reportedly has resulted in decreased plasma levels of phytanic acid and in subsequent clinical improvement (83). Hearing aids may help some of these patients.

In a questionnaire study of 670 probands with retinitis pigmentosa, Boughman found that hearing loss was the most commonly reported extraocular symptom, with severe loss being noted by 10.6% and mild loss by 19.9% of the respondents (83a). The reported losses were not invariably congenital and were described by probands in families showing either a dominant, recessive or X-linked pattern of inheritance. Bieber, et al (83b) further described these 204 RP probands who reported a hearing loss. Of interest is the disparity in the degree of hearing loss in the probands and their affected sibs in the multiplex sibships. In 29 sibships the probands had either mild or severe hearing loss, but the sibs affected with RP had no hearing loss; and in six sibships the probands reported severe loss while the sibs had only a mild loss.

Deafness and Electrocardiographic Abnormalities—The Jervell and Lange-Neilsen Syndrome

In 1957, Jervell and Lange-Neilsen described a Norwegian family in which deaf siblings were subject to recurrent fainting spells, presumably resulting from cardiac arrhythmias, that were associated with a marked prolongation of the QT interval of the ECG (Lange-Neilsen syndrome) (84). Three of the four affected children in the original family died suddenly during fainting attacks, and in subsequently reported cases more than one-half of affected persons died suddenly before the age of 20. The pedigree reported by Jervell and Lange-Neilsen suggested autosomal recessive inheritance.

Similar clinical and ECG findings may be seen in the absence of deafness, often showing an autosomal dominant mode of transmission (85). In addition some reported families have included both deaf and hearing members with the cardiac conduction defect (86). In a recent review, 53 of 203 reported patients with prolonged QT interval were deaf, and the pooled prevalence of the syndrome in several surveys of deaf school children was 14 of 5,077 or 0.28% (87). Despite the relatively low incidence of the syndrome even among the deaf, it is an important condition to consider and exclude in every child with profound sensorineural hearing loss because the cardiac defect can be readily diagnosed and effectively treated either with β-adrenergic blocking agents or by ablation of the left stellate ganglion.

X-Linked Hearing Loss

Profound congenital deafness with an X-linked mode of inheritance is reported in several kindreds. In each one deaf males were born in several generations to normal mothers, some of whom had deaf brothers (88). Affected persons had a 70–100-dB sensorineural hearing loss at all frequencies, but otherwise no physical abnormalities were noted. These cases may represent several types of X-linked deafness. Results of linkage analysis in one family suggest that the responsible gene is not closely linked to the Xg blood group (89).

Moderately severe progressive X-linked deafness with no associated abnormalities is also reported (90). Audiograms showed a 70-dB high frequency loss, with normal hearing in the low frequencies. Vestibular responses were normal. Ten affected males in four generations of a single family were reported to have moderate hearing loss in early childhood, with sufficient hearing for speech development, yet with almost complete loss of hearing by school age (91). This disorder is probably distinct from the X-linked profound congenital sensorineural deafness, in that these persons progressively lost their hearing function.

X-Linked Deafness and Pigmentary Abnormalities

An X-linked form of congenital sensorineural deafness with pigmentary abnormalities was described in a Moroccan Jewish family by Ziprkowski and co-workers (92). Fourteen congenitally profoundly deaf males appeared in three generations. At birth the skin was almost completely albinotic. Pigmentation gradually increased, particularly on the extremities, buttocks and face. Skin changes ultimately were characterized by large leopard-like spots of hypo- and hyperpigmentation. Audiometric testing demonstrated no response at frequencies above 500 Hz. Female heterozygotes had hearing loss detectible on audiometry (93). Caloric tests demonstrated no vestibular responses in three patients, whereas a fourth showed a depressed response that was more marked on the left. The pigmentary skin changes of this syndrome are somewhat similar to those described by Woolf et al in two Hopi Indian brothers (94) with congenital sensorineural deafness, but they differ in that the Hopi Indians had large symmetric areas of pigmentary variegation, including hyperpigmented spots within depigmented areas. In addition the Hopi Indian brothers had normal vestibular responses, whereas the four family members described by Ziprkowski had depressed vestibular responses.

Mixed Hearing Loss and Perilymphatic Gusher
A distinctive form of X-linked hearing loss characterized by moderate-to-severe mixed hearing loss, congenital fixation of the stapes footplate, vestibular hypofunction and perilymphatic gusher is described (2, 95). Recent studies suggest that an abnormal communication between the internal auditory canal and the inner ear vestibule (96) causes the "gusher" encountered during stapedectomy. Audiologic tests on six affected males showed a 60–100-dB pure tone loss and a 20–60-dB bone conduction deficit at all frequencies. In some families female carriers have a variable mixed hearing loss which is more marked in the lower frequencies (97). In other families careful audiologic studies of obligate carriers show no abnormalities. This syndrome may be a relatively common form of hereditary deafness. If prescribed early, hearing aids are valuable in improving sound awareness and speech development of affected persons.

Otopalatodigital Syndrome
Characteristics of the otopalatodigital syndrome include pugilistic facies (hypertelorism, broad nasal root and frontal and occipital bossing), cleft palate, growth retardation, unusual malformations of the hands and feet and mental retardation with conductive hearing loss in some families (5). Thickened ossicles were found in two cases on exploratory tympanotomy (98). Vestibular tests have not been described. The condition is X-linked and although hearing in female heterozygotes is normal, minor skeletal variations have been described (99).

Ocular Pseudoglioma and Deafness
An X-linked recessive disease manifested by lens opacity and proliferating retrolental mass was described in two families by Norrie in 1927 in an article on causes of blindness in children (Norrie's disease) (100). Ocular symptoms begin soon after birth with bilateral retinal vascular proliferation. Subsequent pseudoglioma formation, cataracts, corneal clouding and phthisis bulbi eventually lead to blindness. Progressive sensorineural hearing loss of varying severity is found in about one-third of patients (101). Hearing loss was first noted in childhood or early adult life, and audiometry revealed mild-to-severe sensorineural hearing loss, which was usually symmetrical. Neither vestibular tests nor temporal bone histology findings have been reported. Mental deficiency is common, but may be variable within a single family.

Heterozygous females have no definite abnormalities. Results of genetic linkage studies are consistent with very loose linkage to the glucose-6-phosphate-dehydrogenase locus (102) and have excluded close linkage to the Xg blood group locus (103). Because of the apparent onset after birth and the pathologic similarity of the histologic findings to those seen in retrolental fibroplasia, the mutation in Norrie's disease conceivably renders the developing retina unusually sensitive to ambient levels of oxygen in the same way that the retina of the premature infant is unusually sensitive to high oxygen tensions. If this is true, maintenance in an environment with a low oxygen tension during the neonatal period may provide a rational approach to therapy.

Dystonia and Deafness

Scribanu and Kennedy (104) recently described a progressive familial movement disorder characterized by dystonia and progressive sensorineural hearing loss. There were three affected males in three generations of a single family in a pattern that was consistent with X-linked inheritance, although autosomal dominant inheritance with reduced penetrance could not be ruled out. The 8-year-old male proband had sensorineural hearing loss detected at age 2 years, dysarthria, bizarre posture of the head and neck and hyperactive behavior. His condition gradually deteriorated and despite brief improvement under therapy with 1-DOPA, he died at age 11.

The parents were clinically normal. A maternal uncle was deaf at age 6, developed progressive dystonic posturing and died at age 25. The 26-year-old sister of the proband, who is healthy, has a 6-year-old son with severe bilateral sensorineural hearing loss, confirmed by audiometry, and marked psychomotor retardation. Temporal bone histopathologic examination of the proband revealed degeneration of the sensory epithelium and supporting cells in the basal turn of the cochlea with absence of the organ of Corti.

Although inherited syndromes involving dystonia have been previously described (105), this may be the first reported case of dystonia with deafness.

GENETIC COUNSELING FOR THE DEAF

Perhaps no group could profit more from genetic counseling than the deaf and the parents of the deaf (106). Genetic factors currently account for about one-half of all cases of profound hearing loss, and future advances in decreasing the environmental causes of deafness should increase even more the importance of genetic factors as a cause of deafness (3).

Because of their handicap, the deaf have been largely excluded from the counseling process; as a consequence little is known about the special problems of providing genetic counseling for this group. Even when the problems of communication can be overcome, it may be difficult for deaf couples to achieve a fully adequate understanding of the genetic aspects of deafness because of limitations in their educational background in science. Further, many deaf couples do not consider deafness a handicap and may not be aware that there are benefits to be derived from an accurate genetic diagnosis and counseling, which are unrelated to reproductive decisions. Professional geneticists, who in general have been obsessed with the detection and investigation of rare disorders, must share responsibility for the failure to develop counseling programs in schools for the deaf, where one-half of the students have a genetic disorder. In a recent survey of 1,020 special education programs for hearing-impaired children, only 4.8% stated that genetic counseling was included among the services provided for deaf students and their families (107).

If a precise diagnosis can be made and the mode of inheritance in genetic cases is known, an appropriate risk of recurrence can be provided. However, when a genetic etiology cannot be inferred from either clinical findings or family history, precise counseling may be difficult. In Table 1 empirical risk figures are given for the commonest counseling problems arising in normal by normal, deaf by normal and deaf by deaf matings, respectively. The risk figures are based on recent analyses of several large bodies of pedigree data (12, 107) and were calculated using Bayesian methods.

The commonest counseling problem arises when a deaf child is born to parents with normal hearing. If there are multiple affected siblings or if the parents are consanguineous, recessive inheritance may be assumed with reasonable confidence. However, if there is only one affected child, the proband could have arisen from a sporadic or an environmental cause with a minimal recurrence risk, or the child could represent a chance isolated genetic case, probably recessive, with a 25% recurrence risk. As shown in Table 1 the more normal siblings in a family of this description, the less likely that the case is genetic. Consequently the probability that the next child will be affected is correspondingly lower. A history of deafness in a relative of second degree or less makes it likelier that the index case has genetic deafness and significantly increases the empirical risk for future children.

The risk figures for hearing by hearing matings are based on the analysis of data on 49,765 children in 12,240 families of deaf probands. In 10,509 matings between nonconsanguineous parents with normal hearing and a negative family history of deafness, the estimated proportion of sporadic cases ($\hat{x}$) was 0.605. In 1,391 similar sibships with a positive family history of deafness, the maximum likelihood estimate of $\hat{x}$ was 0.203. In both groups a segregation ratio (p) of 0.25 gave a close fit to the data.

Deaf by hearing matings have a high risk of deaf children only if the affected parent has a dominant form of deafness or if the normal parent is a heterozygous carrier of a recessive gene that the affected parent also carries. As shown in Table 1, the prior probability of a deaf child in matings of this type is only 0.067, and the risk falls to 0.006 following the birth of five normal children. These risk figures are based on a reanalysis by Rose (12) of data on 420 fertile deaf by normal matings originally collected by E. A. Fay (13). The estimated proportion

Table 1. Recurrence Risk of Subsequent Deaf Offspring in Various Matings

Mating Type	Number of Deaf Offspring	\multicolumn{6}{c}{Number of Tested Offspring (S)}
		0	1	2	3	4	5
Normal by normal:							
Positive family history*	1	NA	0.200	0.187	0.172	0.156	0.138
Negative family history†	1	NA	0.098	0.082	0.067	0.054	0.043
Deaf by normal:‡							
All normal children	0	0.067	0.043	0.026	0.016	0.010	0.006
At least 1 deaf child	>0	NA	0.408	0.408	0.408	0.408	0.408
Deaf by deaf:§							
All normal children	0	0.097	0.041	0.029	0.020	0.014	0.010
All deaf children	S	0.097	0.617	0.799	0.918	0.966	0.990
Deaf and normal children	>0,<S	NA	NA	0.325	0.325	0.325	0.325

*$\hat{x}$ = 0.203, p = 0.25.
†$\hat{x}$ = 0.605, p = 0.25.
‡$\hat{h}$ = 0.835, $\hat{p}$ = 0.408.
§$\hat{h}$ = 0.789, $\hat{y}$ = 0.042, $\hat{p}$ = 0.325.
NA = Not applicable.

($\hat{h}$) of matings that could produce only normal offspring was 0.835. The average segregation ratio for deafness ($\hat{p}$) in families that could produce affected offspring was 0.408—a finding in keeping with the reduced penetrance characteristic of many dominant deafness syndromes.

Deaf by deaf matings are more complex. They include matings with a negligible risk of affected offspring; matings between persons who are homozygous for the same form of recessive deafness and who can produce only deaf offspring; and matings that can produce both deaf and hearing offspring. In a reanalysis of data on 1,299 deaf by deaf matings collected by Fay, Rose estimated that the proportion of matings that could produce only normal offspring ($\hat{h}$) was 0.789; the proportion that could produce only deaf offspring ($\hat{y}$) was 0.042; and the proportion that could produce deaf and normal offspring ($1-\hat{h}-\hat{y}$) was 0.169. The maximum likelihood estimate of the segregation ratio in the last-mentioned group was $\hat{p} = 0.325$. As shown in Table 1, the prior probability of an affected child in a deaf by deaf mating is only 0.097. If the first child is normal, the risk is reduced by more than one-half, falling to 0.010 after the birth of four additional normal children. On the other hand, if the first child is deaf, the risk for the next pregnancy increases by more than six-fold and rapidly approaches 1.0 as additional deaf children are born. In matings capable of producing both deaf and normal children, the estimated average empirical risk of recurrence for deafness ($\hat{p}$) was 0.325.

FUTURE PROSPECTS

Meaningful research on the genetic mechanisms in deafness in man is essential not only for the study of genetics itself, but also for the public health aspects of inherited deafness, such as genetic counseling and the prediction of future needs for care and special education of the hard-of-hearing. Moreover the neonatal diagnosis of hearing loss often depends on genetic studies and is important to the early implementation of appropriate treatment or rehabilitative measures, or both.

The prenatal diagnosis of deafness would add a new dimension to genetic counseling. It is already feasible in the X-linked deafness syndromes, in which males at high risk can be detected by prenatal sex determination. Direct visualization of the fetus, which has recently been accomplished using fiber optic devices, may permit detection of those deafness syndromes associated with conspicuous morphologic or pigmentary abnormalities. The development of physiologic testing procedures may also permit the diagnosis of certain types of hereditary deafness, such as detection of a fetus with the Jervell and Lange-Neilsen syndrome by fetal electrocardiography.

The establishment of pedigree linkages through a national registry of hereditary deafness would aid in identifying cases of inherited deafness. Finding a remote genealogic relationship to one or more deaf persons would be invaluable to accurate diagnosis, the first step in treatment and in genetic counseling.

Successful medical intervention requires a proper and careful diagnostic evaluation. Surgical correction has enjoyed considerable success in the treatment of some types of hearing loss, often with predictable and lasting results. The development and application of microelectronic instrumentation for the inner ear holds exciting potential. Early detection of hearing loss with remediation by sound amplification through hearing aids has also been effective in treating many kinds of hearing loss.

Medical intervention can take other forms as well. A diet low in phytanic acid e.g., seems useful in patients with Refsum syndrome. Pharmacologic intervention is useful in treating the cardiac manifestations in the Jervell and Lange-Neilsen syndrome. Eventually other syndromes may be found to be responsive to pharmacologic or dietary manipulation. Clearly, as our understanding of the molecular pathologic ramifications of human deafness becomes sharper, specific therapy for other types of inherited hearing loss will certainly follow.

REFERENCES

1. Miller MH: What's new and important about hearing impairment? *Hospital Trubune* 10:16, 1976
2. Nance WE, McConnell FE: Status and prospects of research in hereditary deafness. *Adv Hum Gen* 4:175, 1973
3. Nance WE, Sweeney A: Genetic factors in deafness in early life. *Otol Clin North Am* 8:19, 1975
4. Konigsmark BW: Hereditary deafness in man. *N Engl J Med* 281:713, 774, 827, 1969
5. Konigsmark BW, Gorlin RJ: Genetic and Metabolic Deafness. Philadelphia, Saunders, 1976
6. Fraser GR: Profound childhood deafness. *J Med Genet* 1:118, 1964
7. Katz J (ed): Handbook of Clinical Audiology. Baltimore, Williams & Wilkins, 1972
8. Brown KS: The genetics of childhood deafness, in McConnell F, Ward PH (eds): Deafness in Childhood. Nashville, Vanderbilt University Press, 1967, p. 177
9. Fraser GR: The Causes of profound Deafness in Childhood. Baltimore, Johns Hopkins, 1976
10. Stevenson HC, Cheeseman EA: Hereditary deaf-mutism with particular reference to Northern Ireland. *Ann Hum Genet* 20:177, 1956
11. Chung CS, Robison OW, Morton NE: A note on deaf mutism. *Ann Hum Genet* 23:357, 1969
12. Rose SP: *Genetic Studies of Profound Prelingual Deafness*. Ph.D. Thesis. Indianapolis, Indiana University, 1975
13. Fay EA: *Marriages of the Deaf in America*. Washington, Volta Bureau, 1898
14. Rowley PT: Familial hearing loss associated with branchial fistulas. *Pediatrics* 44:978, 1969
15. McLaurin JW, Kloepfer H, Laguaite JK, et al: Hereditary branchial anomalies and associated hearing impairment. *Laryngoscope* 76:1277, 1966
16. Fara M, Chlupackova V, Hrivnakova J: Dismorphia oto-facio-cervicalis familiaris. *Acta Chir Plast* 9:255, 1967
17. Jaffe BF: Pinna anomalies associated with congenital conductive hearing loss. *Pediatrics* 57:332, 1976
18. Fourman F, Fourman J: Hereditary deafness in a family with ear pits (fistula auris congenita). *Br Med J* 2:1354, 1955
19. Wildervanck LS: Hereditary malformations of the ear in three generations. *Acta Otolaryngol* 54:553, 1962
20. Melnick M, Bixler D, Nance WE, et al: Familial branchio-oto-renal dysplasia. *Clin Genet* 9:25, 1976
21. Fitch N, Linday JR, Srolovitz H: The temporal bone in the preauricular pit, cervical fistula, hearing loss syndrome. *Ann Otol Rhinol Laryngol* 85:268, 1976
22. Altman F: Problem of so-called congenital atresia of the ear. *Arch Otolaryngol* 50:759, 1949
23. Esher F, Hirt H: Dominant hereditary conductive deafness through lack of incus-stapes junction. *Acta Otolaryngol* 65:25, 1968

24. Wilmot TJ: Hereditary conductive deafness due to incus-stapes abnormalities and associated with pinna deformity. *J Laryngol Otol* 84:469, 1970
25. Mengel MC, Konigsmark BW, Berlin CI, et al: Conductive hearing loss and malformed low-set ears as a possible recessive syndrome. *J Med Genet* 6:14, 1969
26. Ellwood LC, Winter ST, Dar H: Familial microtia with meatal atresia in two sibships. *J Med Genet* 5:289, 1968
27. Konigsmark BW, Nager GT, Haskins HL: Recessive microtia, metal atresia, and hearing loss. *Arch Otolaryngol* 96:105, 1972
28. Dar H, Winter ST: Letter to the editor. *J Med Genet* 10:305, 1973
29. Vanderbilt University Hereditary Deafness Study Group: Dominantly inherited low-frequency hearing loss. *Arch Otolaryngol* 88:242, 1968
30. Konigsmark BW, Mengel MC, Berlin CI: Familial low frequency hearing loss. *Laryngoscope* 81:759, 1971
31. Konigsmark BW, Salman S, Haskins H, et al: Dominant midfrequency hearing loss. *Ann Otol Rhinol Laryngol* 79:42, 1970
32. Mårtensson B: Dominant hereditary nerve deafness. *Acta Otolaryngol* (Stockh) 52:270, 1960
33. Williams F, Roblee LA: Hereditary nerve deafness. *Arch Otolaryngol* 75:69, 1962
34. Paparella MM, Sugiura S, Hoshino T: Familial progressive sensorineural deafness. *Arch Otolaryngol* 90:44, 1969
35. Crowe SJ, Guild SR, Polvogt LM: Observations on pathology of high-tone deafness. *Bull Johns Hopkins Hosp* 54:315, 1934
36. Keleman G, Linthicum FH: Labyrinthine otosclerosis. *Acta Otolaryngol* (Stockh), Suppl 253, 1969, p. 1
37. Larsson A: Otosclerosis, a genetic and clinical study. *Acta Otolaryngol* (Stockh), Suppl 154, 1960, p. 1
38. Morrison AW, Bundy SE: The inheritance of otosclerosis. *J Laryngol Otol* 84:921, 1970
39. Morrison AW: Genetic factors in otosclerosis. *Ann R Col Surg Engl* 41:202, 1967
40. Fowler EP: Otosclerosis in ten pairs of identical twins. *Ann Otol Rhin Laryngol* 67:889, 1958
41. Alport AC: Hereditary familial congenital haemorrhagic nephritis. *Br Med J* 1:504, 1927
42. Cassidy G, Brown K, Cohen M, et al: Hereditary renal dysfunction and deafness. *Pediatrics* 35:967, 1965
43. Miller GW, Joseph DJ, Cozad RL, et al: Alport's syndrome. *Arch Otolaryngol* 92:418, 1970
44. Gregg JB, Becker SF: Concomitant progressive deafness, chronic nephritis, and ocular lens disease. *Arch Ophthalmol* 69:293, 1963
45. Babai F, Bettez P: Lésions auditives du syndrome d'Alport nephrite hématurique héréditaire). *Ann Anat Pathol* (Paris) 13:289, 1968
46. Winter LE, Cram BM, Banovetz JD: Hearing loss in hereditary renal disease. *Arch Otolaryngol* 88:238, 1968
47. Fujita S, Hayden RC Jr: Alport's syndrome. *Arch Otolaryngol* 90:453, 1969
48 Westergaard O, Kluyskens P, John HD: Alport's syndrome. Histopathology of human temporal bones. *Oto Rhino Laryngol* 34:263, 1972
49. Myers GJ, Tyler HR: The etiology of deafness in Alport's syndrome. *Arch Otolaryngol* 96:333, 1972
50. Bergstrom L, Jenkins P, Sando L, et al: Hearing loss in renal disease: Clinical and pathological studies. *Ann Otol Rhinol Laryngol* 82:555, 1973
51. MacNeill E, Shaw RE: Segregation ratios in Alport's syndrome. *J Med Genet* 10:23, 1973
52. Preus M, Fraser FC: Genetics of hereditary nephropathy with deafness (Alport's syndrome). *Clin Genet* 2:331, 1971
53. Mayo O: Alport's syndrome. *J Med Genet* 10:396, 1973
54. Quick CA, Fish A, Brown C: The relationship between cochlea and kidney. *Laryngoscope* 83:1469, 1973
55. Arnold W, Weidauer H: Experimental studies on the pathogenesis of inner ear disturbance in renal disease. *Arch Oto Rhino Laryngol* 211:217, 1975

56. Gorlin RJ, Pindborg JJ, Cohen MM Jr: Syndromes of the Head and Neck, 2d ed. New York, McGraw-Hill, 1976
57. Gorlin RJ, Kietzer G, Wolfson J: Stapes fixation and proximal symphalangism. *Z Kinderheilk* 108:12, 1970
58. Murakami Y: Nievergelt-Pearlman syndrome with impairment of hearing. *J Bone Joint Surg* 57B:367, 1975
59. Wildervanck LS: Dysostosis mandibulofacialis (Franceschetti-Zwahlen) in four generations. *Acta Genet Med* 9:447, 1960
60. Fisch L: Deafness as a part of an hereditary syndrome. *J Laryngol Otol* 73:355, 1959
61. Bergsma DR, Brown KS: White fur, blue eyes, and deafness in the domestic cat. *J Hered* 62:171, 1971
61a. Hageman MJ, Delleman JW: Heterogeneity in Waardenburg syndrome. *Am J Hum Genet* 29:468, 1977
62. Seuanez H, Mañe-Garzon F, Kolski R: Cardio-cutaneous syndrome (the "Leopard" syndrome). Review of the literature and a new family. *Clin Genet* 9:266, 1976
63. Moynahan EJ: Progressive cardiomyopathic lentiginosis: First report of autopsy findings in a recently recognised inheritable disorder (autosomic dominant). *Proc R Soc Med* 63:448, 1970
64. Young DF, Eldridge R, Gardner WJ: Bilateral acoustic neuroma in a large kindred. *JAMA* 214:347, 1970
65. Hicks EP, Camp MB: Hereditary perforating ulcer of the foot. *Lancet* 1:319, 1922
66. Denney-Brown D: Hereditary sensory radicular neuropathy. *J Neurol Neurosurg Psychiatr* 14:237, 1951
67. Hallpike CS: Observations on the structural basis of two rare varieties of hereditary deafness, in de Reuch AVS, Knight J (eds): CIBA Foundation Symposium: Myotatic, Kinesthetic, and Vestibular Mechanisms. Little, Brown, 1967, p. 285
68. Fitzpatrick DB, Hooper RE, Seife B: Hereditary deafness and sensory radicular neuropathy. *Arch Otolaryngol* 102:552, 1976
69. Muckle TJ, Wells M: Urticaria, deafness, and amyloidosis: A new heredofamilial syndrome. *Q J Med* 31:235, 1962
70. Black JT: Amyloidosis, deafness, urticaria, and limb pains: A hereditary syndrome. *Ann Int Med* 70:989, 1969
71. Fraser GR: Association of congenital deafness with goitre (Pendred's syndrome). *Ann Hum Genet* 28:201, 1965
72. Hvidberg-Hansen J, Balslev-Jorgensen M: The inner ear in Pendred's syndrome. *Acta Otolaryngol* (Stockh) 66:129, 1968
73. Vernon M: Usher's syndrome: Deafness and progressive blindness. *J Chron Dis* 22:133, 1969
74. Bossu A, Luypaert RB: The syndrome of Usher. *Ann Oculist* 191:529, 1958
75. Hallgren B: Retinitis pigmentosa combined with congenital deafness: With vestibulocerebellar ataxia and mental abnormality in a proportion of cases. *Acta Psychiatr Scand* 34 (Suppl 138):1, 1959
76. DeHaas EBH, Van Lith G, Rijnder J, et al: Usher's syndrome. *Doc Ophthalmol* 28:166, 1970
77. Holland MG, Cambie E, Kloepfer W: An evaluation of genetic carriers of Usher's syndrome. *Am J Ophthalmol* 74:940, 1972
78. Alstrom CH, Hallgren B, Nilsson LB, et al: Retinal degeneration combined with obesity, diabetes mellitus, and neurogenous deafness. *Acta Psychiatr Neurol Scand* 34 (Suppl 129):1, 1959
79. Garstecki D, Borton T, Stark E, et al: Speech, language, and hearing problems in the Laurence-Moon-Biedl syndrome. *J Speech Hear Disord* 37:407, 1972
80. Edwards JA, Sethi PK, Scoma AJ, et al: A new familial syndrome characterized by pigmentary retinopathy, hypogonadism, mental retardation, nerve deafness and glucose intolerance. *Am J Med* 60:23, 1976
81. Schönenberg H, Frohn K: Das Cockayne-syndrom. *Mschr Kinderkeilk* 117:103, 1969
82. Steinberg D: The metabolic basis of the Refsum syndrome. *Birth Defects* 7:42, 1971

83. Kark AP, Engel WK, Blass JP, et al: Heredopathia atactica polyneuritiformis (Refsum's disease): A second trial of dietary therapy in two patients. *Birth Defects* 7:53, 1971

83a. Boughman JA: Population Genetic Studies of Retinitis Pigmentosa. Ph.D. Thesis, Indianapolis, Indiana University, 1978

83b. Bieber FR, Nance WE, Boughman JA: Genetic and clinical heterogeneity in families with retinitis pigmentosa and hearing loss. *Am J Hum Genet* 29:23A, 1977

84. Jervell A, Lange-Neilsen F: Congenital deafmutism, functional heart disease with prolongation of the Q-T interval and sudden death. *Am Heart J* 54:59, 1957

85. Ward OC: A new familial cardiac syndrome in children. *J Irish Med Assoc* 54:103, 1964

86. Mathews EC, Blount AW, Townsend JI: QT prolongation and ventricular arrythmias with and without deafness in the same family. *Am J Cardiol* 29:702, 1972

87. Schwartz PJ, Periti M, Malliani A: The long Q-T syndrome. *Am Heart J* 89:378, 1975

88. Fraser GR: Sex-linked recessive congenital deafness and the excess of males in profound childhood deafness. *Ann Hum Genet* 29:171, 1965

89. McRae KN, Uchida IA, Lewis M: Sex-linked congenital deafness. *Am J Hum Genet* 21:415, 1969

90. Pelletier LP, Tanguay RB: X-linked recessive inheritance of sensorineural hearing loss expressed during adolescence. *Am J Hum Genet* 27:609, 1975

91. Mohr J, Mageroy K: Sex-linked deafness of a possible new type. *Acta Genet* (Basel) 10:54, 1960

92. Ziprkowski L, Krakowski A, Adam A, et al: Partial albinism and deaf mutism due to a recessive sex-linked gene. *Arch Dermatol* 86:530, 1962

93. Fried K, Feinmesser M, Tsitsianov J: Hearing impairment in female carriers of the sex-linked syndrome of deafness with albinism. *J Med Genet* 6:132, 1969

94. Woolf CM, Dolowitz DA, Aldous HE: Congenital deafness associated with piebaldness. *Arch Otolaryngol* 82:244, 1965

95. Nance WE, Setleff RC, McLeod A, et al: X-linked mixed deafness with congenital fixation of the stapedial footplate and perilymphatic gusher. *Birth Defects* 7(4):64, 1971

96. Glassock ME: The stapes gusher. *Arch Otolaryngol* 98:82, 1973

97. Nance WE, Sweeney A, McLeod AC, et al: Hereditary deafness: A presentation of some recognized types, modes of inheritance and aids in counseling. *South Med Bull* 58:41, 1970

98. Buran DJ, Duvall AJ: The oto-palato-digital (OPD) syndrome. *Arch Otolaryngol* 85:394, 1967

99. Gorlin RJ, Poznanski AK, Hendon I: The oto-palato-digital (OPD) syndrome in females. Heterozygotic expression of an X-linked trait. *Oral Surg* 35:218, 1973

100. Norrie G: Causes of blindness in children. *Acta Ophthalmol* 5:375, 1927

101. Warburg M: Norrie's disease, a congenital progressive oculo-acoustico-cerebral degeneration. *Acta Opthalmol* (Kbh), Suppl 89, 1966, p. 1

102. Nance WE, Hara S, Hansen A, et al: Genetic linkage studies in a Negro kindred with Norrie's disease. *Am J Hum Genet* 21:423, 1969

103. Warburg M, Hauge M, Sanger R: Norrie's disease and the Xg blood group system: Linkage data. *Acta Genet Stat Med* 15:103, 1965

104. Scribanu N, Kennedy C: Familial syndrome with dystonia, neural deafness, and possible intellectual impairment: Clinical course and pathological findings. *Adv Neurol* 14:235, 1976

105. Eldridge R: The torsion dystonias: Literature review and genetic and clinical studies (dystonia musculorum deformans). *Neurol* 20(pt 2) (suppl):1, 1970

106. Nance WE: Genetic counseling for the hearing impaired. *Audiology* 10:222, 1971

107. Nance WE, Rose SP, Conneally PM, et al: Opportunities for genetic counseling through institutional ascertainment of affected probands, in Lubs HA, de la Cruz F (eds): Genetic Counseling. New York, Raven, 1977, p. 307

16
Heritable Diseases of the Kidney

Kenneth D. Gardner, Jr.
Raymond Garrett

The heritable renal diseases together make up a diverse group of structural and functional kidney abnormalities, whose recognized number has increased over the past half-century. The incidence of heritable renal disease has not been established, but a reasonable guess would place it at about 7% of all kidney disease. Polycystic kidney disease alone accounts for more than 5% of end-stage renal disease coming to renal transplantation (1). In some instances familial (and presumed genetic) disease will be shown to be acquired rather than heritable (2). Genetic predisposition to environmental factors, however, may complicate the issue.

No single scheme of classification for these conditions is entirely adequate; nor is it reasonable to list these conditions on the bases of their prevalence or relative importance, since meager reported experience and nosologic confusion make it impossible to estimate their individual frequencies accurately. Because of utilitarian advantages, we prefer the scheme used by Perkoff and Sly to categorize these disorders and have followed it in preparation of this review (3). In that scheme heritable renal diseases are classified as 1) abnormalities of structure (noncystic and cystic); 2) abnormalities of function (proximal tubular and distal tubular); 3) abnormalities of the genitourinary tract; and 4) systemic illnesses with major renal manifestations.

The practicality of recognizing that some renal diseases are inherited lies in a better understanding of pathogenesis, in an increased accuracy of genetic counseling and in the selection of appropriate donors for renal transplantation. The heritable renal diseases classified by their established or suspected Mendelian modes of inheritance include: 1) autosomal dominant (adult polycystic disease; Alport's syndrome; idiopathic Fanconi syndrome; medullary cystic disease; nail-patella syndrome; renal tubular acidosis (distal); urticaria, deafness and amyloidosis; and vasopressin-resistant diabetes insipidus); 2) X-linked dominant (hypophosphatemic rickets and pseudohypoparathyroidism); 3) autosomal recessive (childhood polycystic disease, cystinuria [Types I-III], dibasicaminoaciduria, glucose-galactose malabsorption, Hartnup disease, hypercystinuria, infantile nephrosis, iminoglycinuria, juvenile nephronophthisis, nephritis and hyperprolinemia, renal agenesis [?], renal glycosuria and retinorenal dysplasia); and 4) X-linked recessive (Fabry's disease, oculocerebrorenal syndrome [?], unilateral hydronephrosis [?] and vasopressin-resistant diabetes insipidus).

HERITABLE DIFFUSE NEPHROPATHIES

The heritable disorders of the kidney in which morphologic changes predominate include 1) noncystic (chronic hereditary nephritis [Alport's syndrome], with nerve deafness and ocular abnormalities, without nerve deafness and ocular abnormalities and with polyneuropathy; familial hyperprolinemia; familial Mediterranean fever with renal amyloidosis; and amyloidosis, deafness and urticaria); and 2) cystic (polycystic disease; nephronophthisis—cystic renal medulla complex; medullary sponge kidney; congenital or infantile nephrosis; and miscellaneous, e.g., chromosomal anomalies).

Chronic Hereditary Nephritis

Chronic hereditary nephritis (Alport syndrome) has been recognized for more than a century (4). Well over 100 families are described in the literature to date, and most centers have had experience with multiple kindreds. Most reports describe families of European descent, but the syndrome has been observed in Japanese, Ceylonese and black kindreds (5, 6). One group of workers believes it to be so prevalent that they advise that chronic hereditary nephritis be considered in every patient with unexplained hematuria, proteinuria or renal failure (7).

Early reports of chronic hereditary nephritis have been traced back to the 1870s. That of Dickinson in 1875 is especially charming, for it alludes to family portraits which carried back into the 1500s and exhibited a "clear transparent pallor" among the ancestral complexions of a family in which 11 of 17 individuals had proteinuria and several had died of renal disease (4). Classic reports are those that trace the same kindred (the original "Alport" family) for 66 years (5, 8-12). In that family the disease has now apparently run its course and there is no possibility of new cases appearing (5). A number of families containing large numbers of affected patients are reported (13, 14).

Nephritis, deafness and ocular defects make up the traditional constellation of signs in the Alport syndrome. In affected kindreds, males and females appear to be involved with equal frequency, but the disease is more fulminant in males and most of them die by the fourth decade. Affected female subjects usually live a normal lifespan, although some of them die from renal failure at somewhat later ages than do the men. The earliest and most constant evidence of the renal lesion in the Alport syndrome appears to be hematuria occurring in virtually 100% of affected persons (15, 17). Proteinuria is a far less constant finding, and only rarely has the nephrotic syndrome been reported (7, 18, 19).

Prior to the onset of renal failure, most affected subjects are asymptomatic. Hypertension is not a constant or prominent feature and routine blood studies are nonspecific, although platelet abnormalities are described in two unrelated families, a finding perhaps indicative of genetic heterogeneity (20). The associated deafness is commoner in the male. It is not present at birth, but appears to start gradually during childhood, often paralleling the progressive renal disease. Characteristically, high tones are lost first and often asymmetrically.

Roughly one-half of the patients with end-stage renal disease due to the syndrome also are completely deaf, but no characteristic histopathologic lesion has been described (5). Hyperosteosis of the bony labyrinth is noted in one case and cochlear degeneration in another. Ocular lesions affect a minority (about 15%) of families with the Alport syndrome, and they typically involve the lens or its capsule, with lenticonus being the commonest lesion. Other lenticular defects include ruptured lens capsule, nuclear and subcapsular cataracts and spherophakia.

Pathologic changes in the kidney are nonspecific (15, 20-23). During childhood the kidney may appear normal to light microscopic examination aside from the presence of red blood cells (RBCs) in some tubules. In more advanced cases a greater variety of changes may be seen, ranging from focal through diffuse glomerular involvement, focal proliferative glomerulonephritis and crescent formation. In other cases the features may be more typical of interstitial nephritis. Interstitial foam cells were at one time considered to be diagnostic, but experience shows that foam cells frequently are found in the kidneys of patients with various renal diseases.

The mode of inheritance of the Alport syndrome continues to be the subject of some debate (24, 25). Early reports and several more recent ones conclude that simple autosomal dominant inheritance is operative. However, in some families more abnormal female than male members were noted and the vast majority of children of affected males were affected females (13). Further, male-to-male transmission was infrequently recorded. Two hypotheses and one criticism are brought forth to explain these findings. The disease may be transmitted by partial sex-linked dominant inheritance, although this form of hereditary has no proven precedence in human genetics. Preferential segregation and chromosomal association also may be responsible for the distorted sex ratio observed in some families (24).

Again because preferential segregation is not shown to have a proven precedence in human genetics, this hypothesis remains dubious. Chazan et al point up another possible reason for the apparent excess of affected female members in some of these families (7). In earlier studies one of the criteria used for the diagnosis of hereditary nephritis was isolated pyuria. Because isolated pyuria occurs more frequently in women than in men, the use of this sign for diagnosis may have resulted in the assignment to the affected group of some women who are actually unaffected. A review of 37 families, containing 150 affected males, for ages at death showed two distinct populations (25). In one, affected males died between the ages of 16 and 28 years. In the other, affected males died between 33.5 and 52.5 years. These findings support the existence of genetic heterogeneity even within the classic Alport syndrome.

Variants of Hereditary Chronic Nephritis

In addition to descriptions of the "classic" form of Alport syndrome are other familial reports in which renal disease occurs in the absence of either ocular defects or nerve deafness (5). Chronic hereditary nephritis is also reported with peroneal muscular atrophy (Charcot-Marie-Tooth disease) and with polyneuropathy (20). Families are described in which hematuria appeared to be transmitted as an autosomal dominant trait (26). In these families the hematuria was considered to be benign, inasmuch as no other evidence of nephritis existed, renal function remained normal and renal biopsies in 10 subjects with hematuria showed no changes suggestive of nephritis; one patient showed a moderately proliferative lesion that was thought to be consistent with nephritis of an indeterminant stage.

Progressive nephritis, hearing loss and ocular defects serve to separate the Alport syndrome from nonfamilial cases of benign hematuria with focal glomerulitis in adults. It also must be kept in mind that progressive nephritic renal failure in a family need not imply a heritable disorder. Dodge and his associates found evidence of acute glomerulonephritis in sibling contacts in one-fourth (proven by renal biopsy) to one-half (proven and suspected) of 20 index case sibships (2). A demonstration of progressive nephritis through multiple generations and in individuals who have had no physical contact with one another strengthens a diagnosis of chronic hereditary nephritis in any suspected case.

Familial Hyperprolinemia and Hereditary Nephropathy

Unrelated families are reported in which diffuse chronic nephritis is associated with elevations of proline in the blood (27-29). Of 102 identified individuals in one pedigree, 10 exhibited hematuria or other evidence of renal disease. Affected males transmitted the hematuria trait to four of their seven daughters but to none of their five sons. Affected females transmitted the trait to six of their eight daughters and to three of their seven sons. Similar to the evolution of disease in the Alport syndrome, affected males had severe renal involvement and rarely lived beyond the third decade; affected females suffered a relatively benign form of the disease. Reduced auditory acuity occurred in association with the renal lesion as well as alone, involving nine individuals in the kindred.

In addition to these findings, some members of the family exhibited hyperprolinemia, convulsions, mental retardation, increased sensitivity to photic stimulation on electroencephalography and congenital renal hypoplasia. The most important characteristics differentiating the lesion from the Alport syndrome were the hyperprolinemia and the excessive urinary excretion of proline, hydroxyproline and glycine, the last-mentioned occurring because of a common saturable renal transport mechanism (27). In 1965, an unrelated family with a similar constellation of signs was reported (28). In this kindred the propositus also had suffered mental retardation and had died from uremia and urinary tract infection secondary to congenital renal malformations. Unlike the earlier kindred, however, deafness and seizures were not detected. The propositus from this family was able to metabolize hydroxyproline normally, and hepatic proline oxidase secretion was reduced.

Although such an abnormality was not documented in the earlier family, the similarity of clinical manifestations between the two kindreds led to conclusions that the biochemical abnormalities were identical. In each of these two families renal disease appeared to be transmitted as an autosomal dominant trait from generation to generation. In contrast, hyperprolinemia occurred only in one generation. Thus it appeared to be inherited in a manner different from that of the nephritis. Efron concluded that two distinct genetic defects were segregating in each family, one for hereditary nephritis, the other for proline oxidase deficiency (28).

Hereditary nephritis, hyperprolinemia, neurosensory hearing loss and ichthyosis are reported in one family in which consanguineous mating occurred (30).

Hereditary Renal Amyloidosis

The commonest form of genetically transmitted amyloidosis of the kidney is that which occurs in association with familial Mediterranean fever (FMF), and the kidney is virtually always the organ most seriously affected (31). Sohar and his co-workers, in reporting on 470 cases of FMF in which 125 had amyloidosis, found 68 fatalities (32). In all but one (intestinal malabsorption) death was caused by renal amyloidosis.

These same workers distinguished four stages of renal involvement by amyloid in FMF. In the *preclinical stage* amyloid was present on biopsy (kidney and rectum) in the absence of clinical signs. Amyloid nephropathy becomes clinically evident with the onset of progressive *proteinuria*, which culminates in the *nephrotic stage* and finally, terminal *uremia*. The true incidence of renal involvement in FMF is not established, although most authors place it at around 30% of all cases. Histologic examination shows deposition of amyloid substance, primarily involving glomeruli and renal tubules.

The disease is usually transmitted as an autosomal recessive trait, but exceptions are reported (32). There is a predominance of affected male subjects in reported material (32, 33), a finding for which no obvious explanation is available. Although FMF occurs predominantly among peoples of the Mediterranean area, usually non-Ashkenazic Jews and Armenians, evidently heritable renal amyloidosis has been described in families whose backgrounds are different. Van Allen and his associates, e.g., found eight (and perhaps 12) cases of renal amyloidosis among the members of an American family whose genealogy extended into northern Europe (34). Although only two generations were involved, they concluded that dominant inheritance was most likely. An English family is reported in which fever, deafness, urticaria and amyloidosis affected nine of 18 members of three successive generations (35). Three siblings died of amyloid renal disease after experiencing the nephrotic syndrome. Thus there is evidence that amyloid nephropathy is genetically heterogeneous (36).

POLYCYSTIC KIDNEY DISEASE

Polycystic kidney disease describes two specific conditions in which normal renal architecture is distorted by multiple cysts. Analyses of the ages at onset among patients with polycystic disease indicate a clear-cut bimodality. In some cases, a small minority, a diagnosis is made during the first decade, often at or shortly after birth. In the vast majority of cases, however, the diagnosis is made during the fourth to sixth decades. These forms are designated the *childhood-onset* and *adult-onset* forms of polycystic disease. They are distinguished by several criteria.

Childhood Onset

Childhood polycystic kidney disease (CPKD) is set apart from the adult variety by at least three important criteria: morphology, clinical manifestations and genetics. Exact data as to its incidence are not available. It is reported to occur with a frequency varying from 1 in 6,000 to 1 in 14,000 live births. Dalgaard surveyed five reports and included his own experience to establish a frequency between 1 in 225 and 1 in 450 autopsies (37).

Generally it is assumed that CPKD is inherited as an autosomal recessive trait (38). Blyth and Ockenden suggest that four types can be recognized, including *perinatal, neonatal, infantile* and *juvenile* (39). In all, both renal and hepatic involvement occur and all are inherited as autosomal recessive. Variations in the extent to which either organ is involved and differences in the evolutionary patterns of clinical signs and symptoms distinguish the four types.

In the *perinatal* form, patients have massively enlarged, cystic kidneys at birth. Liver involvement is relatively less dramatic, with proliferation and redundancy of the lining of intrahepatic bile ducts. There is relatively little periportal fibrosis. Death from renal failure occurs within weeks. In the *neonatal* form affected persons have less fulminant renal failure during the first month of life. Approximately 60% of renal substance exhibits cystic nephrons. Hepatic fibrosis is found in addition to the bile duct proliferation. Although death from renal failure is still characteristic, survival for months or several years is possible.

The *infantile* form is characterized by still less renal involvement, with perhaps 25% of nephrons being affected. Periportal fibrosis in the liver is more pronounced and hepatic signs may predominate. Patients have hepatomegaly, splenomegaly and nephromegaly. Death occurs from either chronic renal failure or portal hypertension in the late childhood or early adolescent years.

In the fourth variety, the *juvenile* form, patients during the first five years have hepatosplenomegaly. In the kidneys only about 10% of the nephrons are cystic, and renal failure does not develop. The signs and symptoms of the juvenile form are those of portal hypertension. Surgical intervention in the form of portocaval shunt has prolonged life into the third decade for some persons.

Not all workers accept the existence of four types of CPKD; some of them regard the disease as a single entity following its natural evolution rather than a condition with clear-cut variants (40).

Renal cystic disease and hepatic fibrosis also coexist in a condition known as *congenital hepatic fibrosis*. The hepatic lesion consists of periportal fibrosis, which frequently connects portal triads, and bile duct proliferation. In roughly one-half the cases, renal cystic disease of a mild nature is found. In the remaining 50%, congenital hepatic fibrosis coexists with normal kidneys. Congenital hepatic fibrosis is described in two or more children in a given family, suggesting that it too is transmitted as a recessive trait.

Remember that the morphologically distinct variety of renal cystic disease which characterizes CPKD invariably is associated with some degree of hepatic fibrosis. Whether a spectrum of a single disease or several distinct genetic variants, the hepatic fibrosis increases as the renal cystic lesion regresses. Therefore at one end of the spectrum is marked renal involvement and minimal hepatic fibrosis, leading to death from renal failure at or soon after birth. Near the opposite end of the spectrum is minimal renal involvement and pronounced hepatic fibrosis, leading to death from portal hypertension. If one regards congenital hepatic fibrosis as the far end of the spectrum, cystic renal disease does not invariably accompany hepatic fibrosis. A clearer picture of this constellation of conditions awaits further clinical experience.

Childhood polycystic kidney disease is to be distinguished from *congenital multicystic kidney* and the adult-type of polycystic kidney disease in a child or infant. Congenital multicystic kidney, considered to be the commonest cause of an abdominal mass in childhood, is a form of renal dysplasia. It occurs in association with ureteral obstruction and presents no threat to longevity, provided that the deformity is not bilateral. The anatomic changes characterizing CPKD and distinguishing it from the adult variety include a smooth rather than bosselated renal surface and radially oriented fusiform rather than round cysts on the cut surface. Kidneys the morphology of which is characteristic of the adult form of polycystic kidney disease have been found in children and infants, albeit rarely.

Adult Onset

Adult polycystic kidney disease (APKD) is the most prevalent form of cystic renal disease in man. It affects perhaps 1 in every 500 adults and accounts for renal failure in more than 5% of all patients coming to renal transplantation (1). The pathologic characteristics of APKD have been recognized for almost a century. In contrast to CPKD, the adult form is characterized by an autosomal dominant pattern of inheritance. Credit for recognizing the heritable nature of APKD is given to the British urologist, Cairns, who reported a family of 84 individuals over four generations in 1925 (41).

According to Dalgaard, whose study of 284 patients and their respective families remains a classic contribution 20 years after its publication (37), gene penetrance is about 100% by age 80 years. Within affected kindreds there is a remarkably constant pattern of evolution. Renal failure tends to appear at virtually the same age among affected members of successive generations in any given kindred. Dalgaard found an average standard deviation of ± four years around each family's mean age at onset. A spontaneous mutation rate is calculated as being 6.5×10^{-5} to 12×10^{-5} mutations per gene per generation for APKD (37).

The manner in which heritable influences cause cystic deformity of the kidney in APKD is unknown, as is true for all forms of cystic renal disease. Tubular obstruction by casts, by the deposition of "salts" or by disordered tubular growth were considered (42) until definitive studies indicated that the overwhelming majority of cystic nephrons communicate with both glomerulus and urinary space (43, 44). Possible etiologies include partial rather than complete tubular obstruction or a defect in tubular basement membrane. Whatever the pathogenetic mechanism, it appears to operate over a long time in kidneys that originally seemed grossly normal in both structure and function. The end-stage adult polycystic kidney is dramatically deformed and may achieve the size of a small watermelon without losing its typical kidney shape.

Liver disease also accompanies APKD in about 30% of cases. It takes the form of cyst formation not fibrosis, however, and rarely is it responsible for significant morbidity. In affected subjects cysts in organs other than kidneys and liver occasionally may be found. Berry aneurysms of the cerebral circulation are described in roughly 10% of cases and rarely may be responsible for bringing the disease to initial clinical attention.

Characteristically, flank pain is the earliest symptom of APKD (37). It precedes the development of clinically detectable flank masses by several months or years. Renal function initially may be normal, even if the kidneys are enlarged, but ultimately kidney function fails. Hematuria, urinary tract infection, nephrolithiasis and obstruction are common presenting features; APKD may be manifested as anuric renal failure when strategically located cysts, through enlargement, obstruct outflow from otherwise functioning kidneys.

Instrumentation of the urinary tract should be avoided in making a diagnosis, since polycystic kidneys appear to be unusually susceptible to infection. Therapy is nonspecific. Complications should be treated specifically. Dialysis or renal transplantation prolong life once end-stage disease is reached. Genetic counseling offers the only hope of prevention. Unfortunately affected persons generally become symptomatic only after their childbearing years, thus assuring transmission of the condition.

Nephronophthisis–Cystic Renal Medulla Complex

The nephronophthisis–cystic renal medulla complex takes its name from the appearance of the kidneys, which are bilaterally scarred and shrunken and display a variable number of corticomedullary and medullary cysts. Microscopically, aside from these cysts, the renal substance resembles that of "pyelonephritis," with interstitial infiltration by round cells, periglomerular fibrosis, tubular atrophy and dilatation. Clinically a positive family history for renal disease, progressive renal failure, renal sodium wasting and anemia characterize the complex (45-47).

More than 250 examples of the disease have been reported in the literature through 1975. All of them do not fit exactly the clinicopathologic criteria just described. In a minority of cases, perhaps 15 to 20%, renal cysts are not described and in about 15% of the cases, no positive family history can be obtained (45). Renal sodium wasting, a characteristic of several of the cases described by Strauss in his classic article (46), in fact is documented in fewer than 25 instances (45). Anemia is probably not disproportionate to the degree of renal failure.

Genetic criteria distinguish at least four variants of the nephronophthisis–cystic renal medulla complex. There are reports of isolated cases, of recessively inherited cases in which retinal lesions may be present and of dominantly inherited disease (45). Sporadic cases may represent a new dominant mutation or the recessive transmission of the complex to only children. They may also reflect instances in which medullary cysts occur rather as an acquired than as an inherited condition (48). In most reported cases, autosomal recessive inheritance is most likely. Consanguineous mating has occurred among several families. Renal involvement between the sexes is equal. Some clinicians have elected to retain the title *familial juvenile nephronophthisis* to categorize recessively inherited examples of the complex in which renal disease is the sole feature (38, 45, 49).

An additional 15-20% of reported cases are distinguished by retinal lesions, usually retinitis pigmentosa, while including most or all of the other clinicopathologic characteristics of the complex. Both renal and ocular abnormalities usually occur together, but occasionally either may occur in the absence of the other. Again a recessive mode of inheritance seems likely. This variant of the complex is designated *renal-retinal dysplasia* (50).

A dominant pattern of inheritance is described in about 18% of all published cases. Retinal abnormality has never been observed among these individuals. Onset and death, as in APKD, occur during adult years. This variant is designated *adult-onset medullary cystic disease* (45). The dominantly and recessively inherited variants are distinguished by a difference in their patterns of clinical evolution. Recessively inherited disease appears at a significantly earlier age.

Gardner analyzed ages at onset and death among 91 reported instances of the complex (45). Recessively inherited disease in 34 cases had a mean age at onset of 10 years and a mean age at death of 14.4 years. In 24 cases of dominantly inherited disease, onset and death occurred at ages 28.4 and 31.7 years, respectively. These differences were significant ($P < 0.001$). Of 14 cases with a negative family history, ages at onset and death were intermediate to these figures. Ages at onset and death were not significantly different among cases with proven medullary cysts vs cases in which medullary cysts were not found, giving no credence to an idea that cyst formation in the complex is related to the duration of disease (51).

The discovery of a urinary concentrating defect as the first abnormality of impaired renal function has been described in six instances in which other signs and symptoms of the complex ultimately evolved (45). This observation has led to a suggestion that an inability to maximally concentrate urine may be the earliest clinical sign of involvement in the complex.

It has been suggested that most affected patients with the nephronophthisis-cystic renal medulla complex have red or blond hair (52), but documentation is lacking. Medullary cysts have been found in patients with Laurence-Moon-Biedel syndrome (53) and in an individual with a horseshoe kidney (54). Boichis and his associates described a family with the characteristic clinical picture, but with very unusual pathological findings (55). Instead of medullary cysts they described a renal lesion in which glomerular cysts predominated. The lesion of congenital hepatic fibrosis characterized liver findings in this family, in which the patients were first cousins. Because no other similar families have been reported and because the renal lesion with its "multiple glomerular cysts" is not typical, it is possible these workers have described a new disease comprising nephronophthisis and congenital hepatic fibrosis.

The differential diagnosis of the nephronophthisis-cystic renal medulla complex includes those conditions in which progressive renal failure is associated with a positive family history and shrunken kidneys. Until kidney tissue is available for direct pathological examination, an accurate diagnosis may be impossible. Both closed and open renal biopsy techniques may fail to deliver corticomedullary tissue and consequently the pathologic hallmark of medullary cysts. In some cases renal arteriography with tomography has proven helpful (56).

As to pathogenesis of the complex, nothing is known. In the laboratory it is possible to induce renal cystic disease by the feeding of various chemicals (57).

The resultant gross lesion may resemble the human form of the disease, giving rise to speculation that a circulating nephrotoxin in man may be involved in pathogenesis. However, recurrence of the condition has not been described in transplanted kidneys to date.

Medullary Sponge Kidney

Medullary sponge kidney (MSK) is a relatively common, roentgenographically diagnosed disorder of the intrapyramidal or intrapapillary segments of the collecting tubules, characterized by ectasia sometimes to the point of actual cyst formation (58). More than 600 case reports have appeared in the literature (59). Estimates of its frequency range from 1 in 5,000 to 1 in 20,000 of the general population. MSK is distinguished from the nephronophthisis–cystic renal medulla complex by its tendency to predispose to nephrocalcinosis and infection and by the location and nature of the "cystic" lesions.

In the absence of stone formation, obstruction and infection, MSK carries a benign prognosis and is a nonprogressive lesion (58). Its cause is unknown. By most experienced nephrologists it is considered to be a congenital lesion (58). Recent reports document its familial occurrence; one describes MSK in members of three successive generations (60). The disease has also been reported in association with other heritable disorders, e.g., congenital hemihypertrophy (61). However, the extent to which heritable factors participate in the transmission of MSK has not been established.

Congenital Nephrotic Syndrome

The nephrotic syndrome customarily is considered to be a nonfamilial disorder, since most cases are sporadic. The *congenital nephrotic syndrome* is a heritable type of nephrosis that is relatively rare (62-65). It can be distinguished from other forms of the nephrotic syndrome by its onset at birth or during the first few weeks of life, placental weight in excess of 25% of the birth weight, failure to thrive and resistance to therapy, all of which portend death (66). More than one-half of the reported cases have come from Finland. Norio, in 1969, reviewed them and concluded that the condition is a distinct entity transmitted as an autosomal recessive defect (64). The condition causes marked rise of amniotic fluid and maternal serum α-fetoprotein in early pregnancy so that prenatal diagnosis may be suspected if neural tube defects can be excluded. Dilatation of the renal proximal convoluted tubules may be marked, giving rise to the term *microcystic disease* (67), thereby justifying inclusion of this rare and as yet little understood condition among the heritable forms of renal cystic disease.

Heritable Renal Cysts in Miscellaneous Diseases

Renal cysts have been described in heritable disorders other than those already mentioned. In these disorders the extent to which cyst formation encompasses a significant part of the total syndromes is not established, even though cyst formation in the kidneys may be severe. Included are such conditions as von Hippel-Lindau disease, tuberous sclerosis and the syndromes of Ehlers-Danlos, Meckel, Zellweger and Jeune (49).

HERITABLE FUNCTIONAL NEPHROPATHIES

The traditional concept of discrete renal transport systems for four classes of aminoacids (neutral, dibasic, dicarboxylic and iminoacids) is challenged by recent data.

Aminoacidurias

The description of 5-oxoprolinuria has added support to a new theory on the mechanism of aminoacid transfer involving the γ-glutamyl cycle (68). In theory an increase in the urinary excretion of aminoacids can result from 1) increased production or decreased degradation, resulting in serum concentrations in excess of the tubular reabsorptive maxima ("overflow aminoaciduria"); 2) the production through errant metabolism of a novel aminoacid for which no reabsorptive mechanism was present; or 3) a discrete or generalized defect in nephron transport of these compounds. The most significant of the two latter conditions will be dealt with herein. Two penetrating reviews are available (68, 70).

Cystinuria and Related Disorders

Cystinuria and its variants represent the most important group of aminoacid transport disorders, although genetic studies have been hindered by incomplete assessment of phenotype (Table 1). McKusick lists three separate entities with affected sulfur–dibasic aminoacid transport (71).

Hyperdibasic aminoaciduria is associated with increased urinary excretion of lysine, ornithine and arginine (but no alteration in cystine) with an analogous intestinal (and possibly hepatic and neuronal) transport defect. There are two types and the mutations may be allelic. Type I shows growth retardation and hyperammonemia, and Type II is reflected by mental retardation. The inheritance is autosomal with evidence of both dominance and recessivity. Renal

Table 1. Abnormalities of Renal Function

Proximal tubule:
Cystinuria (Types I-III)
Dibasic aminoaciduria
Glucoglycinuria
Glucose-galactose malabsorption
Hartnup disease
Hypercystinuria
Hypophosphatemic rickets
Idiopathic Fanconi syndrome
Iminoglycinuria
Oculocerebrorenal syndrome
Proximal renal tubular acidosis
Pseudohypoparathyroidism
Renal glycosuria (Types A and B)

Distal tubule:
Distal renal tubular acidosis
Vasopressin-resistant diabetes insipidus

Unspecified:
Familial hypokalemia (Bartter's syndrome)
Hereditary urolithiasis

lithiasis does not appear to be part of the syndrome (70). The lesion clearly shows that cystine transport is separable from the dibasic aminoacid handling.

Another rare experiment of nature confirms the presence of a distinct renal tubular system for disposition of sulfur-containing aminoacids. Cystine (in the absence of lysine, ornithine or arginine) is excreted in excess (*hypercystinuria*). Inability to reabsorb filtered cystine may be responsible for the disease. Only one family study is available, and the natural history of the potential renal dysfunction is inadequately assessed.

The compelling significance of *cystinuria* was exemplified by its inclusion in Garrod's 1908 Croonian Lectures (72). The disease occurs in 1 of 7,000 live births and is autosomal recessive in inheritance; however, at least three controlling alleles have been identified with characteristics described in Table 2. As shown therein, cystine is transported by receptors distinct from those carrying dibasic aminoacids; however, large quantities of cystine, ornithine, lysine and arginine appear in the urine of cystinurics. Paradoxically infusion of lysine can increase clearance of the other aminoacids, as would be expected with competitive inhibition. In-vitro studies show normal cystine transport in isolated intestinal preparations. Finally the clearance of cystine in cystinuria can exceed the inulin clearance; hence a secretory process or "negative reabsorption" must be invoked. Thier and Segal (73) postulate that, distinct from the competitive events on the lumenal border of the tubular cells, a defect in efflux of cystine at the antilumenal surface may be primary and the high intracellular cystine concentration may retard cystine reabsorption and produce net flux into the lumen.

The disease is characterized by nephrolithiasis, with an onset varying from infancy to advanced age. Renal dysfunction from obstruction or secondary infection and ureteral colic, or both, are the only definite maladies suffered as a direct result of the metabolic error, although the relatively short stature of cystinurics is attributed to aminoacid malabsorption (73).

Diagnosis is initially supported by cystine crystals, which are best seen in the acidic and concentrated urine, and confirmed by positive qualitative testing using conventional chromatography. Treatment requires high urine output. Cystine is highly soluble in alkaline urine (pH $\geq$ 8), but this is often unobtainable in practice. Strict compliance to hydration therapy can prevent recurrence of nephrolithiasis and occasionally effect dissolution of formed stones. Progressive renal deterioration while the patient is on this regimen necessitates consideration of solubilizing cystine by administering penicillamine, a drug with a spectrum of serious toxicity (73).

Hartnup Disease

Named after the family first described with this disorder, Hartnup disease is an autosomal recessive mutant in transepithelial transport of monoamino-monocarboxylic aminoacids. Although not common—about 1 in 15,000 in surveys of newborn babies (69)—it derives clinical importance from mimicry of vitamin B-deficiency and partial correctability (74). The pathognomonic clinical triad includes 1) pellagraform rash 2) cerebellar dysfunction and 3) specific aminoaciduria. The rash is erythematous and eczmatoid and is occasionally blistered, although more often dry. It is the most frequent presenting complaint, occurring any time from infancy to prepuberty and primarily over sun-exposed

Table 2. Classification of Cystinuria*

Experimental Observations	Type I	Type II	Type III
Intestine:			
In vitro transport	No transport of cystine lysine, or arginine; normal cysteine transport	No transport of lysine; markedly reduced cystine transport	Transport of cystine reduced but may be normal; lysine variably reduced;
Oral cystine administration	No plasma cystine elevation	No plasma cystine elevation	Slow increase in plasma cystine to normal elevation
Kidney:			
In vitro transport	Normal cystine, cysteine; reduced lysine transport	—†	Normal cystine; reduced lysine transport
Urinary amino acid excretion	Increased cystine, lysine, arginine, and ornithine	Increased cystine, lysine, arginine, ornithine excretion	Increased cystine, lysine, arginine, ornithine excretion
Urinary amino acid excretion in heterozygotes	Normal	Cystine and lysine above normal	Cystine and lysine above normal

*By permission of Stanbury JB, et al (eds): *The Metabolic Basis of Inherited Disease*, 3d ed. New York: McGraw-Hill, 1972, p. 1552).
†No data.

areas. The rash is intermittent with exacerbations during summer months and is identical to that of nicotinic acid deficiency. The CNS is complexly affected. Episodes of cerebellar ataxia, nystagmus and diplopia are very common. A diffuse pain syndrome of undefined origin may appear as well as multifarious psychiatric manifestations. Mental retardation is an inconstant feature and appears to be overestimated in frequency due to the populations studied. Clinical manifestations are provoked by various stresses, e.g., growth, anxiety, fever and inadequate diet. The disorder relents in adulthood unless dietary faddism or economic malnutrition intervene.

Aminoaciduria is constant. The intestinal and renal epithelial cells are equally affected for transport of a group of neutral aminoacids, although a variant with normal intestinal transport exists. Cultured fibroblasts show no defect (70). Most of the symptoms are related to deficient tryptophan absorption and thus defective nicotinamide production. A shortage of this aminoacid is usually circumvented by the normal intestinal absorption of tryptophan-containing dipeptides. However, when normal intake decreases, the available nicotinamide precursors may become acutely insufficient and prompt the episodic expression of a pellagra-like illness. Therapy includes nicotinamide supplements, high protein diet and attempts at decreasing colonic bacteria (75).

Familial Hyperglycinuria and Iminoaciduria

Familial hyperglycinuria and iminoaciduria together represent disordered function of a renal tubular transport system that is shared by glycine and two iminoacids, proline and hydroxyproline (76-78). In one entity, familial renal glycinuria (glucoglycinuria), abnormally large amounts of glycine are excreted in the urine despite normal plasma concentrations. The urinary excretion of other iminoacids is normal. The condition may manifest itself through nephrolithiasis, although surprisingly the stones are composed of calcium oxalate. Their cause is unknown, but it is interesting that oxalic acid can be formed by deamination and oxidation of glycine (78).

Familial iminoglycinuria is transmitted as an autosomal recessive trait. Per se it appears not to cause a characteristic illness. Some obligate heterozygotes exhibit hyperglycinuria, whereas others do not, and some homozygotes have impaired intestinal absorption of glycine and iminoacids; genetic heterogeneity is present (70). At least four mutant alleles seem to be required to account for the observed interfamilial variations (70).

Vitamin D-Resistant Rickets

Vitamin D-resistant rickets (VDRR) is a disorder of calcium-phosphate homeostasis that may progress to radiologic and clinical rickets, which is refractory to conventional dosages of vitamin D. First described by Albright in 1937 (79), the defect has defied any unitary hypothesis. Two lesions are postulated as etiologic: 1) aberrant cellular phosphate transfer, and 2) intestinal malabsorption of calcium. Evidence for the former is that 1) phosphaturia is the only invariant characteristic of the syndrome—it occurs earliest and may portend the development of osteomalacia (80); 2) like other inborn errors in solute transport multiple-cell types may be affected—in this instance decreased renal tubular phosphate reabsorption and subnormal intestinal phosphate uptake coincide

(81); 3) two mechanisms of renal tubular phosphate reabsorption have been defined and the parathyroid hormone (PTH)-modulated transport system appears to be absent in VDRR (81); 4) there is absence of generalized tubular reabsorptive abnormalities associated with secondary hyperparathyroidism (70); 5) the absence of increased cAMP in the urine as would be expected if secondary hyperparathyroidism pertained (although this is disputed) (70); and 6) no form of vitamin D (25 HCC or 1,25 DHCC) in physiologic doses has been able to correct this disorder; hence a primary error in vitamin D metabolism seems unlikely (83).

Some unknown abnormality in the activation of vitamin D may still underlie the malabsorption. Initial reports of hyperplastic parathyroids were supportive of this concept, but serum measurements of parathormone have been normal or only slightly elevated (84), although measurement of supranormal PTH secretion in response to EDTA-induced hypocalcemia (i.e., the response characteristic of a large glandular mass) argues favorably for the concept of chronic secondary hyperparathyroidism (85). That a humoral mechanism (PTH or vitamin-D metabolites) invokes the phosphaturia is consistent with the return and persistence of phosphaturia in the only reported case of renal transplantation in this disorder (86). In addition this syndrome is mimicked by certain tumors and disappears after curative resections, thus providing circumstantial evidence for a circulating factor (87).

Clinically the disorder is suspected by the familial clustering of short stature and rickets. X-linked dominant inheritance is accepted by most workers with generally severer bone disease in the hemizygous male than in the heterozygous female who may show only hypophosphatemia (89). An autosomal dominant form may also exist and thereby explain some of the conflicting biochemical data. There are no other disturbances of renal function in the untreated subject; however, the renal sequelae of hypercalcemia may ensue in those patients on high-dose vitamin-D therapy. The skin may show the rare lesion of granuloma annulare, but pathogenetic association is not firm (88). Radiographically osteomalacia and rickets compose the predominant osseous lesion. Scanning electron microscopy shows decrements of mineral in the circumlacunar areas, prompting a hypothesis that osteocytic inhibition of mineralization may mediate the development of osteomalacia (90).

Hereditary Vitamin D-Dependency Rickets
This entity, also known as hereditary pseudovitamin D-deficiency rickets, or hereditary vitamin D-refractory rickets with aminoaciduria, was differentiated from the better known X-linked vitamin D-resistant rickets by its earlier presentation, severer radiographic abnormalities, profound weakness, generalized aminoaciduria and complete healing with high dosages of vitamin D. The inheritance is autosomal recessive. The etiology was postulated to be a deficient activity of 25-hydroxycholecalciferol-1-hydroxylase, in that vitamin-D absorption is normal, as is the intestinal responsiveness to minute doses of 1,25 dihyroxycholecalciferol (DHCC) in initial reports (91-93). Subsequent contradictory evidence exists, however, that greater-than-normal therapeutic doses of 1,25 DHCC and its synthetic analog, 1 alpha HCC, are required in this disease (84).

These patients have symptoms that resemble severe nutritional rickets, but they fail to show radiologic healing on 2000-4000 IU of vitamin D daily for three months. Early diagnosis is important since the prognosis is excellent in those children who are identified and treated daily with vitamin D in the range of 100,000 IU (92). In addition to the resolution of rickets and increase in growth rate, aminoaciduria may disappear (94). The life expectancy is normal (95), although there are disturbing reports of childhood cirrhosis and hepatoma (96), which may be prevented by the correction of the aminoaciduria. The condition may be heterogenous (97).

Fanconi Syndrome
In its broadest definition the Fanconi syndrome is a nonspecific renal tubular disorder characterized by aminoaciduria, renal glycosuria and hypophosphatemic rickets. The syndrome may be inherited or acquired (only the former is discussed here) and may be observed in both children and in adults.

Among the former, cystine storage disease is a common finding. In contrast, among adults cystinosis does not occur. In recent years it has become fashionable to distinguish between these forms of the Fanconi syndrome. Bickel et al suggested the now generally accepted term *Lignac-Fanconi* disease for the Fanconi syndrome with cystinosis (98). *Adult* or, preferably, *idiopathic Fanconi syndrome* is distinguished by the lack of cystinosis, good prognosis and general lack of advancing morbidity (99). Genetic factors are operative in the transmission of both forms.

Currently it is held that solute transport across the renal tubular membrane is facilitated by genetically determined protein molecules. Defects in these molecules or in the energy requirements needed to drive them are believed to be responsible for the defect in the various transport systems in the Fanconi syndrome. The increased excretion of aminoacids, phosphate and glucose represents a failure of tubular reabsorption.

Lignac-Fanconi Syndrome. The Lignac-Fanconi syndrome is a recessively inherited disease occurring in about 1 in 40,000 births (100). The disease generally is reflected in the first year or two of life by rickets and failure to thrive. It is characterized by the accumulation of cystine crystals in the internal organs (kidney, lymph nodes, bone marrow, cornea and conjunctiva). To consider that cystine accumulation in these locations is of itself damaging probably is an oversimplification since at least nine persons are described in whom neither renal dysfunction nor ocular disease is present despite the fact that cystine crystal accumulation has occurred in the cornea, bone marrow and white blood cells (101).

Familial studies indicate that the Lignac-Fanconi syndrome is transmitted as an autosomal recessive trait. Variability in the severity of symptoms and in the evolution of clinical disease, and a tendency for the condition to follow a similar pattern among all members of a given family, suggest that genetic heterogeneity is present.

The discovery of the Fanconi syndrome in an adult population is an unusual occurrence. Genetically transmitted adult Fanconi syndrome is rarer still. Hunt and his associates in 1956 were able definitely to establish a genetic basis for the disease in only six cases (102). In addition they reported a family in which transmission of the complete syndrome was observed in successive generations, suggesting autosomal dominant inheritance. In other cases recessive inheritance seems likelier. It seems safe to say then that the heritable form of adult Fanconi syndrome with cystinosis represents a genetically heterogeneous group. The exact nature of the genetic defect—or more probably—defects remains unknown. The Lignac-Fanconi and the adult Fanconi syndrome share a common pathologic finding of hypoplasia of the renal proximal tubular cells, producing the characteristic "swan neck" deformity.

Efforts to recalcify bone through calciferol, the correction of acidosis and the maintenance of serum potassium in the normal range constitute appropriate therapy. Renal transplantation can be considered in advanced cases. Briggs and associates observed a recurrence of the Fanconi syndrome after renal transplantation in a 14-year-old male who had no evidence of cystinosis. Their experience, although still an isolated one, raises the possibility of a systemic cause for renal dysfunction in idiopathic Fanconi syndrome (103).

Pseudo—and Pseudopseudohypoparathyroidism

The clinical presentation of hypoparathyroidism in kindreds with certain somatic features suggests the diagnosis of pseudohypoparathyroidism (PH) or Albright's hereditary osteodystrophy. The diagnosis is confirmed by demonstrating normal or high levels of circulating parathyroid hormone (PTH). The disease is inherited by X-linked dominant (104) or autosomal dominant mechanism (105) with sex influenced expressivity. The female-to-male ratio is 2:1. Pseudopseudohypoparathyroidism describes the same hereditary dystrophy, but with normal calcium and phosphate, and really represents the same disease (104).

The excretion of cyclic adenosine monophosphate (cAMP) in the urine defines this condition (106). Initially the urine cAMP did not show an expected rise to infused PTH in PH, and this was offered as a technically easier laboratory diagnosis for PH than measuring tubular reabsorption of phosphate (TRP) (107). Subsequently two reports have identified another defect in which urinary cAMP increased normally, but phosphate clearance was unaffected (108, 109), the so-called PH Type II. The inheritance pattern of this lesion has not been clarified and since it is unaccompanied by bony abnormalities, it is likely distinct from Albright's osteodystrophy. Type I is thought to be anomalous adenyl cyclase or its PTH-specific membrane receptor; in Type II the target response to adenyl cyclase appears to be deficient.

Two observations defy explanation on the basis of an inborn ultrastructural mutation: 1) target organ responsiveness may return in both Type I and Type II PH after vitamin-D therapy (108, 109), and 2) renal tubular adenyl cyclase isolated from a patient with PH is responsive to PTH in vitro.

The patient has a characteristic physiognomy in PH: short stature, mild obesity, rounded facies, short neck and shortened metacarpals and metatarsals. These patients usually have, at age 5 to 10 years, muscle cramps and weakness, tetany, convulsions, cataracts, dental defects and mental retardation. X-ray studies reveal subcutaneous and basal ganglia calcifications. Occasionally the bone retains its responsiveness to PTH so that there is also radiographic evidence of osteopenia and osteitis fibrosa cystica. The chemical profile includes a low level of calcium with elevated levels of phosphate, alkaline phosphatase, PTH and TRP. There is minimal change in the tubular reabsorptive maximum for phosphate (TmPO$_4$) with exogenous PTH infusion.

The treatment is empirical, each patient requiring monitoring of the serum calcium while receiving large doses of vitamin D, 20,000 to 100,000 units daily, or dihydrotachysterol in doses as high as 5 mg daily. Supplemental calcium of 1 to 4 g daily may be required. With adequate therapy the chemical imbalances will be corrected, and the effects of secondary hyperparathyroidism will reverse. Growth retardation remains and its cause is obscure, but it appears not to be the result of inadequate growth hormone or sulfation factor (110).

Renal Glucosuria

Generalized proximal tubular defects, such as the Fanconi syndrome, display glucosuria. There are also other well-defined examples of inborn errors in selective renal glucose transport. *Familial renal glucosuria* is a disorder that is manifested by glucosuria during euglycemia. Other proximal tubular functions and glomerular filtration rates are normal.

Glucose titration curves identify two forms of familial glucosuria: Type A with a low threshold and low tubular maximum (Tm$_g$) and Type B with a low threshold but normal Tm$_g$. Both types are inherited as autosomal recessive disorders with occasional heterozygous expression.

In clinical medicine one of the most perplexing problems is distinguishing this lesion from the occasionally abnormal tubular handling of glucose in the early stage of diabetes mellitus. Although there is no convincing evidence that renal glycosuria is a commonly associated lesion of diabetes, it is recommended that frequent laboratory surveillance be done to exclude the possibility (111).

Glucosuria is also a relatively consistent finding in *glucose-galactose malabsorption* syndrome. This autosomal recessive disorder results from the inability of the intestinal mucosa to transport either glucose or galactose, resulting in life-threatening, neonatal diarrhea (111). Disaccharidase levels are normal. Severe malnutrition and dehydration result, however, if the infant is fed milk or routine formulas.

Scriver considers the epithelial glucose transport system to be bipartite, each subsystem under the control of a different gene with one or more alleles. The first division, termed G$_1$, carries both galactose and glucose; its operation is dysfunctional in both the intestine and kidney in glucose-galactose malabsorption (70). Since it is proposed that G$_1$ accounts for only 30% of the renal glucose-carrying capacity, glucosuria is mild. The G$_2$ system is specific for glucose transport. Mutants at the G$_2$ locus produce familial renal glucosuria and may, as indicated earlier, result in a low Tmg and low threshold, i.e., Type A, or a widened splay, Type B. The G$_2$ system either is an insignificant component in the gut transport of glucose or remains unaffected by the mutation, since patients have no hexose malabsorption. No therapy is indicated for renal glucosuria, and removal from the diet of carbohydrate that contains glucose and galactose prevents the morbidity related to glucose-galactose malabsorption.

Renal Tubular Acidosis

This syndrome represents an inability of the kidneys to excrete the equivalent of the hydrogen ion (H$^+$) derived from normal dietary intake and catabolism. This type of acidosis may be generated by defects in either the reabsorption of filtered bicarbonate, i.e., defective *bicarbonate reclamation,* or an inability to facilitate net H^+ excretion, i.e., defective *bicarbonate regeneration.* These physiologic divisions correlate to some extent with the anatomic division of proximal and distal tubule, respectively. Detailed discussions are available (112-115).

Clinical complications may be predicted, given a knowledge of the pathophysiology of urine acidification. Both the reclamation and regeneration of HCO_3^- is contingent on H^+ secretion, which is linked to sodium (Na^+) reabsorption. Hence an element of Na^+ loss and (1) volume depletion is an expected consequence. Hypokalemia (2) is frequent and occasionally so severe as to produce flaccid paralysis. The calcium carbonate of bones acts as a major buffer in this chronic acidosis and their absorption of H^+ evokes dissolution of bone, hence (3) osteopenia and retarded growth result. The mobilization of calcium from bone causes hypercalciuria, which in concert with an alkaline urine with low citrate content leads to (4) renal stones and (5) nephrocalcinosis. Thus five major presentations of this disorder are easily predicted pathophysiologically. (Fig. 1).

In excess of 80% of filtered bicarbonate is reabsorbed in the proximal tubule. In the normal kidney HCO_3^- reabsorption is increased quantitatively with increments in the filtered load until a threshold plasma HCO_3^- concentration is reached; then HCO_3^- appears in the urine. Bicarbonate reabsorption continues to increase with increasing serum bicarbonate even after the threshold is exceeded, until a P_{HCO_3} is reached, after which all additional HCO_3^- appears in the urine. This is the tubular maximum ($Tm_{HCO_3^-}$).

Normally the threshold and the Tm for HCO_3^- are only a few mEq/liter apart. The difference is called the splay and probably represents nephron heterogeneity for HCO_3-thresholds. A mutant process could derive from abnormally depressed Tm or widened splay, either of which would allow bicarbonaturia at lower plasma HCO_3^-, i.e., HCO_3^- wastage and hyperchloremic metabolic acidosis. Additionally, defective carbonic anhydrase could give rise to proximal or distal renal tubular acidosis (RTA), or both (115, 116).

Finally several investigators emphasize the role of parathyroid hormone as an inhibitor of proximal tubular HCO_3^- reabsorption and "cures" of proximal RTA after parathyroidectomy are reported (88, 117-119).

Clinically one can diagnose proximal RTA by establishing that an acid urine becomes abruptly alkaline as the serum HCO_3^- level is increased, but remains in an acidotic range. Large amounts of HCO_3^- (10 mEq/kg/day) must be given to attempt to correct a moderately severe proximal RTA, whereas once initial acidosis is corrected in distal RTA, the maintenance HCO_3^- dose equals the net acid production of 1-1.5 mEq/kg/day.

Most proximal RTA is acquired (115), and those forms of RTA with an established hereditary pattern generally belong to a group of systemic disorders in which proximal RTA is only one component of the Fanconi syndrome. A primary isolated proximal RTA is described in a number of infants. Unlike most forms of RTA, children with proximal RTA, by and large recover without sequelae. The inheritance pattern, if any, is obscure (113).

This is the classic form of RTA involving an inability to regenerate HCO_3^- consumed during metabolism. No significant HCO_3^- wastage occurs with the pure form. The distal nephron is permanently unable to lower urine pH to less than 5.5, despite very low blood pH. Micropuncture studies show that the defect

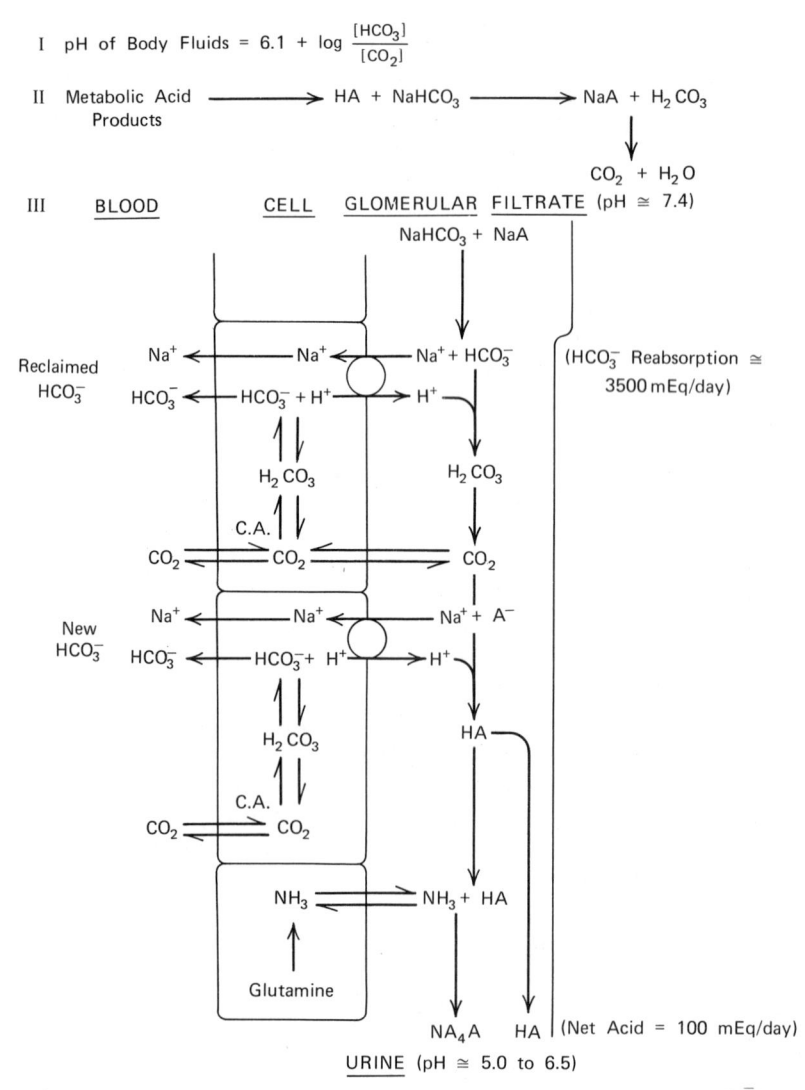

Figure 1. Role of the kidney in regulation of acid-base balance. (Courtesy of JB Stanbury et al (eds): *The Metabolic Basis of Inherited Disease*, 3d ed., New York, McGraw-Hill, 1972.)

actually is in the collecting duct rather than in the distal tubule (115). Unlike proximal RTA, which reaches a steady state when plasma HCO_3^- equals the renal threshold, classic RTA patients have a continuous accumulation in H^+. Hence this group was first described with nephrocalcinosis, kidney stones and azotemia.

Sporadic forms occur, but an autosomal dominant variety (with high penetrance in the female) occurs as early as a few months of age or as late as adulthood. Volume depletion dominates the clinical picture in the infant. Later, growth retardation occurs, followed soon by calcium deposition in the renal parenchyma. There is a transient, sporadic form occurring in infants, and a distal RTA with deficient erythrocyte carbonic anhydrase and associated nerve deafness (120). Many inborn developmental defects are associated with distal RTA. Osteopetrosis, Lowe's syndrome, sickle cell disease and MSK represent a partial list.

Hypercalciuria may underlie the development of RTA in selected families. In a recently reported 64-member kindred (121), 13 showed hypercalciuria, whereas only four could be shown to have a gradient defect in H^+ secretion; six had nephrocalcinosis, two of whom had no evidence of RTA. The authors propose that increased urinary calcium was the primary event. The enhanced sensitivity of these tubules to hypercalciuria is unexplained.

Treatment for RTA should include potassium and HCO_3^- replacement on a permanent basis.

Nephrogenic Diabetes Insipidus

The term nephrogenic diabetes insipidus (NDI) implies renal tubular resistance to circulating antidiuretic hormone (ADH). The defect is unidentified (122, 123). The mode of inheritance is controversial. Sex-linked recessive transmission appears most likely (124). All the cases of NDI in North America may be descendents of the same Scottish clan (125). The male members have severe disease, which develops in infancy. Untreated survivors often show mental and growth retardation, which is generally attributable to episodes of cerebral dehydration and nutritional neglect because of polydipsia. The diagnosis is generally made without difficulty. Guidelines for urinary milliosmolar response to ADH are available (126).

The treatment of NDI deserves emphasis since it is seemingly paradoxical. Daily obligatory urinary losses can be 20 to 30 liters in the adult; therefore it is virtually mandatory to attempt to slow the diuresis. The most important agents in this regard are the diuretics, which reduce the volume of filtrate presented to the dysfunctional collecting duct and thus decrease C_{H_2O} (127-130). The mechanism is increased proximal tubular solute reabsorption stimulated by vascular volume contraction. Dietary sodium also must be restricted and replacement of potassium lost is recommended. When water replacement is impossible orally, intravenous administration of 3% dextrose in water is recommended, since rapid infusion of 5% dextrose in water causes an increment of solute-free water loss and hence intensifies the intracellular dehydration.

Familial Hypokalemia

Hypokalemic metabolic alkalosis, increased concentrations of plasma renin, hyperaldosteronism, juxtaglomerular apparatus (JGA) hyperplasia and normotension characterize this syndrome described by Bartter in two children in 1962 (Bartter syndrome) (131). Subsequent studies favor inheritance as an autosomal recessive trait, but some patients may have acquired disease (132). The nature of the defect has been subject to much debate.

Bartter initially proposed a primary decrease in vascular reactivity to angiotensin, which obligated compensatory increased renin secretion, angiotensin generation and hyperaldosteronism. The accelerated renal H^+ loss mediated the alkalosis (131). Although very appealing, this proposal now appears less cogent. If the vascular volume is acutely restored, the pressor effect of angiotensin returns (132) and if the adrenals are removed (either surgically or chemically), K^+ loss continues (133, 134). Several mechanisms are offered to explain this phenomenon; however, the most parsimonious is that of a single lesion in distal nephron chloride transport (136). The condition is likely heterogeneous, and a subset of patients may have the vascular affliction that Bartter proposed. An excess of renal prostaglandins also has been regarded as etiologic (137).

The clinical spectrum is broad; however, most cases are children with failure to thrive, weakness and polyuria (much of which is attributable to chronic hypokalemia). There is no edema and the blood pressure is normal or low. Occasionally short stature, tetany and erythrocytosis are reported (138).

Most therapeutic regimens contain large dosages of K^+, often in combination with spironolactone in an attempt to delay excretion of this cation although this is only partially successful. Bilateral adrenalectomy and aminoglutethimide are of no value (133, 134). α-Methyldopa (139) and propranolol (140) suppress the augmented renin secretion with only temporary efficacy. Most recently indomethacin has been advocated. Its beneficial effect, coupled with the finding of pretreatment hyperplasia of medullary interstitial cells, reemphasizes a pathogenetic role for prostaglandins (141).

Hereditary Urolithiasis

A number of inherited metabolic abnormalities predispose to the development of urinary tract stones. Best known of these are cystinuria and RTA, which were previously discussed. Several other entities remain.

Hyperoxaluria

Two autosomal recessive mutations in the disposition of oxalate precursors have been defined. *Type I hyperoxaluria* is a deficiency in the enzyme α-ketoglutarate:glyoxalate carboxylase (142). The defect in *Type II hyperoxaluria*, or glyceric aciduria, is not precisely defined. The biochemical pathogenesis has been exhaustively reviewed (143e). Both conditions produce childhood nephrolithiasis, and in the former condition there is generalized oxalate deposition (144). End-stage renal disease occurs by the third decade as a consequence of oxalosis, obstruction and infection, with uremia as the leading cause of death. Significant extrarenal disease may occur, e.g., dysrhythmias due to oxalate deposits in the cardiac conduction system, and acute monoarthritis, resembling gout (143b).

The diagnosis is suspected in children passing pure calcium oxalate stones. The secondary hyperoxalurias should be excluded (143a-c). Confirmation is obtained by the high levels of urinary oxalate excretion per surface area (in the absence of renal failure); oxalosis demonstrated on tissue biopsy, or (for Type I) the assay of low levels of the specific carboligase in kidney or liver tissue (141).

Multiple treatment regimens have been attempted; pyridoxine in high dosages; allopurinol; disulfiram; organic and inorganic phosphates; and thiazides are all of theoretical value, but without documented efficacy. The restriction of dietary calcium and oxalate and insistence on large urine volumes appear justified. The rate of production of oxalate in Type I patients exceeds the clearance by peritoneal dialysis, but hemodialysis can stabilize plasma levels although it is ineffective in removal of deposited oxalate (143e). Renal transplantation is of limited value (pending a means of blocking oxalate production) since the graft is rapidly destroyed (142).

Uricosuria

The metabolic pathway leading to uric acid is regulated by intermediary compounds inhibiting the activity of the rate-limiting enzyme. The enzyme hypoxanthine-guanine phosphoribosyltransferase (HPRTase) is responsible for salvaging the purine compound (guanylic acid) that has this inhibitory function. In the absence of HPRTase, large quantities of substrate descend the pathway to uric acid. Several mutants in the activity of HPRTase are described. X-linked uric aciduria is a result of near-complete deficiency (less than 1% of this enzyme).

The renal dysfunction in this syndrome is a less prominent aspect of the disease, which is dominated by severe neurologic defects manifested as mental and motor retardation and bizarre self-mutilation (Lesch-Nyhan syndrome). Gout occurs. A minority of patients die of uremia, but most of them have renal insufficiency with gouty nephropathy (demonstrated at postmortem). Nephrolithiasis and secondary pyelonephritis are frequent. Assay of HPRTase in unwashed erythrocytes confirms the diagnosis.

Alkaline diuresis and neonatal therapy with allopurinol may obviate the renal compromise; however, the neurologic findings continue unabated. X-linked partial HPRTase deficiency also exists, and instituting the therapy just described is more gratifying in this condition in which the neurologic disease is much less severe, and childhood urolithiasis and gout dominate. These entities have been authoritatively reviewed (145).

Hypercalcuria

Recently it was proposed that a subset of patients with familial RTA in actuality had primary defects in renal tubular calcium reabsorption with subsequent RTA secondary to nephrocalcinosis (121). This intriguing group is discussed under the topic of RTA. In addition to a renal defect, hypercalciuria can result from enhanced calcium absorption or increased endogenous calcium mobilization, or both (146). The first mechanism has yet to be attributed to a genetic cause. Resorptive hypercalciuria occurs in hereditary syndromes that include primary hyperparathyroidism, e.g., Sipple's syndrome.

GENITOURINARY TRACT: HEREDITARY AND CONGENITAL ANOMALIES

In 1967, when Perkoff published his review of the hereditary renal diseases, he devoted nine lines to this subject and cited three references (147). The situation then remains the situation now. Reported data and experience are scarce, and single family experience is the rule at most centers.

In 1968, Winter and associates reported a family in which four female siblings demonstrated varying degrees of renal dysgenesis, vaginal atresia and anomalies of the bones in the middle ear (148). One sibling had trisomy X, which was thought to be coincidental. Autosomal recessive inheritance was suspected. Burger and Smith interpreted evidence gathered from five families as indicative that a ureteral orifice that allowed reflux was congenital and familial and showed no sex-linkage (149). Middleton and his associates describe a family in which three brothers and the maternal grandfather had vesicoureteral reflux, whereas the parents and three daughters were urologically normal (151). An X-linked mode was suspected. Kohn and Borns have reported bilateral and unilateral renal agenesis in the same family (152). Other syndromes in which renal or genitourinary malformations may occur include 1) cryptorchidism (Smith-Lemli-Opitz syndrome, Laurence-Moon-Biedl syndrome, trisomy D, trisomy E and XO Turner syndrome); 2) vaginal atresia; 3) "prune belly" syndrome; 4) tuberous sclerosis; and 5) various intersex states.

The reporting of isolated families and the overall paucity of cases make it difficult to bring any meaningful order to the subclassification of these lesions, especially on a basis of genetic characteristics.

HERITABLE SYSTEMIC DISEASES: RENAL MANIFESTATIONS

A variety of heritable illnesses the principal manifestations of which are extrarenal may also involve the kidney directly. They include 1) cerebrohepatorenal syndrome, 2) Ehlers-Danlos syndrome [cutis hyperelastica], 3) Fabry's disease [angiokeratoma corporis diffusum universale], 4) familial thrombocytopenia, 5) Lowe's syndrome, 6) nail-patella syndrome [arthro-onychodysplasia], 7) sickle cell nephropathy, 8) tuberous sclerosis and 9) Wilson's disease [hepatolenticular degeneration].

The *Ehlers-Danlos syndrome* is characterized by an abnormality in collagen structure permitting the skin to be stretched excessively. Colonic diverticula, diaphragmatic hernia, hyperextensible joints and poor wound healing characteristically coexist. Heterogeneity characterizes the genetic mechanisms, some seven varieties having been recognized. Most common are those transmitted as autosomal dominant traits, but autosomal recessive and X-linked forms have also been identified (153). Polycystic kidney disease (and berry aneurysms) are found in a few patients, leading to speculation that the collagen defect favors renal cyst formation (38).

Fabry's disease, or angiokeratoma corporis diffusum universale, is characterized by small purple papules and macules involving the umbilical, lumbar, scrotal, and other dermal regions. It is transmitted as an X-linked disorder with constant penetrance in the hemizygous male and variable penetrance in the heterozygous female. Deficiency of the enzyme, ceramide trihexosidase, is thought to be responsible for the progressive accumulation of galactosylgalactosylglucosyl ceramide, which gives rise to symptoms in this fatal illness. The accumulation of trihexosyl ceramide in the kidney, including glomerular endothelium and epithelium, leads to renal failure. The condition is discussed in detail elsewhere in this textbook.

Familial thrombocytopenia is a genetically heterogeneous disorder characterized by an inconstantly present bleeding diathesis. It occurs as an autosomal dominant, autosomal recessive or X-linked recessive trait (154). In 1970, Gutenberger et al reported a family of 45 members in which thrombocytopenia, elevated levels of serum immunoglobulin A (IgA) and hematuria without a bleeding tendency were found (154). Renal biopsies in three brothers showed varying degrees of glomerulonephritis. The authors considered both the nephritis and the elevated IgA levels to reflect rather an abnormal immunologic response than an interdependent casual relationship.

The *cerebrohepatorenal,* or *Zellweger,* syndrome is a rare autosomal recessive trait characterized by generalized hypotonia, craniofacial dysmorphia, hepatomegaly and renal cortical cysts (155). Although albuminuria is present, renal function is normal. The number and size of cysts may vary, but they are not a major component of the disorder. Microcysts of the renal cortex also have been described in chromosomal translocation, in congenital cutis laxa and in the syndromes of Goldenhar, Marden-Walker and Majewski (49). Renal dysplasia is reported to accompany the cystic changes in the additional syndromes of Jeune and Meckel (49). Thus there exists a number of disorders in which renal microcystic or dysplastic changes accompany CNS and hepatic or craniofacial abnormalities, or both.

The *oculocerebrorenal,* or *Lowe, syndrome* is characterized by mental retardation, hypotonia, ocular abnormalities, metabolic acidosis, generalized aminoaciduria and, usually, rickets (156). Early reports described its occurrence only in males (156, 157), leading to a conclusion that the syndrome was transmitted as an X-linked recessive trait. The condition is seen in females (158), possibly due to Lyonization or perhaps heterogeneity. The kidney exhibits swelling, irregularity and rounding of its proximal tubular mitochondria. Thickening and splitting of its glomerular basement membranes are seen on electron microscopic examination (159).

The *nail-patella syndrome* (hereditary arthro-onychodysplasia) is a dominantly inherited disorder of connective tissue in which dysplasia of fingernails, webbed elbows, rudimentary patellas and iliac horns constitute a dramatic clinical profile. The nephropathy accompanying the disorder manifests itself anatomically through an irregular thickening and "moth-eaten" appearance of the glomerular basement membrane, fusion of foot processes and inclusion of a fibrillar collagen-like material within the basement membrane itself (160-162). These changes may be accompanied by proteinuria, and renal failure necessitating dialysis or transplanation may ensue. Uranga and his associates found no evidence that the disease was recurring in the kidney of an affected subject one year after renal transplantation (163).

Sickle cell nephropathy encompasses both the functional and the anatomic changes associated with sickle cell anemia and sickle cell trait. A discussion of the generalities of these disorders appears elsewhere. Their renal manifestations are classified as 1) functional (azotemia, hematuria, impaired urinary concentrating ability, incomplete renal tubular acidosis [?], nephrotic syndrome and reduced glomerular filtration [late]; and 2) structural (cortical infarcts, focal mesangial proliferation, glomerular congestion, glomerulosclerosis, hemosiderosis and papillary necrosis).

Tuberous sclerosis is a dominantly inherited disorder in which angiomyolipomas, commonly a part of the complex and cysts may be found in the kidneys. Cysts may enlarge to the point of causing renal failure (49).

Wilson's disease (hepatolenticular degeneration) is a rare autosomal recessive trait in which an accumulation of copper in the kidney leads to renal distal tubular dysfunction, acidosis, aminoaciduria and Fanconi syndrome (164-165). Death generally intervenes as a consequence rather of CNS or hepatic involvement than of renal failure.

REFERENCES

1. Advisory Committee to the Renal Transplant Registry: The eleventh report of the human transplant registry. *JAMA* 226:1197, 1973
2. Dodge WF, Spargo BH, Travis LB: Occurrence of acute glomerulonephritis in sibling con-

tacts of children with sporadic acute glomerulonephritis. *Pediatrics* 40:1029, 1967
3. Perkoff GT, Sly WS: Genetics in renal disease, in Strauss MB, Welt LG (eds): Diseases of the Kidney. Boston, Little Brown, 1971, p. 1215
4. Dickinson WH: Diseases of the Kidney, Vol. 1. London, Longnes, Green & Co., 1875, p. 379
5. Crawfurd MD, Toghill PJ: Alport's syndrome of hereditary nephritis and deafness. *Q J Med New Series* 148:563, 1968
6. Grace SG, Suki WN, Spjut HJ, et al: Hereditary nephritis in the Negro. *Arch Int Med* 125:451, 1970
7. Chazan JA, Zacks J, Cohen JJ, et al: Hereditary nephritis. *Am J Med* 50:764, 1971
8. Guthrie LG: "Idiopathic," or congenital, hereditary and family hematuria. *Lancet* 1:1243, 1902
9. Kendall C, Hertz AC: Hereditary familial nephritis. *Guy's Hosp Rep* 66:137, 1912
10. Hurst AF: Hereditary familial congenital haemorrhagic nephritis occurring in sixteen individuals in three generations. *Guy's Hosp Rep* 73:368, 1923
11. Eason J, Smith GLM: Hereditary and familial nephritis. *Lancet* 2:639, 1924
12. Alport AC: Hereditary familial congenital haemorrhagic nephritis. *Br Med J* 1:504, 1927
13. Perkoff GT, Nugent CA Jr, Dolowitz DA, et al: A follow-up study of hereditary chronic nephritis. *Arch Int Med* 102:733, 1958
14. Shaw RF, Glover RA: Abnormal segregation in hereditary renal disease with deafness. *Am J Hum Genet* 13:89, 1961
15. Royer P: Familial nephropathy with deafness, in Hamburger J, et al (eds): Nephrology. Philadelphia, Saunders, 1966, p. 803
16. Ferguson AC, Rance CP: Hereditary nephropathy with nerve deafness (Alport's syndrome). *Am J Dis Childh* 124:84, 1972
17. Reyersbach GC, Butler AM: Congenital hereditary hematuria. *N Engl J Med* 251:377, 1954
18. Albert MS, Leeming JM, Wigger HJ: Familial nephritis associated with the nephrotic syndrome. *Am J Dis Childh* 117:153, 1969
19. Felts JH: Hereditary nephritis with the nephrotic syndrome. *Arch Int Med* 125:459, 1970
20. Heptinstall RH: Pathology of the Kidney, 2d ed. Boston, Little Brown, 1974, p. 1138
21. Joshi VV: Pathology of hereditary nephritis. *J Clin Pathol* 21:744, 1968
22. Grunfeld JP, Bois EP, Hinglais N: Progressive and nonprogressive hereditary chronic nephritis. *Kidney Int* 4:216, 1973
23. Sherman RL, Churg J, Yudis M: Hereditary nephritis with a characteristic renal lesion. *Am J Med* 56:44, 1974
24. Cohen MM, Cassady G, Hanna BL: A genetic study of hereditary renal dysfunction with associated nerve deafness. *Am J Hum Genet* 13:379, 1961
25. Tishler PV, Rosner B: The genetics of the Alport syndrome. *Birth Defects* 10:93, 1974
26. McConville JM, West CD, McAdams AJ: Familial and nonfamilial benign hematuria. *J Pediatr* 69:207, 1966
27. Schafer IA, Scriver CR, Efron ML: Familial hyperprolinemia, cerebral dysfunction, and renal anomalies occurring in a family with hereditary nephropathy and deafness. *N Engl J Med* 267:51, 1962
28. Efron ML: Familial hyperprolinemia. *N Engl J Med* 272:1243, 1965
29. Kopelman H, Asatoor AM, Milne MD: Hyperprolinaemia and hereditary nephritis. *Lancet* 2:1075, 1964
30. Goyer RA, Reynolds J Jr, Burke J, et al: Hereditary renal disease with neurosensory hearing loss, prolinuria, and ichthyosis. *Am J Med Sci* 256:166, 1968
31. Gafni J, Sohar E, Heller H: The inherited amyloidosis: Their clinical and theoretical significance. *Lancet* 1:71, 1964
32. Sohar E, Gafni J, Pras M, et al: Familial mediterranean fever. *Am J Med* 43:227, 1967
33. Khachadurian AK, Armenian HK: Familial paroxysmal polyserositis (familial mediterranean fever) incidence of amyloidosis and mode of inheritance. *Birth Defects* 10:62, 1974

34. Van Allen MW, Frohlich JA, Davis JR: Inherited predisposition to generalized amyloidosis: Clinical and pathological study of a family with neuropathy, nephropathy, and peptic ulcer. *Neurology* 19:10, 1969
35. Muckle TJ, Wells M: Urticaria, deafness, and amyloidosis: A new heredo-familial syndrome. *Q J Med* 31:235, 1962
36. Schimke RN: The hereditary nephritides. *Birth Defects* 6:12, 1970
37. Dalgaard OZ: Bilateral polycystic disease of the kidneys. A follow-up of two hundred and eighty-four patients and their families. *Acta Med Scand* 158 (Suppl 328), 1957
38. Schimke RN: Genetics in cystic kidney disease, in Gardner KD Jr (ed): Cystic Diseases of the Kidney. New York, Wiley, 1976
39. Blyth H, Ockenden BG: Polycystic disease of the kidneys and liver presenting in childhood. *J Med Genet* 8:257, 1971
40. Lieberman E, Salinas-Madrigal L, Gwinn JL, et al: Infantile polycystic disease of the kidneys and liver: Clinical, pathological, and radiological correlations and comparisons with congenital hepatic fibrosis. *Medicine* 50:277, 1971
41. Cairns HWB: Heredity in polycystic disease of the kidney. *Q J Med* 18:359, 1925
42. Virchow R: Discussion über den Vortrag des Herrn A. Ewald: Zur totalen cystischen Degeneration der Nieren. *Klin Wochenschr* 29:104, 1892
43. Lampert P: Polycystic disease of the kidney: A review. *Arch Pathol* 44:34, 1947
44. Bricker NS, Patton JF: Cystic disease of the kidneys: A study of dynamics and chemical composition of cyst fluid. *Am J Med* 18:207, 1955
45. Gardner KD Jr: Juvenile nephronophthisis and renal medullary cystic disease, in Gardner KD Jr (ed): Cystic Diseases of the Kidney. New York, Wiley, 1976, p. 173
46. Strauss MB: Clinical and pathological aspects of cystic disease of renal medulla. *Ann Int Med* 57:373, 1962
47. Smith CH, Graham JB: Congenital medullary cysts of the kidneys with severe refractory anemia. *Am J Dis Childh* 69:369, 1945
48. Mangos JA, Opitz JM, Lobeck CC, et al: Familial juvenile nephronophthisis. An unrecognized renal disease in the United States. *Pediatrics* 34:337, 1964
49. Bernstein J: A classification of renal cysts, in Gardner KD Jr (ed): Cystic Diseases of the Kidney. New York, Wiley, 1976, p. 26
50. Schimke RN: Hereditary renal-retinal dysplasia. *Ann Int Med* 70:736, 1969
51. Sworn MJ, Eisinger AJ: Medullary cystic disease and juvenile nephronophthisis in separate members of the same family. *Arch Dis Child* 47:278, 1972
52. Rayfield EJ, McDonald FD: Red and blond hair in renal medullary cystic disease. *Arch Int Med* 130:72, 1972
53. Hurley RM, Dery P, Nogrady MB, et al: The renal lesion of the Laurence-Moon-Biedl syndrome. *J Pediatr* 87:206, 1975
54. Whelton A, Ozer FL, Bias W: Renal medullary cystic disease: A family study. *Birth Defects* 10:154, 1974
55. Boichis H, Passwell J, David R, et al: Congenital hepatic fibrosis and nephronophthisis: A family study. *Q J Med* 42:221, 1973
56. Mena E, Bookstein JJ, McDonald FD, et al: Angiographic findings in renal medullary cystic disease. *Radiology* 110:277, 1974
57. Resnick JS, Brown DM, Vernier RL: Normal development and experimental models of cystic renal disease, in Gardner KD Jr (ed): Cystic Diseases of the Kidney. New York, Wiley, 1976
58. Ekstrom T, Engfeldt B, Lagergren C, et al: Medullary Sponge Kidney. Stockholm, Almqvist and Wiksell, 1959
59. Kuiper JJ: Medullary sponge kidney, in Gardner KD Jr (ed): Cystic Diseases of the Kidney. New York, Wiley, 1976, p. 151
60. Kuiper JJ: Medullary sponge kidney in three generations. *NY State J Med* 71:2665, 1971
61. Sprayregen S, Strasberg Z, Naidich TP: Medullary sponge kidney and congenital total hemihypertrophy. *NY State J Med* 73:2768, 1971

62. Hallman N, Hjelt L: Congenital nephrotic syndrome. *J Pediatr* 55:152, 1959
63. Hallman N, Norio R, Kouvalainen K: Main features of the congenital nephrotic syndrome. *Acta Paediatr Scand* 172:75, 1967
64. Norio R: The nephrotic syndrome and heredity. *Hum Hered* 19:113, 1969
65. Bader PI, Grove J, Trygstad CW, et al: Familial nephrotic syndrome. *Am J Med* 56:34, 1974
66. Vernier RL, Brunson J, Good RA: Studies on familial nephrosis. *J Dis Childh* 93:469, 1957
67. Oliver J: Microcystic renal disease and its relation to "infantile nephrosis." *Am J Dis Childh* 100:312, 1960
68. Meister A: The γ-glutamyl cycle: Diseases associated with specific enzyme deficiencies. *Ann Int Med* 81:247, 1974
69. Segal S: Disorders of renal amino acid transport. *N Engl J Med* 294:1044, 1976
70. Scriver CR, Chesney RW, McInnes RR: Genetic aspects of renal tubular transport: Diversity and topology of carriers. *Kidney Int* 9:149, 1976
71. McKusick VA: Mendelian Inheritance in Man: Catalogs of Autosomal Dominant Autosomal Recessive and X-linked Phenotypes. Baltimore, Johns Hopkins Press, 1975
72. Garrod AE: Inborn errors of metabolism. *Lancet* 2:1,73,142,214, 1908
73. Thier SO, Segal S: Cystinuria, in Stanbury JB, et al (eds): Metabolic Basis of Inherited Disease. New York, McGraw-Hill, 1972
74. Jepson JB: Hartnup disease, in Stanbury JB, et al (eds): Metabolic Basis of Inherited Disease. New York, McGraw-Hill, 1972
75. Harrison RJ, Feiwel M: Pellagra caused by isoniazid. *Br Med J* 2:852, 1956
76. Vries de A, Kochwa S, Lazebnik J: Glycinuria, a hereditary disorder associated with nephrolithiasis. *Am J Med* 23:408, 1957
77. Kaser H, Cottier P, Antener I: Glucoglycinuria, a new familial syndrome. *J Pediatr* 61:386, 1962
78. Greene ML, Lietman PS, Rosenberg LE: Familial hyperglycinuria: New defect in renal tubular transport of glycine and iminoacids. *Am J Med* 54:265, 1973
79. Albright F, Butler AM, Bloomberg E: Rickets resistant to vitamin D therapy. *Am J Dis Childh* 54:529, 1937
80. Williams FT, Winters RW: Familial (hereditary) vitamin-D-resistant rickets with hypophosphatemia, in Stanbury JB, et al (eds): Metabolic Basis of Inherited Disease. New York, McGraw-Hill, 1972, p. 1456
81. Short EM, Binder HJ, Rosenberg LE: Familial hypophosphatemic rickets: Defective transport of inorganic phosphate by intestinal mucosa. *Science* 179:700, 1973
82. Glorieux F, Scriver CR: Loss of a parathyroid hormone-sensitive component of phosphate transport in x-linked hypophosphatemia. *Science* 175:997, 1972
83. Hahn TJ, Scharp OR, Halstead LR, et al: Parathyroid hormone status and renal responsiveness in familial hypophosphatemic rickets. *J Clin Endocrinol Metab* 41:926, 1975
84. Balsan S, Garabedian M, Sorgniard R, et al: 1-25 dihydroxy-vitamin D₃ and 1,a-hydroxyvitamin D₃ in children: Biologic and therapeutic effects in nutritional rickets and different types of vitamin D resistance. *Pediatr Res* 9:486, 1975
85. Arnand CD, Glorieux F, Scriver CR: Serum parathyroid hormone in x-linked hypophosphatemia. *Science* 173:845, 1971
86. Reitz RE, Weinstein RL: Parathyroid hormone secretion in familial vitamin-D-resistant rickets. *N Engl J Med* 289:941, 1973
87. Morgan JM, Hawley WL, Chenoweth AI, et al: Renal transplantation in hypophosphatemia with vitamin-D-resistant rickets. *Arch Int Med* 134:549, 1974
88. Salassa RM, Jowsey J, Arnaud CD: Hypophosphatemia osteomalacia associated with "nonendocrine" tumors. *N Engl J Med* 283:65, 1970
89. Steendijk P, Boyde A: Scanning electron microscopic observations on bone from patients with hypophosphatemic (vitamin-D-resistant) rickets. *Calcif Tissue Res* 11:242, 1973
90. Rabinowitz B, Roenigk HH Jr, Schumacher OP: Granuloma annulare and vitamin D-resistant rickets. *Cleve Clin Q* 41(2):75, 1974

91. Lyon MF: Sex chromatin and gene action in the mammalian x-chromosome. *Am J Human Genet* 14:135, 1962
92. Fraser D, Salter RB: The diagnosis and management of the various types of rickets. *Pediatr Clin North Am*, May 1958
93. Fraser D, Kooh SW, Kind HP, et al: Vitamin-D-dependent rickets: An inborn error of vitamin D metabolism. *N Engl J Med* 289:817, 1973
94. Hamilton R, Harrison J, Fraser D, et al: The small intestine in vitamin-D-dependent rickets. *Pediatrics* 45:364, 1970
95. Morris CR, McSherry E, Sebastian A: Modulation of experimental renal dysfunction of hereditary fructose intolerance by circulating parathyroid hormone. *Proc Natl Acad Sci USA* 68:132, 1971
96. Fraser D: Rickets, in Conn D (ed): Current Diagnosis. Philadelphia, Saunders, 1968
97. Prader VA, Illig R, Hierli E: Eine besondere Form der primären vitamin-D-resistenten Rachitis mit Hypocalcamie und autosomaldominantem Erbgang: die hereditäre Pseudomangelrachitis. *Helv Paediatr Acta* 16:452, 1961
98. Bickel H, Baar HS, Astley R, et al: Cystine storage disease with aminoaciduria and dwarfism (Lignac-Fanconi disease). *Acta Paediatr* 42 (Suppl 90):9, 1952
99. Dent CE, Harris H: Hereditary forms of rickets and osteomalacia. *J Bone Joint Surg* 38:204, 1956
100. Bickel H, Smallwood WC, Smellie JM, et al: Clinical description, factual analysis, prognosis and treatment of Lignac-Fanconi disease. *Acta Paediatr* 42 (Suppl 90):27, 1952
101. Schneider JA, Seegmiller JE: Cystinosis and the Fanconi syndrome, in Stanbury JB, et al (eds): The Metabolic Basis of Inherited Disease 3d ed. New York, McGraw-Hill, 1972, p. 1581
102. Hunt DD, Stearns G, McKinley JB, et al: Long-term study of family with Fanconi syndrome without cystinosis (DeToni-Debre-Fanconi syndrome). *Am J Med* 40:492, 1966
103. Briggs WA, Kominami N, Wilson RE: Kidney transplantation in Fanconi syndrome. *N Engl J Med* 286:25, 1972
104. Mann JB, Herman A, Hills AG: Albright's hereditary osteodystrophy comprising pseudohypoparathyroidism and pseudo-pseudohypoparathyroidism. *Ann Int Med* 56:315, 1962
105. Arnstein AR, Frame B, Frost HM, et al: Albright's hereditary osteodystrophy: Report of a family with studies of bone remodeling. *Ann Int Med* 64:996, 1966
106. Rasmussen H: Parathyroid hormone, calcitonin, and the calciferols, in Williams RH (ed): Textbook of Endocrinology. Philadelphia, Saunders, 1974
107. Drezner M, Neelon FA, Lebovitz HE: Pseudohypoparathyroidism type II: Possible defect in reception of cyclic AMP signal. *N Engl J Med* 289:1056, 1973
108. Rodriguez HJ, Villarreal H Jr, Klahr S, et al: Pseudohypoparathyroidism type II: Restoration of normal renal responsiveness to parathyroid hormone by calcium administration. *J Clin Endocrinol Metab* 39:693, 1974
109. Stogmann W, Fischer JA: Pseudohypoparathyroidism: Disappearance of the resistance to parathyroid extract during treatment with vitamin D. *Am J Med* 59:140, 1975
110. Urandivia E, Mataverde A, Cohen MP: Growth hormone secretion and sulfation factor activity in pseudohypoparathyroidism. *J Lab Clin Med* 86:772, 1975
111. Calcagno PL, Hollerman A: Hereditary renal disease, including certain renal tubular disorders, in Rubin MI, Barratt T (eds): Pediatric Nephrology. Baltimore, Williams & Wilkins, 1975, p. 668
112. Seldin DW: Renal tubular acidosis, in Stanbury JB, et al (eds): Metabolic Basis of Inherited Disease. New York, McGraw-Hill, 1972, p. 1548
113. Edelmann CM: Renal tubular acidosis. *Kidney* 1:1, 1971
114. Morris RC Jr, Sebastian A, McSherry E: Renal acidosis. *Kidney Int* 1:322, 1972
115. Steinmetz PR, Al-Awqati Q, Lawton WJ: Specialty rounds: Nephrology rounds, University of Iowa Hospital: Renal tubular acidosis. *Am J Med Sci* 271:41, 1976

116. Donckerwolcke RA, Van Stekelenburg J, Tiddens HA: A case of bicarbonate-losing renal tubular acidosis with defective carbonic anhydrase activity. *Arch Dis Child* 45:769, 1970
117. Massry S, Kurokawa K, Arieff A, et al: Metabolic acidosis of hyperparathyroidism. *Arch Int Med* 134:385, 1974
118. Beck N, Kim KS, Wolak M, et al: Inhibition of carbonic anydrase by parathyroid hormone and cyclic AMP in rat renal cortex in vitro. *J Clin Invest* 55:149, 1975
119. Muldowney FP, Carroll DV, Donohoe JF, et al: Correction of renal bicarbonate wastage by parathyroidectomy. *Q J Med* XL(160):487, 1971
120. Shapira E, Ben-Yoseph Y, Eyal FG, et al: Enzymatically inactive red cell carbonic anhydrase B in a family with renal tubular acidosis. *J Clin Invest* 53:59, 1974
121. Buckalew VM Jr, Purvis ML, Shulman MG, et al: Hereditary renal tubular acidosis. *Medicine* 53:229, 1974
122. Avery S, Clark CM Jr, Trygstad C, et al: Effects of cyclic AMP in antidiuretic hormone-deficient and antidiuretic hormone-resistant diabetes insipidus. *J Clin Invest* 50:3A, 1971
123. Dousa TP: Cellular action of antidiuretic hormone in nephrogenic diabetes insipidus. *Mayo Clin Proc* 49:188, 1974
124. Williams RH, Henry C: Nephrogenic diabetes insipidus: Transmitted by females and appearing during infancy in males. *Ann Int Med* 27:84, 1947
125. Bode HH, Crawford JD: Nephrogenic diabetes insipidus in North America—the Hopewell hypothesis. *N Engl J Med* 280:750, 1969
126. Gardner KD Jr: Diabetes insipidus, in Conn HF (ed): Current Therapy. Philadelphia, Saunders, 1973
127. Earley LE, Orloff J: The mechanism of antidiuresis associated with the administration of hydrochlorothiazide to patients with vasopressin-resistant diabetes insipidus. *J Clin Invest* 41:1988, 1962
128. Brown D, Reynolds JW, Michael AF, et al: The use and mode of action of ethacrynic acid in nephrogenic diabetes insipidus. *Pediatrics* 37:447, 1966
129. Kowarski A, Moshe B, Grossman M, et al: Antidiuretic properties of aldactone (spironolactone) in diabetes insipidus: Studies on the mechanism of antidiuresis. *Bull Johns Hopkins Hosp.* 119:413, 1966
130. Baum NH, Burger R, Carlton C Jr: Nephrogenic diabetes insipidus associated with posterior urethral valves. *Urology* 4:581, 1974
131. Bartter FC, Pronove P, Gill JR Jr, et al: Hyperplasia of the juxtaglomerular complex with hyperaldosteronism and hypokalemic alkalosis: A new syndrome. *Am J Med* 33:811, 1962
132. White MG: Bartter's syndrome: A manifestation of renal tubular defects. *Arch Int Med* 129:41, 1972
133. Trygstad CW, Margos JA, Bloodworth JM Jr, et al: A sibship with Bartter's syndrome: Failure of total adrenalectomy to correct the potassium wasting. *Pediatrics* 44:234, 1969
134. Goodman AD, Vagnucci AH, Hartroft PM: Pathogenesis of Bartter's syndrome. *N Engl J Med* 281:1435, 1969
135. Kurtzman NA, Gutierrez LF: The pathophysiology of Bartter's syndrome. *JAMA* 234:758, 1975
136. Kurtzman NA: Regulation of renal bicarbonate reabsorption by extracellular volume. *J Clin Invest* 49:586, 1970
137. Fichman MP, Telfer N, Zia P, et al: Role of prostaglandins in the pathogenesis of Bartter's syndrome. *Am J Med* 60:785, 1976
138. Erkelens DW, Statius LW, Van Eps LW: Bartter's syndrome and erythrocytosis. *Am J Med* 55:711, 1973
139. Strauss RB: Failure of methyldopa therapy in Bartter's syndrome. *J Pediatr* 85:101, 1974
140. Schwartz G, Cornfield D: Bartter's syndrome: Clinical study of its treatment with salt loading and propranolol. *Clin Nephrol* 4:45, 1975
141. Verberckmoes R, van Damme B, Clement J, et al: Bartter's syndrome with hyperplasia of renomedullary cells: Successful treatment with indomethacin. *Kidney Int* 9:302, 1976

142. Saxon A, Busch GJ, Merrill JP, et al: Renal transplantation in primary hyperoxaluria. *Arch Int Med* 133:464, 1974
143. Hagler L, Herman RH: Oxalate metabolism I-V. *Am J Clin Nutr* 26:758(a), 882(b), 1006(c), 1073(d), 1242(e), 1973
144. Williams HE: Nephrolithiasis. *N Engl J Med* 290:33, 1974
145. Seegmiller JE: Disorders of purine and pyrimidine metabolism, in Bondy PK, Rosenberg LE (eds): Duncan's Diseases of Metabolism. Philadelphia, Saunders, 1974
146. Pak CYC, Ohata M, Lawrence EC, et al: The hypercalciurias: Causes, parathyroid functions, and diagnostic criteria. *J Clin Invest* 54:387, 1974
147. Perkoff GT: The hereditary renal diseases. *N Engl J Med* 277:79, 1967
148. Winter JSD, Kohn G, Mellman WJ, et al: A familial syndrome of renal, genital, and middle ear anomalies. *J Pediatr* 72:88, 1968
149. Burger RH, Smith C: Hereditary and familial vesicoureteral reflux. *J Urology* 106:845, 1971
150. Frye RN, Patel HR, Parsons V: Familial renal tract abnormalities and cortical scarring. *Nephron* 12:188, 1974
151. Middleton GW, Howards SS, Gillenwater JY: Sex-linked familial reflux. *J Urol* 114:36, 1975
152. Kohn G, Borns P: The association of bilateral and unilateral renal agenesis in the same family. *Birth Defects* 10:162, 1974
153. McKusick VA: Heritable Disorders of Connective Tissue, 4th ed. St. Louis, Mosby, 1972, p. 292
154. Gutenberger J, Trygstad CW, Stiehm ER, et al: Familial thrombocytopenia, elevated serum IgA levels and renal disease. *Am J Med* 49:729, 1970
155. Passarge E, McAdams AJ: Cerebro-hepato-renal syndrome: A newly recognized hereditary disorder of multiple congenital defects, including sudanophilic leukodystrophy, cirrhosis of the liver, and polycystic kidneys. *J Pediatr* 71:691, 1967
156. Abbassi V, Lowe CU, Calcagno PL: Oculo-cerebro-renal syndrome. *Am J Dis Child* 115:145, 1968
157. Pallisgaard G, Goldschmidt E: The oculo-cerebro-renal syndrome of Lowe in four generations of one family. *Acta Paediatr Scand* 60:146, 1971
158. Sagel I, Ores RO, Yuceoglu AM: Renal function and morphology in a girl with oculo-cerebro-renal syndrome. *J Pediatr* 77:124, 1970
159. Ores RO: Renal changes in oculo-cerebro-renal syndrome of Lowe. *Arch Pathol* 89:221, 1970
160. Leahy MS: The hereditary nephropathy of osteo-onychodysplasia: Nail-Patella syndrome. *Am J Dis Childh* 112:237, 1966
161. Pozo E dP, Lapp H: Ultrastructure of the kidney in the nephropathy of the nail-patella syndrome. *Am J Clin Pathol* 54:845, 1970
162. Vernier RL, Hoyer JR, Michael AF: The nail-patella syndrome—pathogenesis of the kidney lesion. *Birth Defects* 10:57, 1974
163. Uranga VM, Simmons RL, Hoyer JR, et al: Renal transplantation of the nail-patella syndrome. *Am J Surg* 125:777, 1973
164. Wolff SM: Renal lesions in Wilson's disease. *Lancet* 1:843, 1964
165. Walshe JM: Effect of penicillamine on failure of renal acidification in Wilson's disease. *Lancet* 1:775, 1968

17
Genetic Disorders of the Respiratory System

Enid F. Gilbert
John M. Opitz

Although the field of medical genetics has advanced rapidly in the last decade, the understanding of genetic contributions to respiratory disorders has not matched this overall pace. Cystic fibrosis, although not exclusively a pulmonary disease, has received more attention than most disorders. The understanding of familial emphysema has been advanced with the elucidation of the α-1-antitrypsin protein marker associated with the genetics of the disease. Biochemical genetic understanding, even in these two examples, is incomplete and completely lacking in most respiratory conditions. This chapter delineates those conditions involving the lung, either exclusively or as part of more generalized disorders in which genetic influences are present.

Clearly defined genetic malformations and diseases of the respiratory tract are summarized in Table 1. The incidence and extent of pulmonary involvement in these disorders may vary considerably. In some conditions, such as the G (dysphagia-hypospadias) syndrome, respiratory tract malformations may be an integral part of the disorder, whereas in other respiratory abnormalities they are of lesser importance.

ABNORMALITIES OF NASOPHARYNX, TRACHEA AND BRONCHI

Choanal atresia is an etiologically nonspecific malformation, which may occur alone or as a component manifestation of etiologically specific or idiopathic multiple congenital anomaly (MCA) syndromes. It is observed in an inherited chromosome translocation and in the craniofacial dysostoses of the types Pfeiffer, Crouzon and Apert. Other malformations found in association with choanal atresia include palatal defect, coloboma, tracheoesophageal fistula, congenital heart disease and the Treacher Collins syndrome (1). In about 90% of cases the obstruction is bony and requires surgical intervention. Several instances of af-

The work described in this chapter was supported by USPHS/NIH Grant GM-20130.

Table 1. Genetic Disorders and Pathologic Changes in Respiratory Tract

Condition	Malformation or Pathologic Change	Inheritance*
Achondroplasia	Multiple small supernumerary lobes	AD
α-1-Antitrypsin deficiency	Panacinar emphysema	AR
Asphyxiating thoracic dystrophy (Jeune's syndrome) and short-rib polydactyly syndromes	Severe chondrodystrophy of ribs with small thoracic cage and hypoplastic lungs	AR
Azygous lobe	Septation of right upper lobe of lung by abnormal course of azygous vein	Sometimes AD
C syndrome†	Pulmonary hypoplasia, hypoplasia of right main bronchus	AR
Choanal atresia, isolated	Choanal atresia	Sometimes AD
Chondrodystrophies:		
Achondrogenesis (Types I and II)	Abnormal tracheal and bronchial cartilages	All AR
Chondrodysplasia punctata (Rhizomelic type)		
Ellis Van Creveld syndrome		
Mesomelic dwarfism (Langer type)		
Congenital bronchiectasis (Williams Campbell syndrome)	Deficiency of bronchial cartilage distal to main segmental bronchi	H
Craniofacial dysostosis (Crouzon disease)	Maxillary hypoplasia	AD (many cases)
Cystic fibrosis	Thick mucoid secretions with obstructive emphysema, bronchiectasis and secondary pulmonary infection	AR
Ehlers-Danlos syndrome	Pneumothorax	H with AD, AR and/or XL forms are known
Familial hemorrhagic telangiectasia (Osler-Rendu-Weber syndrome)	Pulmonary arteriovenous fistulas	AD with high penetrance (homozygous form lethal)
G Syndrome†	Failure of closure of laryngotracheal groove, short trachea with high carina, absence of lung lobation, symmetry of bronchi or unilateral bronchial and pulmonary agenesis, tracheoesophageal fistula	AD
Hyaline membrane disease	Hyaline membranes lining respiratory bronchioles and alveolar ducts	Possible genetically determined maternal factor or homozygosity in affected infants

Idiopathic benign recurrent pneumothorax	Unknown	Occasionally AD
Immunologic Deficiency Diseases:		
Combined immunodeficiency with or without adenosine deaminase deficiency	Thicket-like pattern and "tram" lines of long straight thickened bronchial walls	AR
DiGeorge syndrome (developmental field defect with right-sided aortic arch, agenesis of thymus and parathyroids)	Thickening of walls of small bronchi and bronchioli	S
Interstitial pulmonary fibrosis (familial Hamman-Rich syndrome)	Diffuse pulmonary interstitial fibrosis	AD
Kartagener syndrome	Paranasal sinusitis, bronchiectasis	AR with variable expressivity
Lysosomal storage diseases:		
I-cell disease (mucolipidosis II)	Infiltration of tongue, nasopharynx and larynx; lipid granulomas of lungs	AR
Mucopolysaccharidoses	Mucopolysaccharide infiltration of tongue, nasopharynx and larynx; chronic sinus and pulmonary infections	AR except XL for MPS II, (Hunter syndrome)
Nieman-Pick disease	Accumulation of sphingomyelin in alveolar macrophages	AR
Wolman disease	Xanthomatous cells in lung parenchyma	AR
Mandibulofacial dysostosis	Micrognathia, maxillary hypoplasia and (rarely) choanal atresia	AD
Marfan syndrome	Anomalous lobation of lungs, cystic lymphangiectasia, bronchial stenosis, bronchomalacia, bronchogenic cysts, interstitial pulmonary, fibrosis, pectus excavatum and pneumothorax	AD
Neurofibromatosis	Neurofibromata, diffuse interstitial fibrosis, pectus excavatum	AD
Pectus excavatum (isolated)	Pectus excavatum	Occasionally AD
Pierre-Robin anomaly	Micrognathia, cleft palate	H
Potter "syndrome" (renal agenesis),	Hypoplasia of lungs	H
Primary pulmonary hypertension	Persistence of fetal pulmonary vascular pattern, progressive pulmonary arterial hypertensive changes	AR
Pulmonary alveolar microlithiasis	Microliths in alveoli, later interstitial fibrosis	H
Pulmonary edema (acute) of mountaineers	Pulmonary edema at high altitudes	AD

Table 1. (Continued)

Condition	Malformation or Pathologic Change	Inheritance*
Pulmonary hemosiderosis	Hemosiderin deposits in pulmonary macrophages, interstitial fibrosis	S
Rowley-Rosenberg syndrome	Pulmonary hypertension, atelectasis and recurrent pulmonary infections	AR
Rubinstein-Taybi syndrome	Azygous lobe	S(?), M
Scimitar syndrome	Anomalous pulmonary venous connection to IVC, systemic arterial supply to RLL from aorta; pulmonary hypoplasia	Sometimes AD
Stickler syndrome	Micrognathia, cleft palate, myopia and retinal detachment	AD
Tracheobronchomegaly	Dilation of trachea and major bronchi, redundant musculomembranous tissue protrudes between cartilage rings	AR
Tracheoesophageal fistula (isolated)	Tracheoesophageal fistula	Rarely F
Tuberous sclerosis	Hamartoma of lungs, angiomyolipoma, angiofibroma, leiomyoma, leiomyomatosis, bronchogenic cysts and diffuse interstitial fibrosis	AD with variable expressivity
Yellow nail syndrome	Bronchiectasis and recurrent pleural effusions	AD

*AD, autosomal dominant; AR, autosomal recessive; H, heterogeneous; S, sporadic; XL, X-linked recessive; M, multifactorial; and F, familial.

†See text.

fected sibs suggest autosomal recessive inheritance in some families (2-5); an autosomal dominant form is described (6-8). Most cases, however, are sporadic.

Cleft palate is by far the commonest malformation of the respiratory tract. An etiologically nonspecific anomaly, it may occur as an isolated defect or with other malformations in MCA syndromes. Many of these syndromes are reviewed by Gorlin et al (7) and by Goodman and Gorlin (8).

The association of median cleft lip and arrhinencephaly–alobar holoprosencephaly is a nonspecific developmental field complex or defect (DFC), which may also occur by itself or as a component manifestation of such MCA syndromes as the 13 trisomy, 13q- and 18p- syndromes. An autosomal recessive form is known, but the autosomal dominant form of the alobar holoprosencephaly DFC is rare.

Cleft palate associated with micrognathia—the Pierre-Robin anomaly—may be complicated by severe respiratory obstruction when the tongue falls back and obstructs the pharynx. It may be a component manifestation of MCA syndromes or of generalized connective-tissue dysplasias or a symptomatic anomaly complex secondary to prenatal neuromuscular disturbances (e.g., congenital or CNS hypotonias or severe congenital myotonic dystrophy) or oligohydramnios (as in the Potter syndrome).

The association of cleft palate with micrognathia, myopia and facultative retinal detachment is seen in spondyloepiphyseal dysplasia congenita, a rare disorder, and in the Stickler syndrome, a very common disorder; both of them are dominantly inherited. Association with the Stickler syndrome seems to represent the commonest occurrence of autosomal dominant cleft palate (9, 10).

Syndromes associated with isolated cleft palate and their presumed inheritance are shown in Table 2 (10).

The pathogenesis of *laryngeal atresia or stenosis* and *laryngeal* webs is unknown. *Congenital abductor paralysis of the vocal cords* is reported as an X-linked trait (Plott syndrome) (11) and may be a cause of hypoxic brain damage.

A cleft in the posterior wall of the larynx (*laryngotracheoesophageal cleft*) is a rare anomaly and resembles an H-type tracheoesophageal fistula. It appears to result from an arrest in the rostral growth of the tracheoesophageal septum and from dorsal fusion of the cricoid cartilages (1). It is an etiologically nonspecific DFC and may occur with or without associated congenital anomalies.

Failure of closure of the laryngotracheal groove and a characteristic pattern of associated malformations occur in the G (or dysphagia–hypospadias) syndrome. This is a dominantly inherited MCA syndrome with variable penetrance. In one pedigree (12) a mildly affected mother with partial agenesis of one lung gave birth to three severely affected infants. The G syndrome was first described by Opitz et al (13) in four male siblings (the G family) as a syndrome of multiple congenital anomalies, which included a distinctive facial appearance, a defect in deglutition with regurgitation, a hoarse cry, minor anomalies of the ears and hypospadias with descended testes. Pathologic studies (12, 14) in two cases have demonstrated a closure defect of the laryngotracheal groove, a short trachea with high carina, complete absence of lobation of the lungs, asymmetry of the bronchi with pulmonary isomerism in one case and absence of the left lung with a tracheoesophageal fistula in the other case.

Table 2. Syndromes Associated with Isolated Cleft Palate*†

	Presumed Inheritance		
Autosomal dominant	Autosomal Recessive	X-linked	Sporadic Occurrence
Stickler syndrome	Diastrophic dwarfism	Otopalatodigital syndrome	Aglossia-adactylia syndrome[15,16]
Apert syndrome	Smith-Lemli-Opitz (RSH) syndrome	Bruan-type nephrosis[13]	Congenital oral teratoma
Marfan syndrome		Gorlin skeletovascular syndrome[14]	Buccopharyngeal membrane
Mandibulofacial dysostosis	Multiple pterygium syndrome		Oral duplication
Spondyloepiphyseal dysplasia congenita	Stapes fixation and oligodontia		
	Cerebrocostomandibular syndrome		Caudal regression syndrome[17]
Camptodactyly and clubfoot	Chondrodystrophia calcificans congenita		Klippel–Feil syndrome[18]
Larsen syndrome[1]			Oligohydramnios[19]
Wiedemann-Beckwith syndrome[2]	Dubowitz syndrome[9]		Bilateral renal agenesis[20]
Wildervanck syndrome[3]	Brachmann–de Lange syndrome[10]		Various chromosomal syndromes
Chotzen syndrome[4]	Campomelic syndrome[11,12]		
CPLS syndrome[5,6]			
Achondroplasia[7]			
Cleidocranial dysplasia[8]			

*Classification and nomenclature as in Gorlin et al (7) and Bergsma (185) except as referenced below:

1. McFarlane, *Brit J Surg* 34:388, 1947
2. Personal observation
3. Kirkham, *Amer J Ophthalmd* 70:209, 1970
4. Kreiborg et al *Teratology* 6:287, 1972
5. Fuhrmann et al, *Humangenetik* 14:196, 1972
6. Hayward et al, *J Oral Surg* 15:320, 1957
7. Hay, *Am J. Epidemiol* 94:572, 1971
8. Winter, *Am J. Orthodont Oral Surg* 29:61, 1943
9. Grosse et al, *Z. Kinderheilkd* 110:175, 1971
10. Motl et al, *Hum Heredity* 21:1, 1971
11. Maroteaux et al, *Presse Med* 79:1157, 1971
12. Stüve et al, *Lancet* 495, 1971
13. Braun et al, *J Pediat* 60:33, 1962
14. Gorlin et al, *Am J Dis Child* 119:176, 1970
15. Pettersson, *Acta Chir Scand* 122:93, 1961
16. Wehinger, *Z. Kinderheilk.* 108:46, 1970
17. Bailey et al, *Clin Pediatr* 9:668, 1970
18. Cohney, *Plast Reconstr Surg* 31:179, 1963
19. Poswillo, *Br J Surg* 52:902, 1965
20. Potter, *Obstet Gynecol* 25:3, 1965

†Reprinted from (10), courtesy of authors and publisher.

With the exception of tracheoesophageal fistula, malformations of the *trachea and bronchi* are uncommon. *Tracheomalacia* may be a localized or diffuse defect of the cartilaginous rings. Malformations of the saddle cartilage complex and of the trachea, and abnormal tracheal and bronchial cartilages have also been demonstrated in the Ellis van Creveld syndrome, mesomelic dwarfism (Langer type) and diaphragmatic hernia (15).

The incidence of *tracheoesophageal fistula (TEF)* with or without esophageal atresia is 1 in 1,000 to 1 in 2,500 births. Rarely familial, there are at least seven anatomic types. More than 85% of all cases of TEF are of Type A—esophageal atresia with fistula from the trachea or carina to the lower esophageal segment. Type B is the next most common—esophageal atresia with TEF; the other types are rare. The anomaly should be considered in the presence of polyhydramnios (1).

Congenital bronchiectasis (Williams-Campbell syndrome) was first described by Williams and Campbell (16) and later reported by Williams et al (17) in 16 cases as congenital bronchiectasis due to extensive deficiency of bronchial cartilage distal to the main segmental bronchi. The abnormality can best be demonstrated by careful dissection and special staining (18). The compliant bronchial wall collapses during expiration with distal air trapping. Death may occur before 5 years. Survivors suffer from frequent respiratory tract infections and have limited exercise tolerance, but there appears to be some improvement over the years (19).

Bronchiectasis is also a component of the *yellow nail syndrome* (20) with lymphatic and fingernail abnormalities and may be a complication of other genetic diseases, such as cystic fibrosis (21-23) and immunodeficiency diseases (24). The *Kartagener syndrome* (25-27) is an autosomal recessive trait characterized by situs inversus, chronic paranasal sinusitis and absence of frontal sinuses, with or without nasal polyposis and bronchiectasis. The incidence of complete situs inversus is 1 in 10,000, and one-sixth to one-fourth of affected persons have either bronchiectasis or the complete triad of the Kartagener syndrome.

In two extensive reviews Kartagener and Mülly (28) addressed the topics of bronchiectasis in situs inversus (essentially the Kartagener syndrome) and familial occurrence of bronchiectasis. The latter topic is beclouded by difficulties of sorting familial occurrence of purely environmental agents (i.e., pertussis) from genetic disease complicated by chronic infection. Existing evidence is insufficient to clearly identify an autosomal recessive form of bronchiectasis.

Tracheobronchiomegaly (Mounier-Kühn syndrome) seems to have been noted first by Rokitansky; apparently it is an autosomal recessive trait manifested by chronic cough and progressive respiratory impairment with recurrent attacks of "bronchitis" and pneumonia, which begins in infancy or childhood and may lead to death. Radiologically one notes marked dilatation of the trachea and all major bronchi. On bronchographic examination there are deep corrugations produced by the redundant musculomembranous tissue protruding between the cartilaginous rings. This finding has prompted several workers to call the condition trachiectasis with diverticula. Tracheobronchiomegaly is described in a sibship (29).

LUNG MALFORMATIONS

An azygous lobe results from an unusual course of the azygous vein through the substance of the right upper lobe. The medial portion of lung demarcated by this septum is the azygous lobe *(anomalous lobation of the lungs)* (30). Pipkin et al (31) reported autosomal dominant occurrence of an azygous lobe, and at least one other instance of apparent autosomal dominant inheritance of the trait is known. Occurrence in sibs with apparently normal parents may possibly represent instances of autosomal recessive inheritance or incomplete penetrance of a dominant trait. A large number of anatomic variations of lung lobes is reported and reviewed by Lehmann (32), who discusses the genetic implications of several familial observations. Multiple, small supernumerary lobes are common in achondroplasia (33).

Unilateral severe hypoplasia or agenesis of the lung is reported in mother and son with the G syndrome. Other observations of pulmonary agenesis are reviewed by Lehmann (32). The only clear indication of possible autosomal recessive inheritance of such a severe pulmonary defect is the family reported by Brimblecombe (34). The parents were first cousins and had two normal children and two affected girls with agenesis of the right upper and middle lobes. In three instances pulmonary agenesis has been reported in each of twins (35).

Bilateral *pulmonary hypoplasia* occurs in infants with bilateral renal agenesis or polycystic disease (Potter syndrome), the infantile form of which may be recessively inherited. Oligohydramnios is always present, and the infant has a typical facial appearance with micrognathia, flat nose, prominent epicanthic folds, hypertelorism and large posteriorly rotated ears.

A possible instance of autosomal dominant pulmonary hypoplasia was described by Neill et al (36). Opitz and Faith (37) reported an infant with the Goldenhar syndrome with unilobar lungs and hypoplasia of the right lung. The C syndrome of multiple congenital anomalies, first described by Opitz et al (38), is characterized by dyschondroplastic dwarfism; anomalies of the skull, face and ears; postaxial hexadactyly; laxity of skin; rib and sternal deformities; and cutaneous syndactyly of toes. The lungs are described as hypoplastic and immature and are associated in one case with hypoplasia of the right main bronchus (15). A syndrome of camptodactyly, multiple ankyloses, facial anomalies and pulmonary hypoplasia is described by Pena and Shokeir (39).

Familial occurrence of *pulmonary cysts* is reported (40) as well as documentation in ethnic groups (41). These bronchogenic cysts may occur anywhere in the lung, most frequently near the hilum (42). Immediately after birth the cyst is fluid filled, but rapidly becomes distended with air. Bronchogenic cysts may be solitary or related to a diffuse pulmonary cystic disease (43) (multicystic lung disease), which is associated with a high risk for the development of bronchogenic carcinoma (44). Associated malformations are rare, but pulmonary cystic disease is reported in patients with arachnodactyly and tuberous sclerosis.

The relatively high occurrence of (sporadic) pulmonary cystic disease in Yemenite Jews in Israel (41, 45) and in the Maoris of New Zealand (46) suggests that in these groups the diseases may represent either an autosomal recessive trait—perhaps with high early lethality—or a multifactorial trait with "low" developmental threshold. Other forms of pulmonary cystic disease, such as congenital pulmonary lymphangiectasis and congenital cystic adenomatoid malformation of the lung, are not known to have a genetic basis.

Congenital Alveolar Dysplasia is considered to be a primary malformation of the pulmonary alveoli. The histologic pattern suggests a severe retardation in normal alveolar development (47, 48), as the pathologic appearance is that of a relatively immature lung with septal spaces rich in mesenchyme (49). Both lungs may be uniformly involved, or only a portion of a single lobe. It occurs in full-term newborn babies, in contradistinction to the Wilson-Mikity syndrome with which it has a histologic similarity. The condition is described in identical twins (47), and multiple cases in a sibship have also been noted (50).

An autosomal dominant form of *interstitial pulmonary fibrosis*, or the Hamman-Rich syndrome (51-58), is reviewed by Swaye et al (59), who were able to find 51 cases in 14 "family groups" with multiple affected family members (60, 61).

The gross pathologic features of the syndrome include bulky, firm lungs that, on cut surface, have a coarsely nodular appearance due to extensive interstitial fibrosis. Frequently there are porous cysts in the intervening parenchyma, classically described as the "honeycomb lung." The microscopic appearance is characterized by fibrosis of alveolar septa, an interstitial infiltrate of mononuclear cells and cuboidal or columnar metaplasia of alveolar lining cells. The latter may be a precursor of bronchioloalveolar carcinoma. Fibrous obliterative changes in pulmonary arterioles usually supervene with the development of pulmonary hypertension and right-sided heart failure. Congenital cystic disease of the lung with progressive pulmonary fibrosis and carcinomatosis (54) is probably the same entity.

Interstitial pulmonary fibrosis may also be seen in certain heritable disorders, such as tuberous sclerosis (62), the Marfan syndrome (63) and neurofibromatosis (64). A syndrome of pulmonary fibrosis associated with oculocutaneous albinism and a platelet function defect has recently been described in a family with four affected siblings (65).

Tuberous sclerosis (TS) is inherited as an autosomal dominant disorder. Pulmonary involvement occurs in TS in fewer than 1% of cases. It usually occurs with stigmata of the disease that are milder than the cutaneous and cerebral manifestations. Of patients with pulmonary lesions, 60% have normal intelligence, and 84% of these patients are female (62). Both pulmonary and renal involvement tends to develop if the patient survives to middle life. Dyspnea is the commonest presenting symptom, and more than one-half of affected persons have spontaneous pneumothorax, which is the commonest cause of death in patients with TS and pulmonary lesions.

Pulmonary leiomyomas (lymphangiomyomatosis) in association with multiple cutaneous leiomyomas were reported in a case by Freiman and Castleman (66). Although a genetic basis for multiple skin lesions is demonstrated (67), a genetic basis for the associated lung leiomyomas is not established. Lung involvement progresses slowly, with an average survival of four years from the onset of respiratory symptoms. The pulmonary abnormality varies considerably and may include diffuse nodular infiltration with focal emphysema, multiple fibromas, leiomyomatosis, cystic changes or various combinations thereof (68).

The lesion is a proliferation of cords of smooth muscle associated with lymphatic vessels surrounding alveoli (69). Pulmonary glomangiomas have also been described in TS (70). Honeycomb lung is the commonest end-stage lesion.

Pulmonary arteriovenous fistula is an abnormal communication between pulmonary arteries and veins and most commonly involves the lower lobes. It is one of the manifestations of familial hemorrhagic telangiectasia (Osler-Weber-Rendu disease), accounting for 50 to 60% of cases of pulmonary arteriovenous fistula (71). Conversely 25% of patients with hemorrhagic telangiectasia have pulmonary arteriovenous fistulas. This is an autosomal dominant trait with high penetrance; some evidence indicates that the homozygous form is lethal in early life (72). Cyanosis, digital clubbing and polycythemia are frequent, and a bruit is usually audible over the lesion. Cor pulmonale is a rare complication.

Pulmonary artery hypoplasia is described in the Goldenhar (oculoauriculovertebral) syndrome (37, 73).

In the *scimitar syndrome* the venous connections of all lobes of the right lung are anomalous and drain into the inferior vena cava. A "scimitar"-shaped shadow of an anomalous pulmonary vein along the right cardiac border and a shift of the heart to the right are characteristic roentgenologic findings (74, 75). The systemic arteries supply the right lower lobe. Hypoplasia of the lung is an integral part of the syndrome, and hemivertebrae have been observed in some cases. Familial occurrence in a father and daughter is described (36).

Familial occurrence of *primary pulmonary hypertension* is well documented (76). The symptoms may begin in childhood, occasionally neonatally. Although numerous theories of the pathogenesis of this condition are suggested (77), the basic lesion apparently is a persistence of the fetal pulmonary arterial pattern with a prominent muscular media that has failed to involute after birth. Medial hyperplasia and an increase in elastic tissue is evident in the pulmonary arterial vascular tree. Degeneration of elastic tissue ultimately ensues, with the formation of pools of metachromatic amorphous ground substance. All the sequelae of pulmonary hypertension develop, with aneurysmal dilatations of the vessels and angiomatoid lesions (77). A syndrome of primary pulmonary hypertension, congenital heart disease and skeletal anomalies in three generations was described by Kuhn, Schaaf and Wagner (78).

Rowley et al (79) described the findings of growth retardation, poor muscular development, scanty adipose tissue, recurrent pulmonary infections, atelectasis, pulmonary hypertension, cor pulmonale and aminoaciduria as a new, inherited syndrome in three of six siblings. (*Rowley-Rosenberg syndrome*).

Idiopathic benign pneumothorax is a relatively uncommon disorder with a predeliction for young men and a tendency to recur; it has been observed as an apparently dominantly inherited trait in a number of families (80). Its occasional isolated occurrence in sibs with an unremarkable family history may reflect autosomal recessive inheritance in some cases, or more likely, nonpenetrance of dominant cases (81).

Pulmonary alveolar microlithiasis is an unusual pulmonary lesion characterized by sand-like mottling of the lungs on roentgenographic examination. The lungs are pathologically hard, gritty and incompressible. On microscopic examination the alveoli are filled with concentrically laminated calcific deposits. Late in the disease interstitial fibrosis may occur. Genetic heterogeneity of this entity is possible, and autosomal dominant instances have been reported as well as affected sibs with apparently normal parents (82).

The *Pickwickian syndrome* (obesity-hypoventilation syndrome) is characterized by obesity, somnolence, muscular twitching, cyanosis and respiratory dysfunction with poor ventilation and perfusion. All these symptoms result in significant restriction of vital capacity, total lung capacity and increased expiratory reserve volume. The condition is usually sporadic, but one pair of affected sibs of opposite sex has been described, a finding that suggests recessive inheritance (83). The syndrome is a manifestation of obesity and may be seen as an end stage of the Prader-Willi syndrome.

COMPLEX CONDITIONS AND PULMONARY INVOLVEMENT

In certain families an increased susceptibility for *pulmonary neoplasms* is demonstrated (84). An impressive familial aggregation of bronchogenic carcinoma is described by Tokuhita and Lilienfeld (85). The genetic influence here appears to be a modification of tissue vulnerability to excitation, such as the carcinogenic action of cigarette tars.

A high percentage of persons with blood group O has been described in *bronchopneumonia* of childhood (86, 87). The predisposition for the development of *tuberculosis* may be genetically determined (88). Studies show a high rate of concordance in monozygotic twins, whereas dizygotic twins have the same rate of tuberculosis as do non-twin siblings.

Genetic predisposition for the development of *silicosis* has been demonstrated in Sardinian miners (89), a condition that may predispose to tuberculosis.

Familial aggregation of *sarcoidosis* has been clearly shown by Buck and McKusick (90). All of the patients in five families studied in this report as well as in one other report (91) had generalized sarcoidosis. (Whether all sarcoidosis is heritable is moot; however, the possibility of familial aggregation should certainly be investigated when patients with generalized sarcoidosis are encountered.)

The development of nodules in the pulmonary parenchyma varying in diameter from 5 mm to 5 cm, which may occur in cases of rheumatoid arthritis after "exposure to coal dust" (92), appears to have a definite genetic relationship (Caplan syndrome) (93). These lesions can be clearly distinguished from the interstitial pulmonary fibrosis that may accompany rheumatoid arthritis and which is unassociated with coal-dust exposure.

Some cases of *hyaline membrane disease* (respiratory distress syndrome [RDS]) clearly show a familial incidence, and the condition is described in a series of twins. Karpatkin et al (94) observed two siblings, a male and a female, with idiopathic RDS and disseminated intravascular coagulation. A genetically determined maternal factor as well as homozygosity in the affected infants are suggested as possible genetic causes (95). Lankenau has reported the incidence of RDS among full sibs as 12-19%; among the low birth weight (< 2.5 kg) and/or preterm infants (< 37 weeks) in the sibships 32-50%; the empirical recurrence risk in sibs born after the proband for all younger sibs, 17-27%; and 39-67% for low birth weight per preterm younger sibs.

Although usually sporadic, idiopathic *pulmonary hemosiderosis* has been described in families (96).

Acute pulmonary edema of mountaineers was originally described by Fred et al (97) in two physicians who, while skiing at high altitudes, developed acute pulmonary edema. The father of one of these patients died while mountain climbing. Cardiohemodynamic studies in a series of patients with this syndrome disclose pulmonary hypertension, low cardiac output, arterial unsaturation and low pulmonary wedge pressures (98). Pathologic changes are those of severe pulmonary edema at high altitudes (99).

Cystic Fibrosis

Cystic fibrosis (CF) was first reported in 1936 by Fanconi et al (100) and later by Andersen (101) as a distinct clinical entity.

It is now recognized as the commonest single gene disorder involving the respiratory tract, and in many populations as the commonest autosomal recessive trait. Steinberg (102) has described cystic fibrosis as "the most frequent lethal genetic disease among white children."

The disease has its onset either at birth as meconium ileus or early in infancy or childhood, predominant clinical manifestations usually being related to pulmonary symptoms. The disease is characterized by dysfunction of the exocrine and mucus-secreting glands of pancreas, biliary tract, GI tract and respiratory system. The major clinical manifestations are chronic bronchiolar obstruction with pulmonary infection and insufficiency, pancreatic achylia, meconium ileus, steatorrhea, biliary cirrhosis, malnutrition and growth failure. Steatorrhea may lead to deficiency of liposoluble vitamins, particularly of vitamin A and vitamin E (103). Exclusive of meconium ileus, death in patients with cystic fibrosis usually occurs from respiratory failure.

The incidence of cystic fibrosis in the white population may be as high as 1 in every 1,600 live births (104). The incidence is highest in whites and much lower in Orientals and in American blacks (105). Autosomal recessive inheritance is firmly established (105). The gene frequency may be as high as 0.025, and the carrier state in American whites as high as 5% (104). More than 95% of affected males are sterile (106). Females can reproduce, which coupled with improved survival, has increased the carrier rate minimally in terms of the total population (107). The risks of producing a child with cystic fibrosis are shown in Table 3 (104).

Accurate detection of the heterozygote and the diagnosis of CF in the fetus are not possible at present. So far no single biochemical defect has been identified as the primary molecular lesion.

Mangos (108) reported an unidentified substance in the sweat and saliva of patients with CF, which inhibited sodium reabsorption. Rao and Nadler (109) reported a deficiency in trypsin-like activity in saliva of CF patients. It is postulated that these two factors may be related "through the presence of a unique molecule." Other studies demonstrate the disruption of the normal beating of explanted rabbit trachea cilia (110) and oyster ciliary inhibition (111) by serum from CF patients. This ciliary dyskinetic factor is also found in the serum of healthy CF heterozygotes, a finding that suggests a relationship of this factor to

Table 3. Risks of Producing Child with Cystic Fibrosis (CF)*,†

One Parent (CF history)	Other Parent (CF history)	Risk of CF in each Pregnancy
None	None	1/1,600
None	First cousin with CF	1/320
None	Aunt or uncle with CF	1/240
None	Sib with CF	1/120
None	Child by previous marriage with CF	1/80
None	Parent with CF	1/80
None	Has CF	1/40
Sib with CF	Sib with CF	1/9
With CF child	With CF child	1/4

*These data are based on prevalence of cystic fibrosis of 1/1,600 in white population and its mode of inheritance being autosomal recessive with complete penetrance.

†Courtesy of Dr. Barbara H. Bowman and Dr. John A. Mangos, and the New England Journal of Medicine.

the underlying genetic abnormalities. Disturbances of the Na^+ transport in RBCs (112-116) have been suggested, and the serum and saliva from patients with CF inhibit the transport of small organic molecules (117-119). Tissue culture studies demonstrate metachromasia within cultured fibroblasts (120-121), a finding that is shown to be related ultrastructurally to the increase in size and number of lysosomes containing mucopolysaccharides (122-124).

Diagnosis

The most reliable diagnostic test for cystic fibrosis is the demonstration of abnormal sweat electrolytes. A chloride value in excess of 60 mEq per liter is considered to be diagnostic, provided that other causes of high sweat chlorides, such as respiratory allergic states, renal diabetes insipidus and untreated adrenal insufficiency can be excluded (124). Pancreatic function tests are also diagnostically helpful. The trypsin and chymotrypsin content of the stool as measured by the method of Dyck (125), with results less than 20 units/g of stool for trypsin and 30 units/g for chymotrypsin, indicates pancreatic insufficiency (107). The triad of abnormal sweat chlorides, pancreatic insufficiency and pulmonary symptoms is diagnostic for cystic fibrosis.

The radiologic changes of the lung fields vary according to the stage of the disease (126). In earliest infancy there may be severe asphyxiating bronchitis with overaeration of the lungs—perhaps associated with segmental or even lobar collapse. Later a characteristic linear pattern of thickened bronchial walls (127) is superimposed on the normal vascular pattern. This pattern may appear as early as age 6 months and becomes more prominent with time. In a severe or far advanced stage of the disease, peripheral opacities or radiolucencies develop due to small abscesses or purulent bronchiectatic cavities. At this stage there is marked overexpansion of the lungs due to chronic obstructive lung disease, and pneumothorax may occur (128). With further deterioration gross varicose bronchiectasis, enlarged hilar glands, occasional persistent segmental or lobar collapse and even cavitary pneumonic consolidation may develop.

Sinus roentgenograms demonstrate pansinusitis in over 95% of patients with chronic lung involvement (107). Nasal polyposis is present in 15% of cases. The increased frequency of hemolytic *Staphylococcus aureus* in the respiratory tract of CF patients is well established (129-134). In more recent years *Pseudomonas aeruginosa, Hemophilus influenzae, Escherichia coli, Proteus* and other organisms have been incriminated (132-133).

Other Clinical Manifestations

Bone age is retarded in patients with significant pancreatic insufficiency (107). Other clinical manifestations of CF include hypoelectrolytemia due to massive loss of electrolytes through excessive sweating, which is more common when the environmental temperature is high or the patient has a fever (107). Failure to thrive—with hypoalbuminemia, if severe—may lead to generalized edema (107).

The pathologic findings have recently been reviewed by Oppenheimer and Esterly (135). The lungs appear normal in most patients who die of meconium ileus in the neonatal period (135, 136). Acute and chronic bronchiolitis and bronchiectasis appear to be a pathologic sequence in older patients. Patchy atelectasis, particularly of the upper lobes, with areas of emphysema, are apparent in autopsy specimens. Postmortem bronchograms demonstrate that the distortion of pulmonary architecture is confined almost exclusively to the subsegmental bronchi (137). Cor pulmonale with arterial changes of severe pulmonary hypertension are seen in advanced pulmonary disease.

Striking medial hypertrophy of small pulmonary arteries with only rare intimal changes suggests that chronic hypoxia and acidosis are primarily responsible for the contraction and hypertrophy of arterial muscle (138). Squamous metaplasia of glandular epithelium appears to be a manifestation of vitamin-A deficiency.

Therapy for CF consists of dietary and pancreatic replacement and early institution of prophylactic pulmonary therapy. In the United States over the last 30 years, the survival of patients with CF has dramatically improved from death in childhood to an average life expectancy of about 20 years. Patients with meconium ileus make up 7% of CF patients, and they have a much poorer prognosis than have CF patients who have not had meconium ileus. Recent statistics, however, show a 79% five-year survival in meconium ileus (139).

α-1-Antitrypsin Deficiency

Alpha-1-Antitrypsin (α-1-AT) is a glycoprotein protease inhibitor with a molecular weight of about 54,000, found in the serum α-1 globulins and identified by electrophoresis, trypsin inhibitory capacity, phenotyping and radioimmunodiffusion techniques (140-145). It is produced by the liver (146, 147) under the control of a pair of completely penetrant codominant autosomal alleles (148), 24 of which have been identified (149). Protease inhibitor (Pi) designates the alleles of the genes for α-1-AT activity (150). The Pi types may be the result of amino-acid substitutions caused by point mutations (151). There is extensive electrophoretic polymorphism of α-1-AT, and letter designations of the gene products identify the electrophoretic mobilities.

Thus the most frequent type is PiM (the corresponding gene is labeled Pi$_m$), and the fast and slow variants are PiF and PiS, respectively. Most individuals are homozygous for the common Pi$_m$ gene, and therefore have the genotype PiMM.

From 2 to 12% of the population, however, carry variant genes. Autosomal linkage is confirmed between the Gm and Pi loci in man (152). One of the alleles, Pi$_z$, is associated with low levels of serum α-1-AT, and the homozygous individual with a PiZZ genotype usually has less than 15% of the levels of α-1-AT that are present with the PiMM genotype. In one large series in the United States the frequencies of alleles M, S, Z, F and I were respectively 94.8, 3.44, 1.27, 0.27 and 0.12% (153). Pi$_z$ gene frequencies have been determined as 0.0045 in Finns (154), 0.0157 in Norwegians (155) and 0.018 in American whites (156).

The two major diseases associated with α-1-AT deficiency are pulmonary emphysema of early onset and cirrhosis. Eriksson (157) first recognized a deficiency of α-1-AT in a group of patients with chronic obstructive pulmonary disease (COPD). They were characterized by familial incidence, onset at an unusually early age, higher-than-expected percentage of affected females and absence of chronic bronchitis in the early stages of the disease. It is proposed that deficiency of α-1-AT predisposes the affected person to degeneration of connective tissue fibers. These fibers when resynthesized have diminished tensile strength, thereby yielding to pressure, with resultant enlargement of peripheral air spaces (158).

The glycoprotein inhibitor, α-1-AT, also normally inhibits elastase and collagenase, and deficiency of this function may contribute to the destruction of the pulmonary parenchyma (159). The emphysema associated with α-1-AT deficiency is panacinar in distribution and principally involves the lower lobes (160). Smoking appears to accelerate the development of pulmonary emphysema in α-1-AT–deficient individuals. The exact incidence of emphysema in α-1-AT–deficient individuals is not known; however, as many as 60% (159) of homozygous individuals develop COPD. The heterozygous state, although showing a higher risk for the development of emphysema than does the normal, manifests much less emphysema than does homozygous deficiency (161-162).

Although the age of onset is characteristically in early adult life, it has also been reported in childhood (163). Most patients of α-1-AT deficiency are white. One black, a 39-year-old male who had early-onset familial pulmonary emphysema and α-1-AT heterozygous deficiency, is reported (164). Neonatal RDS recently has been shown to be related to α-1-AT (165), which forms part of the pulmonary hyaline membranes. Serum levels of α-1-AT were found to be reduced, but returned to normal after recovery. These infants, however, all had the normal PiMM phenotype.

Hepatic Complications

Deficiency of α-1-AT is also related to chronic liver disease and juvenile cirrhosis (166, 167), which latter is usually fatal at an early age. Of 50 cases with homozygous α-1-AT deficiency and PiZZ phenotypes, 30 had COPD (168). Of the 20 patients without COPD, two died of hepatic cirrhosis and one developed hepatocellular carcinoma (169). Other instances of hepatic carcinoma are reported in α-1-AT deficiency (170). Only in severe deficiency characteristic of the homozygous state does there appear to be a predisposition to cirrhosis (166, 171). Clinical manifestations begin in the first year of life with the development of hepatomegaly and usually jaundice. The disease progresses relentlessly with severe portal hypertension as a result of macronodular and micronodular cirrhosis. The accumulation of α-1-AT can be demonstrated as PAS-positive material in the peripheral portions of the liver lobules and can be more specifically identified by fluorescent techniques (166). It accumulates within the dilated cisternae of the rough endoplasmic reticulum (166). In a prospective study of 200,000 newborn infants followed until age 6 months, 120 were PiZ, 48 PiSZ, two PiZ- and one PiS- (172). Fourteen of the 120 PiZ infants developed prolonged obstructive jaundice; none of the PiSZ infants had clinical liver disease, but 11 had abnormal liver function. The PiZ and PiSZ phenotypes were associated with hepatic dysfunction in the first three months of life.

Both COPD and juvenile cirrhosis appear to have a multifactorial etiology, since some homozygous individuals are protected from the development of either liver or lung disease. A severe deficiency of α-1-AT must combine with other unfavorable factors for overt disease to develop.

METABOLIC DISEASE AND LUNG

Many of the genetic lysosomal storage diseases affect the lung. They are discussed in detail in Chapter 5, "Genetic Metabolic Disorders."

Among *mucopolysaccharidoses* (MPS), respiratory complications have been described in the Hurler syndrome (MPSI) and in the Hunter syndrome (MPSII) (173). The larynx, nasopharynx and tongue may be heavily infiltrated by mucopolysaccharides in these disorders. Respiratory disability is aggravated by kyphoscoliosis, restriction of chest movement, chronic sinus infection and repeated pulmonary infections.

In one of the *lipid storage diseases*, Niemann-Pick disease, the lungs are frequently extensively involved by the accumulation of phospholipid and sphingomyelin in the alveolar macrophages. In Wolman's disease there is multisystem involvement, with xanthomatous cells, particularly in the GI tract. The lungs are also involved, but to a lesser extent (174). The condition is due to deficiency of cholesterol esterase. The lipid accumulated is principally composed of cholesterol and cholesterol esters.

In I-cell disease, a *mucolipid storage disease:* (mucolipidosis II), lipid granulomas may occur in the pulmonary parenchyma as well as extensive mucolipid storage in the tongue, nasopharynx and larynx. Other forms of mucolipid storage diseases have not as yet been associated with mucolipid storage within the lung.

Pulmonary Involvement and Immunodeficiency Diseases

Roentgenographic and histologic changes in the lungs in various heritable forms of severe combined immunodeficiency disease (SCID) have been established by Wolfson and Cross (175). These diseases are discussed in Chapter 4, "Genetic Disorders of the Immune System". Pneumocystis carinii, and less frequently, cytomegalovirus infection may complicate the severer immunodeficiency diseases (176). Common to all SCID are osteoporosis, a unique and characteristic lung pattern and a small or absent thymus gland. In addition SCID patients who have adenosine deaminase deficiency also exhibit bone abnormalities, particularly abnormally thick growth arrest lines, platyspondyly, cupping and flaring of anterior rib ends and a broad, short pelvis (175). Histologic studies (177) show abnormalities of the physeal cartilage with overall thinness and absence of columnization.

Ataxia-telangiectasia is an autosomal recessive trait involving deficient immunity associated with recurrent sinopulmonary infection (178). Telangiectatic lesions are observed in the lungs of these patients.

Chronic granulomatous disease is a disorder in which leukocytes and macrophages cannot adequately kill and dispose of certain bacteria and fungi. It is characterized by fever, systemic granulomatous response, repeated infections, especially draining lymph nodes, and pneumonias leading to death (179, 180). Bronchial and lobar pneumonia resolve slowly and tend to be complicated by pulmonary abscesses and empyema. Encapsulating pneumonias and reticulation of the lungs caused by granulomas are distinctive features.

Deformities of the Chest Wall

Pectus excavatum (funnel chest) may be a dominantly inherited trait; it can occur as an isolated malformation or in association with arachnodactyly or von Recklinghausen's neurofibromatosis (72, 181). The basic defect in pectus excavatum appears to be an abnormal development of the diaphragmatic tendons (182, 183). When the central tendon is abnormally short, the xiphoid and lower sternum are retracted. Shortness of the lateral tendons causes a transverse groove resembling the Harrison's groove of rickets on both sides of the chest.

Respiratory symptoms of pectus excavatum may be due to pulmonary compression or to displacement of the heart and great vessels.

Asphyxiating thoracic dystrophy (ATD) is an autosomal recessive disorder characterized by a severe chondrodystrophy affecting chiefly the ribs and resulting in a small thoracic cage and hypoplastic lungs. It may be lethal. Severe, ATD-like manifestations may also be seen in the Ellis van Creveld syndrome, the short-rib polydactlyly syndromes and rarely in a severely affected infant with achondroplasia (184).

REFERENCES

1. Avery ME: The Lung and Its Disorders in the Newborn Infant, 2 ed. Philadelphia, Saunders, 1968
2. Dirlewanger A: Hereditäres Vorkommen von Choanalatresien. *Pract Otorhinolaryng* 28:211, 1966
3. Fendel K: Zur familiären Häufung der angeborenen Choanalatresie. *Z Laryng Rhinol Otol* 45:67, 1966
4. Grahne B, Kaltiokallio K: Congenital choanal atresia and its heredity. *Acta Otolaryngol* 62:193, 1966
5. Ransome J: Familial incidence of posterior choanal atresia. *J Laryngol* 78:551, 1964
6. Lang J: Ueber Choanalatresie (Heredität derselben). *Mschr Ohrenheilk* 46:970, 1912
7. Gorlin RJ, Cohen MM, Jr, Pindborg JJ: Syndromes of the Head and Neck, 2 ed. McGraw-Hill, 1976

8. Goodman RM, Gorlin RJ: The Face in Genetic Disorders, 2d ed. St Louis, Mosby, 1977
9. Herrmann J, France TD, Opitz JM: The Stickler syndrome. *Birth Defects* 11:203, 1975
10. Herrmann J, France TD, Spranger JW, et al: The Stickler syndrome (hereditary arthroophthalmopathy), *Birth Defects* 11:76, 1975
11. Watters GV, Fitch N: Familial laryngeal abductor paralysis and psychomotor retardation. *Clin Genet* 4:429, 1973
12. Kasner J, Gilbert EF, Viseskul C: Studies of malformation syndromes VII. The G syndrome. Further oserrvations. *Z Kinderheilk* 118:81, 1974
13. Opitz JM, Frias JL, Gutenberger JE, et al: The G syndrome of multiple congenital anomalies. *Birth Defects* 2:95, 1969
14. Gilbert EF, Viseskul C, Mossman HW, et al: The pathologic anatomy of the G syndrome. *Z Kinderheilk* 111:290, 1972
15. Landing BH, Wells TR: *Tracheobronchial anomalies in children,* in Perspectives in Pediatric Pathology. Chicago, Year Book, 1973, p. 1
16. Williams H, Campbell P: Generalized bronchiectasis associated with deficiency of cartilage in the bronchial tree. *Arch Dis Child* 35:182, 1960
17. Williams HW, Landau LI, Phelan PD: Generalized bronchiectasis due to extensive deficiency of bronchial cartilage. *Arch Dis Child* 47:423, 1960
18. Campbell PE: Congenital lobar emphysema: etiological studies, *Aust Paediatr J* 5:226, 1969
19. Mitchell RD, Bury RG: Congenital bronchiectasis due to deficiency of bronchial cartilage (Williams- Campbell syndrome). *J Pediatr* 87:230, 1976
20. Hiller E, Rosenow ECIII, Olsen AM: Pulmonary manifestations of the yellow nail syndrome. *Chest* 61:452, 1972
21. Esterly JR, Oppenheimer EH: Observations in cystic fibrosis of the pancreas. Johns Hopkins Med J 122:94, 1968
22. Esterly JR, Oppenheimer EH: Cystic fibrosis of the pancreas, structural changes in peripheral airways. *Thorax* 23:670, 1968
23. Mearns MB, Hodson CJ, Jackson ADM, et al: Pulmonary resection in cystic fibrosis; results in 23 cases, 1957-1970, *Arch Dis Child* 47:499, 1972
24. McFarlin DE, Strober W, Waldman TA: Ataxia-telangiectasia. *Medicine* 51:281, 1972
25. Gaby P, Dorsti O: Facteurs congénitaux dans la génesè de la bronchite chronique, les bronchiectasies y compris les syndromes de Kartagener et de Mounier-Kühn. *Poumon Coeur* 27:133, 1971
26. Hartline JV, Zelkowitz PS: Kartagener's syndrome in childhood. *Am J Dis Child* 121:349, 1971
27. Kartagener M, Stucki P: Bronchiectasis with situs inversus. *Arch Pediatr* 79:193, 1962
28. Kartagener M, Mülly K: Familiäres Vorkommen von Bronchiectasien. *Schweiz A Tuberk* 13:221, 1956
29. Johnson RF, Green RA: Trachiobronchiomegaly. *Am Rev Resp Dis* 91:35, 1965
30. Glenn WWL, Liebow AA, Lindkog GE: Thoracic and Cardiovascular Surgery. New York, Appleton-Century-Crofts, 1975, p. 1
31. Pipkin SB, Kegel R, Pipkin AC: Inheritance of the accessory lobe of the azygos vein. *J Hered* 43:260, 1952
32. Lehmann W: Krankheiten und Missbildungen der Bronchien und der Lunge, in Becker PE (ed): Humangenetik, Vol. 3. Stuttgart, Thieme, 1972
33. Kissane JM: Pathology of Infancy and Childhood. St. Louis, Mosby, 1975
34. Brimblecombe FSW: Pulmonary agenesis. *Br J Tubercul* 45:7, 1951
35. Maltz DL, Nadas AS: Agenesis of the lung. *Pediatrics* 42:175, 1968
36. Neill CA, Ferencz C, Sabiston DC, et al: The familial occurrence of hypoplastic right lung with systemic arterial supply and venous drainage "Scimitar Syndrome." *Bull Johns Hopkins Hosp* 107:1, 1960
37. Opitz JM, Faith GC: Visceral anomalies in an infant with the Goldenhar syndrome. *Birth Defects* 5:161, 1969

38. Opitz JM, Johnson RC, McCreadie SR, et al: The C syndrome of multiple congenital anomalies. *Birth Defects* 5:161, 1969
39. Pena SDJ, Shokeir MHR: The syndrome of camptodactyly, multiple ankyloses, facial anomalies and pulmonary hypoplasia: A lethal condition. *J Pediatr* 85:373, 1974
40. Adelman AG, Chertkow G, Hayton RC: Familial fibrocystic pulmonary dysplasia: Detailed family study. *Can Med Assn J* 95:603, 1966
41. Baum G, Raoz I, Bubis JJ, et al: Cystic disease of the lung; report of eighty-eight cases with an ethnologic relationship. *Am J Med* 40:578, 1966
42. Sante LR: Cystic disease of the lung. *Radiology* 33:15b, 1939
43. Heller EL, Householder JH, Benshoff AM: Bronchogenic cysts. *Am J Clin Pathol* 23:131, 1953
44. Korol E: The correlation of carcinoma and congenital cystic emphysema of the lungs. Report of ten cases. *Dis Chest* 23:403, 1953
45. Racz I, Baum GL: The relationship of ethnic origin to the prevalence of cystic lung disease in Israel. A preliminary report. *Am Rev Resp Dis* 91:552, 1965
46. Hinds JR: Bronchiectasis in the Maori. *NZ Med J* 57:328, 1958
47. MacMahon EH: Congenital alveolar dysplasia of the lungs: *Am J Pathol* 24:919, 1948
48. Kaufman N, Spiro RK: Congenital alveolar dysplasia of the lungs. *Arch Pathol* 51:434, 1951
49. Esterly JR: Anatomic malformations of the lower respiratory tract. *Birth Defects* 10:217, 1974
50. Liebow AA: Personal communication, June 1971, as cited by Esterly JR: Anatomic malformations of the lower respiratory tract. *Birth Defects* 10:217, 1974
51. Hamman L, Rich A: Fulminating diffuse interstitial fibrosis of the lungs. *Trans Am Clin Climatol Assoc* 51:154, 1935
52. Hamman L, Rich AR: Acute diffuse interstitial fibrosis of the lungs. *Bull Johns Hopkins Hosp* 74:177, 1944
53. Davies GM, Potts MW: Chronic diffuse interstitial pulmonary fibrosis in brothers. *Guy's Hospital Rep* 113:36, 1964
54. McKusick VA, Fisher AM: Congenital cystic disease of the lung with progressive pulmonary fibrosis and carcinomatosis. *Ann Intern Med* 48:774, 1958
55. Bonanni PP, Frymoyer JW, Jacox RF: A family study of idiopathic pulmonary fibrosis. *Am J Med* 39:411, 1965
56. Young WA: Familial fibrocystic pulmonary dysplasia: A new case in a known family. *Can Med Assoc J* 94:1059, 1966
57. Donahue WL, Laski B, Uchida I, et al: Familial fibrocystic pulmonary dysplasia and its relation to the Hamman-Rich syndrome. *Pediatrics* 24:786, 1959
58. Rezek PhR, Talbert WM Jr: Kongenitale (familiäre)zystische Fibrose der Lunge. *Wien Klin Wochenschr* 74:869, 1962
59. Swaye P, van Ordstrand HS, McCormack LJ, et al: Familial Hamman-Rich syndrome; report of 8 cases. *Dis Chest* 55:7, 1969
60. Hughes EW: Familial interstitial pulmonary fibrosis. *Thorax* 19:515, 1964
61. Koch B: Familial fibrocystic pulmonary dysplasia; observations in one family. *Can Med Assoc J* 92:801, 1965
62. Dwyer JM, Hickie JB, Garvan J: Pulmonary tuberous sclerosis; report of three patients and a review of the literature. *Q J Med* 40:115, 1971
63. Lipton RA, Greenwald RA, Seriff N: Pneumothorax and bilateral honey-combed lung in Marfan syndrome; report of a case and review of the pulmonary abnormalities in this disorder. *Am Rev Resp Dis* 104:924, 1971
64. Massaro D, Katz S: Fibrosing alveolitis; its occurrence, roentgenographic and pathologic features in von Recklinghausen's neurofibromatosis. *Am Rev Resp Dis* 93:934, 1966
65. Davies BH, Tuddenham EGD: Familial pulmonary fibrosis associated with oculocutaneous albinism and platelet function defect: A new syndrome *Q J Med* 178:219, 1976
66. Freiman DG, Castleman B: Fluctuating pulmonary shadows for twenty-two years. *N Engl J*

Med 268:550, 1963
67. Kloepfer HW, Krafchuk J, Derbes V, et al: Hereditary multiple leiomyoma of the skin. *Am J Hum Genet* 10:48, 1958
68. Harris JO, Waltuck BL, Swenson EW: The pathophysiology of the lungs in tuberous sclerosis. *Am Rev Resp Dis* 100:379, 1969
69. Vasquez JJ, Fernandez-Cuerro L, Fidalgo B: Lymphangiomyomatosis. *Cancer* 37:2321, 1976
70. Barr HS, Galindo J: Pulmonary glomangiomata and hamartoma in tuberous sclerosis. *Arch Pathol* 78:287, 1964
71. Landing BH: Anomalies of the Respiratory Tract. *Pediatr Clin North Am* 1957, p. 73
72. Snyder LH, Doan CA: Studies in human inheritance. *J Lab Clin Med* 29:1211, 1944
73. Greenwood RD, Sommer A, Wolff G, et al: Cardiovascular malformations in oculo-auriculo-vertebral dysplasia (Goldenhar syndrome). *J Pediatr* 85:816, 1974
74. Mathey J, Galey JJ, Logeais Y, et al: Anomalous pulmonary venous return into inferior vena cava and associated bronchovascular anomalies (the scimitar syndrome). *Thorax* 23:398, 1968
75. Halasz NA: Bronchial and arterial anomalies with drainage of the right lung into the inferior vena cava. *Circulation* 14:826, 1956
76. Robertson B, Rosenhamer G, Lindberg J: Idiopathic pulmonary hypertension in two siblings; clinical, microangiographic and histological observations *Acta Med Scand* 186:569, 1969
77. Wagenvoort CA, Wagenvoort N: Primary pulmonary hypertension; A pathologic study of the lung vessels in 156 clinically diagnosed cases. *Circulation* 42:1163, 1970
78. Kuhn E, Schaaf J, Wagner A: Primary pulmonary hypertension, congenital heart disease and skeletal anomalies in three generations. *Jpn Heart J* 4:205, 1963
79. Rowley PT, Mueller PS, Watkins DM, et al: Familial growth retardation; renal aminoaciduria and cor pulmonale. *Am J Med* 31:187, 1961
80. Ziegler E: Familiärer idiopathischer Spontanpneumothorax. *Helv Paediatr Acta* 16:347, 1961
81. Leites V, Tannenbaum E: Familial spontaneous pneumothorax. *Am Rev Resp Dis* 82:240, 1961
82. Sosman MC, Dodd GD, Jones WD, et al: The familial occurrence of pulmonary alveolar microlithiasis. *Am J Roentgenol* 77:947, 1975
83. Falsetti HL, Hanson JS, Tabakin BS: Obesity-hypoventilation syndrome in siblings. *Am Rev Resp Dis* 90:105, 1964
84. Burch PRJ: Genetic carrier frequency for lung cancer. *Nature* 202:711, 1964
85. Tokuhata GK, Lilienfeld AM: Familial aggregation of lung cancer in humans. *J Natl Cancer Inst* 30:289, 1963
86. Carter CO, Heslop B: ABO blood groups and bronchopneumonia in children. *Br J Prev Soc Med* 11:214, 1957
87. Struthers D: ABO groups in infants and children dying in the west of Scotland (1949-1951). *Br J Soc Med* 5:223, 1951
88. Gedda L: La complexe contribution du genotype à la prédisposition morbide. (Le rôle de la prédiposition dans la tuberculose, la silicose et les maladies allergiques). *Acta Genet Med Gemellol* (Roma) 14:1, 1965
89. Gedda L, Bolognesi M, Bandino R, et al: Richerche di genetica sulla silicosi dei minatori della Sardegna. *Lav Um* 16:555, 1964
90. Buck AA, McKusick VA: Epidemiologic investigations of sarcoidosis III. Serum proteins; syphilis; association with tuberculosis; familial aggregation. *Am J Hyg* 74:174, 1961
91. Quinn KJ: Familial sarcoidosis. *J Irish Med Assoc* 52:161, 1963
92. Caplan A: Certain unusual radiological appearances in the chest of coal miners suffering from rheumatoid arthritis. *Thorax* 8:29, 1953
93. Kantor M, Morrow CS: Caplan's syndrome: A perplexing pneumoconiosis with rheumatoid arthritis. *Am Rev Tuberc* 78:274, 1958
94. Karpatkin M, Sacker I, Ackerman N: Respiratory distress syndrome and disseminated intravascular coagulation in two siblings (letter). *Lancet* 1:102, 1972

95. Lankenau HM: A genetic and statistical study of the respiratory distress syndrome. *Eur J Pediatr* 12:1, 1976
96. Soergel KH, Sommers SC: Idiopathic pulmonary hemosiderosis and related syndromes. *Am J Med* 32:499, 1962
97. Fred HL, Schmidt AM, Bates T, et al: Acute pulmonary edema of altitude. Clinical and physiologic observations. *Circulation* 25:929, 1962
98. Hultgren HN, Lopez CE, Lundberg E, et al: Physiologic studies of pulmonary edema at high altitude. *Circulation* 29:393, 1964
99. Arias-Stella J, Kruger H: Pathology of high altitude pulmonary edema. *Arch Pathol* 76:147, 1963
100. Fanconi G, Uehlinger E, Knauer C: Das Coeliakiesyndrom bei angeborener zystischer Pankreasfibromatose und Bronchiektasien. 86:753, 1936
101. Andersen DH: Cystic fibrosis of pancreas and its relation to celiac disease: Clinical and pathological study. *Am J Dis Childh* 56:344, 1938
102. Steinberg AG, Brown DC: On incidence of cystic fibrosis of pancreas. *Am J Hum Genet* 12:416, 1960
103. Underwood BA, Denning CR, Navab M: Polyunsaturated fatty acids and tocopherol levels in patients with cystic fibrosis. *Ann NY Acad Sci* 203:237, 1972
104. Bowman BH, Mangos JA: Current concepts in genetics. Cystic fibrosis. *N Engl J Med* 294:937, 1976
105. McKusick VA, Mutalik GS: Genetics and pulmonary disease, in Liebow AA, Smith DE (eds): The Lung. International Acadamy of Pathology Monographs, Baltimore, Williams & Wilkins, 1972, p.
106. Denning CF, Sommers SC, Quigley HJ: Infertility in male patients with cystic fibrosis. *Pediatrics* 41:7, 1968
107. Crozier DN: Cystic fibrosis: A not so fatal disease. *Pediatr Clin North Am* 21:935, 1974
108. Mangos JA, McSherry NR, Benke PJ: A sodium transport inhibitory factor in the saliva of patients with cystic fibrosis of the pancreas. *Pediatr Res* 1:436, 1967
109. Rao GJS, Nadler HL: Deficiency of trypsin-like activity in saliva of patients with cystic fibrosis. *J Pediatr* 80:573, 1972
110. Spock A, Heick HMC, Cress H, et al: Abnormal serum factor in patients with cystic fibrosis of the pancreas. *Pediatr Res* 1:173, 1967
111. Bowman BH, Lockhart LH, McCombs ML: Oyster ciliary inhibition by cystic fibrosis factor. *Science* 164:325, 1969
112. Balfe JW, Cole C, Welt LG: Red cell transport defect in patients with cystic fibrosis and in their parents. *Science* 162:689, 1968
113. Horton C, Cole W, Bader H: Depressed (Ca^{++})-transport ATPase in cystic fibrosis erythrocytes. *Biochem Biophys Res Commun* 40:505, 1960
114. Hadden JW, Hansen LG, Shapiro BL, et al: Erythrocyte enigmas in cystic fibrosis. *Proc Soc Exp Biol Med* 142:577, 1973
115. Fitzpatrick D, Landon E, James W: Serum binding of calcium and the red cell membrane in cystic fibrosis. *Nature* 235:173, 1972
116. Feig SA, Segel GB, Kern KA, et al: Erythrocyte transport function in cystic fibrosis. *Pediatr Res* 8:594, 1974
117. Morin CL, Desjeux JF, Authier L: Effect of saliva and serum from patients with cystic fibrosis on intestinal uptake of amino acids in rats. *Biomedicine* 19:133, 1973
118. Taussig LM, Gardner JD: Effects of saliva and plasma from cystic fibrosis patients on membrane transport. *Lancet* 2:1367, 1972
119. Brown GA, Oskin A, Goodchild MC, et al: Inhibition of sugar transport by plasma from cystic fibrosis patients. *Lancet* 2:639, 1971
120. Danes BS, Bearn AG: A genetic cell marker in cystic fibrosis of the pancreas. *Lancet* 1:1061, 1968

121. Danes BS, Bearn AC: Cystic fibrosis of the pancreas; a study in cell culture. *J Exp Med* 129:775, 1969
122. Bartman J, Weismann V, Blanc WA: Ultrastructure of cultivated fibroblasts in cystic fibrosis of the pancreas. *J Pediatr* 76:430, 1970
123. Matalon R, Dorfman A: Acid mucopolysaccharides in cultured human fibroblasts. *Lancet* 2:838, 1966
124. di Sant'Agnese PA: Cystic fibrosis (mucoviscidosis): *Am Fam Physician* 7:102, 1973
125. Dyck WP: Titrimetric measurements of fecal trypsin and chymotrypsin in cystic fibrosis with pancreatic exocrine insufficiency. *Am J Dig Dis* 12:310, 1967
126. Hodson CJ, France NE: Pulmonary changes in CF of the pancreas: A radiopathological study. *Clin Radiol* 13:54, 1962
127. Wolfson JJ: Personal communication, 1976
128. Shwachman J, Holsclaw D: Pulmonary complications of cystic fibrosis. *Minn Med* 52:1521, 1969
129. di Sant'Agnese PA: Pancreas, in Nelson WE (ed): Textbook of Pediatrics, 8th ed. Philadelphia, Saunders, 1964, p. 771
130. Shwachman H: Cystic fibrosis, in Kendig EL Jr (ed): Disorders of the Respiratory Tract in Children. Philadelphia, Saunders, 1967, p. 541
131. Pittman FE, Howe C, Goode L, et al: Phage groups and antibiotic sensitivity of *Staphylococcus aureus* associated with cystic fibrosis of pancreas. *Pediatrics* 24:40, 1959
132. Huang NN, van Loon EL, Sheng KT: Flora of respiratory tract of patients with cystic fibrosis of pancreas. *J Pediatr* 59:512, 1961
133. Iacocca WF, Sibinga MS, Barbero GJ: Respiratory tract bacteriology in cystic fibrosis of pancreas. *J Pediatr* 59:512, 1961
134. Huang NN, Shen KT: Staphylococcal carrier rates of patients with cystic fibrosis and of members of their families. *J Pediatr* 62:36, 1963
135. Oppenheimer EH, Esterly JR: Pathology of cystic fibrosis: Review of the literature and comparison with 146 autopsied cases. *Perspect Pediatr Pathol,* 2:241, 1975
136. Andersen DH: Pathology of cystic fibrosis. *Ann NY Acad Sci* 93:500, 1962
137. Bowden DH, Fischer VW, Wyatt JP: Cor pulmonale in cystic fibrosis: Morphometric analysis. *Am J Med* 38:226, 1965
138. Symchych PA, Blanc WA: Morphometry of pulmonary arterial tree/cor pulmonale in cystic fibrosis. Cystic Fibrosis Club Conference, Atlantic City, 1967, p. 33
139. Robinson MJ, Norman AP: Life tables for cystic fibrosis. *Lancet* 1:962, 1976
140. Kueppers F: Alpha-1-antitrypsin: Physiology, genetics and pathology. *Humangenetik* 11:177, 1971
141. Bundy HF, Mehl JW: Trypsin inhibitors of human serum. *J Biol Chem* 234:1124, 1959
142. Schultze HE, Goellner I, Heide K, et al: Zur Kenntnis der alpha-globuline des menschlichen Normalserum. *Z Naturforsch* 10:463, 1953
143. Shamash Y, Rimon A: The plasmin inhibitors of human plasma. *Biochem Biophys Acta* 121:35, 1955
144. Sun T, Kurtz S, Copeland BE: Alpha-1-antitrypsin deficiency and pulmonary disease. *Am J Clin Pathol* 62:725, 1974
145. Musiani P, Tomasi TB: Isolation, chemical and physical properties of alpha-1-antitrypsin. *Biochemistry* 15:798, 1976
146. Asofsky R, Thorbecke GJ: Sites of formation of immune globulins and of a component of C'3. *J Exp Med* 114:471, 1961
147. Lieberman J, Mittman C, Gordon HW: Alpha-1-antitrypsin in the livers of patients with emphysema. *Science* 175:64, 1972
148. Fagerhol MK: Serum Pi types in Norwegians. *Acta Pathol Microbiol Scand* 70:421, 1967
149. Hoffman JJML, van den Brock WGM: Distribution of alpha-1-antitrypsin phenotypes in two Dutch population groups. *Hum Genet* 32:43, 1976

150. Fagerhol MK, Laurell CB: The polymorphism of "prealbumins" and alpha-1-antitrypsin in human sera. *Clin Chim Acta* 16:199, 1967
151. Fagerhol MK, Laurell CB: The Pi system, inherited variants of serum alpha-1-antitrypsin, *Prog Med Genet* 7:96, 1970
152. Gedde-Dahl T, Fagerhol MK, Cook PJL, et al: Autosomal linkage between the Gm and Pi loci in man. *Ann Hum Genet* 35:393, 1972
153. Dew, TA, Eradio B, Pierce JA: Antitrypsin gene frequencies in healthy people (abstract). *J Clin Invest* 52:23a, 1973
154. Fagerhol MK, Eriksson AW, Monn E: Serum Pi types in some Lappish and Finnish populations. *Hum Hered* 19:360, 1969
155. Fagerhol MK: Serum Pi types in Norwegians. *Acta Pathol Microbiol Scand* 70:421, 1967
156. Kueppers F, Briscoe WA, Bearn AG: Hereditary deficiency of alpha-1-antitrypsin. *Science* 146:1678, 1964
157. Erikkson S: Pulmonary emphysema and alpha-1-antitrypsin deficiency. *Acta Med Scand* 175:197, 1964
158. Hunter CC, Pierce JA, LaBorde JB: Alpha-1-antitrypsin deficiency. *JAMA* 205:93, 1968
159. Kanner RE, Klauber MR, Watanabe S, et al: Pathologic patterns of chronic obstructive pulmonary disease in patients with normal and deficient level of alpha-1-antitrypsin. *Am J Med* 54:706, 1973
160. Mazodier P, Orell SR, Siken L, et al: Deficit constituionnel en alpha$_1$ antitripsine et emphysema panlobulaire. *J Fr Med Chir Thorac* 25:5, 1971
161. Lieberman J: The mechanisms of antitrypsin deficiency and their role in the pathogenesis of pulmonary deficiency. *Pneumonologie* 152, 7, 1975
162. Lieberman J: The association of alpha$_1$ antitrypsin with pulmonary disease, in Junod AF, de Haller R (eds): Lung Metabolism: Proteolysis and antiproteolysis, biochemical pharmacology. New York, Academic, 1975, p. 83
163. Talamo RC, Levison H, Lynch M, et al: Symptomatic pulmonary emphysema in childhood associated with hereditary alpha$_1$-antitrypsin deficiency. *J Pediatr* 79, 20, 1971
164. Tarkoff MP, Kueppers F, Miller WF: Pulmonary emphysema and alpha$_1$-antitrypsin deficiency. *Am J Med*, 45:220, 1968
165. Mathis RD, Freier EF, Hunt CE, et al: Alpha-1-antitrypsin in the respiratory distress syndrome. *N Engl J Med* 288:59, 1973
166. Sharp HL: Alpha-1-antitrypsin deficiency. *Hosp Pract* 6:83, 1971
167. Sharp HL, Bridges RA, Krivit W, et al: Cirrhosis associated with alpha-1-antitrypsin deficiency: A previously unrecognized inherited disorder. *J Lab Clin Med* 73:934, 1969
168. Erikkson S: Studies in alpha-1-antitrypsin deficiency. *Acta Med Scand* 177:1, 1965
169. Ganrot PC, Laurell CB, Eriksson S: Obstructive lung disease and trypsin inhibitors in alpha-1-antitrypsin deficiency. *Scand J Clin Lab Invest* 19:205, 1967
170. Schleissner LA, Cohen AH: Alpha-1-antitrypsin deficiency and hepatic carcinoma, *Am Rev Resp Disease* 3:863, 1975
171. Burke JA, Keesel JL, Blair JO: Alpha-1-antitrypsin deficiency and liver disease in children. *Am J Dis Child* 130:621, 1976
172. Sveger T: Liver disease in alpha-1-antitrypsin deficiency detected by screening of 200,000 infants. *N Engl J Med* 294:1316, 1976
173. Murray JF: Pulmonary disability in the Hurler syndrome (lipochondrodystrophy): A study of two cases. *N Engl J Med* 261:378, 1959
174. Crocker AC, Vawter GF, Neuhauser EBD, et al: Wolman's disease: Three new patients with a recently described lipidosis. *Pediatrics* 35:627, 1962
175. Wolfson JJ, Cross VF: The radiographic findings in forty-nine patients with combined immunodeficiency, in Combined Immulodeficiency Disease and Adenosine Deaminase Deficiency, New York, Academic, 1975
176. Wolfson JJ: Personal communication

177. Alexander WJ, Dunbar JS: Unusual bone changes in thymic alymphoplasia. *Ann Radiol* (Paris) 11:389, 1968
178. Teller WM, Millichap JG: Ataxia-telangiectasia (Louis-Bar syndrome) with prominent sinopulmonary disease. *JAMA* 175:779, 1961
179. Quie PG, Holmes B, Good RA: In-vitro bactericidal capacity of human polymorphonuclear leukocytes: Diminished activity in chronic granulomatous disease of childhood. *J Clin Invest* 46:668, 1967
180. Wolfson JJ, Quie PG, Laxdal SD, et al: Roentgenologic manifestations in children with a genetic defect of polymorphonuclear leukocyte function. *Radiology* 91:37, 1968
181. Jeune M, Beraud C, Mounier-Kühn P: Les anomalies de l'arbre bronchique au cours de la dystrophie thoracique en entonnoir. *Pediatrie* 7:1087, 1952
182. Brodkin HA: Congenital chondrosternal depression (funnel chest). Its treatment by phrenosternolysis and chondrosternoplasty. *Dis Chest* 19:288, 1951
183. Lester CW: Funnel chest and allied deformities of the thoracic cage. *J Thorac Surg* 19:507, 1950
184. Bargman G, Langer LO Jr, Opitz JM: Achondroplasia and thanatophoric dwarfism, in Gardner LI (ed): Endocrine and Genetic Diseases of Childhood, 2d ed. Philadelphia, Saunders, 1975
185. Bergsma D (ed): Birth Defects Atlas and Compendium. National Foundation March of Dimes. Baltimore, Williams & Wilkins Co, 1973, p. 264

18
Heritable Skeletal Dysplasias

Charles I. Scott, Jr.

The understanding, diagnosis and classification of intrinsic bone diseases has undergone considerable change within the past 20 years. Hereditary disorders of the skeleton are relatively rare and any classification of skeletal dysplasias can be only partially complete. The 1969 Paris Nomenclature for Constitutional Disorders of Bone will be used in this chapter (1). Since that conference, a considerable number of newly recognized entities have demonstrated further heterogeneity, especially in the minor categories. Attention here will be given only to the major disorders that are well established. Certain conditions, such as the mucopolysaccharidoses, mucolipidoses and metabolic disorders of calcium and phosphorous, are discussed in other chapters.

The chondrodystrophies are a heterogeneous group of disorders resulting in disproportionately short stature. Clinically the term dwarf has been used for those individuals of disproportionately short stature: the label midget has been reserved for those of proportionate but reduced height. The terms little, short or small people are currently preferred. Most disproportionately short patients have a skeletal dysplasia, whereas those with normal body proportions have nutritional, endocrine, cytogenetic or nonosseous defects. Clinical evaluation of the short individual should include measurements of the circumference of the head, height, weight, span and determination of body segment proportions in addition to a detailed general physical examination. A complete series of skeletal radiographs may then permit recognition of a specific entity, since most disorders cannot be diagnosed on the basis of one or two X-rays of selected body parts.

Review of previously made X-rays is frequently necessary, as the radiographic characteristics of many dysplasias change with time. This is especially true with epiphyseal dysplasias—growth plates fuse and all evidence of disturbed epiphyseal development is submerged. Familial history analysis may be quite useful. Certain disorders, which clinically are indistinguishable, may occur as either autosomal recessive or autosomal dominant traits, and attention to other affected family members can help differentiate these conditions.

This work was supported in part by funds from Grant No. G.M. 19513 from the National Institutes of Health.

SHORT-LIMB DWARFISM RECOGNIZABLE AT BIRTH OR IN FIRST YEAR

Achondroplasia (Chondrodystrophia Fetalis; Chondrodystrophic Dwarfism)

Achondroplasia (Table 1) is recognizable at birth because of disproportionate short stature with a relatively long trunk and shortened extremities, more pronounced in the proximal segments (rhizomelia) (2, 3). There is a disproportionately large head, a prominent or bulging forehead and midface hypoplasia with a depressed nasal bridge and narrow nasal passages. Ligamentous laxity is noteworthy, particularly at the knees, but elbow extension is limited. The hands are short and stubby, wide proximal phalanges preventing full approximation of all fingers (trident hand). Later the abdomen and buttocks become prominent and there may be exaggerated lumbar lordosis and a waddling gait, although kyphosis or gibbus at the thoracolumbar junction is frequent in infancy even prior to ambulation. Head size increases rapidly in infancy, but true hydrocephalus with symptoms of increased intracranial pressure is uncommon and invasive diagnostic techniques are not indicated for large head size alone.

Achievement of developmental milestones is frequently delayed because of the physical difficulties posed by short levers (limbs) and hypotonia, which abates by age 2 years. Otitis media, dental malalignment and obesity are common. The mean adult height for men is 132 cm and for women, 123 cm.

The most serious major complication of adults is spinal cord or nerve root compression, or both usually in the lower thoracic or lumbar region, and surgical decompression may be required. Bracing is of no value for bowleg deformity, which may require osteotomy for control. Normal lifespan can be expected in the absence of serious complications, and intelligence is normal. There is no increased frequency of death in the neonatal period. In the past confusion has arisen over this point because of misdiagnosis.

Major radiographic features include a small foramen magnum and skull base with a bulging neurocranium; progressive narrowing of the interpediculate distances of the lumbar spine in anteroposterior (A-P) view, and short pedicles on lateral projection: a short, broad pelvis with horizontal acetabular margins and narrow, deep greater sciatic notches; disproportionately long fibula, especially proximally, and shortened, relatively wide long bones with mild metaphyseal irregularities.

The trait is an autosomal dominant one, although most cases (85%) are sporadic and thought to result from a new mutation. When both parents are of normal height, the risk of additional achondroplastic children is negligible. When both parents have achondroplasia, 25% of their children will be expected to be homozygous for the achondroplasia gene, and such infants have indeed been reported. All of them have been severely affected, and none has survived more than a few weeks. All achondroplastic women must deliver by caesarean section. Prenatal diagnosis by fetal radiography at 18 weeks' gestation has been accomplished.

Achondrogenesis, Type I (Achondrogenesis, Parenti-Fracarro type)

This disorder is recognizable at birth as short-limbed dwarfism with a hydropic appearance associated with a soft skull that does not appear large when com-

Table 1. Short-Limb Dwarfism Recognizable at Birth or in First Year

Name	Inheritance*	Adult Height (cm)	Life Expectancy*†
Achondroplasia (heterozygous)	AD	122-132	Good unless severe neurologic complications occur
Achondroplasia (homozygous)	AD	——‡	SB, or die within months
Achondrogenesis, Type I	AR	——	SB, or neonatal death
Achondrogenesis, Type II	AR	——	SB, or neonatal death
Grebe disease	AR	99-107	Good unless cardiorespiratory complications occur
Thanatophoric dwarfism	?AR	——	SB, or neonatal death
Chondrodysplasia punctata, Conradi-Hunerman type	AD	151-Normal	SB, or neonatal death; good unless severely affected
Chondrodysplasia punctata, rhizomelic type	AR	——	Most die by age 5 months
Diastrophic dwarfism	AR	86-127	Good unless cardiorespiratory complications occur secondary to kyphoscoliosis
Metatropic dwarfism	AR	109-119	Many die in infancy
Asphyxiating thoracic dysplasia	AR	Near normal	Poor unless mildly affected and survive childhood
Mesomelic dwarfism, Nievergelt type	AD	147	Good
Mesomelic dwarfism, Langer type	AR	122-130	Good
Hypophosphatasia, congenital lethal type	AR	——	SB, or neonatal death
Short rib-polydactyly syndrome, Majewski type	?	——	SB, or neonatal death
Short rib-polydactyly syndrome, Saldino-Noonan type	AR	——	SB, or neonatal death
Campomelic dwarfism	?	——	SB, or neonatal death
Acrodysostosis	?	132-158	Good
Acromesomelic dwarfism	AR	119	Good
Osteogenesis imperfecta, dominant type	AD	Variable	Good beyond infancy
Osteogenesis imperfecta, recessive type	AR	——	SB, or death in months
Metaphyseal chondrodysplasia, McKusick type	AR	107-122	Good unless GI complications or severe varicella occur
Femoral hypoplasia–unusual facies syndrome	?	109-122	Good

*AD, autosomal dominant; AR, autosomal recessive; XR, X-linked recessive.
†SB, stillborn.
‡No data available.

pared with the trunk (4, 5). The neck is extremely short, and there is micromelia, a small barrel-shaped chest and a prominent abdomen. Birth lengths range between 25 and 29 cm, and weight between 800 and 900 g. All patients are stillborn or die in the immediate neonatal period.

The significant radiographic features include variable ossification of the calvarium; and absent or severely retarded ossification of the cervical, thoracic, lumbar and sacral vertebral bodies with ossification centers of the pedicles and neural arches present. There are short, relatively thin ribs with splayed costochondral junctions and multiple rib fractures. The pelvis is poorly ossified and the long bones are extremely short and bowed with wide metaphyses and spur formation. This condition is radiologically distinct from Type II, and yet clinically the two conditions resemble each other closely. Both have been confused with achondroplasia, a fact that helps to account for past suggestions that achondroplasia may be inherited as a recessive trait or that it is lethal in the neonatal period.

Achondrogenesis, Type I is an autosomal recessive trait.

Achondrogenesis, Type II (Achondrogenesis, Langer-Saldino type)

The disorder is recognizable at birth as patients show short-limb dwarfism with a hydropic appearance (6). There is a relatively huge head compared to the short, square-shaped trunk and extremely short extremities held extended from the trunk. Most reported cases have been premature or stillborn infants. Death occurs in six hours or less.

Major radiographic features include marked underossification of the lumbar vertebrae, pubis and ischium, together with absent ossification of the sacrum and normal ossification of the calvarium. The chest is barrel-shaped with short ribs and the tubular bones are short, with flaring and cupping of the metaphyses. The differential diagnosis of this condition includes Type I achondrogenesis, Grebe disease, thanatophoric dwarfism and hypophosphatasia.

Achondrogenesis, Type II is an autosomal recessive trait.

Grebe Disease (Nonlethal Achondrogenesis)

This disorder was described in 1952 by Grebe (7) and later by Quelce-Salgado. Although often included with achondrogenesis, type II, Grebe disease is a distinct entity. At birth affected persons show short-limbed dwarfism with severe hypomelia of both upper and lower extremities, increasing in severity from proximal to distal segments, lower extremities being more severely affected than upper ones. There are extremely short toe-like fingers and short feet in valgus position. The toes may be rudimentary, ball-like structures, and 50% of patients have polydactyly. Adult height ranges from 99 to 104 cm. A high stillborn or neonatal mortality rate is observed in the largest reported series from Brazil, but after childhood the prognosis appears good. Intelligence is normal.

X-ray pictures show a normal skull and axial skeleton. The remaining bones of the extremities show a range of abnormalities from variable hypoplasia to aplasia, and bone age may be retarded.

Grebe disease is an autosomal recessive trait. At least 55 cases have been identified in the literature and of these, 47 belong to a large kindred of inbred Brazilian indians.

Thanatophoric Dwarfism

This form of dwarfism is also detectable at birth. Fetal movements are weak and polyhydramnios has been frequently noted (8). The infant may be stillborn, and viable infants are hypotonic. Short-limb dwarfism is present with a relatively long trunk and a small, pear-shaped thorax, which is narrow in all dimensions. There is a disproportionately large head, frequent hydrocephalus, enlarged fontanelles, open cranial sutures, bulging forehead, prominent eyes and a flat nasal bridge. Excessive skin folds are apparent over the extremely short limbs, which extend from the torso. Death from cardiorespiratory failure intervenes within the first few days unless life is prolonged by mechanical ventilatory assistance. Congenital heart disease is commonly found at autopsy, as well as major brain malformations.

Significant radiographic features include flatness of the vertebral bodies with notch-like ossification defects of the central portion of the upper and lower plates and excessive vertical dimensions of the intervertebral spaces. The ribs are very short with flared anterior ends, and there is a small, narrow thorax. The long bones are very short and broad. In utero diagnosis has been made by fetal radiographs. Cloverleaf skull (Kleeblattschädel anomaly) has been reported with thanatophoric dwarfism.

Autosomal recessive inheritance is suggested, but careful analysis of the reported cases does not support the hypothesis. Most cases are sporadic.

Chondrodysplasia Punctata, Conradi-Hunermann (Conradi Disease; Stippled Epiphyses; Chondrodystrophia Calcificans Congenita)

The major clinical features are usually recognizable at birth; asymmetric shortening of the extremities with leg-length discrepancy is present, together with mild flattening of the nasal bridge and the face, which disappear in early childhood (9). Dry and scaly skin with alopecia is frequent. Joint contractures occur later but are usually mild, although scoliosis is common after age 1 year. Cataracts may not be detected until early childhood, but should be anticipated. Intelligence is normal.

The prognosis is better than with the rhizomelic type; however, severely affected infants are stillborn or die in the newborn period. Less severely affected infants have a relatively good prognosis. Adult height varies between 130 cm and normal. Treatment is directed toward orthopedic correction of leg-length discrepancy and control of scoliosis. Alopecia or sparse and poorly growing hair may be cosmetically managed.

Early in life radiographically detectable stippling of the epiphyses affects primarily the long bones, carpals, tarsals, pelvis and spine—usually in an asymmetric fashion; however, this is not diagnostic. Later in life there may be unequal shortening of the long bones, irregular epiphyses and scoliosis.

The disorder is inherited as an autosomal dominant trait.

Chondrodysplasia Punctata, Rhizomelic Type (Stippled Epiphyses: Chondrodystrophia Calcificans Congenita)

Rhizomelic (short-limb) dwarfism is manifest at birth, together with joint contractures; flat dish-like midface; bilateral cataracts; alopecia or sparse and lusterless hair; and dry, red ichthyosiform skin (9). Most of these infants die during the first year of life, the average age at demise being about 5 months. Spastic paresis, microcephaly and mental retardation are frequent in survivors.

Major radiographic features are symmetric stippling of the epiphyses, primarily about the ends of the shortened humeri and femora, both of which show metaphyseal splaying. Vertical radiolucent areas split the thoracolumbar vertebral bodies into dorsoventral segments. Stippling of the epiphyses is not itself diagnostic, as it may be found in a wide variety of disorders, including the cerebrohepato-renal syndrome, generalized gangliosidosis, Smith-Lemi-Opitz syndrome, anencephaly, trisomy 21, trisomy 18 and cretinism.

Autosomal recessive inheritance has been established.

Diastrophic Dwarfism

This is another form of short-limb dwarfism recognizable at birth (10). Both micromelia and club feet are present. The hands are broad and short with proximally set, hypermobile thumbs held in hitch-hiker position; short, stiff fingers with symphalangism; and fusion of the proximal interphalangeal joints. There are multiple joint contractures, particularly of the hips and knees, which may also subluxate or dislocate. Progressive scoliosis or kyphoscoliosis is present and can become severe with ambulation.

An unusual finding is cystic swelling of the upper portion of the external ear within the first two or three months followed by subsidence and residual cauliflower deformity. Cleft palate is found in one-half the cases. The mean adult height is 112 cm, the range for males being 86-127 cm, and for females, 104-122 cm. Life expectancy is normal unless cardiorespiratory complications occur because of severe kyphoscoliosis. Neurologic problems may develop if cervical kyphosis is marked. Ossicle deformity of the middle ear is reported and can lead to impaired hearing. Degenerative change in hips may cause severe gait disturbances and pain. Intelligence is normal. Any treatment is symptomatic and primarily orthopedic.

Major radiographic features include short, thick long bones with broad flared metaphyses; delayed, irregular and flattened epiphyses; short, wide femoral necks; coxa vara; subluxation or dislocation of the hips; precocious ossification of costal cartilage; small, oval or triangular first metacarpals; irregular deformity of the metacarpals, metatarsals and phalanges; talipes equinovarus; cervical kyphosis and thoraclumbar kyphoscoliosis.

Autosomal recessive inheritance is established.

Metatrophic Dwarfism

The disproportionately long narrow trunk and short extremities in infancy suggests achondroplasia (12, 13). In fact this condition is inappropriately labeled as a hypoplastic form of achondroplasia or chondrodystrophy in the older literature. Rapidly progressive kyphoscoliosis develops by early childhood, resulting in short-trunk dwarfism with a tail-like appendage in the sacral area. The extremities are short with bulbous enlargement of the metaphyses and limited joint movement. Short hands and feet are present with hyperextensibility of finger joints. There is a normal skull and facies and normal intelligence in surviving infants. Many patients die in infancy from respiratory failure secondary to a narrow thorax. The range of adult height of survivors is 109 to 120 cm. Severe scoliosis may cause incapacitation.

Radiographic evaluation shows underossification of the vertebral bodies, which are reduced to narrow tongue-like structures in early infancy and platyspondyly with kyphoscoliosis in later life; exaggerated metaphyseal widening, which produces long bones with a dumb-bell appearance; irregular and deformed epiphyses; and a flared iliac crest, which gives a battle-ax shape to the pelvis.

The condition is an autosomal recessive trait.

Chondroectodermal Dysplasia (Ellis-van Creveld Syndrome)

This is a mesomelic form of short-limb dwarfism, the forearms and shanks being comparatively shorter than the proximal segments of the extremities (14, 15). Polydactyly of the hands is common, with the extra digit located on the ulnar or postaxial side of the hand, and hypoplastic nails are present. Natal teeth are frequently encountered, but later teeth are missing or conical with areas of enamel hypoplasia. Partial midline cleft of the upper lip and multiple buccolabial frenula are frequent. Congenital heart disease—usually an atrioseptal defect—occurs in one-half the cases. Intelligence is normal.

Major radiographic features include mesomelic shortening of the extremities as manifested by progressive distal shortening of the long bones and expanded metaphyses; retarded ossification of the lateral portions of the proximal tibial epiphyses and metaphysis, resulting in knock-kneed deformity; bilateral hexadactyly; progressive distal hypoplasia of the phalages; deformed or fused capitate and hamate carpals; short iliac bones; and hook-like inferior projections from the medial and lateral acetabular margins in infancy. There is progressive normalization of the pelvic appearance later in childhood.

Autosomal recessive inheritance is well-recognized.

Asphyxiating Thoracic Dysplasia (Jeune Syndrome; Thoracic-pelvic-phalangeal Dystrophy; Asphyxiating Thoracic Dystrophy)

This variety of short-limb dwarfism is characterized by an extremely small thorax, which frequently results in respiratory distress, cyanosis or recurrent respiratory infection. The hands and feet are short and broad, occasionally with postaxial polydactyly. There is a variable degree of clinical severity (16, 17). Some patients survive, and they tend to have less and less respiratory difficulty with time. Survivors of childhood may have only minimally reduced adult height. Progressive renal disease may occur as manifested by proteinuria, uremia and hypertension, leading to death from renal failure without dialysis or transplantation. Kidney biopsy shows nonspecific changes.

X-ray evaluation reveals a small thoracic cage in all dimensions, with horizontally orientated ribs showing widened and cupped anterior ends. The iliac bones are short in their cephalocaudal diameter, and the acetabular margins have inferior spurs. The pelvic changes are virtually identical to those of the Ellis-van Creveld syndrome. Metaphyseal irregularities are more obvious in midchildhood. The thorax is relatively large in later childhood and the pelvic appearance becomes normal.

The condition is an autosomal recessive trait.

Mesomelic Dwarfism, Nievergelt Type

The major clinical features of this rare autosomal dominant condition are mesomelic dwarfism with deformity of the lower legs manifested by a visible bony projection along the medial aspect of the shank, a normal head and trunk, limited extension and supination of elbows, club feet and genu valgum (18). There is reduction in adult height to roughly 147 cm. Normal intelligence and lifespan are to be expected.

Radiographically there is marked shortening of the middle segments of the extremities with bizarre triangular or rhomboid deformity of the tibias; short, broad and deformed radius and ulna with radioulnar synostosis; and tarsal synostosis. Skull and spine are normal.

Mesomelic Dwarfism, Langer Type

The extremities in general are short, although the middle segments are strikingly abbreviated with more pronounced shortening in the lower limbs (19-21). There is forearm bowing and normal hands, feet and trunk. Mandibular hypoplasia is common. Adult height varies between 122 and 130 cm. Life expectancy and intelligence are normal.

Mesomelic shortening of the long bones of the extremities; marked hypoplasia of the distal ulna and proximal fibula; and bowing of the radius and tibia are evident on X-ray examination. Other structures are normal.

Langer type mesomelic dwarfism apparently results from homozygosity of the dyschondrosteosis gene, i.e., the severer bony dysplasia behaves like a recessive trait. Careful measurement of the forearm and shank segments of both parents, coupled with radiographs shows mild mesomelia and variable degrees of the Madelung deformity—characteristics that alone may be labeled dyschondrosteosis. Stature may be only minimally reduced in the parents who are regarded as skeletally "normal" on clinical examination.

Hypophosphatasia, Congenital Lethal Type (Neonatal Hypophosphatasia)

There is a globular "boneless" skull and soft skeleton at birth with a short trunk, neck and extremities (22, 23). Congenital angulation deformities of the extremities are present with overlying skin dimples. A low serum alkaline phosphatase level and increased urinary phosphoethanolamine levels are virtually diagnostic. These patients are stillborn or die within the first few hours from respiratory embarrassment secondary to the soft thorax. A few patients survive several weeks or months.

The entire skeleton shows extreme underossification on X-ray examination. Often only the central portion of bones are visible, whereas in other areas ossification is absent. The metaphyses are irregular and spotty, and there are short thin ribs and long bones.

Autosomal recessive inheritance is established. Prenatal diagnosis has been accomplished most reliably by ultrasonography (23a).

Short Rib–Polydactyly Syndrome, Majewski Type

This is a multiple malformation syndrome detectable at birth. Affected individuals show median cleft lip, short nose, malformed low-set ears, polysyndactyly, brachydactyly, short extremities and ribs, narrow chest, large protruding abdomen, hypoplastic epiglottis, congenital heart disease and genitourinary anomalies, including polycystic kidneys (24, 25). Death occurs perinatally secondary to cardiorespiratory insufficiency (24, 25).

Major radiographic features include disproportionately shortened tibias with rounded proximal contours, mild shortening of all long bones, short horizontal ribs and a normal pelvis. Either pre- or postaxial polydactyly, or both, may be present.

Genetic factors are unknown.

Short Rib–Polydactyly Syndrome, Saldino-Noonan Type

These infants are stillborn or die in the newborn period (25, 26). They have an hydropic appearance with postaxial polydactyly, brachydactyly, narrow chest, protuberant abdomen, short ribs, severely shortened and flipper-like limbs, congenital heart disease, genital anomalies, renal malformations including polycystic kidneys, anal atresia and pulmonary hypoplasia.

Horizontally oriented, severely shortened ribs are seen radiographically, together with ossification defects of the vertebral bodies. There is extreme shortness of the long bones, with spurs extending from the medial and lateral segments of the metaphyses. Partial ossification of the intrinsic bones of the hands and feet is evident. The iliac bones are small with horizontal acetabular roofs.

The condition is probably an autosomal recessive trait.

Campomelic Dwarfism

This form of short-limb dwarfism frequently shows forward angulation of the tibias with a cutaneous dimple over the apex of the bend, bowed femurs, bilateral clubfoot deformity and mild bowing of the upper extremities (27, 28). There is a large neurocranium compared to the rather flat face characterized by a prominent forehead, depressed nasal bridge, micrognathia, cleft palate and low-set ears. Stillbirths are frequent, but if the baby is liveborn, severe respiratory distress soon develops secondary to marked hypoplasia of the cartilaginous elements of the tracheobronchial tree. Congenital heart disease may also be present. Most liveborn infants die in the neonatal period; a few survive 5 to 7 months, usually with mental retardation. The oldest living patient was 21 years old and had produced three affected children by two different mates.

Radiographically the calvarium is large relative to face size. There is a small, bell-shaped thorax with thin and wavy ribs and small scapulas. Bowing of the long bones of the upper limbs is mild if present. The hips are dislocated, and the pelvis is underdeveloped and poorly mineralized. The femurs are bowed, and there is anterolateral bowing of the lower third of the tibias with fibular hypoplasia. Autosomal dominant inheritance is implied by an affected mother of three affected children, but normal parents are reported, suggesting autosomal recessive inheritance. Some evidence of genetic heterogeneity exists in that long and short bone varieties of campomelia are reported (28a). Until these matters are clarified, parents who have had an affected child should be given at least a 25% risk of recurrence.

Acrodysostosis

This disorder is characterized by progressive shortening of the limbs from proximal to distal, more marked in the upper extremities; limitation of elbow movement; and short, pudgy and stubby fingers and toes with broad, foreshortened nails. The facies is striking, as the entire nose is flattened, broad and small with anteverted nostrils (pug nose). There is an open-mouth habitus with a long philtrum, maxillary hypoplasia, hypertelorism, epicanthic folds and malocclusion. Recurrent otitis media or hearing loss, or both is common, and mental retardation occurs in 90% of cases. Adult height varies between 132 and 158 cm. Arthritic changes involving the hands, feet and hips may develop in young adulthood, and progressive limitation of movement of the hands, elbows and spine can also occur.

The most striking radiographic abnormalities are in the hands and feet and include severe shortening of metarcarpals and phalanges; and deformed, prematurely fused, cone-shaped epiphyses and small carpal bones with comparable changes in the feet. Minor and inconsistent abnormalities are found in the long bones, particularly the ulnas, radii and femurs. Partially collapsed vertebral bodies and irregularities of the end plates may occur, suggesting juvenile spondylitis. Skull films usually show brachycephaly, and in some children the calvaria are slightly thickened.

The inheritance pattern is not known as all cases are sporadic.

Acromesomelic Dwarfism

This dysplasia was delineated in 1971 and the number of identified cases is too few to permit final conclusions as to phenotype. It is a short-limb dwarfism predominantly involving the forearms, hands and feet (31, 32). The hands are short with stubby fingers and the feet are broad and flat. There is limitation of elbow motion, especially extension, and a mildly shortened trunk. Scaphocephaly may be present although the face is normal, as is intelligence.

A major radiographic feature is marked shortening and lateral bowing of the radius with posterior dislocation of the radial head and even greater shortening and hypoplasia of the ulnar distal end. The long bone epiphyses are generally normal, whereas the epiphyses of the shortened phalanges fuse prematurely. The metacarpals and metatarsals are very short. Minimal changes are noted in the spine. Acromesomelic dwarfism does not have the epiphyseal and metaphyseal changes found in pseudoachondroplasia.

Autosomal recessive inheritance is presumed on the basis of parental consanguinity and affected sibs of normal parents.

Osteogenesis Imperfecta, Dominant Type (Lobstein Disease; Thin Bone Type Osteogenesis Imperfecta; Brittle Bone Disease; Osteogenesis Imperfecta Tarda)

This form of osteogenesis imperfecta has a wide range of expression from severe dwarfism to near-normal stature; hence it may be detected initially at birth or even later in childhood (33-35). The short stature can be related to the degree of deformity of the extremities. The hallmark of the disease is increased bone fragility with fractures due to trivial trauma. Other features include a broad thorax with or without sternal depression or protuberance; hyperextensible joints and ligamentous laxity; kyphoscoliosis; triangular-shaped face; frontal bossing; delayed closure of the wide fontanelles; blue sclera; thin and translucent skin; excessive perspiration; hearing deficits due to otosclerosis after age 20 years; dentogenesis imperfecta with translucent, amber, yellow or blue-gray tooth color and frequent caries; increased capillary fragility; and bleeding tendency or abnormal platelet function tests, or both.

The greatest frequency of fractures occurs between infancy and puberty. Fractures heal in the usual time. Occasionally exuberant callus forms, which can be misinterpreted as sarcomatous degeneration. Surgical scars may stretch and widen. Occasional findings include umbilical and inguinal hernias, joint dislocation, hydrocephalus, spinal cord or nerve root compression, or both, and the floppy mitral value syndrome. Pregnant women with this disorder are best delivered by caesarean section. Prognosis for survival is good once the baby is beyond infancy, although there may be great disability and deformity.

Radiographically there is generalized demineralization of the skeleton with an increased tendency to fractures. Variable degrees of long bone and rib deformity are present and there is biconcave flattening of the vertebrae, producing a codfish-mouth appearance, together with kyphoscoliosis. Delayed ossification of the calvarium is present with frontal and temporal bulging, multiple wormian bones and platybasia.

The tarda form is an autosomal dominant trait, with new mutation thought to be responsible for isolated occurrences. The basic pathogenesis is not known. Numerous substances have been tried as treatment on empirical grounds, but none is established as beneficial. Treatment is primarily orthopedic. In utero diagnosis by fetal radiography is reported.

Osteogenesis Imperfecta, Recessive Type (Vrolik Disease; Thick Bone Type Osteogenesis Imperfecta)

Bizarre, grossly deformed extremities are detectable at birth. The hands and feet are normal although they may be in abnormal positions. There is a large head with soft calvarium and wide fontanelles, shallow orbits, blue sclera and small nose with depressed nasal bridge (36–38). Hypotonia, ligamentous laxity and inguinal hernias are usually present. The patients are generally stillborn or survive only a few days or months. Infrequently survival into late childhood occurs and is associated with dwarfism and severe deformities. This thick bone type of osteogenesis imperfecta is always congenital and is relatively infrequent.

Generalized demineralization with multiple fractures in varying states of callus formation and healing are seen on X-ray examination. The long bones shafts are thick, short and rectangular or ribbon-like. Flattened vertebrae are present, together with an underossified calvarium with multiple wormian bones. Cystic honeycomb-like changes in the long bones may be found in early childhood survivors.

Autosomal recessive inheritance is established.

Metaphyseal Chondrodysplasia, McKusick Type (Cartilage–hair Hypoplasia)

In this disorder, birth length is typically reduced although weight is normal, and the condition is usually not radiographically detectable until age 9-12 months. Features include short-limb dwarfism with a relatively long trunk; short and pudgy hands and feet; generalized ligamentous laxity, which is pronounced in the fingers and toes; normally wide but foreshortened nails; prominent sternum; mild flaring of lower ribs; bowleg deformity, with or without unilateral genu valgum; sparse, light-colored, fragile hair of small caliber; and normal head, face and intelligence. Adult height varies from 107 to 147 cm. Intestinal malabsorption and Hirschsprung disease are reported in affected children. There is an increased—often fatal—susceptibility to varicella. Apparently this is due to a specific defect in T-cell function. Except for these complications, prognosis for survival is good.

The major radiographic findings are irregular metaphyses of the long bones, which are scalloped and sclerotic, often with cystic areas. The changes at the knee are generally severer than they are elsewhere. A relatively long fibula distally produces ankle deformity. The vertebral bodies, although slightly decreased in height, are normal as are the skull and epiphyses.

Autosomal recessive inheritance has been confirmed.

Femoral Hypoplasia–Unusual Facies Syndrome

Relatively few cases of this condition are reported, and all of them are sporadic. Short stature is detectable at birth, primarily due to abbreviated lower extremities. The facies are characteristic with upslanting palpebral fissures, short nose, broad nasal tip and hypoplastic alae nasi, a long philtrum, thin upper lip, micrognathia and cleft palate (41, 42). Clubfoot, shortening of the proximal segment of the arm and limitation of elbow movement are seen. Shortening of the lower limbs may be so great that the tibia clinically appears to articulate with the acetabulum. Cutaneous dimples on the lateral thighs or in the buttocks are associated with the proximal end of the deficient femurs. Walking is delayed, but ultimately these patients ambulate without assistance. Occasionally esotropia, astigmatism, inguinal hernias, cryptorchidism or underdeveloped labia majora are observed. Intelligence is normal.

On X-ray evaluation, proximal focal femoral deficiency varies in severity from hypoplasia to absence, and similar changes are noted in the fibula. Hypoplasia of the humeri is not usually so severe as with the femurs. Radiohumeral synostosis is frequent. Pelvic dysplasia includes the hypoplastic acetabula, a small constricted iliac base and large obturator foramina. Talipes equinovarus may be present.

SHORT-LIMB DWARFISM RECOGNIZABLE IN CHILDHOOD

Hypochondroplasia

Hypochondroplasia resembles achondroplasia and must be distinguished from it, although both are autosomal dominant disorders (2, 43). Both are specific entities on clinical and radiographic grounds. Although the incidence is unknown, it is believed that hypochondroplasia is probably not rare. Many cases are overlooked or mislabeled as being the result of constitutional or familial short stature.

It is recognizable in early childhood by a short stocky habitus with a relatively long trunk and disproportionate shortening of the limbs (Table 2). Circumference of the head is normal as is the face, although the forehead may be prominent. Slight exaggeration of lumbar lordosis with a sacral tilt and protuberant abdomen are common. Mild generalized ligamentous laxity of the joints is present as is mild genu varum, which may be self-correcting with time. Limitation of both elbow extension and supination is seen. The fingers are short, but the hand is not of the trident type. Adult height varies between 132 and 147 cm. No neurologic complications of the lower spinal cord or nerve roots, or both, are reported. The proximal fibula is not elongated and if tibial bowing occurs, it tends to correct spontaneously. Intelligence is normal.

X-ray changes are mild. The pelvis is basically normal, and there are short, broad femoral necks. The skull is normal. Mild narrowing of interpediculate distances in the spine from the first to the fifth lumbar vertebra may be present as well as shortening of the pedicles with normal vertebral body height. The long bones may be slightly abbreviated and square with slight metaphyseal flaring.

Pseudoachondroplasia (Spondyloepiphyseal Dysplasia of the Pseudoachondroplastic Type)

Delay in walking may be the first manifestation of pseudoachondroplasia, and reduced height is usually not appreciated until age 2 to 4 years (44). The body habitus is similar to achondroplasia, but with a normal head and face, i.e., relatively long trunk, rhizomelic shortening of the limbs, exaggerated lumbar lordosis, scoliosis, genu varum or genu valgum or combination of both (tackle deformity), and joint hypermobility. The fingers are short and pudgy, and there is prominent enlargement of wrists. The hands tend to be held in ulnar deviation. The gait is waddling. Joint pain, particularly in the knees and hips occurs because of degenerative arthritic changes, which may be incapacitating; normal intelligence is found and life expectancy is normal. Adult height varies between 75 and 125 cm.

Moderate vertebral platyspondyly with anterior protrusion of the central portion producing a tongue-like structure is seen radiographically, together with biconvex deformity of the upper and lower margins of the vertebrae and scoliosis. The long bones are shortened with irregular, expanded and flared metaphyses, often mushrooming out around the epiphyses—it is especially striking at the knees. There are small, irregular proximal femoral epiphyses in childhood, and marked dysplasia of the femoral heads in adulthood. The iliac, pubic and ischial bones are underdeveloped and irregular.

The condition is genetically heterogeneous. Pseudoachondroplasia can be

Table 2. Short-Limb Dwarfism Recognizable in Childhood

Name	Inheritance*	Adult Height (cm)	Life Expectancy
Hypochondroplasia	AD	132-147	Good
Pseudoachondroplasia	AD, AR	81-130	Good
Parastremmatic dwarfism	AD	109	Good
Metaphyseal chondrodysplasia, Jansen type	AD	102-125	Good
Metaphyseal chondrodysplasia, Schmid type	AD	130-160	Good
Metaphyseal chondrodysplasia, with thymolymphopenia	AR	—†	Poor; death from sepsis usually before 1 year
Multiple epiphyseal dysplasia	AD	137-152	Good
Dyschondrosteosis	AD	152-168	Good
Pycnodysostosis	AR	130-152	Good
Asphyxiating thoracic dystrophy	AR	Near normal	Poor unless mildly affected and survive infancy
Osteogenesis imperfecta, dominant type	AD	Variable	Good beyond infancy

*See footnote, Table 1.
†No data available.

transmitted as an autosomal dominant trait and as an autosomal recessive trait. Variability of physical findings and radiographic expression is considerable in both forms. They cannot be differentiated with complete certainty on clinical or radiologic grounds alone, and an adequate family history may be vital.

Parastremmatic Dwarfism

This condition is named from the Greek word for twisted or distorted limbs (45). Newborn infants affected with the disorder may be considered to have "stiff joints". Between 6 and 12 months, skeletal deformities develop and the full clinical picture is recognizable by age 10 years. Dwarfism is severe. Major features are short neck; increased anteroposterior diameter of the thorax; kyphoscoliosis; asymmetric, bizarre, short lower extremities with bowed and twisted long bones; and genu varum and valgum. The arms are relatively long with short and stubby hands. Multiple large joint contractures are common. The skull is normal. Survival through adulthood occurs, but disability may be severe, with adult height ranging between 89 and 109 cm.

Major radiographic features include generalized decreased bone density; very coarse trabeculations with areas of irregular, dense stippling and streaking, producing a "flocky or wolly" appearance (in the pelvis this appears as a lace-like border of the iliac crest) and a similar wooly appearance of the vertebral bodies, which are also flat. The metaphyses are clear and contain flocky bone, and there are small or unossified femoral heads or necks with severely deformed radiolucent epiphyses. In general there is marked shortening and distortion of the long bones. The flocky appearance disappears by adulthood.

The inheritance pattern is not established, since pedigrees consistent with both autosomal and X-linked dominant inheritance have been described.

Metaphyseal Chondodysplasia, Jansen Type

Metaphyseal chondrodysplasia of the Jansen type is the severest and rarest of the 13 or more forms of metaphyseal chondrodysplasia that have been reported (46-48). It is manifest by short-limb dwarfism, with the forearm and lower extremities being more severely shortened and deformed than the trunk and humeri. The joints are enlarged with limitation of extension of the hips and knees, producing stooped posture. The forehead is prominent and there are unusual facies with supraorbital and frontonasal hyperplasia and micrognathia. The fingers are short, especially the distal phalanges. Adult height ranges between 102 and 125 cm. Asymptomatic hypercalcemia and decreased renal phosphate reabsorption are poorly understood nosologic features. Serum alkaline phosphatase level is normal. Deafness in adulthood is attributed to sclerosis of the petrous bones.

Markedly enlarged, wide, fragmented and irregular metaphyses that enclose areas of radiolucent and dense material are noted radiographically. Essentially normal epiphyses are widely separated from the metaphyses by areas of unossified tissue. Severe shortening, curvature and abnormal alignment of the long bones are seen. Sclerosis of the cranial bones is common.

The trait is an autosomal dominant one.

Metaphyseal Chondrodysplasia, Schmid Type

The initial manifestation may be only bowlegs noted shortly after initial ambulation, and the bowing may continue and become severe (49, 50). If joint alignment is poor, degenerative hip and knee disease produces pain. This chondrodysplasia usually becomes evident at age 18 to 24 months by moderately short stature due to relatively short extremities with prominent, often severe, bowed legs. The wrists are enlarged and patients are often unable to fully extend the fingers. Flaring of the lower rib cage and a waddling gait are also seen. The feet and head are normal and adult height ranges between 130 and 160 cm. Intelligence is normal.

Metaphyseal abnormalities are revealed by X-ray, and they vary from scalloping to gross cupping, fraying, widening and fragmentation. The changes are most frequent at the ankle, knee, wrist, shoulder and hip. Coxa vara is common as is shortening of the long bones and genu varum. There is no involvement or only minor radiographic changes in the hands and feet.

Autosomal dominant inheritance is well recognized. Sporadic cases are related to advanced paternal age. A rare and nearly identical metaphyseal chondrodysplasia is the Spahr-Hartmann type, which is transmitted as an autosomal recessive trait. It cannot be clinically or radiographically distinguished from the Schmid type.

Metaphyseal Chondrodysplasia with Thymolymphopenia

Total birth length in this disease is short primarily because of the short, bowed limbs. Hair is sparse and tends to be lost and the skin is erythematous and dyskeratotic (51). The infants fail to thrive and have recurrent bacterial infections or diarrhea, or both. Lymphopenia, agammaglobulinemia and thymic hypoplasia are present and bone marrow failure may occur. Death usually occurs from bacterial sepsis before the first birthday.

Radiographically the long bones are short and broad, and the metaphyses are wide, slightly irregular and cupped. A peculiar pelvic configuration is primarily due to shortening and hypoplasia of the iliac bones and large sacrosciatic notches.

Autosomal recessive inheritance is established.

Multiple Epiphyseal Dysplasia

The typical child with this condition is first recognized in early childhood because of a waddling gait or slow growth, or both. Other symptoms include difficulty in climbing stairs and running (52-54). Pain and joint stiffness are more common with increasing age. Premature degenerative joint disease may be incapacitating. Many adults, however, either have minimal symptoms or are asymptomatic and aside from shortness of stature, might escape detection, since they may be near-normal in height. Deficiency in the lateral part of the distal tibial ossification center in children causes a sloping, wedge-shaped distal tibial articular surface, an important diagnostic sign when evaluating adults. The head and trunk are normal.

This disorder is often misdiagnosed as bilateral Legg-Perthes disease in childhood. When such a diagnosis is suggested, multiple epiphyseal dysplasia must be considered and a complete skeletal radiographic survey performed. Long-term immobilization used as a therapy for Legg-Perthes disease is not indicated in this disease.

Radiographic features include bilateral, symmetrical irregularity and underdevelopment of the epiphyseal ossification centers, with the severest changes in the hips, knees and ankles. The ossification centers may be mottled and fragmented, but fuse with time, and the adult epiphyses may be nearly normal in size. Contours of the epiphyses in adulthood may be nearly normal or show irregularity and degenerative changes of the articular surfaces. Mild shortening of the long bones develops and the metaphyses occasionally show minimal irregularity and fraying. Vertebral involvement is uncommon, but if present consists primarily of Schmorl's nodes of the end plates and varying degrees of flattening.

Autosomal dominant inheritance is found in the majority of families. Clinically identical disease is reported in a number of families as an autosomal recessive trait. Heterogeneity exists even among the dominant forms: a mild type is referred to as Ribbing disease, a severe type as Fairbank disease. Data are at present insufficient to permit differentiation.

Dyschondrosteosis (Leri-Weill Disease)

This condition is usually recognized in adolescence as a very mild mesomelic type of short stature with greater shortening of the forearms and shanks when related to the upper arm and thighs (21, 55). There is limitation of motion at the elbow and wrist. Madelung deformity produces a bayonet appearance of the wrists, although it is usually asymptomatic. Prognosis is good. Evidence now strongly supports the belief that the Langer type mesomelic dwarfism is the result of homozygosity for the single autosomal dominant gene, which results in dyschondrosteosis, i.e., the heterozygote shows dyschondrosteosis, the homozygote Langer type mesomelic dwarfism.

By X-ray, the ulna, radius, tibia and fibula are short compared to normal length standards and there is lateral bowing of the radii. The humerus and femur are less shortened. Palmar slanting of the radial articular surface at the wrist, dorsal subluxation of the distal ulna and trianular-shape of the carpal bones create the Madelung deformity.

Pycnodysostosis

These patients may come for evaluation because of failure to grow or because of persistence of the open anterior fontanelle (56, 57). Approximately two-thirds of the patients have had fractures, and there may be secondary deformities of the long bones because of this fact. Vertebral segmentation abnormalities and scoliosis may be seen. There is craniofacial disproportion as a result of underdeveloped facial bones, prominence of the frontoparietal bones and an obtuse angle of the mandible, which may be small. There is persistence of the fontanelles even into adulthood. A prominent, parrot-like nose is normally present. Delayed exfoliation of deciduous teeth is common as are malformed and malpositioned permanent teeth. There is a narrow, grooved palate. The sclera may be blue and the skin over the fingers may be wrinkled. Mental retardation is reported in about one-sixth of the cases.

In the past these patients were thought to have osteopetrosis, a condition usually associated with anemia, hepatosplenomegaly, diminished marrow space and deafness. It is believed that the famous French impressionist painter, Toulouse-Lautrec, had pycnodysostosis.

Major radiographic features are generalized increased bone density (osteosclerosis); delayed closure of the cranial sutures; open fontanelles; multiple wormian bones; hypoplastic facial bones and sinuses; obtuse mandibular angle; dental irregularities; tapering of the distal phalanges with aplasia of the ungual tufts; and dysplasia and hypoplasia of the acromial end of the clavicles.

Pycnodysostosis is an autosomal recessive trait.

SHORT STATURE AT BIRTH OR IN FIRST YEAR

Hypophosphatasia Tarda

This disease is usually recognizable at birth (22). Anorexia, vomiting, constipation and failure to thrive often appear between 4 weeks and 6 months of age, together with pneumonia, fever, irritability and occasionally convulsions, polyuria and polydipsia (Table 3). The cranial sutures are widened and there is a bulging fontanelle with prominent scalp veins. Bowlegs with skin dimples may be present. Other variable features are blue sclera, hypercalcemia, nephrocalcinosis and pyuria. Low levels of serum alkaline phosphatase and increased urinary excretion of phosphoethanolamine are found as in the congenital lethal type of hypophosphatasia.

Table 3. Proportionate Short Stature Recognizable at Birth or in First Year

Name	Inheritance*	Adult Height (cm)	Life Expectancy
Hypophosphatasia tarda	AR	Variable	Good
Metaphyseal chondrodysplasia, malabsorption and neutropenia syndrome	AR	Variable	Death may occur in infancy, but most patients survive
Hereditary artho-ophthalmopathy	AD	Near normal	Good
Tricho-rhino-phalangeal dysplasia, Type I	AD, AR	129-163	Good
Osteogenesis imperfecta, dominant type	AD	Variable	Good beyond infancy
Acrodysostosis	?	132-158	Good

*See footnote, Table 1.

Shortening of stature is variable and depends on the severity of the disease. In severely affected infants, death usually intervenes toward the end of the first year of life; however, spontaneous improvement can occur, although motor development, especially walking, is late. Because of premature craniostenosis, increased intracranial pressure and mental retardation are sometimes seen. The deciduous teeth may be shed prematurely without corresponding early eruption of the permanent dentition. Adults may show only premature loss of the secondary dentition, low serum alkaline phosphatase levels and increased urinary excretion of phosphoethanolamine. Others may experience bone pain, mild osteoporosis and fractures or pseudofractures. The heterozygote can be recognized by low serum levels of alkaline phosphatase.

Irregular, frayed, streaked metaphyses with radiolucent defects, extending into the diaphyses and spotty demineralization of epiphyses are seen radiographically. Delayed ossification of the calvarium and skull base and suture widening are common. In later childhood premature cranial suture closure may occur, together with ectopic calcifications. Bowlegs and pseudofractures of the long bones may be present.

All the typical clinical and radiographic changes of the disease are reported in a child with normal alkaline phosphatase activity in the serum and yet with elevated urinary excretion of phosphoethanolamine—"pseudohypophosphatasia."

Like the congenital form, hypophosphatasia tarda is an autosomal recessive trait.

Metaphyseal Chondrodysplasia, Malabsorption and Neutropenia Syndrome (Shwachman Syndrome; Exocrine Pancreatic Insufficiency–Dwarfism Syndrome)

This disease usually develops in infancy because the baby fails to thrive and has diarrhea and symptoms of malabsorption. Normal duodenal secretion viscosity and sweat electrolyte studies rule out cystic fibrosis (58, 59). Leukopenia of variable degree is often accompanied by frequent bacterial infections, and anemia and thrombocytopenia are reported in a few patients. Intestinal lipases apparently are sufficient to prevent overt steatorrhea even though the exocrine pancreas is replaced by fibrous tissue and fat. Treatment is directed to control of infection and pancreatic replacement therapy. Prognosis is variable. Death may occur in infancy; most patients, however, survive.

Major radiographic features include mild to moderate irregularity and dysplasia of the metaphyses of the long bones, particularly at the hips and knees with occasional coxa vara. Mild cupping and irregularities of the ribs may occur.

The inheritance is autosomal recessive.

Hereditary Artho-Ophthalmopathy (Stickler Syndrome)

Ophthalmopathy is the severest complication of this disorder. Myopia begins in early childhood and chorioretinal pigmentary changes are present (60-62). Spontaneous retinal detachment can produce blindness. Skeletal changes are variable and short stature is not a significant feature of the syndrome. There is usually a Marfanoid, slender body habitus, although none of the affected subjects is excessively tall. Prominence of the knee, ankle and wrist joints is present, together with hypotonia; hyperextensibility of the fingers, wrists, elbows and knees; and arthritis and joint pains. The disease may simulate juvenile rheumatoid arthritis in childhood. Midface hypoplasia with a depressed nasal bridge and epicanthic folds are present, and micrognathia, cleft palate, conductive deafness and cataracts are described.

Disproportionately narrow shafts of the long bones relative to their metaphyseal width (overtubulation), mild flattening and irregularity of the epiphyses, especially of the proximal head of the femur and distal portion of the tibia and mild to moderate irregularity and flattening of the vertebrae with mild anterior wedging are seen radiographically. Degenerative changes in the large weight-bearing joints are present in adulthood.

The disorder is an autosomal dominant trait, but there are variable intrafamilial clinical manifestations. It is likely that many of the patients with the "Pierre-Robin syndrome" in fact have the Stickler syndrome.

Trichorhinophalangeal Dysplasia, Type I (Giedion Syndrome)

Recognition of this condition in infancy is possible, although it is more usually detected between 3 and 5 years of age (63, 64). There is scanty, slowly growing scalp hair requiring infrequent cutting, early balding and thinning of the lateral portions of the eyebrows. A prominent, bulbous or pear-shaped nose with a long philtrum and large prominent ears, characterize the facies. Height is variable; most adults have been below the tenth percentile, although a few have been less than 127 cm tall. Asymmetric, short and crooked fingers with essentially normal function are seen. Intelligence is normal.

Cone-shaped epiphyses are found on X-ray examination in the phalanges, particularly the second through fourth fingers and toes. Premature fusion of the epiphyses produces variable shortening and angulation defects of the digits. The metacarpals and metatarsals are short. Changes in the femoral heads simulating aseptic necrosis are frequent.

The condition is an autosomal dominant trait in most families. Rarely it can be transmitted as an autosomal recessive trait. Unfortunately the two forms cannot be distinguished clinically or radiologically.

Trichorhinophalangeal Dysplasia, Type II

Although some growth retardation is apparent in infancy, the radiographic features are not recognizable until age 3 to 4 years (65). Mild short stature and microcephaly with mild to moderate mental retardation are present. There is sparse, slowly growing scalp hair, a large bulbous nose with pudgy nostrils and superior notching defects, protruding ears, prominent philtrum of the lip, micrognathia, maculopapular nevi scattered over the entire body and redundant and lax skin at birth, which disappears in early childhood. Multiple exostoses may be palpable by midchildhood. Mild demineralization and a tendency to fractures are observed. Frequent upper respiratory infections in early childhood abate, and prognosis for survival is good.

Cone-shaped epiphyses of multiple phalanges of the hands and feet are present radiographically. Multiple cartilaginous exostoses, producing irregular expansion of the metaphyses and sessile or pedunculated excrescences appear and may involve all tubular bones, the pelvis or ribs, or both. Aseptic necrosis of the hips may occur.

Genetic factors are unknown. All cases have been sporadic. Multiple exostoses, a different nasal configuration and mental retardation allow this disorder to be differentiated from Type I trichorhinophalangeal dysplasia.

SHORT-TRUNK DWARFISM

Spondyloepiphyseal Dysplasia Congenita

This disorder can be detected at birth as a form of short-trunk dwarfism (Table 4). A broad or barrel-shaped chest with deep Harrison grooves and pectus carinatum are present (66, 67). There is a flat or dish-like facies; occasional cleft palate; hypertelorism; and myopia or retinal detachment, or both, in 50% of cases. A normal head appears to rest directly on the shoulder due to a very short neck. Ligamentous laxity is significant. Marked lumbar lordosis and moderate kyphoscoliosis are present by late childhood or early adulthood. Motor milestone achievement may be delayed. Hypotonia, lax ligaments and odontoid hypoplasia may cause high spinal cord compression symptoms. An early symptom of this impingement may be increased fatigability and decreased stamina. If no neurologic complications occur, prognosis for longevity is good. Adult height varies from 84 to 132 cm.

Major radiographic features include retarded ossification of the pubic bones, femoral heads, knee epiphyses, calcanei and tali in infancy. The vertebral bodies are ovoid or pear-shaped early, becoming flat and irregular later with kyphoscoliosis. Hypoplasia of the odontoid process may be associated with atlantoaxial subluxation or dislocation. Severe coxa vara is present. Shortening of the long bones with rhizomelia is present, but there are minimal changes in the hands and feet.

Autosomal dominant inheritance is recognized although most cases are sporadic, presumably because of new mutation. Paternal age is statistically increased. Therapeutic efforts should be preventive, attention being directed to possible atlantoaxial dislocation and retinal detachment.

Kniest Disease

This disorder is generally manifest at birth as a form of short-trunk dwarfism (68, 69). An unusual facies is present with midface flatness, depressed nasal bridge, protruding eyes with shallow orbits and cleft palate. A hearing deficit is common. Myopia or retinal detachment, or both, occurs in 50% of cases. Skeletal findings include exaggerated lumbar lordosis and kyphoscoliosis, prominent joints with limited motion and contractures, relatively long fingers with bulbous joints and disturbed and difficult gait. Hernias are frequent. Intelligence is normal. Adult height varies greatly (104 to 145 cm) because of the degree of joint contractures and kyphoscoliosis. Premature degenerative arthritis may be incapacitating by late childhood. Deafness may be helped by hearing-aids. Myopia and retinal detachment are major hazards, and periodic life-long ophthalmologic examinations are necessary for early or presymptomatic detection of myopia and retinal degeneration.

X-ray examination reveals generalized platyspondyly with anterior wedging in the lower thoracic and upper lumbar spine and coronal clefts of the lumbar vertebrae. The iliac bones are small with irregular acetabular margins. Femoral head ossification may not be apparent until age 3 years. The long bones are short

Table 4. Short-Trunk Dwarfism

Name	Inheritance*	Adult Height (cm)	Life Expectancy
A†:			
Spondyloepiphyseal dysplasia congenita	AD	84-132	Good unless neurologic complications occur
Kniest disease	AD	104-145	Good unless incapacitated from arthritis
Spondylothoracic dysplasia	AR	?	High mortality rate in first two years of life
B‡:			
Spondyloepiphyseal dysplasia tarda	XR	129-158	Good
Spondylometaphyseal dysplasia, Kozlowski type	AD	130-150	Good
Metatropic dwarfism	AR	109-119	Many die in infancy from respiratory insufficiency
Dyggve-Melchior-Clausen disease	AR	119-130	Good
Spondylocostal dwarfism	AD	129	Good

*See footnote, Table 1.
†Recognizable at birth or in first years.
‡Recognizable in childhood.

with broad, irregular metaphyses, and the epiphyses are irregular and punctate.

The disorder is an autosomal dominant trait.

Spondylothoracic Dysplasia

As the name implies, this is a short-trunk dwarfism due primarily to posterior shortening of the thorax and a marked thoracolumbar lordosis with a barrel-shaped anteriorly bulging chest. The lower anterior ribs may actually infringe on iliac crests. A short and nearly immobile neck is present and the head appears to rest on the shoulder (70, 71). The abdomen is protuberant. The limbs are relatively long, the fingers may reach the knees and span exceeds height. Recurrent respiratory infections presumably are related to the chest deformity or perhaps to pulmonary hypoplasia. There is a high mortality during the first 2 years.

Vertebral X-rays are striking. The entire spine from the cervical region through the sacrum is affected by malformations in the form of hemivertebrae, fused (block) vertebrae, absent vertebrae and sagittal clefting (butterfly) vertebrae. The posterior rib–vertebral articulations are bizarrely approximated, producing a fan-like radiation of the ribs whose number are reduced. The sacrum may not be well mineralized in infancy. Posterior shortening of the spine causes anterior flaring of the chest and rib cage deformity. There are no significant abnormalities of the skull or appendicular skeleton.

Autosomal recessive inheritance is recognized.

Spondyloepiphyseal Dysplasia Tarda

This disease usually becomes apparent between age 10 and 14 years. Since it is an X-linked recessive trait, it affects only male subjects, who show short-trunk dwarfism; a hunched appearance of shoulders; relatively short neck; a broad, barrel-shaped, large chest; and mild pectus carinatum. The extremities are relatively long. The head, hands and feet are normal, as is intelligence. Adult height ranges between 125 and 158 cm. By puberty, premature osteoarthritis may involve the shoulders, spine, hips and knees, causing disability and pain.

Initially the vertebral bodies are mildly flattened on X-ray with a hump-shaped accumulation of bone in the central and posterior portion of the inferior and superior plates, producing a distinctive and diagnostic configuration most noticeable in the lumber spine. The disk spaces are correspondingly narrowed. Premature osteoarthritic changes occur, causing limitation of movement and pain. The hip, knees and shoulders may show mild degenerative changes. The pelvis is small, particularly as compared to the thoracic size.

Although most cases are transmitted as X-linked recessive traits, autosomal dominant and autosomal recessive forms are reported.

Spondylometaphyseal Dysplasia, Kozlowski Type

This condition is usually recognized between 1 and 2 years of age as a form of short-trunk dwarfism with relatively long limbs, short neck, pectus carinatum, kyphoscoliosis and a waddling gait. There is hip, knee and shoulder joint limitation of motion and precocious osteoarthritis. Intelligence is normal (74-76).

There is a characteristic radiographic irregularity and poor ossification of the metaphyses of all long bones, the proximal femoral metaphyses being most severly involved. Normal or minimally deformed epiphyses are present. Generalized platyspondyly with kyphoscoliosis occurs. The iliac bone height is decreased and there is a horizontal acetabular roof.

Autosomal dominant inheritance is reported. Undoubtedly heterogeneity exists in the spondylometaphyseal dysplasia group of disorders. The type described by Kozlowski appears to represent a distinct entity, but there are several other types that are not well delineated.

Dyggve-Melchior-Clausen Disease

This unusual disorder is recognizable from age 6 to 12 months of age (77). It is of moderate severity with a short neck; exaggerated lumbar lordosis; prominence of the interphalangeal joints of the fingers with mild contractures and clawing; and scoliosis. Mental retardation is frequent, but not invariable. Although often confused with the Morquio syndrome, the corneas are clear and, studies of mucopolysaccharide metabolism are normal. Histologic changes indicate that it is not a lysosomal storage disease. It is an autosomal recessive trait.

Spondylocostal Dysplasia

These patients come to medical attention because of short stature noted from early childhood. Slowly progressive decreased range of movement of the entire spine may be associated with multiple episodes of generalized neck and back pain (78). There is moderate shortening of a thick, broad "bull-neck" and rigid positioning of the neck and trunk with limitation of movement in all planes. The chest is barrel-shaped with increased A-P diameter and horizontal grooves of the lower rib cage. The face is normal, as is intelligence. Prognosis for longevity is good.

Major radiographic features include a shortened spine from a decreased number of vertebral segments, although not in the cervical region. Severity of thoracic scoliosis is moderate, and the thoracic, lumbar and sacral spine contains diverse malformations, including congenitally absent or fused vertebrae, hemivertebrae and butterfly vertebrae. Rib numbers are reduced and some are hypoplastic, whereas others are posteriorly fused. The pelvis shows flared ilia. There are no abnormalities of the skull or appendicular skeleton.

The condition is an autosomal dominant trait. Spondylocostal dwarfism must be differentiated from spondyloepiphyseal dysplasia tarda, which has characteristic morphology of the vertebral bodies, but no segmentation anomalies. Spondylothoracic dysplasia is transmitted as an autosomal recessive trait with severe shortening and immobility of the cervical spine, dorsolumbar lordosis and an increased likelihood of death in infancy from respiratory infections.

In summary the skeletal dysplasias present the clinician with a genetically heterogeneous group of disorders manifesting a wide range of clinical features. As treatment and prenatal diagnostic approaches improve, it will become increasingly important to distinguish the form of skeletal dysplasia accurately. This challenge requires a careful and thorough application of available clinical, radiographic and genetic diagnostic measures.

REFERENCES

1. McKusick VA, Scott CI: A nomenclature for constitutional disorders of bone. *J Bone Joint Surg* 53A:978, 1971
2. Scott CI: Achondroplasia and hypochondroplastic dwarfism. *Clin Orthoped* 114:18, 1976
3. Langer LO, Baumann PA, Gorlin RJ: Achondroplasia. *Am J Roentgenol* 100:12, 1967
4. Spranger JW, Langer LO Jr, Wiedemann H-R: Achondrogenesis, Type I, in Bone Dysplasias. Philadelphia, Saunders, 1974, p. 24
5. Houston CS, Awen CF, Kent HP: Fatal neonatal dwarfism. *J Can Assoc Radiol* 23:45, 1972
6. Spranger JW, Langer LO Jr, Wiedemann H-R: Achondrogenesis, Type II, in Bone Dysplasias. Philadelphia, Saunders, 1974, p. 26
7. Romeo G, Zonana J, Lackman RS, et al: Grebe disease and similar forms of severe short-limbed dwarfism. *J. Pediatr* 91:918, 1977
8. Langer LO, Spranger J, Greinacher J, et al: Thanatophoric dwarfism. *Radiology* 92:285, 1969
9. Spranger J, Opitz JM, Bidder U: Heterogeneity of chondrodysplasia punctata. *Humangenetik* 11:190, 1971
10. Walker BA, Scott CI Jr, Hall JG, et al: Diastrophic dwarfism. *Medicine* 51:1, 1972
11. Langer LO: Diastrophic dwarfism in early infancy. *Am J Roentgenol* 93:399, 1965
12. Maroteaux P, Spranger J, Wiedemann H: Metatropischer Zwergwuchs. *Arch Kinderheilk* 173:211, 1966
13. Jenkins P, Smith MB, McKinnel JS: Metatropic dwarfism. *Br J Radiol* 43:561, 1970
14. McKusick VA, Egeland JA, Eldridge R, et al: Dwarfism in the Amish. I. The Ellis-van Creveld syndrome. *Bull Johns Hopkins Hosp* 115:306, 1964
15. LeMarec B, Passarge E, Dellenbach P, et al: Les formes néonatales lithales de la dysplasie chondro-ectodermique: A propos de cinq observations. *Ann Radiol* 16:19, 1973
16. Juene M, Beraud C, Carron R: Dystrophie thoracique asphyxiante de caractère familial. *Arch Fr Pediatr* 12:886, 1955
17. Langer LO: Thoracic-pelvic-phalangeal dystrophy. *Radiology* 91:447, 1968
18. Young LW, Wood BP: Nievergelt syndrome (Mesomelic dwarfism-type Nievergelt). *Birth Defects* 10:81, 1974
19. Silverman FN: Mesomelic dwarfism, in Kaufmann HJ (ed): Progress in Pediatric Radiology. Intrinsic Diseases of Bones. Basel, Karger, 1973, p. 81
20. Langer LO: Mesomelic dwarfism of the hypoplastic ulna, fibula, mandible type. *Radiology* 89:654, 1967
21. Espiritu C, Chen H, Wooley PV Jr,: Probable homozygosity for the dyschondrosteosis genes. *Am J Dis Childh* 129:375, 1975
22. Curranrino G: Hypophosphatasia, in Kaufmann HJ (ed): Progress in Pediatric Radiology. Intrinsic Diseases of Bones. Basel, Karger, 1973, p. 469
23. MacPherson RI, Kroeker M, Houston CS: Hypophosphatasia. *J Can Assoc Radiol* 23:16, 1972
23a. Rudd NL, Miskin M, Hoar D, et al: Prenatal diagnosis of hypophosphatasia. *N Engl J Med* 295:146,148, 1976
24. Majewski F, Pfeiffer RA, Lenz W, et al: Polysyndaktylie, verkürzte Gliedmassen und Genitalfehlbildungen: Kennzeichen eines selbständigen Syndroms. *Z Kinderheilkd* 111:118, 1971
25. Spranger J, Grimm B, Weller M, et al: Short-rib polydactyly (SRP) syndromes, types Majewski and Saldino-Noonan. *Z Kinderheilkd* 116:73, 1974
26. Saldino RM, Noonan CD: Severe thoracic dystrophy with striking micromelia, abnormal osseous development, including the spine, and multiple visceral anomalies. *Am J Roentgenol* 114:257, 1972
27. Maroteaux P, Spranger J, Opitz JM, et al: Le syndrome campomélique. *Presse Med* 79:1157, 1971
28. Schmickel RD, Heidelberger KP, Pozanski AK: The campomelique syndrome. *J Pediatr* 82:299, 1973

28a. Khajavi P, Lachman RS, Rimoin DL, et al: Heterogenity in the campomelic syndromes: Long and short bone varieties. *Birth Defects* 12:83, 1976
29. Maroteaux P, Malamut GL: L'acrodysostose. *Presse Med* 76:2189, 1968
30. Robinow M, Pfeiffer RA, Gorlin RJ, et al: Acrodysostosis. A syndrome of peripheral dysostosis, nasal hypoplasia and mental retardation. *Am J Dis Childh* 121:195, 1971
31. Maroteaux P, Martinelli B, CampaillaE: La nanisme acromesomelique. *Presse Med* 79:1839, 1971
32. Maroteaux P: Acromesomelic dwarfism, in Kaufmann HJ (ed): Progress in Pediatric Radiology. Intrinsic Diseases of Bones. Basel, Karger, 1973, p. 563
33. King JD, Bobechko WP: Osteogenesis imperfecta. *J Bone Joint Surg* 53B:72, 1971
34. Smars G: Osteogenesis Imperfecta in Sweden. Clinical, Genetic, Epidemiological, and Socio-Medical Aspects. Stockholm, Scandinavian University Books, 1961
35. McKusick VA: Osteogenesis imperfecta, in Heritable Disorders of Connective Tissue, 4th ed. St. Louis, Mosby, 1972, p. 390
36. Laplane MR, Lasfargues G, Debray P: Essai de classification genetique des osteogeneses imparfaites. *Presse Med* 67:893, 1959
37. Remigio PA, Grinvalsky HT: Osteogenesis imperfecta congenita. *Am J Dis Childh* 119:524, 1970
38. McKusick VA: Osteogenesis imperfecta, in Heritable Disorders of Connective Tissue, 4th ed. St. Louis, Mosby, 1972, p. 390
39. McKusick VA, Eldridge R, Hostetler JA, et al: Dwarfism in the Amish, II. Cartilage-hair hypoplasia. *Bull Johns Hopkins Hosp* 116:285, 1965
40. Ray HC, Dorst JP: Cartilage-hair hypoplasia, in Kaufmann HJ (ed): Progress in Pediatric Radiology. Intrinsic Diseases of Bones. Basel, Karger, 1973, p. 270
41. Daentl DL, Smith DW, Scott CI, et al: Femoral hypoplasia—unusul facies syndrome. *J Pediatr* 86:107, 1975
42. Holmes LB: Femoral hypoplasia—unusual facies syndrome. *J Pediatr* 87:668, 1975
43. Walker BA, Murdoch JL, McKusick VA, et al: Hypochondroplasia. *Am J Dis Childh* 122:95, 1971
44. Spranger JW, Langer LO Jr, Wiedemann H-R: Pseudoachondroplasia, in Bone Dysplasias. Philadelphia, Saunders, 1974, p. 124
45. Langer LO, Petersen D, Spranger J: An unusual bone dysplasia: Parastremmatic dwarfism. *Am J Roentgenol* 110:550, 1970
46. DeHaas WHD, DeBoer W, Griffioen J: Metaphyseal dysostosis. A late follow-up of the first case. *J Bone Joint Surg* 51B:290, 1969
47. Ozonoff MB: Metaphyseal dysostosis of Jansen. *Radiology* 93:1047, 1969
48. Sutcliffe J, Stanley P: Metaphyseal chondrodysplasias, in Kaufmann HJ (ed): Progress in Pediatric Radiology. Intrinsic Diseases of Bones. Basel, Karger, 1973, p. 250
49. Schmid F: Bietrag zur Dysostosis enchondralis metaphysaria. *Z Kinderheilk* 97:393, 1949
50. Stickler GR, Maher FT, Hunt JC, et al: Familial bone disease resembling rickets (hereditary metaphyseal dysostosis). *Pediatrics* 29:996, 1962
51. Gatti RA, Platt N, Pomerance HH, et al: Hereditary lymphopenic agammaglobulinemia associated with a distinctive form of short-limbed dwarfism and ectodermal dysplasis. *J Pediatr* 75:675, 1969
52. Kozlowski K, Lipska K: Hereditary dysplasia epiphysialis multiplex. *Clin Radiol* 18:330, 1967
53. Jacobs P: Multiple epiphyseal dysplasia, in Kaufmann (ed): Progress in Pediatric Radiology. Intrinsic Diseases of Bones. Basel, Karger, 1973, p. 309
54. Spranger JW, Langer LO Jr, Wiedemann H-R: Multiple epiphyseal dysplasia, in Bone Dysplasias. Philadelphia, Saunders, 1974, p. 10
55. Langer LO: Dyschondrosteosis, a heritable bone dysplasia with characteristic roentgenographic features. *Am J Roentgenol* 45:178, 1965
56. Elmore SM: Pycnodysostosis. A review. *J Bone Joint Surg* 49A:153, 1967

57. Maroteaux P, Faure C: Pycnodysostosis, in Kaufmann HJ (ed): Progress in Pediatric Radiology. Intrinsic Diseases of Bones. Basel, Karger, 1973, p. 403
58. Schwachman H, Diamond LK, Oski FA, et al: Pancreatic insufficiency and bone marrow dysfunction. A new clinical entity. *Pediatrics* 63:835, 1963
59. McLennan TW, Steinbach HL: Schwachmans syndrome: The broad spectrum of bony abnormalities. *Radiology* 112:167, 1974
60. Stickler G, Belau PG, Farrell FJ, et al: Hereditary progressive arthro-ophthalmopathy. *Mayo Clin Proc* 40:433, 1965
61. Stickler GB, Pugh DG: Hereditary progressive artho-ophthalmopathy. II. Additional observations on vertebral abnormalities, a hearing defect, and a report of a similar case. *Mayo Clin Proc* 42:495, 1967
62. Spranger J: Arthroophthalmopathia hereditaria. *Ann Radiol* 11:359, 1968
63. Giedion A: Das tricho-rhino-phalangeale Syndrom. *Helv Paediatr Acta* 21:475, 1966
64. Giedion A, Burdea M, Fruchter H, et al: Autosomal-dominant transmission of the tricho-rhino-phalangeal syndrome. *Helv Paediatr Acta* 28:249, 1973
65. Hall BD, Langer LO, Giedion A, et al: Langer-Giedion syndrome. *Birth Defects* 10:147, 1974
66. Spranger J, Langer LO: Spondyloepiphyseal dysplasia congenita. *Radiology* 94:313, 1970
67. Spranger JW, Langer LO Jr, Wiedemann H-R: Spondyloepiphyseal dysplasia congenita, in Bone Dysplasias. Philadelphia, Saunders, 1974, p. 95
68. Kniest W: Zur Abgrenzung der Dysostosis enchondralis von der chondrodystrophie. *Z Kinderheilk* 70:633, 1952
69. Lachman RS, Rimoin DL, Hollister DW, et al: The Kniest syndrome. *Radiology* 123:805, 1975
70. Moseley JE, Bonforte RJ: Spondylothoracic dysplasia—a syndrome of congenital anomalies. *Am J Roentgenol* 106:166, 1969
71. Pochaczevsky R, Ratner H, Perles D, et al: Spondylothoracic dysplasia. *Radiology* 98:53, 1971
72. Maroteaux P, Lamy M, Bernard J: La dysplasie spondylo-epiphysaire tardive. *Presse Med* 65:1205, 1957
73. Langer LO: Spondyloepiphyseal dysplasia tarda. *Radiology* 82:833, 1964
74. Kozlowski K, Maroteaux P, Spranger J: La dysostose spondylometaphysaire. *Presse Med* 75:2769, 1967
75. LaQuesne GW, Kozlowski K: Spondylometaphyseal dysplasia. *Br J Radiol* 46:685, 1973
76. Sutcliffe J: Metaphyseal dysostosis. *Ann Radiol* 9:215, 1966
77. Dyggve HV, Melchior JC, Clausen J: Morquio-Ullrich's disease. *Arch Dis Child* 37:525, 1962
78. Rimoin DL, Fletcher BD, McKusick VA: Spondylocostal dysplasia. A dominantly inherited form of short-trunked dwarfism. *Am J Med* 45:948, 1968

19
Genetic Disorders of the Skin

Laird G. Jackson

Skin is an organ composed of tissues derived from all three germ layers. The epidermis derives from ectoderm and the dermis from mesoderm, with neural, muscular and hypodermal fat tissue also arising from mesoderm. Only the vascular structures in the skin are in part endodermal. All of the cutaneous appendages and adnexa (nails, hair and sebaceous, apocrine and sweat glands) derive from epidermis and are therefore ectodermal. The surface of the skin is composed of compacted dead, keratinized cells and is marked by fine ridges and furrows, which sometimes follow patterns as on the palms and soles. Hairs are everywhere except on the palms, soles, glans penis, knuckles and dorsal terminal phalanges, the last-mentioned which are covered by the nails. Hairs and nails are essentially strands and plates of horn similar to the skin. The outer portion of the epidermis, the stratum corneum, serves as a protective barrier. Other elements of the skin serve in heat regulation (vascular and glandular loss; epidermal insulation), sense perception (neural) and cleansing (self-renewal and sweat waste removal).

The following discussion of inherited skin disorders is a brief review of a broad topic. The older literature describes many of the inherited disorders in detail, notably Cockayne's 1933 treatise on inherited anomalies of the skin and appendages (1). More recent texts (2) review articles and symposium volumes (3) deal with the subject in detail, particularly newly described individual disorders.

DERMATOGLYPHICS

Dermal ridges occurring in a parallel pattern on the palms and soles begin to appear in the late part of the first trimester of gestation. Their subsequent pattern depends on the shape of the developing hand and foot and their areas of growth stress (4, 5). The lines curve over pads or mounds of tissue as on fingertips and palms between fingers at this early fetal stage. Subsequent resorption of the mound or compensatory adjacent tissue growth results in the dermal markings seen at the infant's birth. These patterns may be studied either by a light and magnifying lens, such as stamp collectors or photographic printers use, or by making prints of the fingers or palms or soles on glossy paper with "dry" ink pads, such as those used in delivery rooms or neonatal units.

The patterns are frequently altered in multiple congenital anomaly syndromes, particularly those of chromosomal origin. For diagnosis the changes in dermal ridge patterns are chiefly useful in the diagnosis of the Down syndrome in which alterations are characteristic. Three easily observable areas of dermatoglyphic patterns on the palms and soles (Fig. 1) give reasonable diagnostic aid (6). *Digital patterns* are on the finger pads and are called whorls, loops or arches. Children with the Down syndrome commonly have loop digital patterns opening to the ulnar side of the finger, i.e., ulnar loops. Triradii occur at the juncture of three sets of converging ridges. On the palm this normally occurs at the base of the palm (t) and at the base of each of the fingers (a-d).

In the Down syndrome a curved pattern on the hypothenar area causes a distal palmar triradius and the angle formed by a line drawn from the atd triradii is increased over normal (Fig. 1). The ball of the foot or *hallucal area* has a pattern of skin markings similar to those on the finger pads and is normally in

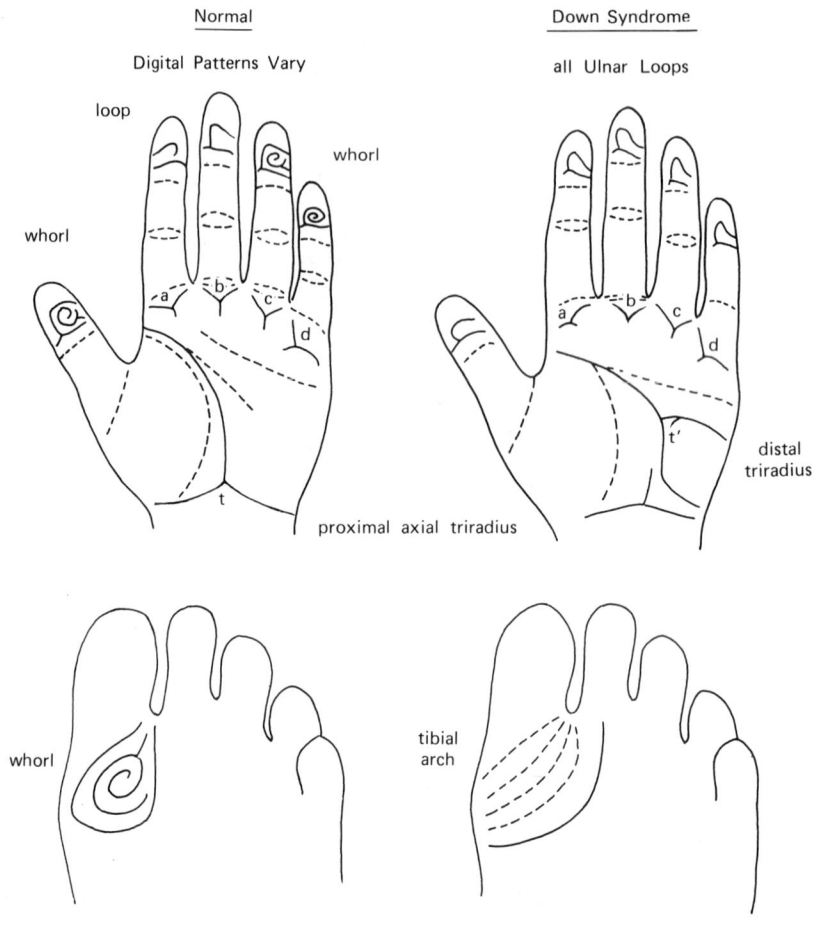

Figure 1. Dermatoglyphic patterns as they appear normally and in the Down syndrome.

the loop or whorl configuration. In Down syndrome individuals the pattern is frequently simple or "open," with an arch pattern concavely opening to the tibial side of the foot, i.e., a tibial arch.

CONGENITAL SKIN DEFECTS

Congenital skin defects are rare anomalies, usually occurring on the scalp but occasionally seen on the limbs, trunk or face (7). Their size varies from pinpoint lesions to 3 inches and larger in diameter. The commonest lesion is seen as a single scalp defect in the parietal area near the midsagittal line. The surface may be raw granulation tissue or a thin membrane. The defect is common in trisomy 13, but both autosomal dominant (8) and recessive cases (9) are reported. The smaller lesions usually heal rapidly, whereas the larger ones may require extensive plastic surgery. A family is reported in which 25 of 103 persons were affected with congenital localized absence of the skin of the lower limbs. This was associated with blistering of the skin and mucous membranes and congenital absence and deformity of the nails (10). Inheritance was of the autosomal dominant type.

EPIDERMOLYSIS BULLOSA

This is a hereditary skin disorder characterized by vesicles and bullae on the skin and (occasionally) the mucous membranes; they occur spontaneously or in response to minimal trauma. The disease assumes several forms, which are usually classified according to clinical severity (11). Both autosomal dominant and recessive genetic transmission have been observed (Table 1) (11). Although 70% of cases have been sporadic, many of them probably represent fresh dominant mutations. The cause of the skin change is not known although relationships to vascular and metabolic abnormalities have been sought, and a lack of elastic tissue has been observed in the dermis of some cases.

The simplex type of epidermolysis bullosa is usually inherited as an autosomal dominant trait and is seen more often in female than in male subjects. The skin lesions are rare in infancy, but develop later in childhood, usually over areas exposed to repeated trauma. This tendency is increased in hot and humid weather and with sweating. Amelioration often occurs at puberty with eventual disappearance of the disorder in adult life. The bullae commonly heal rapidly within a few days, but with scarring. The palms and soles may show hyperhidrosis, and there may be dermatographia. Ordinarily mucous membranes, nails and teeth are not affected in the usual case. The disease has been followed through as many as eight generations of a single family (12).

Epidermolysis bullosa dystrophica may occur in several varieties with dominant or recessive inheritance. The dominant forms are usually less serious than the recessive ones, but manifest themselves at birth after little or no trauma. Healing leaves a thin, atrophic scar. Mucous membranes and nails are occasionally involved. Pubertal improvement does not always occur. In many cases the distinction between the dominant trophic and simplex diseases is obviously sub-

Table 1. Epidermolysis Bullosa

Type (syndrome)	Clinical Features	Inheritance*
Epidermolysis bullosa simplex		
I (Koebner)	Congenital, generalized nonscarring blistering after mechanical trauma	AD
II (Weber-Cockayne)	Congenital blistering limited to palms and soles	AD
III (Ogna)	Congenital generalized blistering with generalized bruising	AD
Edipermolysis bullosa dystrophica		
I (Albopapuloid-Pasini)	Congenital blistering, erythema, scarring, nail dystrophy; localized to limbs with discrete albopapuloid papules (Pasini) present	AD
II (Cockayne-Touraine)	As above, but without papules	AD
III (localized nonlethal)	Congenital, lesions localized to limbs; nonlethal	AR
IV (generalized nonlethal)	Congenital lesions on limbs, trunk, neck; nonlethal	AR
V (sublethal mutilating)	Congenital generalized and persistent blistering with severe skin atrophy, growth retardation and anemia	AR
VI (lethal)	Congenital generalized blistering becoming confluent with pyogenic infection, fever leading to death	AR
VIII (generalized-inverse)	Infancy with transient pyoderma; childhood with keratitis and paronychia; 3-5 years on have periodic truncal blistering; late childhood on have permanent scarring, skin atrophy and blistering	AR
VIII (neurotrophic)	Onset of traumatic blistering in late childhood or adolescence with onset of nail manifestations earlier; diffuse, slowly progressive skin atrophy of hands, feet, elbows, knees; congenital slowly progressive neurogenic hypacusis	AR

*AD, autosomal dominant; AR, autosomal recessive.

tle. In contrast, the recessive forms are frequently severe. The bullae, which appear at birth with or without trauma, are often large and filled with serous or bloody fluid. Healing is accompanied by scarring, pigmentation and contractures. If mucous membranes or respiratory or alimentary tract tissue is involved, the scarring may cause stenosis or soft tissue ankylosis of oropharyngeal structures. Involvement of conjunctiva may cause corneal opacities, and anomalies of teeth, nails and hair may occur. Digits may be involved with formation of a tightly constricting cutaneous envelope (13). Children frequently die of complications and survivors may be physically and mentally retarded.

The most serious form is the lethalis type. There is evidence of in utero involvement with severe deformation of finger- and toenails at birth, and the more severely affected cases are probably aborted or stillborn. Bullae are ubiquitous and accompanied by serous or hemorrhagic exudate; they also ulcerate and become secondarily infected. Necrosis of parts is evident at birth (14). Renal and cardiovascular anomalies are reported, together with these severe skin changes. Treatment is ineffective. The mode of inheritance is usually autosomal recessive.

ICHTHYOSES

The ichthyoses are a group of inherited noninflammatory skin diseases characterized by hyperkeratinization and thickening of the stratum corneum combined with underdevelopment of the stratum granulosum and a normal dermis (Table 2) (15). Recent studies indicate that epidermal turnover is increased (16), allowing for more accurate classification and understanding of this group of disorders. From a pathophysiologic point of view, the keratinocyte is primarily involved in the ichthyoses through a disturbance of timing of its normal function of proliferation, migration through disintegration and desquamation (17). The proteins produced by the keratinocyte are numerous and form a set of helical organized structures known as keratins. The differentiation of the cell takes 14 days and is followed by another 14 day stay in the stratum corneum as a nonviable cell prior to desquamation.

Ordinarily the cells of the stratum corneum are arranged in neat stacks, which apparently contribute to the protective function of the skin by limiting shearing damage (18). This orderliness disappears in ichthyoses as well as in other diseases related to increased epidermal proliferation. The abnormal stacking may promote cell adhesiveness and formation of scales, together with an increase in transepidermal water loss through production of cracks, fissures or incomplete arrangement of the lipid–protein fine structure within the cells of the stratum corneum. Curiously the increase in water loss may act as a feedback mechanism to increase the rate of epidermal proliferation and thereby epidermal thickness.

Biochemical epidermal studies reveal several distinctive structural proteins, and specific protein abnormalities may be associated with specific disorders (19). Moreover injection of radioactive DNA precursors into the skin followed by sequential skin biopsies show differences in the cellular cycle in the various forms of ichthyosis. In general the mitotic rate is slightly decreased in both ichthyosis vulgaris and X-linked ichthyosis. A three- to four-fold increase is seen in lamellar ichthyosis and epidermolytic hyperkeratosis, in that the cell transit time through the epidermis is drastically shortened from the normal 12-14 days to 4-5 days.

Lamellar Ichthyosis

This generalized disorder begins at birth and affects the entire epidermis, including scalp, palms and soles. The newborn baby is frequently covered with a smooth, glistening layer of epidermis, which dries and desquamates over two to three weeks. Large scales then develop, which cover the entire body, including intertriginous areas, axillae, palms and soles. The lips and mucous membranes are spared, but the scalp is covered with thick hyperkeratotic plagues. Hair and nail growth may be increased, decreased or absent. Ectropion is frequent.

Mechanical blockage of eccrine glands leads to impaired sweating and exercise intolerance, whereas transepidermal water loss is increased, often to dehydration. The appearance of the skin at birth has led to the name "collodion baby"

Table 2. Ichthyoses

Type	Clinical Features	Age of Onset	Inheritance*
Ichthyosis vulgaris			
I	Hyperkeratotic scales on trunk and extensor surface of limbs; flexor surfaces, nails, mucous membranes spared	Never congenital, several months to years	AD
II (bullous)	Similar to vulgaris with bullae present	One family reported, onset in childhood	AD
Ichthyosiform erythroderma			
I (bullous Brocq)	Diffuse hyperkeratosis, erythema and erosions with intact bullae	Congenital	AD
II (nonbullous-Brocq)	Diffuse hyperkeratosis, erythema and erosions	Congenital	AR
III	As above with deafness	Congenital?	AR
IV	As above in unilateral pattern with ipsilateral limb malformation or absence	Congenital	AR
V (porcupine man)	Heaped up skin, flexor surfaces involved; palms and soles spared	Childhood	AD
VI	Localized to palms and soles;	Childhood	AD
VII	Linearis circumflexor form; migratory lesions alternating with normal skin; bamboo hair present in form called Netherton syndrome	Congenital	AR
Ichthyosis congenita			
I (lamellar)	Infant covered with smooth glistening epidermis; erythema (collodion baby)	Congenital	AR
II	Infant usually premature and covered with diamond-shaped plaques	Congenital	AR
III	As in Type I with biliary atresia	Congenital	AR
IV	As above with cortical cataract	Congenital	AR
X-linked ichthyosis			
I	Intermediate between congenital and vulgaris types	Congenital to 3 months	XR
II	As above with male hypogonadism	Congenital	XR
III	As type I with hypogonadism, mental retardation and anosmia (may be a variant of II)	Congenital	XR

*AD, autosomal dominant; AR, autosomal recessive; XR, X-linked recessive.

due to the similarity of the skin to celluloid film. It should be noted, however, that other forms may be present in the same manner. The skin may clear completely within three months, and prognosis for an individual case must be based on either observation during the early infant period or the familial history. Inheritance is autosomal recessive, with many consangineous marriages having been reported (20). The frequency of the disorder is estimated at 1 in 300,000 births (21).

Microscopically one sees hyperkeratosis, parakeratosis, acanthosis and keratotic plugging of the hair follicles. The kinetic mechanisms of epidermal proliferation are increased, with disordered stacking of the cells in the stratum corneum (22). Treatment with systemic administration of methotrexate is reportedly successful in some patients (23), and topical application vitamin A is also useful, although irritating (24).

Epidermolytic Hyperkeratosis (Ichthyosiform Erythroderma)

This rare form of ichthyosis is inherited as an autosomal dominant trait and may occur as a generalized or localized disorder. The generalized form is frequently called bullous congenital ichthyosiform erythroderma (25), whereas the localized form is sometimes called ichthyosis hystrix. The disease develops congenitally and is usually recognized in the first week of life. It is marked by generalized hyperkeratosis, erythema and erosions. Intact bullae may be purulent, but they heal without scarring.

The scalp is covered with a thick, greasy crust, which also adheres to the hair shafts in a tube-like fashion although the hair and nails themselves are normal. The palms and soles are affected by uniform hyperkeratosis that will frequently increase with age. There are mild forms that affect primarily the knees, elbows and ankles, and another type is limited to the palms and soles (26). One form with "heaped-up" skin—resulting from variation in scale production along the lines of normal skin markings—is called the "porcupine-man" form (ichthyosis hystrix gravior). Distinctive recessive types are reported, such as generalized nonbullous ichthyosiform erythroderma alone (25), in association with variable corneal involvement and deafness (27) and a severe unilateral form with either ipsilateral limb malformations or absence thereof (28).

The histopathologic features of the skin are distinctive, with a thickened stratum corneum in a "basket-weave" pattern, and a vacuolated reticulated pattern of cells is present in the upper spinous and granular layers. There is some clumping of protein filaments, producing an abnormal periodicity with cross-striations (2). The epidermal proliferation kinetics are increased. There is no safe and effective therapy for the disorder. Topical application of vitamin A affords some relief, and prophylactic treatment with penicillin reduces the frequency of erosions.

Ichthyosis Vulgaris

This is a common form of ichthyosis that begins in childhood and lasts throughout life. Fine white scales, more pronounced on the extensor than the flexor portions of limbs, are characteristic. The antecubital, popliteal and axillary areas are, e.g., clear. Although not usually evident at birth, the skin lesions become manifest early in infancy. Palms and soles are generally fissured. With age there is some improvement. The disorder is inherited as an autosomal dominant trait; the frequency in England is estimated at 1 in 5,300 persons (21).

Light microscopic examination shows mild hyperkeratosis and a thinned to almost absent granular layer. These features should be diagnostic, although mild cases may show essentially normal skin. The kinetics of epidermal proliferation are normal. Treatment is most effective with propylene glycol-containing solutions (29).

X-Linked Ichthyosis

X-linked pedigrees of ichthyosis patients have been long recognized. The form is usually slightly severer than the vulgaris type, but there are still many patients whose disease never prompts them to see a physician. Onset is frequently at birth, the skin peeling and therefore reflecting the "collodion baby" form as in the autosomal recessive disorder.

In addition to the pedigree in distinguishing the X-linked from the vulgaris form, the following features are useful (21): X-linked ichthyosis begins earlier in life, frequently within the first 3 months; flexural areas, neck and ears are involved in X-linked ichthyosis; scales in the X-linked variety are often larger and darker; and dermal atopy and palm and sole fissuring are less frequent in the X-linked form. Confusion and misdiagnosis in the absence of a family history result in misclassification at least 10% of the time.

In England the X-linked gene for this disorder occurs in 1 in 6,190 males (21). It is closely linked to the gene for the Xg^a blood group, but not near the two color blindness loci or that for glucose-6-phosphate dehydrogenase (G-6-PD) (30). Some reports describe affected females, probably explained by the Lyon phenomenon. Both patients and female carriers frequently demonstrate fine, punctate corneal stromal lesions in their second or third decade, which may be useful for genetic counseling (31). Histologically the granular layer is normal in X-linked ichthyosis in contrast to ichthyosis vulgaris. The patients also respond well to treatment with propylene glycol-containing medications.

Harlequin Fetus

This is a severe form of congenital ichthyosis that has frequently been confused in its relationships with the lamellar ichthyosis forms of "collodion baby." The infant is usually premature or small for gestational age and covered with large yellow plaques over the entire body and scalp. The hands and feet may be covered with tight glistening sleeves, and digits may be necrosed. The plaques may be as large as 4-5 cm on a side and have a diamond-like configuration, the latter which gives the appearance of a clown's suit and the name to the condition. The condition is severe and the infant usually dies within the first week of life. Ectropion, which will obliterate the globe, absent pinnae and distortion of the alae nasi are usual findings. Hair, if present, appears between plaques in the fissures formed.

Familial studies suggest that this disorder is separate from the recessive lamellar form and is caused by a distinct autosomal recessive gene. Some evidence also indicates that the harlequin fetus epidermis contains a distinct fibrous protein, and histologic studies suggest that all of the epidermis formed during fetal life is retained by the fetus; hence no squamae should be present in the gastrointestinal (GI) tract or in amniotic fluid, making the prenatal diagnosis of this condition conceivable.

Sjögren-Larsson Syndrome

This autosomal recessive disorder was originally described in northern Sweden in 1956-57 (32). It comprises ichthyosis, spastic diplegia, mental retardation and occasionally retinal degeneration. The almost constant skin changes are usually the initial manifestation. Although affected persons do not have the "colloidion baby" appearance, the skin lesions are frequent at birth—and almost always within the first month. The skin changes resemble those of lamellar ichthyosis. Severe retardation of neuromuscular development, hyperactive deep tendon reflexes and spastic diplegia or paraplegia are found. Ocular involvement is variable, with about 30% of patients having atypical pigmentary retinal degeneration. No treatment is known.

Ichthyosis linearis circumflexa, an autosomal recessive trait, is a rare congenital form of ichthyosis. Characterized by cyclic migratory lesions with double-edged peripheral scales alternating with areas of normal skin (33), it may be combined with a bamboo-like appearance to the hair (trichorrhexis invaginata) and is then called Netherton's syndrome (34).

Miscellaneous Syndromes with Ichthyosis

Ichthyosis is reported as a prominent feature of both the autosomal dominant and recessive forms of the Conradi syndrome (35), a short-limbed form of skeletal dysplasia (see Chapter 18, "Heritable Skeletal Dysplasias"). Ichthyosis is also a frequent feature of the rare autosomal recessive disorder of phytanic acid storage, Refsum disease (36). In addition two forms of ichthyosis are reported in association with hypogonadism in males following an X-linked inheritance pattern. In one family five males in three generations were affected with secondary hypogonadism and congenital ichthyosis (37). The affected males were eunuchoid with a small phallus, testicular atrophy and azospermia. In another kindred of Mexican-American origin, 20 of 114 males were affected with a combination of congenital ichthyosis, hypogonadism, mental retardation and anosmia (38). The prepubertal males had small genitals and cryptorchidism. Postpubertally they had a eunuchoid habitus, absent axillary and pubic hair and a high voice. There was a history of early developmental lag with mental retardation (intelligence quotient of 50-80), and anosmia was demonstrated. Exclusion of close linkage to the Xg[a] locus suggests that the condition was caused by a gene distinct from the classic X-linked ichthyosis gene.

KERATOSIS PALMARIS ET PLANTARIS (TYLOSIS)

Although this disorder is inherited, the skin changes do not occur for months or even years. Typically the thickening of the palms and soles begins in the second six months and is well demarcated from the surrounding normal skin. Severity varies from mild and insignificant to a painful, deforming and crippling disorder. Keratosis may involve fingers, toes and nails, and the hands may be forced into a flexed posture. Thickening in some areas (the heels) may grow to several centimeters with fissuring and a foul odor.

Improvement is common in warm weather with subsequent shedding of the horny epidermis. Most families show an autosomal dominant inheritance (1) with some evidence of lack of penetrance. A number of sporadic cases are likely the result of fresh mutation. Five families are reported in whom late-onset disorder was associated with esophageal carcinoma (39). The morphologic presentation of the dominant types varies, but the genetic distinction is unclear (27). Autosomal recessive types also occur, with the mal de Meleda form being well known for its description in intermarried families on the Dalmatian island of Meleda. Morphologic features are essentially the same as in the dominant type, but the backs of the hands and feet, elbows, knees and forearms may also be involved (40). Recessive forms are reported in association with corneal dystrophy (41) and with periodontopathia (42) (Table 3).

BENIGN FAMILIAL PEMPHIGUS (HAILEY-HAILEY DISEASE)

This disorder is characterized by the recurrent eruption of vesicles and bullae involving predominantly the neck, groin and axilla. It may appear in the first decade, but more often is delayed until middle age. Warm weather excerbates, and cool weather benefits, the eruption. The lesions tend to spread centrifugally with central clearing. The disease runs in cycles with spontaneous clearing, but appears to be precipitated by pyogenic bacteria. Histologic examination shows numerous acantholytic cells and the suprabasal type of blister formation. Antibiotics are effective in the treatment of this disorder, which is inherited as an autosomal dominant trait.

POROKERATOSIS

There may be three distinct hereditary conditions under this title, all of which are inherited as autosomal dominant traits and characterized by keratoatrophoderma. The lesions in the Mibelli type are centrifugally spreading patches

Table 3. Keratosis Palmaris et Plantaris

Type	Clinical Features	Inheritance*
Keratosis palmaris et plantaris familians (tylosis)	Diffuse keratosis of palms and soles (onset in infancy)	AD
Keratosis palmaris et plantaris papulosa	Punctate keratosis of palms and soles	AD
Keratosis palmaris et plantaris striata	Linear keratosis of palms and soles	AD
Keratosis palmaris et plantaris with esophageal cancer	Late-onset keratosis of palms and soles associated with esophageal cancer onset in late adult life	AD
Keratosis palmoplantaris with corneal dystrophy (Richner-Hanhart syndrome)	Herpetiform corneal ulcers, mental retardation and palmoplantar keratosis	AR
Keratosis palmoplantaris with periodontopathia (Papillon-Lefevre syndrome)	Premature loss of milk and permanent teeth, calcification of dura and palmoplantar keratosis	AR
Mal de Meleda	Congenital symmetrical cornification of palms and soles with other ichthyotic skin changes	AR

*See footnote, Table 2.

surrounded by narrow horny ridges with central atrophy. The lesions tend to be crater-like, and epitheliomas have been described. Another variety appears to begin on the palms and soles in the late teens and early 20s, subsequently spreading to other parts of the body. This type is not clearly established as being distinct from the Mibelli or disseminated types. The disseminated superficial actinic type of porokeratosis has lesions occurring almost exclusively on sun-exposed areas (43). The lesions begin to appear in the midteens, and at-risk persons are usually affected by age 30-40 years. This form is apparently commoner than the Mibelli type.

DARRIER-WHITE DISEASE (KERATOSIS FOLLICULARIS)

Keratosis follicularis is characterized by the formation of brownish keratotic papules, which have a greasy appearance and are found in the "seborrheic areas." Lesions may appear on the hands and feet and resemble and overlap with lesions seen in acrokeratosis verruciformis (see subsequent discussion). Bullous lesions may occur, resulting in confusion with Hailey-Hailey disease (44). Nail dystrophy is often seen. Histologically there is 1) mild nonspecific perivascular infiltration in the dermis, 2) dermal villi protruding into the epidermis, 3) suprabasal detachment of the spinal layer, leading to formation of lacunae containing acantholytic cells, 4) distinctive dyskeratotic round epidermal cells in the more superficial epidermis and 5) "grains" in the stratum corneum, which resemble parakeratotic cells in a hyperkeratotic horny layer. Sunlight exposure may exacerbate the lesions, and many cases are benefited by oral administration of vitamin A. Dominant transmission through five generations is reported (45).

ACROKERATOSIS

Two forms of this disorder exist, the commonest being acrokeratosis verruciformis, which may be confused with Darrier-White disease (46). Warty hyperkeratotic lesions are found on the dorsal aspect of hands and feet and on knees and elbows. Autosomal dominant transmission is proved.

The second form is termed acrokeratoelastoidosis and is characterized by lesions primarily on the palms and soles (47). There may be extension to the dorsum of the hands and feet with nodular, yellow, hyperkeratotic lesions. Histologically there is hyperkeratosis combined with disruption of normal organization of the elastic fibers.

ECTODERMAL DYSPLASIA

The ectodermal dysplasias include a large group of syndromes. There are at least three forms with classic findings, two with mild-to-moderate sweating abnormality—the hypohidrotic forms—and one anhidrotic form with severe lack of sweating ability. In addition there are many syndromes with some manifestation of dysplastic development of ectodermal derivatives. Many of these disorders were reviewed by Cockayne (1) and later classified by Freire-Maia (48). The latter author classified the ectodermal dysplasias on the basis of at least two signs related to 1) trichodysplasia, 2) abnormal dentition, 3) onychodysplasia and 4) dyshidrosis. Witkop summarized the findings of Freire-Maia and others and added a new syndrome in his review (49).

It is usual for the manifestations of ectodermal dysplasia to vary widely among affected persons with the same disorder and even within the same family. The mild or hypohidrotic type occurs as both an X-linked recessive and an autosomal recessive type as well as in a milder form found as an autosomal dominant, especially among French Canadians (50). Symptoms include sparse hair, missing and conical teeth, hypoplastic nails and variable difficulty in sweating. The last-mentioned feature may either cause thermal regulatory difficulty or only produce dry skin with relative intolerance to heat. Dry mucous membranes, especially in the nose, may lead to retained, foul secretions. In the autosomal dominant type the effect of the gene may be mild enough to be missed in some members of a family. Neuromuscular development is usually unaffected.

The anhidrotic, X-linked recessive form of ectodermal dysplasia is generally severer than the autosomal dominant type in its effect on heat regulation. The infant with the disorder may be seriously threatened by the inability to control fever. Scanty hair and eyebrows, skin lesions, saddle nose and dry "snuffly" nose are usually noticed at birth. The inability to sweat leads to heat intolerance and absent sebaceous glands produce dry and often red skin.

Absence or hypoplastic development of the breast or nipples may occur rarely. The hair is sparse and thin with absent or underdeveloped eyelashes and eyebrows. In contrast facial hair grows well in the male subject. Atrophy of the nasal and oral mucous membranes may cause loss of sense of smell, atopic rhinitis and hoarseness. The teeth are always reduced in number with conical incisors and relative preservation of the molars. The upper lip is frequently short, the nasal bridge saddle-shaped and the supraorbital ridges are prominent. Mental development may be entirely normal, but there is a 25% chance of mental deficiency (51). The nails are usually normal in appearance.

Of the remaining syndromes listed in the table on pages 582–583, most of them are relatively rare. One condition, the Coffin-Siris syndrome, may be commoner than is realized (52). It consists of mild-to-moderate mental retardation, variable growth retardation, small teeth, mild hypotrichosis of scalp hair and absent-to-hypoplastic fifth finger- and toenails. It appears either to be multifactorial or caused by fresh dominant mutation, as few familial cases have been described.

Van Den Bosch Syndrome

This X-linked disorder is reported in a single family (53). It includes mental deficiency, choroideremia, acrokeratosis verruciformis, anhidrosis and skeletal deformity. It is noted that at least three of these components may be inherited separately as X-linked traits.

Erythrokeratoderma Variabilis

Autosomal dominant pedigrees of this are described in several publications (54). The onset is in early childhood with skin disease marked by hyperkeratosis, hyperpigmentation and hypertrichosis as well as erythema varying by site and time of occurrence. A cardinal feature comprises sharply demarcated areas of erythrokeratoderma from infancy.

A variant form is recently described in two patients with slightly dissimilar features (55), in which the erythrokeratoderma was associated with profound bilateral perceptive deafness, physical growth retardation, developmental retardation, absent deep tendon reflexes in the lower limbs and abnormal EEG findings.

PIGMENTED LESIONS

Pigmented nevi consist of nevus cells or melanocytes that are thought to be of neural crest origin which have migrated to the epidermis early in fetal life. They are so common that it is almost impossible to establish a hereditary pattern. Pigmented nevi or lesions resembling them are seen in the hereditary syndromes of neurofibromatosis, fibrous dysplasia, Peutz-Jeghers syndrome, nevoid basal-cell carcinoma syndrome, melanoblastosis and the "Leopard" syndrome of multiple lentigines, hypertelorism, pulmonic stenosis, cardiac conduction defects, genital anomalies, sensorineural deafness and growth retardation (56). Large hairy moles or nevus pilosis is seen congenitally and may be due to a new dominant mutation, but has never been transmitted within a pedigree. Mongolian spots or collections of pigmented cells in a single patch usually found in the lumbosacral region are another congenital lesion without well-documented genetic pattern. Similarly the common blue nevus, or that occurring in the facial region supplied by the first and second trigeminal branches (nevus of Ota), may frequently be congenital, but is not hereditary.

Incontinentia pigmenti, named by Bloch (57) and Sulzburger (58), is a condition affecting the eye, CNS and teeth, and producing neonatal skin lesions. Vesicles, blisters or papules may be present at birth or (more frequently) develop within minutes or days after birth. Usually these lesions wax and wane for several months before the typical macular, linear or patchy pigmentation occurs on the legs, pelvic area, trunk or other area of the body. The child may have associated defects in CNS development of various types with subsequent developmental delay. There may be growth retardation. Eye lesions occur in about one-third of cases, frequently with mass lesions in the posterior chamber (59). Dentition is delayed with reduced number of teeth and conical formation of the incisors, canines and bicuspids. The disorder is believed to be inherited as an X-linked dominant with variable clinical effect in the female, and severe to lethal intrauterine expression in the males.

Acanthosis nigricans is a skin lesion consisting of a velvet-like thickening of the skin in the intertriginous areas particularly in axilla, neck and groin, with associated dark pigmentation. In extreme cases it may be generalized. It is frequently a warning sign of underlying malignancy, particularly adenocarcinoma in the elderly. Removal of the primary carcinoma results in disappearance of the skin lesion. A benign form exists as an autosomal dominant disorder, several pedigrees having been reported (60).

Dyskeratosis congenita occurs as an X-linked recessive condition with cutaneous pigmentation, dystrophy of the nails, leukoplakia of the oral mucosa, atretic lacrimal ducts with excessive lacrimation, testicular atrophy and frequent thrombocytopenia and anemia. The anemia may be severe and the oral leukoplakia may become malignant. The hematologic symptoms are thought to be manifestations of the Fanconi pancytopenia but this is unlikely.

Urticaria pigmentosa is the dermatologic manifestation of the systemic disease, mastocytosis or mast cell disease. Infiltration of the skin produces a generalized urticarial eruption, which may be present at birth. Although both dominant and recessive types of inheritance are proposed, the genetic pattern of transmission is not clear (61).

Xeroderma pigmentosum, a rare skin disorder predisposing to skin cancer, is discussed in Chapter 3 "Genetics and Cancer."

ALBINISM

Albinism is a hereditary disorder of melanin pigment metabolism occurring in man as well as in many other members of the animal kingdom. The condition was first associated with a possible biochemical defect by the brilliant work of Sir Archibald Garrod in his conception of inborn metabolic errors (62). Recently Witkop has reviewed the subject, adding his extensive experience to the material previously available in the literature (63). Albinos have been noted since ancient times because of their frequently striking appearance, especially among people who are normally darkly pigmented.

The condition results from a failure of normal melanin pigmentation in the melanocyte of the skin. Melanocytes are specialized dendritic cells the precursors of which, the melanoblasts, arise in the neural crest and migrate to peripheral sites (65). The melanocyte is found in mature human subjects in skin (hair bulbs, dermis and dermoepidermal junction), mucous membranes, nervous system (pia arachnoid) and eye (uveal tract and retina). Melanin is synthesized by specialized organelles, the melanosomes, utilizing the enzyme tyrosinase to catalyze the conversion of the amino acid tyrosine to melanin. The entire melanocyte system may be involved as oculocutaneous albinism or only at a particular site as in ocular or cutaneous albinism.

Although albinism in genetic isolates as well as its pedigree pattern indicates an autosomal recessive genetic basis, there has always been a suggestion of genetic heterogeneity. Clinical variation is observed in the degree of hypopigmentation. The statistical incidence of albinism is higher than expected from the observed parental consanguinity if only one gene was responsible (66). Genetically there are reports of normal offspring of two albino parents, strongly implying genetic heterogeneity with phenotypic complementation in the double heterozygote offspring of such a mating.

Now biochemical evidence exists for at least two distinctive forms of autosomal recessive oculocutaneous albinism in man (Table 4) (67-68). Incubation of fresh hair bulbs in a tyrosine solution produces a darkening of the hair bulb in some albinos. The basis for this finding is the presence or absence of tyrosinase. The genetic determination appears to be distinct, and tyrosinase-positive persons do not change to tyrosinase-negative ones or vice versa (63).

Albinism is widespread, the oculocutaneous tyrosinase-negative form occur-

Table 4. Albinism

Type (syndrome)	Clinical Features	Inheritance*
Oculocutaneous albinism:		
Ia (tyrosinase −)	Skin and hair white throughout life; little or no eye pigment with red reflex present and persistent and no pigment visible on transillumination; nystagmus, photophobia and visual defect, severe; strong susceptibility to skin cancer	AR
Ib (Hermansky-Pudlak)	Skin, hair and eyes lack pigment; nystagmus moderately severe; progressive bleeding tendency with bruising, epistaxis leading to death if critical bleeding	AR
IIa (tyrosinase +)	Skin white, but hair changes to yellow, blond or red with age; eye pigment increases with age, red reflex disappears age 5 and transillumination of iris shows "cartwheel" pigment deposits at borders; nystagmus, photophobia, visual defect and skin cancer susceptibility less than Type Ia	AR
IIb (Chédiak-Higashi)	Moderate oculocutaneous albinism with frosted gray hair and variable eye color; nystagmus and photophobia mild; white cell abnormality leads to infection and leukemia-like state	AR
IIc (Cross)	Skin and hair hypopigmentation, microphthalmia, mental retardation; spasticity and athetoid movements	AR
III (tyrosinase variable)	Features similar to Type IIa; red reflex usually persists into adulthood; cartwheel iris pigmentation only in adults; skin photosensitivity is minimal	AR
Ocular albinism	Eye alone is affected with iris accumulating slight pigment with age	XR
Cutaneous albinism:		
I without deafness		
Ia (Piebaldism with white forelock)	Triangular white forelock, forehead and midface; ventral midlimb hypopigmentation	AD
Ib (Menkes')	White, twisted, easily fractured hair; progressive cerebral degeneration with spasticity, seizures, decerebration leading to death; defective copper metabolism	XR
II with deafness		
IIa (Waardenburg)	Appearance of broad midface, white forelock, heterochromia iridis, bilateral nerve deafness	AD
IIb (Ziprkowski-Margolis)	Localized hypopigmentation, heterochromia iridis and nerve deafness in Sephardic Jewish family	XR

*See footnote, Table 2.

ring in roughly 1 in 35,000 persons of black or white origin (4). The oculocutaneous tyrosinase-positive form is more frequent in blacks, the incidence being 1 in 4,000 as opposed to 1 in 60,000 among whites. Ocular albinism has a frequency of 1 in 54,000 in Denmark (64). Several genetic isolates are described with a high frequency of oculocutaneous albinism, most notably the Tule Cuna indians of the San Blas islands off the coast of Panama. The effect on vision is clinically important and accounts for 4-10% of visually handicapped children (63).

Tyrosinase-negative oculocutaneous albinism, an autosomal recessive form, is clinically severer than the tyrosinase-positive type. Skin and hair are milk-white. Eye color in oblique light ranges from translucent gray to blue without evidence of pigment accumulation. No retinal pigment is detectable, and all pigmentary changes remain stable with age. Photophobia and nystagmus are severe, the patient exhibiting severe squint in moderate light. In direct light a red reflex is present in adults and children. Abnormal iris translucency may be present in many tyrosinase-negative white heterozygotes.

In *albinism with hemorrhagic diathesis* (Hermansky-Pudlak syndrome) the patients have a combination of tyrosinase-negative oculocutaneous albinism and a history of progressive bleeding tendency manifested by bruising, repeated epistaxis and hemorrhage following dental extractions (69). Examination of reticulendothelial cells in blood vessels, spleen, liver, lymph nodes and bone marrow showed them to be packed with a lipid-like pigment (70). All patients have a generalized lack of skin and hair pigment. Iris color is light gray to yellow-green, and nystagmus and astigmatism are present. Bleeding time is slighly prolonged, but other routine hematologic studies are usually normal. Recent studies suggest a platelet defect, probably in aggregation (63). Death may intervene from bleeding in a critical area as within the cranium (70). Inheritance is as an autosomal recessive trait.

Tyrosine-positive oculocutaneous albinism is a variant of classic autosomal recessive oculocutaneous albinism with a positive hair bulb test. Clinically the amount of skin, hair and eye pigment varies with race and age. The features overlap with those of the tyrosinase-negative form and normal but lightly pigmented individuals. Tyrosinase-positive albinos tend to have gradual accumulation of pigment with age and frequently have a history of change in eye or hair color. Hair color is usually white in infancy, becoming yellow or even red in adult life. Freckles and pigmented nevi are often seen in these individuals. The red eye reflex is seen in young children, but disappears in older children and adults as retinal and iris pigment accumulates. Although nystagmus and photophobia are present, they are less severe than in the tyrosinase-negative form.

The *Chédiak-Higashi syndrome* is a fatal childhood disease characterized by moderate oculocutaneous albinism and leukocytic abnormalities (63) with a positive hair test. Although melanocytes can form melanin, their melanosomes are abnormal with giant forms and premature degeneration (71). Skin color varies from light cream to slate-gray, and most patients burn easily in sunlight. Hair color ranges from very light blond to brunette and usually has a striking frosted gray sheen. Eye findings are variable, but patients usually have nystagmus and photophobia.

Children have repeated infections with relative lymphocytosis and abnormal giant peroxidase-positive granules in the leucocytes. Younger children may die from these infections or from hemorrhage related to thromocytopenia. Those who survive beyond 5 years develop muscular weakness, decreased deep tendon reflexes, convulsions, cranial and peripheral neuropathy and an abnormal, wide-based gait. A leukemia-like state with cellular infiltration of organs by immature lymphoid cells and histiocytes leads to pseudolymphoma and death in older children. The disease is transmitted as an autosomal recessive genetic condition in man. It also occurs in cattle and Aleutian mink (72).

Three children from a consanguineous family were described with skin and hair hypopigmentation, microphthalmia, mental retardation, spasticity and athetoid movements *(Cross syndrome)* (73). The pigmentary lack and microphthalmia were noted at birth, and the abnormal movements began by age 3 months. Both growth and neuromuscular development were retarded. Hair and skin were white, and the children were very sensitive to ultraviolet light. A high-arched palate, gingival hyperplasia, scoliosis and hyperdevelopment of secondary sex characteristics were also noted. The hair bulb test was weakly positive. The inheritance is assumed to be an autosomal recessive one.

Yellow-mutant albinism resembles tyrosinase-positive albinism clinically, but the hair-bulb test is equivocal—some hair bulbs responding and others not (74). The hair is dead white at birth, but becomes yellow by age 1 year and is a deeper shade than that of the tyrosinase-positive albino. The yellow mutant tans somewhat, and photophobia and nystagmus are less severe than in other forms of oculocutaneous albinism. Inheritance is autosomal recessive.

In *ocular albinism*, an X-linked trait, pigmentation is normal except in the eye. Affected males have hypopigmented irides and retinas, and only the iris accumulates small amounts of pigment with age. Nystagmus and other ocular symptoms occur. Female carriers may have translucent irides and fundic mosaic pigmentation. Linkage between this gene and the one for the Xg^a blood group is established (76). The condition is less frequent than oculocutaneous albinism.

A possible variant was reported by Forsius and Eriksson (77) in which the affected male subjects had protanomalous colorblindness and the female carrier had slightly disturbed color vision and no evidence of mosaic retinal pigmentation. The condition may simply be the simulatneous occurrence of ocular albinism and protanopia in the same kindred (63).

In cutaneous albinism (without deafness), *piebaldism with white forelock* occurs as an autosomal dominant trait and is likely due to incomplete migration of melanoblasts from the neural crest with normal dorsal pigmentation, but ventral albinism (78). The unpigmented areas are congenital, appearing as a triangular white forelock with a posterior apex, possible extension to the forehead and nasal root, a white chin patch and a white area from the nipples to the iliac fossa. The limbs may lack pigment from both forearm to wrist and midthigh to midcalf on the ventral surfaces. Photophobia and ocular symptoms are absent but heterochromia iridis may occur. The condition is rare.

Menkes' syndrome is an X-linked recessive disorder, apparently involving copper metabolism (79) with white, twisted hair that fractures easily (pili torti and trichorrhexis nodosa). In addition the affected males have cerebral degeneration characterized by onset in early infancy with rapid deterioration with seizures, spastic quadriparesis, growth retardation and decerebration leading to death.

In cutaneous albinism (with deafness) the *Waardenburg syndrome* is a well-known autosomal dominant condition with wide clinical variability (80). Lateral displacement of the lacrimal puncta, giving an appearance of a broad nasal root with hypertelorism, is generally present and pathognomonic. In fact the interpupillary distances are normal. Heterochromia iridis occurs in about 25% of cases, and a white forelock with unpigmented spots is present in 20%. Mild-to-severe bilateral nerve deafness is detectable in 17 to 20% of affected persons (81).

The *Ziprkowski-Margolis syndrome*, consisting of cutaneous albinism, localized hyperpigmentation, heterochromia iridis and nerve deafness, was first reported in a large Sephardic Jewish kindred as an X-linked trait (82). Infants are albinotic except for gluteal and scrotal pigmentation. Other hyperpigmented areas develop subsequently. Ocular symptoms are absent, but heterochromia iridis may occur. Congenital total deafness is present in all cases.

VASCULAR LESIONS

Vascular skin lesions, although common congenital abnormalities, are generally not heritable. The common vascular nevus, or strawberry mark, may occur as singular or multiple forms in 12% or more of infants, but is not inherited, and nearly all of them regress spontaneously. Similarly the nevus flammeus, or capillary telangiectatic lesion, which frequently shows a segmental (port-wine nevus) distribution, is nonheritable. Nevus flammeus of the forehead is frequently observed as part of the Rubinstein-Taybi syndrome of peculiar facies, broad thumbs and great toes and mental retardation. Facial port-wine nevus associated with intracranial calcification, seizure and mental retardation form the Sturge-Weber syndrome. Nevus flammeus, or nevus vasculosis combined with enlargement of an extremity, is known as the Klippel-Trenaunay-Weber syndrome. None of these syndromes is inherited. Angiokeratomas occur in X-linked Fabry disease, which is also called angiokeratoma corporis diffusum universale. This disorder of spingolipid metabolism is discussed in Chapter 5, "Genetic Metabolic Disorders."

Lymph vessels may also be involved in congenital vascular lesions. Cystic hygromas of the neck are thought to be associated with failure of normal lymphatic development in that region. Aside from their probable association with the pterygium colli seen in the Turner syndrome, they are nongenetic.

Congenital lymphedema may occur as a hereditary condition. The original description was given in 1892 by Milroy who reported 22 cases of edema in six generations of a kinship of 97 persons (83). In this family the edema appeared congenitally in the majority of cases; it may also appear at puberty in a so-called tardive form. There is disagreement as to whether these two types, sometimes called Milroy (early-onset) and Miege (late-onset) are two distinct conditions or one variable disease. There is speculation that the basic anatomic defect is agenesis of the leg lymphatics, but the explanation is not entirely satisfactory. Although locally disturbing, the edema does not cause a problem with general health or longevity and improves as the patient is at bed rest and frequently with aging. Inheritance appears to be as an autosomal dominant trait with wide clinical variability. Congenital lymphedema may also occur in association with the Turner syndrome or as a sporadic, nonhereditary disorder.

Other vascular telangiectatic lesions are inherited and are often regarded as hamartomatous lesions. Telangiectasias occur in a recessively inherited syndrome, including cerebellar ataxia and immunodeficiency, which is described as ataxia-telangiectasia. Rendu (84) and later Osler (85) described families in which members had recurrent attacks of epistaxis associated with multiple telangiectasias. The lesions were seen on the skin of the face, ears, scalp and digits and were also found in the mucous membranes of nose, mouth, stomach, intestine, kidney, bladder, vagina and uterus. The CNS and reticuloendothelial organs may also be affected. Pulmonary arteriovenous aneurysms may occur and cause circulatory failure (86). The condition, usually given the name Rendu-Osler-Weber syndrome, may cause bleeding from any of the affected sites just listed. It is inherited as an autosomal dominant trait.

Hemangiomatous lesions may occur on cutaneous and intestinal sites in lesions that are 0.5 to 4.0 cm in diameter, and have a blue-black color and soft spongy texture. This has led to their designation as the blue rubber-bleb nevus syndrome (87). At autopsy some patients had malignant lesions; the association of the latter with the syndrome is not clear. The hamartosis syndrome is present at birth and transmitted as a dominant autosomal trait.

Glomus tumors appear congenitally or in the early years as small, pigmented, painful nodules. Usually located on the upper limbs, they may cause vasodilation and temperature elevation of the limb. Pain may be paroxysmal. Histologically the lesions resemble hemangiomas, and the inheritance pattern observed in some families is that of an autosomal dominant trait (88).

Other Skin Syndromes

Several connective-tissue disorders have cutaneous manifestations and are covered elsewhere in this text (see Chapter 6, "Heritable Connective Tissue Disorders."

Cutaneous webs occur in association with other findings in a number of syndromes. The Turner and the Noonan syndromes both feature pterygium colli in a percentage of affected cases. Both of them are associated with short stature and difficulties in sexual development and are discussed under chromosomal and endocrine disorders (see Chapter 2, "Cytogenetic Disorders," and Chapter 9, "Genetics of Endocrine Diseases"). Popliteal pterygia are associated with cleft lip and lip pits in a syndromal presentation as autosomal dominant traits. In addition there is a generalized pterygium syndrome consisting of webbing of the neck with antecubital and popliteal webs occurring in a recessive familial pattern (89). Cryptorchidism was an associated finding in one patient. Finally pterygium colli is seen in some families as an autosomal dominant pattern occurring as an isolated trait (90). Whether this represents a mild expression of the Noonan syndrome or a separate entity is unclear.

Cutis verticis gyrata, a rare skin anomaly, is characterized by furrows of the scalp running over the crown and back of the head. They resemble the convolutions of the brain, hence the name. The hair is sparse, and the lesion may either be associated with numerous causes or occur alone in an apparent dominant pattern.

Generalized *lipodystrophy*, or loss of body fat, has been described extensively in the medical literature. Lipodystrophy, gigantism and endocrinologic abnormalities appear in a syndromal pattern in siblings in a manner consistent with a recessive genetic condition. Hyperlipemia, hepatomegaly, acanthosis nigricans, elevated metabolic rate and nonketotic insulin-resistant diabetes are associated manifestations. Pituitary and adrenal functions, when studied, have been normal. In some cases polycystic ovaries, muscular hypertrophy and mental retardation have occurred. The condition has been reviewed by Seip (91).

An interesting syndrome of *focal dermal hypoplasia* was described by Goltz et al (92) and reviewed by Goltz (93). The syndrome consists of atrophy and linear pigmentation of the skin, herniation of fat through the dermal defects and multiple papillomas of the mucous membrances or skin. Digital anomalies may be associated and include syndactyly, polydactyly, camptodactyly and absence deformities. Hypoplastic teeth and ocular anomalies may also be present. Almost all affected persons have been female with vertical transmission in families, leading to the conclusion that this is caused by an X-linked dominant gene, which is lethal in males.

Amyloidosis may involve skin and other organs. Several types are inherited (Table 5) (see also Chapter 13 Heritable Neurologic Diseases, 16, "Heritable Disorders of the Kidney").

Osteopoikilosis in association with dermatofibrosis lenticularis disseminata is a disorder of no pathologic consequence. It produces "spotted bones" or circumscribed sclerotic areas near the ends of many bones and similarly spotted skin lesions, which are connective-tissue nevi. Families are reported with autosomal dominant inheritance (94).

Cockayne syndrome consists of the infantile presentation of erythema, vesicles and bullae in light-exposed skin areas in association with subsequent growth retardation (95). There is a noticeable loss of subcutaneous fat and a senile appearance. Deafness, dementia and retinal degeneration as well as cryptorchidism and cataracts also occur. Death usually intervenes before age 30 in this autosomal recessive condition.

The *Werner syndrome* is another heritable condition of accelerated aging with striking skin manifestations. There is symmetrical growth retardation, premature graying of hair, atrophy and hyperkeratoses of the skin, hair loss, cataracts and muscular atrophy, and subcutaneous calcification. Premature arteriosclerosis and diabetes mellitus occur. Early death appears to be a feature of the condition, which is inherited as an autosomal recessive disorder (96).

Xanthomatosis is simply the skin manifestation of an underlying systemic disorder. Usually it reflects disordered fat metabolism. The topic is discussed in Chapter 8 "Genetics of Cardiovascular Diseases". The lesions vary considerably in appearance and in mode of onset, from sudden to gradual. They may be flat, planar, nodular (tuberosum) or disseminated or appear on tendons predominately. They may be flesh-colored, brown or yellowish with a smooth and frequently glistening surface.

Lipoid proteinosis is a disorder the earliest manifestation of which is hoarseness in the infant followed by skin lesions during the first year of life (97). The progress of the skin changes is variable. Exposed skin has a waxy appearance with generalized loss of pliability. There are atrophic scars, giving the skin of

Table 5. Genetic Forms of Amyloidosis

Type (syndrome)	Clinical Features	Skin Manifestations	Inheritance*
Type I (Andrade or Portuguese)	Onset between age 20 and 30 of neuropathic symptoms mainly in legs; death follows in 10 yrs	None unless trophic ulcers	AD
Type II (Indiana or Rukavina)	Neuropathic symptoms in upper limbs; onset in 40s with slow progression	None	AD
Type III (cardiac)	One family in Denmark with mother and 7 children showing cardiac amyloidosis with progressive heart failure beginning at age 40	None	AD
Type IV (Iowa or Van Allen)	Neuropathy begins at age 35 with nephropathy following later and causing death about 12 years from onset	None	AD
Type V (Finland or Meretoija)	Corneal lattice dystrophy and cranial neuropathy	None	AD
Type VI (cerebral arterial)	Icelandic kindred with cerebral arterial amyloid	None	AD
Type VII (familial visceral)	Variable age of onset with chronic nephropathy, arterial hypertension and hepatosplenomegaly	None	AD
Type VIII (primary cutaneous)	Onset at puberty of cutaneous amyloid with extension with aging, but no systemic involvement	Lichen amyloidosis	AD
Type IX (Muckle-Wells)	Recurrent fever and itching, progressive perceptive deafness, amyloidosis, especially renal	Urticaria	AD
Cold hypersensitivity	Exposure to cold causes urticarial wheals, pain and swelling of joints, chills and fever; amyloidosis is complication with amyloid nephropathy frequent cause of death	Urticarial wheals	AD

*See footnote, Table 2.

younger patients a pock-marked appearance. Older patients tend to have hypertrophic lesions with consequent nodularity associated with irregular depigmentation. The vermilion and mucosal portions of the lips are irregularly thickened. Patchy alopecia is a regular finding. Neurologic symptoms, including mental retardation, are occasionally found. The disorder appears to be transmitted as an autosomal recessive trait.

The Rothmund-Thomson syndrome is a hereditary dermatosis characterized by atrophy, pigmentation and telangiectasia of the skin, frequently accompanied by juvenile cataract, saddle nose, congenital bone defects, disturbances of hair growth and hypogonadism (98). The basic defect is unknown but, familial studies indicate autosomal recessive inheritance.

The *Bloom syndrome* is a condition with skin manifestations consisting of multiple telangiectasias in association with short stature, mental retardation, chromosomal breakage and rearrangement and subsequent development of leukemia leading to death (99). This autosomal recessive condition appears primarily in Ashkenazic Jews.

Table 6. Dysmorphic Syndromes Affecting Hair Growth

Disorder	Hair Phenotype	Associated Findings	Inheritance*
Chromosomal abnormalities:			
Down syndrome (trisomy 21)	Fine light colored hair with atrophic bulbs	Growth and mental retardation	—†
Chromosome no. 4 short arm deletion	Scalp defects	Cleft lip and palate growth and mental retardation	—
Trisomy 18	Hypertrichosis	Mental retardation, early death	—
Trisomy 13	Scalp defects	Mental retardation, poor survival	—
Ectodermal disorders:			
Marinesco-Sjögren syndrome	Fragile, brittle, rough hair	Congenital cataracts, cerebellar ataxia and mental retardation	AR
Netherton syndrome	Trichorrhexis invaginata	Exfoliative erythroderma and ichthyosis linearis circumflexa	AR
Anhidrotic ectodermal dysplasia	Hypotrichosis	Epidermis, teeth, sweat glands, breast tissue, nails hypodeveloped	XR (AR)
Hidrotic ectodermal dysplasia	Alopecia	Nails and epidermis	AD
Pili torti and nerve deafness	Pili torti	Nerve deafness (maybe variant of pili torti)	AR
Menkes syndrome	Monilethrix	Somatic retardation with focal cerebellar and cerebral degeneration, inconstant amino aciduria	XR
Keratosis follicularis spinulosa decalvans cum ophiasi	Hypotrichia eyebrows, eyelashes, one or more winding bald streaks on scalp	Corneal degeneration	XR
Gingival fibromatosis with hypertrichosis	Extreme hirsutism	Gingival fibromatosis	AD
Atrichia with papular lesions	Complete absence of hair	Papular lesions over most of body	AR
Trichodentoosseous syndrome	Curly hair	Defective enamel, increased bone density	AD
CHANDS (curly hair ankyloblepharon–nail dysplasia	Curly hair	Fusion of eyelids at birth, nail dysplasia	AR
Keratosis palmoplantaris with peridontopathia (Papillon-Lefevre syndrome)	Hypotrichosis	Cystic eyelids, hypodentia, keratosis palmoplantaris	AR
Alopecia universalis, epilepsy, pyorrhea and mental defect	Permanent, universal alopecia	Mental subnormality, psychomotor epilepsy, pyorrhea	AD

Associated with prominent skeletal defects:

Syndrome	Hair	Features	Inheritance
Orofaciodigital syndrome I	Sparse, fragile, rough hair	Cleft palate, syndactyly, mental retardation	XR
Metaphyseal chondrodysplasia, McKusick type (cartilage-hair hypoplasia)	Hairs with small diameter	Dwarfism and immunodeficiency	AR
Apert syndrome	Trichorrhexis	Skeletal deformity including syndactyly and mental retardation	AD
Congenital trichomegaly with retinal pigmentary degenerations, dwarfism and mental retardation	Long eyelashes and eyebrows	Retinal degeneration, mental retardation and dwarfism	AR
Hypotrichosis syndactyly and retinitis pigmentosa	Hypotrichosis	Syndactyly and retinitis pigmentosa	AR
Trichorhinophalangeal syndrome	Thin slow-growing hair	Brachyphalangy, pear-shaped nose, supernumerary incisors	AR
Hallerman-Streiff syndrome	Hypotrichosis	Bird-like facies, dwarfism, microphthalmia and cataracts	AR?
Pierre-Robin (Stickler?)	Fine, light-colored hair	Micrognathia, glossoptosis and cleft palate	AD?
Chondrodysplasia punctata	Spotty alopecia, coarse hair	Calcific cartilaginous deposits, osseous deformities and cataracts	AR, AD
Cockayne syndrome	Fine, sparse hair, premature graying	Dwarfism, retinal degeneration and photosensitivity	AR

*See footnote, Table 2.
†Not applicable.

Table 7. Metabolic Disorders Affecting Hair Growth

Disorder	Hair Phenotype	Involvement	Inheritance*
Argininosuccinicaciduria	Monilethrix, trichorrhexis nodosa, fragile hair	Growth and mental retardation with seizures	AR
Citrullinemia	Fragile hair, atrophic bulb	Growth and mental retardation	AR
Hartnup disease	Fine, fragile hair	Cerebellar ataxia, photosensitivity and mental retardation	AR
Homocystinuria	Sparse, fragile, fine hair	Mental retardation, dislocated lenses and long, thin limbs	AR
Phenylketonuria	Fine, light colored hair	Growth and retardation, eczema and hypopigmentation	AR
Tyrosinemia	Fine, light colored hair	Hepatosplenomegaly, mental retardation and resistant rickets	AR
Congenital adrenal hyperplasia	Excess terminal hair	Adrenogenital syndrome	AR
Erythropoietic protoporphyria	Excess terminal hair on face	Photosensitivity	AD
Mucopolysaccharidoses	Excess terminal and vellus hair	Growth and mental retardation, with systemic symptoms	AR, XR

*See footnote, Table 2.

Pachyonychia congenita is an autosomal dominant disorder characterized by onychogryphosis, hyperkeratosis of palms, soles, knees and elbows, tiny cutaneous horns in many areas and leukoplakia of the oral mucous membranes. Hyperhydrosis of palms and feet is usually present (100).

Several reports implicate hereditary factors in systemic lupus erythematosus and related disorders. Although the reports of familial aggregation-shared antibodies and HLA-associations (101) suggest hereditary factors, the preponderance of nonfamilial cases and evidence for viral etiology suggest nongenetic explanations (102).

Psoriasis is another common disorder occurring in familial forms. Autosomal dominant pedigrees with penetrance reduced to 60% are described (103). However, HLA association and variability of clinical disease strongly suggest for most forms a multifactorial inheritance pattern (104).

Hair

Hereditary disorders affecting only the appearance of the hair or the distribution of normal hair are rare, although either the phenotype or the distribution may be altered in skin diseases or in more generalized disorders. Three types of conditions involving human hair include (105):

1. dysmorphic syndromes with significant hair involvement
2. metabolic disorders affecting hair
3. single gene disorders of hair

Tables 6, 7 and 8 list some of these disorders.

Table 8. Single Gene Disorders of Hair

Type*	Clinical Features	Inheritance†
Hypotrichosis (Marie-Unna type) (106)	Congenital absence of eyebrows, eyelashes and body hair; coarse, wiry childhood hair followed by alopecia	AD
Hypotrichosis (107)	Failure to replace intrauterine hair; no pubertal body hair	AR
Milia-hypotrichosis syndrome (108, 109)	Facial milia with sparse, hypopigmented scalp hair	AD or XD
Aplasia cutis congenita (110)	Skin and scalp defects with underlying skull defect	AD, AR
Baldness (111)	Male pattern baldness	AD (male) AR (female)
Hypertrichosis universale (112)	Excessive body hair; double eyebrows	AD
Pili torti (113)	Twisted, flattened hair shafts; coarse, dry hair breaks easily	AR
Wooly hair (114)	Hair similar to that of Negro in nonblack	AD
Monilethrix (115)	Microscopically beaded hair; breaks at beads	AD
Alopecia (116)	Patchy to complete baldness	Multifactorial

*Numbers in parentheses are reference citations.
†See footnote, Table 2.

REFERENCES

1. Cockayne EA: Inherited Abnormalities of the Skin and Its Appendages. New York, Oxford University Press, 1933
2. Solomon LM, Esterly NB: Neonatal Dermatology, Philadelphia, Saunders, 1973
3. Bergsma D (ed): Skin, Hair and Nails. *Birth Defects* 7: Part 12, 1972
4. Mulvihill J, Smith DW: Genesis of dermatoglyphics. *J Pediatr* 75:578, 1969
5. Popich GA, Smith DW: The genesis and significance of digital and palmar hand creases: Preliminary report. *J Pediatr* 77:1917, 1970
6. Reed TE, Borgaonkar DS, Conneally PM, et al: Dermatoglyphic nomogram for the diagnosis of Down's syndrome. *J Pediatr* 77:1024, 1970
7. Walker JC, Koenig JA, Irwin L, et al: Congenital absence of skin (aplasia cutis congenita). *Plast Resconstruct* Surg. 26:209, 1960
8. Cutlip BD Jr, Cryan DM, Vineyard WR: Congenital scalp defects in mother and child. *Am J Dis Childh* 113:597, 1967
9. Gedda L, Muratore A, Bernardi A: La gangrena asettica della teca cranic come aplasia circoscritta ereditaria del Neonato. *Acta Genet Med* 12:117, 1963
10. Bart BJ: Congenital localized absence of skin, blistering and nail abnormalities, a new syndrome. *Birth Defects* 7:118, 1972
11. Gedde-Dahl T Jr: Epidermolysis Bullosa. A Clinical, Genetic and Epidemiological Study. Baltimore, Johns Hopkins Press, 1971
12. Noojin RO, Reynolds JP, Croom WC: Genetic study of hereditary type of epidermolysis bullosa simplex. *Arch Dermatol Syph* 65:477, 1952
13. McDaniel WH: Epidermolysis bullosa. *Arch Dis Child* 29:334, 1954
14. Walther T: Epidermolysis bullosa hereditaria letalis. A review and report of two new cases. *Ann Pediatr* 180:382, 1953
15. Goldsmith LA: The Ichthyoses, in Steinberg AG, Bearn AG, Motulsley AG, and Childs B. (eds): Progress in Medical Genetics. Philadelphia, Saunders, 1976, p. 185
16. Frost P, Van Scott EJ: Ichthyosiform dermatoses: Classification based on anatomic and biometric observation. *Arch Dermatol* 94:113, 1966
17. Lever WF: Histopathology of the Skin, 4th ed. Philadelphia, Lippincott, 1967
18. Menton DN, Eisen AZ: Structure and organization of mammalian stratum corneum. *J Ultrastruct Res* 35:247
19. Baden HP, Goldsmith LA: The structural proteins of the epidermis. *J Invest Dermatol* 59:66, 1972
20. Nix TE, Kloepfer HW, Derbes VJ: Ichthyosis-lamellar exfoliative type. *Dermatol Trop* 2:142, 1963
21. Wells RS, Kerr CB: Clinical features of autosomal dominant and sex-linked ichthyosis—an English population. *Br Med J* 1:947, 1966
22. Frost P, Weinstein GD, Van Scott EJ: The ichthyosiform dermatoses II. Autoradiographic studies of epidermal proliferation, *J Invest Dermatol* 47:561, 1966
23. Esterly NB, Maxwell E: Nonbullous congenital ichthyosiform erythroderma; treatment with methotrexate. *Pediatrics* 41:120, 1968
24. Frost P, Weinstein GD: Topical administration of vitamin A acid for ichthyosiform dermatoses and psoriasis. *JAMA* 207:1863, 1969
25. Schnyder UW: Inherited ichthyoses *Arch Dermatol* 102:240, 1970
26. Klaus S, Weinstein G, Frost P: Localized epidermolytic hyperkeratosis *Arch Dermatol* 101:272, 1970
27. McKusick VA: Mendelian Inheritance in Man, 4th ed. Baltimore, Johns Hopkins Press, p. 467, 1975
28. Falek A, Heath CW, Ebbin AJ, et al: Unilateral limb and skin deformities with congenital ear disease in two siblings; a lethal syndrome. *J Pediatr* 73:910, 1968
29. Goldsmith LA, Baden HP: Propylene glycol with occlusion in ichthyosis. *JAMA* 220:579, 1972

30. Adam A, Ziprkowski L, Feinstein A, et al: Linkage relations of X-borne ichthyosis to the Xg blood groups and to other markers of the X in Israelis. *Ann Hum Genet* 32:323, 1969
31. Sever RR, Frost R, Weinstein G: Eye changes in ichthyosis. *JAMA* 206:2283, 1968
32. Sjögren T, Larsson, T: Oligophrenia in combination with congenital ichthyosis and spastic disorders. *Acta Psychiatr Neurol Scand.* 32:133, 1957
33. Comel M: Ichthyosis linearis circumflexa. *Dermatologica* 98:133, 1949
34. Netherton EW: A unique case of trichorrhexis nodosa "bamboo hairs." *Arch Dermatol* 78:483, 1958
35. Bodian EL: Skin manifestations of Conradi Disease. *Arch Dermatol* 94:743, 1966
36. Steinberg D, Vroom FQ, Engel WK, et al: Refsum's disease—a recently characterized lipidosis involving the nervous system. *Ann Intern Med* 66:365, 1967
37. Lynch HT, Ozer F, McNutt CW, et al: Secondary male hypoganadism and congenital ichthyosis. Association of two rare genetic diseases. *Am J Hum Genet* 12:440, 1960
38. Perrin JCS, Idemato JY, Sotos JF, et al: X-linked syndrome of congenital ichthyosis, hypogandism, mental retardation and anosmia. *Birth Defects* 12:267, 1976
39. Howel-Evans W, McConnell RB, Clarke CA, et al: Carcinoma of the oesophagus with keratosis palmaris et plantaris (tylosis) a study of two families. *Q J Med* 27:413, 1958
40. Franceschetti A, Reinhart V, Schnyder UW: La maladie de Meleda. *J Genet Hum* 20:267, 1972
41. Hanhart E: Neue Sondefformen von Keratosis palmo-plantaris, U.A. eine regelmässig-dominante mit systematisierten Lipomen, ferner 2 einfachrezessive mit Schwachsinn und Z.T. mit Hornhautveraenderungen des Auges (Ektodermatosyndrom). *Dermatologica* 94:286, 1947
42. Gorlin RJ, Sedano H, Anderson VE: The syndrome of palmar-plantar hyperkeratosis and premature periodontal destruction of the teeth. A clinical and genetic analysis of the Papillon-Lefevre syndrome. *J Pediatr* 65:895, 1964
43. Anderson DE, Hernosky ME: Disseminated superficial actinic porokeratosis. Genetic aspects. *Arch Dermatol* 99:408, 1969
44. Niordson AM, Sylvest B: Bullous dyskeratosis follicularis and acrokeratosis verruciformis. *Arch Dermatol* 92:166, 1965
45. Hitch JM, Callaway JL, Moseley V: Familial Darrier's disease (keratosis follicularis) *South Med J* 34:578, 1941
46. Herndon JH Jr, Wilson JD: Acrokeratosis verruciformis (Hopf) and Darrier's disease. Genetic evidence for a unitary origin. *Arch Dermatol* 93:305, 1966
47. Jung EG: Acrokeratoelastoidosis. *Humangenetik* 17:357, 1973
48. Freire-Maia N: Ectoderma dysplasias. *Hum Hered* 21:309, 1971
49. Witkop CJ, Brearley LJ, Gentry WC: Hypoplastic enamel, onycholysis and hypohidrosis inherited as an autosomal dominant trait. *Oral Surg,* 39;71, 1975
50. Clouston HR: A hereditary ectodermal dystrophy. *Can Med Assoc J* 21:18, 1929
51. Halperin SL, Curtis GM: Anhidrotic ectodermal dysplasia associated with mental deficiency. *Am J Ment Def* 46:459, 1942
52. Coffin GS, Siris E: Mental retardation with absent fifth fingernail and terminal phalanx. *Am J Dis Childh* 119:433, 1970
53. Van den Bosch J: A new syndrome in three generations of a Dutch family. *Opthalmologica* 137:422, 1959
54. Brown J, Kierland RR: Erythrokeratodermia variabilis. Report of three cases and review of the literature. *Arch Dermatol* 93:194, 1966
55. Beare JM, Nevin NC, Froggatt P, et al: Atypical erythrokeratoderma with deafness, physical retardation and peripheral neuropathy. *Br J Dermatol* 87:308, 1972
56. Gorlin RJ: Multiple lentigenes syndrome. *Am J Dis Childh* 117:652, 1969
57. Bloch B: Eigentümiliche bisher nicht beschriebene Pigmentaffection (incontinentia pigmenti). *Schweiz Med Wochnschr* 56:404, 1926
58. Sulzberger MB: Uber eine bisher nicht beschriebene Pigmentanomalie (incontinentia pigmenti). *Arch Dermatol Syph* 154:19, 1928

59. Cole JG, Cole HG: Incontinentia pigmenti associated with changes in the posterior chamber of the eye. *Am J Ophthalmol* 47:321, 1959
60. Curth HO, Aschner BM: Genetic studies on acanthosis nigricans. *Arch Dermatol* 79:55-66, 1959
61. Shaw JM: Genetic aspects of urticaria pigmentosa. *Arch Dermatol* 97:137, 1968
62. Garrod AE: Inborn errors of metabolism, Croonian Lectures, Lecture I. Lancet 2:1, 1908
63. Witkop CJ, Jr: Albinism. *Adv Hum Genet* 2:19, 1971
64. Norn MS: Ocular albinism. *Acta Ophthalmol* 44, 20, 1966
65. Pearson K, Nettleship E, Usher CH: A Monograph on Albinism in Man. Drapers' Co. Res. Memoirs, Biometric Series VI, VIII, and IX; parts I, II, and IV. Dulau, London, 1911
66. Stern C: Principles of Human Genetics, 2d ed., San Francisco, Freeman, 1960, p. 385
67. Kugelman TP, Van Scott EJ: Tyrosinase activity in melanocytes of human albinos, *J Invest Dermatol* 37:73, 1961
68. Witkop CJ Jr, Van Scott EJ, Jacoby GA: Evidence for two forms of autosomal recessive albinism in man, *Proc Second Internat Cong Human Genet* Rome, Institute Gregor Mendel, 1961
69. Hermansky F, Pudlak P: Albinism associated with hemorrhagic diathesis and unusual pigmented reticular cells in the bone marrow. *Blood* 14:162, 1959
70. Bednar B, Hermansky F, Lojda Z: Vascular pseudohemophilia associated with ceroid pigmentophagia duplication in albinos. *Am J Pathol* 45:283, 1964
71. Windhorst DB, Zelickson AS, Good RA: A human pigmentary dilution based on a heritable subcellar structural defect—the Chédiak-Higashi syndrome, *J Invest Dermatol* 50:9, 1968
72. Padgett GA, Leader RW, Gorham JR, et al: The familial occurrence of the Chédiak-Higashi syndrome in mink and cattle. *Genetics* 49:505, 1964
73. Cross HE, McKusick VA, Breen W: A new oculocerebral syndrome with hypopigmentation. *J Pediatr* 70:3, 1967
74. Nance WE, Jackson CE, Witkop CE Jr: Amish albinism: A distinctive autosomal recessive phenotype. *Am J Hum Genet* 22:579, 1970
75. Waardenburg PJ, in Waardenburg PJ, Franschetti A, Klein D (eds): Genetics and Ophhalmology, Vol. 1. Springfield, Ill., Thomas, 1961, p. 704
76. Fialkow PJ, Giblett ER, Motulsky AG: Measurable linkage between ocular albinism and Xg. *Am J Hum Genet* 19:63, 1967
77. Forsius H, Eriksson AW: Ein neues Augensyndrom mit X-chromosomaler Transmission. *Klin Monatsbl Augenheilkd* 144:447, 1964
78. Comings DE, Odland GF: Partial albinism. *JAMA* 195:510, 1966
79. Danks DM, Cartwright E: Menkes kinky hair disease: Further definition of the defect in copper transport. *Science* 179:1140, 1973
80. Waardenburg PJ: A new syndrome combining developmental anomalies of the eyelids, eyebrows, and nose root with congenital deafness. *Am J Hum Genet* 3:195, 1951
81. Partington MW: Waardenburg's syndrome and heterochromia iridum in a deaf school population. *Can Med Assoc J* 90:1008, 1964
82. Ziprkowski L, Krakowski A, Adam A, et al: Partial albinism and deaf mutism—due to a recessive sex-linked gene. *Arch Dermatol* 86:530, 1962
83. Milroy WI: An undescribed variety of hereditary oedema. *NY J Med* 56:87, 1892
84. Rendu M: Epistaxis repetees chez un suget porteur de petits angiomes cutanes et muqueux. *Bull Soc Med Hop* (Paris) 13:731, 1896
85. Osler W: A family form of recurring epistaxis associated with multiple telangiectases of the skin and mucous membranes. *Bull Johns Hopkins Hosp* 11:333, 1901
86. Berquist N: Arteriovenous pulmonary aneurysms in Osler's disease. Report of four cases in the same family. *Acta Med Scand* 171:301, 1962
87. Fine RM, Derbes VJ, Clark WH: Blue rubber-bleb nevus. *Arch Dermatol* 84:802, 1961
88. Gorlin RJ, Fusaro RM, Benton JW: Multiple glomus tumor of the pseudocavernous hemangioma type. *Arch Dermatol* 82:776, 1960

89. Norum RA, James VL, Mabry CC: Pterygium syndrome in three children in a recessive pedigree pattern. *Birth Defects* 5:233, 1969
90. Jackson L: Unpublished observations, 1970
91. Seip M: Generalized lipodystrophy. *Ergeb Inn Med Kinderheilkd* 3:59, 1971
92. Goltz RW, Peterson WC Jr, Gorlin RJ, et al: Focal dermal hypoplasia. *Arch Dermatol* 86:708, 1962
93. Goltz RW, Henderson RR, Hitch JM, et al: Focal dermal hypoplasia syndrome. *Arch Dermatol* 101:1, 1970
94. Schorr WF, Opitz JM, Reyes CN: The connective tissue nevus–osteopoikilosis syndrome. *Arch Dermatol* 106:208, 1972
95. MacDonald WB, Fitch KD, Lewis IC: Cockayne's syndrome: A heredofamilial disorder of growth and development. *Pediatrics* 25:997, 1960
96. Epstein CJ, Martin GM, Schultz AL, et al: Werner's syndrome: A review of its symptomatology, natural history, pathologic features, genetics and relationship to the natural aging process. *Medicine* 45:177, 1966
97. Gordon H, Gordon W, Bertha V, et al: Lipoid proteinosis. *Birth Defects* 7:164, 1971
98. Taylor WB: Rothmund's syndrome–Thomson's syndrome. *Arch Dermatol* 75:236, 1957
99. German J: Bloom's syndrome. I. Genetical and clinical observations in the first twenty-seven patients. *Am J Hum Genet* 21:196, 1969
100. Jackson ADM, Lawler SD: Pachonychia congenita. *Ann Eugen* 16:42, 1951
101. Larsen RA, Godal T: Family studies in systemic lupus erythematosus (SLE). *J Chron Dis* 25:225, 1972
102. Lewis R, Tannenberg W, Smith C, et al: Human systemic lupus erythematosus and C-type RNA viruses. *Clin Res* 22:422A, 1974
103. Abele DC, Dubson RL, Graham JB: Heredity and psoriasis. *Arch Dermatol* 88:38, 1963
104. Watson W, Cann HW, Farber EM, et al: The genetics of psoriasis. *Arch Dermatol* 105:197, 1972
105. Porter PS: The genetics of human hair growth. *Birth Defects* 7:69, 1971
106. Peadry RD, Wells RS: Hereditary hypotrichosis (Marie Unna type). *Trans St. Johns Hosp Dermatol Soc* 57:157, 1971
107. Sly WS, Treister M: Isolated congenital hypotrichosis: Recessive hairlessness in man. Personal communication 1976
108. Parrish JA, Baden HP, Goldsmith LA, et al: Studies of the density and the properties of the hair in a new inherited syndrome of hypotrichosis. *Ann Hum Genet* 35:349, 1972
109. Goldsmith LA, Baden HP: Analysis of genetically determined hair defects. *Birth Defects* 12:86, 1971
110. McKusick VA: Mendelian Inheritance in Man. Baltimore, Johns Hopkins Press, 4th ed. 1975
111. Osborne D: Inheritance of baldness. *J Hered* 7:347, 1916
112. Beighton PH: Congenital hypertrichosis lanuginosa. *Arch Dermantol* 101:669, 1970
113. Gedda L, Cavalier R: Rilievi genetici delle distrafie congenite dei capelli. *Proc Sec Intern Cong Hum Genet* 2:1070, 1963
114. Anderson E: An American pedigree for woolly hair. *J Hered* 27:444, 1936
115. Baker H: An investigation of monilethrix. *Br J Dermatol* 74:24, 1962
116. Lubowe II: The clinical aspects of alopecia areata, totalis and universalis. *Ann NY Acad Sci* 83:1458, 1959

20
Hereditary Defects of Teeth and Oral Structures

Carl J. Witkop, Jr.

More than 600 genetic disorders affect the orofacial structures (1-3). Some of these diseases affect only teeth or the oral tissues; however, in most of them the alterations seen in the orofacial structures are only one manifestation of abnormal processes affecting other systems. This division suggests that those conditions involving only teeth most likely are determined by mutations that regulate functions unique to the highly specialized odontogenic cells, whereas those with general effects result from mutations regulating less differentiated cells or general metabolic or regulatory systems (4). As an aid to differential diagnosis, selected examples, including the more commonly encountered inherited diseases affecting the orofacial region, will be classified on the basis of their local or generalized manifestations. A classification of inherited defects of tooth structure and form, together with the mode of inheritance, is given in Tables 1 through 4. A classification of disorders that can be suspected from the radiographic appearance of the tooth pulp chamber is given in Table 5 (5). Table 6 classifies the ectodermal dysplasia syndromes on the basis of involvement of hair, dentition, nails and sweat glands following the system of Freire-Maia (6-7).

TOOTH ABNORMALITIES WITHOUT GENERALIZED DEFECTS

Defects of Enamel (Amelogenesis Imperfecta)

Inherited defects involving only enamel without alterations in other body tissues or metabolic processes are designated as types of amelogenesis imperfecta (AI) (8, 9). Three general types can be distinguished by their clinicohistologic appearance. Hypoplastic types have thin enamel. Hypomaturation types have enamel of normal thickness and a ground-glass or mottled yellow-white appearance and chips from the dentin. Hypocalcified types have enamel of normal thickness on newly erupted teeth, which is soft, friable and lost rapidly. Teeth of both denti-

This work was supported by NIH Grant DE 03686.

Table 1. Hereditary Defects of Enamel

Defect in enamel primary (amelogenesis imperfecta):
Hypoplastic:
  Pitted (AD)
  Local (AD)
  Smooth (AD, XD)
  Rough (AD)
  Enamel agenesis (AR)
Hypocalcified (AD, AR)
Hypomaturation
  XR form
  Pigmented (AR)
  Snow-capped teeth (AD)
  Enamel opacities, white hypomature enamel (?AR)

Defect in enamel accompanied by generalized disease:
Hypoplasia, hypocalcification, hypomaturation
  Trichodento-osseous syndrome (AD)
  Amelo-onycho-hypohidrotic syndrome (AD)
  Oculodentodigital dysplasia (oculodento-osseous dysplasia) (AD)
  Epidermolysis bullosa dystrophica (AR)
  Focal dermal hypoplasia (XD)
  Mucopolysaccharidosis IV (AR)
  Epilepsy and hypoplastic enamel (?XR)
  XTE syndrome (xeroderma, talipes and enamel defect) (AD)
  Vitamin D-dependent rickets (AR)
  Hypohidrotic ectodermal dysplasia with cleft lip/palate, ocular, genital and digital anomalies (AD?)
  Oculomandibulodyscephaly (NK)
  Marshall syndrome (AD)
  Lesch-Nyhan syndrome (XR)
  Lipoid proteinosis (AR)
  Bird-headed dwarf (AR)
  Tuberous sclerosis (AD)
Associated enamel defect reported, but less frequently:
  Ehlers-Danlos (AD)
  Trisomy 21
  Congenital insensitivity to pain syndrome (AR)
  Chondroectodermal dysplasia (AR)
  Popliteal pterygium syndrome (AD)
  Rieger syndrome (AD)
  Mandibulofacial dysostosis (AD)
  Congenital facial diplegia (sporadic)
  Orodigitofacial dysotosis (OFD 1) (XR)
  Cleidocranial dysplasia (AD)
  Ichthyosis vulgaria (AD)
  Phenylketonuria (AR)
  Infantile allergy syndrome (AR?)
Reported in few cases, questionable association:
  Sjögren-Larsson syndrome (AR)
  Papillon-Lefevre syndrome (AR)
  Neurofibromatosis (AD)
Pigmented enamel
  Progeria (AR)
  Naegeli syndrome (AD)

AD-autosomal dominant; AR-autosomal recessive; XD,XR-X-linked dominant, recessive; NK-inheritance not known; MF-multifactorial.

Table 2. Hereditary Defects of Enamel and Dentin

Defect in enamel and dentin primary in odontogenesis:
Odontodysplasia or ghost teeth (NK)
Trichoonychodental syndrome (AD)
Amelointerradicular dentin dysplasia (Kimura and Nakata) (AR)

Defect in enamel and dentin associated with generalized disorder:
Hypoplasia or hypocalcification, or both
  Pseudohypoparathyoidism (XD)
  Idiopathic juvenile hypoparathyroidism (AR,AD,XD)
  Fanconi syndrome (heterogenous)
  Vitamin D-dependent rickets (AR)
  Other aminoacidurias with nephrotic syndrome (oculocerebral renal syndrome [XR]; Luder-Sheldon syndrome [?AD])
  Ehlers-Danlos syndrome (AD)
  Sickle cell anemia (homozygote shows defect)
  Acrodermatitis enteropathica (AR)
Pigment with or without hypoplasia-hypocalcification
  Erythroblastosis fetalis (multiple causes)
  Erythropoietic porphyria (AR)
  Alkaptonuria (AR)
  Progeria (AR)

Table 3. Hereditary Defects of Dentin and Cementum

Hereditary defects of dentin alone:
Defect in dentin primary in dentinogenesis
  Opalescent dentin (dentinogenesis imperfecta) (AD)
  Radicular dentin dysplasia (dentin dysplasia, Type I) (AD)
  Coronal dentin dysplasia (dentin dysplasia, Type II) (AD)
  Shell teeth (Rushton) (NK)
  Pulpal dysplasia (NK)
Defect in dentin accompanied by generalized disorder
  Dentinogenesis imperfecta (Osteogenesis imperfecta [AD, AR])
  Dentin dysplasia (Dentino-osseous dysplasia [AD], Branchioskeletogenital syndrome [AR], Fibrous dysplasia of dentin [Vickers] [AD?], Calcinosis universalis and tumoral [AD, AR], Ehlers-Danlos syndrome, type unknown [AD?], Growth retardation and dentin dysplasia [NK] and Mucopolysaccharidosis III [AR])
  Interglobular dentin "ricketic dentin" (Vitamin D-resistant rickets [XD], Fanconi syndrome [heterozygous], Vitamin D-dependent rickets [AR], Hypophosphatasia [AR, AD], Pseudohypoparathyroidism [XD], Idiopathic juvenile hypoparathyroidism [AD, AR] and Cystinosis [AR]

Hereditary defects of cementum alone:
Defect in cementum primary in cemetogenesis
  Multiple gigantiform cementoma (AD?)
Defect in cementum associated with generalized disorder
  Cleidocranial dysplasia (AD)
  Paget's disease (AD?)
  Amyotrophic lateral sclerosis (AD, AR?)

Hereditary defects of both dentin and cementum:
Defects in dentin and cementum primary in formation (multiple forms)
Defects in dentin and cementum associated with generalized disorder
  Hypophosphatasia (AR, AD)

Table 4. Hereditary Defects of Tooth Form and Number

Defect primary in tooth germ:
Axial core defects:
   Dens invaginatus and dens in dente (AD, ?MF)
   Dens evaginatus (occlusal tubercle) (AD?)
Taurodontism:
   Isolated taurodontism (?MF)
   With amelogenesis imperfecta (AD)
   Trichodento-osseous syndrome (AD)
   X-aneuploidy syndromes
   Scanty hair, oligodontia and taurodontia (AR)
   Oro-facio-digital syndrome II (AR)
   Microcephalic dwarfism (?AR)
   Trisomy 21
Cynodont teeth with large pulp chambers
   Thistle-shaped pulp chambers (AD)
   Lobodontia (AD)
Hypodontia
   Peg or missing maxillary lateral incisors (AD)
   Missing bicuspids (multiple forms)
   Missing third molars (?MF)
   Missing mandibular lateral incisors (AD)
   Missing maxillary central incisors (AD)
   Missing 8 incisors (AD)
   Missing mandibular incisors and maxillary lateral incisors (XR)
   Multiple teeth (multiple forms)
Supernumerary teeth:
   Mesodens (MF)
   Maxillary lateral incisors (AD)
   Premolars (NK)
   Cuspids (?AD)
   Gemination (AD, MF)

Defect in teeth accompanied by generalized disorder:
Gardner syndrome (AD)
Otodental syndrome (AD)
Short, thin dilacerated roots and short stature (AD)
Hypodontia
   Hypodontia and nail dysgenesis (AD)
   Hypodontia, nail dysgenesis and hypotrichosis (AD)
   Hypodontia and hypotrichosis (AD)
   Hypohidrotic ectodermal dysplasia (XR, AR)
   Chondroectodermal dysplasia (AR)
   Deafness, ectodermal dysplasia, polydactylism and syndactylism (AD)
   Marshall syndrome (AD)
   Charlie M. syndrome (sporadic)
   Rieger syndrome (AD)
   Incontinentia pigmenti (XD)
   Oculomandibulodyscephaly (NK)
   Focal dermal hypoplasia (XD)
   Lipoid proteinosis (AR)

Table 4. (Continued)

    Hurler syndrome (AR)
    Hunter syndrome (XR)
    Trisomy 21
    Cranio-oculodental syndrome (AD)
    Craniofacial dysostosis (AD)
    Cleft palate, stapes fixation and oligodontia (AR)
    Pycnodysostosis (AR)
    Oro-facio-digital syndrome II (AD)
    Mandibulofacial dysostosis (AD)
    Progressive hemifacial atrophy (NK)
    Poikiloderma congenita (Rothmund-Thomson) (AR)
    Otopalatodigital syndrome (XR)
    Progeria (AR)
    Cherubism (AD)
    Cleft lip/palate and cleft lip/palate syndromes (multiple forms)
    Ehlers-Danlos syndrome (AD)
    Sturge-Weber syndrome (sporadic)
    Osteopetrosis (AR)
    Dyskeratosis congenita, infrequent malformed teeth (Zinsser-Engman-Cole) (XL and XR)
    Odontotrichomelic hypohidrotic dysplasia (NK)
    Hypohidrotic ectodermal dysplasia with cleft lip/palate, ocular, genital and digital anomalies (AD?)
    Coffin-Siris syndrome (sporadic)
    Palmoplantar keratosis, hypodontia, hypotrichosis and cysts of eyelids (AR)
    Premolar aplasia, hyperhidrosis and cavities premature (AD)
    Ectrodactyly, ectodermal dysplasia and cleft lip/palate (AD)
    Gorlin-Chaudry-Moss syndrome (Hypertrichosis—missing or bell shaped roots, craniofacial dysostosis, heart, genital and eye abnormalities [AR])
    Hypertrichosis lanuginosa (AD?)
    Melanoleukoderma (AR)
    Otodental syndrome (AD)
Hyperdontia:
    Natal teeth (oculomandibulodyscephaly [NK], pachyonychia congenita, Type II [AD], chondroectodermal dysplasia [AR], cyclopia [AR, ?MF] and osteogenesis imperfecta [AD])
    Permanent dentition (cleidocranial dysostosis [AD], Gardner syndrome [AD], achondroplasia [AD], cleft lip/palate syndromes [multiple forms] and tricho-rhino-phalangeal syndrome [Giedion] [AD])
    Macrodontia (congenital hemihypertrophy [AD], angiosteohypertrophy [NK], taurodontism, chromosomal and polygenic, otodental syndrome [AD], Rothmund-Thompson syndrome [anomalous cusps, AR] and one large maxillary central incisor and growth hormone responsive dwarfism [sporadic])

Table 5. Diseases of Pulp Chambers

Large	Small
Taurodont teeth:	*Dentin dysplasia:*
Racial	Not associated with syndrome
Chromosomal	Radicular (Type I)
Aneuploidy of X-chromosome	Coronal (Type II)
Autosomal (Down syndrom, T-22, 18q+)	Fibrous dysplasia of dentin
Associated with syndromes (trichodento-osseous syndrome and oro-facio-digital syndrome II [Mohr])	Associated with syndromes
	Dentino-osseous dysplasia
	Dysplasia of secondary dentin and growth retardation
Microcephalic dwarfism	Branchio-skeleto-genital syndrome
Scanty hair-oligodontia-taurodontia (SOT syndrome)	Mucopolysaccharidosis III (Sanfilippo syndrome)
	Calcinosis
	Ehlers-Danlos syndrome
Cynodont teeth:	*Dentinogenesis imperfecta:*
Metabolic alterations	Not associated with syndrome
Vitamin D-resistant rickets	Opalescent dentin (Mayflower type)
Vitamin D-dependent rickets	Opalescent dentin (Brandywine type)
Pseudohypoparathyroidism	Associated with syndromes (osteogenesis imperfecta)
Hypophosphatasia	
Alterations in development	Alteration of development (pulpal dysplasia, lobodontia, dens invaginatus and thin, short, dilacerated roots and growth retardation)
Thistle-shaped pulp chambers	
One large central incisor and growth retardation	

tions are affected except in the local hypoplastic form in which only primary teeth may be affected. The prevalence of AI, all forms included, is about 1 in 14,000 (10). Full crown restoration is the treatment of choice for the severe forms of AI; mild forms can be restored using acid-etch resins. For mode of inheritance, see Table 1.

Hypoplastic Types

1. *Pitted hypoplastic AI:* The enamel has small pinpoint-to-pinhead pits, often in rows and columns, particularly on the labiobuccal surfaces (8, 11).
2. *Smooth hypoplastic AI:* The enamel is thin, glossy, smooth and white, and teeth do not meet at contact points. Radiographically teeth may show delayed eruption and partial resorption of crowns prior to eruption (8, 12). Enamel contrasts normally with dentin on radiographs. Histologically enamel rods are broad and short.
3. *Rough hypoplastic AI:* The enamel is thin, rough, granular, very hard and yellow-brown. A thin layer of enamel, which contrasts normally with dentin can be seen radiographically. Histologically enamel rods are distorted and have a gnarled pattern (8, 13).
4. *Local hypoplastic AI:* The hypoplastic defect affects primarily the midthird of the buccal surface of the crown. The defect occurs as a horizontal row of pits or linear depressions. Only primary teeth may be affected. Histologically the

enamel has voids in the rod sheaths, disoriented rods and hypocalcified rods (8, 13-14).

5. *Rough hypoplastic AI* (enamel agenesis): Teeth have essentially no clinically or radiographically detectable enamel. The surface of the crown is rough and yellow-brown. Many permanent teeth fail to erupt, and radiographically teeth retained in the alveolus can be seen to be undergoing resorption. Histologically the only evidence of enamel is a thin layer of amorphous laminated calcified material resembling layers of agate (8, 15).

6. *Smooth hypoplastic AI:* Hemizygous males have thin smooth glossy brown enamel, which contrasts well with dentin on radiographs. Heterozygous females show random patterned vertical bands of normal thickness translucent enamel alternating with bands of hypoplastic enamel, a pattern compatible with lyonization of the X-chromosome (8, 9, 15-19).

Hypomaturation Types

1. *Hypomaturation AI:* Hemizygous males have enamel of normal thickness and the teeth meet proximally. The enamel is a mottled white-yellow-brown color, chips from the dentin and can be penetrated by a sharp dental explorer with firm pressure. Radiographically the enamel has the same radiodensity as dentin. Histologically the outer two-thirds of the enamel has large voids in the rod sheaths, which may contain remnants of esotropic gel. Heterozygous females have vertical bands of opaque white hypomature enamel alternating with bands of normal translucent enamel (8, 20, 21).

2. *Pigmented hypomaturation AI:* Male and female subjects are affected with equal severity. The enamel is of normal thickness, has a white-yellow-brown mottled color and chips from the dentin. Enamel has about the same radiodensity as dentin when observed on radiographs. Histologically voids are seen in both rods and rod sheaths (8, 9, 13, 15, 22)

3. *Snow-capped AI:* Snow-capped teeth have a superficial opaque white appearance affecting the incisal or occlusal one-fourth to one-third of the enamel. The dentition appears to have been dipped in white paint. The defect is not related to the temporal developmental pattern of teeth, but shows an anteroposterior pattern of severity among patients. Histologically the surface enamel in the opaque areas has abnormalities in crystal orientation and voids the enamel rod sheaths (4, 5, 9, 11).

Hypocalcified Types

1. *Hypocalcified AI:* With the exception of snow-capped teeth, this is the commonest form of AI (8, 10, 12). The enamel is of normal thickness on newly erupted teeth. It is yellow-brown to orange-brown, very friable and rapidly lost by attrition, although the enamel at the cervix of the crown is often better calcified. Large amounts of calculus accumulate rapidly on these teeth (13). Radiographically the enamel has a moth-eaten appearance and is less radiodense than dentin. Histologically the enamel matrix is uncalcified and immature and can be retained in decalcified sections (8, 9, 10, 12, 15).

Table 6. Ectodermal Structures Affected in Various Forms of Ectodermal Dysplasia*

Condition	Hair	Dentition	Nails	Sweat	Inheritance Pattern†
Hypohidrotic ectodermal dysplasia, XL	Hypotrichosis	Missing and conical teeth	Infrequently spoon-shaped	Decreased number of sweat glands	XLR
Hypohidrotic ectodermal dysplasia, AR	Hypotrichosis	Missing and conical teeth	Spoon-shaped	Decreased number of sweat glands	AR
Pachyonychia congenita, Type II (Jackson and Lawler)	Dry hair, occasionally alopecia	Natal teeth	Congenital to infantile onset of extremely thick subungual hyperkeratosis	Palmar and plantar hyperhidrosis	AD
Epidermolysis bullosa dystrophica; several genetic types	Hypotrichosis in some patients	General hypoplasia or pitting of enamel; severity related to genetic type; eruption retarded in some patients	Moderate-to-marked dystrophy, depending on genetic type	Palmer hyperhidrosis may be present in some cases	AR and AD (genetic heterogeneity)
Focal dermal hypoplasia	Sparse hair in focal areas of scalp and pubis	Small malformed teeth, hypoplastic enamel, notching of incisors, extra incisors	Dystrophic, thin nails, spoon-shaped or absent in about 50% of patients	Dyshidrosis, hypo- or hyperhidrosis, mostly palmoplantar	Possibly XLD, lethal in male
Dyskeratosis congenita	Loss of cilia secondary to chronic blepharitis and ectropion	Teeth infrequently malformed	Dystrophic with onset late childhood or at puberty; short atrophic nails	Palmoplantar hyperhidrosis with palmar keratosis	XL and AR forms (genetic heterogeneity)
Hypohidrotic ectodermal dysplasia with cleft lip, palate, ocular, genital, and digital anomalies	Stiff, sparse, "steel wool," sparse eyebrows and lashes, no axillary or pubic hair	Missing teeth; short, square incisors and canines; hypoplastic enamel	Small, dysplastic, with soft-tissue tufting	Reduced sweat on iontophoresis; no skin appendages on biopsy; fewer sweat glands; reduced response to pilocarpine and heat; hyperthermia	Possible AD

Ectodermal dysplasia cleft lip and palate, popliteal pterygia	Hypotrichosis	Small and conical teeth	Dysplastic	Hypohidrosis	Possibly AR
Xeroderma, talipes and enamel defect–XTE syndrome	Dry, slow growing, no lashes on lower lids, scanty follicles in body skin	Yellow enamel (probably hypomaturation of enamel)	Small malformed toenails	Reduced number of sweat glands	Probably AR, possibly homozygous AD
Odontotrichomelic hypohidrotic dysplasia	Hypotrichosis	Small, conical and missing teeth	Possibly hypoplastic		Possibly AR
Coffin-Siris syndrome	Hypotrichosis of scalp and hair; hypertrichosis of eyebrows and forehead	Small teeth	Absent to hypoplastic fifth fingernails and toenails; other nails hypoplastic	No‡	Not an inherited disorder
Rothmund-Thomson syndrome	Sparse to alopecia	Anomalous cusps and ridge pattern	Dystrophic in about one-fourth of patients	No	AR
Tooth and nail syndrome	Fine hair in few patients; sparse to absent eyebrows	Missing and conical teeth	Slow nail growth in children; small and frequently spoon-shaped; toenails most severely involved; adult fingernails may appear to be normal	No	AD
Trichodento-osseous syndrome	Tight curly hair	Taurodontism; hypoplastic enamel	Thick and split	No	AD
Incontinentia pigmenti	Alopecia (pseudopelade) in 20 to 25%	Missing, small, conical and impacted teeth	Rarely dystrophic	No	XLD, lethal in male
Palmoplantar keratosis, hypodontia, hypotrichosis and cysts of eyelids	Male type alopecia of scalp, little body hair	Missing teeth	Fragility	No	Possibly AR

Table 6. (Continued)

Condition	Hair	Dentition	Nails	Sweat	Inheritance Pattern†
Premolar aplasia, hyperhidrosis and canities prematura (Böök syndrome)	White hair	Missing premolars	No	Palmoplantar hyperhidrosis	AD
Ectrodactyly, ectodermal dysplasia and cleft lip/palate (EEC syndrome)	Sparse and fine; cilia absent	Small and conical teeth	No	Hypohidrosis (starch test negative)	AD
Pachyonychia congenita, Type I (Jadassohn-Lewandowski)	Dry hair; infrequent alopecia	No	Congenital to infantile onset of extremely thick subungual hyperkeratosis	Palmar and plantar hyperhidrosis	AD
Freire-Maia syndrome	Partial alopecia with follicular hyperkeratosis of scalp	No	Severe onychogryphosis	Hypohidrosis unresponsive to pilocarpine	Inheritance pattern if any is unknown
Hypoplastic enamel, onycholysis and hypohidrosis	No	Hypoplastic-hypocalcified enamel	Onycholysis	Hypofunction of sweat glands	AD
Monilethrix and anodontia	Monilethrix	Conical and missing teeth in few cases	No	No	Possibly AD
Gorlin-Chaudhry-Moss syndrome	Hypertrichosis	Missing; bell-shaped roots; reduced pulp chambers	No	No	AR
Oculomandibulodyscephaly (Hallermann-Streiff syndrome)	Hypotrichosis	Conical, missing, natal and supernumerary teeth; hypoplastic enamel	No	No	Not an inherited disorder
Trichorhinophalangeal syndrome	Thin and slow-growing hair	Supernumerary incisors	No	No	AR and AD forms (genetic heterogeneity)

Syndrome	Hair	Teeth	Nails	Sweating	Inheritance
Oculodento-osseous dysplasia	Dry, short hair in some cases	Hypoplastic enamel	No	No	AD
Hypertrichosis lanuginosa	Facial and body hypertrichosis	Missing teeth	No	No	Possibly AD
Orofaciodigital syndrome	Alopecia and dryness	Missing mandibular lateral incisors	No	No	AD lethal in males
Ungual type ectodermal dysplasia (hidrotic ED)	Alopecia, lanugo or fine, short, dry straight hair	No	Children—convex nails with onycholysis of outer third; adults—longitudinal striations, convex ends and thick horny subungual keratosis	No	AD
Palmoplantar hyperkeratosis and alopecia	Alopecia to hypotrichosis of scalp; alopecia of brows and lashes	No	Small dystrophic with onycholysis	No	AD
Curly hair-ankyloblepharon-nail dysplasia syndrome (CHANDS)	Curly	No	Small dysplastic	No	AR
Focal facial dermal dysplasia	Focal hairless pigmented temporal forehead and chin lesions; may have absent or double row of eyelashes	No	No	Absent sweat glands in focal lesions	AD
Basan syndrome	Sparse body hair, brows and lashes; scalp coarse—sheds 2d decade	? Reported rampant caries	Thick longitudinal ridges	Present but decreased	AD, single palmar flexion crease, thin alae and long philtrum
Ectodermal dysplasia and deafness	No	Missing and conical teeth	Small dystrophic with furrows	Sweat chloride, possibly elevated	AD

Table 6. (Continued)

Condition	Hair	Dentition	Nails	Sweat	Inheritance Pattern†
Chondroectodermal dysplasia (Ellis-van Creveld syndrome)	No	Missing, conical and natal teeth	Hypoplastic spoon-shaped and wrinkled nails	No	AR
Ectodermal dysplasia, deafness and ocular anomalies (Marshall syndrome)	No	Missing, small and conical teeth	No	Mild hypohidrosis	AD
Palmoplantar hyperkeratosis with reticular pigmentation	No	Teeth infrequently involved with yellow flecks in enamel	No	Palmoplantar hypohidrosis with keratosis	AD
Onycholysis	No	No	Onycholysis	No	AD
Onychodystrophy and deafness	No	No	Slow-growing, small dystrophic nails	No	AR
Otodental syndrome	No	Globodontia, large globe-shaped teeth, canines through molars affected	No	No	AD

*Modified from (6).
†AD, autosomal dominant; AR, autosomal recessive; XLD, X-linked dominant; XLR, X-linked recessive.
‡No, not affected.

2. *Hypocalcified AI:* Clinical and radiographic features are similar to those of the dominant type (8).

Defects of Dentin

Radicular dentin dysplasia (dentin dysplasia, Type I, rootless teeth) is an autosomal dominant condition affecting teeth in both dentitions (27, 28). The coronal dentin occlusal to the usual pulp chamber location is histologically normal. The defect is found primarily in the dentin below the cementoenamel junction. Tooth color is normal, but the teeth are frequently malaligned in the arch. Radiographically the pulp chambers are obliterated except for a small demilunar remnant of pulp chamber in the permanent teeth. Roots frequently are distorted, short or absent. Periapical radiolucent cysts are frequent. Histologically dentin forms in the developing dental papillae and is incorporated by the developing root dentin, which creates the appearance of a stream flowing around boulders (29).

Primary teeth affected with *coronal dentin dysplasia* (Type II) are discolored, whereas permanent teeth have normal color. Primary teeth are brown to blue-brown with a translucent opalescent sheen. Because the enamel abraids easily from the defective dentin, primary teeth wear rapidly. Radiographically pulp chambers, which are of normal size in newly erupted teeth, become obliterated with age. This is a feature that differentiates coronal dentin dysplasia from opalescent dentin, in which latter pulp chamber obliteration frequently occurs prior to eruption. Histologically the dentin has an amorphous ground substance practically devoid of tubules. Permanent teeth have flame-shaped pulp chambers, which contain denticles (26, 27). The disorder is an autosomal dominant one.

Opalescent dentin, an autosomal dominant trait, is also called maladie de Capdepont, dentinogenesis imperfecta without osteogenesis imperfecta and dentinogenesis imperfecta, Type II (26, 27). Two types may exist, the Mayflower type (9, 25) and the Brandywine type (dentinogenesis imperfecta, Type III) (25, 31), but clinicohistologic criteria alone are insufficient to distinguish between these types. Opalescent dentin occurs in about 1 in 8,000 persons in the United States. Most families in the United States trace ancestry to Norwegian, Norman-French, Norman-English or Norman-Irish families (9, 25). It is one of the most consistently autosomal dominant traits transmitted, with a low mutation rate and a high degree of penetrance. Teeth have a brown to bluish-brown color and a translucent opalescent sheen. Both dentitions are affected. Primary teeth wear rapidly from attrition.

Radiographically the crowns are bulbous, the roots short and thin and the pulp chambers obliterated. Obliteration of the pulp chamber begins before eruption of teeth, in contrast to coronal dentin dysplasia in which it occurs after eruption. Variations in the radiographic picture occur, however, and teeth with pulp chambers of normal size and shell teeth have been observed, particularly in the primary dentition. Histologic changes involve dentin that shows a large proportion of immature periodic acid–Schiff positive reaction matrix, rich in glycosaminoglycans, short distorted tubules and interglobular calcification.

TOOTH ABNORMALITIES WITH GENERALIZED DEFECTS

Defects of Enamel

Defects in enamel, which are but one feature of a generalized disease, are numerous. Detailed reviews are found elsewhere (8, 9, 15).

Amelo-onychohypohidrotic syndrome is an autosomal dominant trait, consisting of hypocalcified-hypoplastic enamel, onycholysis with subungual hyperkeratosis, seborrheic dermatitis of the scalp and hypofunction of the sweat glands with rough, dry skin (6). Normal numbers of sweat gland openings are found on dermal ridges, but the glands are hyporesponsive to sweating stimuli. Many permanent teeth fail to erupt and radiographically are seen within the alveolus undergoing resorption.

Oculodento-osseous dysplasia (33), an autosomal dominant disorder, is characterized by typical facies showing microphthalmia and microcornia with anomalies of the irides and thin nose with anteverted nostrils; hypoplasia of enamel; bony anomalies of the middle digits of the fifth finger, absence of the middle phalanx of the second through fifth toes, and syndactyly and camptodactyly of the fourth and fifth fingers; and thickened mandibular bone.

The most consistent features of *tuberous sclerosis* are blastomatous lesions of skin, mucosa and visceral organs, epilepsy; and mental retardation. Oral lesions consist of fibroepithelial papillomas and fibroangiomas of oral mucosa, particularly of the gingiva, and micropits of the enamel (34). These latter defects may result from small angiomatous lesions of the enamel organ.

Amelocerebrohypohidrotic syndrome (35) is probably an X-linked disorder. Children develop progressively severe seizures from age 1 to 4 years attended by progressive oligophrenia. Sweating is decreased because of minimal sweat glands, and the sweat contains large amounts of potassium. All teeth of both dentitions have thin hypoplastic enamel except for small islands of enamel at the cervix of the crown.

Epidermolysis bullosa includes at least two autosomal dominant (simplex and dystrophic) and two autosomal recessive (dystrophic and lethal) types. Defects of enamel occur not at all in the dominant simplex type, rarely in the dominant dystrophic type (36), frequently in the recessive dystrophic type (37) and probably always in the lethal form (37). In general, however, the severity of the enamel defect does not parallel that of the skin lesions. Thin hypoplastic enamel or honeycomb-patterned pitted enamel reflect the most frequent lesions. Histologic changes in the enamel (38) show atrophy and metaplasia of ameloblasts accompanying microvesicles in the enamel epithelium.

Mucopolysaccharidosis IV (Morquio syndrome) is the only type of mucopolysaccharidosis yet reported to have a defect of enamel (39). Type III (Sanfilippo syndrome) has a defect of dentin formation, but without changes in the enamel. In Morquio disease teeth of both dentitions have thin enamel of normal radiodensity. They do not meet at contact points, are gray in color and have pitted or dimpled enamel.

Defects of Dentin

Dentino-osseous dysplasia (40) is an autosomal dominant condition in which the teeth in all respects resemble those in radicular dentin dysplasia. In addition the patients have a generalized sclerosis of cortical bone; abnormalities in number and form of carpal bones; medial displacement of thumbs; and normal serum levels of calcium, phosphorus and alkaline phosphatase. Dental radiographic changes are typical of radicular dentin dysplasia.

Branchioskeletogenital syndrome (41), probably an autosomal recessive disorder, is characterized by brachycephalic skull, hypertelorism, nystagmus, ptosis of eyelids, strabismus, hypoplasia of the maxilla with relative mandibular prognathism, high arched palate, cleft palate or bifid uvula, pectus excavatum, vertebral anomalies, penoscrotal hypospadias, oligophrenia, convulsions in childhood, multiple impacted teeth, dentigerous cysts and dentin dysplasia. Radiographic features include multiple impacted and malpositioned teeth with obliteration of pulp chambers, surrounded by radiolucent areas with characteristics of dentigerous cysts. The histologic features of the teeth resemble those of radicular dentin dysplasia (4, 25).

Some forms of *generalized calcinosis* (42) and *Ehlers-Danlos syndrome* (43) have radiographic alterations in dentin that may resemble radicular dentin dysplasia, but are more variable. Some patients show obliteration of pulp chambers in permanent teeth after eruption, with multiple radiodense opacities in the pulp and root canal areas. Not all teeth in any one person show changes; canine, premolar and molar teeth frequently have radiographic alterations. The commonest pattern, however, is one large radiopaque mass at midpoint in the root, giving the impression of a tooth on top of a tooth. The histologic features are not described.

Dysplasia of secondary dentin and growth retardation is possibly an autosomal recessive trait. Birth weight of affected persons is in the lower 10% of normal. During childhood, weight and height continue to remain below the lower 10% for age. The developing and newly erupted teeth are normal radiographically, but gradually the pulp chambers and root canals become obliterated so that by age 10 to 12 years the radiographic appearance of the teeth is similar to those affected with opalescent dentin. Histologically the dentin in crowns and roots is normal. The secondary dentin, which fills in the pulp chamber and root canal, is dysplastic and contains globules of dentin with concentrated tufts of dental tubules.

Mucopolysaccharidosis III (Sanfilippo syndrome) is an autosomal recessive trait resulting from two distinct enzyme deficiencies, one of heparin sulfate sulfatase and the other of N-acetyl-α-D-glucosaminidase. Defects of teeth are described only for the former. It is not known if they are also in the second type. Radiographically the pulp chambers are obliterated. Histologically the dentin of the body of the tooth appears normal, and pulp and root canal areas are filled with a fibrous calcified material in which abundant collagen bundles can be identified. The dental tubules are short, distorted and scanty. The abnormal dentin shows intense periodic acid–Schiff reaction.

Osteogenesis imperfecta is an autosomal dominant disorder, characterized by fragile bones, flaccid ligaments, blue sclera, brown opalescent teeth, otosclerosis and cardiac anomalies. A recessive form is described, but represents a disorder clinically different from typical osteogenesis imperfecta. The frequency of brown opalescent teeth by clinical type and inheritance pattern among 122 propositi of which 100 were sporadic and 22 familial are shown in Table 7 (44).

Table 7. Percentage of Patients with Affected Teeth by Inheritance Pattern and Clinical Type of Osteogenesis Imperfecta*

	Type		
Inheritance	Congenita	Tarda	Total
Dominant	75	12	21
Sporadic	54	43	50
Total	59	19	

*100 sporadic propositi, 22 familial, 102 affected relatives.

Pulpal dysplasia (45) may be an autosomal recessive trait. The patients have short stature and mild mental retardation. Radiographically teeth in both dentitions show large flame-shaped pulp chambers and large root canals, with delayed closure of root apices. The large chambers are filled with multiple pulp stones, which histologically begin as dystrophic calcifications around blood vessels and develop into true denticles in their peripheral aspects. Coronal and radicular dentin is histologically normal.

Vitamin D-resistant rickets (46) is inherited as an X-linked trait. Hemizygous males most frequently have both hypophosphatemia and rickets, whereas heterozygous females most frequently have only hypophosphatemia. Affected males frequently have dental symptoms as the first sign of the disorder (47). Tooth eruption may be delayed. The typical patient has abscesses, fistulas, granulomas or cysts developing periapically on teeth the gross appearance of which is normal. Radiographically the teeth have large pulps with pulp horns extending to the dentinoenamel junction. Histologically teeth show interglobular dentin and delayed closure of root apices (47). Enamel is rarely hypoplastic. In vitamin D-dependent rickets—an autosomal recessive disorder affecting both male and female patients—hypoplasia of enamel accompanies the defect of dentin (5, 25).

Hypophosphatasia (48) may be either an autosomal dominant or an autosomal recessive trait. In both forms the serum levels of alkaline phosphatase are subnormal, and excessive levels of phosphorylethanolamine are found in plasma and urine. Many patients are first detected by spontaneous premature exfoliation of primary teeth or exfoliation following minor trauma due to a lack of cementogenesis of the roots of teeth; this deficiency fails to provide an attachment for periodontal fibers, which normally maintain the tooth in the socket (49). Not all patients with hypophosphatasia have radiographically detectable changes, however. Some of them show large pulp chambers and wide periodontal ligaments (50). Microscopic alterations include absence of cementum on roots, large collagen bundles in dentin, interglobular calcification of dentin and delayed closure of root apices (15).

Defects of Enamel and Dentin

Trichodento-osseous syndrome (51) is an autosomal dominant trait showing hypoplastic-hypomature enamel, taurodont teeth, tightly curled hair in infancy that may persist into adulthood and, in some patients, laminated split nails and sclerosteosis. The enamel of both dentitions is pitted, thin and soft with a white-yellow-brown mottling; pulp horns reach the dentinoenamel junction in a way similar to that seen in vitamin D-resistant rickets, providing the mechanism for microexposure of the dental pulp. The first complaints of more than one-half the patients relate to the sequellae from exposure of dental pulps (52). Dental radiographic features are thin enamel, which contrasts poorly with dentin; teeth with large pulp chambers and pulp horns reaching the dentinoenamel junction; taurodont molars; and periapical radiolucencies characteristic of abscesses, granulomas or cysts. Histologic features of the teeth consist of thin enamel with short rods and included cell remnants, large pulp chambers in anterior and posterior teeth with a taurodontic configuration, high pulp horns reaching the dentinoenamel junction and interglobular calcification of dentin (8).

Pseudohypoparathyroidism (Albright's osteodystrophy) occurs in two forms, Types 1 and II. The former is probably an X-linked trait showing unresponsiveness of target cells in bone and kidney to parathyroid hormone (53). In Type II (54) there is a rise in urinary cyclic adenosine monophosphate (cAMP) following administration of parathyroid hormone, but a phosphate diuresis does not result. The site of the chemical lesion possibly involves an abnormality in cAMP-dependent phosphodiesterase, at a step beyond cAMP. The inheritance of Type II is not clear.

In general patients with pseudohypoparathyroidism have short necks; round facies; thick, stocky builds; short digits (most frequently the fourth metacarpal); extraosseous calcifications (frequently in the basal ganglia and falx cerebri [55]); thin pitted enamel; short wedge-shaped teeth; delayed eruption of teeth (56); and failure to respond to parathyroid hormone. Convulsions from tetany may occur in childhood from low serum calcium levels. Teeth developing subsequent to the onset of refractiveness to parathyroid hormone are defective. In heterozygous females the most frequently affected teeth are the permanent premolars (second and third molars, which are delayed in eruption time, show a gross random pitting of enamel and delayed closure of root apices). In male subjects, hemizygous for the gene, there is thin enamel with small pits, usually involving more of the permanent teeth.

Radiographic changes in dentition show gross, random pitting of enamel, delayed eruption of teeth, short wedge-shaped teeth and large pulp chambers and root canals (57). Histologic features of the teeth are pitted enamel, which is normally calcified, open apices of root canals and interglobular dentin (57).

Vitamin D-dependent rickets (58) is inherited as an autosomal recessive trait in which hypoplasia of enamel affects the incisal and occlusal fractions of the crowns of teeth (5, 8, 25). The children are normal at birth, develop hypocalcemic, hypophosphatemic rickets in the second half of the first year of life and have generalized renal tubular dysfunction. The radiographic and histologic features of the affected teeth are similar to those in vitamin D-resistant rickets with the added features of hypoplasia of incisal and occlusal enamel (5, 8, 25). Patients receiving renal transplants and those developing de Toni-Debre-Fanconi syndrome during the odontogenesis may show similarly affected teeth (13).

Idiopathic hypoparathyroidism is genetically heterogeneous. One autosomal recessive form includes hypoparathyroidism, Addison's disease, mucocutaneous candidiasis and keratoconjunctivitis. The fungal infection, a prominent feature and usually the first sign of the disease, appears during the first six years of life. Some patients have chalky, pitted enamel—frequently with transverse grooves, delayed eruption of teeth, wedge-shaped teeth with short roots and delayed closure of root apices. Histologically the dentin shows extensive areas of interglobular calcification, particularly in the root area.

Developmental Disorders and Tooth Morphology

Among the disorders in which alterations of teeth are prominent are those generally termed ectodermal dysplasias. Freire-Maia (59) has classified this nosologic group according to the forms that demonstrate at least two signs related to 1) trichodysplasia, 2) abnormal dentition, 3) onychodysplasia and 4) dyshidrosis. Table 6 lists various disorders by the Freire-Maia (59) classification.

The dental radiograph, one of the most useful tools for detecting alterations in teeth, can help the practioner to diagnose the patient with a generalized disorder. Among the most efficient criteria is the size of the pulp chamber. Those disorders with alterations in pulp chamber morphology are classified in Table 5 on the basis of pulp chambers being larger or smaller than normal.

Enlarged pulp chambers may occur in teeth of two morphologic classes, namely, taurodontic and cynodontic. *Taurodontism* (bull-shaped teeth) refers particularly to molar teeth that have pulp chambers enlarged at the expense of the body of the tooth, in which the bifurcation or trifurcation of roots are apically placed and the external outline of the tooth has a block configuration (59). *Cynodontism* (dog-like teeth) is the form usually encountered in modern man. The pulp chambers are set relatively high in the tooth, and the roots are relatively long.

Taurodontism is most commonly seen as a racial trait (5), but is also frequent in certain syndromes. Of particular interest is the finding that most patients with poly-X chromosomal syndromes have taurodontic molars (5). Patients with the classic Klinefelter syndrome (XXY), e.g., as well as those with karyotypes XXXY, XXXXY and various mosaics XXY/XXXY; XY/XXYY have taurodontic molars (5, 24, 60, 61). The condition is observed in patients with a female habitus with XXX and XXXX karyotypes (5). Patients with taurodontic teeth, particularly with other signs of X-chromosomal polyploidy should be karyotyped.

REFERENCES

1. Gorlin RJ, Pindborg JJ, Cohen MM Jr: Syndromes of the Head and Neck, 2d ed. New York, McGraw-Hill, 1976
2. Stewart RE, Prescott GH (eds): Oral Facial Genetics. St. Louis, Mosby, 1976
3. Bergsma D (ed): Birth Defects Compendium, 2 ed. White Plains, New York, The National Foundation, 1979
4. Witkop CJ Jr: Genetics. *Schweiz Monatschr Zahnkund* 82:917, 1972
5. Witkop CJ Jr: Clinical aspects of dental anomalies. *Int Dent J* 26:378, 1976
6. Witkop CJ Jr, Brearley LJ, Gentry WC Jr: Hypoplastic enamel, onycholysis, and hypohidrosis inherited as an autosomal dominant trait. A review of ectodermal dysplasia syndromes. *Oral Surg* 39:71, 1975

7. Freire-Maia N: Ectodermal dysplasias. *Hum Hered* 21:309, 1971
8. Witkop CJ Jr, Sauk JJ Jr: Heritable defects of enamel, in Stewart RE Prescott GH (eds): Oral Facial Genetics. St. Louis, Mosby, 1976, p. 151.
9. Witkop CJ Jr, Rao SJ: Inherited defects in tooth structure. *Birth Defects* 7:153, 1971
10. Witkop CJ Jr: Hereditary defects in enamel and dentin. *Acta Genet Stat Med* 7:236, 1957
11. Hals E: Dentin and enamel anomalies. Histologic observation, in Witkop CJ Jr (ed): Genetics and Dental Health. New York, McGraw-Hill, 1962, p. 246
12. Weinmann JP, Svoboda JF, Woods RW: Hereditary disturbances of enamel formation and calcification. *JADA* 32:397, 1945
13. Witkop CJ Jr: Genetic disease of the oral cavity, in Tiecke RW (ed): Oral Pathology. New York, McGraw-Hill, 1965, p. 786
14. Sauk JJ Jr, Vickers RA, Copeland JS, et al: The surface of genetically determined hypoplastic enamel in human teeth. *Oral Surg* 34:60, 1972
15. Witkop CJ Jr, Sauk JJ Jr: Dental and Oral Manifestations of Hereditary Disease. Washington, DC, American Academy of Oral Pathology, 1971
16. Rushton MA: Hereditary enamel defects. *Proc R Soc Med* 57:53, 1964
17. Schulze C: Beitrag zur Frage der angeborenen Schmelzhypoplasie. *Dtsch Zahn-Mund-Kieferhkd* 16:108, 1952
18. Schulze C: Erbbedingte Strukturanomalien menschlicher Zähne. *Acta Genet Stat Med* 7:231, 1957
19. Fishman SL, Fischman IC: Hypoplastic amelogenesis imperfecta: Report of a case. *JADA* 75:929, 1967
20. Witkop CJ Jr: Partial expression of sex-linked recessive amelogenesis imperfecta in females compatible with the Lyon hypothesis. *Oral Surg* 23:174, 1967
21. Sauk JJ Jr, Lyon HW, Witkop CJ Jr: Electron optic microanalysis of two gene products in enamel of females heterozygous for X-linked hypomaturation amelogenesis imperfecta. *Am J Hum Genet* 24:267, 1972
22. Witkop CJ Jr, Kuhlmann W, Sauk JJ Jr: Autosomal recessive pigmented hypomaturation amelogenesis imperfecta. *Oral Surg* 36:367, 1973
23. Giansanti JS: A kindred showing hypocalcified amelogenesis imperfecta: Report of case. *JADA* 86:675, 1973
24. Witkop CJ Jr: Manifestations of genetic diseases in the human pulp. *Oral Surg* 32:278, 1971
25. Witkop CJ Jr: Hereditary defects of dentin. *Dent Clin North Am* 19:25, 1975
26. Bixler D: Heritable disorders affecting dentin, in Stewart RE, Prescott GH (eds): Oral Facial Genetics. St. Louis, Mosby, 1976, p. 227
27. Shields ED, Bixler D, El-Kafrawy AM: Heritable dentine defects: Dentine dysplasia type II. *Arch Oral Biol* 18:543, 1973
28. Ballschmiede (1920): Dissertation, Berlin. Quoted in Herbst EH, Apffelstaedt M: Malformations of the Jaws and Teeth. Fairlawn, N.J., Oxford University Press, 1930
29. Sauk JJ Jr, Lyon HW, Trowbridge H, et al: Electron optic analysis and explanation for the etiology of dentinal dysplasia. *Oral Surg* 33:763, 1972
30. Capdepont C: Dystrophie dentaire non encore decrit a type hereditare et familial. *Rev Stomat* (Paris) 12:550; 13:15, 1905–1906
31. Hodge HC, Finn SB: Hereditary opalescent dentin: A dominant hereditary tooth anomaly in man. *J Hered* 29:359, 1938
32. Hursey RJ, Witkop CJ Jr, Miklashek D, et al: Dentinogenesis imperfecta in a racial isolate with multiple hereditary defects. *Oral Surg* 9:641, 1956
33. Gorlin RJ, Meskin LH, Geme JW: Oculo-dento-digital dysplasia. *J Pediatr* 63:69, 1963
34. Hoff M, van Grunsven MF, Jongebloed WL, et al: Enamel defects associated with tuberous sclerosis: A clinical and scanning-electron-miscroscope study. *Oral Surg* 40:261, 1975
35. Köhlschütter A, Chappuis D, Meier C, et al: Familial epilepsy and yellow teeth—a disease of the CNS associated with enamel hypoplasia. *Helv Paediatr Acta* 29:283, 1974

36. Winter GB, Brook AH: Enamel hypoplasia and anomalies of the enamel. *Dent Clin North Am* 19:3, 1975
37. Rushton MA: Teeth, in Sorsby A (ed): Clinical Genetics. St. Louis, Mosby, 1953, p. 382
38. Gardner DG, Hudson CD: The disturbance in odontogenesis in epidermolysis bullosa hereditaria lethalis. *Oral Surg* 40:483, 1975
39. Levin LS, Jorgenson RJ, Salinas CF: Oral findings in the Morquio syndrome (Mucopolysaccharidosis IV). *Oral Surg* 39:390, 1975
40. Morris ME, Augsburger RH: Dentine dysplasia with sclerotic bone and skeletal anomalies inherited as an autosomal dominant trait, a new syndrome. *Oral Surg* 43:267, 1977
41. Elsahy NI, Waters RW: The branchio-skeleto-genital syndrome: A new hereditary syndrome. *Plast Reconstr Surg* 48:542, 1971
42. Hunter IP, Macdonald DG, Ferguson MM: Developmental abnormalities of the dentine and pulp associated with tumoral calcinosis. *Br Dent J* 133:446, 1973
43. Barabas GM: The Ehlers-Danlos syndrome. *Br Dent J* 126:509, 1969
44. Witkop CJ Jr: Original observations by author, 1977
45. Rao S, Witkop CJ Jr, Yamane GM: Pulpal dysplasia. *Oral Surg* 30:682, 1970
46. Winters RW, Graham JB, Williams TF, et al: A genetic study of familial hypophosphatemia and vitamin D-resistant rickets, with review of the literature. *Medicine* 37:142, 1958
47. Archard HO, Witkop CJ Jr: Hereditary hypophosphatemia (vitamin D-resistant rickets) presenting primary dental manifestations. *Oral Surg* 22:184, 1966
48. Rathbun JC: Hypophosphatasia, a new developmental anomaly. *Am J Dis Childh* 75:822, 1948
49. Bruckner RJ, Rickles NH, Porter DR: Hypophosphatasia with premature shedding of teeth and aplasia of cementum. *Oral Surg* 15:1351, 1962
50. Baer, PN, Brown NC, Hamner JE III: Hypophosphatasia. Report of two cases with dental findings. *Periodontics* 2:209, 1964
51. Robinson GC, Miller JR, Worth HM: Hereditary enamel hypoplasia: Its association with characteristic hair structure. *Pediatrics* 37:498, 1966
52. Lichtenstein J, Warson R, Jorgenson, R, et al: The tricho-dento-osseous (TDO) syndrome. *Am J Hum Genet* 24:569, 1972
53. Aurbach GD, Patts JT Jr, Chase LR, et al: Polypeptide hormones and calcium metabolism. *Ann Intern Med* 70:1243, 1969
54. Rodriguez HV Jr, Klahr S, Slatopolsky E: Pseudohypoparathyroidism type II: Restoration of normal renal responsiveness to parathyroid hormone by calcium administration. *J Clin Endocrinol* 39:693, 1974
55. Albright F, Burnett CH, Smith PH, et al: Pseudohypoparathyroidism—an example of Seabright-Bantam syndrome. *Endocrinology* 30:922, 1942
56. Witkop CJ Jr: Inborn errors of metabolism with particular reference to pseudohypoparathyroidism. *J Dent Res* 45:568, 1966
57. Croft LK, Witkop CJ Jr, Glas JE: Pseudohypoparathyroidism. *Oral Surg* 20:758, 1965
58. Arnaud C, Maijer R, Reade T, et al: Vitamin D dependency: An inherited postnatal syndrome with secondary hyperparathyroidism. *Pediatrics* 46:871, 1970
59. Hamner JE III, Witkop CJ Jr, Metro PS: Taurodontism: Report of a case. *Oral Surg* 18:409, 1964
60. Mednick GA: Two case reports: Taurodontism and taurodontism in Kleinfelter's syndrome. *J Mich Dent Assoc* 55:212, 1973
61. Keeler C: Taurodont molars and shovel incisors in Kleinfelter's syndrome. *J Hered* 64:234, 1973

Part 4

GENETIC COUNSELING

21
The Technique of Genetic Counseling

Robert F. Murray, Jr.

Without equivocation, there is no single perfect technique of genetic counseling, mainly because the currently accepted definition and goals of genetic counseling are so comprehensive that a single approach can hardly encompass the varied needs of people who seek such help. There will always be troublesome situations in which no technique is entirely satisfactory; hopefully, however, with time the patient, persistent and compassionate counselor will help persons or families solve their genetic problems.

DEFINITION AND GOALS

The emphasis in genetic counseling must be on communication, communication about the disease in question, the genetic mechanisms that operate and the family planning options that are available to couples at risk. A critical and often underemphasized element of the process, however, is the necessity of helping the family identify its personal goals and values so that it can make workable decisions. Continuing support for the family at risk must be provided while the members work through their socioeconomic problems, emotionally adjust to the disorder and eventually make decisions. Information, ideas, concepts and options are for ethical reasons best presented in a nondirective fashion, i.e., they must be given in a way so balanced as not to influence the person being counseled in a particular decision. Obviously counselors can never be completely neutral; each counselor must nonetheless strive to achieve this ideal.

Even though there are common goals in all genetic counseling the method of achieving them differs according to 1) the type of disorder, including its psychological and emotional burden; 2) the age of onset of symptoms and eventual death; 3) the risk of recurrence; and 4) the value system and socioeconomic status of the person being counseled ("counselee").

An essential characteristic of the effective genetic counselor is flexibility, i.e.,

This work was supported in part by MCH Project Grant No. 000414-15-0 and NIH Grant HL 15160.

the ability to tailor the approach to counseling a family so that the four factors just cited will be appropriately acknowledged.

INFLUENCES ON COUNSELING

Genetic counseling is a multifaceted process, the success of which hinges on 1) the motivation, emotional state and educational background of the counselee; 2) the background and training of the counselor; and 3) the setting of the counseling session.

The motivation and emotional state of the counselee are probably the most important of the factors listed because the effectiveness of the communication depends on them. A counselee who is interested neither in learning about the disease nor in hearing about the risk of recurrence will hardly be receptive to information of any kind. A counselee who is hostile, seriously depressed or anxious will likely receive information inaccurately and incompletely. In large-scale screening programs for the hemoglobinopathies, e.g., the motivation of at-risk patients is circumvented because counselors personally deliver the test results and advise the families. The problem with this "nonvoluntary counseling" is that most persons (more than 70% in our limited experience) feel that they are not carriers of the sickle cell or other hemoglobinopathy gene. When they receive the news that they carry an atypical or abnormal gene, they generally become significantly anxious and sometimes hostile to the counselor, thereby reducing the effectiveness of the counseling.

In a voluntary program our own studies in hemoglobinopathy counseling show that a healthier response is elicited when a person already recognizes that he or she may be a gene carrier. This same study also demonstrated a positive correlation between educational level and test scores in sickle cell trait counseling (1). This information may not be directly extrapolated to counseling in all circumstances, but it does indicate that at least for *this* condition the educational level of the counselee is an important factor in information acquisition. Most programs tend toward uniformity in the educational level of the clients and their families. Members of the latter are usually middle-class persons with at least a high-school education (2). In Tay-Sachs programs the study population is extremely well educated, the average person having a college or postgraduate education. However, at least one study suggests that the information that even these clients acquired after one counseling session tended to be meager (3).

Some of the variability in the educational and psychological outcome of genetic counseling is almost certainly the result of differences in the counselor's background and training. The traditional role and training of the physician, e.g., may be responsible for the tendency of the medical doctor (MD)-genetic counselor to be more directive than the counselor with a Ph.D. degree (4, 5). On the other hand the counselor with a strong background in behavioral sciences, e.g., the psychologist or social worker, is much likelier to focus on the emotional aspects of counseling. Since facts about genetic disease, hereditary mechanisms and risks of recurrence must be communicated to laymen, the counselor *must* be an effective teacher. Some experienced counselors think that the psychotherapeutic aspect of counseling is much more important than the communication of facts about inherited disease (6, 7). Kallman, who held this view, called counseling "short-term psychotherapy."

The setting in which a counseling session occurs may influence the quality and quantity of communication. The number of people in the counseling session will also influence what happens. In many genetics clinics, counseling is done with several people present, including students and other physicians. The advantage gained by bringing collective knowledge and judgment to bear may be lost when the counselee is ill at ease. The person in charge should question the patient about his feelings privately before the session to minimize embarrassment.

Ideally then effective counseling consists of a combination of effective communication of facts in a setting wherein close attention is paid to the counselee's emotional needs. In some cases, e.g., if the counselee is being counseled only because he or she carries a gene-determining recessive disease (sickle cell anemia or Tay-Sachs Disease), the major counseling emphasis will be on education, especially reproductive information, whereas the only psychological need may be to ease the attendant anxiety when a person learns that he or she carries a mutant gene. On the other hand, some genetic conditions elicit an almost overwhelming psychological impact, requiring intensive or extensive psychotherapeutic interaction, or both, before the counselee or the family can effectively receive the information about the disease, its mode of inheritance or its risk of recurrence. Both high-risk disorders like Duchenne type muscular dystrophy and Huntington's chorea and low-risk congenital malformations like anencephaly and severe cleft lip and palate can create the same kind of psychological stress. The counselor cannot assume that the reactions of the client and his family in any particular case will be appropriate to his (the counselor's) assessment of the genetic burden.

PHASES OF COUNSELING

Four readily discernible—but not mutually exclusive—phases in counseling are characterization, education, evaluation and followup. Each phase is comprehensive and does not necessarily proceed in the order listed. Education, e.g., may have begun before the client arrives for counseling. Evaluation of the educational process and of the counselee's emotional state occurs in all phases of counseling. Other chapters in this source book consider in depth the diagnostic aspects of counseling. The counselor must establish an accurate diagnosis of the disorder and make some calculations about the risk of recurrence. In some cases during the diagnostic workup the counselor may have determined that the condition does not have a significant genetic contribution.

Characterization

To exchange ideas effectively, it helps to know some of the characteristics of the person(s) with whom one is talking. The same observation holds for the counselor and counselee. Some of the important characteristics in counseling are emotional status (e.g., anxiety, hostility or depression); educational level; what is expected from genetic counseling; socioeconomic status; religious beliefs or ethical value systems, or both; cultural values (e.g., family structure and the value of children); and information and attitudes already held about the disorder.

Assessment of the counselee's *emotional status* is listed first because it so strongly influences and shapes the counseling. Some informal evaluation of the client's mental state should be recorded (privately) at least at the beginning and end of each session to promote the counselor's awareness of subtle indicators of emotional distress, rendering counseling more effective. Counselees may not feel comfortable with the counselor in the first session and sublimate feelings of hostility or depression.

Evidence already presented suggests that *educational level* directly correlates with the amount of information retained. Logically, the higher the clients' educational level, the better equipped they should be to assimilate complicated and unfamiliar biologic concepts. This is generally true except for the client's ability to understand and apply the statistical principles of odds or risk of recurrence. Few people, even well-educated ones, are familiar with or understand probability theory. The counselor often must use special techniques to communicate these concepts, examples of which are presented later in this chapter.

Counselees have different expectations from genetic counseling. If counseling is to meet their needs, the counselor must have a firm idea of what is expected from the process. The counselor's best approach is probably to begin the session by asking directly what information is wanted. Often clients are not interested in knowing about the disease or the genetic mechanisms thereof; they want simply to know what will happen to them or whose fault it is that their child has such a grave disease.

Knowledge of the family's *socioeconomic status* enables the counselor to help the parents of a handicapped child plan for his special needs. The counselor will also better appreciate the economic burden imposed on the family by the child and can take specific steps to find public assistance when it is needed.

Religious or ethical value systems, or both, can be critical elements in a family's decision to follow a course of action or to reject an option that might otherwise seem positive. This may be especially true where prenatal diagnosis and the termination of pregnancy is concerned. If the counselor determines at the outset that a family has strong religious beliefs that forbid abortion, even when an abnormal fetus is identified, he or she will not be surprised when this option—attractive to many people—is rejected. Religious beliefs may also strongly bias a family's reaction to the birth of a deformed child. The skilled counselor may elicit feelings of extreme guilt based on the ideas that the mother or father is being punished for past sins.

Cultural attitudes and practices should be considered as well because, like religious values, they may condition the client's responses to reproductive options or the interpretation of medical or genetic information. Considerations like these are vital when one attempts to modify traditional reproductive practices. Studies of the attitudes of black and Hispanic males, e.g., reveal an almost universal rejection of artificial insemination as an alternative reproductive option (7). In these same cultural communities children are highly valued so that abstaining from childbearing can hardly be considered a viable reproductive option for many at-risk couples. In the Hispanic community in particular, sickness is often considered to be a normal part of existence and the threat of a chronically ill child may not seem so serious (8). There is also the rare person whose attitude is that one must give birth to an abnormal child to be considered normal!

Most counselees already have some preconceived ideas about the disorder in question, even if they have never seen an affected person. These notions are often erroneous and if not recognized in advance, they will almost certainly not be corrected.

Education

Providing information to the counselee in clearly understandable language with appropriate illustrations lies at the heart of counseling. If clients do not clearly understand the facts about the disease, including its manifestations, prognosis, genetic mechanisms and risk of recurrence, they can hardly be expected to make an intelligent decision. The educational principles are well known, but difficult to apply. Facts must be presented in easily digestible form and in language appropriate to the counselee. One should have clearly defined learning objectives, i.e., the specific ideas or concepts to be communicated to the counselee should be delineated and laid out in advance. The counselor can keep a checklist to insure that the essential concepts are presented and discussed. He must take special pains not to lecture the counselee; rather the information should be presented in a style as conversational as possible.

One way of presenting information is by using a question-and-answer format. The counselor may ask, "What have you heard about this disease before?," and the counselee may respond with incorrect or incomplete information. The counselor may then proceed from this base to supplement or correct the client's knowledge. If the client has no knowledge of the disorder, the counselor may then ask him or her about a familiar aspect of the condition and proceed from that point. The depth of inquiry and the extent of discussion about the condition will vary according to the background information, level of education and interest of the client. This kind of question-and-answer format is helpful because it stimulates the counselee to begin asking his or her own questions and leads to optimal interaction between counselor and counselee.

In some screening programs counselees are exposed to a fair amount of information prior to testing and mainly need to have information reinforced or expanded. Families seeking counseling because they have an affected child also likely have some knowledge of the condition and almost always have specific questions about prognosis and treatment.

Perhaps the most difficult aspect of counseling is attempting to translate complex biologic and medical principles into language readily understandable by persons with little grasp of the most elementary functions of the human body. This is a challenging task that *can* be accomplished, but only with patience, skill and experience.

Evaluation

To insure that counseling achieves its goals, both formal and informal evaluative methods may be needed. Informal methods rely on a subjective assessment of how the counselor thinks the client is receiving the information by clinical clues to judge the client's emotional state. This approach may be adequate, depending largely on the counselor's experience. It does not, however, permit comparisons between counselors or between counseling techniques. For these and other reasons many workers prefer a more formal method of evaluating counseling. The simplest and most easily administered evaluation is a written short-answer, multiple-choice test of 10-20 questions designed to determine if the counselee has received and can correctly recall information.

Exhibit 1 lists some of the questions included on a short-answer quiz to evaluate communication and retention of information presented in a counseling session for sickle cell trait. The information tested for by each question is also indicated. A more precise estimate of what information is acquired can be obtained if the quiz is given just before and after the session.

Under some circumstances a formal oral evaluation can take the place of a written evaluation. A little thought reveals definite advantages and disadvantages to each approach to evaluation, and one should determine what approach or combination thereof best suits the client's needs, the counseling program and the disease under discussion.

Followup

The counselor must be vitally interested in what the counselees do after counseling. Most follow-up reports have focused almost exclusively on the reproductive outcome and generally showed that couples with a moderate-to-high risk of recurrence (25-100%) of a genetic disorder have fewer offspring than couples with a low genetic risk (10% and below) (9, 10). They also showed that couples at risk to have offspring with genetically determined disorders with a perceived high burden had fewer offspring than couples who are at risk to have children with a perceived low burden (11). These findings are defined in terms of the expected cost of the disease not only in money but also in physical and emotional pain, labor, death and anxiety. The burden of color blindness or postaxial polydactyly is trivial, whereas that related to trisomy 13, anencephaly or Tay-Sachs disease is very heavy. The duration of the burden is also vital, and of course the burden of a given disease may be different to different families. The scope of genetic counseling must go beyond reproductive decision-making. It should also involve helping a family not only adapt to a child with a serious defect or handicap but also to make the best adjustment to the situation.

We have developed a questionnaire for followup of couples at risk to have children with sickle cell disease after they have received comprehensive genetic counseling. Examples of the questions asked of these couples are listed in Exhibit 2. In keeping with the more comprehensive goals of sickle cell counseling, this kind of followup is as concerned with the couple's familial interactions and personal adjustment to the disease as with how much information they still recall from the session. Based on preliminary information from an ongoing evaluation, seemingly the basis of the decision that couples make about reproduction is complex; it involves emotional, cultural, religious and economic factors. All the couples in this evaluation who decided to have children so far, recalled and understood the risks of recurrence in each pregnancy, a finding consistent with

Exhibit 1. Sample Questions From a Written, Multiple Choice Quiz Used to Evaluate Communication and Retention of Information Presented During Counseling for Sickle Cell Trait

The sickle cell gene affects which of the following?
a. the blood type
b. the hemoglobin inside the red blood cell
c. the shape of the blood vessels (arteries or veins)
d. the blood pressure
e. the white blood cells

Sickle cell trait is a mild form of sickle cell anemia?
a. true
b. false
c. don't know

What does malaria have to do with sickle cell trait?
a. nothing at all
b. people with sickle cell trait are more likely to die from malaria
c. sickle cell trait protects people from dying from malaria
d. don't know

Hemoglobin is:
a. a protein that makes the blood red
b. a protein that makes the skin yellow
c. a protein that makes the skin brown
d. the major component of fingernails

You might have a child with sickle cell trait if:
a. you eat certain foods
b. you come in contact with someone with sickle cell trait
c. you or your husband/wife has the trait
d. someone works roots on you when you are pregnant

If both parents have hemoglobin-S trait (AS), what is the chance that their *first* child would have hemoglobin-S disease (SS)?
a. 25% (1 out of 4)
b. 50% (2 out of 4)
c. 100% (4 out of 4)
d. 0% (0 out of 4)

If both parents have hemoglobin-S trait (AS), what is the chance that their *third* child would have hemoglobin-S disease (SS)?
a. 25% (1 out of 4)
b. 50% (2 out of 4)
c. 100% (4 out of 4)
d. 0% (0 out of 4)

If one parent has hemoglobin-S (AS) and the other does not have either hemoglobin-S trait or disease (AS or SS), what is the chance that the couple would have a child with hemoglobin-S disease (SS)?
a. 25% (1 out of 4)
b. 50% (2 out of 4)
c. 100% (4 out of 4)
d. 0% (0 out of 4)

observations of other authors (12). The counselor who follows the comprehensive definition of counseling should see these couples at intervals appropriate to their needs.

THE COUNSELING PROCESS

Intake

Counseling in many settings begins with an intake procedure that can be performed by any one of several health professional workers, often not the MD-or PhD-genetic counselor. Usually a social worker, family worker, psychologist or genetic assistant conducts the intake interview. In this procedure how and why the client arrived at the clinic is determined, and family background and demographic data are collected and recorded. The intake interviewer makes an initial judgment of familial attitudes, conflicts and expectations. It is especially helpful if he or she can detect signs of a serious conflict between husband and wife, especially if it is related to their adjustment to a child with a birth defect.

It is helpful to have available a brochure containing material explaining to the clients both what a genetic clinic is and in outline form how genetic counseling functions. These materials may either be sent to counselees when they are given their clinic appointment or furnished to them after the intake interview. Counselees are encouraged to read the materials and to call and ask questions if anything is unclear. If the intake interview is carried out successfully, the counselor will be well informed about the counselee and the family background, and some of the anxiety and apprehension experienced by most people who attend a genetics clinic for the first time will be allayed.

Initial Counseling Session

The initial counseling session might follow a format that includes 1) an accurate pedigree of at least three generations on both sides of the family, 2) physical examination of the proband and all appropriate persons suspected of being at risk and 3) all laboratory tests indicated by pedigree analysis and physical examination.

Some counseling may take place during the initial evaluation and diagnostic sessions, but formal and intensive counseling really begins after the diagnosis is confirmed. The tone of the formal session is set by reassuring the counselees that genetic counseling is designed to meet their needs. Counselees should be reassured that any questions they have will be answered to the best of the counselor's ability. The counselor should make every effort to create a friendly, concerned and relaxed atmosphere directed to the goal of establishing trust and encouraging and facilitating communication between client and counselor.

The counselor should identify any pressing concerns of the counselee so that they can be dealt with at the outset. If important questions are not discussed before presentation of facts, it is probable that much of the information presented will not be heeded, especially if the concepts are unfamiliar or complex, or both.

As in any sound instructional process, the counselor should begin by delineat-

Exhibit 2. Sample Questions From a Questionnaire Used for the Followup of Couples at Risk to have Children with Sickle Cell Disease After Receiving Comprehensive Genetic Counseling

What is the risk to a couple of having a child with sickle cell disease where each has the trait?
a. 0%
b. 25%
c. 50%
d. 75%
e. 100%
f. don't know

What do you think of this risk? It is
a. high
b. low
c. moderate
d. don't know
e. meaningless

Did the counseling influence your plans for future children?
a. yes
b. no

If yes, in what way?
a. to have fewer children than originally planned
b. to have more children than originally planned
c. still plan to have children but not sure when
d. to have no more children
e. to have no more children as originally planned
f. to adopt children
g. to have prenatal diagnosis with abortion of affected fetus

Has the frequency of your sexual activity changed since you discovered that both of you have sickle cell trait?
a. yes
b. no

Are you currently using a birth control method?
a. yes
b. no

Have there been serious family conflicts and/or problems related to the presence of sickle cell disease in the family?
a. yes
b. no

What advice would you wish to offer trait couples at risk of having children with sickle cell disease out of your personal experience?
a. change lifestyle
b. don't change lifestyle
c. take a chance at having normal children
d. avoid having children
e. marry someone without the trait or the disease
f. other (specify) _____

---

ing the scope of the information to be presented. Areas and depth of discussion will vary, according to what has been learned at the precounseling evaluation. If several professional workers have been involved in the client's precounseling diagnostic and genetic workup, it is probably helpful to hold a case conference to be certain that no facet has been overlooked. This approach is especially important if other counselors are involved in the long-term monitoring.

An often-neglected task in counseling is the presentation and discussion of the diagnosis in *simple nontechnical language*. Most counselees are interested in knowing what tests were performed and why. Illustrative material may be needed to explain the cause of the disorder together with simplified explanations of the pathophysiologic features. The explanation should develop in small, discrete steps with frequent pauses so that the client can ask questions. In both initial session and all subsequent ones the counselor must promote interaction with the counselee. Without overt signs of interaction or feedback, e.g., questions or comments by the counselee, the counselor cannot be sure that effective communication is taking place.

The initial session should probably not be longer than one hour, experience having shown that attention and retention spans diminish after this time. A second counseling session should be scheduled two-four weeks later. For practical reasons it may sometimes be necessary to have a longer session so as to have all the information presented at one sitting. This is the case if, e.g., the client lives far away and cannot return for a follow-up visit conveniently. At the end of the first session the counselor should mention what information he expects to discuss at the next visit. Some workers encourage the client to call in the interim should problems arise, although this practice frequently promotes confusion.

Second Counseling Session

The second session should begin, like the first, by establishing a relaxed atmosphere and rapport. Previous material should be reviewed. The counselor should look for signs of overt or latent depression, hostility or anxiety, emotions that interfere with communication, and deal with them appropriately. When undue tensions are resolved, the counselor can present the genetic mechanisms and the risk of recurrence for the disorder. It is helpful to use audiovisual methods to illustrate genetic mechanisms and (especially) the meaning of recurrence risk.

Flipping coins or throwing specially color coded dice are useful in explaining and illustrating simple Mendelian modes of inheritance. A few well-framed questions about the genetic principles and risk of recurrence that have just been presented will help the counselor decide if the concepts have been understood. An especially sensitive and vital area of counseling is the presentation of reproductive options, since certain of them may upset clients. The reproductive options that will generally be presented to couples at risk are as follows:

1. Accept the risk and take the chance.
2. Adopt.
3. Where indicated undergo prenatal diagnosis and terminate the pregnancies when the fetus is affected.
4. When the disorder is transmitted as an autosomal recessive trait or when an autosomal dominant trait is transmitted by the male, the woman can have artificial insemination by an unaffected donor.
5. When the disorder is transmitted as an autosomal recessive trait or when an autosomal dominant or X-linked dominant or recessive trait is transmitted by the woman, the sperm can be used for artificial insemination of a surrogate mother.
6. Wait, avoid having children and hope for a medical breakthrough in treatment or in prenatal diagnosis when this is not yet available.

7. Conclude that the risk is too great, that other options are not viable and that one or the other partner be sterilized.

Difficulties may arise in discussions of prenatal diagnosis and artificial insemination because of religious or cultural prohibitions against abortion, and most males of all ethnic groups do not accept artificial insemination (7). Further, even though many female clients are willing to consider artificial insemination, 80% of them reject it when they learn that the sperm will be from a donor other than their husband. On the other hand many female clients will find artificial insemination an acceptable option, whereas their husbands reject it. This can lead to a rift in the marital relationship if the wife perceives her husband's attitude as directed primarily toward blocking her desire for maternal self-fulfillment. The feelings of both partners should be thoroughly explored before accepting this option.

Counselors must be especially aware of situations in which the partners are split on the action to take and make it clear to the couple that they will continue to work with them until the conflict is resolved. If the rift is deep, psychiatric consultation may have to be sought, or the couple may be referred to a specialist in marital relations. The birth of a defective child into the family may have served to accelerate an already-eroded relationship.

At the conclusion of this discussion the couple may voluntarily state that they have made a decision about future pregnancies. It is very important to encourage them not to fix this decision in their minds, but rather to go home and discuss it together, with close friends or relatives, or their priest or minister if they have a religious affiliation. An appointment for a third session should be made so that the couple can both discuss their decision with the counselor and consider other questions that have surfaced in the interim.

Care of the Affected Child

Even though the counselor and professional staff members deliver accurate, sensitive, concerned and supportive genetic counseling to marital partners, the latter may be dissatisfied unless time is devoted to discussing and making specific plans for the existing affected child. The question of his or her treatment and long-term management may emerge in the intake interview, the diagnostic evaluation or the initial counseling session, especially if other specialists who have seen the child have given the parents the opportunity to discuss this aspect of the child's problem. Even when the physician has outlined the child's management and projected his developmental program, the parents still need to be directed to specific professional agencies or institutions where they can find the required assistance.

PSYCHOSOCIAL CONSIDERATIONS

At each stage of counseling psychological and sociological considerations obviously must remain prominent in the counselor's mind. Counseling is expected not only to inform and instruct but also to help families deal with an affected child. Parents who learn that they are carriers of mutant genes have to cope with this fact. (14) They may also be confronted with society's negative attitudes toward handicapped children.

When recurrence risks are presented to counselees, counselors should determine their psychological state since it can significantly influence their interpretation of the risk figures (15). Clients who are emotionally positive may interpret a 1 in 4 or 25% risk as "good news," whereas to a depressed individual a 1 in 20 or 5% risk (a low genetic risk) is unacceptable. Parents of genetically handicapped children may experience other basic emotions, including denial, guilt, grief and mourning and the psychology of defectiveness (16).

Denial

Denial, a psychological defense mechanism that can be self-defeating, interferes with the understanding and acceptance of the genetic recurrence risk to the extent that the parents may have a second affected child. Denial is particularly likely to operate when a child with an inherited or genetically determined defect is not grossly deformed. That parents deny that their child is genetically handicapped is not necessarily a function of their intelligence. A report from one counseling clinic reveals that 40% of couples waited two or more years after an affected child was born before they actually sought genetic counseling. A second abnormal child was born to 15% of the members of this group in the interim (16). Because denial can be a long-term and persistent reaction to the handicapped child, the counselor who recognizes this phenomenon may be more direct in his approach, especially if definite positive reproductive alternatives exist.

Grief

Parents do not mourn because of the birth of a handicapped or sick child, but because of the loss of the perfect child whom they expected (15). This emotional stress can be especially strong in the mother who provides day-to-day care for the abnormal child. Not only depression, but also hostility, sometimes accompanies grief, which may be exaggerated by society's rejection of the child. Since grieving is continuous or repetitive, repeated sessions are necessary. A feeling of almost absolute trust in the counselor must be cultivated.

If the counselor lacks the background to treat this problem or if the grief seems especially severe, psychiatric consultation may be required. The parents must eventually come to accept the reality of the situation to gain enough insight to enable them to make effective use of the information provided in counseling.

Irrational Guilt

Parents of children with genetically determined defects often experience deep-seated guilt. Some of them are certain that they are being punished for something they did or perhaps did not do. Deeply religious persons who interpret the Bible literally are especially liable to these feelings. They are loathe to accept the idea that the birth of their abnormal child was a chance happening. Parents who experience a second abnormal birth in the family may experience "cosmic guilt," in which they perceive themselves as having been singled out for special punishment. These feelings are accentuated if the parents had ambivalence about conceiving the child in the first place.

Such guilt feelings can become overwhelming in families in whom the mother or father, or both, already suffer neurotic tendencies or if they are having an especially difficult time with the child. No matter how the counselor tries to work with the parents to relieve the guilt, it may be fruitless. This is yet another instance in which psychiatric consultation may be necessary, if genetic counseling is to be successful. Irrational guilt can inappropriately motivate parents to devote an inordinate amount of energy and familial resources to the care of the defective child. In an extreme case healthy siblings may be virtually ignored to the extent that they can be physically and emotionally damaged. If this kind of behavior is suspected, a home visit by the counselor or a specialist in family dynamics is indicated.

Stigmatization

Inherited or genetically determined disorders are often viewed with considerable disdain in a sociocultural sense because parents transmit them to their children. The patient affected by a genetically determined disease may develop ideas of unworthiness or inadequacy because the effect of the genetic defect is extrapolated to the whole person (15). These feelings exist even when the gene does not produce obvious external abnormalities (e.g., hemophilia). Feelings of worthlessness may also be incorporated by the parents of a child affected by an autosomal recessive disorder or by the mother of a child with the Down syndrome.

If latent feelings of unworthiness already exist, the knowledge that they carry a mutant gene or that they may have transmitted a chromosomal anomaly can result in feelings of stigmatization in the parents of an affected child. The potential for these feelings is so strong in many persons that they refuse to be tested to confirm that they are mutant gene carriers. If the counselor is aware that such feelings exist even in carrier parents, they are much likelier to be discovered and placed in the proper perspective so as not to interfere with the counseling process.

Genetic Counseling Team

The methodology of genetic counseling and the technique outlined in this chapter clearly require the skills and knowledge of several professional workers, a minimal list of whom may include:

1. One or more medical specialist or subspecialists;
2. the medical (physician) geneticist who may (or may not) be the immediate genetic counselor;
3. a social worker (genetic assistant);
4. a nurse (coordinator or public health);
5. a psychologist or psychiatrist;
6. a child development specialist or vocational rehabilitation specialist;
7. religious counselor.

Persons in categories 1 through 4 are likely to be involved routinely in genetic

Table 1. Team Approach to Genetic Counseling

Procedure	Medical Professional(s)
Diagnostic workup	MD, specialist or subspecialist
Intake interview (collecting background data); psychosocial background	Social worker or genetic assistant
Confirmation of diagnosis	MD-geneticist
Pedigree and pedigree analysis	MD-geneticist or genetic assistant
Genetic counseling	MD-geneticist, genetic assistant, psychologist or psychiatrist
Follow-up counseling	Genetic assistant, MD-geneticist or psychiatrist

Medical followup:
  Artificial insemination
  Prenatal diagnosis
	MD-specialist or subspecialist

Psychosocial followup:
  Adoption
  Placement of affected children
	Social worker or genetic assistant

counseling for serious genetic diseases. Persons in categories 5, 6 and 7 on the other hand, although not routinely involved in counseling, may be asked to help in the management of difficult cases (Table 1). Experienced genetic counselors (18) advocate the team or group approach to counseling because it makes likelier the optimal performance of all functions and activities of genetic counseling. A major problem with this system is how to provide the finances for and smoothly coordinate the services of these professional workers. The central figure in counseling may be a physician-counselor with the job of making sure that all professional services and skills are brought to bear in each case. Alternatively, the MD-geneticist may serve as the director of services, whereas the actual counseling is done by a specially trained genetic counselor or assistant.

The reason advanced for having a non-MD health investigator do the counseling is that physicians without special psychiatric training tend to be directive and lack the sensitivity of other specially trained health-care workers. It may be too that clients are likelier to be open and communicative with health-care staff members other than physicians.

A major advantage of the group or team system of counseling is that it promotes better quality control. When limited facilities preclude the development of a counseling team, some areas have developed satellite counseling clinics, which appropriate members of a medical center genetic team to visit on a regular basis. This should help ensure the widest possible availability of quality counseling services.

REFERENCES

1. Murray RF, Bolden R, Headings VE, et al: Information transfer in genetic counseling for sickle cell trait. *Am J Hum Genet* 26:63A, 1974
2. Carter CO: Comments on genetic counseling, in Proceedings of the Third International Congress of Human Genetics, Baltimore, Johns Hopkins Press, 1967

3. Childs B, Gordis L, Kaback MM, et al: Tay-Sachs screening: Motives for participating and knowledge of genetics and probability. *Am J Hum Genet* 28:537, 1976
4. Sorenson JR: Counselors: Self portrait. *Gen Counseling* 1:31, 1973
5. Headings VE: Associations between type of health profession and judgements about prevention of sickling disorders. *J Med Educ* 51:682, 1976
6. Money J: Counseling in genetics and applied behavior genetics, in Schaie KW, Anderson VE, McClearn GE, Money J, (eds): Developmental Human Behavior Genetics, Lexington, Lexington Books, 1975
7. Buckhout R: The war on people: A scenerio for population control. Collected papers of Department of Psychology, California State College at Hayward, 1971
8. Warwick DP, Williamson N: Population policy and Spanish-surname Americans, in Veatch RM (ed): Population Policy and Ethics: The American Experience. New York, Irvington, NY, 1977
9. Carter CO, Evans K, Fraser-Roberts JA, et al: Genetic clinic: A follow-up. Lancet 1:281, 1971
10. Stevenson AC: Frequency of congenital and hereditary disease, with special reference to mutation. *Br Med Bull* 17:254, 1961
11. Murphy EA: Probabilities in genetic counseling, in Contemporary Genetic Counseling. Birth Defects Original Art. Ser. IX:19-33, The National Foundation–March of Dimes, White Plains, NY, 1973
12. Emery AE: Genetic counseling. *Br Med J* 3:219, 1975
13. Falek A, Britten S: Phases in coping. The hypothesis and its implications. *Soc Biol* 21:1, 1974
14. Pearn JH: Patient's subjective interpretation of risks offered in genetic counseling. *J Med Genet* 10:129, 1973
15. Shore MF: Psychological issues in counseling the genetically handicapped, in Birch C, Abrecht M. (eds): Genetics and the Quality of Life. Elmsford, Pergammon, NY, 1975
16. Emery AEH, Watt MS, Clack ER: Social effects of genetic counseling. *Br Med J* 1:724, 1973
17. Epstein CJ: Who should do genetic counseling and under what circumstances? in Contemporary Genetic Counseling, Birth Defects Original Article Series, Vol IX, No. 4, The National Foundation–March of Dimes, 1973
18. Hsia YE: Choosing my children's genes: Genetic counseling, in Lipkin M Jr, Rowley PT (eds): Genetic Responsibility: On Choosing Our Children's Genes. New York, Plenum, 1974, p. 43

22
Prenatal Diagnosis

Laird G. Jackson

One of the most significant advances in medicine—and surely an outstanding advance in clinical genetics—is the development of prenatal diagnosis. This tool allows one to know, before birth, exactly what the risk of certain birth defects will be in the pregnancy at risk. Within the limits of the procedure, risks of 50%, 25% or 5% are transformed into 0 or 100% for all detectable chromosomal disorders and an increasing number of biochemical genetic diseases. The techniques of prenatal diagnosis are varied (1, 2), but the commonest is the use of midtrimester amniocentesis and analysis of amniotic fluid or amniotic fluid cells in tissue culture for chromosomal or enzymatic properties.

The technique of midtrimester transabdominal amniocentesis is a safe, reliable procedure, which to date has been used in several thousand at-risk pregnancies (3, 5). It is best performed with ultrasound guidance at about 15 weeks' gestation or 17 weeks from onset of the last menses. Ultrasound itself may be used for fetal imaging and diagnosis of certain defects (6). Other radiologic imaging techniques are less useful. Finally, direct fetoplacental visualization by fetoscopy may be applicable, especially for fetal blood aspiration and testing (7, 8).

INDICATIONS FOR PRENATAL DIAGNOSIS

Prenatal diagnosis requires an accurate diagnosis of the condition for which the fetus is at risk, preferably before onset of pregnancy. Determination of the heterozygous carrier state for recessive genes, diagnosis from family studies and stillborn or neonatal autopsy examination, among others, is discussed elsewhere in this textbook; in general, however, attempts to establish or verify genetic diagnoses during pregnancy for subsequent prenatal diagnostic attempts are complicated by both the physiologic changes and the time constraints of the advancing pregnancy. Assuming that a proper diagnosis is known, several groups of patients qualify for prenatal diagnosis. Currently the commonest indication is that of advanced maternal age and the risk of a trisomic chromosomal condition. Table 1 gives the age-related risk for Down syndrome, the most frequent autosomal trisomy, from maternal age 35 through 40 (9). Amniocentesis is generally offered at maternal age 35 and older.

Table 1. Age-Related Risk for the Down Syndrome*

Age/Risk	Age/Risk	Age/Risk
34: 1 in 527	37: 1 in 266	40: 1 in 106
35: 1 in 413	38: 1 in 183	41–45: 1 in 50
36: 1 in 355	39: 1 in 135	over 45: 1 in 25

*Ages 34–45 years.

A second group at risk for chromosomal trisomies are those women who have previously given birth to a trisomic child. This experience raises their risk to roughly 1-1.5% for a repeat trisomy, regardless of age (12). Other groups in which diagnostic amniocentesis for chromosomal analysis should be offered include any couple in whom either parent is a known carrier of a balanced chromosomal rearrangement. The risk of an unbalanced fetal karyotype is about 5-15%, but this probably varies with the specific rearrangement.

Couples who are both heterozygous carriers of an autosomal recessive genetic trait that is expressed in tissue-cultured cells have a 25% risk of a homozygous affected fetus in each pregnancy. The list of such conditions expands rapidly enough to make listing them individually pointless. The general conditions amenable to this approach include 1) mucopolysaccharidoses, 2) mucolipidoses, 3) sphingolipidoses, 4) the lysosomal-storage forms of carbohydrate metabolic disorders, 5) aminoacidopathies with cultured cell expression of a known enzymatic defect and 6) a growing list of heterogeneously classified disorders. For the possible availability of such tests for specific diseases, consult appropriate sections in this source book or inquire at the nearest genetic center. (Genetic centers are generally located in university hospital teaching centers, and a listing of them may be obtained from the National Foundation–March of Dimes.)

Women who are either known or suspected carriers of an X-linked recessive disorder may wish to utilize prenatal diagnosis, for they have a potential risk of 50% with each male birth of having an affected son. Carrier tests for X-linked conditions are difficult and in many cases not completely reliable, so many women may choose to consider themselves carriers when test results are in doubt. Only a few X-linked conditions (MPS II, Lesch-Nyhan syndrome) may be diagnosed by enzymatic analysis. Duchenne type muscular dystrophy may be diagnosed by fetal blood aspiration and serum creatine phosphokinase determination (11). Other mothers may choose to have the fetal sex identified by chromosomal study or testosterone determination from amniotic fluid (12), or both, and elect to bear only girls—who will be unaffected.

Couples who have had a previous child with a neural-tube defect (spina bifida or anencephaly, e.g.) may also be advised. An open neural-tube defect can be signaled by elevated levels of amniotic fluid α-fetoprotein (13, 14).

In addition to amniotic fluid and amniotic fluid cell analysis, fetoscopy offers the opportunity for fetal blood aspiration and analysis of blood and serum components. So far hemoglobin structural variants (15) and synthesis defects (thalassemia) (16) have been diagnosed as well as X-linked muscular dystrophy (11). Recently other molecular biological tools, some of which do not require cell culture or fetal blood aspiration, have been used for the antenatal diagnosis of HbS or α-thalassemia problems (16a,b). Undoubtedly future research will expand the use of this approach.

Ultrasound imaging offers various diagnostic approaches in this area. Diagnosis of neural-tube defects by visualization of subnormal head size (17) or protruding myelomeningocele has been accomplished (18), and ultrasound examination during amniocentesis for α-fetoprotein analysis should complement this assay. Renal imaging is used to detect enlarged cystic kidneys (19) or renal agenesis (20). Hydrocephalus may be recognized, and early visualization of expanding intracranial fluid spaces is possible with new equipment (21). Skeletal dysplasias with prenatal onset of reliable skeletal changes may be diagnosed by this technique (22), and this area will probably expand rapidly.

COUNSELING IN PRENATAL DIAGNOSIS

Prior to the application of any of these techniques, each couple should receive appropriate, informative genetic counseling. The latter shou,d define the use of the procedure, its risks and complications, the time required for establishing a fetal diagnosis (as long as 3-5 weeks for the use of cultured cells), the risk of culture failure or failure to obtain a clear-cut diagnosis because of difficulty in interpreting test results and even the small risk of diagnostic error (probably due to interpretative error). Each couple should be aware that therapeutic abortion for a diagnosis of an affected fetus is the only alternative to delivery of the affected child (with one exception to date, namely, treatment of methylmalonic acidemia with vitamin B_{12}) (23). This is not to say that prenatal diagnosis is predicated on the couples' agreement to abort an affected fetus, only that the partners should understand that there is a real risk that they may have to make a decision about abortion after prenatal diagnosis.

Following completion of tests, results are reported to both the referring physician and the couple. A positive diagnosis of an affected child should be reported in another counseling session with both parents, if possible. Decisions at this time are as difficult as those following diagnosis of a liveborn child with a genetic handicap, and similar psychologic reactions occur (24). The reactions of both parents—but especially the mother—following abortion for a genetic diagnosis also require supportive counseling (25).

Properly applied, prenatal diagnosis with attendant genetic counseling can be a powerful preventive medical tool. In the more personal sense it frequently allows at-risk couples to have healthy children when they might otherwise forfeit the opportunity.

REFERENCES

1. Milunsky A: The Prevention of Genetic Disease and Mental Retardation. Philadelphia, Saunders, 1975
2. Menutti M: Perinatal medicine, in Schwartz R, Bolognese R (eds): Clinical Perinatology. Baltimore, Williams & Wilkins, 1977
3. NICHD Amniocentesis Registry Symposium Report, *JAMA* 236:1471, 1976
4. The Canadian Registry for prenatal diagnosis of genetic diseases: Preliminary report. *MRC* (Canada) *Prenatal Diagnosis Newsletter* 5:1, 1976

5. Galjaard H: European experience with prenatal diagnosis of congenital disease: A survey of 6121 cases. *Cytogen Cell Genet* 16:453, 1976
6. Santo-Ramos R, Duenholter JH: Diagnosis of congenital fetal abnormalities by sonography. *Obstet Gynecol* 45:279, 1975
7. Valenti C: Endoamnioscopy and fetal biopsy: A new technique. *Am J Obstet Gynecol* 114:561, 1972
8. Kaback MM, Valenti C (eds): Intrauterine Fetal Visualization: A Multidisciplinary Approach. New York, American Elsevier, 1976
9. Hook EB: Estimates of maternal age-specific risks of a Down-syndrome birth in women aged 34-41. *Lancet* 2:33, 1976
10. Mikkelsen M, Stene J: Genetic counseling in Down's syndrome. *Hum Hered* 20:457, 1970
11. Mahoney MJ, Haseltine FP, Hobbins JC, et al: Prenatal diagnosis of Duchenne's muscular dystrophy. *N Engl J Med* 297:968, 1977
12. Menutti MT, Wu CH, Mellman WJ, et al: Amniotic fluid testosterone and follicle stimulating hormone levels as indicators of fetal sex during pregnancy. *Am J Med Genet* 1:211, 1977
13. Brock DJH: The prenatal diagnosis of neural tube defects. *Obstet Gynecol Surv* 31:32, 1976
14. Cowchock FS: Use of alpha-fetoprotein in prenatal diagnosis. *Clin Obstet Gynecol* 19:871, 1976
15. Kan YW, Golbus MS, Trecartin R: Prenatal diagnosis of sickle cell anemia. *N Engl J Med* 294:1039, 1976
16. Kan YW, Golbus MS, Trecartin RF, et al: Prenatal diagnosis of β-thalassemia and sickle cell anemia: Experience with 24 cases. *Lancet* 1:269, 1977
16a. Kan YW and Dozy AM: Antenatal diagnosis of sickel-cell anemia by D.N.A. analysis of amniotic-fluid cells. *Lancet* II: 910, 1978
16b. Wong V, Ma HK, Todd D, Golbus MS, Dozy AM, Kan YW: Diagnosis of homozygous α-thalassemia in cultured amniotic-fluid fibroblasts. *N Eng J Med* 298:669, 1978
17. Campbell S, Johnstone FD, Holt EM, et al: Anencephaly: Early ultrasonic diagnosis and active management. *Lancet* 2:1126, 1972
18. Campbell S, Pryse-Davies J, Coltart TM, et al: Ultrasound in the early diagnosis of spina bifida. *Lancet* 1:1065, 1975
19. Garrett WJ, Greenwald G, Robinson DE: Prenatal diagnosis of fetal polycystic kidney by ultrasound. *Aust NZ Obstet Gynecol* 10:7, 1970
20. Kaffe S, Godmilow L, Walker B, et al: Prenatal diagnosis of bilateral renal agenesis. *Obstet Gynecol* 49:478, 1977
21. Rose J: Fetal abnormalities diagnosed by ultrasound in the second and third trimester, in Saunders R, James AE Jr (eds): Ultrasonography in Obstetrics and Gynecology. New York, Appleton-Century-Crofts, 1977, p. 158
22. Rudd NL, Miskin M, Hoar DI, et al: Prenatal diagnosis of hypophosphatasia, *N Engl J Med* 295:146, 1976
23. Ampola MG, Mahoney MJ, Nakamura E, et al: Prenatal therapy of a patient with vitamin B-12 responsive methylmalonic acidemia. *N Engl J Med* 293:313, 1975
24. Jackson LG: Prenatal genetic counseling. *Prim Care* 3:710, 1976
25. Blumberg BD, Golbus MS, Hanson KH: The psychological sequelae of abortion performed for a genetic indication. *Am J Obstet Gynecol* 122:799, 1975

Glossary

Acrocentric. A chromosome with the centromere near one end, so that one arm is very short.

Allele. (allelomorph) An alternative form of a gene occupying the same locus on homologous chromosomes. Alleles segregate at meiosis, and a child normally receives only one of each pair of alleles from each parent. See also *Multiple alleles* and *Isoalleles*.

Allograft. A tissue graft from a donor of one genotype to a host of another, host and donor being members of the same species.

Allotypes. Genetically determined differences in antigens, e.g., the Gm and Inv variants, which are detected by their antigenic properties.

Amino acid. Small molecules that are the building blocks of proteins. There are 20 amino acids that commonly make up proteins. All amino acids have the same general structure with one acidic (carboxyl) and one alkaline (amino) end, but they differ in the side (R) groups.

Aminoacidurias. Diseases in which there is abnormally high urinary excretion of one or more amino acids.

Amniocentesis. A clinical procedure by which a few milliliters of the amniotic fluid surrounding the fetus are withdrawn. The fluid and fetal cells contained in the fluid may then be subjected to tests for various genetic diseases.

Anaphase. The phase of mitosis or meiosis at which the chromosomes leave the equatorial plate and pass to the poles of the cell.

Aneuploidy. A karyotypic abnormality resulting from extra chromosomes or absence of chromosomes, so that the karyotype is neither haploid nor an exact multiple thereof.

Anticipation. The apparent tendency of certain diseases to appear at earlier-onset ages and with increasing severity in successive generations. It is an artifact of ascertainment.

Anticodon. A triplet of nucleotides specific to each t-RNA, corresponding to a particular amino acid and complementary to the codon (the triplet of nucleotides in m-RNA).

Antigen. A molecule that stimulates production of specific antibodies.

Ascertainment. The method of selection of families for inclusion in a genetic study.

Association. Two or more phenotypic characteristics in members of a kindred occurring together more often than expected by chance. See *Linkage*.

Assortment. The random distribution to the gametes of different combinations of chromosomes. At anaphase of the first meiotic division, one member of each pair passes to each pole, and the gametes thus contain one chromosome of each type; the origin of this chromosome, however, may be either paternal or maternal. Thus nonallelic genes assort independently to the gametes. The exception is linked genes.

Auto-antibodies. Antibodies against components of the self.

Autoimmune disease. A disease presumably resulting from abnormal reactivity of either or both the cellular and humoral components of the immune system against one's own tissues or organs.

Autoradiography. A technique in which a living cell incorporates a radioactive label, and a radiograph of it is then made. It is used particularly in cytogenetics to study DNA synthesis in chromosomes labeled with tritiated thymidine.

Autosome. Any chromosome other than the sex chromosomes. Man has 22 pairs.

Backcross. A cross between a heterozygote (Aa) and a corresponding homozygote (AA or aa).

Bacteriophage or phage. A virus parasitic to a bacterium.

Balanced polymorphism. A polymorphism that is stable (tends to remain unchanged with time) and is probably maintained by selective advantage of the heterozygote over both homozygotes.

Banding. The techniques of staining chromosomes in a characteristic pattern of cross bands, thus allowing individual identification of each chromosome pair. Giemsa banding (G banding) and quinacrine fluorescence banding (Q banding) are the best known banding tecnhiques.

Barr body (sex chromatin). A mass of chromatin in the nucleus of resting cells, resulting from inactivation of an X-chromosome. A cell ordinarily contains a number of Barr bodies equal to the number of X-chromosomes minus one.

Base. A substance with alkaline reaction. Bases forming nucleic acids are purines and pyrimidines.

Base pairing. In the DNA double helix, the coupling of adenine with thymine and of guanine with cytosine; in RNA, adenine pairs with uracil. The specificity of base pairing is fundamental to DNA replication and to its transcription into RNA for protein synthesis.

Base substitution. Replacement of one base for another in the DNA molecule.

Bimodal distribution. A distribution characterized by two modes (frequency peaks).

Carcinogen. A chemical substance or physical agent (such as radiation) that causes cancer.

Carrier. An individual who is heterozygous for a normal gene and an abnormal gene, which is not expressed phenotypically, although it may be detectable by appropriate laboratory tests.

Centric fusion. Union of the long arms of two acrocentric chromosomes at the centromere. See also *Robertsonian translocation.*

Centriole. The cellular organelle that migrates to opposite poles of the cell in meiosis and mitosis, thus insuring (through the spindle apparatus) the separation of each replicated chromosome and the equal partition of the chromosomes among the daughter cells.

Centromere. The small mass of heterochromatin within a chromosome by which both the chromatids are held together and the chromosome is attached to the spindle. Also called the primary constriction.

Chiasma. Literally, a "cross." The term refers to the crossing of chromatid strands of homologous chromosomes, seen at diplotene of the meiotic division. Chiasmas either result in or are evidence of interchanges of chromosomal material between members of a chromosome pair. See also *Crossover.*

Chimera. An individual whose cells are not all of the same genotype. Dizygotic twins are occasionally chimeras. See *Mosaic,* which is distinct from a chimera.

Chromatid. One of two parallel strands of chromosome held together by the centromere and seen during prophase and metaphase. A single-stranded chromosome replicates during the DNA synthesis stage of the cell cycle and is then composed of two chromatids until the next mitotic division, at which phase each chromatid becomes a chromosome of a daughter cell.

Chromatin. The basic substance of the chromosome, including both proteins and DNA. The term essentially means "chromosome material."

Chromosomal aberrations. Karyotypic alterations involving whole chromosomes or portions of them, sufficiently large to be detectable through light microscopic examination.

Chromosome. A thread-like body found in the nucleus of a cell and containing the genes. Under the light microscope the chromosomes are not visible during interphase.

Cistron. A chromosomal segment corresponding to the gene regarded as a functional unit; defined operationally by complementation tests. Independent mutants of the same cistron should not complement or should do so less efficiently than mutants of different cistrons.

Clone. A line of cells derived from a single cell (usually presumed to contain the same genetic information). The term may be applied also to multicellular organisms that can propagate by vegetative (nonsexual) reproduction.

Codominant. Pertaining to the traits determined by both alleles of a pair expressed in the heterozygote.

Codon. A triplet of nucleotides in m-RNA, coding for an amino acid.

Complement. A series of serum proteins that under appropriate conditions cause antibody-coated cells to lyse.

Complementation. The production of normal offspring from a mating between two homozygotes affected by a similar defect, showing that the parents actually suffered from different genetic lesions.

Compound. A genotype in which either the two alleles are different mutations from the wild type or is derived from an individual who carries two different mutant alleles at a locus.

Concordance. Similarity between individuals for an "all-or-none" trait. Usually the term is applied in twin research; the comparison of concordance of monozygotic (MZ) and dizygotic (DZ) twins is a test for heritability.

Congenital defect. A defect present at birth; it may be determined genetically or by external influences during intrauterine life.

Congenital trait. A trait present at birth, not necessarily genetic.

Consanguinity. A characteristic of two or more individuals if they have a common recent ancestor (usually not more remote than three or four generations).

Coupling. The situation characteristic of dominant alleles at two different loci in a heterozygote if they are located on the same chromosome. Thus the double heterozygote AB/ab is said to be in coupling, whereas Ab/aB is said to be in repulsion.

Crossover. The process of exchange of genetic information between two homologous chromosomes, presumed to occur through breakage of both chromosomes at homologous sites followed by reunion after exchange.

Cytogenetics. The study of the relationship of the microscopic appearance of chromosomes and their behavior during cell division to the genotype and phenotype of the individual.

Deletion. At the molecular level, the removal of one or more bases from a DNA sequence. At the cytologic level the absence of a segment of a chromosome (also known as a deficiency).

Demography. The study of populations and how they reproduce, grow, survive and die.

Deoxyribonucleic acid (DNA). A polymer of nucleotides in which the sugar residue is deoxyribose; DNA is found primarily in the double-helical conformation.

Dermatoglyphics. The patterns of the ridged skin of the palms, fingers, soles and toes.

Dicentric. A chromosome with two centromeres; it can arise because of crossing-over in a heterozygote for a paracentric inversion.

Diploid. The number of chromosomes in most somatic cells, which is double the number in the gametes. In man, the diploid chromosome number is 46.

Diplotene. Late meiotic prophase; chiasmas are visible at this stage.

Discontinuous variation. Divergence of a trait occurring so that the trait can be classified into two or more easily distinguishable, clearly separated categories.

Discordant. A term often used in twin studies to describe a twin pair in which one member shows a certain trait and the other does not.

Dizygotic. Type of fraternal twins produced by two separate ova, separately fertilized. The genotypes of dizygous twins are thus no more related than those of any two sibs.

DNA (deoxyribonucleic acid). The nucleic acid of the chromosomes, which carries the genetic code.

Dominant. Pertaining to a trait that is expressed in the heterozygote. Modifications of this definition are discussed in the text.

Dosage compensation. The mechanism by which genes on the X-chromosome have the same physiologic effect in females (who have two X-chromosomes) as in males (who have only one).

Duplication. Presence of two copies of a chromosomal segment, usually in the same chromosome—sometimes in immediate sequence (tandem duplication) or elsewhere in the same or other chromosomes.

Dysgenic. Due to, or determining, an increase in the frequency of deleterious genes. Dysgenic effects may be spontaneous, or they may result from medical or social interventions that improve the fitness of the handicapped. See also *Eugenic*.

Egg. The female gamete.

Electrophoresis. A technique for separating molecules, particularly proteins, according to the overall electric charge of the molecules.

Empirical risk. An estimate of the risk that a trait will recur in a family, based on experience rather than on knowledge of the mechanism by which the trait is produced.

Enzyme. A protein that catalyzes a specific chemical reaction.

Euchromatin. Most of the chromosomal material, which stains uniformly. See *Heterochromatin*.

Eugenics. A program of decreasing the frequency of deleterious genes in a human population (negative eugenics) or increasing that of advantageous genes (positive eugenics) through artificial selection against the genetically handicapped or in favor of the types considered especially desirable.

Feedback inhibition. Inhibition of a sequence of reactions by the end product of those reactions.

Fibroblast. A type of spindle-shaped cell grown in cell culture, especially from skin biopsies or amniotic fluid cells, often used in genetic studies.

Forme fruste. An expression of a genetic trait so mild as to be of no clinical significance.

Founder effect. An expression emphasizing the drift effect (on gene frequencies) that results when a new population is founded by a small group of people selected from an old population.

Frameshift mutation. A mutation resulting from the insertion or deletion in DNA of one or more bases (but not a multiple of three), thus changing the boundaries of the codons and altering the whole string of amino acids subsequent to the insertion or deletion.

Gamete. The haploid cell generated by meiosis that may fuse with another appropriate gamete to form a zygote. In a bisexual species a gamete is either male (sperm) or female (egg).

Gametic selection. Differential survival (and/or capacity to fertilize) of sperm or egg cells.

Gene. A segment of chromosome with a detectable function. Used as a synonym for locus or cistron, and sometimes for allele.

Gene-dosage effect. The relation between the number of functional copies of a given gene present and the level of their activity (enzymatic activity, e.g.).

Gene flow. The exchange of genes at a low rate (in one or both directions) between two initially different populations. Because the rate of exchange is low, the groups may retain their identity.

Gene frequency. The proportion (of all alleles at a locus) in which a given allele is found in the people forming a specified population.

Genetic code. The code that relates nucleotide sequences in nucleic acids to amino-acid sequences. Each triplet of nucleotides designates a particular amino acid; thus the genetic code allows both the translation of information stored in DNA and the use of that information in protein synthesis.

Genetic drift. Random fluctuation of gene frequencies in small populations.

Genetic lethal. A genetically determined trait in which affected persons do not reproduce.

Genetic marker. A trait that can be used as a genetic tracer in studies of cell lines, persons, families and populations if it 1) is genetically determined, 2) can be accurately classified and 3) has both a simple unequivocal pattern of inheritance and heritable variations common enough to allow it to be classified as a genetic polymorphism.

Genetic screening. A systematic testing of persons designed to detect potential genetic handicaps in them or in their progeny, particularly those who may respond to treatment or prophylaxis.

Genetic trait. A trait determined by genes, not necessarily congenital.

Genome. The total of genetic material in a cell.

Genotype. The genetic constitution of a person at one or more loci.

Genotype frequency. The relative proportion of a particular genotype in a population.

Germ cells. Gametes, or their precursors.

Germ line. The line of cells that produces gametes.

Haploid. The chromosome number of a normal gamete containing only one chromosome of each type. In man the haploid number (n) is 23.

Hardy-Weinberg law. A rule for predicting genotypic frequencies on the basis of gene frequencies, under the assumption either of random mating in the absence of selection or some other function disturbing population equilibrium.

Hemizygous. A term applicable to the genes on the X-chromosome in a male. Since males have only one X, they are hemizygous (not homozygous or heterozygous) with respect to X-linked genes.

Heterochromatin. Chromosomal material staining differently from the rest of the chromosomal material in the cell, e.g., the Barr body. See *Euchromatin*.

Heterogametic. The sex that produces gametes of two types: in man this is the male, who produces X-bearing and Y-bearing sperm.

Heterogeneity. The phenomenon by which a certain phenotype (or closely similar phenotypes) can be produced by different genetic mechanisms.

Heterokaryon. A cell having two different types of nuclei, as a result of fusion of two cell types without nuclear fusion.

Heterozygous. Having different alleles at a given locus on homologous chromosomes.

Histocompatibility. Capacity to accept a tissue or organ graft.

Homogametic. Referring to the sex whose gametes all carry the same sex chromosome. In the human, females are homogametic because all eggs carry an X-chromosome.

Homograft. A tissue graft between two members of the same species. Normally a homograft is rejected, but if the donor and host are isogenic it is accepted.

Homologous chromosomes. A "matched pair" of chromosomes, one from each parent, having the same gene loci in the same order; "homologs" is the abbreviated form of the term.

Homozygous. Having the same allele at a given locus on homologous chromosomes.

Hybrid. The progeny resulting from a cross between different parental stocks.

Inborn errors of metabolism. Inherited disorders that can be explained as genetic blocks in specific metabolic pathways, usually due to recessive alleles determining a decreased activity (or absence) of a specific enzyme. Examples include albinism and alkaptonuria.

Inbreeding. The mating of consanguineous (closely related) individuals. Their progeny are said to be inbred.

Incidence. The relative frequency of a trait or disease at birth.

Incompatibility. Antigens in a fetus (or graft, or donated blood) that can evoke an immune response in the mother (or graft recipient, or blood recipient).

Inducer. A molecule that binds to a repressor, preventing it from blocking transcription.

Inversion. A chromosomal aberration that arises when two breaks occur in the same chromosome, and the region between the breaks is reinserted after a 180° rotation.

Isoalleles. Allelic genes that are "normal" and which can be distinguished from one another only by their differing phenotypic expression when in combination with a dominant mutant allele.

Isochromosome. An abnormal chromosome with two arms of equal length and bearing the same loci in reverse sequence, formed rather by transverse than by logitudinal division of the centromere.

Isograft. A tissue graft between two persons with identical genotypes.

Isolate. A subpopulation in which mating take place exclusively with other members of the same subpopulation.

Isozymes. Enzymes performing similar functions (in the same or different organs). They may be in the same person and are usually differentiated by electrophoresis.

Karyotype. The chromosome set. The term is often used for photomicrographs of chromosomes arranged in a standard classification.

Kindred. An extended family group.

Lethal. An allele that kills all carriers (or only homozygotes in the case of a recessive lethal) before reproductive age and usually in the first years of life.

Linkage. Two or more loci on a single chromosome, causing a tendency for alleles at the linked loci to be inherited together. Linkage is observed only when the loci are sufficiently close to one another; crossing-over can lead to random assortment of loci that are far apart on the same chromosome.

Linkage group. A group of linked loci. A linkage group must correspond to the chromosome on which the genes are located.

Locus (plural, loci). Position of a gene on a chromosome. Used as a synonym for gene or cistron, but not for allele.

Lyon hypothesis (lyonization). Random and fixed inactivation of one X-chromosome in the somatic cells of female mammals during early embryonic life, which leads to dosage compensation, heterozygote variability and female mosaicism.

Manifesting heterozygote. A female heterozygous for an X-linked disorder, in whom, because of lyonization, the trait is expressed clinically with about the same degree of severity as in hemizygous affected males.

Mapping a chromosome. Determining the position and order of gene loci on a chromosome, especially by analyzing the frequency of recombination between the loci.

Meiosis. The special type of cell division occurring in the germ cells by which gametes containing the haploid chromosome number are produced from diploid cells. Two meiotic division occur, the first and second (meiosis I and meiosis II). Reduction in number takes place during meiosis I. See also *Mitosis.*

Messenger RNA (m-RNA). Ribonucleic acid that is synthesized using one strand of DNA as a template, and which is then used to direct protein synthesis.

Metacentric. A chromosome with the centromere near the middle of the chromosome arm(s).

Metaphase. The phase of mitosis or meiosis in which the condensed chromosomes attached to the spindle fibers line up on an equatorial plane between the two poles of the cell.

Missense mutation. The substitution of one amino acid for another, resulting from the substitution of one base for another in the DNA molecule.

Mitosis. Somatic cell division resulting in the formation of two cells, each with the same chromosomal complement as the parent cell. See also *Meiosis.*

Mitotic cycle. The cycle of a cell between two successive mitoses, in which four periods are distinguished: G_1, S (DNA synthesis), G_2 and mitosis.

Mode. A peak in a frequency distribution.

Monogenic. Referring to a single-gene model that postulates that one gene determines a trait.

Monosomy. A condition in which one chromosome of one pair is missing, as in 45, X Turner syndrome. Partial monosomy may occur.

Monozygous (MZ) twins. Twins developing from a single zygote that divides to give rise to two complete embryos. Monozygous twins have identical genotypes.

Morgan. A unit of recombination; a centimorgan indicates 1% of recombination; 1 morgan equals 100 centimorgans. One chiasma corresponds on the average to 50 centimorgans.

Mosaic. A person or tissue with at least two cell lines differing in genotype or karyotype, derived from a single zygote.

Mosaicism. The presence in a person of two or more cell genotypes arising by mutation or by chromosomal aberration (including nondisjunction).

Multifactorial trait. A trait the phenotypic expression of which is influenced by the cumulative effects of both genes and significant environmental factors.

Multiple alleles. The possibility of more than two different alleles at a locus in the population as a whole, even though only two alleles at a locus can be present in any one normal person.

Mutagen. A physical or chemical agent that increases the mutation rate.

Mutant. A gene in which a mutation has occurred or an individual carrying the gene.

Mutation. A sudden heritable change in the genetic material.

Mutation rate. The frequency of mutations per generation at a given locus.

Natural selection. The process by which changes occur spontaneously in the proportions of genetic types within populations of a living organism, due to differences in the (Darwinian) fitnesses of these types in the existing environment.

Nondisjunction. Failure of homologous chromosomes to separate during the first stage of meiosis (primary nondisjunction) or of chromatids to separate in the second division of meisosis (secondary nondisjunction). Nondisjunction can also occur in mitosis.

Nonsense mutation. A mutation involving a base change in DNA that converts a triplet specifying an amino acid into one specifying chain termination of a polypeptide chain.

Nucleic acids. Deoxyribonucleic acid (DNA) and ribonucleic acid (RNA).

Nucleolus. A body in the nucleus involved in r-RNA synthesis; it is usually associated with secondary constrictions of certain chromosomes.

Nucleoside. A purine or pyrimidine base attached to a sugar (deoxyribose or ribose); it is a nucleotide without the phosphate group.

Nucleotide. A monomer consisting of a base, a pentose sugar and a phosphate group. A nucleic acid molecule consists of many such nucleotides, held together by sugar-phosphate bonds.

Oncogenic. Having a tendency to "cause" cancer.

Operator. A genetic region that, by interaction with a repressor, can prevent transcription of the DNA in a specific set of adjacent genes.

Operon. A postulated unit of gene action, consisting of an operator and the closely linked structural gene (s) whose action it controls.

p. The short arm of a chromosome (from the French *petit*); often used to indicate the frequency of the commoner allele of a pair. See also *q*.

Pachytene. The stage of meiotic prophase at which chromosomes are fully paired.

Paracentric inversion. An inversion in which both breaks occur on the same arm of the chromosome (i.e., same side of the centromere).

Pedigree. A diagram delineating the genetic relationships of members of a family for more than two or more generations.

Penetrance. The term applied when the frequency of expression of a genotype is less than 100%. In a person with a genotype that characteristically produces an abnormal phenotype but who is phenotypically normal, the trait is said to be nonpenetrant.

Peptide bond. A covalent bond formed between the amino group of one amino acid and the carboxyl group of another.

Pericentric inversion. A reversal in which the breaks occur on opposite sides of the centromere.

Phenocopy. A phenotype that is usually determined by one specific genotype, produced instead by the interaction of an environmental factor with a different genotype.

Phenotype. The entire physical, biochemical and physiologic nature of a person, as determined by his genotype and the environment in which he develops; or in a more limited sense the expression of some particular gene or genes, as classified in some specific way.

Philadelphia chromosome. The deleted chromosome no. 22 typically occurring in a proportion of bone-marrow cells in patients with chronic myelogenous leukemia.

Pleiotropy. If a single gene or gene pair produces multiple effects, it exhibits pleiotropy (or pleiotropic effects).

Polar bodies. Products of meiosis in the female subject. Of the four daughter cells, three are polar bodies and one is an egg. The polar bodies are smaller than the egg and are unable to form a zygote.

Polygenic. Determined by many genes at different loci, with small additive effects. Aslo termed quantitative. To be distinguished from multifactorial in which environmental as well as genetic factors may be involved.

Polymerase. An enzyme that catalyzes polymerization; DNA and RNA polymerases catalyze the formation of the nucleic acids from the nucleotide constituents on the basis of a single-stranded DNA template. Reverse transcriptase catalyzes the formation of DNA on the basis of an RNA template.

Polymorphism. Two or more genetically determined alternative phenotypes in a population, in frequencies too great to be maintained by mutation alone, e.g., the rarer of two alleles having a population frequency of at least 0.01.

Polynucleotide. A polymer formed by the covalent linkage of nucleotides; the phosphate group of one nucleotide is bonded to the sugar of the next.

Polypeptide. A chain of amino acids held together by peptide bonds between the amino group of one and the carboxyl group of an adjoining amino acid. A protein molecule may be composed of a single polypeptide chain or of two or more identical or different polypeptides.

Polyploid. Having a simple multiple (greater than two) of the haploid number of chromosomes.

Prenatal diagnosis. Determination of the fetal sex, karyotype or phenotype, usually before the 20th week of gestation. Various techniques, especially amniocentesis and cell culture, are employed.

Prevalance. The relative frequency of a trait or disease in the members of a specified population.

Proband (propositus, or index case). The affected person who first draws the attention of the genetic researcher to a particular family.

Promoter. A region in the operon between the operator and the structural genes; RNA polymerase attaches to the promoter.

Prophase. The first stage of cell division, during which the chromosomes become visible as discrete structures and subsequently thicken and shorten. Prophase of the first meiotic division is further characterized by pairing (synapsis) of homologous chromosomes.

Propositus. The family member who first draws attention to a pedigree of a given trait. Also called index case or proband.

Purines. A class of organic compounds to which the nucleic acid bases adenine and guanine belong.

Pyrimidines. A class of organic compounds to which the nucleic acid bases cytosine, thymine, and uracil belong.

q. The long arm of a chromosome; often used to indicate the frequency of the rarer allele of a pair. See also *p*.

Races. Subdivisions of a species, recognizably different from one another.

Random mating. A situation in which the genetic character being studied has no influence on the choice of a mate.

Recessive. An allele that causes a phenotypic effect different from that of other alleles (dominant) only when present in the homozygous state.

Reciprocal translocation. A translocation in which breaks occur in two different chromosomes, and the resulting fragments are exchanged.

Recombination. The formation of new combinations of linked genes by crossing-over between their loci.

Reduction division. The first meiotic division, so-called because at this stage the chromosome number per cell is reduced from diploid to haploid.

Regulator genes. Genes controlling the rate of production of the product of other genes by synthesis of a substance that inhibits the action of an operator.

Repressor. A molecule that binds to the operator to prevent attachment of the RNA polymerase to the promoter.

Repulsion. The occurrence of two dominant alleles at two different loci when they are located on homologous chromosomes. Thus the double heterozygote Ab/aB is said to be in repulsion. See also *Coupling*.

Ribonucleic acid (RNA). A polynucleotide similar to DNA, but with ribose as the sugar and uracil the pyrimidine component instead of thymine. See also *Messenger RNA, Ribosomal RNA*, and *Transfer RNA*.

Ribosomal RNA (r-RNA). A form of RNA that is a major constituent of ribosomes; the largest portion of RNA in a cell consists of r-RNA.

Ribosomes. Small particles, made of r-RNA and proteins, that are the site of protein synthesis.

Ring chromosome. A chromosome formed by breaks at the two extremities and rejoining of the broken ends; acentric fragments are excluded in the process.

Robertsonian translocation. The fusion of two acrocentric chromosomes at the centromeres.

Satellite. (chromosomal satellite). A small mass of condensed chromatin attached to the short arm of each chromatid of a human acrocentric chromosome by a relatively uncondensed stalk (secondary constriction).

Segregation. In genetics, the separation of allelic genes at meiosis. Since allelic genes occupy the same locus on homologous chromosomes, they pass to different gametes, i.e., they segregate.

Selection. 1. In population genetics, the operation of forces that determines the relative fitness of a genotype in the population. 2. The manner in which kindreds are chosen for study, i.e., ascertainment.

Sex chromatin. A chromatin mass in the nucleus of interphase cells of females of most mammalian species, including man. It represents a single X-chromosome, which is inactive in the metabolism of the cell. Normal female subjects have sex chromatin and thus are chromatin-positive; normal males lack it and thus are chromatin-negative. Synonym: Barr body.

Sex chromosomes. Chromosomes responsible for sex determination. (in human beings XX in females, XY in males).

Sex-limited. A trait expressed in only one sex although the gene determining it is not X-linked.

Sex linkage. X-linkage.

Sibs (siblings or sibship). Offspring of the same parental combination. Brothers and sisters of a proband.

Somatic cell hybrid. A uninuclear cell resulting from the fusion of two somatic cells in culture.

Somatic cells. Cells that are not part of the germ line.

Species. A set of persons who can interbreed and produce fertile progeny.

Sperm (spermatozoon). The male gamete.

Spindle apparatus. The bundle of fibers that joins the centrioles to the centromeres and ensures accurate partitioning of the genetic material during cell division.

Structural gene. A gene coding for a polypeptide. Sometimes used to mean genes other than regulatory genes or genes coding for r-RNA, t-RNA and m-RNA; those regulatory genes that code for a repressor, however, code for a polypeptide.

Suppressor. A mutation or allele that suppresses the phenotypic action of an allele at another locus.

Telcentric. A chromosome with the centromere located at one end.

Telophase. The stage of cell division that begins when the daughter chromosomes reach the poles of the dividing cell and lasts until the two daughter cells take on the appearance of interphase cells.

Teratogen. An agent that produces, or raises the incidence of, congenital malformations.

Teratogenic. Causing a serious deformity in fetal development.

Tetraploid. Having four times the haploid complement of chromosomes.

Transcription. The formation of an m-RNA strand complementary to one strand of DNA.

Transfer RNA (t-RNA). Ribonucleic acid with a partially double-stranded structure that functions as an agent of transfer of information between m-RNA and the protein being synthesized. One end of the molecule is a triplet (anticodon) complementary to the m-RNA triplet (codon). The other end of the t-RNA holds the amino acid specified by the codon.

Translation. The generation of a polypeptide chain on the basis of the information contained in an m-RNA chain.

Translocation. A chromosomal abnormality in which a chromosome (or portion thereof) becomes attached to another chromosome. A balanced translocation exists when an individual (or gamete) with a translocation carries the same number of copies of each locus as exist in the normal diploid or haploid genome.

Triplet. A sequence of three nucleotides in a polynucleotide. Each triplet (except for nonsense or chain-end triplets) codes for a particular amino acid.

Triploid. A cell with three times the normal haploid chromosome number, or an individual made up of such cells.

Triradius. In dermatoglyphics, a point from which the dermal ridges course in three directions at angles of roughly 120°.

Trisomy. Three chromosomes of one type in a person. The commonest in man is trisomy 21 (Down syndrome).

Unequal crossing-over. A situation that results when breaks leading to crossing-over occur in nonhomologous regions of the two chromosomes that exchange.

Wild-type allele. The allele that is most frequent in natural populations; usually indicated by the symbol +. This term cannot be applied for a locus that is recognizably polymorphic.

X-inactivation. The genetic inactivation of all X-chromosomes in excess of one, taking place on a random basis in each cell at an early stage in embryogenesis.

X-linkage. Genes on the X-chromosome or traits determined by such genes.

Y-linkage. Genes on the Y-chromosome or traits determined by such genes.

Zygote. The primordial cell of a new organism, formed by the fusion of an egg and a sperm.

Index

Abetalipoproteinemia, 363, 401
ABO blood groups, 16
Abortion
  genetic counseling, 600, 607, 615
  spontaneous, 40, 50
Acanthocytosis, 363
Acanthosis nigricans, 557, 564
Acatalasia, 245-46, 248
Accelerated aging, 564
Acetazolamide treatment, for periodic paralysis, 381
Acetophenetidin-induced methemoglobinemia, 252
N-Acetyl galactosamine 4-sulfate sulfatase deficiency, 174
α-N-Acetyl glucosaminidase deficiency, 173
N-Acetyl hexosamine 6-sulfate sulfatase, 173
β-D-N Acetyl hexosaminidase, 160-62
Acetyl transferase, 249
Achondrogenesis, 520-23
Achondroplasia, 383, 502, 511, 520-21
Acid-base balance, kidney's regulation of, 482
Acidic glycosaminoglycans, 158
Acid maltase deficiency, 188-89
Acidosis, 202, 480-83
Acid phosphatase deficiency, 181-82
Acoustic neuroma, 450
Acrocentric chromosome, 35-36
Acrocephalopolysyndactyly, 393
Acrocephalosyndactyly (Apert syndrome), 393, 567
Acrodysostosis, 528
Acrokeratoelastoidosis, 555
Acrokeratosis, 555
Acrokeratosis verruciformis, 555-56
Acromegaly, 288
Acromesomelic dwarfism, 528-29
ACTH, inability to respond to, 301
Acute glomerulonephritis, 466
Acute intermittent porphyria, 213, 215-16
Acute lymphatic leukemia, 87

Acute lymphoblastic leukemia, 114
Acute myelogenous leukemia, 72, 113-14
Acute nonlymphocytic leukemia, 115
Acute pulmonary edema of mountaineers, 506
Addison's disease, 302-3, 309, 591
Adenine, 4-7
  metabolic disorders, 220
Adenine phosphoribosyl transferase deficiency, 220
Adenocarcinoma, 89
  acanthosis nigricans, 557
  duodenal ulcer, 335
Adenoma
  duodenal ulcers, 334
  hyperparathyroidism, 298
Adenosine deaminase
  deficiency, 131, 146, 510
  increase, 220
Adenosine triphosphate, 354-55
  deficiency, 358
Adeylate kinase deficiency, 357-58
Adrenal aplasia, 299
Adrenal corticoid biosynthesis, defects of, 303
Adrenal gland disorders, 299-305
Adrenal hyperplasia, 303-4, 312
Adrenal hypoplasia, 299
Adrenal medullary hyperfunction, 305
Adrenocortical hyperfunction, 303-5
Adrenocortical hypofunction, 299-303
Adrenocortical tumor, 97, 99, 304
Adrenoleukodystrophy, 302
Adult-onset medullary cystic disease, 471
Adult-onset ophthalmoplegia, 425
Adult polycystic kidney disease, 469-70
Adverse drug reactions, 245. See also Pharmacogenetics
Afibrinogenemia, 367
Agammaglobulinemia, 533
Aganglionosis, 337-38

Aggressive behavior, in XYY males, 51
ALA synthetase, 208, 212-13
Albinism, 22, 108-10, 419, 424, 427, 430, 503, 558-62
Albright's hereditary osteodystrophy, 297-98, 479
  calcification of basal ganglia, 396
  tooth abnormalities, 591
Alcohol
  and cardiovascular maldevelopment, 278
  and hyperlipoproteinemia, 282
  metabolism, 258-59
Alder anomaly, 364
Aldolase A, 356
Aldosterone
  deficiency, 301, 304
  increase, 305
Alimentary tract neoplasia, in Bloom syndrome, 73
Alkaline phosphatase excretion, 535-36
Alkaptonuria, 196, 237, 419
ALL, see Acute lymphoblastic leukemia
Allele, 10-11
Allopurinol therapy
  gout, 219
  Lesch-Nyhan syndrome, 218
  orotic aciduria, 358
  xanthinuria, 220
Alopecia, 565-67, 569
Alpers disease, 393-94
Alpha thalessemia minor, 352
Alport syndrome, 447-48, 464-66
ALS, see Amyotrophic lateral sclerosis
Alstrom syndrome, 321, 452
Altered affinity hemoglobin variant, 351
Alveolar malformations, 503-4
Alzheimer disease, 394
Amaurosis, 434
Amelocerebrohypohidrotic syndrome, 588
Amelogenesis imperfecta, 575-76, 580-81, 587
Amelo-onychohypohidrotic

631

syndrome, 588
Amenorrhea, in autoimmune ovarian disease, 309
Amino acids
  of collagen, 229
  metabolic abnormalities, 194-207, 345
  ocular findings in disorders of, 419
  prenatal diagnosis, 614
  screening, 207
Aminoaciduria, 206-7, 473-74
α-Aminoadipic aciduria, 204
γ-Aminobutyric aciduria, 206
δ-Aminolevulinic acid synthetase, 208, 212-13
Aminopterin, 30
β-Aminosobutyric acid, 206
AML, see Acute myelogenous leukemia
Amniocentesis, 29, 55, 76 77, 613
Amniotic fluid, 37, 40. See also Amniocentesis
Amphetamines, 32
Amylo-1, 6-glucosidase, 189
Amyloidosis, 405, 451, 464, 467
  tables, 407-8, 565
Amyotrophic lateral sclerosis, 402
Amyotrophy of the hands, 403
Anal atresia, 336, 527
Anal skin carcinoma, 74
Anaphase, in mitosis, 42
Anaphase lag, 42
Anderson's disease, 188
Androgen
  abnormal action of, 315-17
  deficiency, 305
  excessive production, 303
  insensitivity, 317
Androgenic drugs, 312
Androgen-producing maternal tumors, 312
Androgen receptor, 311
Anemia, 354-58
  from drug-sensitive hemoglobins, 256
  in dyskeratosis congenita, 558
  in G-6-PD deficiency, 253
  in kidney disease, 470
  macrocytic, 358-60
Anemia-triphalangeal thumb syndrome, 361
Anencephaly, 286, 392, 524
  prenatal diagnosis, 29, 614
  recurrence risk, 28
Anesthesia, malignant hyperthermia following, 258, 383
Aneuploidy, 10, 72, 115
Angioedema, 145
Angiokeratoma, 175, 562
Angiokeratoma corporis diffusum universale, see Fabry disease
Angiomyolipoma, 487
Angle-closure glaucoma, 431
Anhidrosis, in hereditary sensory neuropathy, 405, 409
Anhidrotic ectodermal dysplasia, 555-56
Aniridia, 90, 94, 426-27

Anisocoria, 430
Ankyloses, multiple, 502
Ankylosing spondylitis, 124, 338
Annular pancreas, 336
Anomalous lobation of the lungs, 502
Anomalous trichromatism, 437
Anophthalmia, 426
Anorchia, 305, 313
Anosmia, 307, 553
Anotia, 445
Anterior chamber cleavage anomalies, 430
Anterior horn cell degeneration, 402
Anterior pituitary, 285-89
Anterior polar cataract, 432
Anterior uveitis, 124
Antibodies, 17, 126
  deficiencies, 139, 146
  fetal synthesis, 122
Antibody-forming cells, 121-22
Anticipation, 20
Anticonvulsants, 278
Antidepressants, 259
Antidiuretic hormone, 289-90, 483
Antiendotoxin antibody deficiency, 138
Antigens, 124
  HLA system, 86, 319
  tumor cells, 87
Antipyrine accumulation, 260, 262-63
α-1-Antitrypsin deficiency, 344, 508-10
α-1-Antitrypsin protein marker, 495
Anuric renal failure, 470
Aortic dilation, in Marfan syndrome, 232
Apert syndrome, 393, 567
APKD, see Adult polycystic kidney disease
Aplasia cutis congenita, 569
Aplastic anemia, 361-62
Apnea, in succinylocholine sensitivity, 249
Apple peel jejunal atresia, 335
Aqueductal stenosis, 392
Arachnodactyly, 502, 511
Areflexia, in Bassen-Kornzweig syndrome, 401
Arginase deficiency, 198
Argininosuccinate synthetase defects, 197
Argininosuccinic aciduria, 198, 568
Arnold, Matthew, 279
Aromatic amino acid metabolism, defects in, 194-96
Arrhinecephaly, 499
Arteriosclerosis, 564
Arthritis, 237, 539
Arthrogryposis multiplex congenita, 376, 379
Arthro-onychodysplasia, 486-87
Arthro-ophthalmopathy, 537
Arthropathy, secondary, 238
Artificial insemination, 600, 607

Aryl hydrocarbon hydroxylase, 116
Arylsulfatase, 166
Arylsulfatase B, 174
Asexual ateliosis, 287
N-Aspartyl-β-glucosaminidase deficiency, 182
Aspartylglycosaminuria, 182
Asphyxiating thoracic dysplasia, see Jeune syndrome
Asphyxiating thoracic dystrophy, 511
Association, 11, 16
Ataxia
  in basal ganglia calcification, 396
  and hypogonadism, 307
  spinocerebellar, 397-401
  vestibulocerebellar, 401
Ataxia-telangiectasia, 75, 135, 309, 511, 563
  gastric cancer, 103
  glucose intolerance, 320-21
  leukemia and lymphoma, 113
  neoplasm-development risk, 88
  virus-associated malignancies, 87
Atherosclerosis, 180, 279
Athyrotic cretinism, 290
Atopic dermatitis, 146
Atopic diseases, immunodeficiency and, 145-46
ATP, see Adenosine triphosphate
Atransferrinemia, 361
Atrial septal defect, 306
Atrioventricular canal defect, 271-73
Atrophic stria vascularis, 447, 450
Attitude tremor, 396
Atypical alcohol dehydrogenase, 258-59
Atypical pseudocholinesterase, 250
Auditory canal, 444, 449
Autoimmune disorders
  Addison's disease, 302
  corticotropin deficiency, 299
  diabetes, 321
  gastric cancer susceptibility, 103
  hemolytic anemia, 362
  hepatitis, 346
  hypoparathyroidism, 296
  juvenile-onset diabetes, 319-20
  lymphocytic thyroiditis, 294
  ovarian disease, 309
  pernicious anemia, 359
  thyroid diseases, 294
Autonomic nervous system manifestations, in familial dysautonomia, 393
Autoradiography, 36
Autosomal dominant inheritance, 17-20
  anterior chamber cleavage malformations, 430
  breast cancer, 101
  cancers, table of, 91-93
  cardiovascular diseases, 274, 280
  cataracts, 423-33

Index

complement system deficiencies, 143
connective-tissue disorders, 238-39, 421
corneal dystrophies, 428-29
craniofacial dysplasias, 423
cutaneous malignancies, 109
ectodermal dysplasias, 582-86
endocrine diseases, 288-90, 295-98, 306
familial spastic paraplegias, 404
gastrointestinal disorders, 104-5, 331-32
genodermatoses, 106
hamartoses, 421
hearing loss, 444-51
iris anomalies, 429
kidney diseases, 463, 473, 477, 479, 486
malabsorption syndromes, 339
microphthalmia, 426
muscular dystrophies, 374
myotonias, 379
neuronal ceroid lipofuscinosis, 183
pain insensitivity and indifference disorders, 405
pharmacogenetic diseases, 246-47
polyneuropathies, 406-8
porphyrias, 211
proportionate short stature, 536
respiratory tract disorders, 496-98
retinal abnormalities, 435-36
short-limb dwarfism, 521, 532
short-trunk dwarfism, 539
skeletal dysplasias, 422
skin disorders, 423, 548, 550, 554, 559, 565-69
spinal muscular atrophy, 403
spinocerebellar ataxias, 398-400
sporadic cases, 25
tooth defects, 576-79
Autosomal gene, 17
abnormalities, 52-62
Autosomal recessive inheritance, 20-21
amino acid metabolism defects, 195, 201
aspartylglycosaminuria, 182
cancers, table of, 91-94
cardiovascular diseases, 273, 280
carriers, prenatal diagnosis of, 614
cataracts, 432-33
coagulation deficiency states, 366
complement system deficiencies, 143
connective-tissue disorders, 238-39, 421
corneal dystrophies, 428-29
craniofacial dysplasias, 423
cutaneous malignancies, 109
ectodermal dysplasias, 582-86
endocrine diseases, 286-87, 289, 299, 301, 303, 321
fructose metabolism disorders, 191-92
galactose metabolism disorders, 192
gastrointestinal disorders, 104-5, 332-33
genodermatoses, 106
glucose metabolism disorders, 185
glycogen synthetase deficiency, 191
glycolytic enzyme deficiency, 356
glycosphingolipidoses, 158
hearing loss, 444-46, 451-53
immunodeficiency syndromes, 129
iris anomalies, 429
kidney diseases, 463, 467-68, 470, 472-74, 476-78, 480, 483-86
lysosomal-storage diseases, 179-80
malabsorption syndromes, 339
microphthalmia, 426
mucosulfatidosis, 174
muscular dystrophies, 374
myotonias, 379
neuronal ceroid lipofuscinosis, 183
pain insensitivity and indifference disorders, 405
pentosuria, 193
phagocytic-bactericidal system disorders, 138
pharmacogenetic diseases, 246-47
phosphorylase-kinase deficiency, 190
polyneuropathies, 406-8
porphyrias, 211
proportionate short stature, 536
respiratory tract disorders, 496-98
retinal abnormalities, 435-36
short-limb dwarfism, 521, 532
short-trunk dwarfism, 539
skeletal dysplasias, 422
skin disorders, 423, 548, 550, 554, 559, 566-69
spinal muscular atrophy, 403
spinocerebellar ataxias, 398-400
sporadic cases, 25
sulfur amino acid metabolism disorders, 200
tooth defects, 576-79
urea cycle enzyme defects, 196
xanthinuria, 220
6-Azauridine therapy, 358
Azotemia, 482, 487

Bactericidal activity, *see* Microbicidal activity
Banding, chromosomal, 36-38, 59
Bardet-Biedl syndrome, 393, 452
Barr body, *see* Sex chromatin
Bartter syndrome, 483-84
Basal-cell carcinoma, 76, 108-10
Basal-cell nevus syndrome, 426
Basal ganglia calcification, 396, 479
Basal ganglia disorders, 395-97
Bassen-Kornzweig syndrome, 363, 401
Batten-Mayou disease (Batten-Spielmeyer-Vogt syndrome), *see* Neuronal ceroid lipofuscinosis
Bayesian analysis, 23-25
Becker dystrophy, 375
Beckwith syndrome, 304
Beckwith-Wiedemann syndrome, 94, 96, 288
Behavioral disorder, in Lesch-Nyhan syndrome, 217
Behr's syndrome, 438
Benign familial pemphigus, 554
Berry aneurysms, 470, 486
Beta, low-density, lipoproteins (LDL), 280
Bicarbonate reclamation, defective, 481
Bicarbonate regeneration, defective, 481
Biemond syndrome, 409
Bilateral breast cancer, 100
Bilateral carotid body tumor, 99
Bilateral rudimentary testes, 305
Bile duct disorders, 341
Bile duct proliferation, 468
Bile excretion, defective, 343
Bile salts, 343
Biliary atresia, 341
Bilirubin, 342-43
Bishydroxycoumarin, 251-52, 260, 262
Blackfan Diamond syndrome, 361
Bleeding tendency
in albinism, 560
in Lobstein disease, 529
see also Hemophilia
Blepharophimosis, 424
Blepharoptosis, 425
Blindness, 421, 434
in metabolic disorders, 416
Usher syndrome, 451
Blood-brain barrier, 185
Blood group
genetic factors and drug reactions, 259
genetic marker, 8
Blood group A, 346
Blood group O, 346, 505
Bloom syndrome, 73-74, 565
Bount's disease, 239
Blue nevus, 557, 563
Blue sclera, 529, 535
B-lymphocyte (B-cell), 122-23, 125-26, 128-33
Bone abnormalities
fractures, in Lobstein disease, 529
in SCID, 510-11
see also Skeletal disorders
Bone marrow cells, 37
Bone marrow transplant, in SCID, 146
Boveri, Theodor, 85
Bowlegs, 527-28, 535
Brachydactyly, 527
Brain stem disorders, 401-2

Branched-chain amino acid metabolism defects, 201-3, 365
Brancher disease, 189
Branchioskeletogenital syndrome, 589
Brandywine-type dentinogenesis imperfecta, 587
Breast cancer, 99-102
  and Klinefelter syndrome, 112
Breast self-examination, 100
Breast tumors, maternal, and sarcomas of children, 96
Brittle bone disease, 529
Broad-beta disease, 281-82
Bronchi, abnormalities of, 495-501
Bronchiectasis, 501, 507-8
Bronchogenic carcinoma, 505
Bronchogenic cyst, 502
Bronchopneumonia, 505
Buccal mucosa cells, 44-45
  in Klinefelter syndrome, 50
  in Turner syndrome, 48-49
Bulbar paralysis of childhood, 402
Buphthalmos, 427, 431
Burkitt lymphoma, 73, 86-87
Bursa-equivalent lymphocyte system, 121-23. *See also* B-lymphocyte
Bursa of Fabricius, 121
BW15 antigen, 319
Byler disease, 343

Calicification of the adrenals, 302
Calcification of the basal ganglia, 396, 479
Calcitonin, in thyroid carcinoma patients, 97-98
Calcium, serum, low levels in pseudohypoparathyroidism, 480
Calcium leak, of RBC membrane, 363
Calcium metabolism disorders, 396
  in malignant hyperthermia, 383
Calcium-phosphate homeostasis, disorder of, 476-77
Calcium reabsorption defects, 485
Campomelic dwarfism, 527-28
Camptodactyly, 502
Cancer, 85-120. *See also* specific cancers, and specific organs and tissues
Cancer families (cancer fraternities), 89-90
Candidiasis, 591-92
  in hypoparathyroidism, 296
  in myeloperoxidase deficiency, 142
  in T-cell system disorders, 133-34
Caplan syndrome, 505
Carbamylphosphatase synthetase, 196
Carbohydrate chain alteration, in inclusion-cell disease, 178
Carbohydrate diet, in treatment of acute intermittent porphyria, 214

Carbohydrate metabolism disorders, 185-94
  liver involvement, 345
  ocular findings in, table, 420
Carcinoma
  in Fanconi syndrome, 74
  in MEN I, 97
  of stomach, 346
Cardiovascular disorders, 269-84
  and deafness, 452-53
  in Leopard syndrome, 450
  in Marfan syndrome, 232-33
  in muscular dystrophy, 374
  in Noonan syndrome, 306
  in trisomy-13 syndrome, 57
Cardiovascular teratogens, table, 278
Carnitine deficiency, 382
Carnitine palmityltransferase, defect of, 381
Carnosinase deficiency, 205
Carnosine, 205
Carpenter syndrome, 308, 393
Carriers, genetic, 21-22
  Beckwith-Wiedemann syndrome, 289
  genetic counseling, 598-99, 609
  hemophilia, 366-67
  melanoma, 110
  MPS II, 172
  muscular dystrophy, 375
  ocular albinism, 561
  prenatal diagnosis, 614
  retinoblastoma, 95
  Wilms' tumor, 90
Cartilage-hair hypoplasia, 135-36, 530
Catalase deficiency, 245
Cataracts, 418, 420-24, 432
  in Alport syndrome, 465
  in cerebrotendinous xanthomatosis, 180
  in chondrodysplasia punctata, 523-24
  in galactose metabolism disorders, 192
  and hearing loss, 454
  in Marinesco-Sjögren syndrome, 401
  and muscular dystrophy, 376
  in pseudohypoparathyroidism, 479
  in Rothmund-Thomson syndrome, 565
  and skin disorders, 564
  tables of, 432-33
Catecholamine-secreting neural tumor, 95, 305
Cat's eye syndrome, 336
Cattle, mannosidosis in, 177
Cauliflower ear, 524
Cebocephaly, 286
Celiac disease, 124, 338-39, 346
Cell-culture techniques, 37, 40
Cell-mediated immune defects, in gastric cancer families, 103
Cell-specific antigen, 87
Cellular defects, in phagocytic bactericidal system disorders, 139

Cellular origin of cancer, studies of, 86-87
Cementum, tooth, defects of, 577
Central core diseases, 384
Central nervous sytem disorders
  adrenoleukodystrophy, 302
  Hartnup disease, 476
  incontinentia pigmenti, 557
  malformations, 392-95
  muscular dystrophy, 374
  neuronal ceroid lipofuscinosis, 182-83
  red cell metabolism errors, 355
  trisomy, 13, 57
  virilizing adrenocortical tumors, 304
Centromere, 35
Centronuclear myopathy, 385
Ceramidase deficiency, 167
Ceramide accumulation, 167
Ceramide degradation, 158
Ceramide β-galactosidase deficiency, 167
Ceramide trihexosidase deficiency, 486
Cerebellum, disorders of the, 397-401
  ataxia associated with hearing loss, 452
  dysfunction in Hartnup disease 474
  hemangioblastoma in von Hippel-Lindau disease, 394
  histologic changes in cerebrotendinous xanthomatosis, 181
  micropolygyria in muscular dystrophy, 376
Cerebral gigantism, 288
Cerebral hemispheres, disorders of, 392-95
  demyelination, 302
  micropolygyria in muscular dystrophy, 376
Cerebral palsy, in Lesch-Nyhan syndrome, 217
Cerebrohepatorenal syndrome (Zellweger syndrome), 486-87
  pyloric stenosis, 337
  renal cysts, 472
  stippling of the epiphyses, 524
Cerebrosides degradation, 158
Cerebrotendinous xanthomatosis, 180-81
Ceroid, 183
Ceroid lipofuscinosis, 417
Ceruloplasmin deficiency, 343, 382
CESD, *see* Cholesteryl ester storage disease
C1 esterase inhibitor deficiency, 145
CGD, *see* Chronic granulomatous disease of childhood
Chain-terminating codon, point mutation in, 9
Charcot-Marie-Tooth syndrome, 406, 466
Chédiak-Higashi syndrome, 139-40, 142, 364, 559-61

Chemotaxis, disorders of, 128, 136-39
  in hyper-IgE syndrome, 134
  Wiskott-Aldrich syndrome, 135
Cherry-red macula, 158, 160, 164, 417-18
Cherry-red spot-myoclonus syndrome, 177
Chest wall deformities, 511
Chiari-Frommel syndrome, 288
Chief-cell hyperplasia of the parathyroids, 298
Childhood cancers, two-hit model for oncogenesis, 88
Childhood polycystic kidney disease, 468-69
Chimerism, 318
Chlorpromazine accumulation, 260
Chlorpromazine treatment, in intermittent porphyria, 214
Choanal atresia, 495
Cholestanol accumulation, 181
Cholestasis, 343
Cholesterol, 279-81
Cholesterol esterase deficiency, 510
Cholesteryl ester storage disease, 180
Cholestyramine, 343
Chondrodysplasia punctata, 567
  Conradi-Hunermann type, 523, 553
  rhizomelic type, 524
Chondrodystrophic dwarfism, 520
Chondrodystrophic myotonia, 380
Chondrodystrophies, 519-44
Chondroectodermal dysplasia, see Ellis-van Creveld syndrome
Chondroitin sulfate, 168
  in Gm$_2$ gangliosidosis, 161
  in Maroteaux-Lamy syndrome, 174
  in Morquio syndrome, 173
Chondrosarcoma, 112
Chorea
  in basal ganglia calcification, 396
  hereditary nonprogressive, 397
  see also Huntington chorea
Choreoathetosis, 397
  in basal ganglia calcification, 396
  in Lesch-Nyhan syndrome, 217
Choroidal anomalies, 434-37
Choroideremia, 436-37
Chromatid, 35
Chromatin, sex, see Sex chromatin
Chromosome, 9-10
  genetic maps, 11-15
  linkage of gene loci, 10-17
Chromosome aberrations and disorders, 42-44
  in congenital heart disease, 271-72
  in GI tract disorders, 344-45
  incidence, 40-41
  and neoplasia, 71-76

ocular manifestations in, table, 424
  see also Chromosome deletion; Chromosome inversion; Chromosome translocation; Monosomy; Mosaicism; Trisomy
Chromosome analysis, for Down syndrome, 55
Chromosome deletion, 10, 42-43
  autosomal abnormalities, 52-53, 64-71
  in chronic myelogenous leukemia, 72
  and embryonic sexual differentiation, 311
  facial clefts, 333
  growth hormone deficiency, 288
  hemoglobin chains, 352
  polycythemia vera, 73
  primary hypogonadism, 308
  Turner syndrome, 48
Chromosome identification, 35
Chromosome imbalance, mechanisms of, 40-44
Chromosome insertion, 352
Chromosome instability
  in Bloom syndrome, 73
  in Fanconi pancytopenia, 361
Chromosome inversion, 42-43
  trisomy, 18, 57
Chromosome 4p-syndrome, see Wolf-Hirschhorn syndrome
Chromosome 5p-syndrome, see Cri du chat syndrome
Chromosome 9p-syndrome, 67-68
Chromosome 13q-syndrome, 67-69
  retinoblastoma, 95
Chromosome 18p-syndrome, 68-69, 288
Chromosome 18q-syndrome, 69-70
Chromosome 21q-syndrome, 70-71
Chromosome 22q-syndrome, 70-71
Chromosome translocation, 10, 42, 44
  in acute leukemia, 114-15
  in chronic myelogenous leukemia, 72
  in Down syndrome, 53-55
  and embryonic sexual differentiation, 311
  in trisomy 13, 57
  in trisomy 18, 57
Chronic autoimmune hepatitis, 346
Chronic familial neutropenia, 364
Chronic glomerulonephritis, 144
Chronic granulomatous disease, 140-41, 355, 511
Chronic hereditary nephritis, see Alport syndrome
Chronic lymphocytic leukemia, 73
Chronic mucocutaneous candidiasis, 134

Chronic myelogenous leukemia, 72, 86-87, 113-15
Chylomicrons, 280
Ciliary dyskinetic factor, in cystic fibrosis, 506
Cinchophen, 252
C3 inhibitor deficiency, 145
Cirrhosis, 509
  antitrypsin deficiency, 344
  in brancher disease, 189
  in cystic fibrosis, 341
  neonatal, 342
  in vitamin D-dependency rickets, 478
Citrullinemia, 197, 568
Clear-cell hyperplasia, in hyperparathyroidism, 298
Clear vacuoles, in Hurler syndrome, 169
Cleft defect
  growth hormone deficiency, 286
  see also Cleft lip; Cleft palate
Cleft lip, 333-34, 499
  recurrence risk, 28, 334
Cleft palate, 333-34, 499
  in branchioskeletogenital syndrome, 589
  in diastrophic dwarfism, 524
  in Kallman syndrome, 307
  recurrence risk, 28, 334
  syndromes associated with, table, 500
Cleft scrotum, 336
Clones
  in ataxia-telangiectasia, 75
  in xeroderma pigmentosum, 76
Cloverleaf skull, 523
Clubfoot
  in diastrophic dwarfism, 524
  in familial spastic paraplegia, 403
  recurrence risk, 28
CML, see Chronic myelogenous leukemia
Coagulation abnormalities, 366-67
Cochlear disorders, 444, 447-48, 450
Cockayne syndrome, 452, 564, 567
Codon, 4-5
  in point mutation, 7
Cofactors, in collagen synthesis, 230
Coffin-Siris syndrome, 556
Colitis, 338
Collagen, 229-30
  abnormalities in Ehlers-Danlos syndromes, 235, 486
Collagenase inhibition, 509
Collodion baby, 549, 552
Coloboma, 421, 424, 426, 434-35, 437
Colon, cancer of the, 101-3, 107-8, 340
Colon diverticula, 336
Color vision, 437
Combined hyperlipemia, 282
Common variable hypogammaglobulinemia, 132-33

636   Index

Complementation, in Gm₁ gangliosidosis, 160
Complement system, 124, 128
  disorders and deficiencies, 137-39, 142-46
Complete heart block, 272
Complex genetic disorders, and immunodeficiency, 135-36
Complicated microphthalmia, 426
Concanavalin A, T-cells' response to, 123
Conditional probability, 24
Conductive hearing loss, 444, 446-49
Cone dystrophies, 424
Congenital adrenal hyperplasia, 303-4, 312
Congenital alveolar dysplasia, 503
Congenital amaurosis, 434
Congenital anemia, 361
Congenital bronchiectasis, 501
Congenital cataracts, 433
Congenital conditions, 17
Congenital dislocation of hip, recurrence risk, 28
Congenital dyserythropoietic anemia, 360
Congenital glaucoma, 431
Congenital heart disease, 271-79
  chromosomal aberrations, table, 272
  Ellis-van Creveld syndrome, 525
  etiologic basis of, table, 271
  and primary pulmonary hypertension, 504
  recurrence risk, 275-77
  in short rib-polydactyly syndrome, 527
  in thanatophoric dwarfism, 523
  in Turner syndrome, 47
Congenital hemihypertrophy, 472
Congenital hepatic fibrosis, 468
Congenital hyperthyroidism, 295
Congenital hypertrophic pyloric stenosis, 337
Congenital hypogammaglobulinemia, 131-32
Congenital hypoparathyroidism, 296
Congenital hypothyroidism, 290-92
Congenital ichthyosis, 307
Congenital intestinal aganglionosis, see Hirschsprung disease
Congenital lymphedema, 562-63
Congenital malformations, 26
  recurrence risk, 28
Congenital multicystic disease, 469
Congenital muscular dystrophy, 376
Congenital nephrotic syndrome, 472
Congenital neutropenia with eosinophilia, 364
Congenital nonspherocytic hemolytic anemias, 354, 356
Congenital nystagmus, and duodenal ulcers, 335
Congenital ophthalmoplegias, 425
Congenital ptosis, 425
Congenital renal hypoplasia, 466
Congenital retinoschisis, 433
Congenital skin defects, 547
Congenital subluxation of the lens, 431
Connective tissue disorders, 229-44
  and complement deficiencies, 144
  homocystinuria 198
  ocular findings in, table, 421
Connective tissue fibers, degeneration of, 509
Connective tissue matrix, 229
Conradi disease, 523, 553
Consanguinity, 21
Constitutional infantile panmyelopathy, see Fanconi syndrome
Convulsions
  in branchioskeletogenital syndrome, 589
  in chronic nephritis, 466
  in pseudohypoparathyroidism, 479, 591
Copper metabolism disorders
  in liver disorders, 343
  in Menkes' syndrome, 237, 382, 561
  in Wilson's disease, 396, 487
Coproporphyria, 214, 216
Coproporphyrinogen, 208
Coproporphyrinogen oxidase deficiency, 216
Cori disease, 188
Corneal disorders, 424, 426-29
  clouding, and hearing loss, 454
  keratosis palmaris et plantaris, 554
  metabolic disorders, table, 416-20
Cornea plana, 427
Coronal dentin dysplasia, 587
Coronary heart disease, 279
Cornelia de Lange syndrome, 336-37
Corpus callosum, agenesis of the, 392
Corticosteroid treatment, in glaucoma, 430
Corticotropin deficiency, 299
Corticotropin releasing factor (CRF), deficiency, 299
Cortisol deficiency, 304, 312
Cosynthetase (CoS) deficiency, 212
Coumarin, 30, 32, 251-53
Counseling, see Genetic counseling
Coupled phase, of mutant gene loci, 11, 16
Cowden's disease, 101
Coxsackie virus B, 278
CPKD, see Childhood polycystic kidney disease
Craniofacial dysostosis, 393, 449
Craniofacial dysplasias, ocular findings in, table, 423
Craniometaphyseal dysplasia, 448
Craniostenosis, 393
Creatine phosphokinase (CPK), 375
Cretinism
  goitrous, 451
  stippling of epiphyses, 524
Creutzfeldt-Jakob disease, 391
Cri du chat syndrome, 65-67
Crigler-Najjar syndrome, 342-43
Crossing over, 11
Cross-linking, in connective tissue formation, 230-31
Cross syndrome, 559, 561
Crouzon syndrome, 393, 449
Cryptorchidism, 112
  in Cockayne syndrome, 564
  and ear malformations, 446
  in Noonan's syndrome, 306
C syndrome, 502
Curvilinear bodies, in neuronal ceroid lipofuscinosis, 183
Cushing syndrome, 288, 304
Cutaneous horns, in pachyonychia congenita, 569
Cutaneous neoplasms, 103, 108-10
Cutaneous webs, 563
Cutis laxa, 235
Cutis verticis gyrata, 563
Cyanosis, 351
Cyclic AMP
  in pseudohypoparathyrodism, 297
  urinary excretion, 479
  in vitamin D-resistant rickets, 477
Cyclic AMP-dependent kinase, abnormalities in, 190
Cyclic neutropenia, 364
Cyclohydrolase deficiency, 360
Cyclopia, 286
Cyclopropane, and malignant hyperthermia, 258
Cynodontism, 592
Cynthiana variant, of atypical pseudocholinesterase, 251
Cystathionine synthetase deficiency, 198, 237
Cystathionuria, 200-1
Cystic fibrosis, 338, 340-41, 495, 506-8
  risks of producing a child with, table, 507
Cystic hygromas of the neck, 562
Cystic kidneys, prenatal diagnosis, 615. See also Polycystic kidney disease; Renal cysts
Cystic liver disease, 342
Cystine storage disease, 478
Cystine transport system, 474
Cystinosis, 295, 419, 424
Cystinuria, 206, 473-75
Cytochrome enzyme system, 17
Cytogenetic disorders, 35-84
Cytomegalovirus, 275, 510
Cytoplasmic membraneous bodies, 156

Cytosine, 4-7
Cytosol receptor for dihydrotestosterone, deficiency of,
  316
Cytotoxicity responses, in
  lymphocytes, 128

Darrier-White disease, 555
Deafness, see Hearing loss
Degenerative diseases, 391
Degranulation, in phagocytosis,
  140
Dejerine-Sottas disease, 406
Deletion mutation, 9. See also
  Chromosome deletion
Dementia, 394-96, 564
Demyelination, 165-66
Dental abnormalities, see Tooth
  abnormalities
Dentigerous cyst, in branchio-
  skeletogenital syndrome, 589
Dentin, defects of, 577, 587-92
Dentino-osseous dysplasia, 588-89
Dentogenesis imperfecta, 236, 529
Deoxyribonucleic acid, see DNA
Deoxyribose, 4
Depression, 259
Dermatan, defective catabolism of,
  161
Dermatan sulfate, 168-69, 172-74
Dermatitis herpetiformis, 124
Dermatoglyphics, 545-47
Dermatomyositis, 144
De Sanctis-Cacchione syndrome,
  76, 108
17, 20-Desmolase deficiency,
  314-15
20, 22-Desmolase deficiency, 314
Desmosine, 231
Desoxycorticosterone, excessive
  production of, 303
Deuteranomaly, 437
Deuteranopia, 437
Dexamethasone hypertension, 257
Dextroamphetamine, and con-
  genital heart disease, 278
Diabetes, in hypertriglyceridemia,
  282
Diabetes insipidus, 289-90
Diabetes mellitus, 289-90, 318-21
  in Werner syndrome, 564
Diabetic ketoacidosis, defective
  phagocytosis in, 140
Diastrophic dwarfism, 524
Diazepam, 218
Diazoxide, 188
Dibucaine, 249
Dibucaine number (DN), 249-50
Dichromatic vision, 437
Dicoumarol resistance, 253
DiGeorge syndrome, 134, 296
Digital patterns, 546
Dihydrofolate reductase
  deficiency, 360
Dihydropteridine reductase
  deficiency, 195
Dihydrotestosterone, 311
5-α-Dihydrotestosterone
  deficiency, 315-16

Dimethyl adipimidate, 363
Diphenylhydantoin, 30, 32
  deficient parahydroxylation of,
  251
  treatment for myotonia, 380
  treatment for paroxysmal
  choreoathetosis, 397
2, 3 Diphosphoglycerate, 354
2, 3 Diphosphoglycerate phosphatase, 356
Diploid number of chromosomes,
  9
Diplopia, in ataxia, 401
Distal muscular dystrophy, 377
Distichiasis, 424
Diverticula, 336
DNA, 4-6
  repair mechanism, 76, 108
  synthesis by T-cells, 127
DNA-dependent DNA synthetase,
  4
DNA-dependent RNA synthetase,
  4
Döhle bodies, 364
Dominant inheritance, 17-20, 23
Dosage compensation, 22
Double fertilization, in
  chimerism, 318
Double helix, of DNA, 5
Down syndrome (Trisomy 21), 35,
  52-55
  aganglionosis, 338
  and autoimmune thyroid
    disease, 294
  cataracts, 432
  congenital heart disease, 271-72
  dermatoglyphics, 546-47
  duodenal obstruction, 336
  GI tract malformation, 345
  hair, 566
  incidence, 41
  leukemia, 71, 73
  maternal age-related risk, 614
  prenatal diagnosis, 77
  pyloric stenosis, 337
  stippling of epiphyses, 524
Doyne's choroiditis, 437
Drug disposition, genetic control
  of, 260-63
Drugs, 245. See also Pharmacogenetics
Drusen, 437
Duane's phenomenon, 425
Duarte variant, in galactosemia,
  193
Dubin-Johnson syndrome, 343
Duchenne type muscular
  dystrophy, 22, 374-75
  prenatal diagnosis, 614
Duncan's disease, 136
Duodenal atresia, 335-36
Duodenal ulcers, 334-35, 346
Dw2 (HL-A7a) antigen, 124
Dwarfism, 519-44
  and cartilage-hair hypoplasia,
    135-36
  dyschondroplastic, 502
  mesomelic, 501
  Winchester syndrome, 237

Dyggve-Melchior-Clausen
  disease, 541
Dysautonomia, 393, 409
Dyschondroplastic dwarfism, 502
Dyschondrosteosis, 526, 534-35
Dysgammaglobulinemic
  syndromes, 125
Dysgerminoma, 111, 309, 313
Dyshidrosis, 556, 592
Dyskeratosis congenita, 557-58
Dysphagia-hypospadias
  syndrome, 499
Dysprothrombinemia, 367
Dysrhythmias, in hyperoxaluria,
  484
Dystonia, 396-97, 455
Dystopia canthorum, 449

Ear, structural development of,
  444
Ear malformations, 445-46
  in diastrophic dwarfism, 524
E-B virus, 86
Ectodermal disorders affecting
  hair
  growth, 566
Ectodermal dysplasias, 555-57,
  582-86, 592
Ectodermal neoplasma, in xero-
  derma pigmentosum, 76
Ectopic hormones, secretion by
  medullary thyroid
  carcinoma, 98
Ectopic lenses, 430-31
  in Marfan syndrome, 232-33
Ectopic pupils, 430-31
Ectropion, 549, 552
Edema, 145
Edwards syndrome, see Trisomy 18
Effector cells, 123
Ehlers-Danlos syndromes, 233-35,
  486
  GI hemorrhage, 335
  renal cysts, 472
  retinal detachment, 433
  tooth abnormalities, 589
Elastase inhibition, 509
Elastin, 230
Elastic fibers, 229-31
Electrocardiographic abnormal-
  ities, associated with deafness,
  452-53
Electromyogram examination,
  375, 380-81
Electrophoresis, 279-80
Electroretinograph, 434
Elliptocytosis, 362
Ellis-van Creveld syndrome, 501,
  511, 525
Embryo
  hemoglobin, 349
  immune system, development
    of, 122
  sexual differentiation, 309, 311
Embryonal tumors, 90, 94-96
Emery-Dreifuss dystrophy, 375
Emphysema, 495, 509
Enamel, tooth, defects of, 575-77,
  580-81, 587-88, 590-92

Encephalocele, 392
Endocrine diseases, 285-329
  neoplasms, 91, 96-99
Endocytosis, 156-57
Endoglycosidases, 168
Enolase, 356
Enteritis, 338
Environmental factors, 26
  congenital heart disease, 274
  effect on fetus, 29-32
  gastrointestinal cancer, 102
  lung tumors, 115
  skin neoplasms, 108
Enzyme replacement therapy, 146, 183-85
Enzyme variations, 8
Eosinophilia, 364
Epicanthus, 424
Epidermolysis bullosa, 547-49, 588
  peptic ulcers, 335
Epidermolytic hyperkeratosis, 551
Epilepsy, myoclonus, 383
Epilepsy-like dyspnea, 394
Epimerase deficiency, 193
Epiphyseal dysplasias, 519, 534
Epiphyses, stippling of, 523-24
Epstein-Barr virus, 86
Erythrocytes, see Red blood cells
Erythrokeratoderma variabilis, 556-57
Erythropoietic porphyrias, 210-12
Erythropoietic protoporphyria, 212
Erythropoietin, 366
Esophageal atresia and stenosis, 335-36, 501
Esophageal cancer, 102-3
  Fanconi syndrome, 74
  keratosis palmaris et plantaris, 554
Esophageal ulceration, 335
Esotropia, 415
Essential hypertension, 270
Essential tremor, 335, 396
Estriol levels, maternal, 300
Ethanol metabolism, 258-59
Ether, and malignant hyperthermia, 258
Exocrine pancreatic insufficiency-dwarfism syndrome, 536-37
Exocytosis, 156
Exoenzyme, 156
Exomphalos-macroglossia-gigantism syndrome, 136
Expressivity, 19-20
External auditory canal, 444, 446
External ear disorders, 445-46
  cystic swelling, 524
External ophthalmoplegia, 402, 425
External rectus muscle, paresis of, 425
Extraosseous calcification, 591
Eye disorders, 415-41
  albinism, 559-61
  and deafness, 454
  developmental, 426
  Ehlers-Danlos syndrome, 235
  incontinentia pigmenti, 557
  in kidney diseases, 464-65, 471
  Marfan syndrome, 232
  ocular muscular dystrophy, 377-78
  steroid application, response of ocular pressure to, 257
  in Stickler syndrome, 537
  Weill-Marchesani syndrome, 238
Eyelashes, 424-25
Eyelids, 424

Fabry disease, 162-63, 406, 409, 418, 486, 562
Facial anomalies, and lung malformations, 502
Facial clefts, 333-34, 425-26, 434
Facial diplegia, 385, 402
Facioscapulohumeral dystrophy, 376
Factor VIII system, in coagulation abnormalities, 366-67
Fairbank disease, 534
Familial calcification of basal ganglia, 396
Familial cancer, 89-116
  simply inherited neoplasms, table, 91-94
Familial diseases, 17
Familial dysautonomia, 338, 393, 409
Familial eosinophilia, 364
Familial exudative vitreoretinopathy, 434
Familial foveal retinoschisis, 433
Familial glucocorticoid deficiency, 301
Familial hemorrhagic telangiectasia, see Osler-Rendu-Weber syndrome
Familial hyperkalemic periodic paralysis, 380
Familial hypokalemic periodic paralysis, 380-81
Familial juvenile nephronophthisis, 470
Familial Mediterranean fever, 464, 467
Familial paroxysmal choreoathetosis, 397
Familial renal glucosuria, 480
Familial spastic paraplegia, 403-4
Familial thrombocytopenia, 486
Family planning, 29-32
Fanconi syndrome (Fanconi pancytopenia), 74-75, 86-87, 191, 288, 361-62, 364-65, 478-79
Farber's disease, 167, 418
Fasciculations, facial, in striatonigral degeneration, 397
Fatal infantile agrulocytosis, 364
Fat metabolism, disordered, 564
Favism, 357. See also Glucose 6-phosphate dehydrogenase deficiency
Favre syndrome, 434
Fazio-Londe disease, 402
Fc-receptor-bearing cells, assay for, 125
Fecal coproporphyrin, 212, 214, 216
Fecal protoporphyrin, 214
Female hypogonadism, 308-9
Female pseudohermaphroditism, 310, 311-12
Femoral hypoplasia-unusual facies syndrome, 530
Ferrochelatase deficiency, 212
Fertile eunuch syndrome, 307-8
Fetal alcohol syndrome, 29, 31-32
Fetal liver-cell transplants, in treatment for SCID, 146
Fetal Warfarin syndrome, 30, 32
α-Fetoprotein, 29, 392, 472, 614-15
Fetoscopy, 613-14
Fetus
  effect of drugs on, 29-32, 312
  electrocardiography, 457
  hemoglobin, 57, 349
  immune system, 122
  see also Prenatal diagnosis
Fibrils, in collagen synthesis, 230
Fibrinase deficiency, 367
Fibroblasts, culture, 40
Fibrodysplasia ossificans progressiva, 238
Fibrous dysplasia, pigmented nevi in, 557
Fingernails, see Nails
Fingerprint body myopathy, 385
Flank pain, in polycystic kidney disease, 470
Fletcher factor deficiency, 367
Flood factor deficiency, 367
Fluorescent Q banding, 45
Fluoride, pseudocholinesterase inhibition, 249
Foam cells, 465
Focal dermal hypoplasia, 564
  ocular changes in, 421
Folate metabolism, abnormal, 359-60
Folic acid, 200, 359
Folic acid antagonists, 278
Follicle-stimulating hormone deficiency, 306
Fontanelles, open, 535
Forbes-Albright syndrome, 288
Forbes disease, 188
Forme fruste, 391
Formiminotransferase deficiency, 360
Fractures, bone, in Lobstein disease, 529
Frame shift mutation, 9, 352
Fructokinase deficiency, 191
Fructose 1,6-diphosphatase deficiency, 192
Fructose metabolism, disorders of, 191-92
Fructosemia (fructose intolerance), 191-92
Fructose 1-phosphate adolase defect, 191
Fructosuria, 191
Fucosidase deficiency, 175
Fucosidosis, 175, 177, 416

Fundus of the eye abnormalities of, 434-35
Fungal infections
 in hypoparathyroidism, 296, 591-92
 in myeloperoxidase deficiency, 142
 in T-cell system disorders, 133-34
Funnel chest, 511
Fusion genes, 353

GAG, see Glycosaminoglycans
Galactokinase deficiency, 192
Galactose metabolism, disorders of, 192-93
Galactosemia, 192-93
Galactose-1-phosphate uridyl transferase, 192
α-Galactosidase, 162
β-Galactosidase, 158, 160
Galactosuria, 192
Galactosylceramide, 167
Gamma-globulin therapy, 146
Ganglion cell deficiency, 337-38, 447
Ganglioside degradation, 158
Gangliosidosis, 158-62, 167, 417, 524
Gastric ulcers, 334
Gastrointestinal bleeding, 335
Gastrointestinal cancer, 93, 102-8, 340, 346
 and ABO blood groups, 16
Gastrointestinal tract disorders, 331-47
Gaucher cells, 163
Gaucher disease, see Glucosylceramidosis
G-deletion syndromes, 71
Gene, 4-10
 deficiency, 16
 duplication, 16
 in families, 17-23
 linkage of loci, 10-17
Gene penetrance, see Penetrance
Genetic carriers, see Carriers, genetic
Genetic code, 4, 7
 hemoglobin beta chain, 9
Genetic counseling, 29, 597-611
 for congenital heart disease, 275
 for deaf, 455-57
 for endocrine diseases, 285
 in prenatal diagnosis, 615
Genetic disease, 17
Genetic-environmental interaction, see Environmental factors; Multifactorial inheritance
Genetic heterogeneity, 25
 acatalasia, 248
 albinism, 558
 Alport syndrome, 465
 drug response, 259-63
 Fanconi syndrome, 479
 galactosemia, 193
 gout, 219
 Marfan syndrome, 233

MPS IV, 173
 succinylocholine sensitivity, 250
 xeroderma pigmentosum, 76
Genetic map, 11-15
 nonalpha hemoglobin chain, 350
Genetic marker, 8
 cancer origin studies, 87
 GI disorders, 345-46
Genetics, history of, 3-4
 cytogenetics, 35
 immunodeficiency syndromes, 121
 neoplasia, 85
Genetics glossary, 617-30
Genitalia, ambiguous, 309
 in adrenal hyperplasia, 303
 in Wilms' tumor patients, 90
 see also Hermaphroditism; Pseudohermaphroditism
Genitourinary tract disorders, 485-86
 and adrenocortical disorders, 303-4
 cancer, 93, 110-12
Genodermatoses, 106
Gestation, prolonged, and placental steroid sulfatase deficiency, 301
G group chromosome, in leukemic cells, 114
Giant cell hepatitis, 342
Gibbus, in achondroplasia, 520
Giedon syndrome, 537-38
Giemsa stain, 36-38
Gigantism, 288
Gigantoblast, 360
Gilbert disease, 342
Glanzman-Naegli thrombasthenia, 365
Glaucoma, 421, 424
 in anterior chamber cleavage anomalies, 430-31
 in microcornea, 427
 response of ocular pressure to steroids, 257
Gliadin fraction, of gluten, 338
Glioblastoma, 98
Glioma, 98
Globoid cell leukodystrophy, 166-67
Globoside, 161-62
Glomerular involvement, in Alport syndrome, 465
Glomerulonephritis, 144, 486
Glomus tumors, 99, 563
Glossary, 617-30
Glucagonoma, 97
Glucocerebroside β-glucosidase, 163
Glucocorticoid deficiency, 301
Glucocorticoid suppressible aldosteronism, 305
Glucoglycinuria, 207, 476
Glucose-galactose malabsorption syndrome, 480
Glucose metabolism, abnormalities in, 185-91

Glucose 6-phosphatase deficiency, 185
Glucose 6-phosphate dehydrogenase deficiency, 142, 253-56, 354-57
 genetic marker, 87
Glucose 6-phosphate dehydrogenase variants, 356-57
Glucose tolerance abnormalities
 in acute intermittent porphyria, 213
 genetic syndromes associated with, 320-21
 in hyperlipoproteinemias, 282
β-Glucosidase deficiency, 188-89
Glucosuria, 480
Glucosylceramidosis (Gaucher disease), 163-64, 363
β-Glucuronidase deficiency, 174
Glucuronyl transferase, 342
 deficiency, 343
Glutaric aciduria, 204
Glutathione, 255, 354, 357
Glutathione peroxidase deficiency, 357
Glutathione reductase deficiency, 357
Glutathione synthetase abnormality, 205, 357
Gluten intolerance, 338-39, 346
 HL-A8 associated with, 124
Glyceraldehyde 3-phosphate dehydrogenase, 356
Glyceric aciduria, 484
Glycine, of collagen, 229
Glycine excretion, 466
Glycine metabolism defects, 203-4
Glycinuria, 476
Glycogen storage diseases, 185-90, 381
Glycogen synthetase deficiency, 191
Glycolipid storage, in inclusion-cell disease, 178
Glycolytic enzymes, abnormalities of, 356
Glycophorin, in RBC membrane, 362
Glycoprotein microfibrils, 230
Glycoprotein protease inhibitor, 508
Glycosaminoglycans (GAG)
 catabolism defects, 167-68
 storage, 174, 178
Glycosphingolipidoses, 157-67
 biochemical aspects, table, 154
Glycosphingolipids, 157
Glycosyltransferase deficiency, 167
GM$_1$ gangliosidosis, 158-60, 417
GM$_2$ gangliosidosis, 160-62, 417
GM$_3$ gangliosidosis, 167
Goiter, 292-93, 295
 Pendred syndrome, 451
 PTC tasting ability, 257
Goldenhar syndrome, 487, 502, 504
Gonadoblastoma, 313
Gonadotropin, 287

deficiencies, 300, 305-7
Gonadotropin-releasing hormone deficiency, 306
Gonads, disorders of the, 305-9, 313-14
  tumors, 110-12, 313
  Turner syndrome, neoplasia in, 49-50, 72
Gout, 219, 485
Granuloma annulare, 477
Graves' disease, 294-95
Grebe disease, 522-23
Ground substance, of connective tissue, 229, 231
Growth acceleration, 288
Growth hormone deficiency, 286-88
Growth hormone receptor deficiency, 287
Growth retardation
  and dysplasia of secondary dentin, 589
  in hypothyroidism, 292
  in kidney diseases, 481
  in osteogenesis imperfecta, 236
  in pseudohypoparathyroidism, 480
  in Werner syndrome, 564
G (dysphagia-hypospadias) syndrome, 499
Guanine, 4-7
Gynecomastia, 309-10
Gyrate atrophy, 437

Hailey-Hailey disease, 554
Hair, disorders affecting the, 566-69
  albinism, 559-61
  ectodermal dysplasia, 556, 582-85
  Menkes' syndrome, 237
Hair bulb test, in albinism, 558, 561
Hallerman-Streiff syndrome, 567
Hallervorden-Spatz syndrome, 395-97
Hallucal area, of the foot, 546
Halothane, 258, 262
Hamartomatous defects, 304, 394, 563
  ocular findings in, table, 421
Hamman-Rich syndrome, 503
Hands, anomalies of the, 336
Haploid number of chromosomes, 9
Haplotype, 240
Hardy-Weinberg law, 21
Harelip, 307
Harlequin fetus, 552
Hartnup disease, 207, 474, 476
Hashimoto's thyroiditis, 294
Hay fever, 124
H-2 chromosomal complex, 124
Hearing loss, 443-61
  and albinism, 562
  in Alport syndrome, 464-65
  in Cockayne syndrome, 564
  and hypoparathyroidism, 296
  incidence of, 444

in Kniest disease, 539
in metaphyseal chondrodysplasia, 533
in osteogenesis imperfecta, 236
in Pendred syndrome, 293
in pseudoglioma, 434
recurrence risks, table, 456
in Wolfram syndrome, 289
Heat-labile opsonins, 139
Heat-stable opsonins, 139
Heinz bodies, 352, 358
Helper cells, 123
Hemangioblastoma, cerebellar, 394
Hemangioma, 365, 563
Hematology, 349-71
Hematopoietic neoplasms, 92
Hematuria, 464, 466, 470, 487
Heme biosynthesis defects, 208
Hemeralopia, 434
Hemihypertrophy, 472
  and hepatoblastoma, 96
  and adrenocortical tumors, 304
  in Wilms' tumor patients, 94
Hemizygous inheritance, 10, 22
Hemochromatosis, 344, 361
  pituitary insufficiency in, 288
Hemoglobin
  and drug sensitivity, 256
  genetic counseling, 598
  and pituitary insufficiency, 288
Hemoglobin A, 349-50
Hemoglobin A_2, 17, 349-50
Hemoglobin anti-Lepore, 16, 353
Hemoglobin Barts, 352
Hemoglobin C, 7, 349, 351
Hemoglobin Capetown, 351
Hemoglobin chains, 16-17, 352
  alpha chain, 10, 16, 349-52
  beta chain, 9-10, 17, 256, 349-51, 365
  delta chain, 17, 350
  epsilon chain, 349
  gamma chain, 17, 349-50
  nonalpha chains, 17
  zeta chain, 349
Hemoglobin C-Harlem, 350
Hemoglobin Chesapeake, 351
Hemoglobin Constant Spring, 9, 352
Hemoglobin E, 8, 10-11, 349
Hemoglobin F, 349-50
Hemoglobin G, 10
Hemoglobin Gower 1, 349
Hemoglobin Gower 2, 349
Hemoglobin H, 256, 352
Hemoglobin Hammersmith, 352
Hemoglobin Kansas, 351
Hemoglobin Kenya, 350, 353
Hemoglobin Köln, 351
Hemoglobin-Korle Bu, 351
Hemoglobin Lepore, 16, 350, 353
Hemoglobin M, 351
Hemoglobin McKees-Rock, 352
Hemoglobin Miyada, 353
Hemoglobin P-Congo, 353
Hemoglobin Portland, 349
Hemoglobin Rainier, 351
Hemoglobin S, 10-11, 349, 353-54

Hemoglobin Santa Ana, 352
Hemoglobin Shepard's Bush, 352
Hemoglobin tetramers, table, 350
Hemoglobin variants, 349-54
  prenatal diagnosis, 614
  table, 351
Hemoglobin Zürich, 256, 351
Hemolysis, 190, 252-56
Hemolytic anemia, 340, 354, 362
Hemophilia, 335, 366
Hemorrhagic telangiectasia, see Osler-Rendu-Weber syndrome
Hemosiderosis, 361
HEMPAS anemia, 360
Heparan sulfate, 168-69, 172, 174
Hepatic carcinoma, 188, 509
Hepatic fibrosis, 468-69, 471
Hepatic porphyrias, 212-13
Hepatitis, 346
  neonatal, 342-44
Hepatobiliary disease, 341-42
Hepatoblastoma, 96
Hepatolenticular degeneration, see Wilson disease
Hepatoma, 344, 478
Hepatomegaly
  alpha-1-antitrypsin deficiency, 509
  fructosemia, 191
  glycogen storage diseases, 185, 189
  lipodystrophy, 564
  polycystic kidney disease, 468
Hepatosplenomegaly, 281, 468
Hereditary benign erythroreticulosis, 360
Hereditary cancer, 89-116
  table, 91-94
Hereditary hemorrhagic telangiectasia, see Osler-Rendu-Weber syndrome
Hereditary persistence of fetal hemoglobin, 352
Hereditary nonprogressive chorea, 397
Hereditary pancreatitis, 340
Hereditary polyneuropathies, 406-8, 409
Hereditary sensory neuropathy, 405
Heritability, 26
Hermansky-Pudlak syndrome, 559-60
Hermaphroditism, 309, 310, 317-18
Hernia, in Kniest disease, 539
Heterochromia irides, 430, 449, 562
Heteromorphisms, karotypic, 40
Heterozygous genes, 10
  carriers, 21
Hexokinase deficiency, 354
β-Hexosaminidase, 160-62
High carbohydrate diet, in treatment of acute intermittent porphyria, 214
High-frequency hearing loss, 447
Hirschsprung disease, 337, 530

recurrence risk, 28
Histadase deficiency, 205
Histidinemia, 204-5
Histocompatibility genes, 86
Histocytosis X, 288
HL-A5 antigen, 124
HL-A7a antigen, *see* DW2 antigen
HL-A8 antigen, 124
HL-A12 antigen, 124
HL-A13 antigen, 124
HLA-B1 antigen, 339
HLA-B8 antigen
  celiac disease, 346
  juvenile-onset diabetes, 319
  malabsorption disorders, 339
  pernicious anemia, 359
  thyroid disorders, 294
HLA-B27 antigen, 346
HLA-D antigen, 124, 128
HLA system, 86
  connective-tissue disorders, 240-41
  hay fever, 124
Hodgkin's disease, 86
Holoprosencephaly, 286, 392
Holt-Oram syndrome, 272
Homocystinuria, 198-99, 237, 360, 419, 431, 568
Homogentisic acid oxidase deficiency, 196, 237
Homologous chromosomes, 10, 16
Homozygous genes, 10, 17
Honeycomb lung, 503-4
Horner syndrome, 95
Host-defense mechanisms, 125-28
Human leukocyte antigen system, *see* HLA system
Hunter syndrome, 172, 416, 510
Huntington chorea, 17, 391, 394
Hurler syndrome, 169-70, 172, 416, 510
Hyaline membrane disease, 505, 509
Hyaluronic acid, 179
H-Y antigen, 311, 318
Hydralazine metabolism, 249
Hydration therapy, in cystinuria, 474
Hydroa aestivale, 210
Hydrocephalus, 392
  prenatal diagnosis, 615
Hydrogen peroxide, 140-41, 293
Hydrolytic enzymes, 153
Hydrops fetalis, 352
25-Hydroxycholecalciferol-1-hydroxylase deficiency, 477
11-Hydroxylase defects, 312
11-β-Hydroxylase deficiency, 303
17-Hydroxylase deficiency, 304
17-α-Hydroxylase deficiency, 314
21-Hydroxylase defects, 303, 312
p-Hydroxylphenylpyruvate hydroxylase, 195
3-Hydroxy-3 methylglutaryl coenzyme A reductase, 280-81
Hydroxyproline, 203-4, 229, 466
Hydroxyprolinemia, 203
Hydroxyproline oxidase defect, 203

3-β-Hydroxysteroid dehydrogenase defects, 304, 312, 314
Hyperalaninemia, 206
Hyperalanuria, 206
Hyperammonemia, 196-98
Hyper-β-alaninemia, 206
Hyperbetalipoproteinemia, 281
Hyperbilirubinemia, 342-43
Hypercalcemia, 298-99, 477, 533, 535
Hypercalcuria, 481, 483, 485
Hypercholesterolemia, 213, 281
Hypercystinuria, 474
Hyperdibasic aminoaciduria, 207, 473-74
Hyperglycinemia, 203-4, 365
Hyperglycinuria, 476
Hyperhydrosis, 569
Hyper-IgE, 134-35, 139
Hyper-IgM, 133
Hyperinsulinism, 96
Hyperkalemic periodic paralysis, 380-81
Hyperkeratosis, 569
Hyperlipidemia, 279-83
Hyperlipoproteinemia, 279-83
Hyperlysinemia, 431
Hyperlysinuria, 204
Hypermetropia, 427, 431
Hypermobile thumbs, 524
Hyperornithinemia, 197, 419
Hyperoxaluria, 206, 484
Hyperparathyroidism, 98, 298, 340, 477, 485
Hyperpituitarism, 288-89
Hyperplastic myelinopathy, 405
Hyperplastic vitreous, 424
Hyperprolinemia, 464, 466-67
Hypersensitivity skin test, 128
Hypertelorism, 392, 589
Hypertension, 270, 504, 509
  in adrenocortical disorders, 303, 305
  in 17-α-hydroxylase deficiency, 314
Hyperthermia, malignant, 258, 383-84
Hyperthyroidism, 295
Hypertrichosis, 210
Hypertriglyceridemia, 282
Hypertrophic neuropathy of Dejerine-Sottas, 409
Hypertrophic subaortic stenosis, 272
Hyperuricemia, 282
Hypervalinemia, 201-2
Hypervitaminosis A, 298
Hypoadrenalism, 286
Hypoaldosteronism, 301
Hypobetalipoproteinemia, 363
Hypocalcemia, 477
Hypocalcified types of amelogenesis imperfecta, 575, 576, 581, 587
Hypochondroplasia, 531
Hypochromic anemia, 358, 360-61, 383
Hypogammaglobulinemia, 125, 131-33

  in infants, 123
  neoplasm development risk, 88, 113
Hypogenitalism, in Laurence-Moon syndrome, 452
Hypoglycemia, 185, 191
  and neuroblastoma, 95
  and pancreatic tumor, 96
Hypogonadism, 305-6
  and ear malformations, 446
  and ichthyosis, 553
  and muscular dystrophy, 376
  in Rothmund-Thomson syndrome, 565
Hypohidrotic ectodermal dysplasia, 555-56
Hypokalemia, 481, 483-84
Hypokalemic alkalosis, 314
Hypokalemic periodic paralysis, 380-81
Hypomagnesemia, 296
Hypomaturation types of amelogenesis imperfecta, 575, 576, 581
Hypoparathyroidism, 295-98
  calcification of basal ganglia, 396
  tooth abnormalities, 591
Hypopharyngeal cancer, 346
Hypophosphatasia, 298, 590
Hypophosphatasia, congenital lethal type, 526-27
Hypophosphatasia tarda, 535-36
Hypophosphatemia, 477
Hypopituitarism, 286-88
Hypoplastic teeth, 564
Hypoplastic types of amelogenesis imperfecta, 575, 576, 580-81
Hypoprothrombinemia, 335, 367
Hypospadia, 304
Hypothalamus disorders, 286, 307, 309
Hypothyroidism, 290-95, 298
Hypotonia, in Leigh disease, 401
Hypotrichosis, 569
Hypoventilation, 505
Hypoxanthineguanine phosphoribosyltransferase deficiency, 133, 218-19, 485

I-cell disease, *see* Inclusion-cell disease
Ichthyosiform erythroderma, 551
Ichthyosis, 301, 425, 549-53
  and hypogonadism, 307
  in mucosulfatidosis, 174
Ichthyosis hystrix, 551
Ichthyosis linearis circumflexa, 553
Ichthyosis vulgaris, 551
Idiopathic Addison's disease, 302-3
Idiopathic benign pneumothorax, 504
Idiopathic Fanconi syndrome, 478
Idiopathic hypoparathyroidism, 296, 591-92
Idiopathic hypopituitarism, 286
Idiopathic seizures, 395

α-L-Iduronidase deficiency, 169, 172
L-Idurosulfate sulfatase deficiency, 172
Ileal transport system, 359
Iminoaciduria, 476
Iminoglycinuria, 203, 206
Imipramine, 259
Immobiliization, prolonged, as cause of hypercalcemia, 298
Immune factors, in thyroid maldevelopment, 291
Immune response, genetic components of, 124
Immune response (Ir) genes, 124, 240-41, 319
Immune system, disorders of, 121-52
Immunodeficiency syndromes, 128-36
 and atopic diseases, 145-46
 lung involvement, 510-11
 relationship to cancers, 87-89, 92, 103, 113
 therapeutic considerations, 146
Immunoglobulin, 122-23, 133
Immunoglobulin A (IgA)
 deficiencies, 103, 113, 133, 135
 elevation, 486
 fetal synthesis, 122
 neonatal synthesis, 123
Immunoglobulin E (IgE)
 elevation, 134-35, 146
 hyper-IgE syndrome, 139
 ragweed pollinosis, 124
 wheal and flare skin responses, 126
Immunoglobulin G (IgG)
 in complement pathway, 142
 fetal synthesis, 122
 neonatal synthesis, 122-23
 ragweed pollinosis, 124
Immunoglobulin M (IgM), 122
 in complement pathway, 142
 deficiency, 88
 elevation, 133
 neonatal synthesis, 123
Immunologic anamnesis, 136
Imperforate anus, 336
Inbreeding, 21
Inclusion-cell (I-cell) disease, 178, 510
Incomplete testicular feminization, 316-17
Incontinentia pigmenti, 557
 ocular changes, 421
Incudostapedial joint, absence of, 446
Incus, malformed, 446
Independent assortment of genes, 10
Indiana variant, galactosemia, 193
Indifference to pain, 404-5, 409
Inducer molecules, 6
Infantile amaurotic idiocy, see Tay-Sachs disease
Infectious mononucleosis, 136
Imflammatory bowel disease, 338
Influenza vaccine, 124

Inner ear abnormalities, 446
Insensitivity to pain, 404-5, 409
Insulin receptor abnormalities, 321
Internal ophthalmoplegia, 425
Intersex states, 486
Interstitial pulmonary fibrosis, 116, 503
Intestinal cancer, 103, 107, 340
Intestinal malabsorption, 530
Intestinal motility, impaired, 337-38
Intestinal obstruction, 335-38
Intestinal pseudoobstruction, 338
Intraocular inflammation, 424
Intraocular pressure, 257, 430-31
 metabolic disorders, table, 416-20
Intrauterine diagnosis, see Prenatal diagnosis
Intrinsic factor (IF) deficiency, 359
Inversion, chromosomal, see Chromosome inversion
Iodide organification defect, 292-93
Iodide trapping defect, 292
Iodine deficiency, in Pendred syndrome, 451
Iodotyrosine coupling defect, 293
Iodotyrosine deiodination defect, 293
Iridodenesis, 232
Iris, disorders of the, 427, 429, 430
Iron, excess body, 361
Isochromosomes
 in primary hypogonadism, 308
 in Turner syndrome, 47-48
Isolated cleft defect, 286
Isolated LH deficiency, 307-8
Isoleucine metabolism defects, 201
Isoniazid, slow inactivation of 248-49
Isovaleric acidemia, 202

Jansen type metaphyseal chondrodysplasia, 533
Jaundice, 343
 neonatal, 342
Jejunal atresia, 335
Jervell and Lange-Neilsen syndrome, 452-53, 457-58
Jeune syndrome, 472, 487, 525-26
Johnson-Blizzard syndrome, 292
Joint disorders
 in Conradi disease, 523
 in diastrophic dwarfism, 524
 in Ehlers-Danlos syndromes, 233
 in Winchester syndrome, 238
Joint probability, in Bayesian analysis, 24
Juvenile-onset diabetes, 318-20
Juvenile osteochondroses, 239

Kallman syndrome, 307
Kartagener syndrome, 501
Kassbach-Merritt syndrome, 365
Keratins, 549
Keratin sulfate, 168, 173

Keratoatrophoderma, 554
Keratoconus, 427
Keratosis follicularis, 555
Keratosis palmaris et plantaris, 553-54
Kernicterus, 343, 397
Ketoacidosis, diabetic, 140
α-Ketoadipic aciduria, 204
α-Ketoglutarate: glyoxalate carboxylase deficiency, 484
17-Ketosteroid reductase deficiency, 315
Ketotic hyperglycinemia, 202
Keyhole limpet hemocyanin, 126
Kidney disorders, 463-93
 in Jeune syndrome, 525
 oxate metabolism defects, 206
 prenatal diagnosis, 615
 in trisomy 13, 57
 and Wilms' tumor, 94
 see also entries under Renal
Kidney stones, 481-82
 in APRT deficiency, 220
Kleeblattschädel anomaly, 523
Klinefelter syndrome, 35, 50-51
 androgen deficiency, 305
 breast carcinoma, 112
 gynecomastia, 309
 incidence, 41, 45
Klippel-Trenaunay-Weber syndrome, 562
Kniest disease, 539-40
Knudson, A.G., 88
Kocher-Debre-Semelaigne syndrome, 292
Kostmann syndrome, 364
Kozlowski type spondylometaphyseal dysplasia, 540-41
Krabbe's disease, see Globoid cell leukodystrophy
Kyphosis
 in achondroplasia, 520
 in diastrophic dwarfism, 524

Lacrimal apparatus, disorders of, 425-26, 558
 and ear malformations, 446
Lactate dehydrogenase, 356
Lactate metabolism defects, 206
Lactic acidosis, 361
Lactoferrin, 141
Lactosylceramidosis, 163
Lafora-body disease, 395
Lamellar bodies, 161-62
Lamellar (zonular) cataracts, 432
Lamellar ichthyosis, 549-51
Landing's disease, 158-60, 417
Lange-Neilsen syndrome, 452-53
Langer-Saldino achondrogenesis, 522
Langer type mesomelic dwarfism, 526, 535
Laryngeal atresia, 499
Laryngeal webs, 499
Laryngotracheoesophageal cleft, 499
Late infantile amaurotic idiocy, see Neuronal ceroid lipofuscinosis

Laurence-Moon syndrome, 321, 393, 452
Laurence-Moon-Biedl syndrome, 308, 471, 485
Lazy leukocyte syndrome, 139, 364
Lead intoxication, 402
Leber's optic atrophy, 438
Lecithin-cholesterol acyltransferase deficiency, 363
Legg-Calvé-Perthes disease, 239, 534
Leigh disease, 401
Leiomyoma, 503
Lens defects, 430-32
  in Alport syndrome, 465
  in Marfan syndrome, 232-33
  metabolic disorders, 207, 416-20
  in Weill-Marchesani syndrome, 238
Lenticonus, 465
Lentigenes, 308, 421, 450
Leopard syndrome, 421, 449-50, 557
Leri-Weill disease, 534-35
Lesch-Nyhan syndrome, 133, 217-18, 485
Leucine metabolism defects, 201
Leukemia, 85-87, 113-15
  in Bloom syndrome, 73, 565
  chromosomal abnormalities, 71-74
  in Down syndrome, 52, 71
  in Fanconi syndrome, 74
  and hypercalcemia, 298
  in Wilms' tumor patients, 95
Leukocyte antigens, 124
Leukocytic abnormalities, 363-64
  in Chédiak-Higashi syndrome, 560-61
  in phagocytic-bactericidal system disorders, 138-39
Leukocytic peroxide, 183
Leukopenia, 359, 536
Leukoplakia, oral, 557-58, 569
Leydig cells, 307, 314
Ligamentous laxity, 520, 538
Lignac-Fanconi syndrome, 478-79
Limb-girdle dystrophy, 377
Lingual thyroid, 291
Linkage, 10-17
  in myotonia, 380
  in Norrie's disease, 454
  in ocular albinism, 561
  protease inhibitor genes, 509
Linkage disequilibrium, 240-41
Lipid metabolic defects, 363, 381-82
Lipid storage disorders, 179-81
  lung involvement, 510
  Wolman disease, 302
Lipoate acetyltransferase deficiency, 193
Lipochrome histiocytosis syndrome, 142
Lipodystrophy, 564
Lipofuscin, 183
Lipogranulomatosis, 167, 418
Lipoid adrenal hyperplasis, 304, 314

Lipoid proteinosis, 418, 564-65
Lipomide dehydrogenase deficiency, 193
Lipoprotein abnormalities, 279-83, 420
Lipoprotein lipase deficiency, 280
Lip pits, 333
Lithium, in congenital heart disease, 278
Lithium carbonate, 259
Lithium chloride, 275
Liver disorders, 341
  antitrypsin deficiency, 509-10
  metabolic lysosomal disorders, 345
  phosphorylase deficiency, 190
  in polycystic kidney disease, 468, 470
  see also entries beginning with Hepatic and Hepato
Lobstein disease, 529
Lobular carcinoma of the breast, 101
Long bone sarcoma, 95
Los Angeles variant, galactosemia, 193
Louis-Bar syndrome, see Ataxia-telangiectasia
Lowe syndrome, 308, 483, 486-87
Low frequency hearing loss, 447
Lung disorders
  cancer, 93, 115-16
  malformations, 502-5
  see also entries beginning with Pulmonary
Lupus erythematosus, 569
Luteinizing hormone deficiency, 306-8
Lymphangiomyomatosis, 503
Lymphatic neoplasms, 86, 92, 113
Lymphedema, 562-63
Lymphocyte ecto-5'-nucleotidase deficiency, 131
Lymphocytes, 37, 40, 121-23: See also B-lymphocyte; T-lymphocyte
Lymphocytic leukemia, chronic, see Chronic lymphocytic leukemia
Lymphocytic thyroiditis, 294
Lymphoid tissue, in host-defense system, 125
Lymphokines, 123, 126-27, 138-39, 146
Lymphoma, 86, 113
Lymphopenia, 533
Lymphoreticular malignancies, 131
Lymph vessels, in vascular skin lesions, 562
Lyon hypothesis, 22-23, 44
Lysine metabolism, defects in, 204-5
Lysinuria, 204
Lysosomal acid lipase-cholesteryl esterase deficiency, 180
Lysosomal hydrolase elevation, 178
Lysosomal storage disorders, 153-79

liver involvement, 345
prenatal diagnosis, 614
therapeutic aspects, 183-85
Lysosome, 153
Lysozyme, 141
Lysyl hydroxylase, reduced activity, 235
Lysyl oxidase, 230
  deficiencies, 235, 237

Mabry-type dystrophy, 375
McArdle's disease, 188-89
McKusick type metaphyseal chondrodysplasia, 530, 567
Macrocytic anemia, 358-60
Macrophage, 123
Macrophage migration inhibitory factor (MIF), 127
Macular defects of the retina, 434-35
Madelung deformity, 526, 534
Majewski syndrome, 487, 527
Major histocompatibility complex, 86, 240
Malabsorption syndromes, 338-39
  in metaphyseal chondrodysplasia, 536-37
Maladie de Capdepont, 587
Malaria, 356
Mal de Meleda, 554
Male hypogonadism, 305-6
Male-pattern baldness, 569
Male pseudohermaphroditism, 310, 313-17
Males, X-linked recessive inheritance in, 22
Malignant hyperthermia, 258, 383-84
Malignant melanoma, 110
Mammography, 100
Mandibular hypoplasia, 526
Mandibulofacial dysostosis, 424, 449
α-D-Mannosidase, 177
Mannosidosis, 175, 177
Maple syrup urine disease, 201
Marcus-Gunn phenomenon, 425
Marden-Walker syndrome, 487
Marfan syndrome, 19-20, 26, 232-33, 431, 503
Marie-Unna type hypotrichosis, 569
Marinesco-Sjögren syndrome, 306, 401, 566
Maroteaux-Lamy syndrome, 171, 173-74, 416
Masculinizing maternal tumors, 312
Mast cell disease (mastocytosis), 558
Maternal age, and risk of trisomic chromosomal condition, 52-53, 55, 57, 77, 613-14
Maternal drug ingestion, and fetal dysmorphogenesis, 29-32
Maternal tumors, masculinizing, 312
Maturity-onset diabetes, 320
Mayflower type dentinogenesis

imperfecta, 587
May-Hegglin anomaly, 364
MEA, see Multiple endocrine adenomatosis
Meckel diverticulum, 336
Meckel syndrome, 392, 472, 487
Meconium ileus, 338
Mediterranean G-6-PD variants, 253-54
Medullary sponge kidney, 472, 483
Medullary thyroid carcinoma, 97-98
Medulloblastoma, 110
Megacolon, 337-38
Megaloblastic anemia, 200, 358-60
Megalocornea, 427
Meiosis, 10, 16, 40
Meiotic nondisjunction, 40-41
Melanin metabolism disorders, 558
Melanoblastosis, 557
Melanocyte, 558
Melanoma, 110
Melanosome, 558
Membrane stabilizer, 380
MEN, see Multiple endocrine neoplasia
Mendelian inheritance (single-gene inheritance), 17
  congenital heart defects, 273
  hair disorders, 569
Mendel's laws, 10
Meningioma, 73, 98
Meningomyelocele, 392
Menkes' syndrome, 237, 382, 420, 559, 561
Menstrual cycle, and acute intermittent porphyria, 213
Mental retardation, 57-58, 65-66
  acrodysostosis, 528
  anal imperforate, 336
  aspartylglycosaminuria, 182
  basal ganglia calcification, 396
  Bloom syndrome, 565
  chronic nephritis, 466
  Down syndrome, 52
  Dyggve-Melchior-Clausen disease, 541
  galactosemia, 192
  Hartnup disease, 476
  and hearing loss, 452
  ichthyosis, 553
  Klinefelter syndrome, 50
  Laurence-Moon syndrome, 452
  Lesch-Nyhan syndrome, 217
  muscular dystrophies, 374, 376
  myotonia, 379
  PKU, 194
  poly-X female, 51
  pseudoglioma, 434
  pseudohypoparathyroidism, 479
  pycnodysostosis, 535
  Sanfilippo syndrome, 172
  Sjögren-Larsson syndrome, 552
  Turner syndrome, 47
  uricosuria, 485
β-Mercaptolactate-cysteine disulfiduria, 201

Mesodermal neoplasms, 76, 91-92
Mesomelic dwarfism, 501, 525-26, 534-35
Messenger RNA, 4-6
Metabolic disorders, 94, 153-228
  albinism, 558-62
  autosomal recessive inheritance, 20
  eye abnormalities, 415-20, 431, 433
  folate metabolism, 359-60
  lipid metabolism, 363
  liver involvement, 342-45
  missense mutation, errors due to, 8
  in myopathies, 381-83
  in neurological diseases, 396, 401
  pulmonary complications, 510-11
  red blood cells, 354-57
  vitamin-dependent errors, 261
Metacentric chromosome, 35-36
Metachromasia, 169, 507
Metachromatic leukodystrophy, 165-66, 418
Metal metabolism disorders, ocular findings in, 420
Metanephric blastema, 90
Metaphase of mitosis, 35, 37
Metaphyseal chondrodysplasia, 298, 530, 533-34, 536-37
Metatrophic dwarfism, 524-25
Methemoglobinemia, 252, 351
Methionine, of collagen, 229
Methionine adenoxyltransferase defect, 200
Methoxyflurane, and malignant hyperthermia, 258
β-Methylcrotonylglycinuria, 202
$N^{5,10}$-Methylenetetrahydrofolate reductase deficiency, 200, 360
N-Methylglycine, 204
α-Methyl-β-hydroxybutyric acidemia, 202
Methylmalonic acidemia, 615
Methylmalonic aciduria, 203, 359
N^5-Methyltetrahydrofolate homocysteine-methionine transferase deficiency, 200
Meulengracht disease, 342
MHC, see Major histocompatibility complex
Mibelli type porokeratosis, 554-55
Microbicidal activity, disorders of, 138, 140-42
Microcephaly, 392-93
Microcornea, 427
Microcystic disease, 472
Microdactyly, 238
Micrognathia, 499
Micromelia, 522, 524
Microphallus, 286
Microphthalmia, 426
Microtia, 446
Middle ear, 444
  abnormalities, 446-47, 524
Middle frequency hearing loss, 447

Midget, 519
Miege type lymphedema, 562
Miescher's elastoma, 233
MIF, see Müllerian-inhibitory factor
Migraine, 395
Milia-hypotrichosis syndrome, 569
Milroy type lymphedema, 562
Mineral metabolism disorders, ocular findings in, 420
Missense mutation, 7-8
Mitochondrial inclusion bodies, 361
Mitochondrial myopathies, 382-83
Mitosis, 42
Mitotic nondisjunction, 42
Mitral regurgitation, in Marfan syndrome, 232
Mixed hyperlipemia, 282
Mixed leukocyte reaction, 128
Mixoploidy, see Mosaicism
Möbius syndrome, 402
Mohr syndrome, 448
Mongolian spot, 557
Monilethrix, 569
Monoamine oxidase (MAO) inhibitor, 259
Monoamino-monocarboxylic aminoacid transport, 474
Monoarthritis, 484
Monochromatism, 437
Mononuclear phagocytes, 141
Monosomy, 40, 64-71, 73
  See also Chromosome deletion; Partial monosomy
Monosomy-X, in Turner syndrome, 46-48
Monozygotic twins, childhood leukemia in, 113
Morquio syndrome, 171, 173, 416, 588
Mosaicism, 42
  Down syndrome, 52-53
  gonadal tumors, 111
  hermaphroditism, 318
  hypogonadism, 308
  Klinefelter syndrome, 50
  polyploidy, 62-63
  trisomy 8, 58-59
  trisomy 13, 57
  trisomy 18, 57
  Turner syndrome, 49-50
Mother pedigrees, in breast cancer, 100
Mounier-Kühn syndrome, 501
Mouse
  immune responses, 124
  neoplasms, 85-86
MPS, see Mucopolysaccharidoses
MSK, see Medullary sponge kidney
Mucocutaneous candidiasis, 591-92
Mucolipidoses, 174-79
  lung involvement, 510
  ocular findings in, 417
  prenatal diagnosis, 614

Index

Mucopolysaccharides, *see* Acidic glycosaminoglycans
Mucopolysaccharidoses, 167-74
  biochemical aspects, 155
  liver involvement, 345
  ocular findings in, 416
  prenatal diagnosis, 614
  respiratory complications, 510
  tooth abnormalities, 588-89
Mucosulfatidosis, 166, 174
Mucous membrane disorders, 547-48
  in ectodermal dysplasias, 556
Müllerian-inhibitory factor (MIF), 311, 315
Multicore disease, 384
Multicystic lung disease, 502
Multifactorial inheritance, 26, 269
  congenital heart disease, 275
  endocrine diseases, 287, 294
  hyperlipoproteinemias, 280
  refractive powers of the cornea, 427
Multilamellar cytosomes, 183
Multiple endocrine adenomatosis (MEA), 334-35
Multiple endocrine neoplasia (MEN), 96-98, 288, 298, 304
Multiple epiphyseal dysplasia, 239, 534
Multiple exostosis, 112
Multiple intestinal atresia, 335
Multiple lentigenes, 308, 421, 450
Multiple mucosal neuroma syndrome, 98
Multiple neuroma syndrome, ocular changes in, 421
Multiple primary malignancies, 89, 108
Multiple sclerosis, 124
Multiple-system neoplasms, table, 92
Muscle disorders, 373-89
  in acid maltase deficiency, 189
  and hypochromic anemia, 361
  in hypothyroidism, 292
  phosphorylase deficiency, 189
  rigidity, in malignant hyperthermia, 258
  spasticity, in Lesch-Nyhan syndrome, 217
  spinal muscular atrophies, 402-3
Muscular dystrophies, 22, 373-78
  and intestinal disorders, 338
  Pelger-Hüet gene, 364
  prenatal diagnosis, 614
  ptosis, 425
Mutation, 6-9
  delayed, in Beckwith-Wiedemann syndrome, 289
  enzymes affected by, 157
  gene duplication, 16
  hemoglobin chains, 10, 350-52, 365
  Knudson's theory of oncogenesis, 88
Myelin figures, 156
Myelogenous leukemia, *see* Acute myelogenous leukemia; Chronic myelogenous leukemia
Myelomeningocele, 615
Myeloperoxidase, 141
  deficiency, 142
Myoclonus epilepsy, 383
Myofibrils, in central core disease, 384
Myoglobinuria, 189, 381, 383
Myopathies, 373. *See also* Muscle disorders
Myopia, 422, 427, 431
  and cleft palate, 499
  in Kniest disease, 539
  in Marfan syndrome, 232
  and retinal detachment, 433
  in spondyloepiphyseal dysplasia, 538
  in Stickler syndrome, 537
  in Weill-Marchesani syndrome, 238
Myotonia, 378-80
Myotonia congenita, 378-79, 384
Myotonic dystrophy, 338, 379-80
Myotubular myopathy, 385

NADPII, *see* Nicotinamide-adenine dinucleotide phosphate
Nail-patella syndrome, 486-87
Nails, disorders of, 547-48
  in ectodermal dysplasias, 556, 582-86
  yellow nail syndrome, 501
Nasopharyngeal cancer, 346
  E-B virus, 86
Nasopharynx abnormalities, 495-501
Natal teeth, 525
Natural selection, 9
NCL, *see* Neuronal ceroid lipofuscinosis
NDI, *see* Nephrogenic diabetes insipidus
Negro variant, galactosemia, 193
Neisseria infections, 145
Nemaline myopathy, 384-85
Neonatal cholestasis, 343
Neonatal cirrhosis, 342
Neonatal giant cell hepatitis, 342
Neonatal Graves' disease, 295
Neonatal hepatitis, 342-44
Neonatal hypophosphatasia, 526-27
Neonatal IgG synthesis, 122-23
Neonatal jaundice, 342
Neonatal primary hyperparathyroidism, 298
Neoplasms
  chromosomal abnormalities and, 71-76
  see also Cancer; Tumor; and specific organs and tissues
Nephritis, 447-48, 464-66
Nephroblastoma, *see* Wilms' tumor
Nephrocalcinosis, 472, 481-83, 535
Nephrogenic diabetes insipidus, 290, 483
Nephrolithiasis, 206, 470, 474, 476, 484-85
Nephromegaly, 468
Nephronophthisis-cystic renal medulla complex, 470-72
Nephropathic cystinosis, 295
Nephrotic syndrome, 472
Nephrotoxin, 472
Nerve deafness, *see* Sensorineural hearing loss
Nerve root compression, 520
Nervous system, neoplasms of the, 91
Netherton syndrome, 553, 566
Neural crest tissue, 95-96, 98
Neural lesion, hearing loss from, 444
Neural-tube defects, 392
  prenatal diagnosis, 29, 614-15
  recurrence risk, 28
Neuraminidase deficiency, 177-78
Neuroblastoma, 88, 95-96, 305
Neurofibromatosis, 99, 335, 394, 450, 503, 557
Neurologic disorders, 391-414
  cataracts in, 433
Neuroma, 98
Neuronal ceroid lipofuscinosis, 161, 182-83, 293
Neurosecretory cells, reduction of, 289
Neutropenia, 139, 363-65, 536
Neutrophilia, 141, 364
Nevi, 557, 562-63
Nevoid basal-cell carcinoma, 108-10, 421, 557
Nevus flammeus, 562
Nevus of Ota, 557
Nezelof syndrome, 134
Nicotinamide-adenine dinucleotide phosphate (NADPH), 255, 354, 356-57
Nicotinamide deficiency, 476
Nicotine, effect on fetus, 32
Niemann-Pick disease, 163-65, 418, 510
Nievergelt type mesomelic dwarfism, 526
Night-blindness, 427, 435
Nitrazepam, 249
Nitrous oxide, and malignant hyperthermia, 258
Noack syndrome, 393
Nomenclature, in chromosomal analysis, 36, 39
Nonallele, 10-11
Nonketotic insulin-resistant diabetes, 564
Nonlethal achondrogenesis, 522-23
Nonsecretors of ABO blood groups, 346
Nonsense mutations, 8
Noonan syndrome, 277, 306, 309, 563
Normokalemic periodic paralysis, 381
Norrie's disease, 434, 454

Nortriptyline accumulation, 260, 262-63
Norum disease, 363
Nuclear cataracts, 432
Nucleic acid, 4. *See also* DNA; RNA
Nucleoside phosphorylase deficiency, 134
Nucleotide, 4
Nystagmus, 415, 425
  and albinism, 560-61
  in branchioskeletogenital syndrome, 589
  in cerebellar ataxias, 401
  and hearing loss, 452

Obesity, 308, 321, 505
Ochronosis, 196, 237
Ocular disorders, *see* Eye disorders
Ocular phakoma, 394
Ocular pseudoglioma, 454
Oculocerebrorenal syndrome, *see* Lowe syndrome
Oculocutaneous albinism, 503, 558-60
Oculodento-osseous dysplasia, 588
Oculopharyngeal muscular dystrophy, 377-78
Oligophrenia, 403, 589
Onychodysplasia, 556, 592
Oogenesis, 40-41
Opalescent dentin, 587
Open-angle glaucoma, 430-31
Open meningomyelocele, 29
Ophthalmopathies, *see* Eye disorders
Ophthalmoplegia, 377, 385, 402, 418, 425
Opsonins, 136, 139
Opsonization, disorders of, 138-40, 144
Optic atrophy, 289
Optic cleft, 426
Optic nerve problems, 437-38
  in metabolic disorders, 416-20
Orbital hypotelorism, 286
Orofacial digital II syndrome, 448
Orofacial structures, disorders affecting, 575-94
Organ of Corti, absence of, 449
Organophosphorus compounds, pseudocholinesterase inhibition, 249
Ornithine transcarbamylase deficiency, 196
Ornithine transcarboxylase deficiency, 358
Orotic aciduria, 358
Osler-Rendu-Weber syndrome, 335, 342, 421, 504, 563
Osteitis fibrosa cystica, 479
Osteoarthritis, 540
Osteochondritis dissecans, 239
Osteochondroma, 112
Osteochondroses, juvenile, 239
Osteogenesis imperfecta, 235-36, 529-30
  tooth abnormalities, 589-90

Osteogenic sarcoma, 96
Osteomalacia, 476-77
Osteopenia, 479, 481
Osteopetrosis, 482-83, 535
Osteopoikilosis, 564
Ostitis fibrosa, 297
Otopalatodigital syndrome, 454
Otosclerosis, 447, 529
Ovalocytosis, 362
Ovarian failure, 309
Ovarian neoplasms, 101, 112
Oxalosis, 484
Oxate metabolism defects, 206
Oxyhemoglobin, 351
Oxytocin secretion, 289

Pachyonychia congenita, 569
Paget's disease, 112
Pain, insensitivity and indifference to, 404-5, 409
Palilalia, 396
Pancreatic cholera syndrome, 97
Pancreatic hypofunction, 318
Pancreatic insufficiency, 340-41, 507-8, 535
Pancreatic tumors, 96-97, 99
Pancreatitis, 281-82
Pancytopenia, 74
Panhypopituitarism, 287, 292
Pansinusitis, 507
Papilloma, 564
Para-aminosalicylic acid acetylation, 249
Paracentric inversion, 42-43
Parafollicular cell, thyroid, 97
Paraganglioma, 99
Paralysis
  in Fazio-Londe disease, 402
  periodic, 380-81
Paramyotonia congenita, 380
Paraplegia, spastic, in Laurence-Moon syndrome, 452
Parastremmatic dwarfism, 532
Parathyroid glands, disorders of 295-99
  and kidney diseases, 479
  tumors, 96-98
  in vitamin D-resistant rickets, 477
  see also Hyperparathyroidism; Hypoparathyroidism
Parathyroid hormone receptors, defect of, 297
Parenti-Fracarro achondrogenesis, 520-22
Parietal cell antibodies, 103
Parkinson disease, 395-96
Partial androgen insensitivity, 317
Partial monosomy, 40, 42, 52. *See also* Chromosome deletion
Partial monosomy-X, 47-49
Partial trisomy, 40, 53, 57-59
Patau syndrome, *see* Trisomy 13
Patent ductus arteriosus, 272-74
Pectus excavatum, 511
Pedigree studies, 17-19, 22-23
  Bayesian analysis, 25
  cancer, 89-90, 100, 115
Pelger-Hüet anomaly, 364

Pellagraform rash, 474
Pendred syndrome, 293, 451
Penetrance, 19
  adult polycystic kidney disease, 469
  colorectum cancer, 107
  hyperparathyroidism, 98
  juvenile-onset diabetes, 319
  MEN, 97-98
  multiple exostosis, 112
  neuroblastoma, 96
  pheochromocytoma, 99
  retinoblastoma, 95
  Wilms' tumor, 90
Penicillamine, 474
Pentosuria, 193-94
Peptic ulcer, 16, 334-35, 346
Pericentric inversion, 42-43
Perilymphatic gusher, 454
Perinatal anoxia, 397
Periodic paralyses, 380-81
Periodontopathia, in keratosis palmaris et plantaris, 554
Peripheral nerves, disorders of, 404-5, 409, 452
Peripheral pulmonary artery stenosis, 274
Periportal fibrosis, 468
Permeability defect, of RBC membrane, 362
Pernicious anemia, 346, 359
Peroneal muscular atrophy, 409
Peroxidase abnormalities, 292-93
  in neuronal ceroid lipofuscinosis, 183
Peroxide reduction, by glutathione, 357
Pes cavus, 403
Peutz-Jeghers syndrome, 335, 557
Phacomatoses, 91
Phagocytic-bactericidal system, disorders of, 136-42
Pharmacogenetics, 245-66
Phenacetin-induced methemoglobinemia, 252
Phenelzine metabolism, 249
Phenobarbital, 343
Phenocopy, 25
Phenothiazine, 214, 252, 397
Phenotype, 17
Phenylalanine hydroxylase, 195
Phenylalanine transaminase, 195
Phenylbutazone, 252, 260, 262
Phenylketonuria, *see* PKU
Phenylthiocarbamide nontasters, 291
Phenylthiocarbamide tasters, 256-57
Phenyramidol, 252
Pheochromocytoma, 88, 97-99, 305, 394
Philadelphia chromosome, 72, 114
Phosphaturia, 476-77
Phosphoenolpyruvate carboxykinase defect, 194
Phosphoethanolamine excretion, 535-36
Phosphofructokinase deficiency, 190, 355

Index 647

Phosphogluconate dehydrogenase deficiency, 357
Phosphoglycerokinase deficiency, 355
Phosphoribosyl-1-pyrophosphate, 218-20
Phosphoribosyl-1-pyrophosphate synthetase, 219
Phosphorus metabolism, disordered, 396
Phosphorylase kinase abnormalities, 190
Photophobia, 424
  in albinism, 560-61
Photosensitivity, in porphyrias, 209-10, 214
Phthisis bulbi, 454
Physiologic neonatal jaundice, 342
Phytanic acid elevation, 452
Phytohemagglutinin, 123
Pick presenile dementia, 394
Pickwickian syndrome, 505
Piebaldism with white forelock, 559, 561
Pierre-Robin syndrome, 499, 537, 567
Pigmentary anomalies
  and deafness, 449-50, 453-54
  of iris, 427, 430
  in von Recklinghausen disease, 394
Pigmented nevi, 557
Pigmented skin lesions, 557-58
Pigmenturia, 383
Pili torti, 237, 561, 566, 569
Pinna (auricle), 444-45
Pipecolic aciduria, 204
Pituitary defects, 285-90
  and corticotropin deficiency, 299
  and gonadotropin deficiency, 307
  tumors, 97
PKU (phenylketonuria), 24, 26, 194, 568
Placental chorionic gonadotropin, 311
Placental steroid sulfatase deficiency, 300-1
Plasma-cell leukemia, 73
Plasma thromboplastin antecedent deficiency, 367
Plasma transfusion, in immunologic deficiency disease, 146
Platelet disorders, 335, 364-65, 503 529
Pleiotropism, 20, 391
Pneumocystis carinii, 510
Pneumothorax, 503-4
Point mutation, 6-9
  hemoglobin chains, 350-52, 365
Poisson distribution, in tumor-development predictions, 88
Pokeweed mitogen (PWM), 123, 126
Poland anomaly, 361
Polycystic kidney disease, 342, 463, 467-72, 486, 527

Polycystic ovaries, 564
Polycythemia, 73, 351, 365-66
Polydactyly, 525, 527
Polygenic inheritance, 26-29
Polyhydramnios, 523
Polymorphism, 10
Polymyoclonia, 95
Polyneuritis, INH-induced, 249
Polyneuropathy, 406-9, 464
Polyostotic fibrous dysplasia, 112
Polyploidy, 10, 40, 42, 53, 62-64
Polypoid diseases, and colorectum cancers, 107
Polyposis coli syndromes, 340
Polysome, 5
Polysomy-X, 41, 45, 51
Polysomy-Y, 45
Polysyndactyly, 527
Pompe's disease, 188
Popliteal pterygia, 563
Population genetics, 8
Porcupine man, 551
Porokeratosis, 108-9, 554-55
Porphobilinogen, 208, 210, 213-14
Porphyria, 211, 407, 409
Porphyria cutanea tarda, 217
Porphyrin metabolism, disorders of, 208-17
Portacaval shunt, 281
Portal hypertension, 509
Port-wine nevus, 562
Posterior pituitary, 289-90
Posterior polar cataracts, 432
Posterior probability, 24
Postlysosome, 156
Postmenopausal breast cancer, 100-1
Potassium, in periodic paralyses, 381
Potter syndrome, 499, 502
Prader-Willi syndrome, 308, 321 505
Prebeta, very low-density lipoproteins (VLDL), 280
Precocious pseudopuberty, 303
Precocious puberty, 290-91
Prednisolone, 182
Pregnancy
  cholestasis, 343
  placental steriod sulfatase deficiency, 301
  see also Fetus; Prenatal diagnosis
Pregnanediol, in breast milk, 343
Premature termination codon, Hb chains, 352
Premenopausal breast cancer, 100-1
Prenatal diagnosis, 29-32, 76-77, 613-16
  achondroplasia, 520
  acid phosphotase deficiency, 181
  CNS malformations, 392
  deafness, 457
  Down syndrome, 55
  glucosylceramidosis, 164
  glycogenosis, 188
  Gm_2 gangliosidosis, 161

harlequin fetus, 552
Hurler syndrome, 169
hypophosphatasia, 527
inclusion-cell disease, 178
metachromatic leukodystrophy, 166
mucolipidosis IV, 179
myotonia, 380
placental steroid sulfatase deficiency, 301
Sanfilippo syndrome, 173
thyroid disorders, 293
xeroderma pigmentosum, 108
Primaquine sensitivity, see Glucose 6-phosphate dehydrogenase deficiency
Primary corneal dystrophies, 427
Primary glaucoma, 430-31
Primary hyperplastic vitreous, 424
Primary hypogonadism, 305-6, 308
Primary lysosomes, 153, 156
Primary microcornea, 427
Primary pulmonary hypertension, 504
Primary siderocytic anemia, 360-61
Prior probability, 24
Probability
  Bayesian theorem, 23-24
  multifactorial inheritance, 26-29
Procainamide accumulation, 260
Procainamide acetylation, 249
Procollagen, 230
Procollagen peptidase, 230, 235
Progestin, 312
Progestogen/estrogen, and cardiovascular maldevelopment, 278
Progressive cerebral degeneration in infancy, 393-94
Progressive intrahepatic cholestasis, 343
Progressive myoclonic epilepsy, 395
Prolactin deficiency, 298
Proline, 229
  excretion, 466
  metabolic defects, 203-4
Proline oxidase abnormalities, 203
Prolonged immobilization, and hypercalcemia, 298
Properdin, 143
Propionic acidemia, 202-3
Propranolol accumulation, 260
Prostaglandins, 484
Protanomaly, 437
Protanopia, 437
Protease inhibitor, 344, 508
Protein synthesis, 5-6
Proteinuria, 359, 464, 467
Prothrombin, 367
Protoporphyrin, 214
Protoporphyrinogen, 208, 212
Prune belly syndrome, 486
Pseudoachondroplasia, 531-32
Pseudocholinesterase, 249

Pseudoglioma, 434, 454
Pseudohermaphroditism, 303, 310-17
  in Wilms' tumor patients, 90
Pseudo-Hurler polydystrophy, 175, 179
Pseudohypertrophy, muscular, 374
Pseydohypoparathyroidism, 297-98, 479-80, 591
Pseudohypophosphatasia, 536
Pseudoidiopathic hypoparathyroidism, 296
Pseudolymphoma, 561
Pseudominant inheritance, 21
Pseudopseudohypoparathyroidism, 479-80
Pseudovaginal perineoscrotal hypospadia, 315
Pseudovitamin D-deficiency rickets, 477-78
Pseudoxanthoma elasticum, 236, 335, 421
Psoriasis, 124, 569
Psychomotor retardation, in Leigh disease, 401
Psychotherapeutic aspects of genetic counseling, 598-99
Pterygia, 333, 563
Ptosis, 378, 425, 589
Pug nose, 528
Pulmonary agenesis, 502
Pulmonary alveolar microlithiasis, 504-5
Pulmonary arteriovenous fistula, 504
Pulmonary cysts, 502-3
Pulmonary edema, 506
Pulmonary emphysema, 509
Pulmonary glomangioma, 504
Pulmonary hemosiderosis, 506
Pulmonary hypoplasia, 502, 504, 527
Pulmonary malformations, 502-5
Pulmonary neoplasms, 93, 115-16, 505
Pulmonic stenosis, 277, 306
Pulp chamber defects, 580, 590
Pupillary abnormalities, 430
Pure gonadal dysgenesis, 308, 313
Purine metabolism disorders, 133, 217-20, 357-58
Pycnodysostosis, 535
Pyelonephritis, 485
Pygmies, unresponsiveness to growth hormone, 287
Pyloric stenosis, 28, 337
Pyogenic infections, 131
Pyridoxine (vitamin B_6)
  in cystathionine synthetase deficiency, 198, 200-1
  deficiency of, in oxalate overproduction, 206
  in hypochromic anemia, 360-61
Pyrimidine metabolism disorders, 357-58
Pyrimidine-5-nucleosidase deficiency, 358
Pyroglutamic acid, 205

Pyruvate carboxylase deficiency, 193
Pyruvate decarboxylase, 193, 382
Pyruvate dehydrogenase, 193, 382
Pyruvate kinase deficiency, 193, 354-55
Pyruvate metabolism disorders, 193-94
  in Leigh disease, 401
Pyuria, 465, 535

Quadriradial chromosome rearrangement, in Bloom syndrome, 73
Quinacrine staining, 36-38, 45

Radiation therapy
  and basal-cell carcinoma, 110
  and hyperparathyroidism, 298
Radicular dentin dysplasia, 587
Radioactive iodine, and congenital hyperparathyroidism, 291
Radium, and skin cancer, 108
Ragweed pollinosis, 124
Receptor proteins, in liver cells, 342
Recessive inheritance, 17, 20-23
Reciprocal chromosome translocation, 44
Recombination, 11, 16-17
Rectum, cancers of the, 107
Red blood cells
  enzymopathies, 190, 356
  genetic markers, 8
  G-6-PD deficiency, 254-56
  membrane alterations, 362-63
  metabolic errors, 354-57
  polycythemia, 366
Reducing body myopathy, 385
5 α-Reductase, 311, 315
Red urine, 210, 212
Refractive errors of the eye, 426-27
Refsum disease, 406, 409, 420, 452, 553
Regional enteritis, 338
Registration peptides, in collagen synthesis, 230
Regulatory genes, 6
Reifenstein syndrome, 317
Reilly bodies, 364
Reiter's disease, 124
Renal abnormalities and diseases, for those not listed below, see Kidney disorders
Renal agenesis, 485
  prenatal diagnosis, 615
Renal aminoacidurias, 206-7
Renal amyloidosis, 451, 464, 467
Renal anomalies, and ear malformations, 446
Renal cell tumors, 112, 394
Renal cysts, 467-72, 486-87
  prenatal diagnosis, 615
Renal dysgenesis, 485
Renal function abnormalities, table of, 473
Renal prostaglandin excess, 484
Renal-retinal dysplasia, 471
Renal transport system, 473-85

Renal tubular acidosis, 480-83
Rendu-Osler-Weber syndrome, see Osler-Rendu-Weber syndrome
Rennes variant, galactosemia, 193
Replicase, 4
Repressor, 6
Repulsion, genetic, 11, 16
Respiratory distress syndrome (RDS), see Hyaline membrane disease
Respiratory system disorders, 495-518
Reticular dysgenesis, 129, 363-64
Reticuloendothelial system neoplasia, 75
Retinal abnormalities, 434-37
  and kidney disease, 471
  in metabolic disorders, 416-20
Retinal angioma, 394
Retinal aplasia, 434
Retinal degeneration
  in Alstrom syndrome, 452
  in Cockayne syndrome, 564
  in familial spastic paraplegia, 403
  in metabolic disorders, 416, 419-20
  in Sjögren-Larsson syndrome, 553
Retinal detachment, 421-22, 424, 433-34, 436
  and cleft palate, 499
  in Kniest disease, 539
  in Marfan syndrome, 232
  and myopia, 427
  in spondyloepiphyseal dysplasia, 538
  in Stickler syndrome, 537
Retinitis pigmentosa, 434, 436-37
  and hearing loss, 451-52
  and kidney disease, 471
Retinoblastoma, 68, 72, 95, 424
  risk of development, 94
  theories of oncogenesis, 88-89
Retinoschisis, 433
Retrolental fibroplasia, 424, 434, 454
Reverse T3 (3,3′,5′-triiodothyronine), 294
Reverse-transcriptase, 4
Reye syndrome, 344
Rhabdomyosarcoma, 96
Rheumatic fever, 269-70
Rheumatoid arthritis, 505
Rhizomelia, 520, 524
Ribbing disease, 534
Ribnucleic acid, see RNA
Ribosomes, 5
Richner-Hanhart syndrome, 196
Rickets, 476-78
  tooth abnormalities, 590-91
Rieger syndrome, 288
Riley-Day-syndrome, 338, 393, 409
Ring chromosomes, 42-43
  hypogonadism, 308
  hypoparathyroidism, 296
  Turner syndrome, 48-49
RNA, 4-6

RNA-dependent DNA synthetase, 4
Robertsonian chromosome translocation, 44, 53-54, 57, 77
Rootless teeth, 587
Rosette-forming T-cells, 123, 126
Rothmund-Thomson syndrome, 565
Rotor syndrome, 343
Roussy-Lévy syndrome, 406, 409
Rowley-Rosenberg syndrome, 504
Rubber-bleb nevus, 563
Rubella, 17, 31, 133
  congenital heart disease, 274-75 278
  deafness, 444
  pyloric stenosis, 337
Rubinstein-Taybi syndrome, 562

Saccharopinuria, 204
Sachs type Ehler-Danlos syndrome, 233-34
Saddle nose, 556, 565
Saldino-Noonan type short rib-polydactyly syndrome, 527
Salt loss, 303-4, 312, 314
Sandhoff disease, 161, 417
Sanfilippo syndrome, 171-73, 416 589
Sarcoidosis, 505
Sarcoma, 96
Sarcomeres, in central core disease, 384
Sarcoplasmic reticulum, 383
Sarcosine, 204
Sarcotubular myopathy, 385
Scapulperoneal syndrome, 376-77
Scheie syndrome, 172, 416
Schizophrenia-like episodes, in metabolic disorders, 200
Schmidt syndrome, 302
Schmid type metaphyseal chondrodysplasia, 533
Schwartz-Jampel syndrome, 380
SCID, see Severe combined immunodeficiency
Scimitar syndrome, 504
Scleroderma, 116, 145
Sclerotylosis, 103
Scoliosis, 523-24
  recurrence risk, 28
Sea-blue histiocyte syndrome, 418
Secondary dentin, dysplasia of, 589
Secondary hypogonadism, 305-6
Secretor gene, 380
Segregation of allelic genes, 10
Seip-Barnardinelli syndrome, 321
Seizures, 394-96, 434
Self-mutilation, in Lesch-Nyhan syndrome, 217
Sella turcica, 287
Senile dementia, 394
Sensorineural hearing loss, 289, 296, 444, 447-48, 450-55
Sensory radicular neuropathy, 450
Septo-optic dysplasia, 286
Serine metabolism defects, 206
Serologically defined (SD) leukocyte antigen, 124

Sertoli cell, 311
Serum proteins, as genetic markers, 8
Severe combined immunodeficiency (SCID), 129-30, 143-44
  lung involvement, 510
  treatment, 146
Sex chromatin, 22, 35, 44-46
  See also X-chromatin; Y-chromatin
Sex chromosomes, 44-51, 110-11, 309, 311. See also X-chromosome; Y-chromosome
Sex identification, of fetus, 614
Sexual ateliosis, 286
Sexual differentiation, disorders of, 309-18
Short-limb dwarfism, 520-35
  and cartilage-hair hypoplasia, 135-36
Short-rib polydactyly syndromes, 511, 527
Short stature, 519
  in Bloom syndrome, 565
  and vitamin D-resistant rickets, 477
  see also Dwarfism
Short-trunk dwarfism, 538-41
Shwachman syndrome, 364, 536-37
Sickle cell disease, 353
  opsonins, 139-40
  renal tubular acidosis, 483
  retinal detachment, 433
  vascular thrombosis, 20
Sickle cell nephropathy, 486-87
Sickle cell trait, genetic counseling for, 598, 602-3, 605
Sideroachrestic (sideroblastic) anemia, see Hypochromic anemia
Silent gene, in succinylcholine sensitivity, 250
Silicosis, 505
Simple chromosome translocation, 44
Single gene inheritance, see Mendelian inheritance
Sipple syndrome, 97, 485
Sirenomelia, 336
Sister chromatid exchange
  in Bloom syndrome, 73
  in xeroderma pigmentosum, 76
Sister pedigrees, in breast cancer, 100
Situs inversus, 501
Sjögren-Larsson syndrome, 393, 552-53
Skeletal disorders, 519-44
  and deafness, 448-49
  and hair growth disorders, 567
  in juvenile osteochondroses, 239
  in Marfan syndrome, 232
  neoplasms, 93, 112
  ocular findings in, 422, 433
  osteogenesis imperfecta, 235-36
  prenatal diagnosis, 615
  and primary pulmonary hypertension, 504

Skeletal muscle disorders, 373-89
Skin disorders, 545-73
  candidiasis, in T-cell disorders, 134
  cutis laxa, 235
  ectodermal dysplasias, 582-86
  in Ehlers-Danlos syndromes, 233-34
  Fabry's disease, 486
  neoplasms, 93, 103, 106, 108-10
  ocular findings in, 423, 433
  in porphyrias, 210, 212, 214, 217
  in pseudoxanthoma elasticum, 236
  in von Recklinghausen disease, 394
  wheal and flare responses, 126
Sleep disturbance, and duodenal ulcers, 335
Sly syndrome, 171, 174
SMA, see Spinal muscular atrophies
Small eye, 424, 431
Smith-Lemli-Opitz syndrome, 337, 485, 524
Smoking and heredity, synergistic effect of, in lung cancer, 115
Snow-capped teeth, 581
Sodium loss, in renal tubular acidosis, 481
Sodium transport disturbances, in cystic fibrosis, 507
Soft tissue ankylosis, in epidermolysis bullosa, 548
Somatomedin deficiencies, 287
Somnolence, in Pickwickian syndrome, 505
Sotos syndrome, 288
Spahr-Hartmann type metaphyseal chondrodysplasia, 533
Spasticity
  in Laurence-Moon syndrome, 452
  in Sjögren-Larsson syndrome, 522-53
  in striatonigral degeneration, 397
Spectrin, in RBC membrane, 362
Spherocytosis, 362
Spherophakia, 465
Sphingolipid, 159
Sphingolipidoses
  liver involvement, 345
  ocular findings in, 417-18
  prenatal diagnosis, 614
Sphingomyelinase deficiency, 163-64
Sphingomyelin degradation, in glycosphingolipidoses, 158
Sphingomyelin lipidosis (Niemann-Pick disease), 163-65, 418, 510
Spina bifida
  prenatal diagnosis, 614
  recurrence risk, 28
Spinal cord compression, 520, 538
Spinal cord disorders, 402-3
Spinal muscular atrophies, 402-3

650 Index

Spinocerebellar ataxias, 397-401
Splenomegaly, 362, 468
Spondylocostal dysplasia, 541
Spondyloepiphyseal dysplasia, 499, 531-32, 538-40
Spondylometaphyseal dysplasia, 540-41
Spondylothoracic dysplasia, 540-41
Spontaneous abortion, 40, 50
Sporadic cases, 24-26, 66
Spotted bones, 564
Squamous cell carcinoma, 76, 106, 108-9
Squamous epithelioma of Ferguson-Smith, 335
Staining techniques, for chromosome identification, 36-38
Stapes abnormalities, 446-47, 449, 454
Steely hair disease, see Menkes' syndrome
Stem cells, 121, 123, 129-31, 364
Steroids
  biosynthesis, errors of, 299
  metabolism, defective, 343
  response of ocular pressure to applications of, 257-58
Steroid sulfatase deficiency, placental, 300-1
Stickler syndrome, 434, 499, 537
Stomach cancer, 16, 102-3, 346
Stomatocytosis, 363
Storage disorders, see Cystine storage disease; Glycogen storage diseases; Lipid storage disorders; Lysosomal storage disorders; Mucopolysaccharidoses
Storage pool disease, 365
Strabismus, 415, 421, 425, 589
Stratum corneum, 545, 549
Strawberry mark, 562
Streak gonads, 308-9, 311, 313
Streptococcal antigens, 124
Streptococcal infections, in rheumatic fever, 270
Striatonigral degeneration, 397
Structural gene, 6
  mutations, 248-49
Sturge-Weber syndrome, 421, 562
Subacute combined immunodeficiency, neoplasm development risk, 88
Subacute necrotizing encephalomyelopathy, 401
Subacute sclerosing panencephalitis, 391
Subcutaneous calcification, 564
Subcutaneous fat necrosis, 298
Subluxation of the lens, 431
Submetacentric chromosome, 35-36
Succinylcholine, 249-51
  and malignant hyperthermia, 258
Sulfamethazine metabolism, 249
Sulfamidase deficiency, 173

Sulfanilamide acetylation, 249
Sulfatide degradation, in glycosphingolipidoses, 158
Sulfatide lipidosis, 165-66, 418
Sulfhydryl groups, 256
Sulfite oxidase deficiency, 201, 419, 431
Sulfur amino acid metabolism disorders, 198-203
  in lens defects, 431
Sulfur-dibasic amino acid transport, 473
Sunlight, and skin cancer, 108, 110
Suppressor cells, 123, 132
Suture cataracts, 432
SV-40 virus, 86
Swan neck deformity, 479
Sweat electrolytes, abnormalities of, in cystic fibrosis, 507
Sweating abnormalities, 555-56, 582-86, 588
Switch mechanism, in hemoglobin chain synthesis, 352
Swyer syndrome, 313
Symphalangism, 448-49
Symptomatic seizures, 394
Synapsis, 16
Systemic diseases, renal manifestations, 486-87
Systemic lupus erythematosus, 569
Systemic lupus erythematosus-like syndrome, in complement deficiencies, 143-44

Talipes, see Clubfoot
Tangier disease, 363, 407, 420
TAR syndrome, see Thrombocytopenia-absent radius syndrome
Taurodontism, 592
Tay-Sachs disease, 160, 418
  genetic counseling, 598
TBG, see Thyroxine-binding globulin
Telangiectasias, 504, 562-63, 565. See also Ataxia-telangiectasia
Temporal bone abnormalities, 446-47, 450
Tendons, abnormal development of, 511
Tendon xanthoma, 180-81
Teratology, 29-32
  cardiovascular teratogens, 277-78
Testes
  atrophy, in dyskeratosis congenita, 558
  defects of, 305
  embryonic development, 311
  neoplasms, 112
Testicular feminization syndrome, 316
Testicular hormone synthesis, defects in, 304, 314-15
Testosterone
  defects in synthesis, 312, 314-15
  fetal secretion, 311
Tetany, in pseudohypoparathyroidism, 479, 591

Tetracaine hydrolyzation, 250
Tetrahexosyl ceramide (globoside), 161-62
Tetrahydrobiopterin, 195
Tetraploidy, 40, 42, 62
Tetrapyrrole structure, 209-10
Thalassemia, 352-53
  prenatal diagnosis, 614
Thalidomide, 29, 275, 278, 337
Thanatophoric dwarfism, 523
Thermal sensitivity, of RBC membrane, 363
Thermography, 100
Thiamine-responsive megaloblastic anemia, 360
Thick bone type osteogenesis imperfecta, 529-30
Thomsen disease (myotonia congenita), 378-79, 384
Thoracic-pelvic-phalangeal dystrophy, see Jeune syndrome
Thrombocytopenia, 359, 364-65
  Chédiak-Higashi syndrome, 561
  familial, 486
  GI hemorrhage, 335
Thrombocytopenia-absent radius (TAR) syndrome, 361, 365
Thrombocytopenic thrombopathy, 365
Thrombopathies, 365
Thrombopoietin deficiency, 365
Thrombotic thrombocytopenia, 365
Thumb abnormalities, 361
Thymine, 4-7
Thymolymphopenia, 533
Thymosin, 146
Thymus, influence on stem cells, 121
Thymus-dependent lymphocyte system, 123, 146
Thyroglobulin synthesis, defective, 293
Thyroid gland disorders, 290-95
  and PTC tasting ability, 256-57
  tumors, 97-99
Thyroid hormone unresponsiveness, 292
Thyroid-stimulating hormone deficiency, 287, 292
Thyrotoxicosis, 294, 298
Thyrotropin deficiency, see Thyroid-stimulating hormone deficiency
Thyroxine-binding globulin (TBG), 295
Thyroxine synthesis, defective, 292-93
T-lymphocyte (T-cell), 123-24, 126-31, 133-34
  in ataxia-telangiectasia, 135
  in atopic dermatitis, 146
  in metaphyseal chondrodysplasia, 530
Tooth abnormalities, 575-94
  acatalasia, 245, 248
  in ectodermal dysplasias, 556, 582-86

in erthropoietic porphyria, 212
growth hormone deficiency, 286
incontinentia pigmenti, 557
in Lobstein disease, 529
in osteogenesis imperfecta, 236
in pseudohypoparathyroidism, 479
in pycnodysostosis,, 535
Torre's syndrome, 107
Torsion dystonia, 397
Toxocara canis cyst, 424
Trabecular pigmentary changes, in glaucoma, 431
Trachea, abnormalities of the, 495-501
Tracheobronchiomegaly, 501
Tracheoesophageal fistula, 501
Tracheomalacia, 501
Transcobalamin, 356, 359
Transcriptase, 4
Transfer factor, 146
Transfer RNA, 5-6
Translation, 5-6
Translocation, chromosomal, see Chromosome translocation
Transplantation therapy, in immunologic deficiency disease, 146
Transport proteins, 359
Treacher Collins syndrome, 449
Tremor, 396
Trichodento-osseous syndrome, 590-91
Trichodysplasia, 556, 592
Trichopoliodystrophy, see Menke's syndrome
Trichorhinophalangeal dysplasia, 537-38
Trichorrhexis nodosa, 561
Trichromatic vision, 437
Triglycerides, 180, 281-82
Trihexosyl ceramidosis, see Fabry disease
Trimethadione (tridione), 30, 32
Triosephosphate isomerase deficiency, 190, 355
Triphalangeal thumbs, 361
Triple helix, in collagen alpha chain, 229
Triplet code, 4, 7
Triploidy, 40, 42, 62-63, 65
Trisomy, 40-41, 52-64
  congenital heart disease, 271-72
  GI tract malformations, 344-45
  ocular findings in, 424
  prenatal diagnosis, 613-14
  solid tumors, 73
Trisomy D, 485
Trisomy E, 485
Trisomy X, 485
Trisomy 1q, 61
Trisomy 2p, 61
Trisomy 2q, 61
Trisomy 3q, 62
Trisomy 4, 336
Trisomy 5p, 62
Trisomy 7q, 63
Trisomy 8, 58-59, 73
Trisomy 9, 59-60

Trisomy 10q, 63
Trisomy 13, 57-58
  facial clefts, 333
  GI tract malformations, 344-45
  hair growth disorders, 566
  primitive hemoglobin, 349
  retinoblastoma, 72
  skin defects, 547
Trisomy 14, 73
Trisomy 14q, 64
Trisomy 18, 55-57
  GI tract malformations, 345
  hair growth disorders, 566
  pyloric stenosis, 337
  stippling of epiphyses, 524
  Wilms' tumor, 95
Trisomy 21, see Down syndrome
Trisomy 22, 336
Trisomy 22q, 64
Tristichiasis, 425
Tritanopia, 437
Tropocollagen, 230
Tropoelastin, 230
Tryptophan deficiency, 229, 476
TSH deficiency, see Thyroid-stimulating hormone deficiency
Tuberculosis, 505
Tuberous sclerosis, 394, 486-87, 588
  ocular changes in, 421
  pulmonary involvement, 502-3
  renal cysts, 472
Tuftsin, 140
Tumors, 73. See also specific tumors, and specific organs and tissues
Turner syndrome, 22, 35, 46-50, 111
  congenital heart disease, 271-72
  congenital lymphedema, 563
  GI hemorrhage, 335
  genitourinary malformations, 486
  gonadal neoplasia, 72
  incidence, 41, 45
  and lymphocytic thyroiditis, 294
  primary hypogonadism, 308
  pterygium colli, 563
  rectal hemorrhage, 345
Twins
  childhood leukemia in, 113
  hypothyroidism, 291
  metabolic differences, 260-63
Two-component model for development of immune system, 121-22
Two-hit model for human oncogenesis, 88
Tylosis, 102-3, 553-54
Tympanic cavity, see Middle ear
Tympanic membrane, 444
Tyrosinase, 558
Tyrosinase-negative oculo-cutaneous albinism, 558, 560
Tyrosinase-positive oculo-cutaneous albinism, 558, 560
Tyrosine elevation, 195
Tyrosinemia, 195-96, 419, 568
Tyrosinosis, 196

Ulcer, gastrointestinal, 334-35, 346
Ulcerative colitis, 338
Ullrich syndrome, see Noonan syndrome
Ultracentrifugation, in study of lipoprotein abnormality, 279-80
Ultrasound imaging, in prenatal diagnosis, 613, 615
Ultraviolet radiation, 76, 108, 110
  response of porphyrins to, 209
Undritz anomaly, 364
Unequal chromatids, 16
Unifactorial inheritance, see Mendelian inheritance
Unstable hemoglobin variants, 351-52
Unverricht-Lundborg's myo-clonic epilepsy, 395
Uracil, 5-7
Urbach-Wiethe disease, 418
Urea cycle enzyme defects, 196-98
Uremia, 467, 484
Uric aciduria, 485
Uricosuria, 484-85
Urinary concentrating defect, 471
Urogenital disorders, see Genitourinary tract disorders; Kidney disorders
Urolithiasis, 484-85
Uroporphyrin, 217
Uroporphyrinogen, 208, 214
Uroporphyrinogen-I-synthetase (URO-S), 213
Uroporphyrinogen III cosynthetase deficiency, 212
Urticaria, 451, 464, 558
Usher syndrome, 451-52
Uterine-hernia syndrome, 315

Vaginal atresia, 485-86
Valine metabolism defects, 201
Van den Bosch syndrome, 556
Variegate porphyria, 214
Vascular neoplasms, 93
Vascular skin lesions, 562-63
Vasoactive intestinal peptide, 97
Vasopressin deficiency, 289-90
VATER association, 336
Ventricular septal defect, 271-74, 336
Vesicoureteral reflux, 485
Vestibular hypofunction, 454
Vestibulocerebellar ataxia, 401
Virilization, 303-4, 311-12
Viruses
  cardiovascular teratogens, 278
  neoplasms, 85-86
Vision disorders of, 415. See also Blindness; Eye disorders
Vitamin A treatment, for skin disorders, 551, 555
Vitamin B6, see Pyridoxine
Vitamin B12
  defective absorption of, 200, 203, 358-59
  deficiency, in Pelger-Huët anomaly, 364
Vitamin C deficiency, 195

Vitamin D-dependent rickets, 477-78
  tooth abnormalities, 591
Vitamin-dependent genetic diseases, 259-60
  metabolic errors, 261
Vitamin D excess, in hypercalcemia, 298
Vitamin D-resistant rickets, 476-77
  tooth abnormalities, 590
Vitamin K deficiency, 252
Vitreous abnormalities, 433-34
Vitreous veil, 433
Vocal cord paralysis, 499
von Gierke's disease, 188
von Hippel-Lindau disease, 99, 112, 394, 472
von Recklinghausen disease, 394, 511
von Willebrand disease, 365-67
Vrolik disease, 529-30

Waardenburg syndrome, 424, 430, 449, 559, 562
Wagner's disease, 433-34
W-16 antigen, 124
W-17 antigen, 124
W-27 antigen, 124
Warfarin, 30, 32, 252-53
Weill-Marchesani syndrome, 238, 431
Werner syndrome, 564
Wheal and flare skin responses, 126
White blood cell abnormalities, see Leukocytic abnormalities
White forelock, in albinism, 561-62
Williams-Campbell syndrome, 501
Williams syndrome, 299
Wilms' tumor, 90, 94-95
  and aniridia, 427
  two-hit model of oncogenesis, 88
  in XY male pseudohermaphroditism, 317
Wilson disease (hepatolenticular degeneration), 343, 396-97, 420, 486-87
Wilson-Mikity syndrome, 503
Winchester syndrome, 237-38
Wiskott-Aldrich syndrome, 88, 113, 135, 365
Wolffian duct virilization, 311
Wolf-Hirschhorn syndrome (4p-syndrome), 65-66, 333, 566
Wolfram syndrome, 289
Wolman's syndrome, 179-80, 302, 510
Woolly hair, 569

Xanthine oxidase deficiency, 219
Xanthinuria, 219-20
Xanthomatosis, 279, 281, 564
X-chromatin, 44-46
X-chromosome, 9
  monosomy, 64
  poly-X females, 51
  see also X-linked conditions; X-linked dominant inheritance; X-linked recessive inheritance
Xeroderma pigmentosum, 75-76, 108-10, 426
X-linked conditions, 17
  carriers, prenatal diagnosis, 614
  connective-tissue disorders, 238
  endocrine disorders, 311
  erythrocytic enzyme disorders, 357
  familial spastic paraplegias, 404
  hearing loss, 444, 453-55
  ichthyosis, 552
  immunodeficiency syndromes, 129
  Klinefelter syndrome, 50-51
  Lesch-Nyhan syndrome, 218
  ornithine transcarbamylase deficiency, 196
  phagocytic-bactericidal system disorders, 138
  pharmacogenetic diseases, 246
  sporadic cases, 25
  Turner syndrome, 46-49
X-linked dominant inheritance, 23
  cancer, 93
  cardiovascular abnormalities, 274
  endocrine diseases, 290, 295, 297
  hair disorders, 569
  hamartoses, 421
  kidney diseases, 463, 465-66, 469, 477, 479, 482
  pharmacogenetic diseases, 246
  skeletal dysplasias, 422
  skin disorders, 423
  tooth defects, 576-78
X-linked recessive inheritance, 22-23
  cancer, 92-94
  cardiovascular abnormalities, 274
  cataracts, 432-33
  coagulation deficiency states, 366
  connective-tissue diseases, 421
  craniofacial dysplasias, 423
  endocrine diseases, 296-97, 299, 302, 307
  glycolytic enzyme deficiencies, 356

gout, 219
granulomatous disease, 141
hearing loss, 445, 454
Hunter syndrome, 172
hypochromic anemia, 360
hypogammaglobulinemia, 131
immunodeficiency with hyper-IgM, 133
kidney diseases, 464, 483, 486-87
malabsorption syndromes, 339
microphthalmia, 426
muscular dystrophies, 374
phosphorylase-kinase deficiency, 190
polyneuropathies, 406
retinal abnormalities, 435-36
SCID, 130
short-trunk dwarfism, 539
skin disorders, 423, 550, 559, 566-68
T-cell disorders, 134
tooth defects, 576-79
trihexosyl ceramidosis, 162
Wiskott-Aldrich syndrome, 135
X-rays, and carcinoma, 108
XX gonadal dysgenesis, 308
XXXXY syndrome, 271-72
XY gonadal agenesis, 305-6, 313-14
XY gonadal dysgenesis, 111, 313
ɪ-Xylulose dehydrogenase deficiency, 193
XY Turner phenotype, see Noonan syndrome
XYY male, 41, 51

Y-chromatin, 44-46
Y-chromosome, 9, 17
  fluorescence, 45
Yellow-mutant albinism, 561
Yellow nail syndrome, 501
Y-linked conditions
  gonadal tumor abnormalities, 110-11
  Klinefelter syndrome, 50-51
  sexual differentiation, 311
Y protein, in liver cells, 342

Z discs, in central core disease, 384
Zebra bodies, 156, 169
Zellweger syndrome, see Cerebrohepatorenal syndrome
Ziprkowski-Margolis syndrome, 559, 562
Zollinger-Ellison syndrome, 96, 334-35
Z protein, in liver cells, 342
Zygotes, fusion of, in chimerism, 318

WICHITA CLINIC

LIBRARY

WICHITA, KANSAS